HEAT TRANSFER
A Basic Approach

M. NECATI ÖZIŞIK

Professor, Mechanical and Aerospace Engineering
North Carolina State University

McGraw-Hill Book Company

New York St. Louis San Francisco Auckland Bogotá Hamburg
London Madrid Mexico Montreal New Dehli
Panama Paris São Paulo Singapore Sydney Tokyo Toronto

HEAT TRANSFER
A Basic Approach
INTERNATIONAL EDITION 1985

6 7 8 9 0 KHL 9 4 3 2 1

This book was set in Times Roman.
The editors were Anne Murphy and Madelaine Eichberg.
The production supervisor was Leroy A. Young.

Library of Congress Cataloging in Publication Data

Özisik, M. Necati.
Heat transfer.

Includes bibliographical references and indexes.
1. Heat—Transmission. I. Title.
TJ260.096 1985 621.402'2 83-20369
ISBN 0-07-047982-8

When ordering this title use ISBN 0-07-066460-9

Printed in Singapore.

To GÜL and HAKAN

CONTENTS

PREFACE

The field of heat transfer is so wide and diversified that an orderly presentation of scientific facts is essential for effective teaching of this subject. In our teaching, we should place emphasis not only to the transmission of the knowledge but also to the laying of a strong foundation on which future knowledge can readily be accumulated and the acquired knowledge can be fully utilized for useful purposes.

This book, although based on my 1977 book entitled *Basic Heat Transfer*, is actually completely rewritten and reorganized, first by pedagogical considerations, and second by providing a large number of fully worked out illustrative examples, a large number of problems at the end of each chapter, summary tables for ready reference, improved heat transfer charts and correlations, and comprehensive physical property tables to help to increase its usefulness in practical applications.

The role of an introductory text on the subject is to establish the guidelines for the transmission of the knowledge and serve as the catalizer in the interaction between the teaching and learning processes. Therefore, it is not only the knowledge contained in a book but also its organization which influences the effectiveness of teaching. These principles have been the basic guidelines in the preparation of this undergraduate text for the teaching of heat transfer in engineering schools for the mechanical, chemical, and nuclear engineering students.

Our primary goal is not only to transmit the knowledge effectively but also to provide a good understanding of the physical aspects of the subject matter and to develop the necessary skills and background for the handling of related heat transfer problems to be encountered later in the professional career. To achieve such an objective, the fundamentals are systematically developed, the physical significance of the developments are emphasized, and applications to the solution of practical problems are illustrated with ample examples in each chapter.

There is sufficient material in this book to meet individual course objectives at different levels, in both the junior and senior years. The spectrum of its possible uses may range from a one-semester basic heat transfer course to a sequence of two-quarter or two-semester courses spread over the junior and/or senior years. When used in a sequence course, conduction, finite differences, and radiation should preferably be covered first, followed by convection, boiling and condensation, and heat exchangers.

The book can also serve as a source of ready reference for engineering graduates and industry. A background in differential equations at the sophomore level and some familiarity with fluid mechanics are all needed for following this book.

Chapter 1 introduces the basic concepts and gives a bird's-eye view of the mechanisms of heat transfer. A discussion of units and conversion factors is also presented. Chapter 2 builds up the necessary background for the understanding of the physical significance of the heat conduction equation and its boundary conditions. Emphasis is placed on the development of necessary skills needed for the mathematical formulation of practical heat conduction problems. This matter is illustrated with numerous representative examples. Chapter 3 provides an introduction to the solving of heat conduction problems and developing analytic expressions for the determination of temperature distribution and heat flow in solids. Only the one-dimensional steady-state heat conduction problems are considered. The thermal-resistance concept is introduced for use in the determination of heat transfer through composite layers. The analysis and application of heat flow through fins are presented. In Chapter 4 the concept of transient heat flow is introduced through the use of the lumped system analysis because of its simplicity. To develop a better understanding of the significance of temperature transients with time and position, the temperature response and heat transfer charts are introduced before presenting an analysis of transient conduction. The use of these charts for predicting temperature transients in solids having shapes such as a slab, cylinder, and sphere is illustrated with numerous examples. As a follow-up to this approach, the use of conduction shape factors is also discussed to predict steady-state heat flow in solids having complicated configurations. To give some idea of the analytic methods of solving transient heat conduction problems, the method of separation of variables is considered for the solution of one-dimensional transient heat conduction in a slab geometry. Chapter 5 presents the fundamentals of finite-difference methods for the solution of both steady and time-dependent heat conduction problems. The application is illustrated with numerous examples and a computer program is given for solving the resulting finite-difference equations.

Chapters 6 through 9 are devoted to heat transfer in forced and free convection. Chapter 6 prepares the necessary background for the understanding of the physical significance of various concepts and fundamental definitions associated with the study of convection. The equations of motion and energy are introduced for the case of two-dimensional constant-property incompressible flow. The physical significance of various terms in these equations is discussed, and the use of these equations in the formulation of forced-convection problems is illustrated with examples. Chapter 7 presents forced-convection inside ducts. To illustrate the use of the equations of motion and energy in the determination of friction factor and heat transfer coefficients, simple forced-convection problems are solved and temperature and velocity distributions are established. Such elementary analysis of forced convection provides a good insight into the role of fluid flow in heat transfer. In addition, it helps toward better understanding of the physical significance of analytic and empirical correlations of friction factor and heat transfer coefficient for complicated situations. Chapter 8 deals with forced convection over bodies. To illustrate the use of boundary layer equations in the prediction of drag and heat transfer coefficients, the integral method of analysis is applied to develop analytic expressions for the drag and heat transfer coefficients for laminar flow over a flat plate. Various correlations are then presented for flow over bodies having other geometrics. In Chapter 9 the principles of free convection are discussed. An approximate boundary layer analysis is presented to illustrate the use of the energy equation to predict the heat transfer coefficient for free convection from a vertical plate. Correlations of free convection for other configurations are then presented and their application is illustrated with examples.

Chapter 10 presents the fundamentals of boiling and condensation. Various regimes of boiling and condensation are discussed and the heat transfer correlations associated with them are presented.

Chapter 11 is devoted to the thermal analysis of heat exchangers. Various types of heat exchangers are discussed, and the use of LMTD and E-NTU methods for the sizing of heat exchangers are illustrated with representative examples.

Heat transfer by radiation is covered in Chapters 12 and 13. Chapter 12 deals with radiation exchange among surfaces separated by a nonparticipating medium. The absorption, emission, and reflection of radiation by real surfaces are discussed and the concept of blackbody radiation is developed. The analysis of radiation exchange among surfaces is introduced first by using the network method, because it provides a good insight into the physical nature of the problem. However, the method is not so practicable when the system involves more than two arbitrarily oriented surfaces. Therefore, a relatively straightforward radiosity-matrix method, capable of handling radiation problems involving any number of surfaces with no additional complexity, is then presented. Chapter 13 considers radiation transfer inside a semitransparent, absorbing, emitting medium, and radiation from hot gases.

Finally, in Chapter 14 the analysis of mass transfer is presented with an analogy of heat transfer by diffusion and forced convection.

The SI (Système Internationale) system of units is used throughout this book. Comprehensive conversion factors and physical property tables are presented in the Appendix. To provide some feel for the relative magnitude of physical properties in the SI and Btu units, some property tables are also included in both system of units.

There are more than 170 fully worked-out examples to illustrate the application of the basic theory and concepts. Over 800 problems, arranged in the same order as the material presented in the text, are included at the ends of the chapters with answers provided for some of the representative ones. A summary of fundamental equations are tabulated at the end of each chapter for ready reference.

This book is the outcome of many years of experience in teaching and writing of textbooks in heat transfer at various levels. It is hoped that it will be helpful to improve the effectiveness of teaching and learning of the subject of heat transfer. I am indebted to many of my colleagues and students for their valuable suggestions toward achieving the objectives stated previously. I wish to thank Dr. S. Kakaç for providing useful comments on the text and Y. Cengel for thoroughly reading the manuscript.

M. NECATI ÖZIŞIK

HEAT TRANSFER
A Basic Approach

CHAPTER
ONE

INTRODUCTION AND CONCEPTS

The concept of *energy* is used in thermodynamics to specify the state of a system. It is a well-known fact that energy is neither created nor destroyed but only changed from one form to another. The science of *thermodynamics* deals with the relation between heat and other forms of energy, but the science of *heat transfer* is concerned with the analysis of the rate of heat transfer taking place in a system. The energy transfer by heat flow cannot be measured directly, but the concept has physical meaning because it is related to the measurable quantity called *temperature*. It has long been established by observations that when there is temperature difference in a system, heat flows from the region of high temperature to that of low temperature. Since heat flow takes place whenever there is a temperature gradient in a system, a knowledge of the temperature distribution in a system is essential in heat transfer studies. Once the temperature distribution is known, a quantity of practical interest, the *heat flux*, which is the amount of heat transfer per unit area per unit time, is readily determined from the law relating the heat flux to the temperature gradient.

The problem of determining temperature distribution and heat flow is of interest in many branches of science and engineering. In the design of heat exchangers such as boilers, condensers, radiators, etc., for example, heat transfer analysis is essential for sizing such equipment. In the design of nuclear-reactor cores, a thorough heat transfer analysis of fuel elements is important for proper sizing of fuel elements to prevent burnout. In aerospace technology, the temperature distribution and heat transfer problems are crucial because of weight limitations and safety considerations. In heating and air conditioning applications for buildings, a proper heat transfer analysis is necessary to estimate the amount of insulation needed to prevent excessive heat losses or gains.

In the studies of heat transfer, it is customary to consider three distinct modes of heat transfer: *conduction*, *convection*, and *radiation*. In reality, temperature distribution in a medium is controlled by the combined effects of these three modes of heat transfer; therefore, it is not actually possible to isolate entirely one mode from interactions with the other modes. However, for simplicity in the analysis, one can consider, for example, conduction separately whenever heat transfer by convection and radiation is negligible. With this qualification, we present below a brief qualitative description of these three distinct modes of heat transfer; they are studied in great detail in the following chapters.

1-1 CONDUCTION

Conduction is the mode of heat transfer in which energy exchange takes place from the region of high temperature to that of low temperature by the kinetic motion or direct impact of molecules, as in the case of fluid at rest, and by the drift of electrons, as in the case of metals. In a solid which is a good electric conductor, a large number of free electrons move about in the lattice; hence materials that are good electric conductors are generally good heat conductors (i.e., copper, silver, etc.).

The empirical law of heat conduction based on experimental observations originates from Biot but is generally named after the French mathematical physicist Joseph Fourier [1]* who used it in his analytic theory of heat. This law states that the rate of heat flow by conduction in a given direction is proportional to the area normal to the direction of heat flow and to the gradient of temperature in that direction. For heat flow in the x direction, for example, the Fourier law is given as

$$\boxed{Q_x = -kA\frac{dT}{dx}} \quad \text{W} \tag{1-1a}$$

or

$$\boxed{q_x = \frac{Q_x}{A} = -k\frac{dT}{dx}} \quad \text{W/m}^2 \tag{1-1b}$$

where Q_x is the rate of heat flow through area A in the positive x direction and q_x is called the *heat flux* in the positive x direction. The proportionality constant k is called the *thermal conductivity* of the material and is a positive quantity. If temperature decreases in the positive x direction, then dT/dx is negative; hence q_x (or Q_x) becomes a positive quantity because of the presence of the negative sign in Eqs. (1-1a) and (1-1b). Therefore, the minus sign is included in Eqs. (1-1a) and (1-1b) to ensure that q_x *(or* Q_x*) is a positive quantity when the heat flow is in the positive x direction.* Conversely, when the right-hand side of Eqs. (1-1a) and (1-1b) is negative, the heat flow is in the negative x direction.

The thermal conductivity k in Eqs. (1-1a) and (1-1b) must have the dimensions W/(m · °C) or J/(m · s · °C) if the equations are dimensionally correct. There is a wide difference in the range of thermal conductivities of various engineering

* Bracketed numbers indicate references at the end of the chapter.

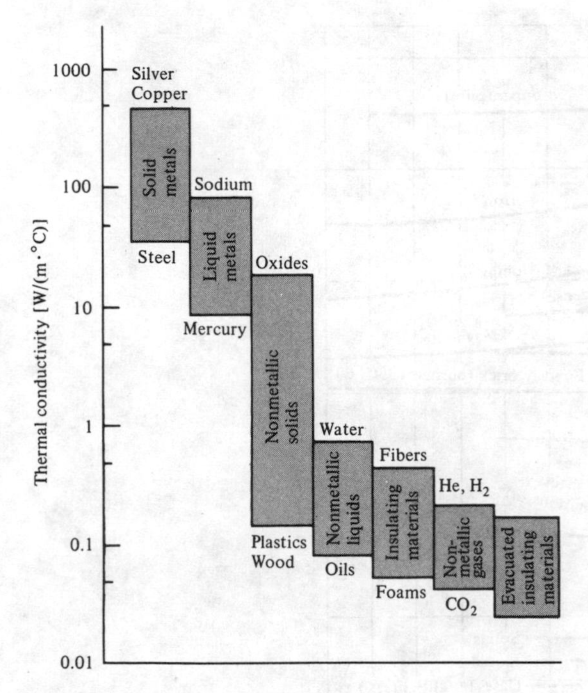

Figure 1-1 Typical range of thermal conductivity of various materials.

materials, as illustrated in Fig. 1-1. Between gases and highly conducting metals, such as copper or silver, k varies by a factor of about 10^4. Thus, in Fig. 1-1 the highest value is for highly conducting pure metals, and the lowest value is for gases and vapors, excluding the evacuated insulating systems. The nonmetallic solids and liquids have thermal conductivities that lie between them. Metallic single crystals are exceptions, which may have very high thermal conductivities; for example, with copper crystals, values of 8000 W/(m · °C) and even higher are possible.

Thermal conductivity also varies with temperature. This variation, for some materials over certain temperature ranges, is small enough to be neglected; but for many cases the variation of k with temperature is significant. Especially at very low temperatures k varies rapidly with temperature; for example, the thermal conductivities of copper, aluminum, or silver reach values 50 to 100 times those that occur at room temperature. Figure 1-2 illustrates how the thermal conductivity of some engineering materials varies with temperature. Actual values of thermal conductivity of various materials are given in App. B, and a comprehensive compilation of thermal conductivities of materials can be found in Refs. 2 to 5.

Example 1-1 Determine the heat flux q and the heat transfer rate across an iron plate with area $A = 0.5\ \text{m}^2$ and thickness $L = 0.02$ m [$k = 70$ W/(m · °C)] when one of its surfaces is maintained at $T_1 = 60°\text{C}$ and the other at $T_2 = 20°\text{C}$.

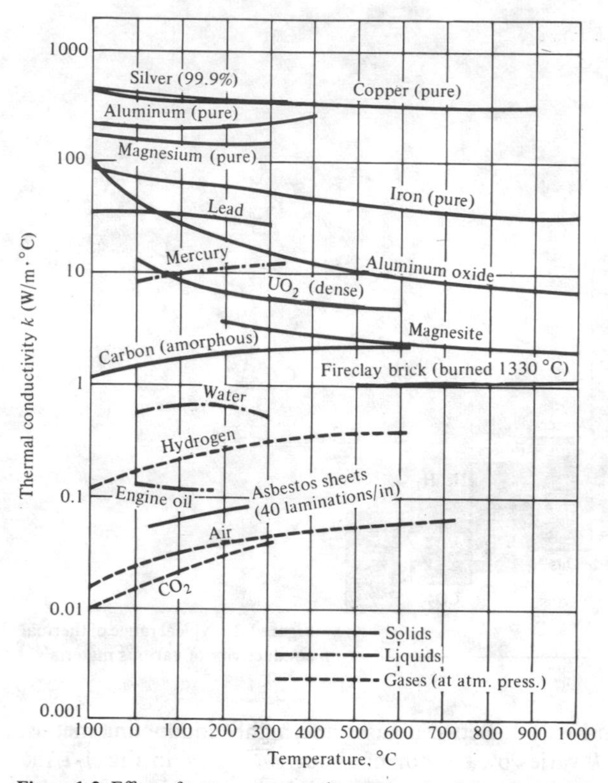

Figure 1-2 Effect of temperature on thermal conductivity of materials.

SOLUTION In this problem the temperature gradient dT/dx is constant; hence the temperature distribution $T(x)$ through the plate is linear, as illustrated in Fig. 1-3. Then the heat flux q is determined by applying Eq. (1-1*b*) as

$$q = -k \frac{dT(x)}{dx}$$

$$= -k \frac{T_2 - T_1}{L} = -70 \frac{20 - 60}{0.02} = 140 \text{ kW/m}^2$$

Thus the heat flow is in the *positive x direction* since the result is a *positive quantity*. The heat flow rate Q through an area $A = 0.5$ m^2 is computed by applying Eq. (1-1*a*):

$$Q = Aq = 0.5 \times 140 = 70 \text{ kW}$$

Example 1-2 The heat flow rate through a wood board $L = 2$ cm thick for a temperature difference of $\Delta T = 25$°C between the two surfaces is 150 W/m^2. Calculate the thermal conductivity of the wood.

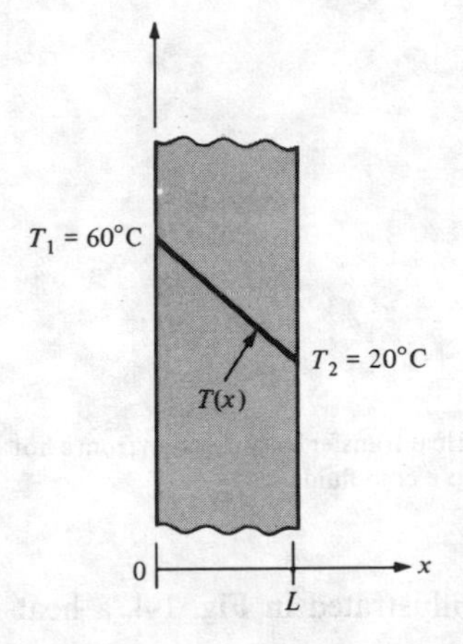

Figure 1-3 Heat conduction through a slab.

SOLUTION Equation (1-1*b*) is applied as follows:

$$q = k\frac{\Delta T}{L}$$

$$150 = k\frac{25}{0.02}$$

$$k = 0.12 \text{ W/(m} \cdot {}^\circ\text{C)}$$

1-2 CONVECTION

When fluid flows over a solid body or inside a channel while temperatures of the fluid and the solid surface are different, heat transfer between the fluid and the solid surface takes place as a consequence of the motion of fluid relative to the surface; this mechanism of heat transfer is called *convection.* If the fluid motion is artificially induced, say with a pump or a fan that forces the fluid flow over the surface, the heat transfer is said to be by *forced convection.* If the fluid motion is set up by buoyancy effects resulting from density difference caused by temperature difference in the fluid, the heat transfer is said to be by *free* (or *natural*) *convection.* For example, a hot plate vertically suspended in stagnant cool air causes a motion in the air layer adjacent to the plate surface because the temperature gradient in the air gives rise to a density gradient, which in turn sets up the air motion. As the temperature field in the fluid is influenced by the fluid motion, the determination of temperature distribution and of heat transfer in convection for most practical situations is a complicated matter. In engineering applications, to simplify the heat transfer calculations between a hot surface at T_w and a cold

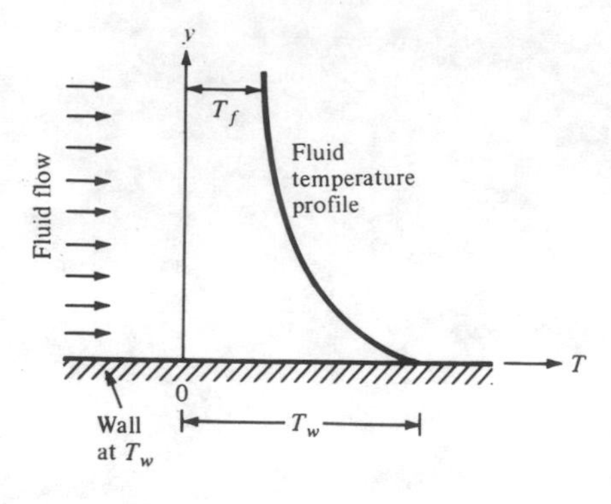

Figure 1-4 Heat transfer by convection from a hot wall at T_w to a cold fluid.

fluid flowing over it at a bulk temperature T_f as illustrated in Fig. 1-4, a heat transfer coefficient h is defined as

$$q = h(T_w - T_f) \tag{1-2a}$$

where q is the heat flux (in watts per square meter) from the hot wall to the cold fluid. Alternatively, for heat transfer from the hot fluid to the cold wall, Eq. (1-2*a*) is written as

$$q = h(T_f - T_w) \tag{1-2b}$$

where q represents the heat flux from the hot fluid to the cold wall. Historically, the form given by Eq. (1-2*a*) was first used as a law of cooling as heat is removed from a body to a liquid flowing over it, and it is generally referred to as "Newton's law of cooling." If the heat flux in Eqs. (1-2*a*) and (1-2*b*) is given in watts per square meter and the temperatures are in degrees Celsius (or kelvins), then the heat transfer coefficient h in Eqs. (1-2*a*) and (1-2*b*) must have the dimensions $W/(m^2 \cdot °C)$ if the equations are dimensionally correct.

The heat transfer coefficient h varies with the type of flow (i.e., laminar or turbulent), the geometry of the body and flow passage area, the physical properties of the fluid, the average temperature, and the position along the surface of the body. It also depends on whether the mechanism of heat transfer is by forced convection (i.e., the fluid motion is caused by a pump or a blower) or by natural convection (i.e., the fluid motion is caused by the buoyancy). When h varies with the position along the surface of the body, for convenience in engineering applications, its average value h_m over the surface is considered instead of its local value h. Equations (1-2*a*) and (1-2*b*) are also applicable for such cases by merely replacing h by h_m; then q represents the average value of the heat flux over the region considered.

The heat transfer coefficient can be determined analytically for flow over bodies having a simple geometry such as a flat plate or flow inside a circular tube.

Table 1-1 Typical values of the convective heat transfer coefficient h

Type of flow	h, W/(m$^2 \cdot$ °C)
Free convection, $\Delta T = 25$°C	
• 0.25-m vertical plate in:	
Atmospheric air	5
Engine oil	37
Water	440
• 0.02-m-OD* horizontal cylinder in:	
Atmospheric air	8
Engine oil	62
Water	741
• 0.02-m-diameter sphere in:	
Atmospheric air	9
Engine oil	60
Water	606
Forced convection	
• Atmospheric air at 25°C with $U_\infty = 10$ m/s over a flat plate:	
$L = 0.1$ m	39
$L = 0.5$ m	17
• Flow at 5 m/s across 1-cm-OD cylinder of:	
Atmospheric air	85
Engine oil	1,800
• Water at 1 kg/s inside 2.5-cm-ID† tube	10,500
Boiling of water at 1 atm	
• Pool boiling in a container	3,000
• Pool boiling at peak heat flux	35,000
• Film boiling	300
Condensation of steam at 1 atm	
• Film condensation on horizontal tubes	9,000–25,000
• Film condensation on vertical surfaces	4,000–11,000
• Dropwise condensation	60,000–120,000

* OD = outer diameter.
† ID = inner diameter.

For flow over bodies having complex configurations, the experimental approach is used to determine h. There is a wide difference in the range of the values of the heat transfer coefficient for various applications. Table 1-1 lists typical values of h encountered in some applications.

Example 1-3 An electrically heated plate dissipates heat by convection at a rate of $q = 8000$ W/m^2 into the ambient air at $T_f = 25$°C. If the surface of the hot plate is at $T_w = 125$°C, calculate the heat transfer coefficient for convection between the plate and the air.

SOLUTION Heat is being transferred from the plate to the fluid, so Eq. (1-2a) is applied:

$$q = h(T_w - T_f)$$
$$8000 = h(125 - 25)$$
$$h = 80 \text{ W/(m}^2 \cdot {}^\circ\text{C)}$$

Example 1-4 Hot air at $T_f = 150°C$ flows over a flat plate maintained at $T_w = 50°C$. The forced convection heat transfer coefficient is $h = 75$ W/(m^2 · °C). Calculate the heat transfer rate into the plate through an area of $A = 2$ m^2.

SOLUTION For heat transfer from the hot fluid to the plate, Eq. (1-2a) is applied:

$$q = h(T_f - T_w)$$
$$q = 75(150 - 50) = 7.5 \times 10^3 \text{ W/m}^2$$
$$Q = qA = (7.5 \times 10^3)(2) = 15 \text{ kW}$$

1-3 RADIATION

All bodies continuously emit energy because of their temperature, and the energy thus emitted is called *thermal radiation*. The radiation energy emitted by a body is transmitted in the space in the form of electromagnetic waves according to Maxwell's classic electromagnetic wave theory or in the form of discrete photons according to Planck's hypothesis. Both concepts have been utilized in the investigation of radiative-heat transfer. The emission or absorption of radiation energy by a body is a bulk process; that is, radiation originating from the interior of the body is emitted through the surface. Conversely, radiation incident on the surface of a body penetrates to the depths of the medium where it is attenuated. When a large proportion of the incident radiation is attenuated within a very short distance from the surface, we may speak of radiation as being absorbed or emitted by the surface. For example, thermal radiation incident on a metal surface is attenuated within a distance of a few angstroms from the surface; hence metals are opaque to thermal radiation.

The solar radiation incident on a body of water is gradually attenuated by water as the beam penetrates to the depths of water. Similarly, the solar radiation incident on a sheet of glass is partially absorbed and partially reflected, and the remaining is transmitted. Therefore, water and glass are considered semitransparent to the solar radiation.

It is only in a *vacuum* that radiation propagates with no attenuation at all. Also the atmospheric air contained in a room is considered transparent to thermal radiation for all practical purposes, because the attenuation of radiation by air is insignificant unless the air layer is several kilometers thick. However, gases such as carbon dioxide, carbon monoxide, water vapor, and ammonia absorb thermal radiation over certain wavelength bands; therefore they are semitransparent to thermal radiation.

It is apparent from the previous discussion that a body at a temperature T emits radiation owing to its temperature; also a body absorbs radiation incident on it. Here we briefly discuss the emission and absorption of radiation by a body.

Emission of Radiation

The maximum radiation flux emitted by a body at temperature T is given by the *Stefan–Boltzmann law*

$$\boxed{E_b = \sigma T^4} \qquad \text{W/m}^2 \tag{1-3}$$

where T is the *absolute temperature* in kelvins, σ is the *Stefan–Boltzmann constant* [$\sigma = 5.6697 \times 10^{-8}$ W/(m$^2 \cdot$ K^4)], and E_b is called the *blackbody emissive power*.

Only an *ideal radiator* or the so-called *blackbody* can emit radiation flux according to Eq. (1-3). The radiation flux emitted by a *real body* at an absolute temperature T is always less than that of the blackbody emissive power E_b; it is given by

$$\boxed{q = \varepsilon E_b = \varepsilon \sigma T^4} \tag{1-4}$$

where the *emissivity* ε lies between zero and unity; for all real bodies it is always less than unity.

Figure 1-5 shows a plot of the blackbody emissive power E_b defined by Eq. (1-3) versus the absolute temperature. The radiation flux emitted rapidly increases with rising temperature. For example, the emissive power increases from $E_b = 461$ W/m^2 at room temperature $T = 300$ K to $E_b = 3562$ W/m^2 at $T = 500$ K and $E_b = 56{,}700$ W/m^2 at $T = 1000$ K.

Absorption of Radiation

If a radiation flux q_{inc} is incident on a blackbody, it is completely absorbed by the blackbody. However, if the radiation flux q_{inc} is incident on a real body, then the energy absorbed q_{abs} by the body is given by

$$q_{abs} = \alpha q_{inc} \tag{1-5}$$

where the *absorptivity* α lies between zero and unity; for all real bodies it is always less than unity.

The absorptivity α of a body is generally different from its emissivity ε. However, in many practical applications, to simplify the analysis, α is assumed to equal ε. Such matters are discussed in depth and actual values of emissivity of various surfaces given in Chap. 12.

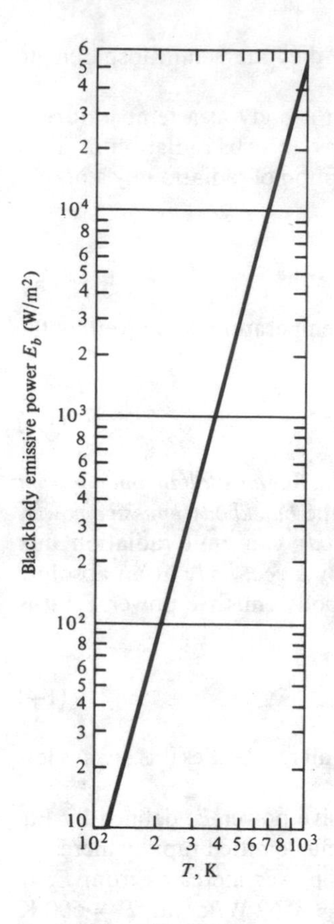

Figure 1-5 Blackbody emissive power $E_b = \sigma T^4$.

Radiation Exchange

When two bodies at different temperatures "see" each other, heat is exchanged between them by radiation. If the intervening medium is filled with a substance such as air which is transparent to radiation, the radiation emitted from one body travels through the intervening medium with no attenuation and reaches the other body, and vice versa. Then the hot body experiences a net heat loss, and the cold body a net heat gain, as a result of the radiation heat exchange. The analysis of *radiation heat exchange among surfaces* is generally a complicated matter and is dealt with in Chap. 12. Here we examine some very special cases with illustrative examples.

Figure 1-6 shows a small, hot, opaque plate of surface area A_1 and emissivity ε_1 that is maintained at an absolute temperature T_1 and exposed to a large surrounding area A_2 (i.e., $A_1/A_2 \to 0$) at an absolute temperature T_2. The space between them contains air which is transparent to thermal radiation. The radiation energy emitted by the surface A_1 is given by

$$A_1 \varepsilon_1 \sigma T_1^4$$

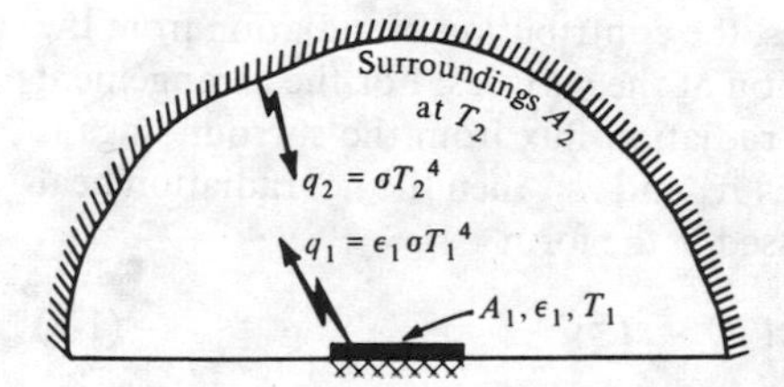

Figure 1-6 Radiation exchange between a surface A_1 and its surroundings.

The large surrounding area can be approximated as a blackbody in relation to the small surface A_1. Then the radiation flux emitted by the surrounding area is σT_2^4, which is also the radiation flux incident on the surface A_1. Hence, the radiation energy absorbed by the surface A_1 is

$$A_1 \alpha_1 \sigma T_2^4$$

The *net radiation loss* at the surface A_1 is the difference between the energy emitted and the energy absorbed:

$$Q_1 = A_1 \varepsilon_1 \sigma T_1^4 - A_1 \alpha_1 \sigma T_2^4 \tag{1-6a}$$

For $\varepsilon_1 = \alpha_1$, this result simplifies to

$$\boxed{Q_1 = A_1 \varepsilon_1 \sigma (T_1^4 - T_2^4)} \tag{1-6b}$$

which provides the expression for calculating the radiation heat exchange between a small surface element A_1 and its surroundings at T_2. Clearly, the positive value of Q_1 implies heat loss from the surface A_1, and the negative value implies heat gain.

We now consider two finite surfaces A_1 and A_2 as illustrated in Fig. 1-7. The surfaces are maintained at absolute temperatures T_1 and T_2, respectively, and have emissivities ε_1 and ε_2. The physical situation implies that part of the radiation leaving surface A_1 reaches surface A_2 while the remaining is lost to the surroundings. Similar considerations apply for the radiation leaving surface A_2. The analysis of radiation heat exchange between the two surfaces for such a case should include

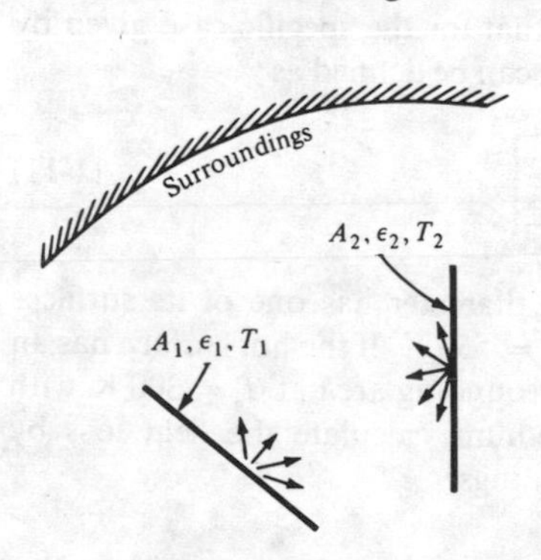

Figure 1-7 Radiation exchange between surfaces A_1 and A_2.

the effects of the orientation of the surfaces, the contribution of radiation from the surroundings, and the reflection of radiation at the surfaces. For the arrangement shown in Fig. 1-7, if we assume that the radiation flux from the surroundings is negligible compared to those from surfaces A_1 and A_2, then the net radiation heat transfer Q_1 at the surface A_1 can be expressed in the form

$$Q_1 = F_1 A_1 \sigma (T_1^4 - T_2^4) \tag{1-7}$$

where F_1 is a factor that includes the effects of the orientation of the surfaces and their emissivities. The determination of this factor is a complicated matter, and the analysis of radiation problems of this type is the subject of Chap. 12.

Radiation Heat Transfer Coefficient

To simplify the heat transfer calculations, it may be possible, *under very restrictive conditions*, to define a radiation heat transfer coefficient h_r, analogous to the convection heat transfer coefficient, as

$$\boxed{q_1 = h_r (T_1 - T_2)} \tag{1-8}$$

This concept can be applied to the result given by Eq. (1-6*b*) as now described.

Equation (1-6*b*) is written as

$$Q = A_1 \varepsilon_1 \sigma (T_1^2 + T_2^2)(T_1 + T_2)(T_1 - T_2) \tag{1-9a}$$

If $|T_1 - T_2| \ll T_1$, this result is linearized as

$$Q_1 \cong A_1 \varepsilon_1 \sigma 4 T_1^3 (T_1 - T_2) \tag{1-9b}$$

or

$$\boxed{q_1 \equiv \frac{Q_1}{A_1} = (4 T_1^3 \varepsilon_1 \sigma)(T_1 - T_2)} \tag{1-10}$$

A comparison of Eqs. (1-8) and (1-10) reveals that for the specific case given by Eq. (1-6*b*), a radiation heat transfer coefficient h_r can be defined as

$$\boxed{h_r \equiv 4 T_1^3 \varepsilon_1 \sigma} \tag{1-11}$$

Example 1-5 A heated plate of $D = 0.2$ m diameter has one of its surfaces insulated, and the other is maintained at $T_w = 550$ K. If the hot surface has an emissivity $\varepsilon_w = 0.9$ and is exposed to a surrounding area at $T_s = 300$ K with atmospheric air being the intervening medium, calculate the heat loss by radiation from the hot plate to the surroundings.

SOLUTION Assuming $\varepsilon_1 = \alpha_1$, we can apply Eq. (1-6b):

$$Q_w = A_w \varepsilon_w \sigma (T_w^4 - T_s^4)$$
$$= \left[\frac{\pi}{4}(0.2)^2\right](0.9)(5.67 \times 10^{-8})[(5.5)^4 - 3^4] \times 10^8$$
$$= 134.5 \text{ W}$$

Example 1-6 A small hot surface at temperature $T_1 = 430$ K having an emissivity $\varepsilon_1 = 0.8$ dissipates heat by radiation into a surrounding area at $T_2 = 400$ K. If this radiation transfer process is characterized by a radiation heat transfer coefficient h_r calculate the value of h_r.

SOLUTION For this particular case, the requirement $T_1 - T_2 \ll T_1$ is satisfied. Then Eq. (1-11) is applied as follows:

$$h_r = 4T_1^3 \varepsilon_1 \sigma$$
$$= 4[(4.3)^3 \times 10^6](0.8)(5.67 \times 10^{-8}) = 14.4 \text{ W/(m}^2 \cdot {}^\circ\text{C)}$$

1-4 COMBINED HEAT TRANSFER MECHANISM

So far we have considered the heat transfer mechanism, conduction, convection, and radiation separately. In many practical situations heat transfer from a surface takes place simultaneously by convection to the ambient air and by radiation to the surroundings. Figure 1-8 illustrates a small plate of area A and emissivity ε that is maintained at T_w and exchanges energy by convection with a fluid at T_∞ with a heat transfer coefficient h_c and by radiation with the surroundings at T_s. The heat loss per unit area of the plate, by the combined mechanism of convection and radiation, is given by

$$q_w = h_c(T_w - T_\infty) + \varepsilon\sigma(T_w^4 - T_s^4) \tag{1-12}$$

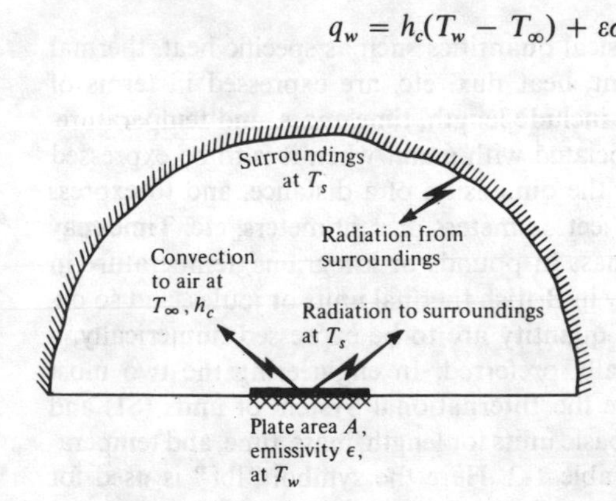

Figure 1-8 Simultaneous convection and radiation from a plate.

If $|T_w - T_s| \ll T_w$, the second term can be linearized. We obtain

$$q_w = h_c(T_w - T_\infty) + h_r(T_w - T_s) \tag{1-13a}$$

where

$$h_r \equiv 4\varepsilon\sigma T_w^3 \tag{1-13b}$$

Example 1-7 A small, thin metal plate of area A m^2 is kept insulated on one side and exposed to the sun on the other side. The plate absorbs solar energy at a rate of 500 W/m^2 and dissipates it by convection into the ambient air at $T_\infty = 300$ K with a convection heat transfer coefficient $h_c = 20$ W/(m$^2 \cdot$ °C) and by radiation into a surrounding area which may be assumed to be a blackbody at $T_{sky} = 280$ K. The emissivity of the surface is $\varepsilon = 0.9$. Determine the equilibrium temperature of the plate.

SOLUTION The energy balance per unit area of the exposed surface is written as

$$500 = 20(T_w - 300) + 0.9 \times 5.67 \times 10^{-8}\left[\left(\frac{T_w}{100}\right)^4 - (2.8)^4\right]10^8$$

or

$$25 = T_w - 300 + 0.255\left(\frac{T_w}{100}\right)^4 - 15.68$$

or

$$T_w = 340.68 - 0.255\left(\frac{T_w}{100}\right)^4$$

The solution of this equation by trial and error yields the plate temperature as

$$T_w = 315.5 \text{ K}$$

1-5 UNITS, DIMENSIONS, AND CONVERSION FACTORS

In the field of heat transfer, the physical quantities such as specific heat, thermal conductivity, heat transfer coefficient, heat flux, etc. are expressed in terms of a few fundamental *dimensions* which include length, time, mass, and temperature, and each of these dimensions is associated with a *unit* when it is to be expressed numerically. For example, length is the dimension of a distance, and to express it numerically one may use units of feet or meters or centimeters, etc. Time may be measured in hours or seconds, mass in pounds or kilograms, temperature in degrees Fahrenheit or Celsius, energy in British thermal units or joules, and so on. When the dimensions of a physical quantity are to be expressed numerically, a consistent system of units is generally preferred. In engineering the two most commonly used systems of units are the International System of units (SI) and the English engineering system. The basic units for length, mass, time, and temperature for each system are listed in Table 1-2. Here the symbol "lbf" is used for

Table 1-2 Systems of units

Quantity	SI	English engineering system
Length	m	ft
Mass	kg	lb
Time	s	s
Temperature	K	R
Force	N	lbf
Energy	J or N · m	Btu or ft · lbf

pound-force to distinguish it from the symbol "lb" commonly used for *pound-mass*, but there is no such misunderstanding in SI because the *kilogram* is the unit of *mass* and the *newton* is the unit of *force*. The physical significance of the force units newton and lbf is better envisioned by considering Newton's second law of motion, written as

$$\text{Force} = \frac{1}{g_c} \times \text{mass} \times \text{acceleration} \tag{1-14}$$

where g_c is the *gravitational conversion-factor constant*. The pound-force lbf is defined as the force that acts on the mass of one pound at a point on the earth where the magnitude of the gravitational acceleration is $g = 32.174$ ft/s^2. Then, in the English engineering system, Eq. (1-14) becomes

$$1\ \text{lbf} = \frac{1}{g_c} \times 1\ \text{lb} \times 32.174\ \text{ft/s}^2 \tag{1-15}$$

According to this relation, one pound of force (that is, 1 lbf) will accelerate one pound of mass (that is, 1 lb) 32.174 ft/s^2; or 1 lbf is equal to 32.174 ft · lb/s^2. The conversion factor g_c in the English engineering system is obtained from this relation as

$$g_c = 32.174\ \text{lb} \cdot \text{ft/(lbf} \cdot \text{s}^2) \tag{1-16}$$

Note that the gravitational acceleration g and the gravitational conversion factor g_c are not similar quantities; g_c is constant, but g depends on the location and on the altitude.

In SI, Eq. (1-14) becomes

$$1\ \text{N} = \frac{1}{g_c} \times 1\ \text{kg} \times 1\ \text{m/s}^2 \tag{1-17}$$

Clearly, in SI, 1 N is a force that will accelerate a 1-kg mass 1 m/s^2, or a 1-N force is equal to 1 kg · m/s^2. The conversion factor g_c in SI becomes

$$g_c = 1\ \text{kg} \cdot \text{m/(N} \cdot \text{s}^2) = 1 \tag{1-18}$$

since
$$1 \text{ N} = 1 \text{ kg} \cdot \text{m/s}^2 \tag{1-19}$$

Therefore, in SI, g_c is not needed.

Energy is measured in Btu or ft · lbf in the English engineering system whereas it is measured in joules (J) or newton-meters (N · m) in SI. Note that 1 J = 1 N · m and 1 J = 1 kg · m²/s² since 1 N = 1 kg · m/s².

Power is measured in Btu/h or ft · lbf/s in the English engineering system and in watts (W) or kilowatts (kW) or J/s in SI. Note that

$$1 \text{ kW} = 1000 \text{ W} \qquad \text{and} \qquad 1 \text{ W} = 1 \text{ J/s} = 1 \text{ N} \cdot \text{m/s} = 1 \text{ kg} \cdot \text{m}^2/\text{s}^3 \tag{1-20}$$

Pressure is measured in lbf/in² in the English engineering system and in bars or N/m² in SI. Note that

$$1 \text{ bar} = 10^5 \text{ N/m}^2 = 10^5 \text{ kg/(m} \cdot \text{s}^2)$$

and

$$1 \text{ atm} = 0.98066 \text{ bar}$$

In SI, when the size of units becomes too large or too small, multiples in powers of 10 are formed with certain prefixes. The important ones are listed in Table 1-3. For example,

$$1000 \text{ W} = 1 \text{ kW (kilowatt)}$$
$$1{,}000{,}000 \text{ W} = 1 \text{ MW (megawatt)}$$
$$1{,}000{,}000 \text{ N} = 1 \text{ MN (meganewton)}$$
$$1000 \text{ m} = 1 \text{ km (kilometer)}$$
$$10^{-2} \text{ m} = 1 \text{ cm (centimeter)}$$

and so forth.

A comprehensive table of conversion factors useful in heat transfer calculations is presented in App. A.

Example 1-8 Convert the heat transfer coefficient h = 20 Btu/(h · ft² · °F) to the units of J/(s · m² · °C) or W/(m² · °C).

Table 1-3 Prefixes for multiplying factors

10^{-12}	= pico (p)	10	= deka (da)
10^{-9}	= nano (n)	10^2	= hecto (h)
10^{-6}	= micro (μ)	10^3	= kilo (k)
10^{-3}	= milli (m)	10^6	= mega (M)
10^{-2}	= centi (c)	10^9	= giga (G)
10^{-1}	= deci (d)	10^{12}	= tera (T)

SOLUTION From Table A1, #9, App. A we have 1 Btu/(h · ft² · °F) = 5.677 W/(m² · °C), which is written as

$$5.677 \frac{\text{W/(m}^2 \cdot {}^\circ\text{C)}}{\text{Btu/(h} \cdot \text{ft}^2 \cdot {}^\circ\text{F)}}$$

Then the conversion is performed as

$$h = 20 \text{ Btu/(h} \cdot \text{ft}^2 \cdot \text{F)} = [20 \text{ Btu/(h} \cdot \text{ft}^2 \cdot {}^\circ\text{F)}]\left[5.677 \frac{\text{W/(m}^2 \cdot {}^\circ\text{C)}}{\text{Btu/(h} \cdot \text{ft}^2 \cdot {}^\circ\text{F)}}\right]$$

$$\cong 113.6 \text{ W/(m}^2 \cdot {}^\circ\text{C)} = 113.6 \text{ J/(s} \cdot \text{m}^2 \cdot {}^\circ\text{C)}$$

since

$$1 \text{ W} = 1 \text{ J/s}$$

1-6 SUMMARY OF BASIC RELATIONS

We summarize in Table 1-4 the basic relations given in this chapter.

Table 1-4 Summary of basic relations

Equation number	Relation	Remarks
	Conduction:	
(1-1*b*)	$q_x = -k \frac{dT}{dx}$ W/m²	Conduction heat flux in the x direction
	Convection:	
(1-2*a*)	$q = h(T_w - T_f)$ W/m²	Convection heat flux from the wall surface to the fluid
	Radiation:	
(1-3)	$E_b = \sigma T^4$ W/m²	Blackbody emissive power
(1-4)	$q = \varepsilon \sigma T^4$ W/m²	Radiation flux emitted by a real body
(1-6*b*)	$Q_1 = A_1 \varepsilon_1 \sigma (T_1^4 - T_2^4)$ W for $\frac{A_1}{A_2} \to 0$	Net radiation loss from surface A_1 at T_1 to a very large medium A_2 at T_2
(1-8)	$q_1 = h_r(T_1 - T_2)$ W/m² where $h_r = 4\varepsilon\sigma T_1^3$	Energy transfer by radiation for $\lvert T_1 - T_2 \rvert \ll T_1$
(1-13*a*)	$q_w = h_c(T_w - T_\infty) + h_r(T_w - T_s)$ W/m² where $h_r = 4\varepsilon\sigma T_w^3$	Energy transfer by combined convection and radiation for $\lvert T_w - T_s \rvert \ll T_w$

PROBLEMS

Conduction

1-1 A temperature difference of 500°C is applied across a fireclay brick 10 cm thick with thermal conductivity 1.0 W/(m · °C). Determine the heat transfer rate per square meter area.

Answer: 5 kW/m^2 or kJ/(m^2 · s)

1-2 A temperature difference of 100°C is applied across a corkboard 5 cm thick with thermal conductivity 0.04 W/(m · °C). Determine the heat transfer rate across a 3-m^2 area per hour.

1-3 A fiber glass insulating board of thermal conductivity 0.05 W/(m · °C) is to be used to limit the heat losses to 80 W/m^2 for a temperature difference of 160°C across the board. Determine the thickness of the insulating board.

Answer: 0.1 m

1-4 Glass wool of thermal conductivity 0.038 W/(m · °C) is to be used to insulate an ice box. If the maximum heat loss should not exceed 45 W/m^2 for a temperature of 40°C across the walls of the ice box, determine the thickness of the insulation.

Answer: 3.4 cm

1-5 A brick wall 15 cm thick with thermal conductivity 1.2 W/(m · °C) is maintained at 30°C at one face and 230°C at the other face. Determine the heat transfer rate across the 4-m^2 surface area of the wall.

Answer: 6.4 kW

1-6 Two large plates, one at 50°C and the other at 200°C, are 8 cm apart. If the space between them is filled by loosely packed rock wool of thermal conductivity 0.08 W/(m · °C), calculate the heat transfer rate across the plates per 1-m^2 area.

1-7 The heat flow rate across an insulating material of thickness 3 cm with thermal conductivity 0.1 W/(m · °C) is 250 W/m^2. If the hot surface temperature is 175°C, what is the temperature of the cold surface?

Answer: 100°C

1-8 A 25-cm-thick concrete wall has a surface area of 40 m^2. The inner surface of the wall is at 20°C, and the outer surface is at −10°C. Determine the rate of heat loss through the wall if the thermal conductivity is 0.75 W/(m · °C).

Answer: 3.6 kW or kJ/s

1-9 The heat flow rate through a 4-cm-thick wood board for a temperature difference of 25°C between the inner outer surfaces is 75 W/m^2. What is the thermal conductivity of the wood?

Answer: 0.12 W/(m · °C)

1-10 The inside and outside surface temperatures of a window glass are 20 and −12°C, respectively. If the glass is 80 cm by 40 cm, is 1.6 cm thick, and has thermal conductivity 0.78 W/(m · °C), determine the heat loss through the glass over 3 h.

Answer: 5391 kJ

1-11 Two plates, one at a uniform temperature of 300°C and the other at 100°C, are separated by a 2-cm-thick asbestos-cement board of thermal conductivity 0.70 W/(m · °C). Determine the rate of heat transfer across the layer per 1-m^2 surface.

1-12 By conduction 2000 W is transferred through a 0.5-m^2 section of a 4-cm-thick insulating material. Determine the temperature difference across the insulating layer if the thermal conductivity is 0.2 W/(m · °C).

Answer: 800°C

Convection

1-13 Water at a mean temperature of 20°C flows over a flat plate at 80°C. If the heat transfer coefficient is 200 W/(m^2 · C), determine the heat transfer per square meter of the plate over 5 h.

Answer: 216 MJ

1-14 A large surface at 50°C is exposed to air at 20°C. If the heat transfer coefficient between the surface and the air is 15 W/(m^2 · °C), determine the heat transferred from 5 m^2 of the surface over 7 h.

1-15 Air at 150°C flows over a flat plate which is maintained at 50°C. If the heat transfer coefficient for forced convection is 300 W/(m^2 · °C), determine the heat transfer to the plate through 2 m^2 over $1\frac{1}{2}$ h.

Answer: 648 MJ

1-16 A 25-cm-diameter sphere at 120°C is suspended in air at 20°C. If the natural convection heat transfer between the sphere and the air is 15 W/(m^2 · C), determine the rate of heat loss from the sphere.

Answer: 294.5 W

1-17 A fluid at 10°C flows over a 2.5-cm-OD and 2-m-long tube whose surface is maintained at 100°C. If the heat transfer coefficient between the tube and the air is 300 W/(m^2 · C), determine the rate of heat transfer from the tube to the air.

1-18 Pressurized water at 50°C flows inside a 5-cm-diameter, 1-m-long tube with surface temperature maintained at 130°C. If the heat transfer coefficient between the water and the tube is $h = 2000$ W/(m · °C), determine the heat transfer rate from the tube to the water.

Answer: 25.13 kW

1-19 Heat is supplied to a plate from its back surface at a rate of 500 W/m^2 and is removed from its front surface by air flow at 30°C. If the heat transfer coefficient between the air and the plate surface is $h = 20$ W/(m^2 · °C), what is the temperature of the front surface of the plate?

1-20 The inside surface of an insulating layer is at 270°C, and the outside surface is dissipating heat by convection into air at 20°C. The insulation layer is 4 cm thick and has thermal conductivity 1.2 W/(m · °C). What is the minimum value of the heat transfer coefficient at the outside surface if the outside surface temperature should not exceed 70°C?

Answer: 120 W/(m^2 · °C)

1-21 A 10-cm-diameter sphere is heated internally with a 100-W electric heater. The sphere dissipates heat by convection from its outer surface to the ambient air. Calculate the heat transfer coefficient for convection from the sphere if the temperature difference between the sphere surface and the ambient air is 50°C.

1-22 A thin metallic plate is insulated at the back surface and is exposed to the sun at the front surface. The front surface absorbs the solar radiation of 900 W/m^2 and dissipates it mainly by convection to the ambient air at 25°C. If the heat transfer coefficient between the plate and the air is 15 W/(m^2 · °C), what is the temperature of the plate?

Answer: 85°C

Radiation

1-23 A thin metal plate 0.1 m by 0.1 m is placed in a large evacuated container whose walls are kept at 300 K. The bottom surface of the plate is insulated, and the top surface is maintained at 500 K as a result of electric heating. If the emissivity of the surface of the plate is $\varepsilon = 0.8$, what is the rate of heat exchange between the plate and the walls of the container? Take $\sigma = 5.67 \times 10^{-8}$ W/(m^2 · K^4).

Answer: 24.7 W

1-24 Two large parallel plates, one at a uniform temperature 500 K and the other at 1000 K, are separated by a nonparticipating gas. Assuming that the surfaces of the plates are perfect emitters and that the convection is negligible, determine the rate of heat exchange between the surfaces per square meter.

1-25 A sphere 10 cm in diameter is suspended inside a large evacuated chamber whose walls are kept at 300 K. If the surface of the sphere has emissivity $\varepsilon = 0.8$ and is maintained at 500 K, determine the rate of heat loss from the sphere to the walls of the chamber.

Answer: 77.52 W

1-26 Two very large, perfectly black parallel plates, one maintained at 1200 K and the other at 600 K, exchange heat by radiation (i.e., convection is negligible). Determine the heat transfer rate per 1-m^2 surface.

1-27 One surface of a thin plate is exposed to a uniform heat flux of 500 W/m^2, and the other side dissipates heat by radiation to an environment at $-10°C$. Determine the temperature of the plate. Assume blackbody conditions for radiation.

Answer: $T_1 = 341.5$ K

1-28 A thin metal sheet separates two large parallel plates, one at a uniform temperature of 1000 K and the other at 400 K. Blackbody conditions can be assumed for all surfaces, and heat transfer can be assumed to be by radiation only. Determine the temperature of the separating sheet.

Combined heat transfer mechanism

1-29 A flat plate has one surface insulated and the other surface exposed to the sun. The exposed surface absorbs the solar radiation at a rate of 800 W/m^2 and dissipates it by both convection and radiation into the ambient air at 300 K. If the emissivity of the surface is $\varepsilon = 0.9$ and the convection heat transfer coefficient between the plate and air is 12 $W/(m^2 \cdot °C)$, determine the temperature of the plate.

Answer: 342.5 K

1-30 A thin plate is exposed to an infrared radiation flux of 1500 W/m^2 on one surface while the other surface is kept insulated. The exposed surface absorbs 90 percent of the incident radiation flux and dissipates it by convection and radiation into the ambient air at 300 K. If the heat transfer coefficient for convection between the surface and the ambient air is 15 $W/(m^2 \cdot °C)$, determine the temperature of the plate. Take the emissivity of plate as $\varepsilon = 0.9$.

1-31 A thin plate 50 cm by 50 cm is subjected to 400 W of heating on one surface and dissipates the heat by combined convection and radiation from the other surface into the ambient air at 290 K. If the surface of the plate has an emissivity $\varepsilon = 0.9$ and the heat transfer coefficient between the surface and the ambient air is 15 $W/(m^2 \cdot °C)$, determine the temperature of the plate.

Answer: 362.2 K

1-32 The solar radiation incident on the outside surface of an aluminum shading device is 1000 W/m^2. Aluminum absorbs 12 percent of the incident solar energy and dissipates it by convection from the back surface and by combined convection and radiation from the outside surface. The emissivity of the aluminum is 0.10, the convection heat transfer coefficient is 15 $W/(m^2 \cdot °C)$ for both surfaces, and the ambient temperature can be taken 20°C for both convection and radiation. Determine the temperature of the shade.

1-33 One surface of a thin metal sheet receives radiation from a large plate at 700°C, while the other surface dissipates heat by convection to a coolant fluid at 20°C. The surfaces can be considered as a perfect absorber and a perfect emitter for radiation. The heat transfer coefficient for convection between the surface and the fluid is 120 $W/(m^2 \cdot °C)$. Determine the temperature of the plate.

Answer: 638 K

1-34 Heat is lost by both convection and radiation from a 2-m-long uninsulated portion of 5-cm-OD hot water pipe into an external environment at 0°C. The convection heat transfer coefficient is 20 $W/(m^2 \cdot °C)$, and for radiation calculations blackbody conditions can be assumed. Determine the rate of heat loss from the uninsulated portion of the pipe for a wall temperature of $T_w = 125°C$.

Conversion factors

1-35 Convert the heat transfer coefficient $h = 50$ $W/(m^2 \cdot °C)$ to $Btu/(h \cdot ft^2 \cdot °F)$.

Answer: 8.807 $Btu/(h \cdot ft^2 \cdot °F)$

1-36 Derive the conversion factor for converting the thermal conductivity from $Btu/[(h \cdot ft^2)(°F/in)]$ to $W/(m \cdot °C)$.

1-37 Derive the following conversion factors:

$$\frac{Btu/(h \cdot ft^2 \cdot °F)}{4.882\ kcal/(m^2 \cdot h \cdot °C)}, \qquad \frac{Btu/(h \cdot ft \cdot °F)}{0.0173\ W/(cm \cdot °C)}$$

1-38 Convert the heat transfer coefficient $h = 100$ Btu/(h · ft^2 · °F) to W/(m^2 · °C).
Answer: 567.7 W/(m^2 · °C)

1-39 Convert the generation rate $g = 50$ Btu/(h · ft^3) to W/m^3.
Answer: 517.5 W/m^3

1-40 Convert the following from English units to SI units:

$C_p = 0.25$ Btu/(lb · °F)	to	W · s/(kg · °C)
$C_p = 0.25$ Btu/(lb · °F)	to	kJ/(kg · °C)
$k = 0.0263$ Btu/(h · ft · °F)	to	W/(m · °C)
$\mu = 0.072$ lb/(ft · h)	to	kg/(m · s)
$\rho = 62.54$ lb/ft^3	to	kg/m^3
$\nu = 1.92$ ft^2/h	to	m^2/s
$\alpha = 2.80$ ft^2/h	to	m^2/s

1-41 Perform the following conversions:

500 Btu	to	kJ
1000 Btu	to	kcal
50 Btu/(h · °F)	to	W/°C
25 Btu/(lb · °F)	to	J/(kg · °C)
5 Btu/(lb · °F)	to	kcal/(kg · °C)

REFERENCES

1. Fourier, J. B.: *Theorie analytique de la chaleur*, Paris, 1822. (English translation by A. Freeman, Dover Publications Inc., New York, 1955.)
2. Powell, R. W., C. Y. Ho, and P. E. Liley: *Thermal Conductivity of Selected Materials.* NSRDS-NBS 8, U.S. Department of Commerce, National Bureau of Standards, 1966.
3. Touloukian, Y. S., R. W. Powell, C. Y. Ho, and P. G. Klemens: *Thermophysical Properties of Matter*, vol. 1, *Thermal Conductivity—Metallic Elements and Alloys*; vol. 2, *Thermal Conductivity—Nonmetallic Solids*, IFI/Plenum Data Corporation, New York, 1970.
4. Touloukian, Y. S., P. E. Liley, and S. C. Saxena: *Thermophysical Properties of Matter*, vol. 3, *Thermal Conductivity—Nonmetallic Liquids and Gases*, IFI/Plenum Data Corporation, New York, 1970.
5. Ho, C. Y., R. W. Powell, and P. E. Liley: *Thermal Conductivity of Elements*, vol. 1, First supplement to *Journal of Physical and Chemical Reference Data* (1972), American Chemical Society, Washington.

SHORT BIBLIOGRAPHY OF TEXTBOOKS IN HEAT TRANSFER

A vast amount of information has accumulated in the heat transfer area, and the literature continues to grow. The following short bibliography of books, selected from the wealth of material available in the literature, may serve as a guide to newcomers to the heat transfer area.

General, basic heat transfer

Bayley, F. J., M. J. Owen, and A. B. Turner: *Heat Transfer*, Barnes & Noble, New York, 1972.
Chapman, Alan J.: *Heat Transfer*, Macmillan, New York, 1967.
Gebhart, B.: *Heat Transfer*, McGraw-Hill, New York, 1971.
Grassmann, Peter: *Physikalische Grundlagen der Verfahrenstechnik*, Saverlander, Aarau, 1982.
Gröber, H., S. Erk, and U. Grigull: *Fundamentals of Heat Transfer*, McGraw-Hill, New York, 1961.
Holman, J. P.: *Heat Transfer*, McGraw-Hill, New York, 1981.
Incropera, Frank P., and David P. Dewitt: *Fundamentals of Heat Transfer*, Wiley, New York, 1981.
Kreith, F.: *Principles of Heat Transfer*, Intext, New York, 1973.
——— and W. Z. Black: *Basic Heat Transfer*, Harper & Row, New York, 1979.
Lienhard, John H.: *A Heat Transfer Textbook*, Prentice-Hall, Englewood Cliffs, N.J., 1981.
Özişik, M. N.: *Basic Heat Transfer*, McGraw-Hill, New York, 1977.
Thomas, Lindon D.: *Fundamentals of Heat Transfer*, Prentice-Hall, Englewood Cliffs, N.J., 1980.
Wolf, Helmut: *Heat Transfer*, Harper & Row, London, 1983.

Conduction

Arpaci, V. S.: *Conduction Heat Transfer*, Addison-Wesley, Reading, Mass., 1966.
Carslaw, H. S., and J. C. Jaeger: *Conduction of Heat in Solids*, Oxford University Press, London, 1959.
Kakaç, S., and Y. Yener: *Heat Conduction*, Middle East Technical University, Ankara, Turkey, 1979.
Mikhailov, M. D., and M. N. Özişik: *Unified Analysis and Solutions of Heat and Mass Diffusion*, Wiley, New York, 1984.
Myers, Glen E.: *Analytical Methods in Conduction Heat Transfer*, McGraw-Hill, New York, 1971.
Özişik, M. N.: *Boundary Value Problems of Heat Conduction*, International Textbook Company, Scranton, Pa., 1968.
———: *Heat Conduction*, Wiley, New York, 1980.
Schneider, P. J.: *Conduction Heat Transfer*, Addison-Wesley, Reading, Mass., 1955.

Convection

Arpaci, V. S., and P. S. Larsen: *Convection Heat Transfer*, Prentice-Hall, Englewood Cliffs, N.J., 1984.
Cebeci, T., and P. Bradshaw: *Physical and Computational Aspects of Convective Heat Transfer*, Springer, New York, 1984.
Eckert, E. R. G., and R. M. Drake: *Analysis of Heat and Mass Transfer*, McGraw-Hill, New York, 1972.
Kakaç, S., and Y. Yener: *Convective Heat Transfer*, Middle East Technical University, Ankara, Turkey, 1980.
Kays, W. M., and M. E. Crowford: *Convective Heat and Mass Transfer*, McGraw-Hill, New York, 1980.
Rohsenow, W. M., and H. Choi: *Heat, Mass and Momentum Transfer*, Prentice-Hall, Englewood Cliffs, N.J., 1961.

Radiation

Edwards, D. K.: *Radiation Heat Transfer Notes*, Hemisphere, New York, 1981.
Hottel, H. C., and A. F. Sarofim: *Radiative Heat Transfer*, McGraw-Hill, New York, 1967.
Love, T. J.: *Radiative Heat Transfer*, Merrill, Columbus, Ohio, 1968.
Özişik, M. N.: *Radiative Transfer*, Wiley, New York, 1973.
Siegel, R., and J. R. Howell: *Thermal Radiative Heat Transfer*, McGraw-Hill, New York, 1972.
Sparrow, E. M., and R. D. Cess: *Radiation Heat Transfer*, Hemisphere, New York, 1978.

Heat exchangers

Fraas, A. P., and M. N. Özişik: *Heat Exchanger Design*, Wiley, New York, 1965.
Kays, W. M., and A. L. London: *Compact Heat Exchangers*, McGraw-Hill, New York, 1958.
Walker, G.: *Industrial Heat Exchangers—A Basic Guide*, Hemisphere, New York, 1982.

CHAPTER

TWO

CONDUCTION—BASIC EQUATIONS

This chapter is devoted to the derivation of the basic equations and the appropriate boundary conditions that govern the temperature distribution in solids. The one-dimensional heat conduction equation is presented in the rectangular, cylindrical, and spherical coordinate systems. The objective of this chapter is to provide a good understanding of the heat conduction equation and the boundary conditions for use in the mathematical formulation of heat conduction problems.

2-1 ONE-DIMENSIONAL HEAT CONDUCTION EQUATION

The temperature distribution in solids can be determined from the solution of the *heat conduction equation* subject to a set of appropriate boundary and initial conditions. For the thermal analysis of bodies having shapes such as a slab, rectangle, or parallelepiped, the heat conduction equation given in the rectangular coordinate system is needed. However, to analyze heat conduction for bodies having shapes of a cylinder and a sphere, the heat conduction equation should be given in the cylindrical and the spherical coordinate systems, respectively. In the analytical solution of heat conduction problems, we use different coordinate systems to ensure that the boundary surfaces of the region coincide with the coordinate surfaces. For example, in the cylindrical coordinate system, one of the coordinate surfaces is a cylinder; hence this coordinate surface coincides with the cylindrical surface of a body in the form of a cylinder, and so forth.

In this section we derive the one-dimensional, time-dependent heat conduction equation in the rectangular, cylindrical, and spherical coordinate systems. Such a derivation helps one understand the physical significance of various terms in the heat conduction equation.

We consider a solid whose temperature $T(x, t)$ depends on time and varies in only one direction, say, along the x coordinate. Here we assume that the x axis in the rectangular coordinate system refers to the usual x axis; but if the cylindrical or the spherical coordinate system is considered, it refers to the radial coordinate r. This matter will be clear later in the analysis.

As discussed in Chap. 1, when the temperature varies in a given direction, say, along the x axis, then there is a heat flow along the x axis given by the Fourier law in the form

$$q = -k \frac{\partial T(x, t)}{\partial x} \qquad \text{W/m}^2 \tag{2-1}$$

For generality in the analysis, we assume that there is also an external source within the medium generating energy at a specified rate of $g \equiv g(x, t)$ W/m^3. In practice, such an energy source may be due to nuclear fission, as in the case of fuel elements for nuclear reactors; or to some chemical reaction taking place within the solid; or to the disintegration of radioactive elements present in the solid, as in nuclear waste; or to the attenuation of gamma rays penetrating the body; or to the passage of electric current through the solid; and so forth.

To derive the one-dimensional heat conduction equation, we consider a volume element of thickness Δx and having an area A normal to the coordinate axis x, as illustrated in Fig. 2-1. The energy balance equation for this volume element is stated as

$$\underset{\text{I}}{\begin{pmatrix}\text{Net rate of}\\ \text{heat gain by}\\ \text{conduction}\end{pmatrix}} + \underset{\text{II}}{\begin{pmatrix}\text{rate of}\\ \text{energy}\\ \text{generation}\end{pmatrix}} = \underset{\text{III}}{\begin{pmatrix}\text{rate of}\\ \text{increase of}\\ \text{internal energy}\end{pmatrix}} \tag{2-2}$$

Each of the terms I, II, and III in this equation is evaluated as described below.

Let q be the heat flux at the location x in the positive x direction at the surface A of the element. Then the rate of heat flow into the element through the surface A by conduction at the location x is written as

$$[Aq]_x$$

Similarly, the rate of heat flowing out of the element by conduction at the location $x + \Delta x$ is written as

$$[Aq]_{x+\Delta x}$$

Then the net rate of heat gain by the element by conduction is the difference between these two quantities:

$$\text{I} \equiv [Aq]_x - [Aq]_{x+\Delta x} \tag{2-3a}$$

The rate of energy generation in the element having a volume $A\,\Delta x$ is given by

$$\text{II} \equiv A\,\Delta x\, g \tag{2-3b}$$

since $g \equiv g(x, t)$ is the energy generation per unit volume.

The rate of increase of internal energy of the volume element resulting from the change of temperature with time is written as

$$\text{III} \equiv A\,\Delta x\, \rho c_p \frac{\partial T(x, t)}{\partial t} \tag{2-3c}$$

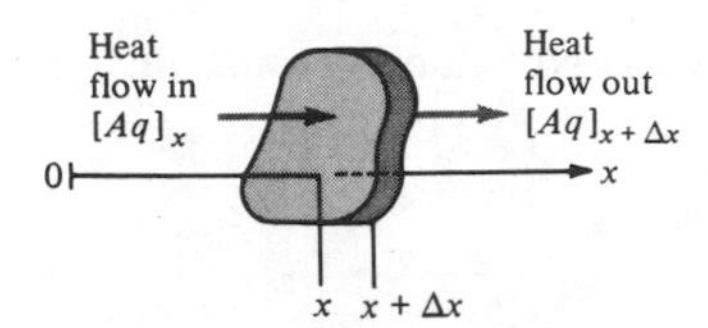

Figure 2-1 Nomenclature for the derivation of one-dimensional heat conduction equation.

since for solids and liquids $c_p = c_v$. Various quantities appearing in Eqs. (2-3*a*) to (2-3*c*) are defined as

c_p = specific heat of material, J/(kg · °C)

g = energy generation rate per unit volume, W/m^3

q = conduction heat flux in the x direction, W/m^2

t = time, s

ρ = density of material, kg/m^3

Equations (2-3*a*) to (2-3*c*) are introduced into Eq. (2-2), and the result is rearranged in the form

$$-\frac{1}{A}\frac{[Aq]_{x+\Delta x} - [Aq]_x}{\Delta x} + g = \rho c_p \frac{\partial T(x, t)}{\partial t} \tag{2-4a}$$

As $\Delta x \to 0$, the first term on the left-hand side, by definition, becomes the derivative of $[Aq]$ with respect to x, and Eq. (2-4*a*) is written as

$$-\frac{1}{A}\frac{\partial}{\partial x}(Aq) + g = \rho c_p \frac{\partial T(x, t)}{\partial t} \tag{2-4b}$$

The heat flux q, given by Eq. (2-1), is now introduced into Eq. (2-4*b*). We obtain

$$\frac{1}{A}\frac{\partial}{\partial x}\left(Ak\frac{\partial T}{\partial x}\right) + g = \rho c_p \frac{\partial T(x, t)}{\partial t} \tag{2-5}$$

So far our analysis has been general, so we do not need to specify a particular coordinate system; but from now on we need to know the dependence of the area A on the coordinate axis x in order to complete the derivation of the heat conduction equation. Therefore, we consider this matter for the rectangular, cylindrical, and spherical coordinate systems.

Rectangular Coordinates

The area A does not vary with x, hence it is considered constant and cancels. Then Eq. (2-5) reduces to

$$\boxed{\frac{\partial}{\partial x}\left(k\frac{\partial T}{\partial x}\right) + g = \rho c_p \frac{\partial T(x, t)}{\partial t}} \tag{2-6a}$$

which is the one-dimensional, time-dependent heat conduction equation in the rectangular coordinate system.

Cylindrical Coordinates

In cylindrical coordinates, it is customary to denote the radial variable by r instead of x. Therefore in Eq. (2-5) we replace x by r and note that the area A is proportional to r. Then Eq. (2-5) takes the form

$$\boxed{\frac{1}{r}\frac{\partial}{\partial r}\left(rk\frac{\partial T}{\partial r}\right) + g = \rho c_p \frac{\partial T(r, t)}{\partial t}} \tag{2-6b}$$

which is the one-dimensional, time-dependent heat conduction equation in the cylindrical coordinate system.

Spherical Coordinates

In the spherical coordinate system, it is also customary to denote the radial variable by r instead of x. Therefore in Eq. (2-5) we replace x by r and note that the area A is proportional to r^2. Then Eq. (2-5) takes the form

$$\boxed{\frac{1}{r^2}\frac{\partial}{\partial r}\left(r^2 k\frac{\partial T}{\partial r}\right) + g = \rho c_p \frac{\partial T(r, t)}{\partial t}} \tag{2-6c}$$

which is the one-dimensional, time-dependent heat conduction equation in the spherical coordinate system.

A Compact Equation

The one-dimensional, time-dependent heat conduction equation in the rectangular, cylindrical, and spherical coordinate systems given above by Eqs. (2-6a) to (2-6c) can be written more compactly in the form of a single equation as

$$\boxed{\frac{1}{r^n}\frac{\partial}{\partial r}\left(r^n k\frac{\partial T}{\partial r}\right) + g = \rho c_p \frac{\partial T}{\partial t}} \tag{2-7}$$

where

$$n = \begin{cases} 0 & \text{for rectangular coordinates} \\ 1 & \text{for cylindrical coordinates} \\ 2 & \text{for spherical coordinates} \end{cases}$$

And in the rectangular coordinate system, it is customary to replace the r variable by the x variable.

Special Cases

Several special cases of Eq. (2-7) are of practical interest.

For constant thermal conductivity k, Eq. (2-7) simplifies to

$$\frac{1}{r^n}\frac{\partial}{\partial r}\left(r^n \frac{\partial T}{\partial r}\right) + \frac{1}{k} g = \frac{1}{\alpha}\frac{\partial T}{\partial t} \tag{2-8a}$$

where

$$\alpha \equiv \frac{k}{\rho c_p} = \text{thermal diffusivity of material, m}^2/\text{s} \tag{2-8b}$$

For steady-state heat conduction with energy sources within the medium, Eq. (2-7) becomes

$$\frac{1}{r^n}\frac{d}{dr}\left(r^n k \frac{dT}{dr}\right) + g = 0 \tag{2-9a}$$

and for the case of constant thermal conductivity, this result reduces to

$$\frac{1}{r^n}\frac{d}{dr}\left(r^n \frac{dT}{dr}\right) + \frac{1}{k} g = 0 \tag{2-9b}$$

For steady-state heat conduction with no energy sources within the medium, Eq. (2-7) simplifies to

$$\frac{d}{dr}\left(r^n k \frac{dT}{dr}\right) = 0 \tag{2-10a}$$

and for constant k, this result reduces to

$$\frac{d}{dr}\left(r^n \frac{dT}{dr}\right) = 0 \tag{2-10b}$$

In the preceding relations n is defined as previously:

$$n = \begin{cases} 0 & \text{for rectangular coordinates} \\ 1 & \text{for cylindrical coordinates} \\ 2 & \text{for spherical coordinates} \end{cases}$$

and for rectangular coordinates it is customary to replace r by x.

Example 2-1 Write the heat conduction equation for one-dimensional, steady-state heat flow in a solid having a constant k and a constant rate of energy generation g_0 W/m^3 within the medium for (*a*) a slab, (*b*) a cylinder, and (*c*) a sphere.

SOLUTION The results are immediately obtainable from Eq. (2-9*b*):

(*a*) By setting $n = 0$ and $r \equiv x$,

$$\frac{d^2T}{dx^2} + \frac{1}{k} g_0 = 0$$

(*b*) By setting $n = 1$,

$$\frac{1}{r}\frac{d}{dr}\left(r\frac{dT}{dr}\right) + \frac{1}{k} g_0 = 0$$

(*c*) By setting $n = 2$,

$$\frac{1}{r^2}\frac{d}{dr}\left(r^2\frac{dT}{dr}\right) + \frac{1}{k} g_0 = 0$$

Table 2-1 Thermal diffusivity of typical materials

	Average temperature, °C	Diffusivity $\alpha \times 10^6$ m^2/s
Metals		
Aluminum	0	85.9
Copper	0	114.1
Gold	20	120.8
Iron, pure	0	18.1
Cast iron ($c \cong 4\%$)	20	17.0
Lead	21	25.5
Mercury	0	4.44
Nickel	0	15.5
Silver	0	170.4
Steel, mild	0	12.4
Tungsten	0	61.7
Zinc	0	41.3
Nonmetals		
Asbestos	0	0.258
Brick, fireclay	204	0.516
Cork, ground	38	0.155
Glass, Pyrex		0.594
Granite	0	1.291
Ice	0	1.187
Oak, across grain	29	0.160
Pine, across grain	29	0.152
Quartz sand, dry		0.206
Rubber, soft		0.077
Water	0	0.129

Table 2-2 Effect of thermal diffusivity on the rate of heat propagation

Material		Silver	Copper	Steel	Glass
α	$10^6 \times m^2/s$	170	103	12.9	0.59
Time		9.5 min	16.5 min	2.2 h	2.00 days

Thermal Diffusivity

The time-dependent heat conduction equation for constant k contains a quantity α, called the thermal diffusivity, as defined by Eq. (2-8*b*). It is instructive to discuss the physical significance of α. Table 2-1 lists thermal diffusivity of typical materials. There are orders of magnitude of difference in the values of thermal diffusivity of different materials. For example, α varies from about 114×10^{-6} m²/s for copper to 0.15×10^{-6} m²/s for ground cork.

The physical significance of thermal diffusivity is associated with the propagation of heat into the medium during changes of temperature with time. The higher the thermal diffusivity, the faster the propagation of heat into the medium. Consider, for example, a semi-infinite medium extending from $x = 0$ to $x \to \infty$ and initially at a uniform temperature $T_0 = 100°C$. Suddenly the temperature of the surface at $x = 0$ is lowered to 0°C and maintained at that temperature. The temperature in the interior of the solid will vary continuously with position and time. Table 2-2 lists the time required for the temperature to be lowered to $\frac{1}{2}T_0 = 50°C$ at a location 30 cm from the boundary surface for materials having different thermal diffusivities. It is apparent from this table that the larger the thermal diffusivity, the less time is required for heat to penetrate into the solid.

2-2 THREE-DIMENSIONAL HEAT CONDUCTION EQUATION

In Sec. 2-1 we derived the one-dimensional, time-dependent heat conduction equation. By following a similar approach but allowing for heat conduction in the three directions, the general equation can be derived also. Such derivations are explained in various texts on heat conduction [1–5]. Here we directly present the resulting equations in the rectangular, cylindrical, and spherical coordinate systems for the case of *constant thermal conductivity*.

For the rectangular coordinate system (x, y, z), the heat conduction equation becomes

$$\boxed{\frac{\partial^2 T}{\partial x^2} + \frac{\partial^2 T}{\partial y^2} + \frac{\partial^2 T}{\partial z^2} + \frac{1}{k} g = \frac{1}{\alpha} \frac{\partial T}{\partial t}} \tag{2-11a}$$

where $T \equiv T(x, y, z, t)$.

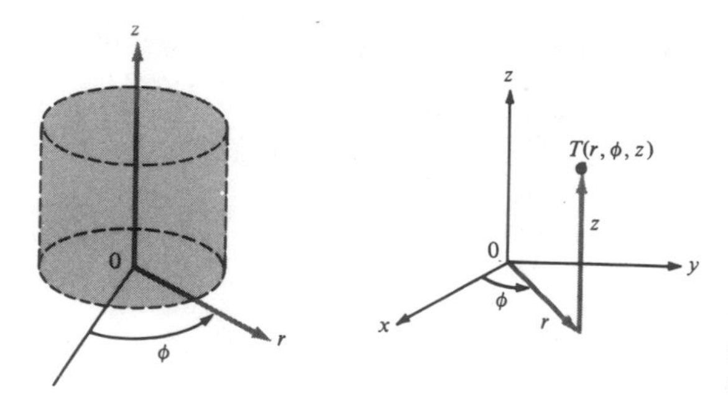

Figure 2-2 Cylindrical coordinate system (r, ϕ, z).

For the *cylindrical coordinate system* (r, ϕ, z), illustrated in Fig. 2-2, it is given by

$$\boxed{\frac{1}{r}\frac{\partial}{\partial r}\left(r\frac{\partial T}{\partial r}\right)+\frac{1}{r^2}\frac{\partial^2 T}{\partial \phi^2}+\frac{\partial^2 T}{\partial z^2}+\frac{g}{k}=\frac{1}{\alpha}\frac{\partial T}{\partial t}} \tag{2-11b}$$

where $T \equiv T(r, \phi, z, t)$.

Finally, for the *spherical coordinate system* (r, ϕ, θ), illustrated in Fig. 2-3, it is given by

$$\boxed{\frac{1}{r^2}\frac{\partial}{\partial r}\left(r^2\frac{\partial T}{\partial r}\right)+\frac{1}{r^2\sin\theta}\frac{\partial}{\partial \theta}\left(\sin\theta\frac{\partial T}{\partial \theta}\right)+\frac{1}{r^2\sin^2\theta}\frac{\partial^2 T}{\partial \phi^2}+\frac{g}{k}=\frac{1}{\alpha}\frac{\partial T}{\partial t}} \tag{2-11c}$$

where $T \equiv T(r, \phi, \theta, t)$.

Clearly, for the special case of temperature depending on time and only the space variable r (or x in the rectangular coordinates), Eqs. (2-11a) to (2-11c) reduce to Eq. (2-8).

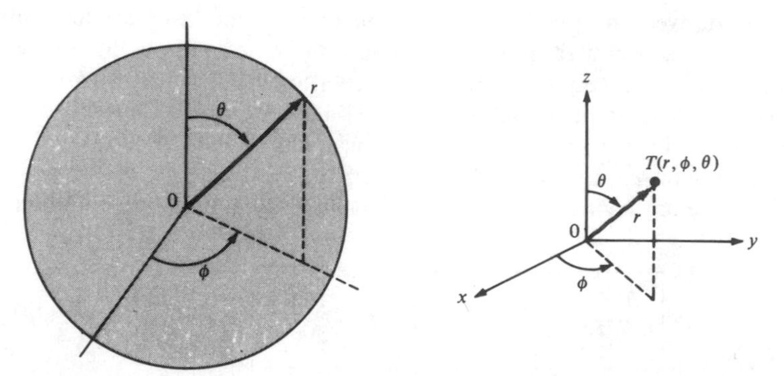

Figre 2-3 Spherical coordinate system (r, ϕ, θ).

Example 2-2 By simplifying the three-dimensional equation (2-11*c*), obtain the heat conduction equation for a sphere in which there is energy generation and the temperature varies with time and the radial coordinate r only.

SOLUTION Since the temperature depends on r and t only, we have $T \equiv T(r, t)$. Then the derivatives with respect to θ and ϕ vanish, and Eq. (2-11*c*) reduces to

$$\frac{1}{r^2}\frac{\partial}{\partial r}\left(r^2 \frac{\partial T}{\partial r}\right) + \frac{1}{k}g = \frac{1}{\alpha}\frac{\partial T}{\partial t}$$

which is the same as Eq. (2-8*a*) for $n = 2$.

2-3 BOUNDARY CONDITIONS

Appropriate boundary and initial conditions are needed for the analysis of heat conduction problems. The initial condition specifies the distribution of temperature at the origin of the time coordinate, $t = 0$. The boundary conditions specify the thermal condition at the boundary surfaces of the region. For example, at a given boundary surface, the distribution of temperature may be prescribed, or the distribution of the heat flux may be specified, or there may be heat transfer by convection into the ambient fluid at a specified temperature with a known heat transfer coefficient.

Therefore, in the analysis of heat conduction problems, such physical boundary conditions should be represented with appropriate mathematical expressions. We now discuss the mathematical representation of three commonly used, different types of boundary conditions, namely, the *prescribed temperature, prescribed heat flux, and convection boundary conditions.*

Prescribed Temperature Boundary Condition (B.C. First Kind)

There are numerous applications in which the temperature of the boundary surface is considered known. For example, a boundary surface in contact with melting ice is said to be maintained at a uniform temperature 0°C, or the distribution of temperature at the boundary surface may be known as a function of time.

Consider a plate of thickness L as illustrated in Fig. 2-4. Suppose the boundary surface at $x = 0$ is maintained at a uniform temperature T_1 and that at $x = L$ at a uniform temperature T_2. The plate is said to be subjected to prescribed temperature boundary conditions at both surfaces, and these boundary conditions are written as

$$T(x, t)|_{x=0} \equiv T(0, t) = T_1 \qquad \text{(2-12}a\text{)}$$

$$T(x, t)|_{x=L} \equiv T(L, t) = T_2 \qquad \text{(2-12}b\text{)}$$

In more general cases, the distribution of temperature at the boundary surface may be specified as a function of position and time. When the value of temperature

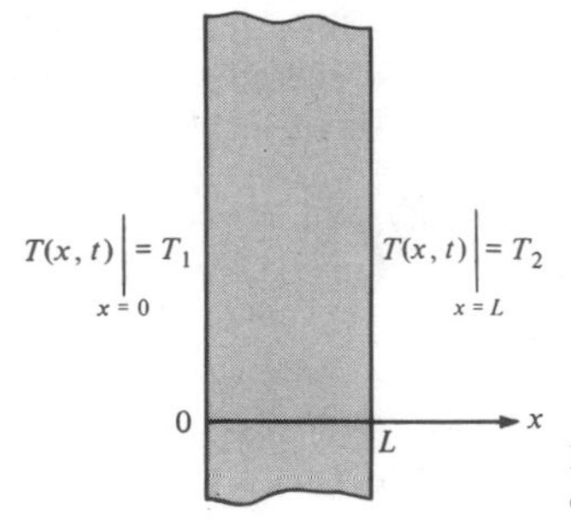

Figure 2-4 Prescribed temperature at the boundary (boundary condition of the *first kind*).

is prescribed at the boundary surface, the boundary condition is said to be of the *first kind*.

Similar considerations are applicable for boundary conditions at the surfaces of a cylinder and sphere.

Prescribed Heat Flux Boundary Condition (B.C. Second Kind)

In some situations, the rate of heat supply to a boundary surface is considered known. For example, on an electrically heated surface the rate of heat flow entering the solid is known. Such boundary conditions are called *prescribed* heat flux boundary conditions.

Consider a plate of thickness L as illustrated in Fig. 2-5*a*. Suppose there is a heat supply into the medium at a rate of q_0 W/m^2 through the boundary surface at $x = 0$ and another heat supply into the medium at a rate of q_L W/m^2 through the boundary surface at $x = L$. The mathematical representation of such boundary conditions is now described.

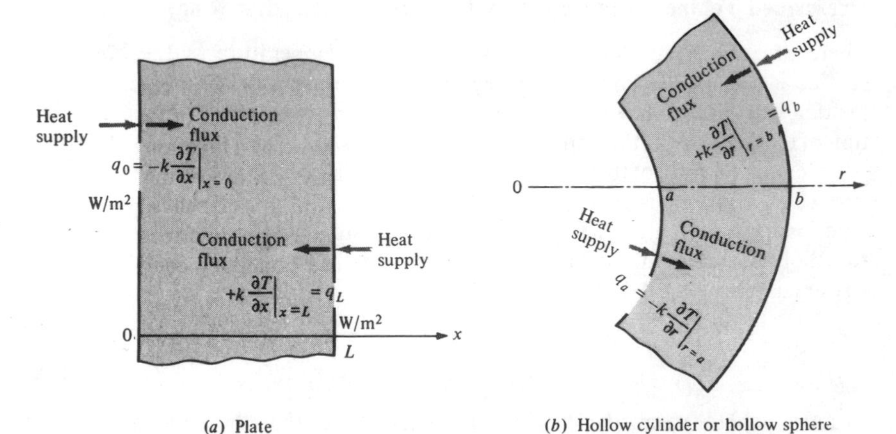

Figure 2-5 Prescribed heat flux at the boundaries (boundary condition of the *second kind*).

At the boundary surface $x = 0$, the external heat supply q_0 W/m^2 is equated to the conduction heat flux into the solid:

$$-k \left. \frac{\partial T}{\partial x} \right|_{x=0} = q_0 \tag{2-13a}$$

Similarly, at the boundary surface $x = L$, the external heat supply q_L W/m^2 is equated to the conduction heat flux into the solid:

$$+k \left. \frac{\partial T}{\partial x} \right|_{x=L} = q_L \tag{2-13b}$$

Equations (2-13*a*) and (2-13*b*) are the mathematical representation of prescribed heat flux boundary conditions. In these equations a positive value for q_0 and q_L implies heat supply into the medium. Conversely, a negative value for q_0 and q_L implies heat removal from the medium.

When the heat flux is prescribed at a boundary surface, the boundary condition is said to be of the *second kind*.

The results given by Eqs. (2-13*a*) and (2-13*b*) are developed with reference to a slab geometry. Similar results are applicable at the boundary surfaces of a cylinder or sphere as illustrated in Fig. 2-5*b* for a hollow cylinder or sphere. For such cases the coordinate x is replaced by the radial variable r.

Convection Boundary Condition (B.C. Third Kind)

In most practical applications, heat transfer at the boundary surface is by convection with a known heat transfer coefficient h into an ambient fluid at a prescribed temperature. We consider again a plate of thickness L as illustrated in Fig. 2-6*a*. Suppose a fluid at a temperature T_1 with a heat transfer coefficient h_1 flows over the surface of the plate at $x = 0$. The mathematical formulation of this convection boundary condition is obtained by considering an energy balance at the surface $x = 0$ stated as

$$\begin{pmatrix}\text{Convection heat flux}\\ \text{from the fluid at } T_1\\ \text{to the surface at } x = 0\end{pmatrix} = \begin{pmatrix}\text{conduction heat}\\ \text{flux from the surface}\\ \text{at } x = 0 \text{ into the plate}\end{pmatrix}$$

or

$$h_1[T_1 - T(x, t)|_{x=0}] = -k \left. \frac{\partial T(x, t)}{\partial x} \right|_{x=0} \tag{2-14a}$$

which is the convection boundary condition at the surface $x = 0$.

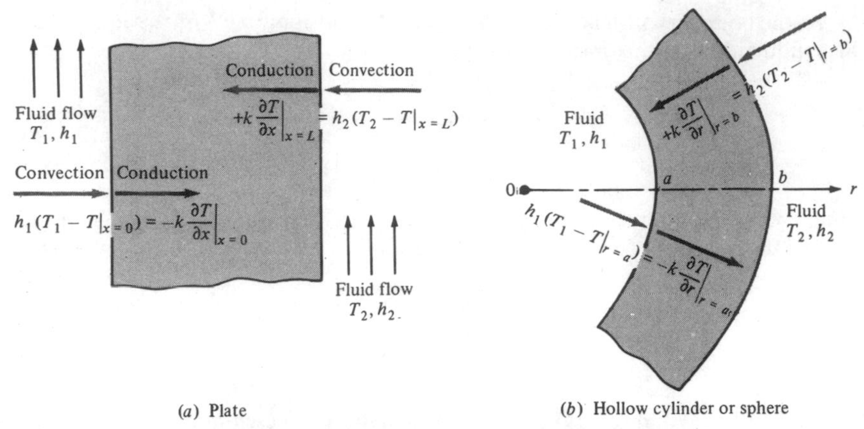

(*a*) Plate (*b*) Hollow cylinder or sphere

Figure 2-6 Convection at the boundaries (boundary condition of the *third kind*).

If a fluid at a temperature T_2 with a heat transfer coefficient h_2 flows over the surface at $x = L$, the energy balance for this surface is stated as

$$\begin{pmatrix}\text{Convection heat flux} \\ \text{from the fluid at } T_2 \\ \text{to the surface at } x = L\end{pmatrix} = \begin{pmatrix}\text{conduction heat flux} \\ \text{from the surface at} \\ x = L \text{ into the plate}\end{pmatrix}$$

or

$$h_2[T_2 - T(x, t)|_{x=L}] = +k \frac{\partial T(x, t)}{\partial x}\Big|_{x=L} \tag{2-14b}$$

which is the convection boundary condition at the surface $x = L$.

Similar expressions are applicable for convection boundary conditions at the surfaces of a hollow cylinder or sphere, as illustrated in Fig. 2-6*b*. Such boundary conditions are called of the *third kind*.

Summary of Boundary Conditions

Of the three different types of boundary conditions considered, the convection boundary condition is the most general one, and the remaining two are obtainable from this boundary condition as special cases.

Consider the convection boundary conditions for a plate given by Eqs. (2-14*a*) and (2-14*b*) rewritten in the form

$$\left[-k \frac{\partial T}{\partial x} + h_1 T\right]_{x=0} = h_1 T_1 \qquad \text{at } x = 0 \tag{2-15a}$$

$$\left[+k \frac{\partial T}{\partial x} + h_2 T\right]_{x=L} = h_2 T_2 \qquad \text{at } x = L \tag{2-15b}$$

Clearly, the prescribed temperature boundary conditions (2-12*a*) and (2-12*b*) are obtainable from these results by dividing both sides of Eqs. (2-15*a*) by h_1 and of Eq. (2-15*b*) by h_2 and then letting $h_1 \to \infty$ and $h_2 \to \infty$.

Similarly, the prescribed heat flux boundary conditions Eqs. (2-13*a*) and (2-13*b*) are also obtainable from Eqs. (2-15*a*) and (2-15*b*) by setting

$$h_1 T_1 \equiv q_0 \qquad \text{and} \qquad h_2 T_2 \equiv q_L$$

on the right-hand sides of Eqs. (2-15*a*) and (2-15*b*) and then letting $h_1 = h_2 = 0$ on the left-hand sides.

Finally, the convection boundary conditions for a hollow cylinder or sphere having an inside radius a and an outside radius b are written, according to Fig. 2-6*b*, in the form

$$\left[-k\frac{\partial T}{\partial r} + h_1 T\right]_{r=a} = h_1 T_1 \qquad \text{at } r = a \tag{2-16a}$$

$$\left[+k\frac{\partial T}{\partial r} + h_2 T\right]_{r=b} = h_2 T_2 \qquad \text{at } r = b \tag{2-16b}$$

Example 2-3 Consider one-dimensional, steady-state heat conduction in a plate with constant thermal conductivity in a region $0 \le x \le L$. Energy is generated in the medium at a rate of $g_0 e^{-\beta x}$ W/m^3, while the boundary surfaces at $x = 0$ are kept insulated and at $x = L$ dissipate heat by convection into a medium at temperature T_∞ with a heat transfer coefficient h W/(m$^2 \cdot$°C). Write the mathematical formulation of this heat conduction problem.

Solution The heat conduction equation is immediately available from Eq. (2-6*a*) by setting in that equation $g = g_0 e^{-\beta x}$ and considering k constant. The boundary condition at $x = 0$ is obtainable from Eq. (2-13*a*) by setting $q_0 = 0$ and that at $x = L$ from Eq. (2-14*b*). Then the mathematical formulation is given by

$$\frac{d^2 T(x)}{dx^2} + \frac{1}{k} g_0 e^{-\beta x} = 0 \qquad \text{in } 0 < x < L$$

$$\frac{dT(x)}{dx} = 0 \qquad \text{at } x = 0$$

$$k\frac{dT(x)}{dx} + hT(x) = hT_\infty \quad \text{at } x = L$$

Example 2-4 Consider one-dimensional, steady-state heat conduction in a hollow cylinder with constant thermal conductivity in the region $a \le r \le b$. Heat is generated in the cylinder at a rate of g_0 W/m^3, while heat is dissipated by convection into fluids flowing inside and outside the cylindrical tube.

Heat transfer coefficients for the inside and outside fluids are h_a and h_b, respectively, and temperatures of the inside and outside fluids are T_a and T_b, respectively. Write the mathematical formulation of this heat conduction problem.

SOLUTION The heat conduction equation is obtainable from Eq. (2-9*d*) by setting $n = 1$ and $g = g_0$. The boundary conditions at $r = a$ and $r = b$ are convective boundary conditions and can be written according to Eqs. (2-16). Then the formulation of the problem becomes

$$\frac{1}{r}\frac{d}{dr}\left(r\frac{dT}{dr}\right) + \frac{g_0}{k} = 0 \qquad \text{in } a < r < b$$

$$-k\frac{dT}{dr} + h_a T = h_a T_a \qquad \text{at } r = a$$

$$k\frac{dT}{dr} + h_b T = h_b T_b \qquad \text{at } r = b$$

where $T \equiv T(r)$.

Example 2-5 Write the mathematical formulation of one-dimensional, steady-state heat conduction for a hollow sphere with constant thermal conductivity in the region $a \le r \le b$, when heat is supplied to the sphere at a rate of q_0 W/m^2 from the boundary surface at $r = a$ and dissipated by convection from the boundary surface at $r = b$ into a medium at zero temperature with a heat transfer coefficient h.

SOLUTION The heat conduction equation is obtained from Eq. (2-10*b*) by setting $n = 2$. Then, the mathematical formulation becomes

$$\frac{d}{dr}\left(r^2\frac{dT}{dr}\right) = 0 \qquad \text{in } a < r < b$$

$$-k\frac{dT}{dr} = q_0 \qquad \text{at } r = a$$

$$k\frac{dT}{dr} + hT = 0 \qquad \text{at } r = b$$

where $T \equiv T(r)$.

2-4 SUMMARY OF BASIC EQUATIONS

We summarize in Table 2-3 the basic equations given in this chapter.

Table 2-3 Summary of basic equations

Equation number	Equation	Remarks
(2-6*a*)	$\frac{\partial}{\partial x}\left(k\frac{\partial T}{\partial x}\right) + g = \rho c_p \frac{\partial T}{\partial t}$	One-dimensional, time-dependent heat conduction equation in rectangular coordinates
(2-6*b*)	$\frac{1}{r}\frac{\partial}{\partial r}\left(rk\frac{\partial T}{\partial r}\right) + g = \rho c_p \frac{\partial T}{\partial t}$	One-dimensional, time-dependent heat conduction equation in cylindrical coordinates
(2-6*c*)	$\frac{1}{r^2}\frac{\partial}{\partial r}\left(r^2 k\frac{\partial T}{\partial r}\right) + g = \rho c_p \frac{\partial T}{\partial t}$	One-dimensional, time-dependent heat conduction equation in spherical coordinates
(2-7)	$\frac{1}{r^n}\frac{\partial}{\partial r}\left(r^n k\frac{\partial T}{\partial r}\right) + g = \rho c_p \frac{\partial T}{\partial t}$ where $n = \begin{cases} 0 & \text{rectangular coordinates} \\ 1 & \text{cylindrical coordinates} \\ 2 & \text{spherical coordinates} \end{cases}$	Equations (2-6*a*) to (2-6*c*) expressed compactly as a single equation
(2-8*a*)	$\frac{1}{r^n}\frac{\partial}{\partial r}\left(r^n \frac{\partial T}{\partial r}\right) + \frac{1}{k} g = \frac{1}{\alpha}\frac{\partial T}{\partial t}$	Constant thermal conductivity
(2-9*a*)	$\frac{1}{r^n}\frac{d}{dr}\left(r^n \frac{dT}{dr}\right) + \frac{1}{k} g = 0$	Constant thermal conductivity, steady state
(2-11*a*)	$\frac{\partial^2 T}{\partial x^2} + \frac{\partial^2 T}{\partial y^2} + \frac{\partial^2 T}{\partial z^2} + \frac{1}{k} g = \frac{1}{\alpha}\frac{\partial T}{\partial t}$	Three-dimensional, time-dependent heat conduction equation with constant k, in rectangular coordinates
(2-11*b*)	$\frac{1}{r}\frac{\partial}{\partial r}\left(r\frac{\partial T}{\partial r}\right) + \frac{1}{r^2}\frac{\partial^2 T}{\partial \phi^2} + \frac{\partial^2 T}{\partial z^2} + \frac{1}{k} g = \frac{1}{\alpha}\frac{\partial T}{\partial t}$	Three-dimensional, time-dependent heat conduction equation with constant k, in cylindrical coordinates
(2-11*c*)	$\frac{1}{r^2}\frac{\partial}{\partial r}\left(r^2\frac{\partial T}{\partial r}\right) + \frac{1}{r^2 \sin\theta}\frac{\partial}{\partial \theta}\left(\sin\theta\frac{\partial T}{\partial \theta}\right) + \frac{1}{r^2 \sin^2\theta}\frac{\partial^2 T}{\partial \phi^2} + \frac{1}{k} g = \frac{1}{\alpha}\frac{\partial T}{\partial t}$	Three-dimensional, time-dependent heat conduction equation with constant k, in spherical coordinates
(2-13*a*)	$-k\left.\frac{\partial T}{\partial x}\right\|_{x=0} = q_0$	Prescribed heat flux boundary conditions at $x = 0$ and
(2-13*b*)	$k\left.\frac{\partial T}{\partial x}\right\|_{x=L} = q_L$	at $x = L$
(2-15*a*)	$\left[-k\frac{\partial T}{\partial x} + h_1 T\right]_{x=0} = h_1 T_1$	Convection boundary conditions at $x = 0$ and
(2-15*b*)	$\left[+k\frac{\partial T}{\partial x} + h_2 T\right]_{x=L} = h_2 T_2$	at $x = L$

PROBLEMS

Derivations of heat conduction equation

2-1 By writing an energy balance for a differential volume element, derive the one-dimensional, time-dependent heat conduction equation with internal energy generation and variable thermal conductivity in the rectangular coordinate system for the x variable.

2-2 By writing an energy balance for a differential cylindrical volume element in the r variable, derive the one-dimensional, time-dependent heat conduction equation with internal heat generation and variable thermal conductivity in the cylindrical coordinate system for the r variable.

2-3 By writing an energy balance for a differential spherical volume element in the r variable, derive the one-dimensional, time-dependent heat conduction equation with internal heat generation and constant thermal conductivity in the spherical coordinate system for the r variable.

2-4 By simplifying the three-dimensional heat conduction equations obtain the one-dimensional, steady-state heat conduction equation with heat generation and constant thermal conductivity for

(*a*) Rectangular coordinates in the x variable
(*b*) Cylindrical coordinates in the r variable
(*c*) Spherical coordinates in the r variable

2-5 By simplifying the three-dimensional heat conduction equation in the rectangular coordinate system, obtain the two-dimensional, steady-state heat conduction equation in rectangular coordinates for the x and y variables.

2-6 By simplifying the three-dimensional heat conduction equation in the cylindrical coordinate system, obtain the two-dimensional, steady-state heat conduction equation with heat generation in the cylindrical coordinate system for the r and z variables.

2-7 By simplifying the three-dimensional heat conduction equation for the spherical coordinate system, obtain the one-dimensional, time-dependent heat conduction equation with heat generation in the variables r and t in the spherical coordinate system.

2-8 Write the one-dimensional, steady-state heat conduction equation for constant thermal conductivity and constant heat generation rate for

(*a*) Rectangular coordinates in the x variable
(*b*) Cylindrical coordinates in the r variable
(*c*) Spherical coordinates in the r variable

2-9 Write the two-dimensional steady-state heat conduction equation in the x and y variables in the rectangular coordinate system for temperature-dependent thermal conductivity and heat generation within the medium.

2-10 Write the one-dimensional, steady-state heat conduction equation for temperature-dependent thermal conductivity and space-dependent heat generation for

(*a*) Rectangular coordinates in the r variable
(*b*) Cylindrical coordinates in the r variable
(*c*) Spherical coordinates in the r variable

2-11 Write the one-dimensional, time-dependent heat conduction equation with temperature-dependent thermal conductivity and heat generation in the spherical coordinate system in the r variable.

Formulation of boundary conditions

2-12 A plane wall, confined to the region $0 \leq x \leq L$, is subjected to a heat supply at a rate of q_0 W/m^2 at the boundary surface $x = 0$ and dissipates heat by convection with a heat transfer coefficient h_∞ W/(m$^2 \cdot$ °C) into the ambient air at temperature T_∞ from the boundary surface at $x = L$. Write the boundary conditions at $x = 0$ and $x = L$.

2-13 Consider a cylindrical wall with inside radius r_1 and outside radius r_2. The inside surface is heated uniformly at a rate of q_1 W/m^2, and the outside surface dissipates heat by convection with a heat transfer coefficient h_2 W/(m$^2\cdot$°C) into the ambient air at zero temperature. Write the boundary conditions for the boundaries at $r = r_1$ and $r = r_2$.

2-14 Write the mathematical formulation of the boundary conditions for heat conduction in a rectangular region $0 \leq x \leq a, 0 \leq y \leq b$ for

(*a*) Boundary at $x = 0$: heat removed at a constant rate of q_0 W/m^2

(*b*) Boundary at $x = a$: heat dissipated by convection with a heat transfer coefficient h_a into the ambient air at constant temperature T_a

(*c*) Boundary at $y = 0$: maintained at a constant temperature T_0

(*d*) Boundary at $y = b$: heat supplied into the medium at a rate of q_b W/m^2

2-15 A spherical shell has inside radius $r = r_1$ and outside radius $r = r_2$. At the inside surface it is electrically heated at a rate of q_1 W/m^2, and at the outside surface it dissipates heat by convection with a heat transfer coefficient h_2 into the ambient air at temperature $T_{\infty 2}$. Write the boundary conditions.

2-16 Consider a tube of inside radius $r = r_1$ and outside radius $r = r_2$. A hot gas at a temperature T_2 flows on the outside, and a cold fluid at a temperature T_1 flows inside. If the heat transfer coefficients for the inside and outside are h_1 and h_2, respectively, write the boundary conditions for the inside and outside surfaces of the tube.

2-17 The inside surface at $z = 0$ of a plane wall is heated by a hot gas flowing at temperature T_1, and the outside surface at $z = L$ is cooled by a cold gas flowing at temperature T_2. The heat transfer coefficients for the hot side and cold side are h_1 and h_2, respectively. Write the boundary conditions for the inside and outside surfaces.

2-18 A copper bar of radius $r = a$, which is heated by the passage of electric current, dissipates heat by convection with a heat transfer coefficient h_∞ from its outer surface at $r = a$ into the ambient air at temperature T_∞. Write the appropriate boundary condition at the outer surface $r = a$.

2-19 Consider a solid cylinder of radius $r = b$ and height $z = c$. The boundary surface at $r = b$ is subjected to a uniform heating electrically at a rate of q_0 W/m^2, the boundary surface at $z = 0$ is insulated, and the boundary surface at $z = c$ dissipates heat by convection into an ambient at zero temperature with a heat transfer coefficient h. Write the boundary conditions.

2-20 Consider a plane wall of thickness L. The boundary at $x = 0$ is subjected to forced convection with a heat transfer coefficient h into an ambient at temperature T_∞. The boundary at $x = L$ is heated by solar radiation flux at a rate of q_0 W/m^2. Write the boundary conditions.

2-21 Consider a hollow cylinder with inside radius r_1, outside radius r_2, and height H, as illustrated in the accompanying figure. The boundary conditions for each of the boundary surfaces are as stated:

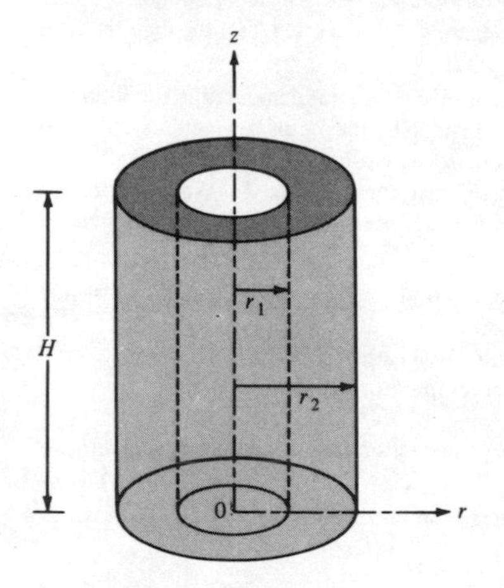

Figure P2-21

(*a*) The inner boundary surface at $r = r_1$ is heated uniformly with an electric heater at a constant rate of q_0 W/m^2.

(*b*) The outer boundary surface at $r = r_2$ dissipates heat by convection, with a heat transfer coefficient h_2, into an ambient at constant temperature T_∞.

(*c*) The lower surface at $z = 0$ is kept insulated.

(*d*) The upper boundary surface at $z = H$ dissipates heat by convection with a heat transfer coefficient h_3 into an ambient at temperature T_∞.

Write the mathematical formulation of each boundary condition.

Formulation of heat conduction problems

2-22 Heat is generated at a constant rate of g_0 W/m^3 in a copper rod of radius $r = a$ by the passage of electric current. The heat is dissipated by convection from the boundary surface at $r = a$ into the ambient air at temperature T_∞ with a heat transfer coefficient h_∞. Write the mathematical formulation of this heat conduction problem for the determination of the *one-dimensional, steady-state temperature distribution* $T(r)$ within the rod.

2-23 A plane wall of thickness L and with constant thermal properties is initially at a uniform temperature T_i. At time $t = 0$, the surface at $x = L$ is subjected to heating by the flow of a hot gas at temperature T_∞, while the other surface at $x = 0$ is kept insulated. The heat transfer coefficient between the hot gas and the surface is h_∞. There is no heat generation within the plate. Write the mathematical formulation of this transient heat conduction problem in order to determine the *one-dimensional, time-dependent temperature distribution* $T(x, t)$ within the plate for $t > 0$.

2-24 A solid bar of radius $r = b$ is initially at a uniform temperature T_0. For times $t > 0$, it is cooled by convection from its surface at $r = b$ into the ambient air at temperature T_∞ with a heat transfer coefficient h_∞. Write the mathematical formulation of this heat conduction problem for the determination of *one-dimensional, time-dependent temperature distribution* $T(r, t)$ within the rod.

2-25 A plane wall of thickness L is exposed to a uniform heat flux of q_0 W/m^2 on one side and dissipates heat by convection with a heat transfer coefficient h_∞ into the ambient air temperature T_∞ on the other side. Write the mathematical formulation of this problem for the determination of *one-dimensional, steady-state temperature distribution* $T(x)$ within the wall.

2-26 A thick-walled circular tube has inside radius r_1 and outside radius r_2. A hot gas at temperature T_1 flows inside the tube, and a cold gas at temperature T_2 flows outside. The thermal conductivity k of the tube is considered constant. The heat transfer coefficients for flow inside and outside the tube are specified as h_1 and h_2, respectively. Write the mathematical formulation of this heat conduction problem in order to determine the *one-dimensional, steady-state temperature distribution* $T(r)$ through the tube wall.

2-27 Consider a solid cylinder of radius $r = b$ and the height $z = H$. Heat is generated in the solid at a rate of g_0 W/m^3. The boundary surface at $z = 0$ is kept insulated; the boundary surface at $z = H$ dissipates heat by convection into a medium at temperature T_∞ with a heat transfer coefficient h. The cylindrical boundary surface at $r = b$ is maintained at a uniform temperature T_0. Write the mathematical formulation of this problem for the determination of *two-dimensional, steady-state temperature distribution* $T(r, z)$ within the cylinder.

2-28 Consider the two-dimensional, steady-state heat conduction problem for a rectangular region $0 \le x \le a$, $0 \le y \le b$ for the following boundary conditions:

(*a*) The boundary surface at $x = 0$ is electrically heated at a rate of q_0 W/m^2.

(*b*) The boundary surface at $x = a$ is kept at constant temperature T_0.

(*c*) The boundary surface at $y = 0$ is insulated.

(*d*) The boundary surface at $y = b$ dissipates heat by convection into a medium at temperature T_∞ with a heat transfer coefficient h.

The thermal conductivity of the solid is constant, and there is no heat generation in the medium. Write the mathematical formulation of this problem for the determination of *two-dimensional, steady-state temperature distribution* $T(x, y)$ within the region.

2-29 A copper bar of radius b is initially at a uniform temperature T_0. The heating of the rod begins at time $t = 0$ by the passage of electric current which generates heat throughout the rod at a constant rate of g_0 W/m^3. The rod dissipates heat by convection from its surface at $r = b$, with a heat transfer coefficient h, into the ambient air at temperature T_∞. Assuming that the thermal conductivity k of the rod is constant and that the problem can be treated as one-dimensional transient heat conduction in the r variable, write the mathematical formulation for the determination of the *one-dimensional, time-dependent temperature distribution* $T(r, t)$ within the solid for $t > 0$.

2-30 Consider a hollow sphere of inside radius $r = a$ and outside radius $r = b$. The inside surface is uniformly heated electrically at a rate of q_0 W/m^2, and the outside surface dissipates heat by convection with a heat transfer coefficient h into an ambient at a constant temperature T_∞. Write the mathematical formulation of this heat conduction problem for the determination of the *one-dimensional steady-state temperature distribution* $T(r)$ in the sphere.

2-31 Consider a long tube of inside radius $r = a$ and outside radius $r = b$. Heat is generated in the tube at a constant rate of g_0 W/m^3 by the passage of electric current. The inside surface is kept insulated, and the outside surface is dissipating heat by convection into an ambient at temperature T_∞ with a heat transfer coefficient h. Write the mathematical formulation of this heat conduction problem for the determination of the *one-dimensional, steady-state temperature distribution* $T(r)$ in the solid.

REFERENCES

1. Schneider, P. J.: *Conduction Heat Transfer*, Addison-Wesley, Reading, Mass., 1955.
2. Carslaw, H. S., and J. C. Jaeger: *Conduction of Heat in Solids*, 2d ed., Oxford University Press, London, 1959.
3. Özişik, M. N.: *Boundary Value Problems of Heat Conduction*, International Textbook, Scranton, Pa., 1968.
4. Myers, G. M.: *Analytical Methods in Conduction Heat Transfer*, McGraw-Hill, New York, 1971.
5. Özişik, M. N.: *Heat Conduction*, Wiley, New York, 1980.

CHAPTER

THREE

ONE-DIMENSIONAL, STEADY-STATE HEAT CONDUCTION

In this chapter we present the application of one-dimensional, steady-state heat conduction problems to determine temperature distribution and heat flow in solids having geometries in the form of a slab (i.e., plane wall), cylinder, and sphere. The one-dimensional, steady-state condition implies that the temperature gradients exist along one coordinate direction only and that temperature within the solid does not vary with time. Situations involving different types of boundary conditions, with and without internal energy generation within the solid, are examined. The concept of *thermal resistance*, analogous to electric resistance, is introduced to solve one-dimensional, steady-state heat conduction with no internal energy generation. The thermal resistance concept is also utilized to determine one-dimensional, steady-state heat flow through layers of slabs, cylinders, or spheres.

Heat flow problems involving the determination of critical thickness of insulation and the use of extended surfaces (i.e., fins) are examined.

3-1 THE SLAB

Consider a slab (i.e., plane wall) of thickness L as illustrated in Fig. 3-1. The plate is sufficiently large in the y and z directions in comparison to its thickness L to ensure that the temperature gradients in the y and z directions are negligible compared with that in the x direction. The temperature within the solid does not vary with time. Then the temperature distribution $T(x)$ within the solid is governed by the one-dimensional, steady-state heat conduction equation. For the case of constant thermal conductivity k and with an energy generation at a rate of $g(x)$ W/m^3, the heat conduction equation is given by

$$\boxed{\frac{d^2T(x)}{dx^2} + \frac{1}{k}g(x) = 0} \tag{3-1}$$

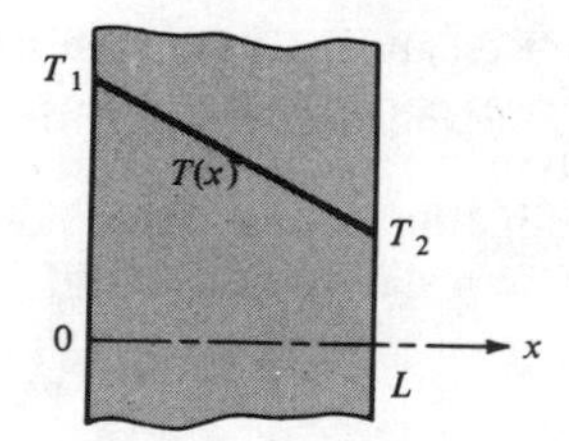

Figure 3-1 One-dimensional steady-state heat flow through a slab and the equivalent thermal resistance concept.

which is valid over the domain of the slab confined to the region $0 < x < L$. Once the temperature distribution $T(x)$ in the slab is established from the solution of this equation, the heat flux $q(x)$ anywhere in the slab is determined from the definition

$$\boxed{q(x) = -k\frac{dT(x)}{dx}} \qquad \text{W/m}^2 \tag{3-2}$$

In this section we are concerned with the solution of the heat conduction equation (3-1) with or without energy generation for a slab of thickness L and the determination of the temperature distribution $T(x)$ within the slab. To illustrate the general procedure, we consider Eq. (3-1) for the case of constant energy generation g_0 and write it in the form

$$\frac{d^2T(x)}{dx^2} = -\frac{g_0}{k} \tag{3-3}$$

The first and the second integrations of this equation give, respectively,

$$\frac{dT(x)}{dx} = -\frac{g_0}{k}x + C_1 \tag{3-4a}$$

$$T(x) = -\frac{g_0}{2k}x^2 + C_1x + C_2 \tag{3-4b}$$

where C_1 and C_2 are the integration constants.

Clearly, two boundary conditions are needed for the determination of these constants. For a slab in the region $0 \le x \le L$, one boundary condition is assigned for the surface at $x = 0$ and another for the surface at $x = L$. These boundary conditions can be a prescribed temperature, a prescribed heat flux, or a convection boundary condition. The case in which both boundary surfaces are subjected to prescribed heat flux is not considered. To illustrate the difficulties associated with such a case, we consider a slab in which energy is generated while both boundary surfaces are kept insulated. Then the steady-state condition can never be established, because the energy generated in the medium cannot escape from the slab since both boundaries are insulated and the temperature rises continuously. Therefore, in the analysis of the problem, we avoid such a situation.

We now illustrate with examples the determination of one-dimensional, steady-state temperature distribution and heat flow in a slab subject to different types of boundary conditions.

Example 3-1 Consider a slab of thickness L as illustrated in Fig. 3-1. The boundary surfaces at $x = 0$ and $x = L$ are maintained at constant, but different, temperatures T_1 and T_2, respectively. There is no energy generation in the solid, and the thermal conductivity k is constant.

Develop an expression for the temperature distribution $T(x)$ in the slab.

Develop an expression for heat flow Q through an area A of the slab.

SOLUTION The mathematical formulation of this heat conduction problem is given by

$$\frac{d^2T(x)}{dx^2} = 0 \qquad \text{in } 0 < x < L$$

$$T(x) = T_1 \qquad \text{at } x = 0$$

$$T(x) = T_2 \qquad \text{at } x = L$$

Two integrations of the differential equation gives

$$T(x) = C_1 x + C_2$$

The application of the boundary condition at $x = 0$ gives

$$C_2 = T_1$$

and the application of the boundary condition at $x = L$ results in

$$C_1 = \frac{T_2 - T_1}{L}$$

Introducing these constants into the above expressions for $T(x)$, we determine the temperature distribution $T(x)$:

$$T(x) = (T_2 - T_1)\frac{x}{L} + T_1$$

This result shows that for one-dimensional, steady-state heat conduction through a slab having a constant thermal conductivity and no energy generation, the temperature $T(x)$ varies linearly with x.

The heat flux q through the slab is determined by differentiating the above solution for $T(x)$ with respect to x and applying the definition of the heat flux given by Eq. (3-2). We find

$$q = k\frac{T_1 - T_2}{L} \qquad \text{W/m}^2$$

When $T_1 > T_2$, the right-hand side of this expression is positive, hence the heat flow is in the positive x direction.

The heat flow rate Q through an area A of the slab normal to the direction of heat flow is

$$Q = Aq = Ak \frac{T_1 - T_2}{L} \qquad \text{W}$$

This result is now rearranged in the form

$$\boxed{Q = \frac{T_1 - T_2}{R}} \tag{3-5a}$$

where

$$\boxed{R = \frac{L}{Ak}} \tag{3-5b}$$

Here R is called the *thermal resistance of the slab* for heat flow through an area A across a temperature potential $T_1 - T_2$. The concept is analogous to electric resistance in Ohm's law and is illustrated in Fig. 3-1.

Example 3-2 Consider a slab of thickness L as illustrated in Fig. 3-2. A fluid at a temperature $T_{\infty 1}$ with a heat transfer coefficient h_1 flows over the surface at $x = 0$, and another fluid at a temperature $T_{\infty 2}$ with a heat transfer coefficient h_2 flows over the surface at $x = L$ of the plate.

Develop an expression for the heat flow Q through an area A of the plate.

Calculate the heat transfer rate through $A = 1\ \text{m}^2$ of the slab for $T_{\infty 1} = 130°\text{C}$, $T_{\infty 1} = 30°\text{C}$, $h_{\infty 1} = 250\ \text{W/(m}^2 \cdot °\text{C)}$, $h_{\infty 2} = 500\ \text{W/(m}^2 \cdot °\text{C)}$, $L = 4\ \text{cm}$, $k = 20\ \text{W/(m} \cdot °\text{C)}$.

SOLUTION Since there is *no energy generation* in the medium and only the heat flow rate through the plate is required, it is more convenient to utilize

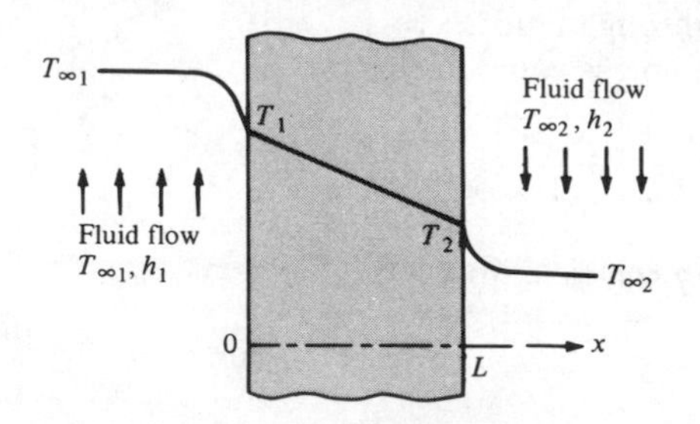

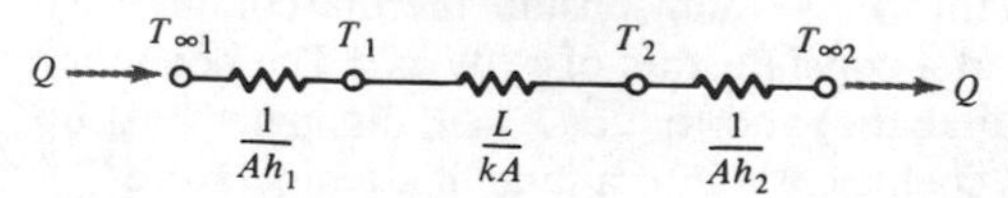

Figure 3-2 Thermal resistance concept for heat flow through a slab with convection at both surfaces.

the thermal resistance concept for the solution of this problem. Refer to the nomenclature shown in Fig. 3-2. The heat flow Q is by convection from fluid 1 to the surface of the plate at $x = 0$, by conduction through the plate, and by convection from the surface at $x = L$ to fluid 2. With this consideration we write

$$Q = Ah_1(T_{\infty 1} - T_1) = Ak\frac{T_1 - T_2}{L} = Ah_2(T_2 - T_{\infty 2})$$

This result is rearranged in the form

$$Q = \frac{T_{\infty 1} - T_1}{1/(Ah_1)} = \frac{T_1 - T_2}{L/(Ak)} = \frac{T_2 - T_{\infty 2}}{1/(Ah_2)}$$

This is analogous to Ohm's law, with each term in the denominator representing the thermal resistance to heat flow of that particular layer.

By adding the numerators and the denominators of this equality, we obtain

$$Q = \frac{T_{\infty 1} - T_{\infty 2}}{1/(Ah_1) + L/(Ak) + 1/(Ah_2)}$$

The heat transfer rate Q through the area A of the slab can be calculated from this expression, since all the quantities on the right-hand side are known. This result can be written more compactly in the form

$$Q = \frac{T_{\infty 1} - T_{\infty 2}}{R_{tot}}$$

where the total thermal resistance to heat flow R_{tot} is defined as

$$R_{tot} = \frac{1}{Ah_1} + \frac{L}{Ak} + \frac{1}{Ah_2}$$

Here $1/(Ah_1)$ is the thermal resistance for convection through fluid 1, $L/(Ak)$ is the thermal resistance for conduction through the slab, and $1/(Ah_2)$ is the thermal resistance for convection through fluid 2.

The numerical part of this example is computed as

$$R_{tot} = \frac{1}{1 \times 250} + \frac{0.04}{1 \times 20} + \frac{1}{1 \times 500} = 8 \times 10^{-3}$$

$$Q = \frac{130 - 30}{8 \times 10^{-3}} = 12{,}500 \text{ W} = 12.5 \text{ kW}$$

Example 3-3 Consider a slab of thickness L and constant thermal conductivity k in which energy is generated at a constant rate of g_0 W/m^3. The boundary surface at $x = 0$ is insulated (adiabatic) and that at $x = L$ dissipates heat by convection with a heat transfer coefficient h into a fluid at a temperature T_∞.

Develop expressions for the temperature $T(x)$ and the heat flux $q(x)$ in the slab.

Calculate the temperatures at the surfaces $x = 0$ and $x = L$ under the following conditions: $L = 1$ cm, $k = 20$ W/(m · °C), $g_0 = 8 \times 10^7$ W/m³, $h = 4000$ W/(m² · °C), and $T_\infty = 100$°C.

SOLUTION There is energy generation in the medium, hence the thermal resistance concept cannot be utilized for this problem. Therefore, the heat conduction equation should be solved to determine the temperature distribution.

The mathematical formulation of this heat conduction problem is given by

$$\frac{d^2T(x)}{dx^2} + \frac{g_0}{k} = 0 \qquad \text{in } 0 < x < L$$

$$\frac{dT(x)}{dx} = 0 \qquad \text{at } x = 0$$

$$k\frac{dT(x)}{dx} + hT = hT_\infty \qquad \text{at } x = L$$

The first integration of the differential equation gives

$$\frac{dT(x)}{dx} = -\frac{g_0}{k}x + C_1$$

and the application of the boundary condition at $x = 0$ yields

$$C_1 = 0$$

A second integration with $C_1 = 0$ results in

$$T(x) = -\frac{g_0}{2k}x^2 + C_2$$

and the application of the boundary condition at $x = L$ gives

$$-g_0 L + h\left(-\frac{g_0 L^2}{2k} + C_2\right) = hT_\infty$$

or

$$C_2 = \frac{g_0 L^2}{2k} + \frac{g_0 L}{h} + T_\infty$$

Then the temperature distribution in the slab becomes

$$T(x) = \frac{g_0 L^2}{2k}\left[1 - \left(\frac{x}{L}\right)^2\right] + \frac{g_0 L}{h} + T_\infty$$

The physical significance of each term in this solution is as follows: The first term on the right-hand side is due to the energy generation in the solid. The second term is due to the presence of a finite heat transfer coefficient at the surface. For $h \to \infty$, this term vanishes, and the boundary surface at $x = L$ is at the temperature T_∞.

The expression for the heat flux anywhere in the medium is determined from its definition as

$$q(x) = -k\frac{dT(x)}{dx} = g_0 x$$

Finally, by using the numerical values given above, the temperatures at the boundary surfaces $x = 0$ and $x = L$ are, respectively,

$$T(0) = \frac{8 \times 10^7 \times (0.01)^2}{2 \times 20} + \frac{8 \times 10^7 \times 0.01}{4000} + 100 = 500°\text{C}$$

$$T(L) = \frac{8 \times 10^7 \times 0.01}{4000} + 100 = 300°\text{C}$$

Example 3-4 An iron plate of thickness L with thermal conductivity k is subjected to a constant, uniform heat flux q_0 W/m^2 at the boundary surface at $x = 0$. From the other boundary surface at $x = L$, heat is dissipated by convection into a fluid at temperature T_∞ with a heat transfer coefficient h. Figure 3-3 shows the geometry and the nomenclature.

Develop expressions for the determination of the surface temperatures T_1 and T_2 at the surfaces $x = 0$ and $x = L$, respectively.

Calculate the surface temperatures T_1 and T_2 for $L = 2$ cm, $k = 20$ W/(m · °C), $q_0 = 10^5$ W/m^2, $T_\infty = 50$°C, and $h = 500$ W/(m^2 · °C).

SOLUTION Since there is no energy generation in the medium and only the temperatures of the boundary surfaces are required, it is more convenient to utilize the thermal resistance concept as illustrated in Fig. 3-3 to solve this problem.

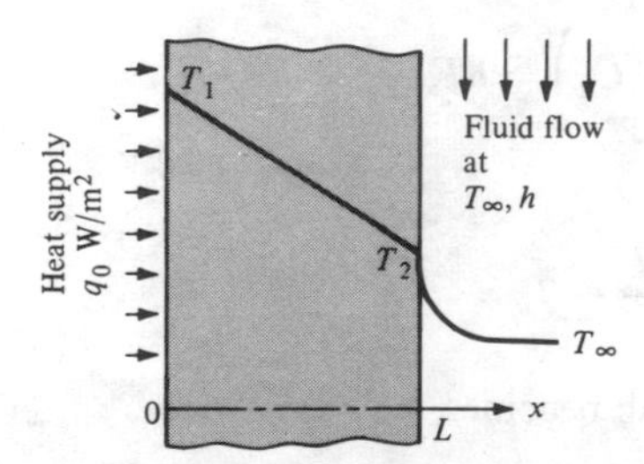

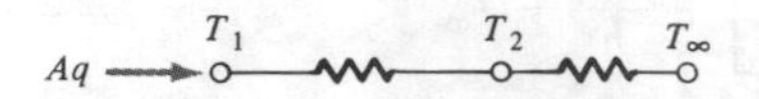

Figure 3-3 Nomenclature and thermal resistance concept for Example 3-4.

Using the nomenclature in this figure, we immediately apply the thermal resistance concept:

$$Aq_0 = \frac{T_1 - T_2}{L/(Ak)} = \frac{T_2 - T_\infty}{1/(Ah)} = \frac{T_1 - T_\infty}{L/(Ak) + 1/(Ah)}$$

Here the last expression is obtained by adding the numerators and the denominators of the second and third expressions. Clearly, in these equations the area A cancels, and the surface temperatures T_1 and T_2 become the two unknowns.

By equating the first and the last expression, T_1 is found:

$$T_1 = \left(\frac{L}{k} + \frac{1}{h}\right) q_0 + T_\infty$$

and by equating the first and the third expressions, T_2 is found:

$$T_2 = \frac{q_0}{h} + T_\infty$$

For the numerical part of this example, the temperatures T_1 and T_2 are calculated by introducing the numerical values of various quantities in the above results. We find

$$T_1 = \left(\frac{0.02}{20} + \frac{1}{500}\right) \times 10^5 + 50 = 350°\text{C}$$

$$T_2 = \frac{10^5}{500} + 50 = 250°\text{C}$$

3-2 THE CYLINDER

The problems of one-dimensional, radial heat flow in solids having a geometry in the form of a cylinder have numerous engineering applications. Heat removal from a cylindrical fuel element of a nuclear reactor by the coolant fluid, heat dissipation from a current-carrying wire, and heat flow across a thick-walled circular tube are typical examples.

Suppose there is an energy generation in the solid at a rate of $g(r)$ W/m^3, and the thermal conductivity is considered constant. The temperature distribution $T(r)$ in the solid is governed by the following heat conduction equation:

$$\boxed{\frac{1}{r}\frac{d}{dr}\left(r\frac{dT}{dr}\right) + \frac{1}{k}g(r) = 0} \qquad (3\text{-}6)$$

which is valid over the cylindrical region. Once the temperature distribution $T(r)$ in the cylinder is established from the solution of this equation, the radial heat flux $q(r)$ anywhere in the solid is determined from the definition

$$q(r) = -k \frac{dT(r)}{dr} \qquad \text{W/m}^2 \tag{3-7}$$

To determine the temperature distribution $T(r)$ in the region, Eq. (3-6) is integrated, and the resulting integration constants are determined from the application of the boundary conditions for the problem.

To illustrate the general procedure, we consider Eq. (3-6) for constant energy generation g_0 and write it in the form

$$\frac{d}{dr}\left(r \frac{dT}{dr}\right) = -\frac{g_0}{k} r \tag{3-8}$$

where g_0 = constant. The first and the second integrations of Eq. (3-8) give, respectively,

$$\frac{dT(r)}{dr} = -\frac{g_0}{2k} r + \frac{C_1}{r} \tag{3-9}$$

$$T(r) = -\frac{g_0}{4k} r^2 + C_1 \ln r + C_2 \tag{3-10}$$

Clearly, two boundary conditions are needed to determine the two integration constants C_1 and C_2. In the case of a hollow cylinder illustrated in Fig. 3-4*b*, the boundary conditions at the inner and outer surfaces can be a prescribed temperature, a prescribed heat flux, or a convection boundary condition. The case in which both boundary surfaces are subjected to prescribed heat flux is not considered in this book, because it requires special consideration. The implications of such a situation can be envisioned if we consider a hollow cylinder in which energy is generated while both boundary surfaces are kept insulated. The energy

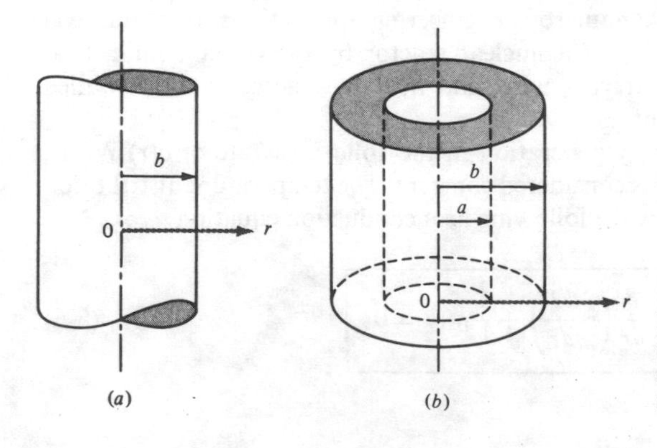

Figure 3-4 Coordinates for one-dimensional heat conduction in a cylinder. (*a*) Solid cylinder, (*b*) hollow cylinder.

generated has no way to escape from the medium because the boundaries are insulated; as a result, the problem has no steady-state solution.

In the case of a solid cylinder, illustrated in Fig. 3-4*a*, a boundary condition can be specified for the outer surface, but another boundary condition is needed at the center of the cylinder. Such a boundary condition is specified from the physical consideration of the temperature distribution in the cylinder. Physically meaningful solution requires that the temperature not be infinite at $r = 0$; this condition is satisfied if

$$T(r) = \text{finite} \qquad \text{at } r = 0 \tag{3-11a}$$

In one-dimensional, radial heat flow in a solid cylinder, the temperature is symmetric about the center of the cylinder. With this consideration, an alternative boundary condition at $r = 0$ can be given as

$$\frac{dT(r)}{dr} = 0 \qquad \text{at } r = 0 \tag{3-11b}$$

Either of the boundary conditions, Eq. (3-11*a*) or Eq. (3-11*b*), at the center of the cylinder can be used, and they yield the same result.

We now illustrate with examples the determination of temperature distribution and heat flow in cylindrical bodies.

Example 3-5 Consider a solid cylinder of radius $r = b$ in which energy is generated at a constant rate of g_0 W/m^3, while the boundary surface at $r = b$ is maintained at a constant temperature T_2.

Develop an expression for the one-dimensional, radial, steady-state temperature distribution $T(r)$ and the heat flux $q(r)$.

Calculate the center temperature $T(0)$ and the heat flux at the boundary surface $r = b$ for $b = 1$ cm, $g_0 = 2 \times 10^8$ W/m^3, $k = 20$ W/(m · °C), and $T_2 = 100$°C.

SOLUTION The mathematical formulation of this problem is given as

$$\frac{1}{r}\frac{d}{dr}\left[r\frac{dT(r)}{dr}\right] + \frac{g_0}{k} = 0 \qquad \text{in } 0 < r < b$$

$$\frac{dT(r)}{dr} = 0 \qquad \text{at } r = 0$$

$$T(r) = T_2 \qquad \text{at } r = b$$

The first and the second integrations of this differential equation give, respectively,

$$\frac{dT(r)}{dr} = -\frac{g_0}{2k}r + \frac{C_1}{r}$$

$$T(r) = -\frac{g_0}{4k}r^2 + C_1 \ln r + C_2$$

The application of the boundary condition at $r = 0$ to the first equation gives

$$C_1 = 0$$

The same result also could be obtained by the application of the alternative form of the boundary condition given by $T(r) =$ finite at $r = 0$ to the second equation. The constant C_2 is determined by the application of the boundary condition at $r = b$ to the second of the above equations. We find

$$C_2 = \frac{g_0 b^2}{4k} + T_2$$

Then the temperature distribution in the cylinder becomes

$$T(r) = \frac{g_0 b^2}{4k}\left[1 - \left(\frac{r}{b}\right)^2\right] + T_2$$

and the heat flux $q(r)$ anywhere in the medium is determined from its definition as

$$q(r) = -k\frac{dT(r)}{dr} = \frac{g_0 r}{2}$$

The center temperature $T(0)$ is calculated from the above expression for $T(r)$ by setting $r = 0$:

$$T(0) = \frac{2 \times 10^8 \times (0.01)^2}{4 \times 20} + 100 = 350°\text{C}$$

The surface heat flux is calculated from the expression for $q(r)$ by setting $r = b$:

$$q(b) = \frac{g_0 b}{2} = \frac{2 \times 10^8 \times 0.01}{2} = 10^6 \text{ W/m}^2$$

Example 3-6 The inner surface at $r = a$ and the outer surface at $r = b$ of a hollow cylinder are maintained at uniform temperatures T_1 and T_2, respectively. The thermal conductivity k of the solid is constant.

Develop an expression for the one-dimensional, steady-state temperature distribution $T(r)$ in the cylinder.

Develop an expression for the radial heat flow rate Q through the cylinder over a length H.

Develop an expression for the thermal resistance of a hollow cylinder of length H.

SOLUTION The mathematical formulation of this problem is given as

$$\frac{d}{dr}\left[r\frac{dT(r)}{dr}\right] = 0 \qquad \text{in } a < r < b$$

$$T(r) = T_1 \qquad \text{at } r = a$$

$$T(r) = T_2 \qquad \text{at } r = b$$

The first and the second integrations of the differential equation give, respectively,

$$\frac{dT(r)}{dr} = \frac{C_1}{r}$$

$$T(r) = C_1 \ln r + C_2$$

The boundary conditions at $r = a$ and $r = b$ are applied to the equation for $T(r)$ to give

$$T_1 = C_1 \ln a + C_2$$

$$T_2 = C_1 \ln b + C_2$$

A simultaneous solution of these two equations gives

$$C_1 = \frac{T_2 - T_1}{\ln (b/a)}$$

$$C_2 = T_1 - (T_2 - T_1)\frac{\ln a}{\ln (b/a)}$$

Introducing these coefficients into the above equation for $T(r)$, we obtain

$$\frac{T(r) - T_1}{T_2 - T_1} = \frac{\ln (r/a)}{\ln (b/a)}$$

The heat flow rate Q over a length H of the cylinder is determined from

$$Q = q(r) \cdot \text{area} = -k\frac{dT(r)}{dr} 2\pi rH$$

$$= -k2\pi HC_1$$

since $dT(r)/dr = (1/r)C_1$. Now, introducing C_1 into the expression Q, we find

$$Q = \frac{2\pi kH}{\ln (b/a)}(T_1 - T_2)$$

This expression for Q is now rearranged in the form

$$\boxed{Q = \frac{T_1 - T_2}{R}} \tag{3-12a}$$

where

$$\boxed{R = \frac{\ln (b/a)}{2\pi kH}} \tag{3-12b}$$

Equation (3-12b) can also be rearranged in the form

$$R = \frac{\ln (b/a)}{2\pi Hk} = \frac{(b - a) \ln [2\pi bH/(2\pi aH)]}{(b - a)2\pi Hk}$$

or

$$\boxed{R = \frac{t}{kA_m}} \tag{3-12c}$$

where

$$\boxed{A_m = \frac{A_1 - A_0}{\ln (A_1/A_0)}} \tag{3-12d}$$

when $A_0 = 2\pi aH$ = area of inner surface of cylinder

$A_1 = 2\pi bH$ = area of outer surface of cylinder

A_m = logarithmic mean area

$t = b - a$ = thickness of cylinder

Here, R, as defined above, is called the *thermal resistance* for a hollow cylinder.

Example 3-7 A hollow cylinder with inner radius $r = a$ and outer radius $r = b$ is heated at the inner surface at a rate of q_0 W/m^2 and dissipates heat by convection from the outer surface into a fluid at temperature T_∞ with a heat transfer coefficient h. There is no energy generation, and the thermal conductivity of the solid is assumed to be constant.

Develop expressions for the determination of the temperatures T_1 and T_2 of the inner and outer surfaces of the cylinder.

Calculate the surface temperatures T_1 and T_2 for $a = 3$ cm, $b = 5$ cm, $h = 400$ W/(m$^2 \cdot$°C), $T_\infty = 100$°C, $k = 15$ W/(m $\cdot$ °C), and $q_0 = 10^5$ W/m^2.

SOLUTION Since there is no energy generation in the medium, it is more convenient to solve this problem by utilizing the thermal resistance approach. Figure 3-5 illustrates various thermal resistances in the path of heat flow. We therefore immediately write

$$2\pi aHq_0 = \frac{T_1 - T_2}{\ln (b/a)/(2\pi kH)} = \frac{T_2 - T_\infty}{1/(2\pi bHh)}$$

$$= \frac{T_1 - T_\infty}{\ln (b/a)/(2\pi kH) + 1/(2\pi bHh)}$$

Here H is the length of the cylinder, and the last expression is obtained by adding the numerators and the denominators of the second and third expressions. Clearly, in these expressions $2\pi H$ cancels, and the surface temperatures T_1 and T_2 are the two unknown quantities.

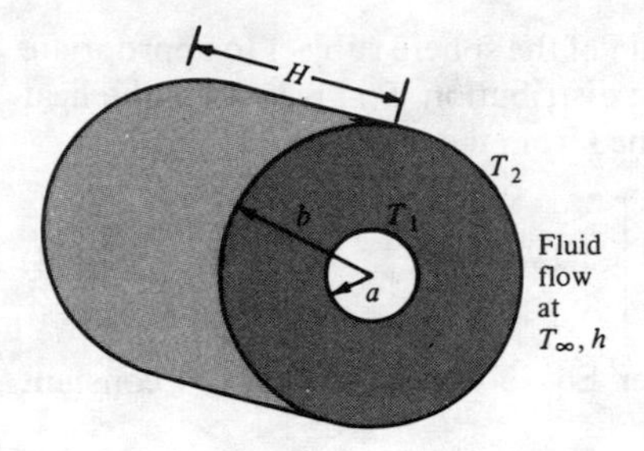

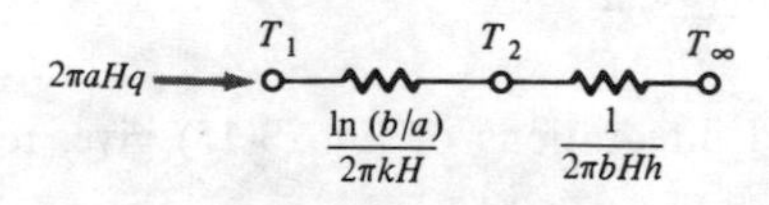

Figure 3-5 Nomenclature for thermal resistance concept for Example 3-7.

By equating the first and the last expressions, T_1 is determined to be

$$T_1 = \left(\frac{a}{k}\ln\frac{b}{a} + \frac{a}{bh}\right)q_0 + T_\infty$$

and equating the first and third expressions, we find

$$T_2 = \frac{a}{bh}q_0 + T_\infty$$

By introducing the numerical values given above into these solutions, we find

$$T_1 = \left(\frac{0.03}{15}\ln\frac{5}{3} + \frac{0.03}{0.05 \times 400}\right) \times 10^5 + 100 = 352.2°\text{C}$$

$$T_2 = \frac{0.03}{0.05 \times 400} \times 10^5 + 100 = 250°\text{C}$$

3-3 THE SPHERE

The one-dimensional, steady-state temperature distribution $T(r)$ in a sphere in which energy is generated at a rate of $g(r)$ W/m^3 is governed by the heat conduction equation

$$\boxed{\frac{1}{r^2}\frac{d}{dr}\left(r^2\frac{dT}{dr}\right) + \frac{1}{k}g(r) = 0} \tag{3-13}$$

This equation must be solved over the domain of the sphere subject to appropriate boundary conditions. Once the temperature distribution $T(r)$ is known, the heat flux $q(r)$ anywhere in the medium is determined from the definition

$$\boxed{q(r) = -k\frac{dT(r)}{dr}} \qquad \text{W/m}^2 \tag{3-14}$$

To illustrate the procedure, we consider Eq. (3-13) for the case of constant energy generation g_0 and write it in the form

$$\frac{d}{dr}\left(r^2\frac{dT}{dr}\right) = -\frac{g_0}{k}r^2 \tag{3-15}$$

where g_0 = constant. The first and second integrations of Eq. (3-15) give, respectively,

$$\frac{dT(r)}{dr} = -\frac{g_0}{3k}r + \frac{C_1}{r^2} \tag{3-16}$$

$$T(r) = -\frac{g_0}{6k}r^2 - \frac{C_1}{r} + C_2 \tag{3-17}$$

Two boundary conditions are needed to determine the two integration constants C_1 and C_2. In the case of a hollow sphere, illustrated in Fig. 3-6*b*, the boundary conditions at the inner and outer surfaces can be a prescribed temperature, a prescribed heat flux, or a convection boundary condition. The case in which both boundary surfaces are subjected to prescribed heat flux is not considered for the reasons stated previously in connection with the slab and the cylinder.

In the case of a solid sphere, illustrated in Fig. 3-6*a*, a boundary condition can be specified for the outer surface, but another boundary condition is needed at the center of the sphere. As discussed in connection with the solid cylinder, the boundary condition at the center of the sphere can be taken as

$$T(r) = \text{finite} \qquad \text{at } r = 0 \tag{3-18a}$$

or

$$\frac{dT(r)}{dr} = 0 \qquad \text{at } r = 0 \tag{3-18b}$$

Both boundary conditions lead to the same result.

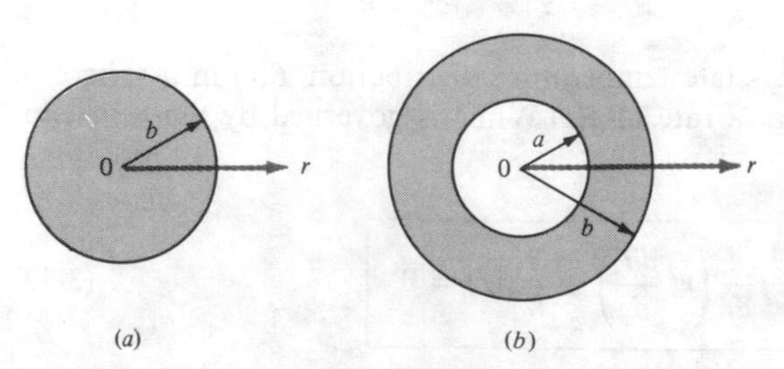

Figure 3-6 Coordinates for one-dimensional heat conduction in a sphere. (*a*) Solid sphere. (*b*) hollow sphere.

We now illustrate with examples the application of heat conduction in a sphere.

Example 3-8 The inner surface at $r = a$ and the outer surface at $r = b$ of a hollow sphere are maintained at uniform temperatures T_1 and T_2, respectively. The thermal conductivity k of the solid is constant.

Develop an expression for the one-dimensional, steady-state temperature distribution $T(r)$ in the sphere.

Develop an expression for the radial heat flow rate Q through the hollow sphere.

Develop an expression for the thermal resistance of the hollow sphere.

SOLUTION The mathematical formulation of this problem is given as

$$\frac{d}{dr}\left[r_2 \frac{dT(r)}{dr}\right] = 0 \qquad \text{in } a < r < b$$

$$T(r) = T_1 \qquad \text{at } r = a$$

$$T(r) = T_2 \qquad \text{at } r = b$$

The first and second integrations of the differential equation give, respectively,

$$\frac{dT(r)}{dr} = \frac{C_1}{r^2}$$

$$T(r) = -\frac{C_1}{r} + C_2$$

The boundary conditions at $r = a$ and $r = b$ are applied to the equation for $T(r)$ to give

$$T_1 = -\frac{C_1}{a} + C_2 \qquad \text{and} \qquad T_2 = -\frac{C_1}{b} + C_2$$

A simultaneous solution of these two equations for constants C_1 and C_2 gives

$$C_1 = -\frac{ab}{b-a}(T_1 - T_2) \qquad \text{and} \qquad C_2 = \frac{bT_2 - aT_1}{b-a}$$

Then the temperature distribution $T(r)$ becomes

$$T(r) = \frac{a}{r} \cdot \frac{b-r}{b-a} \cdot T_1 + \frac{b}{r} \cdot \frac{r-a}{b-a} \cdot T_2$$

The heat flow rate Q through the hollow sphere is determined from

$$Q = (4\pi r^2)\left[-k \frac{dT(r)}{dr}\right] = (4\pi r^2)\left(-k \frac{C_1}{r^2}\right) = -4\pi k C_1$$

When C_1 is substituted, we obtain

$$Q = 4\pi k \frac{ab}{b-a}(T_1 - T_2)$$

This result is now rearranged in the form

$$\boxed{Q = \frac{T_1 - T_2}{R}} \tag{3-19a}$$

where

$$\boxed{R = \frac{b-a}{4\pi kab}} \tag{3-19b}$$

and R is called the *thermal resistance for a hollow sphere.*

Example 3-9 A hollow sphere of inside radius $r = a$ and outside radius $r = b$ is electrically heated at the inner surface at a constant rate of q_0 W/m^2. At the outer surface it dissipates heat by convection into a fluid at temperature T_∞ with a heat transfer coefficient h. The thermal conductivity k of the solid is constant.

Develop expressions for the determination of the inner and outer surface temperatures T_1 and T_2 of the sphere.

Calculate the inner and outer surface temperatures for $a = 3$ cm, $b = 5$ cm, $h = 400$ W/(m$^2 \cdot$ °C), $T_\infty = 100$°C, $k = 15$ W/(m · °C), and $q_0 = 10^5$ W/m^2.

SOLUTION This problem can be readily solved by the thermal resistance concept since there is no energy generated in the medium. Figure 3-7 illustrates the thermal resistance network for this problem. We therefore write

$$4\pi a^2 q_0 = \frac{T_1 - T_2}{(b-a)/(4\pi kab)} = \frac{T_2 - T_\infty}{1/(4\pi b^2 h)}$$

$$= \frac{T_1 - T_\infty}{(b-a)/(4\pi kab) + 1/(4\pi b^2 h)}$$

In these equations, T_1 and T_2 are the only unknowns; the last expression is obtained by adding the numerators and denominators of the second and third expressions.

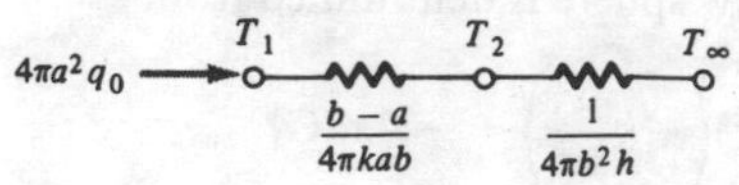

$4\pi a^2 q_0$ → T_1 — $\frac{b-a}{4\pi kab}$ — T_2 — $\frac{1}{4\pi b^2 h}$ — T_∞

Figure 3-7 Thermal resistance network for Example 3-9.

By equating the first and last expressions, T_1 is found:

$$T_1 = \left[\frac{a(b-a)}{bk} + \left(\frac{a}{b}\right)^2 \frac{1}{h}\right] q_0 + T_\infty$$

and by equating the first and third expressions, T_2 is found:

$$T_2 = \left(\frac{a}{b}\right)^2 \frac{1}{h} q_0 + T_\infty$$

By introducing the numerical values given above, we find the surface temperatures T_1 and T_2:

$$T_1 = \left[\frac{0.03 \times 0.02}{0.05 \times 15} + \left(\frac{3}{5}\right)^2 \frac{1}{400}\right] \times 10^5 + 100 = 270°\text{C}$$

$$T_2 = (\tfrac{3}{5})^2 \tfrac{1}{400} \times 10^5 + 100 = 190°\text{C}$$

3-4 COMPOSITE MEDIUM

In many engineering applications, heat transfer takes place through a medium composed of several different layers, each having different thermal conductivity. Consider, for example, a hot fluid flowing inside a tube covered with a uniform layer of thermal insulation. The thermal conductivities of the tube metal and of insulation are different; hence the heat transfer problem from the hot fluid to the colder, outer environment involves conduction through a composite medium consisting of two parallel concentric cylinders. The thermal resistance concept discussed earlier is applied now to the prediction of one-dimensional, steady-state heat transfer rate by conduction through a composite structure.

Composite Slab

Consider a composite wall consisting of three parallel layers in perfect thermal contact as illustrated in Fig. 3-8. Consider the heat flow rate Q through an area A of the slab. The equivalent thermal resistance network is also shown in this figure. By applying the thermal resistance concept, we immediately write

$$Q = \frac{T_a - T_0}{R_a} = \frac{T_0 - T_1}{R_1} = \frac{T_1 - T_2}{R_2} = \frac{T_2 - T_3}{R_3} = \frac{T_3 - T_b}{R_b} \qquad (3\text{-}20)$$

where various thermal resistances are defined as

$$R_a = \frac{1}{Ah_a} \qquad R_1 = \frac{L_1}{Ak_1} \qquad R_2 = \frac{L_2}{Ak_2} \qquad R_3 = \frac{L_3}{Ak_3} \qquad \text{and} \qquad R_b = \frac{1}{Ah_b} \qquad (3\text{-}21)$$

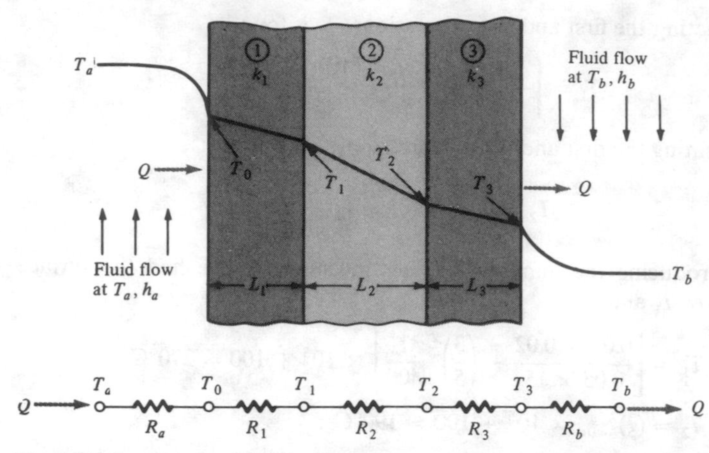

Figure 3-8 A composite of three walls in series paths and the equivalent thermal resistance network.

By summing the numerators and the denominators of the individual ratios in Eq. (3-20), we obtain

$$Q = \frac{T_a - T_b}{R} \quad \text{W} \tag{3-22a}$$

where

$$R = R_a + R_1 + R_2 + R_3 + R_b \tag{3-22b}$$

Here R is the *total thermal resistance* in the path of heat flow through an area A from temperature T_a to T_b, and various individual resistances are defined by Eq. (3-21). The composite wall arrangement shown in Fig. 3-8 is a series arrangement because the individual thermal resistances are connected in series.

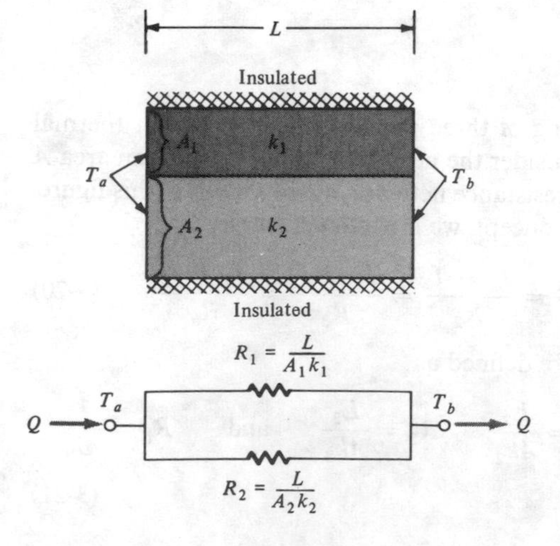

Figure 3-9 A composite of two materials in parallel paths and the equivalent thermal resistance network.

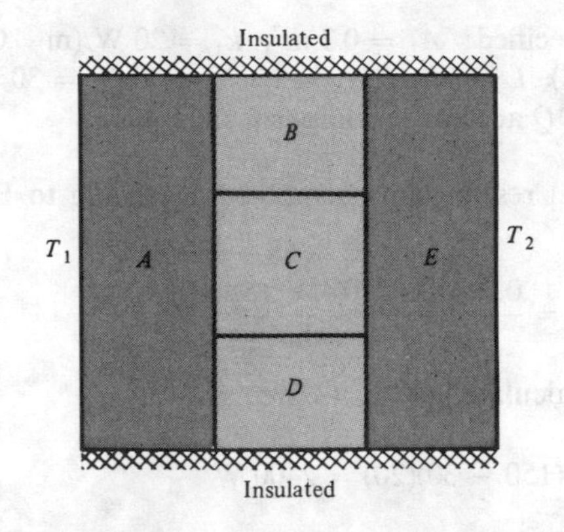

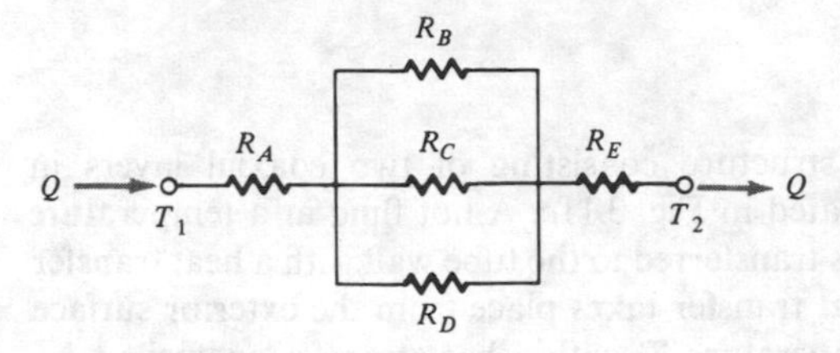

Figure 3-10 Equivalent thermal resistance network approximating heat flow through a composite wall as one-dimensional heat flow.

Figure 3-9 shows a composite of two materials combined in parallel paths with the ends maintained at uniform temperatures T_a and T_b. The equivalent thermal resistance network is also shown in this figure. The total heat transfer rate Q through this parallel arrangement is given by

$$Q = \frac{T_a - T_b}{R} \tag{3-23a}$$

where the equivalent parallel resistance is

$$\frac{1}{R} = \frac{1}{R_1} + \frac{1}{R_2} = \frac{A_1 k_1}{L} + \frac{A_2 k_2}{L} \tag{3-23b}$$

Figure 3-10 illustrates a composite of several different materials arranged in parallel and series paths. An equivalent thermal resistance network shown in this figure is based on the assumption that the heat flow path can be approximated as one-dimensional. In reality, the heat flow through such a system is two-dimensional. But if all the materials have the same thermal conductivity, the heat flow is truly one-dimensional, and the analysis becomes exact. Therefore, if the thermal conductivities of the materials in the composite do not differ significantly, the approximation of the problem as one-dimensional may be reasonable. Otherwise, a two-dimensional analysis is necessary.

Example 3-10 Consider the composite of two materials combined in parallel paths with the ends maintained at uniform temperatures as illustrated in Fig.

3-9. Various quantities are specified: $A_1 = 0.2\ \text{m}^2$, $k_1 = 20\ \text{W/(m}\cdot{}^\circ\text{C)}$, $A_2 = 0.4\ \text{m}^2$, $k_2 = 15\ \text{W/(m}\cdot{}^\circ\text{C)}$, $L = 0.5$ m, $T_a = 150^\circ\text{C}$, and $T_b = 30^\circ\text{C}$. Calculate the rate of heat transfer Q across the composite medium.

SOLUTION The equivalent parallel resistance is computed according to Eq. (3-23*b*) as

$$\frac{1}{R} = \frac{A_1 k_1}{L} + \frac{A_2 k_2}{L} = \frac{0.2 \times 20}{0.5} + \frac{0.4 \times 15}{0.5} = 20$$

Then the heat transfer rate Q is calculated by Eq. (3-23*a*) as

$$Q = \frac{T_a - T_b}{R} = (150 - 30)(20) = 2400\ \text{W}$$

Composite Coaxial Cylinders

Consider a composite cylinder structure consisting of two coaxial layers in perfect thermal contact, as illustrated in Fig. 3.11*a*. A hot fluid at a temperature T_a flows inside the tube, and heat is transferred to the tube wall with a heat transfer coefficient h_a. On the outside, heat transfer takes place from the exterior surface of the tube to a cold fluid at temperature T_b with a heat transfer coefficient h_b. The total heat transfer rate Q from the hot to the cold fluid over the length H of the cylindrical structure is the same through each layer and is given by

$$Q = \frac{T_a - T_0}{R_a} = \frac{T_0 - T_1}{R_1} = \frac{T_1 - T_2}{R_2} = \frac{T_2 - T_b}{R_b} \tag{3-24}$$

where various thermal resistances are defined as

$$R_a = \frac{1}{2\pi r_0 H h_a} \qquad R_1 = \frac{1}{2\pi H k_1} \ln\frac{r_1}{r_0}$$

$$R_2 = \frac{1}{2\pi H k_2} \ln\frac{r_2}{r_1} \qquad R_b = \frac{1}{2\pi r_2 H h_b} \tag{3-25}$$

Here the thermal resistances R_1 and R_2 for conduction through a cylindrical layer are written in accordance with the thermal resistance expression given by Eq. (3-12*b*).

Summing the numerators and denominators of the individual ratios in Eq. (3-24), we write

$$Q = \frac{T_a - T_b}{R} \quad \text{W} \tag{3-26a}$$

where

$$R = R_a + R_1 + R_2 + R_b \tag{3-26b}$$

Here R is the *total thermal resistance* in the path of heat flow through the composite cylinder from T_a to T_b.

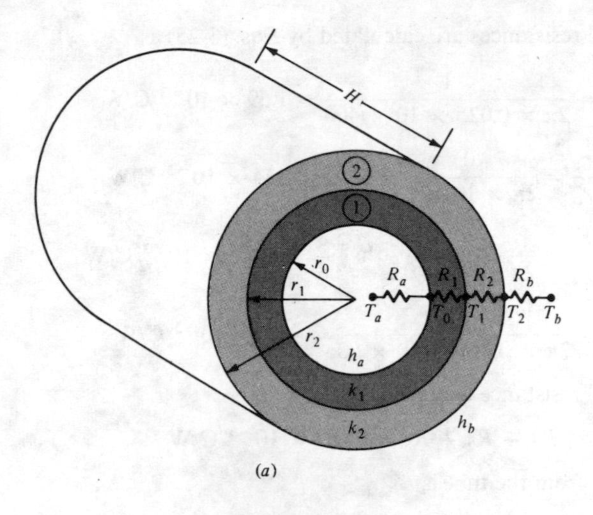

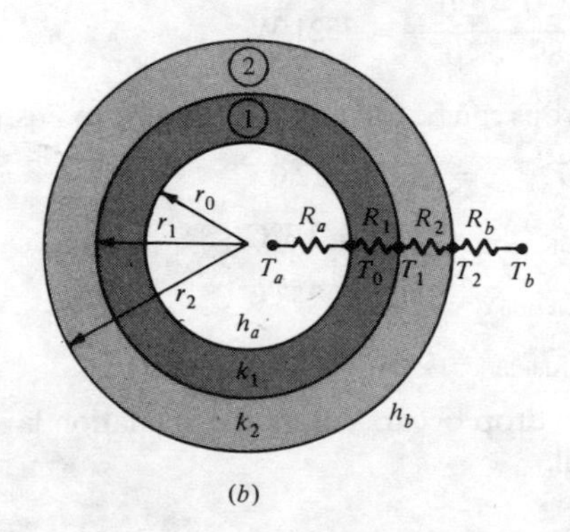

Figure 3-11 Thermal resistances for radial heat flow through a hollow composite cylinder and sphere with layers in perfect thermal contact. (*a*) Coaxial cylinder, (*b*) concentric spheres.

Example 3-11 A steel tube with 5-cm ID, 7.6-cm OD, and $k = 15$ W/(m · °C) is covered with an insulative covering of thickness $t = 2$ cm and $k = 0.2$ W/(m · °C). A hot gas at $T_a = 330$°C, $h_a = 400$ W/(m^2 · °C) flows inside the tube. The outer surface of the insulation is exposed to cooler air at $T_b = 30$°C with $h_b = 60$ W/(m^2 · °C).

Calculate the heat loss from the tube to the air for $H = 10$ m of the tube.

Calculate the temperature drops resulting from the thermal resistances of the hot gas flow, the steel tube, the insulation layer, and the outside air.

SOLUTION The radial heat flow through the tube is given by Eq. (3-26) as

$$Q = \frac{T_a - T_b}{R_a + R_1 + R_2 + R_b} \quad \text{W}$$

where various thermal resistances are calculated by Eqs. (3-25) as

$$R_a = \frac{1}{2\pi r_0 H h_a} = \frac{1}{2\pi \times 0.025 \times 10 \times 400} = 1.59 \times 10^{-3}\text{°C/W}$$

$$R_1 = \frac{1}{2\pi H k_1} \ln \frac{r_1}{r_0} = \frac{1}{2\pi \times 10 \times 15} \ln \frac{3.8}{2.5} = 0.44 \times 10^{-3}\text{°C/W}$$

$$R_2 = \frac{1}{2\pi H k_2} \ln \frac{r_2}{r_1} = \frac{1}{2\pi \times 10 \times 0.2} \ln \frac{5.8}{3.8} = 33.65 \times 10^{-3}\text{°C/W}$$

$$R_b = \frac{1}{2\pi r_2 H h_b} = \frac{1}{2\pi \times 0.058 \times 10 \times 60} = 4.21 \times 10^{-3}\text{°C/W}$$

Then the total thermal resistance becomes

$$R = R_a + R_1 + R_2 + R_b = 39.89 \times 10^{-3}\text{°C/W}$$

and the total heat loss from the tube is

$$Q = \frac{330 - 30}{39.89 \times 10^{-3}} = 7521 \text{ W}$$

Given Q, various temperature drops can be calculated according to Eqs. (3-24):

$$\Delta T_{\text{hot gas}} = QR_a = 12.0\text{°C}$$

$$\Delta T_{\text{tube wall}} = QR_1 = 3.3\text{°C}$$

$$\Delta T_{\text{insulation}} = QR_2 = 253.0\text{°C}$$

$$\Delta T_{\text{outside air}} = QR_b = 31.7\text{°C}$$

Clearly, the largest temperature drop occurs across the insulation layer and the smallest across the tube wall.

Composite Concentric Spheres

Figure 3-11*b* shows a composite sphere consisting of two concentric layers. The interior and exterior surfaces are subjected to heat exchange by convection with fluids at constant temperatures T_a and T_b, with heat transfer coefficients h_a and h_b, respectively. The total radial heat flow Q through the sphere is given by

$$Q = \frac{T_a - T_0}{R_a} = \frac{T_0 - T_1}{R_1} = \frac{T_1 - T_2}{R_2} = \frac{T_2 - T_b}{R_b} \tag{3-27}$$

where various thermal resistances are defined as

$$R_a = \frac{1}{4\pi r_0^2 h_a} \qquad R_1 = \frac{1}{4\pi k_1} \frac{r_1 - r_0}{r_1 r_0}$$

$$R_2 = \frac{1}{4\pi k_2} \frac{r_2 - r_1}{r_2 r_1} \qquad R_b = \frac{1}{4\pi r_2^2 h_b} \tag{3-28}$$

Here the thermal resistances R_1 and R_2 for conduction through a hollow sphere are written according to Eq. (3-19*b*).

Summing the numerators and the denominators of the individual ratios in Eq. (3-27), we write

$$Q = \frac{T_a - T_b}{R} \quad \text{W} \tag{3-29a}$$

where

$$R = R_a + R_1 + R_2 + R_b \tag{3-29b}$$

where R is the *total thermal resistance* for the system.

3-5 THERMAL CONTACT RESISTANCE

Consider one-dimensional heat flow through a composite medium consisting of two bars brought into contact with lateral surfaces insulated. As illustrated in Fig. 3-12, the temperature profile through the solids experiences a sudden drop across the interface between the two materials. This temperature drop across the interface is the result of *thermal contact resistance.*

The physical significance of thermal contact resistance is better envisioned by examining an enlarged view of an interface as shown in Fig. 3-12. The direct contact between the solid surfaces takes place at a limited number of spots, and the voids between them usually are filled with air or the surrounding fluid. Heat transfer through the fluid filling the voids is mainly by conduction, because there is no convection in such a thin layer of fluid and the radiation effects are negligible

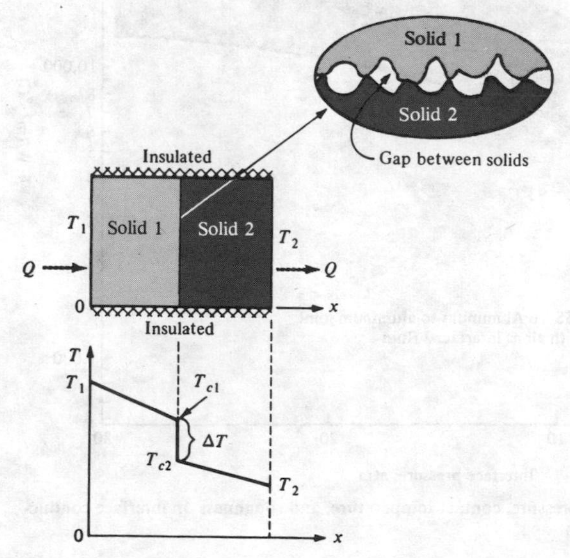

Figure 3-12 Temperature drop across a contact resistance.

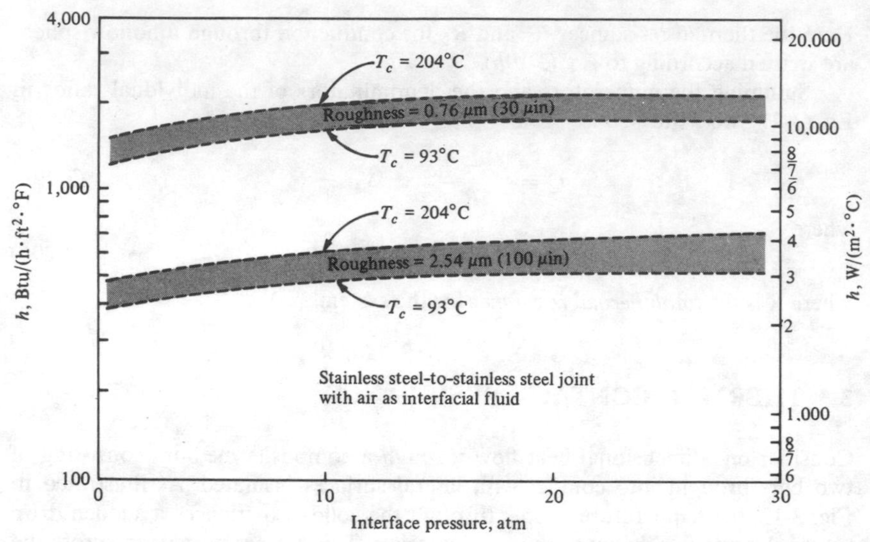

Figure 3-13(*a*) Effects of interface pressure, contact temperature, and roughness on interface conductance *h*.

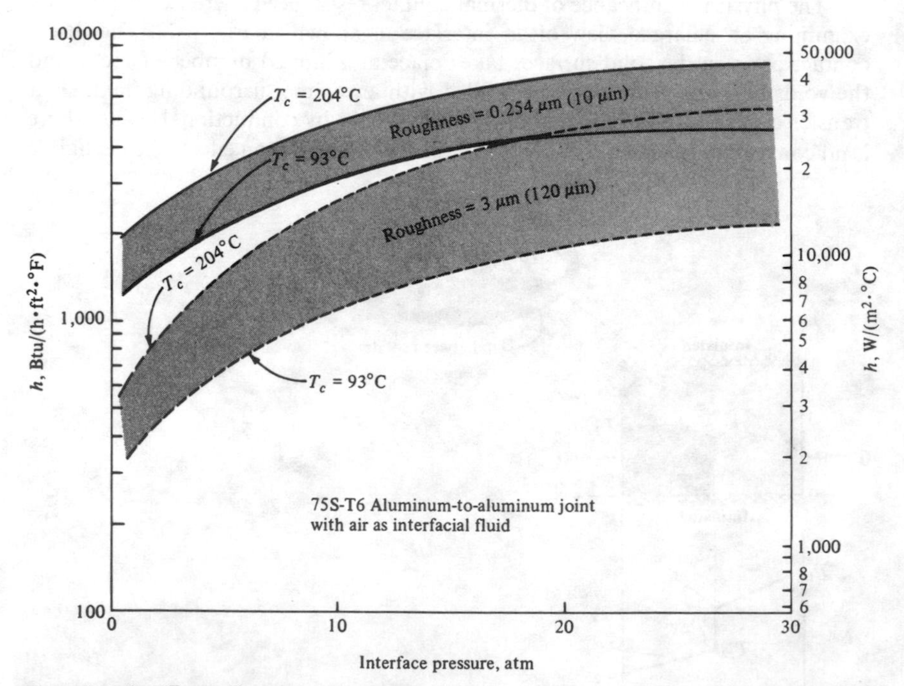

Figure 3-13(*b*) Effects of interface pressure, contact temperature, and roughness on interface conductance *h*. (*Based on data from Ref.* 2.)

Table 3-1 Thermal contact conductance in a vacuum

Type of interface	Contact temperature, °C	Contact roughness, μm	Contact pressure, atm	h_c, $W/(m^2 \cdot °C)$
304 Stainless	25	0.25×0.38	0.6 to 75	300 to 11,000
steel	25	1.1×1.3	0.7 to 75	200 to 2100
6061-T6				
Aluminum	25	0.3	0.7 to 75	1500 to 32,000
Copper	45	0.2	0.7 to 75	6500 to 14,000
Magnesium	40	0.3	0.7 to 75	11,000 to 26,500

Based on data in Ref. 6.

at normal temperatures. Then heat transfer across the interface takes place entirely by conduction through both the thin layer filling the voids and the spots in direct metal-to-metal contact. If the thermal conductivity of the fluid is less than that of the solids, the interface acts as a resistance to heat flow; this resistance is referred to as the *thermal contact resistance.*

Considerable amounts of experimental and theoretical work have been reported in the literature on the prediction of thermal contact resistance [1–12]. It appears that the reliable results for practical use are still those that have been determined experimentally. To illustrate the effects of various parameters such as the surface roughness, the interface temperature, the interface pressure, and the type of material, we present in Fig. 3-13*a* and *b* the *interface thermal contact conductance h* for stainless steel–to–stainless steel and aluminum–to–aluminum joints. The results on these figures show that interface conductance increases with increasing interface pressure, increasing interface temperature, and decreasing surface roughness. The interface conductance is higher with a softer material (aluminum) than with a harder material (stainless steel).

The smoothness of the surface is another factor that affects contact conductance; a joint with a superior surface finish may exhibit lower contact conductance owing to waviness. The adverse effect of waviness can be overcome by introducing between the surfaces an interface shim from a soft material such as lead.

Contact conductance also is reduced with a decrease in the ambient air pressure, because the effective thermal conductance of the gas entrapped in the interface is lowered. To illustrate this effect, Table 3-1 shows appropriate ranges of contact conductance under vacuum conditions.

Example 3-12 Consider one-dimensional, steady-state heat flow along two stainless-steel bars, each of diameter $D = 2$ cm, length $L = 3$ cm and pressed together with a pressure of 10 atm. The surface has a roughness of about 2.5 μm. An overall temperature difference of $\Delta T = 100°C$ is applied across the bars. The interface temperature is about 90°C.

Calculate the heat flow rate along the bars.

Calculate the temperature drop at the interface.

SOLUTION The total heat flow rate Q along the bars is determined by the application of the thermal resistance concept as

$$Q = \frac{\Delta T}{R_1 + R_c + R_1} \quad \text{W}$$

where $\Delta T = 100°\text{C}$. The thermal resistance R_1 for each of the steel bars is determined as

$$R_1 = \frac{L}{kA} = \frac{0.03}{20 \times \pi/4 \times (0.02)^2} = 4.775$$

The contact conductance of the interface is obtained from Fig. 3-13*a* as $h_c = 3000 \text{ W/(m}^2 \cdot °\text{C)}$. Then the thermal contact resistance of the interface becomes

$$R_c = \frac{1}{hA} = \frac{1}{3000 \times \pi/4 \times (0.02)^2} = 1.061$$

So the heat flow rate is

$$Q = \frac{100}{4.775 + 1.061 + 4.775} = 9.42 \text{ W}$$

The temperature drop across the interface becomes

$$\Delta T_c = \frac{R_c}{R_1 + R_c + R_1} \Delta T = \frac{1.061}{10.61} \times 100 = 10°\text{C}$$

3-6 CRITICAL THICKNESS OF INSULATION

Consider a small-diameter tube, cable, or wire the outside surface of which has approximately constant temperature and dissipates heat by convection into the surrounding air. Suppose the surface is covered with a layer of insulation. In some situations the addition of insulation increases the heat loss until a critical thickness of insulation at which the energy loss becomes maximum. Further addition of insulation beyond the critical thickness starts to decrease the energy loss. Therefore, this critical thickness can be used to increase the cooling of a cable, wire, or tube. However, if the insulation is added to reduce the heat loss from a tube, it is essential that the thickness of the insulation added be larger than the critical thickness of the insulation. We now examine the critical thickness of insulation for a cylinder and a sphere.

Cylinder

To develop a relation for the critical thickness of insulation, we consider a circular tube of radius r_i maintained at a uniform temperature T_i and covered with a layer of insulation of radius r_o, as illustrated in Fig. 3-14. Heat is dissipated by convection from the outside surface of the insulation into an ambient at temperature T_∞ with a heat transfer coefficient h_o.

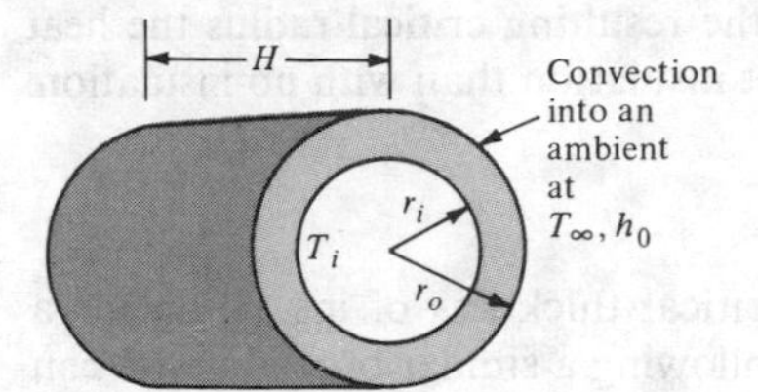

Figure 3-14 Nomenclature for the critical radius of insulation for a tube.

The rate of heat loss Q from the tube is given by

$$Q = \frac{T_i - T_\infty}{R_{\text{ins}} + R_o} \quad \text{W} \tag{3-30}$$

If H is the tube length and k is the thermal conductivity of the insulation, the thermal resistances R_{ins} and R_o of the insulation and the convection at the outer surface are

$$R_{\text{ins}} = \frac{1}{2\pi kH} \ln \frac{r_o}{r_i} \quad \text{and} \quad R_o = \frac{1}{2\pi r_o H h_0} \tag{3-31}$$

Now, it is assumed that T_i, T_∞, k, H, h_0, and r_i remain constant and r_o is allowed to vary (i.e., $r_o \geq r_i$). We note that as r_o increases, the thermal resistance R_o decreases but R_{ins} increases. Therefore, Q may have a maximum for a certain value of $r_o \equiv r_{oc}$. This *critical value of the radius* r_{oc} is determined by differentiating Eq. (3-30) with respect to r_o and setting the resulting expression equal to zero:

$$\frac{dQ}{dr_o} = -\frac{2\pi kH(T_i - T_\infty)}{[\ln (r_o/r_i) + k/(h_0 r_o)]^2} \left(\frac{1}{r_o} - \frac{k}{h_0 r_o^2}\right) = 0 \tag{3-32}$$

The solution of Eq. (3-32) for r_o gives the *critical radius* r_{oc} of insulation at which the heat transfer rate is a maximum; we find

$$\boxed{r_{oc} = \frac{k}{h_0}} \tag{3-33}$$

In practice, the physical significance of this result is as follows: If the radius is greater than the critical radius defined by Eq. (3-33), any addition of insulation on the tube surface decreases the heat loss, as one expects. But if the radius is less than the critical radius, as in small-diameter tubes, cables, or wires, the heat loss will increase continuously with the addition of insulation until the radius of the outer surface of the insulation equals the critical radius. The heat loss becomes maximum at the critical thickness of insulation and begins to decrease with the addition of insulation beyond the critical radius.

There are numerous practical applications of the critical radius of insulation. In electric wires or cables, the critical thickness of coating can be utilized to achieve a maximum amount of cooling. If the insulation on a steam pipe is wetted, the thermal conductivity of the insulation increases, which in turn increases the critical radius. Then it is possible that with the resulting critical radius the heat loss from the pipe will become larger with wet insulation than with no insulation.

Sphere

In the previous analysis we discussed the critical thickness of insulation for a cylindrical body. In the case of a sphere, by following a similar procedure we can show that the critical radius of insulation becomes

$$\boxed{r_{oc} = \frac{2k}{h_0}} \tag{3-34}$$

where k is the thermal conductivity of insulation.

Effects of Radiation

The results given above for the critical radius do not include the effects of thermal radiation. Suppose the heat transfer coefficient h_0 at the outer surface of insulation is approximated by the sum of a convection component h_c and a radiation component h_r in the form

$$h_0 = h_c + h_r \tag{3-35}$$

Then the critical radius given by Eqs. (3-33) and (3-34) becomes, respectively,

$$r_{oc} = \frac{k}{h_c + h_r} \qquad \text{for a cylinder} \tag{3-36a}$$

$$r_{oc} = \frac{2k}{h_c + h_r} \qquad \text{for a sphere} \tag{3-36b}$$

Example 3-13 A tube with OD of $D = 2$ cm is maintained at a uniform temperature and is covered with an insulative tube covering [$k = 0.18$ W/(m · °C)] in order to reduce the heat loss. Heat is dissipated from the outer surface of the cover by natural convection with $h_0 = 12$ W/(m^2 · °C) into the ambient air at constant temperature. Determine the critical thickness of the insulation. Calculate the ratio of the heat loss from the tube with insulation to that without any insulation for (1) the thickness of insulation equal to that at the critical thickness and (2) the thickness of insulation 2.5 cm thicker than the critical thickness.

SOLUTION The critical radius r_{oc} of insulation is determined by Eq. (3-32):

$$r_{oc} = \frac{k}{h_0} = \frac{0.18}{12} = 0.015 \text{ m} = 1.5 \text{ cm}$$

Then the critical thickness of insulation is 0.5 cm.

The heat losses from the tube with and without insulation are respectively,

$$Q_{\text{with}} = 2\pi r_o H h_0 \,\Delta T \,\frac{1}{1 + (r_o h_0/k)\ln(r_o/r_i)}$$

$$Q_{\text{without}} = 2\pi r_i H h_0 \,\Delta T$$

where H is the tube length and r_i is the tube radius. Then the heat-loss ratio becomes

$$\frac{Q_{\text{with}}}{Q_{\text{without}}} = \frac{r_o}{r_i}\left(1 + \frac{r_o h_0}{k}\ln\frac{r_o}{r_i}\right)^{-1} \tag{3-37a}$$

For $r_o = r_{oc}$, this ratio reduces to

$$\frac{Q_{\text{with}}}{Q_{\text{without}}} = \frac{r_{oc}}{r_i}\left(1 + \ln\frac{r_{oc}}{r_i}\right)^{-1} \tag{3-37b}$$

since $r_{oc} h_0/k = 1$.

For the critical radius $r_{oc} = 1.5$ cm, this result gives

$$\frac{Q_{\text{with}}}{Q_{\text{without}}} = \frac{1.5}{1}(1 + \ln 1.5)^{-1} = 1.067$$

which shows that the heat loss is increased about 7 percent despite the fact that there is an insulation of thickness 0.5 cm. If another layer of 2.5-cm-thick insulation is added, we have $r_o = 1.5 + 2.5 = 4$ cm, and the heat-loss ratio is

$$\frac{Q_{\text{with}}}{Q_{\text{without}}} = \frac{4}{1}\left(1 + \frac{0.04 \times 12}{0.18}\ln 4\right)^{-1} = 0.85$$

This result implies that with a 3-cm-thick insulation layer the heat loss is reduced by about 15 percent.

3-7 FINNED SURFACES

Heat transfer by convection between a surface and the fluid surrounding it can be increased by attaching to the surface thin strips of metal called *fins*. A large variety of fin geometries are manufactured for heat transfer applications; Fig. 3-15 shows typical examples.

When heat transfer takes place by convection from both interior and exterior surfaces of a tube or a plate, generally fins are used on the surface where the heat transfer coefficient is low. For example, in a car radiator the outer surface of the tubes is finned because the heat transfer coefficient for air at the outer surface is much smaller than that for water flow at the inner surface. The problem of determination of heat flow through a fin requires a knowledge of temperature

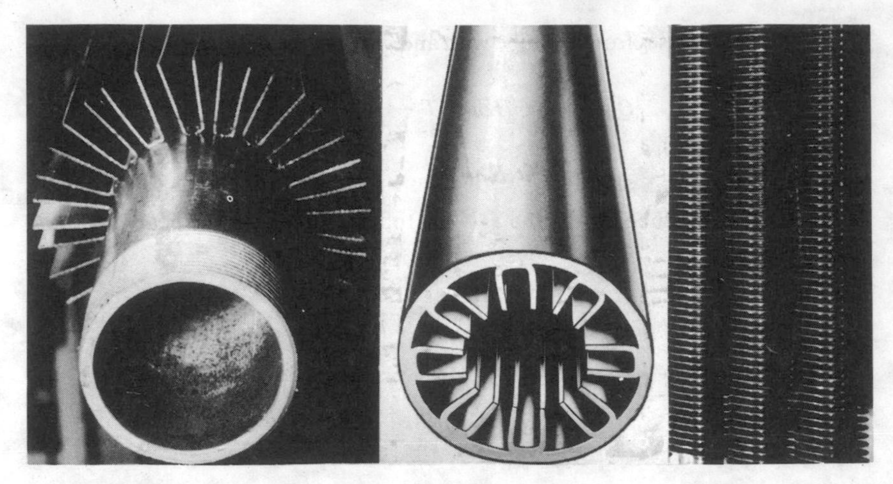

Figure 3-15 Types of finned tubing. (*From Brown Fintube Co.*)

distribution in the fin. In this section we analyze temperature distribution and heat flow through a fin having a simple geometry. The objective is to provide the reader a better insight to the physical significance of various parameters that affect heat transfer through a finned surface.

To determine the temperature distribution through a fin, we must develop the governing energy equation by performing an energy balance on a differential volume element in the fin. Figure 3-16 illustrates the geometry, the coordinates, and the nomenclature for the development of the one-dimensional, steady-state energy equation for fins of *uniform cross section.* We consider a small volume

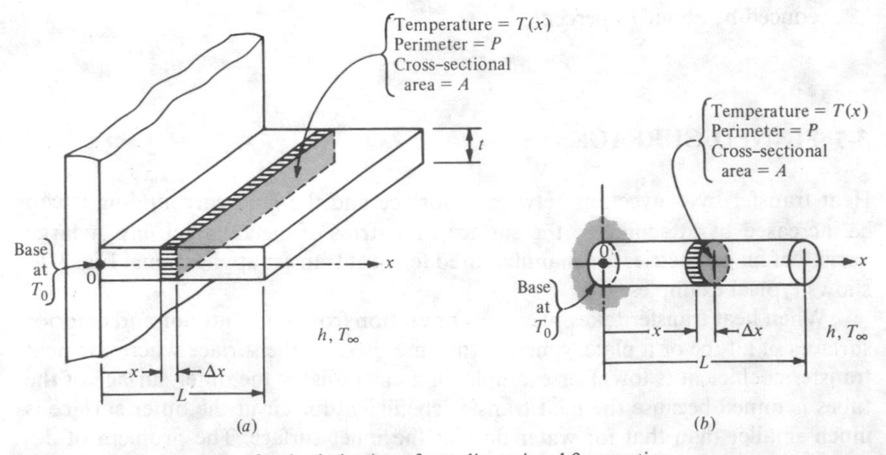

Figure 3-16 Nomenclature for the derivation of one-dimensional fin equation.

element Δx and write the statement of the steady-state energy balance for this volume element as

$$\underbrace{\begin{pmatrix}\text{Net rate of heat gain}\\ \text{by conduction in}\\ x \text{ direction into}\\ \text{volume element } \Delta x\end{pmatrix}}_{\text{I}} + \underbrace{\begin{pmatrix}\text{net rate of heat gain}\\ \text{by convection through}\\ \text{lateral surfaces into}\\ \text{volume element } \Delta x\end{pmatrix}}_{\text{II}} = 0 \tag{3-38}$$

The net heat gain by conduction is given by

$$\text{I} = -\frac{d(qA)}{dx}\,\Delta x = kA\,\frac{d^2T(x)}{dx^2}\,\Delta x \tag{3-39a}$$

and the net heat gain by convection by

$$\text{II} = h[T_\infty - T(x)]P\,\Delta x \tag{3-39b}$$

where the cross-sectional area A, the perimeter P, the heat transfer coefficient h, and the thermal conductivity of the fin material k are constant. Equations (3-39) are introduced into Eq. (3-38), Δx is canceled, and the result is rearranged to yield

$$\frac{d^2T(x)}{dx^2} - \frac{hP}{Ak}[T(x) - T_\infty] = 0 \tag{3-40}$$

This result is written more compactly in the form

$$\boxed{\frac{d^2\theta(x)}{dx^2} - m^2\theta(x) = 0} \tag{3-41a}$$

where

$$m^2 \equiv \frac{hP}{Ak} \qquad \theta(x) = T(x) - T_\infty \tag{3-41b}$$

Equation (3-41*a*) is called the *one-dimensional fin equation for fins of uniform cross section.* The solution of this ordinary differential equation subject to appropriate boundary conditions at the two ends of the fin gives the temperature distribution in the fin. Once the temperature distribution is known, the heat flow through the fin is readily determined.

Equation (3-41*a*) is a linear, homogeneous, second-order ordinary differential equation with constant coefficients. Its general solution may be taken in the form

$$\theta(x) = C_1e^{-mx} + C_2e^{mx} \tag{3-42}$$

where the two constants are determined from the two boundary conditions specified for the fin problem. The solution given by Eq. (3-42) is more convenient to use in the solution of the fin equation (3-41*a*) for the *long fin.* Soon this matter will be clear in the solution of fin problems.

Recalling the fact that hyperbolic sine and hyperbolic cosine can be constructed by the combination of e^{-mx} and e^{mx}, we know it is possible to express the solution (3-42) in the following alternative forms:

$$\theta(x) = C_1 \cosh mx + C_2 \sinh mx \tag{3-43a}$$

or

$$\theta(x) = C_1 \cosh m(L - x) + C_2 \sinh m(L - x) \tag{3-43b}$$

The solution given by Eqs. (3-43) is more convenient to use for the analysis of fins of finite length, as soon will be apparent.

The temperature distribution $\theta(x)$ in a fin having a uniform cross section can be determined from Eq. (3-42) or (3-43) if the integration constants C_1 and C_2 are determined by the application of the two boundary conditions for the problem, one specified for the *fin base* and the other for the *fin tip*. Customarily, the temperature at the fin base $x = 0$ is considered known; that is,

$$\theta(0) = T_0 - T_\infty \equiv \theta_0 \tag{3-44}$$

where T_0 is the fin base temperature. Several different physical situations are possible at the fin tip $x = L$; any one of the following conditions may be considered: *long fin, negligible heat loss from the fin tip,* and *convection at the fin tip.* To illustrate the physical significance of such conditions and the corresponding heat transfer results, we present the solution of the fin problem for each of these three cases.

Long Fin

For a sufficiently long fin, it is reasonable to assume that the temperature at the fin tip approaches the temperature T_∞ of the surrounding fluid. With this consideration the mathematical formulation of the fin problem becomes

$$\frac{d^2\theta(x)}{dx^2} - m^2\theta(x) = 0 \qquad \text{in } x > 0 \tag{3-45a}$$

$$\theta(x) = T_0 - T_\infty \equiv \theta_0 \qquad \text{at } x = 0 \tag{3-45b}$$

$$\theta(x) \to 0 \qquad \text{as } x \to \infty \tag{3-45c}$$

where $m^2 \equiv Ph/(Ak)$. The solution is taken in the form given by Eq. (3-42):

$$\theta(x) = C_1 e^{-mx} + C_2 e^{mx} \tag{3-46}$$

The boundary condition (3-45*c*) requires that $C_2 = 0$, and the application of the boundary condition (3-45*b*) gives $C_1 = \theta_0$. Then the solution becomes

$$\boxed{\frac{\theta(x)}{\theta_0} = \frac{T(x) - T_\infty}{T_0 - T_\infty} = e^{-mx}} \tag{3-47}$$

which is the simplest analysis of the fin problem.

Now, since the temperature distribution is known, the heat flow through the fin is determined either by integrating the convective heat transfer over the entire fin surface according to the relation

$$Q = \int_{x=0}^{L} hP\theta(x)\, dx \tag{3-48a}$$

or by evaluating the conductive heat flow at the fin base according to the relation

$$Q = -Ak \left. \frac{d\theta(x)}{dx} \right|_{x=0} \tag{3-48b}$$

Equations (3-48*a*) and (3-48*b*) give identical results since heat flow through the lateral surfaces by convection is equal to heat flow at the fin base by conduction. Here we prefer to use the relation given by Eq. (3-48*b*). By substituting $\theta(x)$ from Eq. (3-47) into Eq. (3-48*b*), the heat flow rate through the fin becomes

$$\boxed{Q = Ak\theta_0 m = \theta_0 \sqrt{PhkA}} \qquad \text{W} \tag{3-49}$$

since $m = \sqrt{Ph/(kA)}$.

Fins with Negligible Heat Loss at the Tip

The heat transfer area at the fin tip is generally small compared with the lateral area of the fin for heat transfer. For such situations the heat loss from the fin tip is negligible compared with that from the lateral surfaces, and the boundary condition at the fin tip characterizing this situation is taken as $d\theta/dx = 0$ at $x = L$. Then the mathematical formulation of the fin problem becomes

$$\frac{d^2\theta(x)}{dx^2} - m^2\theta(x) = 0 \qquad \text{in } 0 \le x \le L \tag{3-50a}$$

$$\theta(x) = T_0 - T_\infty \equiv \theta_0 \qquad \text{at } x = 0 \tag{3-50b}$$

$$\frac{d\theta(x)}{dx} = 0 \qquad \text{at } x = L \tag{3-50c}$$

Here we choose the solution in the form given by Eq. (3-43*b*):

$$\theta(x) = C_1 \cosh m(L - x) + C_2 \sinh m(L - x) \tag{3-51}$$

The reason for this choice is that the solution (3.51) is in such a form that one of the integration constants is immediately eliminated by the application of one of the boundary conditions. Indeed, the boundary condition (3-50*c*) requires that $C_2 = 0$; then the application of the boundary condition (3-50*b*) gives $C_1 = \theta_0/\cosh mL$, and the solution becomes

$$\boxed{\frac{\theta(x)}{\theta_0} = \frac{T(x) - T_\infty}{T_0 - T_\infty} = \frac{\cosh m(L - x)}{\cosh mL}} \tag{3-52}$$

If we had performed this analysis by choosing $\theta(x)$ in the form given by Eq. (3-42), it would have required rather lengthy manipulations to bring the solution into the form given by Eq. (3-52).

The heat flow rate Q through the fin is now determined by introducing the solution Eq. (3-52) into Eq. (3-48*b*). We obtain

$$\boxed{Q = Ak\theta_0 m \tanh mL = \theta_0\sqrt{PhkA}\tanh mL} \tag{3-53}$$

We note that this result reduces to that given by Eq. (3-49) for the long fin since $\tanh mL \rightarrow 1$ for sufficiently large values of mL (i.e., $\tanh mL$ is equal to 0.7616, 0.964, and 0.995, respectively, for $mL = 1$, 2, and 3). Also, Eq. (3-52) reduces to Eq. (3.47) for sufficiently large values of mL.

Fins with Convection at the Tip

A physically more realistic boundary condition at the fin tip is the one that includes heat transfer by convection between the fin tip and the surrounding fluid. Then the mathematical formulation of the heat conduction problem becomes

$$\frac{d^2\theta(x)}{dx^2} - m^2\theta(x) = 0 \qquad \text{in } 0 \leq x \leq L \tag{3-54a}$$

$$\theta(x) = T_0 - T_\infty \equiv \theta_0 \qquad \text{at } x = 0 \tag{3-54b}$$

$$k\frac{d\theta(x)}{dx} + h_e\theta(x) = 0 \qquad \text{at } x = L \tag{3-54c}$$

where k is the thermal conductivity of the fin and h_e is the heat transfer coefficient between the fin tip and the surrounding fluid.

The solution is chosen in the form given by Eq. (3-43*b*):

$$\theta(x) = C_1 \cosh m(L - x) + C_2 \sinh m(L - x) \tag{3-55}$$

The application of the boundary conditions (3-54*b*) and (3-54*c*), respectively, gives

$$\theta_0 = C_1 \cosh mL + C_2 \sinh mL \tag{3-56a}$$

and

$$-kC_2 m + h_e C_1 = 0 \tag{3-56b}$$

since

$$\left.\frac{d\theta}{dx}\right|_{x=L} = [-C_1 m \sinh m(L - x) - C_2 m \cosh m(L - x)]_{x=L} = -C_2 m$$

When C_1 and C_2 are determined from Eqs. (3-56) and introduced into Eq. (3-55), the temperature distribution in the fin becomes

$$\boxed{\frac{\theta(x)}{\theta_0} = \frac{T(x) - T_\infty}{T_0 - T_\infty} = \frac{\cosh m(L-x) + (h_e/mk)\sinh m(L-x)}{\cosh mL + (h_e/mk)\sinh mL}} \tag{3-57}$$

The heat flow rate through the fin is obtained by introducing this result into Eq. (3-48*b*). We obtain

$$\boxed{Q = \theta_0\sqrt{PhkA}\left[\frac{\sinh mL + (h_e/mk)\cosh mL}{\cosh mL + (h_e/mk)\sinh mL}\right]} \tag{3-58}$$

Clearly, the temperature distribution and the heat flow given by Eqs. (3-57) and (3-58) are for more general situations than those derived under the assumptions of *long fin* and *fin with negligible heat loss from the fin tip*. For example, $h_e = 0$ corresponds to no heat loss from the fin tip. Indeed, for $h_e = 0$, Eqs. (3-57) and (3-58), respectively, reduce to the results given by Eqs. (3-52) and (3-53).

Example 3-14 A steel rod of diameter $D = 2$ cm, length $L = 25$ cm, and thermal conductivity $k = 50$ W/(m · °C) is exposed to ambient air at $T_\infty = 20$°C with a heat transfer coefficient $h = 64$ W/(m^2 · °C). If one end of the rod is maintained at a temperature of 120°C, calculate the heat loss from the rod.

SOLUTION In this problem, the condition for the other end of the rod is not specified explicitly. By considering the length-to-diameter ratio, it appears that the long fin assumption may be applicable, and we can use the simplest analysis to solve this problem. First we compute the value of the parameter m:

$$m^2 = \frac{hP}{Ak} = \frac{h\pi D}{(\pi/4)D^2 k} = \frac{4h}{kD} = \frac{4 \times 64}{50 \times 0.02} = 4 \times 64$$

$$m = 16 \qquad \text{and} \qquad mL = 16 \times 0.25 = 4$$

Therefore Eq. (3-49) can be used to calculate the heat flow rate:

$$Q = \theta_0\sqrt{PAkh} = \theta_0\sqrt{(\pi D)\left(\frac{\pi}{4}D^2\right)kh}$$

$$= (120 - 20)\frac{\pi}{2}\sqrt{(0.02)^3 \times 50 \times 64} = 25.1 \text{ W}$$

If Eq. (3-53) were used, the difference in the result would be a factor of $\tanh mL = \tanh 4 = 0.999$, which is negligible.

Fin Efficiency

In the preceding analysis, we considered only fins of uniform cross section. In numerous applications fins of variable cross section are used. The determination of temperature distribution, and hence the heat flow for such cases, is rather involved and beyond the scope of this book. However, heat transfer analysis has been performed for a variety of fin geometries [12–14], and the results have been presented in terms of a parameter called *fin efficiency* η defined by

$$\eta = \frac{\text{actual heat transfer through fin}}{\text{ideal heat transfer through fin if entire fin surface were at fin base temperature } T_0} = \frac{Q_{\text{fin}}}{Q_{\text{ideal}}} \tag{3-59}$$

Here Q_{ideal} is given by

$$Q_{\text{ideal}} = a_f h \theta_0 \tag{3-60a}$$

where a_f = surface area of fin
h = heat transfer coefficient
$\theta_0 = T_0 - T_\infty$

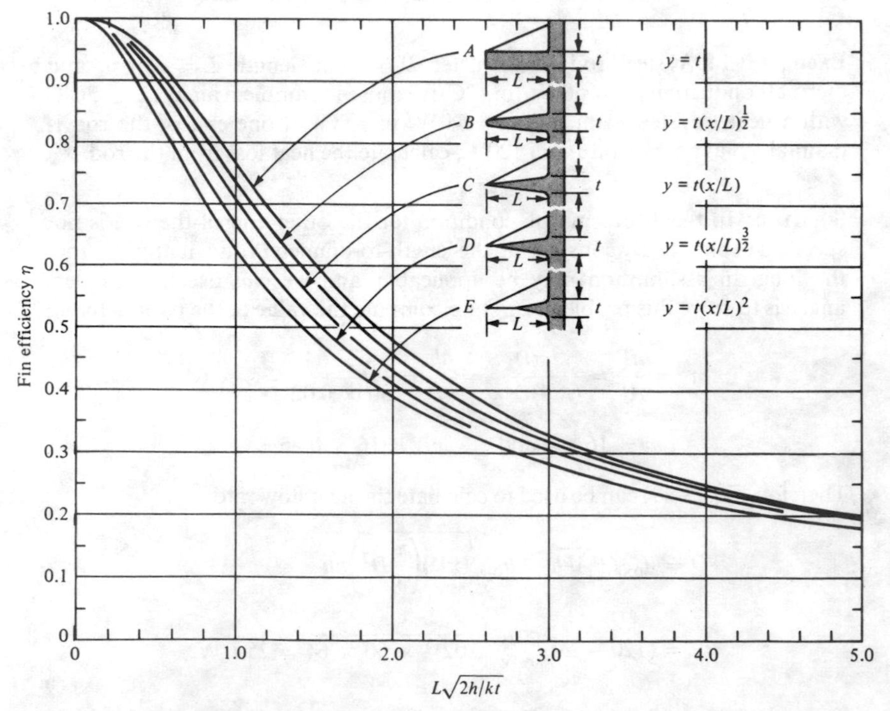

Figure 3-17 Efficiency of axial fins where the fin thickness y varies with the distance x from the root of the fin where $y = t$. (*From Gardner* [14].)

Thus, if the fin efficiency η is known, the heat transfer Q through the fin is determined from

$$Q_{\text{fin}} = \eta Q_{\text{ideal}} = \eta a_f h\theta_0 \tag{3-60b}$$

Figures 3-17 and 3-18 show the fin efficiency η plotted against the parameter $L\sqrt{2h/(kt)}$ for typical fin geometries. Figure 3-17 gives the efficiency of axial fins where the fin thickness y may vary with the distance x from the fin base when the thickness is t. Figure 3-18 is the fin efficiency for circular disk fins of constant thickness.

In practical applications a finned heat transfer surface is composed of the *fin surfaces* and the *unfinned portion*. Then the total heat transfer Q_{total} from such a surface is obtained by summing the heat transfer through the fins and the unfinned portion as

$$\begin{aligned} Q_{\text{total}} &= Q_{\text{fin}} + Q_{\text{unfinned}} \\ &= \eta a_f h\theta_0 + (a - a_f)h\theta_0 \end{aligned} \tag{3-61}$$

where a = total heat transfer area (i.e., fin surface + unfinned surface)
a_f = heat transfer area of fins only

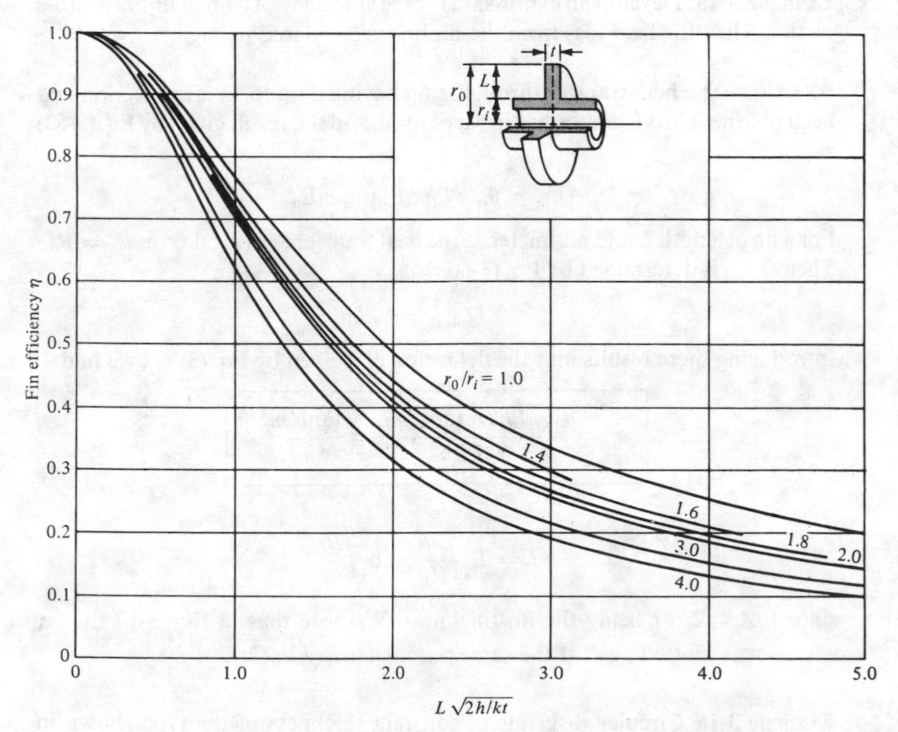

Figure 3-18 Efficiency of circular disk fins of constant thickness. (*From Gardner* [14].)

Equation (3-61) can be written more compactly as

$$Q_{\text{total}} = [\eta\beta + (1 - \beta)]ah\theta_0 \equiv \eta' ah\theta_0 \tag{3-62a}$$

where

$$\eta' \equiv \beta\eta + 1 - \beta = \text{area-weighted fin efficiency} \tag{3-62b}$$

$$\beta = \frac{a_f}{a} \tag{3-62c}$$

Although the addition of fins on a surface increases the surface area for heat transfer, it also increases the thermal resistance over the portion of the surface where the fins are attached. Therefore, there may be situations in which the addition of fins does not improve heat transfer. As a practical guide, the ratio $Pk/(Ah)$ *should be much larger than unity to justify the use of fins.* In the case of plate fins, for example, $P/A \cong 2/t$; then $Pk/(Ah)$ becomes $[2(k/t)]/h$, which implies that internal conductance of the fin should be much greater than the heat transfer coefficient for the fins to improve the heat transfer rate.

Example 3-15 Develop an expression for the efficiency of a fin of uniform cross section when the heat loss from the fin tip is considered negligible.

SOLUTION The heat transfer through a fin having a uniform cross section and negligible heat loss from the fin tip was solved, and the result given by Eq. (3-53) as

$$Q_{\text{fin}} = \theta_0 \sqrt{PhkA} \tanh mL$$

For a fin of length L and perimeter P, the heat transfer area is taken as $a_f = PL$. Then Q_{ideal} is determined by Eq. (3-60a) as

$$Q_{\text{ideal}} = PLh\theta_0$$

Introducing these results into the definition of η given by Eq. (3-59), we find

$$\eta = \frac{\theta_0 \sqrt{PhkA} \tanh mL}{\theta_0 PLh} = \frac{\tanh mL}{mL}$$

where

$$mL = L\sqrt{\frac{Ph}{Ak}} = L\sqrt{\frac{2h}{kt}}$$

since $P/A \cong 2/t$, t being the fin thickness. We note that in Fig. 3-17 the fin efficiency is plotted against the same parameter $L\sqrt{2h/(kt)}$.

Example 3-16 Circular disk fins of constant thickness of the type shown in Fig. 3-18 are attached on a 2.5-cm-OD tube with a spacing of 100 fins per

1-m length of tube. Fins are made of aluminum, $k = 160$ W/(m · °C), thickness $t = 1$ mm, and length $L = 1$ cm. The tube wall is maintained at $T_0 = 170$ °C, and the heat is dissipated by convection into the ambient air at $T_\infty = 30$°C with a heat transfer coefficient $h = 200$ W/(m^2 · °C). The radiation effects are considered negligible.

Calculate the heat loss to the ambient air per 1-m length of tube.

Compare this heat loss with that if there were no fins on the tube.

SOLUTION To determine the fin efficiency η from Fig. 3-18, the following parameters are calculated:

$$L\sqrt{\frac{2h}{kt}} = 1 \times 10^{-2}\sqrt{\frac{2 \times 200}{160 \times 10^{-3}}} = 0.5$$

$$\frac{r_o}{r_i} = \frac{1.25 + 1}{1.25} = 1.8$$

Then the fin efficiency is determined as

$$\eta \cong 0.9$$

We now compute β, the ratio of the heat transfer area for the fin to the total heat transfer area, by considering a 1-cm length (i.e., one cycle of fin) along the tube:

Fin surface per cm of tube length

$$= 2\pi(r_o^2 - r_i^2) = 2\pi[(2.25)^2 - (1.25)^2]$$
$$= 21.99 \text{ cm}^2$$

Total heat transfer surface per cm of tube length

$$= 2\pi(r_o^2 - r_i^2) + 2\pi r_i(1 - t)$$
$$= 2\pi[(2.25)^2 - (1.25)^2] + 2\pi(1.25)(1 - 0.1)$$
$$= 29.06 \text{ cm}^2$$

Then

$$\beta = \frac{21.99}{29.06} = 0.757$$

$$\eta' = \beta\eta + 1 - \beta = 0.757(0.9) + 0.243 = 0.924$$

The total heat transfer surface a per 1-m length of the tube is

$$a = 29.06 \times 100 \text{ cm}^2 = 0.29 \text{ m}^2$$

The total heat loss per 1-m length of tube is determined according to Eq. (3-62a):

$$Q = \eta' a h \theta_0 = 0.924 \times 0.29 \times 200(170 - 30) = 7503 \text{ W}$$

The heat loss per 1-m length of tube with no fins is

$$Q_{\text{nofin}} = 2\pi r_i h\theta_0 = 2\pi \times 0.0125 \times 200 \times (170 - 30) = 2199 \text{ W}$$

Clearly, the addition of fins increased the heat dissipation by a factor of about 3.4.

3-8 TEMPERATURE-DEPENDENT $k(T)$

In many applications the thermal conductivity of the solid varies with temperature very rapidly, or the temperature differences become so large that substantial variation occurs in the thermal conductivity. In such situations it may be necessary to include in the analysis the variation of thermal conductivity with temperature. In general, the solution of heat conduction problems with variable thermal conductivity is a complicated matter; however, in the case of one-dimensional, steady-state heat conduction in a solid subject to prescribed temperature boundary conditions, the effects of variable thermal conductivity on heat flow can be included in the analysis in a straightforward manner as described below.

Slab

Consider a slab in the region $0 \leq x \leq L$ having boundary surfaces at $x = 0$ and $x = L$ kept at uniform temperatures T_0 and T_1, respectively. The thermal conductivity $k(T)$ of the material depends strongly on temperature, and the variation of $k(T)$ with temperature is assumed to be specified. The rate of heat flow through the slab is determined by including in the analysis the temperature dependence of the thermal conductivity.

The mathematical formulation of the problem is given as

$$\frac{d}{dx}\left[k(T)\frac{dT}{dx}\right] = 0 \qquad \text{in } 0 \leq x \leq L \tag{3-63a}$$

$$T = T_0 \qquad \text{at } x = 0 \tag{3-63b}$$

$$T = T_1 \qquad \text{at } x = L \tag{3-63c}$$

The heat flow rate Q through an area A of the slab is determined from the definition

$$Q = Aq = -Ak(T)\frac{dT}{dx} \tag{3-64}$$

The integration of Eq. (3-63a) yields

$$k(T)\frac{dT}{dx} = C \tag{3-65}$$

Introducing Eq. (3-65) into Eq. (3-64), we obtain

$$Q = -AC \tag{3-66}$$

Clearly, the problem of determining the heat flow Q through the slab is reduced now to the determination of the integration constant C. This can be done readily by rearranging Eq. (3-65) in the form

$$k(T)\,dT = C\,dx \tag{3-67}$$

and integrating both sides of this equation over the limits for T from T_0 to T_1 and for x from 0 to L as specified by boundary conditions (3-63*b*) and (3-63*c*). We obtain

$$\int_{T_0}^{T_1} k(T)\,dT = C\int_0^L dx = CL \tag{3-68a}$$

or

$$C = \frac{1}{L}\int_{T_0}^{T_1} k(T)\,dT = -\frac{1}{L}\int_{T_1}^{T_0} k(T)\,dT \tag{3-68b}$$

The substitution of C into Eq. (3-66) gives the heat flow rate through the slab as

$$\boxed{Q = \frac{A}{L}\int_{T_1}^{T_0} k(T)\,dT} \tag{3-69}$$

Once the thermal conductivity $k(T)$ is specified as a function of temperature T, the integral on the right-hand side of this equation can be evaluated, and the heat flow rate Q through the slab is determined. This matter is illustrated with an example.

Example 3-17 In the slab problem defined by Eqs. (3-63), the thermal conductivity $k(T)$ of the material varies with temperature linearly in the form

$$k(T) = k_0(1 + \beta T)$$

where k_0 is a constant thermal conductivity and the constant β is called the *temperature coefficient of thermal conductivity*.

Develop an expression for the rate of heat flow Q through an area A of the slab.

If the slab is $L = 0.18$ m thick and the surfaces at $x = 0$ and $x = L$ are maintained at $T_0 = 500$ K and $T = 300$ K, respectively, calculate the heat flow rate per 1 m^2 of the slab for

$$k_0 = 0.2\ \text{W/(m}\cdot{}^\circ\text{C)} \qquad \beta = 2 \times 10^{-3}\ \text{K}^{-1}$$

SOLUTION The heat flow rate through the slab is determined by introducing the expression $k(T) = k_0(1 + \beta T)$ into Eq. (3-69) and performing the integration. We obtain

$$Q = \frac{Ak_0}{L}\left[(T_0 - T_1) + \tfrac{1}{2}\beta(T_0^2 - T_1^2)\right]$$

$$= Ak_0\left(1 + \beta\frac{T_0 + T_1}{2}\right)\frac{T_0 - T_1}{L}$$

$$= Ak_m\frac{T_0 - T_1}{L} \tag{3-70a}$$

where the *mean thermal conductivity* k_m is defined as

$$k_m \equiv k_0\left(1 + \beta\frac{T_0 + T_1}{2}\right) \tag{3-70b}$$

Equations (3-70) are utilized now to calculate the heat flow rate as

$$k_m = 0.2\left[1 + (2 \times 10^{-3})\frac{500 + 300}{2}\right] = 0.36\ \mathrm{W/(m \cdot {}^\circ C)}$$

$$Q = Ak_m\frac{T_0 - T_1}{L} = (1 \times 0.36)\frac{500 - 300}{0.18} = 400\ \mathrm{W}$$

We note that for the linear variation of k with temperature the heat flow rate through the slab can be calculated by the simple relation given by Eq. (3-70) in which the thermal conductivity k_m is evaluated at the arithmetic mean of the boundary surface temperatures, that is, $(T_0 + T_1)/2$.

Hollow Cylinder

Consider a hollow cylinder in the region $a \le r \le b$ with boundary surfaces at $r = a$ and $r = b$ kept at uniform temperatures T_0 and T_1, respectively. The thermal conductivity $k(T)$ of the material depends strongly on temperatures, and the variation of $k(T)$ with temperature is considered specified. The rate of heat transfer through a length H of the cylinder is determined by including in the analysis the variation of thermal conductivity with temperature.

The mathematical formulation of the problem is given as

$$\frac{d}{dr}\left[rk(T)\frac{dT}{dr}\right] = 0 \qquad \text{in } a \le r \le b \tag{3-71a}$$

$$T = T_0 \qquad \text{at } r = a \tag{3-71b}$$

$$T = T_1 \qquad \text{at } r = b \tag{3-71c}$$

The heat flow rate Q through a length H of the cylinder is determined from its definition:

$$Q = 2\pi rHq = -2\pi rH\left[k(T)\frac{dT}{dr}\right] \tag{3-72}$$

The integration of Eq. (3-71*a*) yields

$$rk(T)\frac{dT}{dr} = C \tag{3-73}$$

Introducing Eq. (3-73) into Eq. (3-72), we obtain

$$Q = -2\pi HC \tag{3-74}$$

Hence the problem of determining Q is reduced to the determination of the integration constant C. This can be done readily by rearranging Eq. (3-73) in the form

$$k(T)\,dT = C\,\frac{dr}{r} \tag{3-75}$$

and integrating both sides over the limits for T from T_0 to T_1 and for r from a to b as specified by the boundary conditions Eqs. (3-71*b*) and (3-71*c*). We obtain

$$\int_{T_0}^{T_1} k(T)\,dT = C\ln\frac{b}{a} \tag{3-76a}$$

or

$$C = -\frac{1}{\ln(b/a)}\int_{T_1}^{T_0} k(T)\,dT \tag{3-76b}$$

The substitution of Eq. (3-76*b*) into Eq. (3-74) gives the heat flow rate Q through the slab as

$$\boxed{Q = \frac{2\pi H}{\ln(b/a)}\int_{T_1}^{T_0} k(T)\,dT} \tag{3-77}$$

Once the thermal conductivity is specified as a function of temperature, the integral in this equation can be evaluated and the heat transfer rate through the cylinder determined.

3-9 SUMMARY OF BASIC RELATIONS

We summarize in Table 3-2 the basic relations given in this chapter.

Table 3-2 Basic relations

Equation number	Relation	Remark
	Thermal resistance concept:	
(3-5*a*)	$Q = \frac{T_1 - T_2}{R}$	Thermal resistance concept for heat flow

Table 3-2 (*continued*)

Equation number	Relation	Remark
(3-5*b*)	$R = \dfrac{L}{Ak}$	Thermal resistance for a slab
(3-12*b*)	$R = \dfrac{\ln (b/a)}{2\pi k H}$	Thermal resistance for a hollow cylinder
(3-19*b*)	$R = \dfrac{b - a}{4\pi kab}$	Thermal resistance for a hollow sphere
	Critical radius:	
(3-33)	$r_{oc} = \dfrac{k}{h_0}$	Critical radius of insulation for a cylinder
(3-34)	$r_{oc} = \dfrac{2k}{h_0}$	Critical radius of insulation for a sphere
	Fins of uniform cross section:	
(3-41*a*)	$\dfrac{d^2\theta(x)}{dx^2} - m^2\theta(x) = 0$	The fin equation
(3-41*b*)	$m^2 = \dfrac{hP}{Ak}$	
	Solution for a long fin:	
(3-47)	$\dfrac{\theta(x)}{\theta_0} = e^{-mx}$	Temperature profile
(3-49)	$Q = \theta_0\sqrt{PhkA}$	Heat flow rate
	Solution for a fin with negligible heat loss at the fin tip:	
(3-52)	$\dfrac{\theta(x)}{\theta_0} = \dfrac{\cosh m(L - x)}{\cosh mL}$	Temperature profile
(3-53)	$Q = \theta_0\sqrt{PhkA}\tanh mL$	Heat flow rate
	Solution for a fin with convection at the fin tip:	
(3-57)	$\dfrac{\theta(x)}{\theta_0} = \dfrac{\cosh m(L - x) + B \sinh m(L - x)}{\cosh mL + B \sinh mL}$	Temperature profile
(3-58)	$Q = \theta_0\sqrt{PhkA}\left(\dfrac{\sinh mL + B \cosh mL}{\cosh mL + B \sinh mL}\right)$	Heat flow rate
	where $B \equiv \dfrac{h_e}{mk}$	
	Temperature-dependent $k(T)$:	
(3-69)	$Q = \dfrac{A}{L}\displaystyle\int_{T_1}^{T_2} k(T)\, dT$	Heat flow rate for a slab
(3-77)	$Q = \dfrac{2\pi H}{\ln (b/a)}\displaystyle\int_{T_1}^{T_2} k(T)\, dT$	Heat flow rate for a hollow cylinder

PROBLEMS

Slab

3-1 Consider a slab of thickness $L = 0.20$ m and thermal conductivity $k = 40$ W/(m · °C) with no internal heat generation. The boundary surface at $x = 0$ is kept at uniform temperature T_1, and the boundary surface at $x = L$ is kept at uniform temperature T_2. Determine the heat flux across this slab under steady-state conditions for each of the cases shown:

T_1, °C	100	−20	−40
T_2, °C	0	40	−10

Also specify the direction of the heat flow for each case.

Answer: 20 kW/m², −12 kW/m², −6 kW/m²

3-2 Consider a slab of thickness $L = 0.25$ m. One surface is kept at 100°C and the other surface at 0°C. Determine the heat flux across this slab if the slab is made from (*a*) pure copper, (*b*) pure aluminum, (*c*) pure iron, (*d*) building brick, (*e*) cement, and (*f*) loosely packed asbestos.

3-3 Determine the steady-state heat flux through a 0.20-m-thick brick wall [$k = 0.69$ W/(m · °C)] with one surface at 30°C and the other at −20°C.

Answer: 172.5 W/m²

3-4 A pressure vessel for a nuclear reactor is approximated as a large flat plate of thickness L. The inside surface of the plate at $x = 0$ is insulated, the outside surface at $x = L$ is maintained at a uniform temperature T_2, and the gamma-ray heating of the plate can be represented as a heat generation term in the form

$$g(x) = g_0 e^{-\gamma x} \quad \text{W/m}^3$$

where g_0 and γ are constants and x is measured from the insulated inside surface.

(*a*) Develop an expression for the temperature distribution in the plate.

(*b*) Develop an expression for the temperature at the insulated surface (that is, $x = 0$) of the plate.

(*c*) Develop an expression for the heat flux at the outer surface $x = L$.

3-5 What is the total rate of heat loss per hour from a 0.5 m by 0.5 m by 1 m container having 5-cm-thick walls made of an insulated material of thermal conductivity $k = 0.04$ W/(m · °C) for a temperature difference of 30°C between the inside and outside? Neglect the thermal resistances for the heat transfer coefficient at the inside and outside surfaces.

Answer: 60 W

3-6 Derive an expression for one-dimensional, steady-state temperature distribution $T(x)$ in a slab of thickness L, with no heat generation, when the boundary surface at $x = 0$ is kept at a uniform temperature T_0 and the boundary surface at $x = L$ dissipates heat by convection with a heat transfer coefficient h into the ambient air at temperature T_∞. Assume constant thermal conductivity.

3-7 Derive an expression for the one-dimensional, steady-state temperature distribution $T(x)$ in a slab of thickness L for the following conditions: Heat is generated in the slab at a constant rate of g_0 W/m³, the boundary surface at $x = 0$ is kept insulated, and the boundary surface at $x = L$ is kept at zero temperature. Assume constant thermal conductivity. Give the relation for the temperature of the insulated boundary.

Calculate the temperature of the insulated surface for $k = 40$ W/(m · °C), $g_0 = 10^6$ W/m³, $L = 0.1$ m.

Answer: 125°C

3-8 A plane wall of thickness L, thermal conductivity k has its surface at $x = 0$ insulated, and the other surface at $x = L$ is kept at zero temperature. Heat is generated within the wall at a rate of

$$g(x) = g_0 \cos \frac{\pi x}{2L} \qquad \text{W/m}^3$$

where g_0 is the heat generation rate per unit volume at $x = 0$.

(*a*) Develop an expression for the one-dimensional, steady-state temperature distribution in the wall.

(*b*) Develop an expression for the temperature of the insulated surface.

3-9 A solid cylindrical rod of diameter $D = 1$ cm and length $L = 15$ cm is *thermally insulated on its cylindrical surface*. One of the end surfaces is kept at 0°C and the other at 200°C. Determine the heat flow rate through this rod if the rod is made of (*a*) pure copper, (*b*) pure iron, and (*c*) portland cement.

3-10 A large window glass $L = 0.5$ cm thick with thermal conductivity $k = 0.78$ W/(m · °C) is exposed to warm air at $T_i = 25$°C over its inner surface, and the heat transfer coefficient for the inside air is $h_i = 15$ W/(m^2 · °C). The outside air is at $T_0 = -15$°C, and the heat transfer coefficient associated with the outside surface is $h_0 = 50$ W/(m^2 · °C). What are the temperatures of the inner and outer surfaces of the glass?

Answer: $T_1 = -3.7$°C, $T_2 = -6.4$°C

3-11 Determine the relation for the steady-state temperature distribution in a slab $0 \leq x \leq L$ in which heat is being generated at a rate of $g(x) = g_0 x^2$ W/m^3, while the boundary surface at $x = 0$ is insulated and that at $x = L$ is maintained at zero temperature. Give the relation for the temperature of the insulated surface at $x = 0$.

3-12 A steel plate of thickness $L = 5$ cm and thermal conductivity $k = 20$ W/(m · °C) is subjected to a uniform heat flux $q = 600$ W/m^2 on one of its surfaces and dissipates heat by convection with a heat transfer coefficient $h_\infty = 80$ W/(m^2 · °C) from the other surface into the ambient air at $T_\infty = 25$°C. What is the temperature of the surface dissipating heat by convection?

Answer: 32.5°C

3-13 A wall of thickness L is irradiated with gamma rays at the surface $x = 0$. The gamma-ray attenuation with the wall can be regarded as a volumetric heat generation at a rate

$$g(x) = g_0 e^{-\beta x} \qquad \text{W/m}^3$$

where g_0 and β are constants. The boundary surface at $x = 0$ is insulated, and the boundary surface at $x = L$ dissipates heat by convection with a heat transfer coefficient h into the ambient air at temperature T_∞.

Develop an expression for the temperature distribution in the wall and the temperature of the insulated surface.

3-14 A concrete wall of thickness $L_1 = 15$ cm has thermal conductivity $k_1 = 0.76$ W/(m · °C). The inside surface is exposed to air at $T_i = 20$°C, and the outside surface to air at $T_0 = -20$°C. The heat transfer coefficients for the inside and outside surfaces are $h_i = 10$ W/(m^2 · °C) and $h_o = 40$ W/(m^2 · °C). Determine the rate of heat loss per square meter of wall surface.

3-15 A slab has a thickness L and thermal conductivity k, and its boundary surface at $x = 0$ is kept at temperature T_1 and at $x = L$ at temperature T_2. Show that the heat transfer Q through an area A of the slab is given by

$$Q = \frac{T_1 - T_2}{R} \qquad \text{W}$$

where the thermal resistance R for the slab is

$$R = \frac{L}{Ak}$$

Cylinder

3-16 Determine the one-dimensional, steady-state temperature distribution $T(r)$ in a solid cylinder of radius $r = b$ for the following conditions: Heat is generated in the cylinder at a constant rate of

g_0 W/m^3, and the boundary surface at $r = b$ is kept at zero temperature. Assume constant thermal conductivity. Also, give the expressions defining the temperature at the center of the cylinder and the heat flux at the surface $r = b$.

3-17 Repeat Problem 3-16 for the case when the heat generation rate $g(r)$ within the cylinder is a function of the radial position in the form

$$g(r) = g_0\left(1 - \frac{r}{b}\right) \qquad \text{W/m}^3$$

where g_0 is a constant.

3-18 A long cylindrical rod of radius $b = 5$ cm, $k = 10$ W/(m · °C) contains radioactive material which generates heat uniformly within the cylinder at a constant rate of $g_0 = 3 \times 10^5$ W/m^3. The rod is cooled by convection from its cylindrical surface into the ambient air at $T_\infty = 50$°C with a heat transfer coefficient $h_\infty = 60$ W/(m^2 · °C). Determine the temperatures at the center and the outer surface of this cylindrical rod.

Answer: 193.7°C, 175°C

3-19 A cylindrical insulation for a steam pipe has an inside radius $r_i = 6$ cm, outside radius $r_o = 8$ cm, and a thermal conductivity $k = 0.5$ W/(m · °C). The inside surface of the insulation is at a temperature $T_i = 430$°C, and the outside surface at $T_o = 30$°C. Determine the heat loss per 1-m length of this insulation.

Answer: 4368 W/m

3-20 A thick cylindrical pipe of thermal conductivity k W/(m^2 · °C) is subjected to a uniform constant temperature T_i at the inner surface $r = r_i$ and to zero temperature at the outer surface $r = r_o$. There is no heat generation. Determine the steady-state temperature distribution $T(r)$ in the cylinder. Obtain the relation for the heat flux at the outer boundary surface $r = r_o$.

3-21 A hollow cylindrical fuel element of inner radius r_i and outer radius r_o is heated uniformly within the entire volume at a constant rate of g_0 W/m^3 as a result of the disintegration of some radioactive material. The inner and outer surfaces of the element are at zero temperature, and the thermal conductivity of the cylinder is constant. Determine the relation for the one-dimensional, steady-state temperature distribution $T(r)$ within the cylinder.

3-22 A pipe of inner radius r_i, outer radius r_o, and thermal conductivity k is heated by passing an electric current through the pipe. The passage of current heats the tube at a constant rate of g_0 W/m^3. The outer surface of the pipe is insulated, and the inner surface is considered to be at zero temperature. Develop an expression for the steady-state temperature distribution $T(r)$ within the pipe and for the temperature of the outer insulated surface.

3-23 A long, hollow cylinder with constant thermal conductivity has a constant heat supply at a rate of q_i W/m^2 at the inner boundary surface at $r = r_i$, while the heat is dissipated from the outer boundary surface at $r = r_o$ by convection into a fluid at zero temperature with a heat transfer coefficient h. Develop an expression for the steady-state temperature distribution $T(r)$ in the cylinder.

3-24 Develop an expression for the steady-state temperature distribution $T(r)$ in a long, hollow cylinder, $a \le r \le b$, in which heat is generated at a rate of

$$g(r) = g_0(1 + Ar) \qquad \text{W/m}^3$$

where g_0 and A are constants, while the boundary surfaces at $r = a$ and $r = b$ are kept at zero temperature.

3-25 An electric resistance wire of radius $a = 1 \times 10^{-3}$ m, with thermal conductivity $k = 25$ W/(m · °C) is heated by the passage of electric current which generates heat within the wire at a constant rate of $g_0 = 2 \times 10^9$ W/m^3. Determine the centerline temperature rise above the surface temperature of the wire if the surface is maintained at a constant temperature.

Answer: 20°C

3-26 A 3-mm-diameter chrome-nickel wire of thermal conductivity $k = 20$ W/(m · °C) is heated electrically by the passage of electric current which generates heat within the wire at a constant rate of

$g_0 = 10^9$ W/m^3. If the outer surface of the wire is maintained at 100°C, determine the center temperature of the wire.

3-27 Heat is generated at a constant rate of $g_0 = 4 \times 10^8$ W/m^3 in a copper rod of radius $r = 0.5$ cm and thermal conductivity $k = 386$ W/(m · °C). The rod is cooled by convection from its cylindrical surface into an ambient at 30°C with a heat transfer coefficient $h = 2000$ W/(m · °C). Determine the surface temperature of the rod.

Answer: 530°C

3-28 In a cylindrical fuel element for a gas-cooled nuclear reactor, the heat generation rate within the fuel element due to fission can be approximated by the relation

$$g(r) = g_0\left[1 - \left(\frac{r}{b}\right)^2\right] \qquad \text{W/m}^3$$

where b is the radius of the fuel element and g_0 is constant. The boundary surface at $r = b$ is maintained at a uniform temperature T_0.

(*a*) Assuming one-dimensional, steady-state heat flow, develop a relation for the temperature drop from the centerline to the surface of the fuel element.

(*b*) For radius $b = 2$ cm, the thermal conductivity $k = 10$ W/(m · °C) and $g_0 = 4 \times 10^7$ W/m^3, calculate the temperature drop from the centerline to the surface.

Answer: 300°C

3-29 Consider a long, hollow cylinder of inner radius $r = a$ and outer radius $r = b$. The inner surface is heated uniformly at a constant rate of q_0 W/m^2, and the outer surface is maintained at zero temperature.

(*a*) Develop an expression for the steady-state temperature distribution.

(*b*) Calculate the temperature of the inner surface for $k = 50$ W/(m · °C), $a = 10$ cm, $b = 20$ cm, and $q_0 = 1.16 \times 10^5$ W/m^2.

3-30 Consider a long, hollow cylinder of inside radius $r = a$ and outside radius $r = b$. The outer surface is heated uniformly at a constant rate of q_0 W/m^2 while the inner surface is maintained at zero temperature.

(*a*) Develop an expression for the steady-state temperature distribution $T(r)$ in the cylinder.

(*b*) Calculate the temperature of the outer surface for $k = 75$ W/(m · °C), $a = 5$ cm, $b = 15$ cm, and $q_0 = 10^5$ W/m^2.

3-31 A hollow cylinder of inner radius r_1, outer radius r_2, and length H has a constant thermal conductivity k. If the inner surface is kept at temperature T_1 and the outer surface at temperature T_2, show that the total heat transfer rate Q through the cylinder is given by

$$Q = \frac{T_1 - T_2}{R} \qquad \text{W}$$

where the thermal resistance R of the cylinder is

$$R = \frac{\ln (r_2/r_1)}{2\pi Hk}$$

3-32 A 10-cm-OD steam pipe maintained at $T_i = 130$°C is covered with asbestos insulation $L = 3$ cm thick [$k = 0.1$ W/(m · °C)]. The ambient air temperature is $T_\infty = 30$°C, and the heat transfer coefficient for convection at the outer surface of the absestos insulation is $h = 25$ W/(m^2 · °C). By using the thermal resistance concept calculate the rate of heat loss from the pipe per 1-m length of pipe.

Answer: 120.8 W/m

3-33 A 2.5-cm-OD pipe carrying oil at $T_i = 150$°C is exposed to an ambient at $T_i = 25$°C with a convection heat transfer coefficient $h = 50$ W/(m^2 · C). Calculate the thickness of the asbestos insulation [$k = 0.1$ W/(m · °C)] required to reduce the heat loss from the pipe by 50 percent.

Answer: 0.23 cm

Sphere

3-34 Determine the steady-state temperature distribution $T(r)$ in a solid sphere of radius $r = b$ in which heat is generated at a uniform rate of g_0 W/m^3 and heat is dissipated from the surface by convection into a medium at a temperature T_∞ with a heat transfer coefficient h. Determine the expressions for the temperature at the center of the sphere and the heat flux at the boundary surface.

3-35 Develop an expression for the steady-state temperature distribution $T(r)$ in a hollow sphere, $a \leq r \leq b$, in which heat is generated at a uniform rate of g_0 W/m^3 and the boundary surfaces at $r = a$ and $r = b$ are kept at the same uniform temperature T_0.

3-36 Develop an expression for the steady-state temperature distribution $T(r)$ in a solid sphere of radius $r = b$ in which heat is generated at a rate of

$$g(r) = g_0\left(1 - \frac{r}{b}\right) \qquad \text{W/m}^3$$

where g_0 is a constant and the boundary surface at $r = b$ is maintained at a uniform temperature T_0.

3-37 Determine the steady-state temperature distribution $T(r)$ in a hollow sphere in which heat is generated at a constant rate of g_0 W/m^3, while the inner surface $t = a$ is insulated and the outer surface $r = b$ is kept at zero temperature. Also give the expressions for the temperature of the insulated inner surface and the heat flux at the outer surface.

3-38 The inner surface of a hollow sphere of radius $r = r_i$ is subjected to a uniform heat flux q_i W/m^2, and the outer surface of radius $r = r_o$ is kept at zero temperature. There is no heat generation within the cylinder, and the thermal conductivity k is constant. Develop an expression for the steady-state temperature distribution $T(r)$ within the sphere and for the temperature of the inner surface.

If the inner and outer radii are $r_i = 4$ and $r_o = 8$ cm, respectively, and the thermal conductivity is $k = 50$ W/(m · °C), determine the heat flux q_i required at the inner surface in order to maintain the inner surface at 100°C, while the outer surface is at 0°C.

Answer: 250 kW/m^2

3-39 Determine the heat tansfer rate through a spherical copper shell of thermal conductivity $k = 386$ W/(m · °C), inner radius $r_i = 2$ cm, and outer radius $r_o = 6$ cm if the inner surface is kept at $T_i = 200$°C and the outer surface at $T_o = 100$°C.

Answer: 14.55 kW

3-40 A solid sphere of radius $b = 5$ cm and thermal conductivity $k = 20$ W/(m · °C) is heated uniformly throughout the volume at a rate of 2000 W/m^3, and heat is dissipated by convection from its outer surface into ambient air at $T_\infty = 25$°C with a heat transfer coefficient $h = 20$ W/(m^2 · °C). Determine the steady-state temperature at the center and the outer surface of the sphere.

3-41 Consider an aluminum hollow sphere of inside radius $r_1 = 2$ cm, outside radius $r_2 = 6$ cm, and thermal conductivity $k = 200$ W/(m · °C). The inside surface is kept at a uniform temperature $T_i =$ 100°C, and the outside surface dissipates heat by convection with a heat transfer coefficient $h_\infty = 80$ W/(m^2 · °C) into the ambient air at temperature $T_\infty = 20$°C. Determine the outside surface temperature of the sphere and the rate of heat transfer from the sphere.

Answer: 96.3°C, 276.3 W

3-42 A hollow sphere, $r_1 \leq r \leq r_2$, is kept at a uniform temperature T_1 at the inner surface $r = r_1$ and at temperature T_2 at the outer surface $r = r_2$. Show that the total heat transfer rate Q through the sphere is given by

$$Q = \frac{T_1 - T_2}{R} \qquad \text{W}$$

where the thermal resistance R of the sphere is

$$R = \frac{r_2 - r_1}{kA_m} \qquad A_m = \sqrt{(4\pi r_1^2)(4\pi r_2^2)} \equiv \sqrt{A_1 A_2}$$

and k is the thermal conductivity.

3-43 Consider a hollow sphere of inner radius $r = a$ and outer radius $r = b$. The inner surface is heated uniformly at a constant rate of q_0 W/m^2 with an electric heater while the outer surface is maintained at zero temperature.

(*a*) Develop an expression for the steady-state temperature distribution $T(r)$ in the sphere.

(*b*) Calculate the temperature of the inner boundary surface for $a = 10$ cm, $b = 20$ cm, $k = 40$ W/(m · °C), and $q_0 = 1.6 \times 10^5$ W/m^2.

Answer: 200°C

3-44 A hollow sphere of inside radius $r = a$ and outside radius $r = b$ is heated uniformly at a rate of q_0 W/m^2 at the inner surface. At the outer surface, heat is dissipated by convection into an ambient at zero temperature with a heat transfer coefficient h.

(*a*) Develop an expression for the steady-state temperature distribution $T(r)$ in the sphere.

(*b*) Calculate the temperature of the inner surface for $k = 50$ W/(m · °C), $a = 8$ m, $b = 16$ cm, $h = 100$ W/(m^2 · °C), and $q_0 = 10^5$ W/m^2.

3-45 A sphere of outside diameter $D = 10$ cm maintained at a uniform temperature $T_i = 225$°C is exposed to an ambient at $T_\infty = 25$°C with a heat transfer coefficient for convection of $h = 50$ W/(m^2 · °C). Calculate the thickness of the insulation [$k = 0.08$ W/(m · °C)] required to reduce the rate of heat loss by 83 percent while the sphere is maintained at $T_i = 225$°C.

Answer: 0.5 cm

3-46 The inside and outside radii of a hollow aluminum sphere are $r_1 = 5$ and $r_2 = 10$ cm, respectively. The inside surface is maintained at $T_1 = 330$°C by heating electrically, while heat is dissipated from the outer surface by convection, with a heat transfer coefficient $h = 100$ W/(m^2 · °C) into an ambient at $T_\infty = 30$°C. By using the thermal resistance concept, calculate the heat transfer rate across the sphere [$k = 204$ W/(m · °C)].

Composite medium

3-47 A composite wall consisting of three different layers in perfect thermal contact is shown in the sketch below. The outer surfaces to the left and right are kept at temperatures $T_1 = 400$°C and $T_2 = 50$°C, respectively. The thickness L_i and the thermal conductivity k_i for $i = 1, 2, 3$ of each layer are also

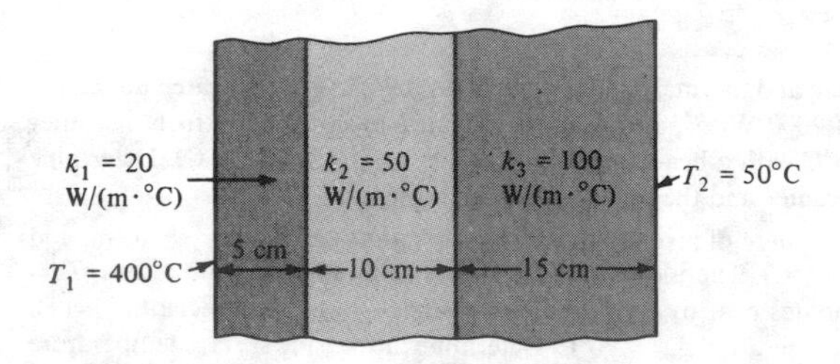

Figure P3-47

specified. Determine the heat transfer rate per square meter across this composite layer by assuming one-dimensional heat flow and using the thermal resistance concept.

Answer: 58.3 kW/m^2

3-48 The wall for an industrial furnace consists of a fireclay brick of thickness $L_1 = 0.20$ m of thermal conductivity $k_1 = 1.0$ W/(m · °C), which is covered on the outer surface with a layer of insulating material of thickness $L_2 = 0.03$ m of thermal conductivity $k_2 = 0.05$ W/(m · °C). If the inner surface of the wall is at temperature $T_i = 830$°C and the outer surface at $T_o = 30$°C, determine the heat transfer rate per square meter of the furnace wall. Assume one-dimensional heat flow, and use the thermal resistance concept.

Answer: 1 kW/m^2

3-49 An industrial furnace is made of fireclay brick of thickness $L_1 = 0.25$ m and thermal conductivity $k_1 = 1.0$ W/(m · °C). The outside surface is to be insulated with material of thermal conductivity $k_2 = 0.05$ W/(m · °C). Determine the thickness of the insulation layer in order to limit the heat loss from the furnace wall to $q = 1000$ W/m^2 when the inside surface of the wall is at 1030°C and the outside surface at 30°C.

Answer: 3.75 cm

3-50 A composite wall consists of an iron plate of thickness $L_1 = 3$ cm and thermal conductivity $k = 60$ W/(m · °C), an asbestos layer of thickness $L_2 = 0.5$ cm and thermal conductivity $k_2 = 0.2$ W/(m · °C), and an insulation of thickness $L_3 = 4$ cm with thermal conductivity $k_3 = 0.05$ W/(m · °C). Calculate the heat flow across this composite wall per square meter of the surface for a temperature difference of 400°C between the inner and outer surfaces.

Answer: 484.6 W/m^2

3-51 A Thermopane window consists of two 6-mm-thick glasses separated by a stagnant air space of thickness 6 mm. The thermal conductivity of the glass is $k_g = 0.78$ W/(m · °C), and that of air can be taken as $k_a = 0.025$ W/(m · °C). The convection heat transfer coefficients for the inside and outside air are $h_i = 10$ and $h_o = 60$ W/(m^2 · °C). Determine the rate of heat loss per square meter of the glass surface for a temperature difference of 40°C between the inside and outside air. Compare this result with the heat loss if the window had a single glass of thickness $L = 6$ mm only instead of the Thermopane.

3-52 Consider a composite plane wall consisting of two layers of different materials A and B, as illustrated below. Layer A is in contact with a hot fluid at $T_a = 250$°C for which the heat transfer coefficient is $h_a = 15$ W/(m^2 · °C), layer B is in contact with a cold fluid at $T_b = 50$°C for which the heat transfer

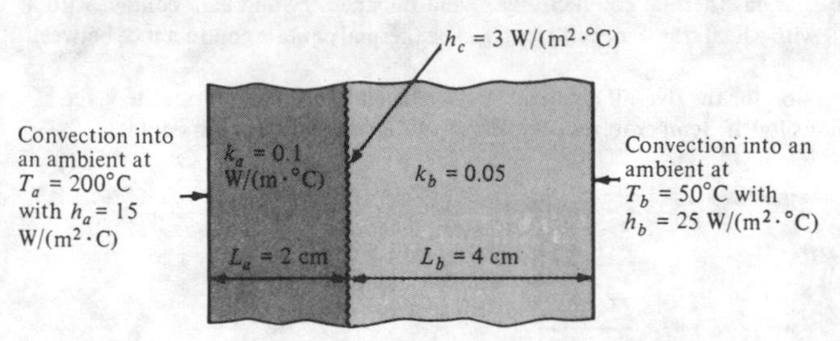

Figure P3-52

coefficient is $h_b = 25$ W/(m^2 · °C), and the contact conductance between layers A and B is $h_c = 3$ W/(m^2 · °C). Layer A is $L_a = 2$ cm thick and has a thermal conductivity $k_a = 0.1$ W/(m · °C), and layer B is $L_b = 4$ cm thick and has a thermal conductivity $k_b = 0.05$ W/(m · °C). Determine the heat transfer rate through this composite wall per square meter of the surface, and make a sketch of the temperature distribution.

Answer: 138.9 W/m^2

3-53 A composite wall consisting of four different materials is shown in the sketch below. Since the

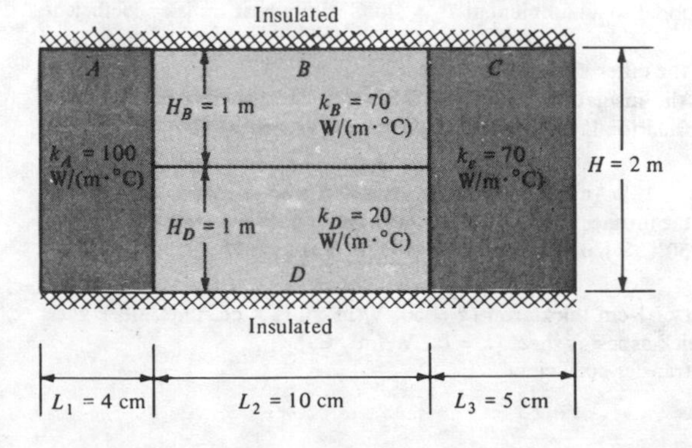

Figure P3-53

upper and lower surfaces are insulated, the heat flow can be assumed to be one-dimensional. The dimensions and the thermal conductivities of each layer are also listed. Using the thermal resistance concept, determine the heat flow rate per square meter of the exposed surface for a temperature difference of $\Delta T = 300°C$ between the two outer surfaces.

3-54 A composite slab consisting of three layers in perfect thermal contact, shown in the accompanying figure, is maintained at uniform temperatures T_0 and T_3 at the outer surfaces. Here k_1, k_2, and k_3 are the thermal conductivities, and L_1, L_2, and L_3 are the thicknesses of the layers. Develop the relations for the determination of the interface temperatures T_1 and T_2.

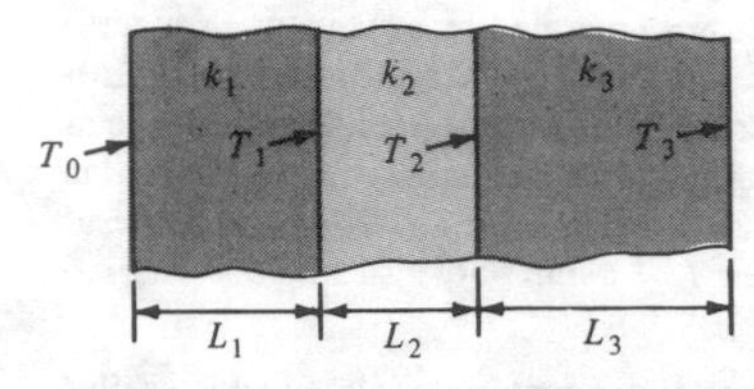

Figure P3-54

3-55 Consider a composite slab consisting of two layers A and B as illustrated below. Layer A has thermal conductivity k_a and thickness L_a and is in contact with a hot fluid at temperature T_a with heat transfer coefficient h_a. Layer B has thermal conductivity k_b and thickness L_b and is in contact with a cold fluid at temperature T_b with a heat transfer coefficient h_b. The thermal contact conductance between layers A and B is h_c.

(*a*) Develop an expression for the overall heat transfer coefficient U for this composite layer.

(*b*) Develop expressions for the temperatures of layers A and B where they are in contact.

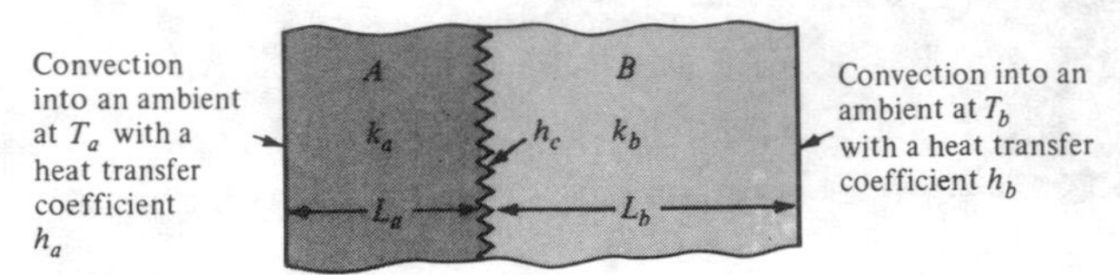

Figure P3-55

3-56 Two 416 stainless-steel plates each $L = 1$ cm thick are bolted together. The thermal conductivity of the stainless steel can be taken as $k = 20$ W/(m · °C). The contact pressure is approximately 10 atm. Calculate the heat transfer rate across the plates for a temperature difference of $\Delta T = 100°C$ per square meter of the surface, assuming that the contact conductance is $h_c = 4000$ W/(m^2 · °C).

Answer: 80 kW/m^2

3-57 The inside surface of a brick wall $L = 10$ cm thick [$k_b = 1$ W/(m · °C)] is at temperature $T_i = 930°C$, and the outer surface is exposed to an ambient at $T_\infty = 30°C$ with a heat transfer coefficient $h_0 = 20$ W/(m^2 · °C).

(*a*) What is the temperature of the outer surface?

(*b*) Calculate the thickness of the insulation layer [$k = 0.1$ W/(m · °C)] needed on the outer surface such that the surface of the insulation layer exposed to air will not exceed 90°C.

Answer: 330°C, 6 cm

3-58 A 0.5-m-thick furnace wall [$k = 1$ W/(m · °C)] is to be insulated with a material [$k = 0.08$ W/(m · °C)]. The temperature inside the furnace is 1350°C. If the heat loss should not exceed 750 W/m^2 for the outer surface temperature of 50°C, what thickness of insulation is required?

Answer: 9.9 cm

3-59 A composite wall is composed of 1-cm-thick iron [$k = 60$ W/(m · °C)], 4-cm-thick fiber glass [$k = 0.04$ W/(m · °C)], and 0.4-cm-thick asbestos sheet [$k = 0.2$ W/(m · °C)].

(*a*) Determine the overall heat transfer coefficient.

(*b*) Determine the heat transfer rate per square meter through this composite wall for a temperature difference of 400°C.

3-60 A composite wall consists of a 10-cm-thick layer of building brick of thermal conductivity $k = 0.7$ W/(m · °C) and 3-cm-thick plaster of thermal conductivity $k = 0.5$ W/(m · °C). An insulating material of thermal conductivity $k = 0.08$ is to be added to reduce the heat transfer through the wall by 70 percent. Determine the thickness of the insulating layer.

Answer: 3.8 cm

3-61 A wall is constructed of 10-cm-thick common brick [$k = 0.69$ W/(m · °C)], 1.5-cm-thick fiber insulating board [$k = 0.048$ W/(m · °C)], followed by a 5-cm layer of glass wool ($k = 0.038$) and 1.5-cm-thick insulating board [$k = 0.048$ W/(m · °C)]. The heat transfer coefficient for both outside and inside is 12 W/(m² · °C). Determine the overall heat transfer coefficient U.

Answer: 0.44 W/(m² · °C)

3-62 Sketch the electric network analog for one-dimensional heat flow through the composite wall shown below.

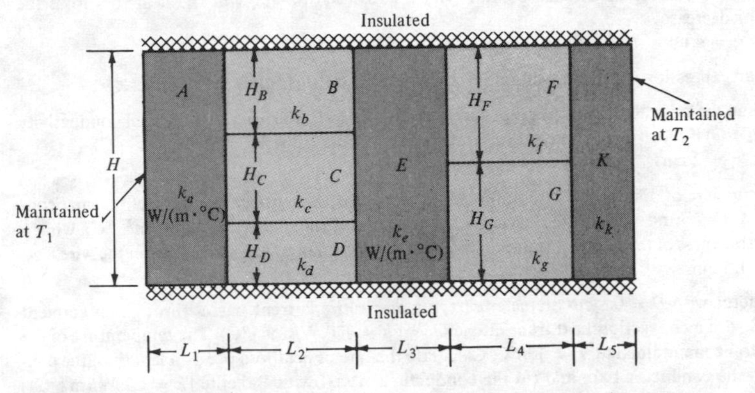

Figure P3-62

3-63 A steam pipe of outside radius $r_1 = 4$ cm is covered with a layer of asbestos insulation of thickness $L_1 = 1$ cm, thermal conductivity $k_1 = 0.15$ W/(m · °C) which is covered in turn with a fiber glass insulation of thickness $L_2 = 3$ cm and thermal conductivity $k_2 = 0.05$ W/(m · °C). The surface of the steam pipe is at temperature 330°C, and the outside surface of the fiber glass insulation is at 30°C.

(*a*) Determine the interface temperature between the asbestos and fiber glass insulations.

(*b*) Determine the heat transfer rate per 1-m length of pipe.

Answer: 173.1 W/m, 289°C

3-64 A steam pipe of outside radius $r_1 = 4$ cm is covered with a layer of asbestos cement of thickness $L_1 = 1.5$ cm of thermal conductivity $k_1 = 0.2$ W/(m · °C) which is covered in turn with glass wool of thickness $L_2 = 5$ cm and thermal conductivity $k_2 = 0.038$ W/(m · °C). If the temperature of the outer surface of the pipe is $T_1 = 300$°C and that of the outer surface of the glass wool is 40°C, determine the heat loss per 1-m length of pipe. Assume one-dimensional heat flow, and use the thermal resistance concept. Also give the overall heat transfer coefficient U_o based on the outside surface of the insulation.

3-65 Consider a hollow steel sphere of inside radius $r_1 = 10$ cm and outside radius $r_2 = 20$ cm. The thermal conductivity of the steel is $k = 10$ W/(m · °C). The inside surface is maintained at a uniform temperature of $T_1 = 230$°C, and the outside surface dissipates heat by convection with a heat transfer coefficient $h = 20$ W/(m² · °C) into an ambient at $T_\infty = 30$°C. Determine the thickness of asbestos insulation [$k_i = 0.5$ W/(m · °C)] required to reduce the heat loss by 50 percent.

Answer: 5.8 cm

Critical radius of insulation

3-66 A tube of outside diameter $D = 2.5$ cm is to be insulated with a layer of asbestos of thermal conductivity $k = 0.2$ W/(m · °C). The convection heat transfer coefficient from the surface of the asbestos to the ambient air is $h_\infty = 12$ W/(m · °C). Calculate the critical radius of the insulation. If an asbestos later 3 mm thick were added on the tube, would the heat transfer be increased or decreased from the tube?

Answer: $r_c = 1.67$ cm

3-67 A wire of diameter $D = 2$ mm is heated as a result of passing electric current through it. The wire dissipates heat by convection with a heat transfer coefficient $h = 125$ W/(m^2 · °C) into the ambient air. If the wire is covered with an insulation of thickness 0.2 mm of thermal conductivity $k = 0.175$ W/(m · °C), will the heat from the wire increase or decrease?

3-68 A copper rod of diameter $D = 5$ mm is heated by the passage of an electric current. The surface of the rod is maintained at a temperature of 175°C while it is dissipating heat by convection into an ambient at $T_\infty = 25$°C with a heat transfer coefficient $h_\infty = 150$ W/(m^2 · °C). If the rod is to be covered with a 1-mm-thick coating of thermal conductivity $k = 0.6$ W/(m · °C), will the heat loss from the rod increase or decrease?

Answer: $r_c = 4$ mm; the heat loss increases.

3-69 Derive an expression for the critical radius appropriate for the insulation of a sphere.

3-70 Determine the critical radius for a pipe covered with a layer of asbestos of thermal conductivity $k = 0.2$ W/(m · °C) if the external convection heat transfer coefficient is $h_\infty = 10$ W/(m^2 · °C).

Answer: $r_c = 2$ cm

3-71 An electric wire of diameter $D = 3$ mm is to be covered with a rubber insulator of thermal conductivity $k = 0.15$ W/(m^2 · °C). If the external heat transfer coefficient is $h_\infty = 50$ W/(m^2 · °C), what is the optimum thickness of the rubber insulation that will produce maximum heat loss from the wire.

Answer: 1.5 mm

3-72 A conductor with $D = 0.8$ cm diameter carrying an electric current passes through an ambient at $T_\infty = 30$°C with a convection heat transfer coefficient $h = 120$ W/(m^2 · °C). The temperature of the conductor is to be maintained at $T_i = 130$°C. Calculate the rate of heat loss per 1-m length of the conductor for (*a*) the conductor bare and (*b*) the conductor covered with Bakelite [$k = 1.2$ W/(m · °C)] with radius corresponding to the critical radius of the insulator.

3-73 An electrically heated sphere with diameter $D = 6$ cm is exposed to an ambient at $T_\infty = 25$°C with a convection heat transfer coefficient $h = 20$ W/(m^2 · °C). The surface of the sphere is to be maintained at $T_i = 125$°C. Calculate the rate of heat loss from the sphere for (*a*) the sphere bare (i.e., uninsulated) and (*b*) the sphere covered with an insulation [$k = 1.0$ W/(m · °C)] having a radius corresponding to the critical radius of the sphere.

Answer: 22.6 W, 44.4 W

Fins

3-74 Heat is generated at a constant rate of g W/m^3 in a thin circular rod of length L and diameter D by the passage of electric current. The two ends at $x = 0$ and $x = L$ are kept at constant temperatures T_0 and zero, respectively, while the heat is dissipated from the lateral surfaces by convection into a medium at zero temperature with a heat transfer coefficient h.

(*a*) Derive the one-dimensional, steady-state energy equation for the determination of temperature distribution $T(x)$ in the rod.

(*b*) By solving this differential equation, develop an expression for the temperature distribution $T(x)$ in the rod.

3-75 Derive the differential equation governing the temperature distribution $T(x)$ in a thin fin of uniform cross section by assuming that the thermal conductivity k, the heat transfer coefficient h, and the ambient temperature T_∞ remain constant.

3-76 A circular rod fin of diameter D, length L, and thermal conductivity k is exposed to an ambient fluid at temperature T_∞ and having a heat transfer coefficient h. The fin base is maintained at temperature T_0, and the heat loss from the fin tip can be considered negligible compared with that from the lateral surfaces of the fin.

(*a*) Using the fin equation derived in Chap. 3, develop an expression for temperature distribution $T(x)$ in the fin.

(*b*) Develop an expression for the heat transfer rate through the fin.

(*c*) Develop an expression for the fin efficiency η.

3-77 Copper-plate fins of rectangular cross section having thickness $t = 1$ mm, height $L = 10$ mm, and thermal conductivity $k = 380$ W/(m · °C) are attached to a plane wall maintained at a temperature $T_0 = 230$°C. Fins dissipate heat by convection into ambient air at $T = 30$°C with a heat transfer coefficient $h = 40$ W/(m^2 · °C). Fins are spaced at 8 mm (that is, 125 fins per meter). Assume negligible heat loss from the fin tip.

(*a*) Determine the fin efficiency.

(*b*) Determine the area-weighted fin efficiency.

(*c*) Determine the net rate of heat transfer per square meter of plane wall surface.

(*d*) What would be the heat transfer rate from the plane wall if there were no fins attached?

Answer: (*c*) 26.87 kW/m^2, (*d*) 8 kW/m^2

3-78 Circular aluminum disk fins of constant rectangular profile are attached to a tube having $D = 2.5$ cm outside diameter with a spacing of 8 mm (that is, 125 fins per 1-m tube length). Fins have a thickness $t = 1$ mm, height $L = 15$ mm, and thermal conductivity $k = 200$ W/(m · °C). The tube wall is maintained at a temperature $T_0 = 190$°C, and the fins dissipate heat by convection into ambient air at $T_\infty = 40$°C with a heat transfer coefficient $h_\infty = 80$ W/(m^2 · °C).

(*a*) Determine the fin efficiency.

(*b*) Determine the area-weighted fin efficiency.

(*c*) Determine the net heat loss per 1-m length of tube.

(*d*) What would be the heat loss per 1-m length of tube if no fins were attached?

Answer: (*c*) 6.4 kW/m, (*d*) 0.94 kW/m

3-79 Aluminum fins of rectangular profile are attached on a plane wall with 5-mm spacing. The fins have thickness $t = 1$ mm, length $L = 10$ mm, and thermal conductivity $k = 200$ W/(m · °C). The wall is maintained at a temperature $T_0 = 200$°C, and the fins dissipate heat by convection into the ambient air at $T_\infty = 40$°C with heat transfer coefficient $h_\infty = 50$ W/(m^2 · °C).

(*a*) Determine the fin efficiency.

(*b*) Determine the area-weighted fin efficiency.

(*c*) Determine the heat loss per square meter of wall surface.

Answer: (*c*) 37.8 kW/m^2

3-80 Aluminum fins of triangular profile are attached on a plane wall with 5-mm spacing (that is, 200 fins per meter). The fin base is $t = 2$ mm thick, the fin height is $L = 8$ mm, and the thermal conductivity of the fin material is $k = 200$ W/(m · °C). The plane wall is maintained at $T_0 = 240$°C, and the fins dissipate heat by convection into the ambient air at $T_\infty = 40$°C with heat transfer coefficient $h_\infty = 50$ W/(m · °C).

(*a*) Determine the fin efficiency.

(*b*) Determine the area-weighted fin efficiency.

(*c*) Determine the heat loss from the wall per square meter of wall surface.

3-81 Circular aluminum fins of constant rectangular profile are attached to a tube of outside diameter $D = 5$ cm. The fins have thickness $t = 2$ mm, height $L = 15$ mm, thermal conductivity $k = 200$ W/(m · °C), and spacing 8 mm (that is, 125 fins per 1-m tube length). The tube surface is maintained at a uniform temperature $T_0 = 200$°C, and the fins dissipate heat by convection into the ambient air at $T_\infty = 30$°C with a heat transfer coefficient $h_\infty = 50$ W/(m^2 · °C). Determine the net heat transfer per 1-m length of tube.

Answer: 7.45 kW/m

3-82 Consider a tube having fins attached to both inside and outside surfaces. The thermal resistance of the tube material is negligible compared with those for flow on the outside and inside. Various quantities are given:

	Outside	Inside
Temperatures of fluids	T_o	T_i
Heat transfer coefficients	h_o	h_i
Fin surface area	a_{fo}	a_{fi}
Total heat transfer area	a_{to}	a_{ti}
Tube surface areas without fins	a_o	a_i
Fin efficiency	η_o	η_i

Write the expressions for the heat transfer rate through the tube for the following cases:

(*a*) Fins are attached on both sides of the tube surface.

(*b*) Fins are attached on the outside surface of the tube only.

(*c*) No fins are attached to the tube surfaces.

3-83 Consider a long, slender copper rod of diameter $D = 1$ cm and thermal conductivity $k = 380$ W/(m · °C), with one end thermally attached to a wall at 200°C. Heat is dissipated from the rod by convection with a heat transfer coefficient $h_\infty = 15$ W/(m^2 · °C). Determine the heat transfer rate from the rod into the surrounding air at $T_\infty = 30$°C.

Answer: 20.16 W

3-84 A thin rod of uniform cross section A, length L, and thermal conductivity k is thermally attached from its ends to two walls which are maintained at temperatures T_1 and T_2. The rod is dissipating heat from its lateral surface by convection into the ambient air at T_∞ with a heat transfer coefficient h_∞.

(*a*) Derive the differential equation governing the one-dimensional, steady-state temperature distribution $T(x)$ in the rod.

(*b*) By solving this equation, develop an expression for the temperature distribution $T(x)$ in the rod.

(*c*) Develop an expression for the heat loss from the rod into the ambient air.

3-85 Consider a steel rod of length $L = 50$ cm, diameter $D = 2$ cm, and thermal conductivity $k = 55$ W/(m · °C). One end of this rod is thermally attached to a hot surface maintained at $T_1 = 150$°C, and the other end is attached to a cold surface maintained at $T_2 = 50$°C. The rod dissipates heat by convection into the ambient air at temperature $T_\infty = 20$°C with a heat transfer coefficient $h_\infty = 15$ W/(m^2 · °C). Determine the heat loss from the rod into the ambient air. What fraction of this heat loss is from the surface maintained at $T_1 = 150$°C?

Answer: 19.4 W, 84.4%

3-86 A very long, slender brass rod of diameter $D = 2$ cm and thermal conductivity $k = 60$ W/(m · °C) is thermally attached at one end to a large, hot surface maintained at $T_0 = 200$°C. The rod dissipates heat by convection with a heat transfer coefficient $h_\infty = 25$ W/(m^2 · °C) into an ambient at $T_\infty = 30$°C.

(*a*) Calculate the heat loss through the rod into the ambient air.

(*b*) If a copper rod [$k = 386$ W/(m · °C)] of identical size were used in place of the brass rod, what would be the resulting change in the heat loss through the rod?

3-87 An iron rod of length $L = 30$ cm, diameter $D = 1$ cm, and thermal conductivity $k = 65$ W/(m · °C) is attached horizontally to a large tank at temperature $T_0 = 200$°C. The rod is dissipating heat by convection into the ambient air at $T_\infty = 20$°C with a heat transfer coefficient $h_\infty = 15$ W/(m^2 · °C). What is the temperature of the rod 10 and 20 cm from the tank?

Answer: 90.1°C, 50.1°C

3-88 Two very long, slender rods of the same diameter are given. One rod is of aluminum and has a thermal conductivity $k_1 = 200$ W/(m · °C), but the thermal conductivity k_2 of the other rod is not known. To determine the thermal conductivity of the other rod, one end of each rod is thermally attached to a metal surface which is maintained at a constant temperature T_0. Both rods are losing heat by convection, with a heat transfer coefficient h_∞ into the ambient air at T_∞. The surface temperature of each rod is measured at various distances from the hot base surface. The temperature of the aluminum rod at $x_1 = 40$ cm from the base is the same as the that of the rod of unknown thermal conductivity at $x_2 = 20$ cm from the base. Determine the thermal conductivity k_2 of the second rod.

Answer: 50 W/(m · °C)

3-89 For a fin with convection at the fin tip, the temperature distribution in the fin is given by

$$\frac{T(x) - T_\infty}{T_0 - T_\infty} = \frac{\cosh m(L - x) + (h_e/mk) \sinh m(L - x)}{\cosh mL + (h_e/mk) \sinh mL}$$

Using this result, develop an expression for heat transfer through the fin.

Temperature-dependent $k(T)$ and variable cross-sectional area

3-90 The lateral surface of a rod is perfectly insulated while its ends at $x = 0$ and $x = L$ are kept at constant temperatures T_1 and T_2, respectively. The cross-sectional area A of the rod is uniform, and the thermal conductivity of the material varies with temperature as

$$k(T) = k_0(1 + \alpha T)$$

where k_0 and α are constants.

(*a*) Develop an expression for the heat flow rate Q through the rod.

(*b*) Develop an expression for the thermal resistance of the rod.

3-91 Thermal conductivity of a plane wall varies with temperature according to the relation

$$k(T) = k_0(1 + \beta T^2)$$

where k_0 and β are constants.

(*a*) Develop an expression for the heat flow through the slab per unit area if the surfaces at $x = 0$ and $x = L$ are maintained at uniform temperatures T_1 and T_2, respectively.

(*b*) Develop a relation for the thermal resistance of the wall if the heat transfer surface is A.

(*c*) Calculate the heat transfer rate through $A = 0.1$ m^2 of the plate for $T_1 = 200$°C, $T_2 = 0$°C, $L = 0.4$ m, $k_0 = 60$ W/(m · °C), and $\beta = 0.25 \times 10^{-4}$ °C^{-2}.

3-92 The two faces of a slab at $x = 0$ and $x = L$ are kept at uniform temperatures T_1 and T_2, respectively, and the thermal conductivity of the material depends on temperature in the form

$$k(T) = k_0(T^2 - T_0^2)$$

where k_0 and T_0 are constants.

(*a*) Find an expression for the heat flow rate per unit area of the slab.

(*b*) Find an expression for the average thermal conductivity k_m for the slab.

3-93 A hollow cylinder, $a \le r \le b$, has its boundary surfaces at $r = a$ and $r = b$ maintained at uniform temperatures T_1 and T_2, respectively. The thermal conductivity varies with temperature in the form $k(T) = k_0(1 + \beta T)$, where k_0 and β are constants.

(*a*) Develop a relation for the heat flow through the cylinder per unit length of cylinder.

(*b*) Develop an expression for the thermal resistance per unit length of cylinder.

3-94 A hollow cylinder, $a \le r \le b$, has its boundary surfaces at $r = a$ and $r = b$ maintained at temperatures T_1 and T_2, respectively. The thermal conductivity of the material varies with temperature in the form $k(T) = k_0(1 + \beta T^2)$. Develop an expression for the heat flow through a unit length of cylinder.

3-95 The inner and outer radii of a hollow cylinder are 5 and 10 cm, respectively. The inside surface is maintained at 300°C, and the outside surface at 100°C. The thermal conductivity varies with temperature over the range of $100 < T < 300$°C as $k(T) = 0.5(1 + 10^{-3}T)$, where T is in degrees Celsius. Determine the heat flow rate per 1-m length of cylinder.

Answer: 1.088 kW/m

3-96 The inside and outside surfaces of a hollow sphere, $a \le r \le b$, at $r = a$ and $r = b$ are maintained at uniform temperatures T_1 and T_2, respectively. The thermal conductivity varies with temperature as

$$k(T) = k_0(1 + \alpha T + \beta T^2)$$

(*a*) Develop an expression for the total heat flow rate Q through the sphere.
(*b*) Develop a relation for the thermal resistance of the hollow sphere.

3-97 Consider a slab of thickness L in which heat is generated at a constant rate of g_0 W/m^3. The boundary surfaces at $x = 0$ and $x = L$ of the slab are maintained at temperatures T_1 and T_2, respectively. The thermal conductivity of the slab varies with temperature as

$$k(T) = k_0(1 + \beta T)$$

where k_0 and β are constants. Develop an expression for the heat flux $q(x)$ in the slab.

3-98 The thermal conductivity of a plane wall of thickness L varies with temperature as

$$k(T) = k_0(1 + \beta T^3)$$

where k_0 and β are constants. The boundary surfaces at $x = 0$ and $x = L$ are maintained at temperatures T_1 and T_2, respectively.

(*a*) Develop an expression for the heat flow rate across an area A of the plate.
(*b*) Calculate the heat transfer rate for $A = 0.1$ m^2, $L = 0.4$ m, $T_1 = 200$°C, $T_2 = 0$°C, $k_0 = 60$ W/(m · °C), and $\beta = 0.25 \times 10^{-4}$°C^{-3}, and compare this result with that obtained in Problem 3-91.

REFERENCES

1. Cetinkale, T. N., and M. Fishenden: Thermal Conductance of Metal Surfaces in Contact. General Discussion on Heat Transfer. Conference of Institution of Mechanical Engineers (London) and ASME, pp. 271–275, 1951.
2. Barzelay, M. E., K. N. Tong, and G. F. Holloway: "Effect of Pressure on Thermal Conductance of Contact Joints," *NACA Tech. Note* 3295, May 1955.
3. Fried, E.: Study of Interface Thermal Contact Conductance, *General Electric Co. 64SD652*, May 1964 (also *ARS* Paper No. 1690-61, April 1961).
4. Fried, E., and F. A. Costello: "Interface Thermal Contact Resistance Problem in Space Vehicles," *ARS Journal*, **32**:237–243 (1962).
5. Fenech, H., and W. H. Rohsenow: "Prediction of Thermal Conductance of Metallic Surfaces in Contact," *J. Heat Transfer, Trans. ASME*, **85**: 15–24 (1963).
6. Atkins, H. L., and E. Fried: "Thermal Interface Conductance in a Vacuum," *AIAA Paper* No. 64–253, 1964.
7. Clausing, A. M., and B. T. Chao: "Thermal Contact Resistance in a Vacuum Environment," *J. Heat Transfer*, **87**:243–265 (1965).
8. Fried, E.: "Study of Interface Thermal Contact Conductance," *General Electric Co. 65SD4395*, March 1965.
9. Moore, C. J., Jr., H. A. Blum, and H. Atkins: "Subject Classification Bibliography for Thermal Contact Resistance Studies," *ASME Paper* 68-*WA*/*HT*-18, December 1968.
10. Cooper, M. G., B. B. Mikic, and M. M. Yovanovich: "Thermal Contact Conductances," *Int. J. Heat Mass Transfer*, **12**:279–300 (1969).
11. Clausing, A. M.: "Heat Transfer at the Interface of Dissimilar Metals—The Influence of Thermal Strain," *Int. J. Heat Mass Transfer*, **9**:791–801 (1966).
12. Veziroglu, T. N.: "Correlation of Thermal Contact Conductance Experimental Results," *Prog. Astron. Aero.*, **20**, Academic Press, Inc., New York, 1967.
13. Harper, W. P., and D. R. Brown: "Mathematical Equations for Heat Conduction in the Fins of Air-Cooled Engines." *NACA Rep.* 158, 1922.
14. Gardner, K. A.: "Efficiency of Extended Surfaces," *Trans. ASME*, **67**:621–631 (1945).
15. Kern, D. Q., and A. D. Kraus: *Extended Surface Heat Transfer*, McGraw-Hill, New York, 1972.

CHAPTER

FOUR

TRANSIENT CONDUCTION AND USE OF TEMPERATURE CHARTS

If the surface temperature of a solid body is suddenly altered, the temperature within the body begins to change over time. It will take some time before the steady-state temperature distribution is reached. The determination of the temperature distribution within the solid during temperature transients is a more complicated matter because temperature varies with both position and time [1–9]. In many practical applications, the variation of temperature with position is negligible during the transients, hence the temperature is considered to vary with time only. The analysis of heat transfer under such an assumption is called the *lumped system analysis*, and since the temperature is a function of time only, the analysis becomes very simple. Therefore, in this chapter we begin with the lumped system analysis of transient heat conduction problems.

The use of *transient-temperature charts* is illustrated for solving simple transient heat conduction in a slab, cylinder, and sphere in which temperature varies with both time and position. The method of *product solution* is described for the analysis of multidimensional transient heat conduction. The use of *conduction shape factors* is discussed for predicting the steady-state heat flow in two-dimensional heat flow systems.

Finally, the analytic solution of transient heat conduction by the method of separation of variables is presented for the case of the slab geometry, and the use of *tabulated solutions* is introduced.

4-1 LUMPED SYSTEM ANALYSIS

Consider that a solid of arbitrary shape, volume V, total surface area A, thermal conductivity k_s, density ρ, specific heat c_p, at a uniform temperature T_0 is suddenly immersed at the time $t = 0$ in a well-stirred fluid which is kept at a uniform temperature T_∞. Figure 4-1 illustrates the considered heat transfer system. Heat transfer between the solid and liquid takes place by convection with a heat transfer

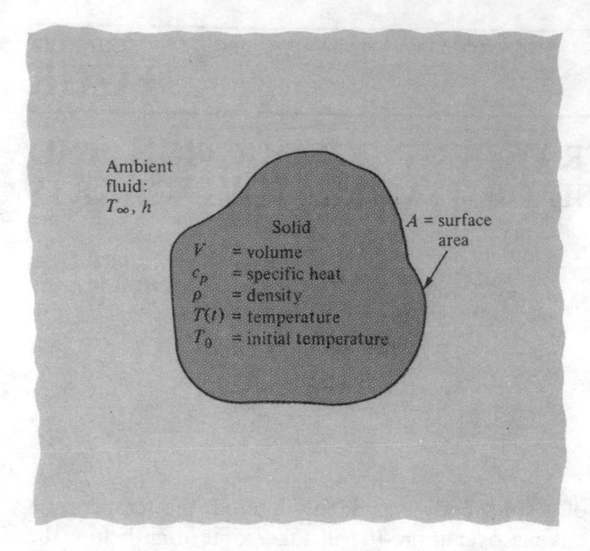

Figure 4-1 Nomenclature for lumped system analysis of transient heat flow.

coefficient h. It is assumed that the temperature distribution within the solid at any instant is sufficiently uniform that the temperature of the solid can be considered to be a function of time only, that is, $T(t)$. The energy equation for heat transfer in the solid may be stated as

$$\begin{pmatrix}\text{Rate of heat flow into the}\\ \text{solid of volume } V \text{ through}\\ \text{boundary surfaces } A\end{pmatrix} = \begin{pmatrix}\text{rate of increase of}\\ \text{internal energy of the solid}\\ \text{of volume } V\end{pmatrix} \tag{4-1}$$

By writing the appropriate mathematical expressions for each of these terms, Eq. (4-1) becomes

$$Ah[T_\infty - T(t)] = \rho c_p V \frac{dT(t)}{dt}$$

or

$$\frac{dT(t)}{dt} + \frac{Ah}{\rho c_p V}[T(t) - T_\infty] = 0 \qquad \text{for } t > 0 \tag{4-2a}$$

subject to the initial condition

$$T(t) = T_0 \qquad \text{for } t = 0 \tag{4-2b}$$

For convenience in the analysis, a new temperature $\theta(t)$ is defined as

$$\theta(t) \equiv T(t) - T_\infty \tag{4-3}$$

Then Eqs. (4-2) become

$$\frac{d\theta(t)}{dt} + m\theta(t) = 0 \qquad \text{for } t > 0 \tag{4-4a}$$

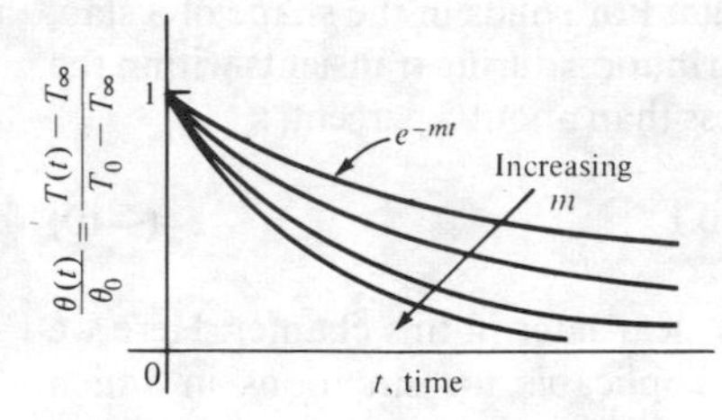

Figure 4-2 Dimensionless temperature $\theta(t)/\theta_0$ as a function of time.

and

$$\theta(t) = T_0 - T_\infty \equiv \theta_0 \qquad \text{for } t = 0 \tag{4-4b}$$

where we have defined

$$m \equiv \frac{Ah}{\rho c_p V} \tag{4-5}$$

Equation (4-4*a*) is an ordinary differential equation for the temperature $\theta(t)$, and its general solution is given as

$$\theta(t) = Ce^{-mt} \tag{4-6}$$

The application of the initial condition (4-4*b*) gives the integration constant as $C = \theta_0$. Then the temperature of the solid as a function of time is given as

$$\boxed{\frac{\theta(t)}{\theta_0} = \frac{T(t) - T_\infty}{T_0 - T_\infty} = e^{-mt}} \tag{4-7}$$

Figure 4-2 shows a plot of the dimensionless temperature given by Eq. (4-7) as a function of time. The temperature decays with time exponentially, and the shape of the curve is determined by the value of the exponent m. Here m has the dimension of $(\text{time})^{-1}$. Clearly, the curves in Fig. 4-2 become steeper as the value of m increases. That is, any increase in m will cause the solid to respond more quickly to a change in the ambient temperature. An examination of the parameters in the definition of m reveals that increasing the surface area for a given volume and the heat transfer coefficient increases m. Increasing the density, specific heat, or volume decreases m.

To establish some criteria under which the temperature distribution can be considered uniform within the solid, and hence the lumped system analysis becomes applicable, we define a *characteristic length* L_s as

$$L_s = \frac{V}{A} \tag{4-8}$$

and the Biot number Bi as

$$\boxed{\text{Bi} = \frac{hL_s}{k_s}} \tag{4-9}$$

where k_s is the thermal conductivity of the solid. For solids in the shape of a slab, long cylinder, and sphere, the temperature distribution during transients within the solid at any instant is uniform, with an error less than about 5 percent, if

$$\text{Bi} = \frac{hL_s}{k_s} < 0.1 \tag{4-10}$$

This matter is discussed further and becomes clear later in this chapter. Here we assume that the lumped system analysis is applicable for situations in which Bi < 0.1.

The physical significance of the Biot number is better envisioned if it is arranged in the form

$$\text{Bi} = \frac{h}{k_s/L_s} \tag{4-11}$$

which is the ratio of the heat transfer coefficient for convection at the surface of the solid to the specific conductance of the solid. Hence, the assumption of uniform temperature within the solid is valid if the specific conductance of the solid is much larger than the heat transfer coefficient for convection.

Example 4-1 An aluminum plate [$k = 160$ W/(m · °C), $\rho = 2790$ kg/m^3, $c_p = 0.88$ kJ/(kg · °C)] of thickness $L = 3$ cm and at a uniform temperature of $T_0 = 225$°C is suddenly immersed at time $t = 0$ in a well-stirred fluid maintained at a constant temperature $T_\infty = 25$°C. The heat transfer coefficient between the plate and the fluid is $h = 320$ W/(m^2 · °C). Determine the time required for the center of the plate to reach 50°C.

SOLUTION This problem can be solved by the lumped system analysis with sufficient accuracy if Bi < 0.1. Therefore, we need to check the magnitude of the Biot number. The characteristic dimension L_s is determined as

$$L_s = \frac{\text{volume}}{\text{area}} \cong \frac{LA}{2A} = \frac{L}{2} = 1.5 \text{ cm}$$

Then the Biot number becomes

$$\text{Bi} = \frac{hL_s}{k} = \frac{320 \times 1.5 \times 10^{-2}}{160} = 3 \times 10^{-2}$$

which is less than 0.1, hence the lumped system analysis is applicable. From Eq. (4-7) we have

$$\frac{T(t) - T_\infty}{T_0 - T_\infty} = e^{-mt}$$

where

$$T(t) = 50°\text{C} \qquad T_\infty = 25°\text{C} \qquad T_0 = 225°\text{C}$$

and

$$m = \frac{hA}{\rho c_p V} \cong \frac{h}{\rho c_p L_s} = \frac{320}{2790 \times 880 \times 1.5 \times 10^{-2}} = 0.00869 \text{ s}^{-1}$$

Then

$$\frac{50 - 25}{225 - 25} = e^{-0.00869t} \qquad \text{or} \qquad 0.125 = e^{-0.00869t}$$

$$0.00869t = 2.079$$

$$t \cong 239 \text{ s} \cong 4 \text{ min}$$

Example 4-2 The temperature of a gas stream is measured with a thermocouple. The junction may be approximated as a sphere of diameter $D = 1$ mm, $k = 25$ W/(m · °C), $\rho = 8400$ kg/m^3, and $c_p = 0.4$ kJ/(kg · °C). The heat transfer coefficient between the junction and the gas stream is $h = 560$ W/(m^2 · °C). How long will it take for the thermocouple to record 99 percent of the applied temperature difference?

SOLUTION The characteristic dimension L_s is

$$L_s = \frac{V}{A} = \frac{\frac{4}{3}\pi r^3}{4\pi r^2} = \frac{r}{3} = \frac{D}{6} = \frac{1}{6} \text{ mm}$$

The Biot number becomes

$$\text{Bi} = \frac{hL_s}{k} = \frac{560}{25}\frac{10^{-3}}{6} \cong 3.7 \times 10^{-3}$$

hence the lumped system analysis is applicable. From Eq. (4-7) we have

$$\frac{T(t) - T_\infty}{T_0 - T_\infty} = e^{-mt}$$

When the temperature reaches 99 percent of the applied temperature difference, we have

$$\tfrac{1}{100} = e^{-mt}$$

or

$$e^{mt} = 100$$

or

$$mt = 4.6$$

The value of m is

$$m = \frac{hA}{\rho c_p V} = \frac{h}{\rho c_p L_s} = \frac{560 \times 6}{8400 \times 400 \times 10^{-3}} = 1 \text{ s}^{-1}$$

Then

$$t = 4.6 \text{ s}$$

Mixed Boundary Condition

In the previous discussion we considered a situation in which all boundary surfaces of the region were subjected to convection. The method is also applicable when part of the boundary surface is subjected to convection and the remainder is subjected to prescribed heat flux as now illustrated.

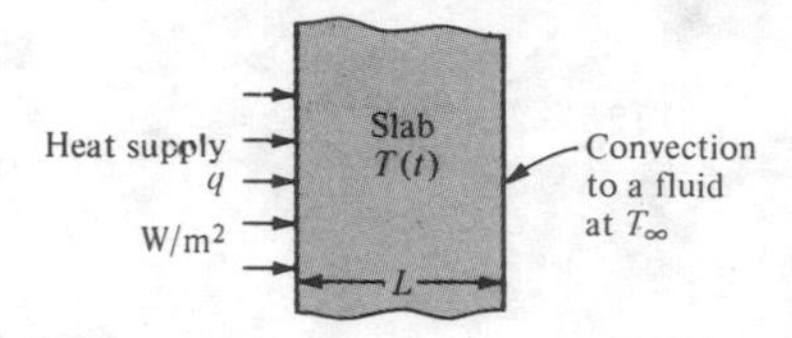

Figure 4-3 Nomenclature for lumped analysis of transient heat flow in a slab.

Consider a slab of thickness L, initially at a uniform temperature T_0. For times $t > 0$, heat is supplied to the slab from one of its boundary surfaces at a constant rate of q W/m² while heat is dissipated by convection from the other boundary surface into a medium at a uniform temperature T_∞ with a heat transfer coefficient h. Figure 4-3 shows the geometry and the boundary conditions for the problem.

We assume an equal area A for heat transfer on both sides of the plate. The application of the energy balance equation (4-1) for this particular case gives

$$Aq + Ah[T_\infty - T(t)] = \rho c_p AL \frac{dT(t)}{dt}$$

or

$$q + h[T_\infty - T(t)] = \rho c_p L \frac{dT(t)}{dt} \qquad \text{for } t > 0 \tag{4-12a}$$

with the initial condition

$$T(t) = T_0 \qquad \text{for } t = 0 \tag{4-12b}$$

For convenience in the analysis, a new temperature $\theta(t)$ is defined as

$$\theta(t) = T(t) - T_\infty \tag{4-13}$$

Then Eqs. (4-12) become

$$\frac{d\theta(t)}{dt} + m\theta(t) = Q \qquad \text{for } t > 0 \tag{4-14a}$$

and

$$\theta(t) = T_0 - T_\infty \equiv \theta_0 \qquad \text{for } t = 0 \tag{4-14b}$$

where we have defined

$$m \equiv \frac{h}{\rho c_p L} \qquad \text{and} \qquad Q \equiv \frac{q}{\rho c_p L} \tag{4-15}$$

The solution of Eq. (4-14a) is written as a sum of the solution of the homogeneous part of Eq. (4-14a) and a particular solution in the form

$$\theta(t) = Ce^{-mt} + \theta_p \tag{4-16}$$

where C is the integration constant. The particular solution θ_p is given by

$$\theta_p = \frac{Q}{m} \tag{4-17}$$

By combining Eqs. (4-16) and (4-17) we obtain

$$\theta(t) = Ce^{-mt} + \frac{Q}{m} \tag{4-18}$$

The integration constant C is determined by the application of the initial condition (4-14*b*) as

$$\theta_0 = C + \frac{Q}{m} \tag{4-19}$$

Substitution of Eq. (4-19) into (4-18) gives the solution of this heat transfer problem:

$$\theta(t) = \theta_0 e^{-mt} + (1 - e^{-mt})\frac{Q}{m}$$

$$\boxed{\theta(t) = \theta_0 e^{-mt} + (1 - e^{-mt})\frac{q}{h}} \tag{4-20}$$

For $t \to \infty$ this solution simplifies to

$$\theta(\infty) = \frac{Q}{m} = \frac{q}{h} \tag{4-21}$$

which is the steady-state temperature in the slab.

Example 4-3 A household electric iron has a steel base [$\rho = 7840\ \text{kg/m}^3$, $c_p = 450\ \text{J/(kg} \cdot {^\circ}\text{C)}$, and $k = 70\ \text{W/(m} \cdot {^\circ}\text{C)}$] which weighs $M = 1$ kg. The base has an ironing surface of $A = 0.025\ \text{m}^2$ and is heated from the other surface with a 250-W heating element. Initially the iron is at a uniform temperature of $T_i = 20°\text{C}$. Suddenly the heating starts, and the iron dissipates heat by convection from the ironing surface into an ambient at $T_\infty = 20°\text{C}$ with a heat transfer coefficient $h = 50\ \text{W/(m}^2 \cdot {^\circ}\text{C)}$.

Calculate the temperature of the iron $t = 5$ min after the start of heating.

What would the equilibrium temperature of the iron be if the control did not switch off the current?

SOLUTION The thickness L of the base is determined to be

$$L = \frac{M}{A\rho} = \frac{1}{0.025 \times 7840} = 0.51 \times 10^{-2}\ \text{m}$$

The Biot number becomes

$$\text{Bi} = \frac{hL}{k} = \frac{50 \times 0.51 \times 10^{-2}}{70} = 0.36 \times 10^{-2}$$

hence the lumped system analysis is applicable. From Eq. (4-20) we have

$$\theta(t) = \theta_0 e^{-mt} + (1 - e^{-mt})\frac{q}{h}$$

where $\theta_0 = T_i - T_\infty = 20 - 20 = 0$

$$q = \frac{250}{0.025} = 10{,}000 \text{ W/m}^2$$

$$h = 50 \text{ W/(m}^2 \cdot {}^\circ\text{C)}$$

$$m = \frac{h}{\rho c_p L} = \frac{50}{7840 \times 450 \times 0.51 \times 10^{-2}} = 0.278 \times 10^{-2}$$

Then the temperature at the end of $t = 5$ min is determined as

$$mt = 0.278 \times 10^{-2} \times 5 \times 60 = 0.834$$

$$\theta(t) = (1 - e^{-0.834})\frac{10{,}000}{50} = 113^\circ\text{C}$$

$$T(t) = \theta(t) + T_\infty = 113 + 20 = 133^\circ\text{C}$$

The equilibrium temperature becomes

$$\theta(\infty) = \frac{10{,}000}{50} = 200^\circ\text{C}$$

$$T(\infty) = \theta(\infty) + T_\infty = 200 + 20 = 220^\circ\text{C}$$

4-2 SLAB—Use of Transient-Temperature Charts

In many situations the temperature gradients within the solid are no longer negligible, hence the lumped system analysis is no longer applicable. Then the analysis of heat conduction problems involves the determination of the temperature distribution within the solid as a function of both time and position, and it is a complicated matter. Various methods of analysis for solving such problems are discussed in several texts [1–5] on advanced treatment of heat conduction. Simple problems, such as one-dimensional, time-dependent heat conduction in a slab with no internal energy generation, can be solved readily by the method of *separation of variables*, as described later in this chapter. In addition, the temperature distribution for such situations has been calculated, and the results presented in the form of transient-temperature charts in several places [6–8]. We now present transient-temperature and heat flow charts and discuss their physical significance and use.

Consider a slab (i.e., a plane wall) of thickness $2L$ that is confined to the region $-L \leq x \leq L$. Initially the slab is at a uniform temperature T_i. Suddenly, at $t = 0$, both boundary surfaces of the slab are subjected to convection with a heat transfer

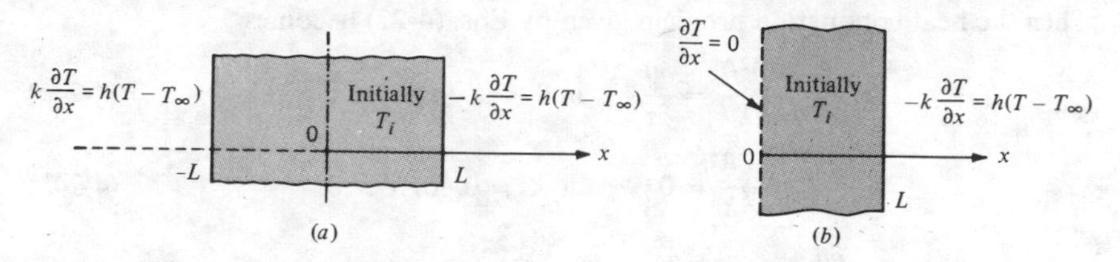

Figure 4-4 Geometry, coordinates, and boundary conditions for transient heat conduction in a slab.

coefficient h into ambients at temperature T_∞ and are maintained so for times $t > 0$. Figure 4-4a shows the geometry, coordinates, and boundary conditions for this particular problem. However, in this problem there is both geometrical and thermal symmetry about the $x = 0$ plane, so we need to consider the heat conduction problem for only half the region, say, $0 \le x \le L$. With this consideration the heat conduction problem for the slab of thickness $2L$ confined to the region $-L \le x \le L$ as illustrated in Fig. 4-4a is equivalent to that for the slab of thickness L confined to the region $0 \le x \le L$ as illustrated in Fig. 4-4b. Then the mathematical formulation of this time-dependent heat conduction problem for the geometry and boundary conditions shown in Fig. 4-4b is given as

$$\frac{\partial^2 T}{\partial x^2} = \frac{1}{\alpha}\frac{\partial T}{\partial t} \qquad \text{in } 0 < x < L, \text{ for } t > 0 \tag{4-22a}$$

$$\frac{\partial T}{\partial x} = 0 \qquad \text{at } x = 0, \text{ for } t > 0 \tag{4-22b}$$

$$k\frac{\partial T}{\partial x} + hT = hT_\infty \qquad \text{at } x = L, \text{ for } t > 0 \tag{4-22c}$$

$$T = T_i \qquad \text{for } t = 0, \text{ in } 0 \le x \le L \tag{4-22d}$$

Dimensionless Equations

The transient heat conduction problem given by Eqs. (4-22) can be expressed in the dimensionless form by introducing the following dimensionless variables:

$$\theta = \frac{T(x, t) - T_\infty}{T_i - T_\infty} = \text{dimensionless temperature} \tag{4-23a}$$

$$X = \frac{x}{L} = \text{dimensionless coordinate} \tag{4-23b}$$

$$\text{Bi} = \frac{hL}{k} = \text{Biot number} \tag{4-23c}$$

$$\tau = \frac{\alpha t}{L^2} = \text{dimensionless time, or Fourier number} \tag{4-23d}$$

Then the heat conduction problem given by Eqs. (4-22) becomes

$$\frac{\partial^2 \theta}{\partial X^2} = \frac{\partial \theta}{\partial \tau} \qquad \text{in } 0 < X < 1, \text{ for } \tau > 0 \tag{4-24a}$$

$$\frac{\partial \theta}{\partial X} = 0 \qquad \text{at } X = 0, \text{ for } \tau > 0 \tag{4-24b}$$

$$\frac{\partial \theta}{\partial X} + \text{Bi}\,\theta = 0 \qquad \text{at } X = 1, \text{ for } \tau > 0 \tag{4-24c}$$

$$\theta = 1 \qquad \text{in } 0 \le X \le 1, \text{ for } \tau = 0 \tag{4-24d}$$

The physical significance of the dimensionless time τ or the Fourier number is better envisioned if Eq. (4-23*d*) is rearranged in the form

$$\tau = \frac{\alpha t}{L^2} = \frac{k(1/L)L^2}{\rho c_p L^3/t} = \frac{\text{rate of heat conduction across } L \text{ in volume } L^3,\ \text{W/°C}}{\text{rate of heat storage in volume } L^3,\ \text{W/°C}} \tag{4-25a}$$

Thus, the Fourier number is a measure of the rate of heat conduction in comparison with the rate of heat storage in a given volume element. Therefore, the larger the Fourier number, the deeper the penetration of heat into a solid over a given time.

The physical significance of the Biot number is better understood if Eq. (4-23*c*) is rearranged in the form

$$\text{Bi} = \frac{hL}{k} = \frac{h}{k/L} = \frac{\text{heat transfer coefficient at the surface of solid}}{\text{internal conductance of solid across length } L} \tag{4-25b}$$

That is, the Biot number is the ratio of the heat transfer coefficient to the unit conductance of a solid over the characteristic dimension.

Comparing the heat conduction problems given by Eqs. (4-22) and (4-24), we conclude that the number of independent parameters that affect the temperature distribution in the solid is reduced significantly when the problem is expressed in the dimensionless form. In the problem given by Eqs. (4-22), the temperature depends on the following eight physical parameters:

$$x, t, L, k, \alpha, h, T_i, T_\infty$$

However, in the dimensionless problem given by Eqs. (4-24), the temperature depends on only the following three dimensionless parameters:

$$X, \text{Bi}, \text{and } \tau \tag{4-26}$$

Clearly, by expressing the problem in the dimensionless form, the number of parameters affecting the temperature distribution is significantly reduced. Therefore, it becomes feasible to solve such a problem once and for all and present the results in the form of charts for ready reference.

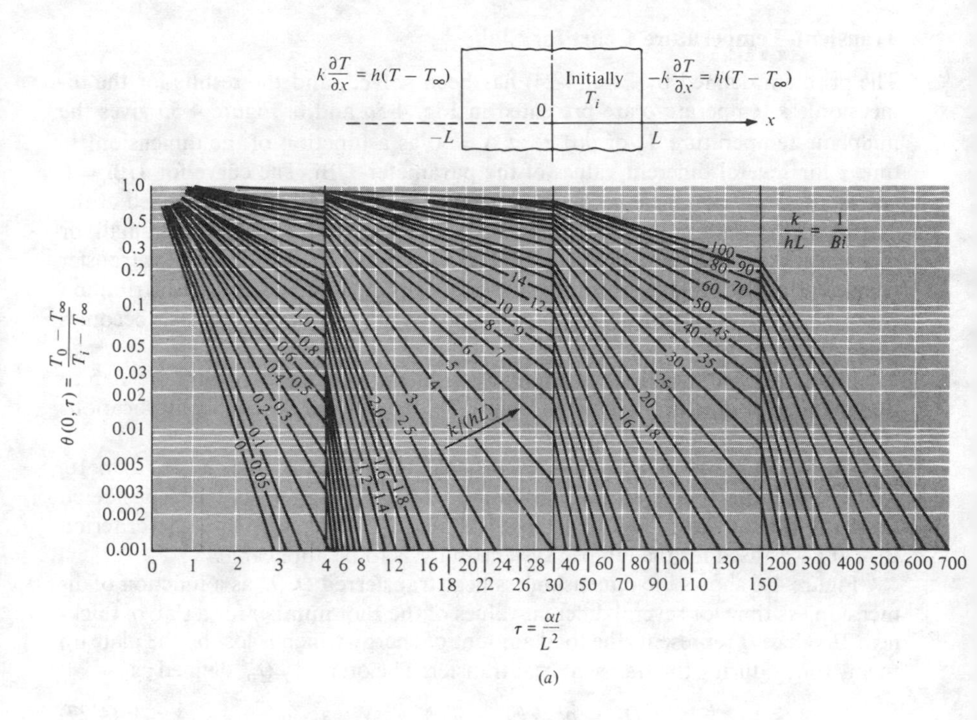

(a)

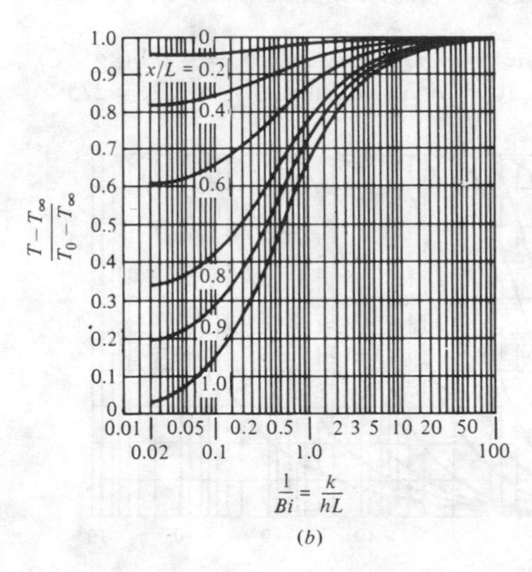

(b)

Figure 4-5 Transient-temperature chart for a slab of thickness $2L$ subjected to convection at both boundary surfaces. (*a*) Temperature T_0 at the center plane, $x = 0$; (*b*) position correction for use with part (*a*). (*From Heisler* [6].)

Transient-Temperature Chart for Slab

The problem defined by Eqs. (4-24) has been solved, and the results for the dimensionless temperature are presented in Fig. 4-5*a* and *b*. Figure 4-5*a* gives the midplane temperature T_0 or $\theta(0, \tau)$ at $X = 0$ as a function of the dimensionless time τ for several different values of the parameter 1/Bi. The curve for 1/Bi = 0 corresponds to the case $h \to \infty$, or the surfaces of the plate are maintained at the ambient temperature T_∞. For large values of 1/Bi, the Biot number is small, or the internal conductance of the solid is large in comparison with the heat transfer coefficient at the surface. This, in turn, implies that the temperature distribution within the solid is sufficiently uniform, hence the lumped system analysis becomes applicable.

Figure 4-5*b* relates the temperatures at different locations within the slab to the midplane temperature T_0. If we know T_0, temperatures at different locations within the slab can be determined.

An examination of Fig. 4-5*b* reveals that for values of 1/Bi larger than 10, or Bi < 0.1, the temperature distribution within the slab may be considered uniform with an error less than about 5 percent. We recall that the criterion Bi < 0.1 was used for the lumped system analysis to be applicable.

Figure 4-6 shows the dimensionless heat transferred Q/Q_0 as a function of dimensionless time for several different values of the Biot number for a slab of thickness 2*L*. Here *Q* represents the total amount of energy which is lost by the plate up to any time *t* during the transient heat transfer. The quantity Q_0, defined as

$$Q_0 = \rho c_p V(T_i - T_\infty) \qquad \text{W} \cdot \text{s} \tag{4-27}$$

represents the initial internal energy of the slab relative to the ambient temperature.

Example 4-4 A 5-cm-thick iron plate [$k = 60$ W/(m · °C), $c_p = 460$ J/(kg · °C), $\rho = 7850$ kg/m^3, and $\alpha = 1.6 \times 10^{-5}$ m^2/s] is initially at $T_i = 225$°C.

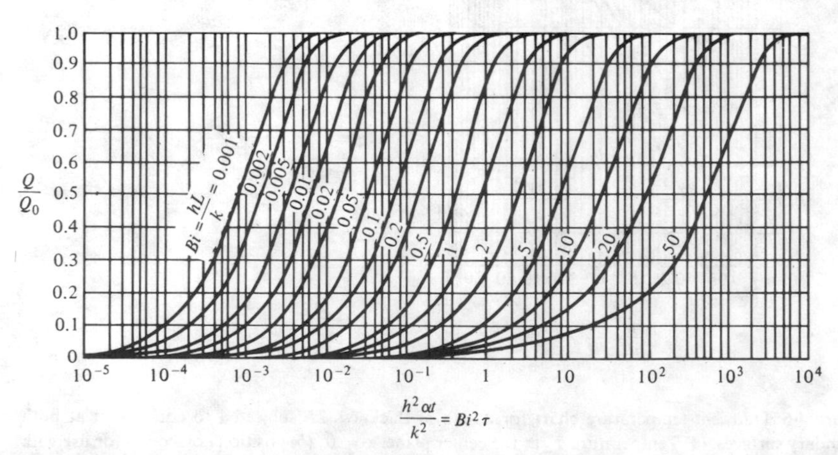

Figure 4-6 Dimensionless heat transferred Q/Q_0 for a slab of thickness 2*L*. (*From Ref.* 8.)

Suddenly, both surfaces are exposed to an ambient at $T_\infty = 25°C$ with a heat transfer coefficient $h = 500$ W/(m^2 · °C).

Calculate the center temperature at $t = 2$ min after the start of the cooling.

Calculate the temperature at a depth 1.0 cm from the surface at $t = 2$ min after the start of the cooling.

Calculate the energy removed from the plate per square meter during this time.

SOLUTION The transient-temperature charts may be used to solve this problem since the lumped system analysis is not applicable. We have

$$2L = 5 \times 10^{-2} \text{ m} \qquad \text{or} \qquad L = 2.5 \times 10^{-2} \text{ m}$$

$$\tau = \frac{\alpha t}{L^2} = \frac{(1.6 \times 10^{-5})(2 \times 60)}{(2.5)^2 \times 10^{-4}} = 3.1$$

$$\frac{1}{\text{Bi}} = \frac{k}{hL} = \frac{60}{500 \times 2.5 \times 10^{-2}} = 4.8$$

$$\text{Bi} = 0.21$$

Then, from Fig. 4-5*a* for $\tau = 3.1$ and $1/\text{Bi} = 4.8$, the center temperature T_0 is

$$\theta(0, \tau) = \frac{T_0 - T_\infty}{T_i - T_\infty} = 0.58$$

$$T_0 = T_\infty + (T_i - T_\infty)(0.58)$$

$$= 25 + 200(0.58) = 141°C$$

The temperature 1.0 cm from the surface is determined as

$$\frac{x}{L} = \frac{2.5 - 1}{2.5} = 0.6$$

For $1/\text{Bi} = 4.8$ and $x/L = 0.6$, from Fig. 4-5*b* we have

$$\frac{T - T_\infty}{T_0 - T_\infty} \cong 0.95$$

$$T = T_\infty + (T_0 - T_\infty)(0.95)$$

$$= 25 + (141 - 25)(0.95) = 135.2°C$$

The heat loss from the plate per square meter (including both sides) during the transients up to $t = 2$ min is determined as follows: From Fig. 4-6, for $\text{Bi} = 0.21$ and $\text{Bi}^2 \cdot \tau = 0.21^2 \times 3 \cdot 1 = 0.137$ we find

$$\frac{Q}{Q_0} = 0.45$$

where Q_0, by Eq. (4-27), is

$$\begin{aligned} Q_0 &= \rho c_p(2L)(A)(T_i - T_\infty) \\ &= (7850)(460)(5 \times 10^{-2})(1)(225 - 25) \\ &= 35.33 \times 10^6 \text{ J} \end{aligned}$$

Then the heat loss from the slab per square meter in 2 min becomes

$$Q = 0.45Q_0 = 15.9 \times 10^6 \text{ J}$$

4-3 LONG CYLINDER AND SPHERE—Use of Transient-Temperature Charts

The dimensionless transient-temperature distribution and the heat transfer results similar to those given by Figs. 4-5 and 4-6 also can be developed for a long cylinder and sphere.

Transient-Temperature Chart for Long Cylinder

Consider one-dimensional, transient heat conduction in a long cylinder of radius b, which is initially at a uniform temperature T_i. Suddenly, at time $t = 0$, the boundary surface at $r = b$ is subjected to convection with a heat transfer coefficient h into an ambient at temperature T_∞ and maintained so for $t > 0$. The mathematical formulation of this heat conduction problem is given in the dimensionless form as

$$\frac{1}{R}\frac{\partial}{\partial R}\left(R\frac{\partial \theta}{\partial R}\right) = \frac{\partial \theta}{\partial \tau} \qquad \text{in } 0 < R < 1, \text{ for } \tau > 0 \qquad (4\text{-}28a)$$

$$\frac{\partial \theta}{\partial R} = 0 \qquad \text{at } R = 0, \text{ for } \tau > 0 \qquad (4\text{-}28b)$$

$$\frac{\partial \theta}{\partial R} + \text{Bi}\,\theta = 0 \qquad \text{at } R = 1, \text{ for } \tau > 0 \qquad (4\text{-}28c)$$

$$\theta = 1 \qquad \text{in } 0 \le R \le 1, \text{ for } \tau = 0 \qquad (4\text{-}28d)$$

where various dimensionless quantities are defined as follows:

$$\text{Bi} = \frac{hb}{k} = \text{Biot number} \qquad (4\text{-}29a)$$

$$\tau = \frac{\alpha t}{b^2} = \text{dimensionless time, or Fourier number} \qquad (4\text{-}29b)$$

$$\theta = \frac{T(r, t) - T_\infty}{T_i - T_\infty} = \text{dimensionless temperature} \qquad (4\text{-}29c)$$

$$R = \frac{r}{b} = \text{dimensionless radial coordinate} \qquad (4\text{-}29d)$$

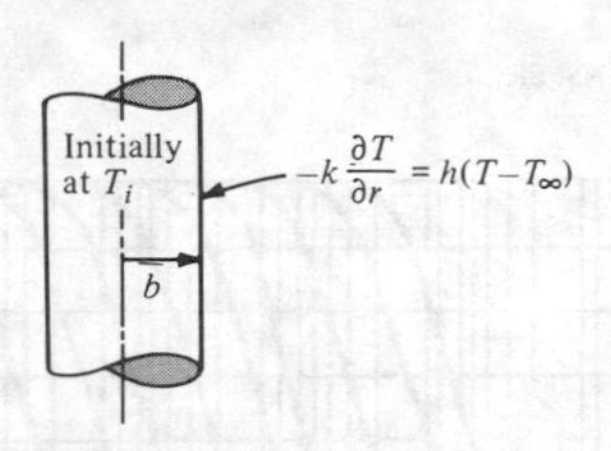

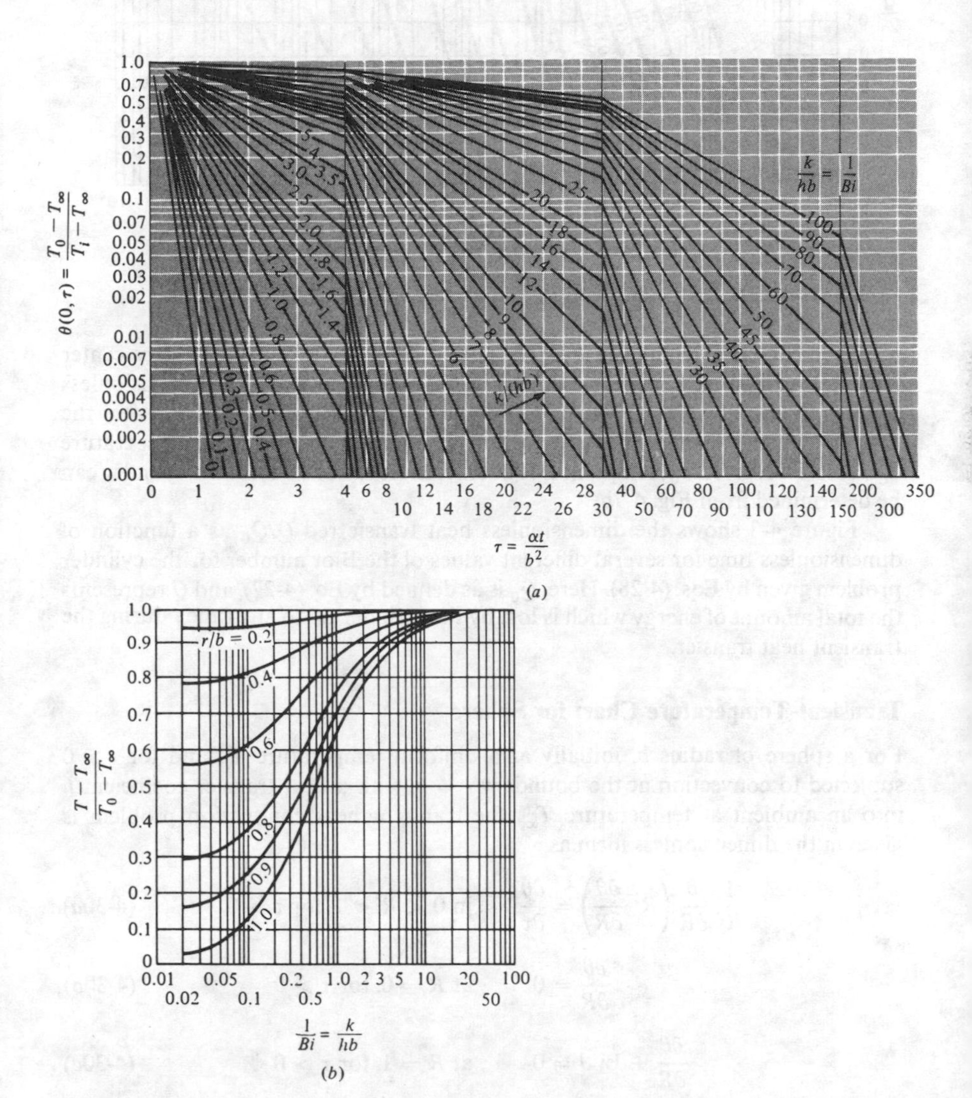

Figure 4-7 Transient-temperature chart for a long solid cylinder of radius $r = b$ subjected to convection at the boundary surface $r = b$. (*a*) Temperature T_0 at the axis of the cylinder; (*b*) position correction for use with part (*a*). (*From Heisler* [6].)

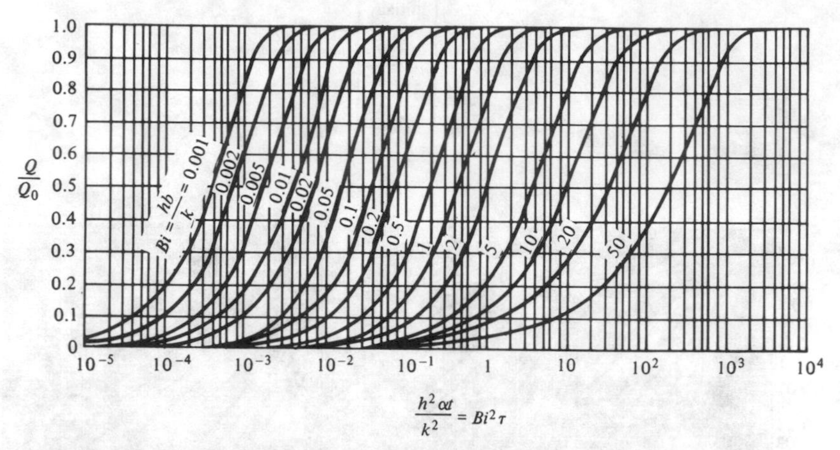

Figure 4-8 Dimensionless heat transferred Q/Q_0 for a long cylinder of radius b. (*From Ref.* 8.)

The problem given by Eq. (4-28) has been solved, and the results for the center temperature T_0 or $\theta(0, \tau)$ are shown in Fig. 4-7*a* as a function of the dimensionless time τ for several different values of the parameter 1/Bi. Figure 4-7*b* relates the temperatures at different locations within the cylinder to the center temperature T_0. Therefore, given T_0, temperatures at different locations within the cylinder can be determined from Fig. 4-7*b*.

Figure 4-8 shows the dimensionless heat transferred Q/Q_0 as a function of dimensionless time for several different values of the Biot number for the cylinder problem given by Eqs. (4-28). Here, Q_0 is as defined by Eq. (4-27), and Q represents the total amount of energy which is lost by the cylinder up to any time t during the transient heat transfer.

Transient-Temperature Chart for Sphere

For a sphere of radius b, initially at a uniform temperature T_i, and for $t > 0$ subjected to convection at the boundary $r = b$, with a heat transfer coefficient h into an ambient at temperature T_∞, the transient heat conduction problem is given in the dimensionless form as

$$\frac{1}{R^2}\frac{\partial}{\partial R}\left(R^2\frac{\partial \theta}{\partial R}\right) = \frac{\partial \theta}{\partial \tau} \quad \text{in } 0 < R < 1, \text{ for } \tau > 0 \tag{4-30a}$$

$$\frac{\partial \theta}{\partial R} = 0 \quad \text{at } R = 0, \text{ for } \tau > 0 \tag{4-30b}$$

$$\frac{\partial \theta}{\partial R} + \text{Bi}\,\theta = 0 \quad \text{at } R = 1, \text{ for } \tau > 0 \tag{4-30c}$$

$$\theta = 1 \quad \text{in } 0 \le R \le 1, \text{ for } \tau = 0 \tag{4-30d}$$

Here, the dimensionless parameters Bi, τ, θ, and R are as defined by Eqs. (4-29).

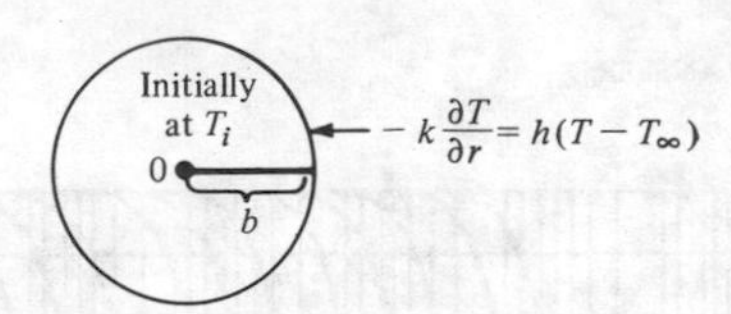

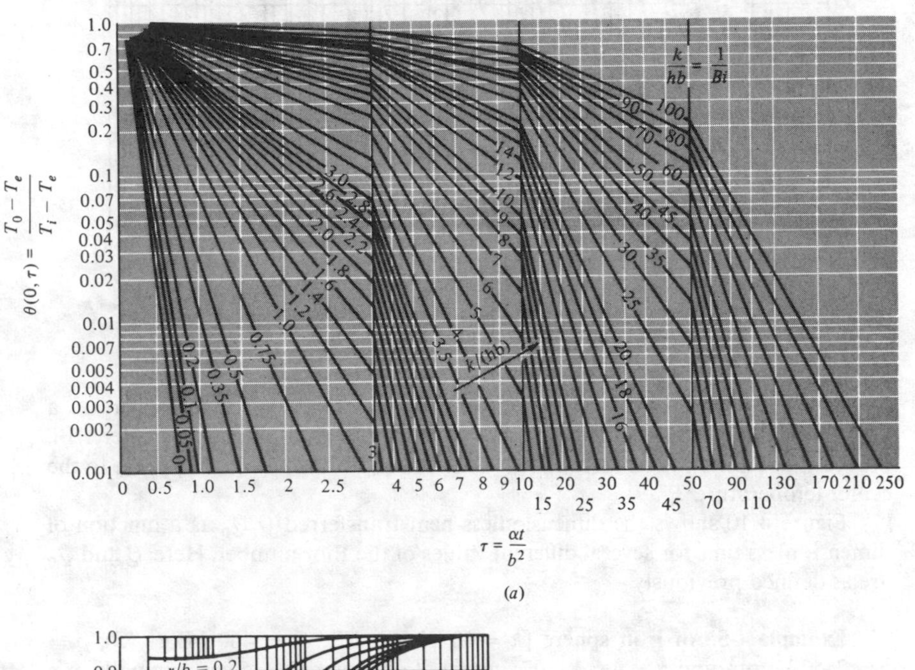

(a)

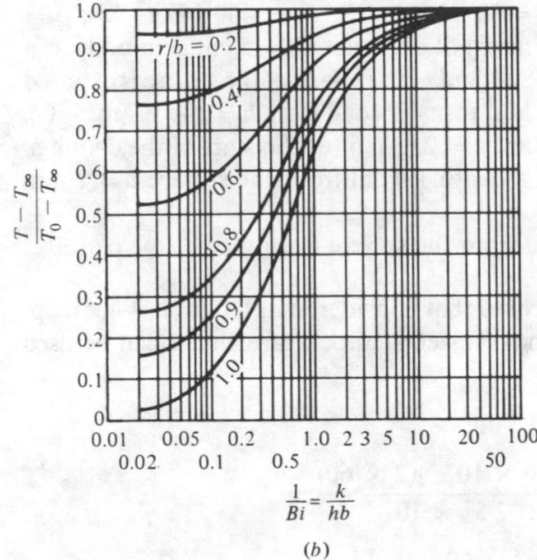

(b)

Figure 4-9 Transient-temperature chart for a solid sphere of radius $r = b$ subjected to convection at the boundary surface $r = b$. (*a*) Temperature T_0 at the center of the sphere; (*b*) position correction for use with part (*a*). (*From Heisler* [6].)

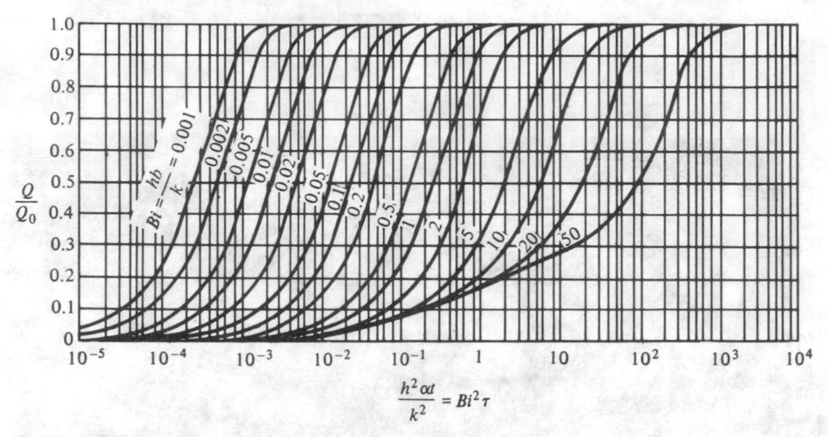

Figure 4-10 Dimensionless heat transferred Q/Q_0 for a sphere of radius b. (*From Ref.* 8.)

Figure 4-9*a* shows the center temperature T_0 or $\theta(0, \tau)$ for the sphere as a function of dimensionless time τ for several different values of the parameter 1/Bi. Figure 4-9*b* relates the temperatures at different locations within the sphere to the center temperature T_0.

Figure 4-10 shows the dimensionless heat transferred Q/Q_0 as a function of dimensionless time for several different values of the Biot number. Here, Q and Q_0 are as defined previously.

Example 4-5 An iron sphere [$k = 60$ W/(m · °C), $c_p = 460$ J/(kg · °C), $\rho = 7850$ kg/m³, and $\alpha = 1.6 \times 10^{-5}$ m²/s] of diameter $D = 5$ cm is initially at a uniform temperature $T_i = 225$°C. Suddenly the surface of the sphere is exposed to an ambient at $T_\infty = 25$°C with a heat transfer coefficient $h = 500$ W/(m² · °C).

Calculate the center temperature $t = 2$ min after the start of the cooling.

Calculate the temperature at a depth 1.0 cm from the surface $t = 2$ min after the start of the cooling.

Calculate the energy removed from the sphere during this time period.

SOLUTION This problem is identical to that considered in Example 4-4, except here the geometry is a sphere. The transient-temperature charts can be used to solve this problem. We have

$$b = 2.5 \times 10^{-2} \text{ m}$$

$$\tau = \frac{\alpha t}{b^2} = \frac{(1.6 \times 10^{-5})(2 \times 60)}{2.5^2 \times 10^{-4}} = 3.1$$

$$\frac{1}{\text{Bi}} = \frac{k}{hb} = \frac{60}{500 \times 2.5 \times 10^{-2}} = 4.8$$

$$\text{Bi} = 0.21$$

Then, from Fig. 4-9*a*, for $\tau = 3.1$ and $1/\text{Bi} = 4.8$, the center temperature T_0 is determined as

$$\theta(0, \tau) = \frac{T_0 - T_\infty}{T_i - T_\infty} = 0.18$$

$$T_0 = T_\infty + (T_i - T_\infty)(0.18)$$

$$= 25 + 200(0.18) = 61°\text{C}$$

The temperature 1.0 cm from the surface is determined as

$$\frac{r}{b} = \frac{2.5 - 1}{2.5} = 0.6$$

From Fig. 4-9*b*, for $1/\text{Bi} = 4.8$ and $r/b = 0.6$, we have

$$\frac{T - T_\infty}{T_0 - T_\infty} \simeq 0.95$$

$$T = T_\infty + (T_0 - T_\infty)(0.95)$$

$$= 25 + (61 - 25)(0.95) = 59.2°\text{C}$$

The heat loss from the sphere during transients up to $t = 2$ min is determined as follows: From Fig. 4-10, for $\text{Bi} = 0.21$, $\text{Bi}^2\tau = 0.137$, we find

$$\frac{Q}{Q_0} = 0.8$$

where Q_0 is given by Eq. (4-27) as

$$Q_0 = \rho c_p(\tfrac{4}{3}\pi b^3)(T_i - T_\infty)$$

$$= 7850(460)(\tfrac{4}{3}\pi \times 2.5^3 \times 10^{-6})(225 - 25)$$

$$= 47{,}268 \text{ J}$$

Then the heat loss from the sphere becomes

$$Q = 37{,}814 \text{ J}$$

The boundary and initial conditions for this problem as well as the physical properties are the same as those for the slab problem considered in Example 4-4. In addition, the half-thickness L of the slab is the same as the radius b of the sphere. A comparison of the center temperatures T_0 for the two examples reveals that at the end of 2 min $T_0 = 141°\text{C}$ for the slab whereas $T_0 = 61°\text{C}$ for the sphere. That is, the sphere loses heat at a much faster rate than the slab. This conclusion is also apparent from the comparison of the fractional heat loss Q/Q_0, which is 0.45 for the slab and 0.8 for the sphere.

4-4 SEMI-INFINITE SOLID—Use of Transient-Temperature Charts

The one-dimensional, transient heat conduction in a semi-infinite solid in rectangular coordinates with no energy generation in the medium is another example of simple geometry for which the temperature distribution in the solid as a function of time and position can be presented in graphical form.

Consider a semi-infinite solid, confined to the domain $0 \leq x < \infty$, which is initially at a uniform temperature T_i, and there is no internal energy generation. Then the temperature transient within the solid can be initiated by a change in the thermal condition at the boundary surface $x = 0$, since the solid extends to infinity. Figure 4-11 illustrates three different possibilities at the boundary surface $x = 0$ to initiate the transients within the solid. The physical significance of these three situations is as follows:

1. In Fig. 4-11*a*, at $t = 0$ the surface temperature is suddenly changed to T_0 and is maintained at that temperature for $t > 0$. The applied temperature T_0 may be higher or lower than the initial temperature T_i of the medium.
2. In Fig. 4-11*b*, at $t = 0$ a constant heat flux q_0 W/m^2 is imposed on the boundary surface at $x = 0$ and maintained for $t > 0$.
3. In Fig. 4-11*c*, at $t = 0$ the boundary surface at $x = 0$ is subjected to convection with a fluid at temperature T_∞ and with a heat transfer coefficient h. This condition is maintained for all $t > 0$. Here, the fluid temperature T_∞ may be higher or lower than the initial temperature T_i of the solid.

Clearly, the temperature response within the solid will be different for each of these three cases.

The semi-infinite medium transient heat conduction problems have numerous practical applications in engineering. Consider, for example, temperature transients in a slab of finite thickness, initiated by a sudden change in the thermal condition at the boundary surface. At very early times, the temperature transients near the boundary surface behave similar to those of the semi-infinite medium, because some time is required for the heat to penetrate the slab before the other boundary condition begins to influence the transients.

The transient heat conduction problems for the three cases illustrated in Fig. 4-11 have been solved, and analytical expressions are available for the temperature

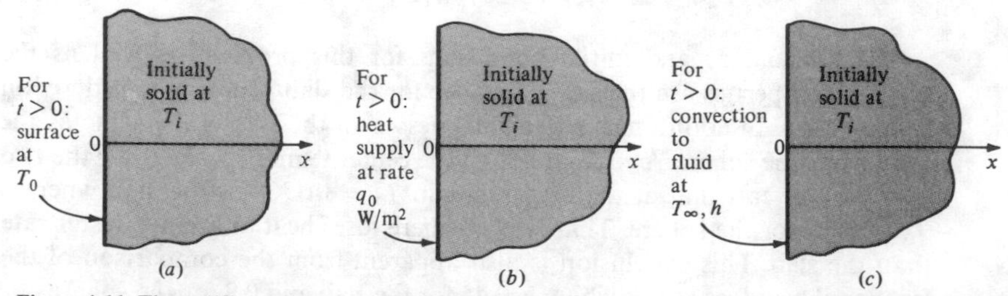

Figure 4-11 Three different boundary conditions for transient heat conduction in a semi-infinite solid.

distribution in the solid as a function of time and position. We now examine the results obtained from such solutions for each case.

Case 1 The solid is initially at a uniform temperature T_i, and for $t > 0$ the boundary surface at $x = 0$ is maintained at a constant temperature T_0, as illustrated in Fig. 4-11*a*. For this particular case the dimensionless temperature

$$\theta(x, t) = \frac{T(x, t) - T_0}{T_i - T_0}$$

is plotted against the dimensionless parameter $\xi = x/(2\sqrt{\alpha t})$, as shown in Fig. 4-12. The physical significance of this graph is as follows: For a given value of x, the graph represents the variation in temperature with time at that particular location x. Conversely, for a given value of t, the graph represents the variation of temperature with position within the solid at that particular time t.

In engineering applications, the heat flux at the boundary surface $x = 0$ is also of interest. For the case illustrated in Fig. 4-11*a*, the analytic expression for the heat flux at the boundary surface $x = 0$ is given by

$$\boxed{q_s(t) = \frac{k(T_0 - T_i)}{\sqrt{\pi \alpha t}}} \qquad \text{W/m}^2 \tag{4-31}$$

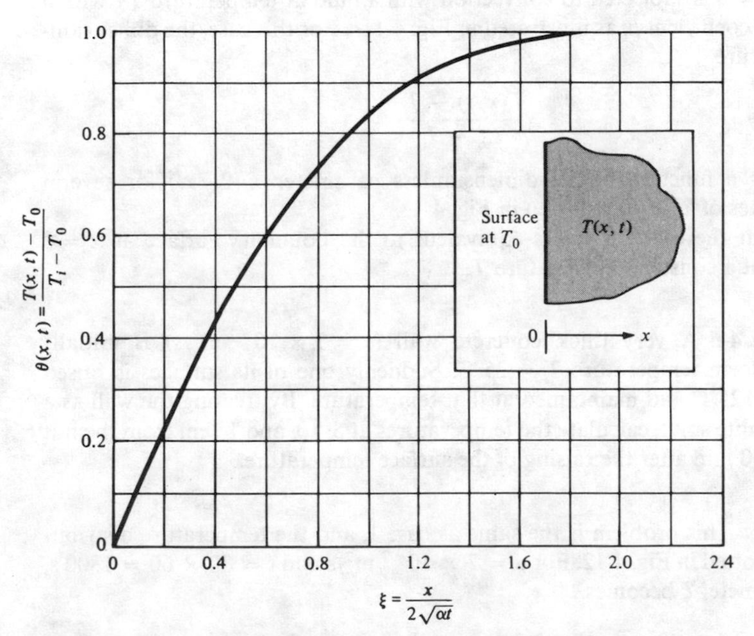

Figure 4-12 Temperature distribution $T(x, t)$ in a semi-infinite solid which is initially at T_i; for $t > 0$ the surface at $x = 0$ is maintained at T_0.

Clearly, when $q_s(t)$ is a positive quantity, the heat flow is into the medium, and vice versa.

Case 2 The solid is initially at a uniform temperature T_i, and for $t > 0$ the boundary surface at $x = 0$ is subjected to a constant heat flux q_0 W/m^2, as illustrated in Fig. 4-11*b*. For this case, the temperature distribution $T(x, t)$ within the solid is given by

$$T(x, t) = T_i + \frac{2q_0}{k}(\alpha t)^{1/2}\left[\frac{1}{\sqrt{\pi}}e^{-\xi^2} + \xi\,\mathrm{erf}(\xi) - \xi\right] \tag{4-32a}$$

where

$$\xi = \frac{x}{2\sqrt{\alpha t}} \tag{4-32b}$$

Here, the function erf(ξ) is called the *error function* of arguments ξ, and its values are tabulated in App. D, Table D-1.

Case 3 The solid is initially at a uniform temperature T_i, and for $t > 0$ the boundary surface at $x = 0$ is subjected to convection with a fluid at temperature T_∞ with a heat transfer coefficient h, as illustrated in Fig. 4-11*c*. For this case, the dimensionless temperature

$$\frac{T(x, t) - T_i}{T_\infty - T_i}$$

is plotted as a function of the dimensionless parameter $x/(2\sqrt{\alpha t})$ for several different values of $h\sqrt{\alpha t}/k$, as given in Fig. 4-13.

Note that the case $h \to \infty$ is equivalent to the boundary surface at $x = 0$ maintained at a constant temperature T_∞.

Example 4-6 A very thick concrete wall ($\alpha = 7 \times 10^{-7}$ m^2/s) is initially at a uniform temperature $T_i = 25$°C. Suddenly one of its surfaces is raised to $T_0 = 125$°C and maintained at that temperature. By treating the wall as a semi-infinite solid, calculate the temperatures at 5, 10, and 15 cm from the hot surface 30 min after the raising of the surface temperature.

SOLUTION This problem is the same as case 1, and the temperature distribution is plotted in Fig. 4-12. For $\alpha = 7 \times 10^{-7}$ m^2/s and $t = 30 \times 60 = 1800$ s, the parameter ξ becomes

$$\xi = \frac{x}{2\sqrt{\alpha t}} = \frac{x}{2\sqrt{7 \times 10^{-7} \times 1800}} = 14x$$

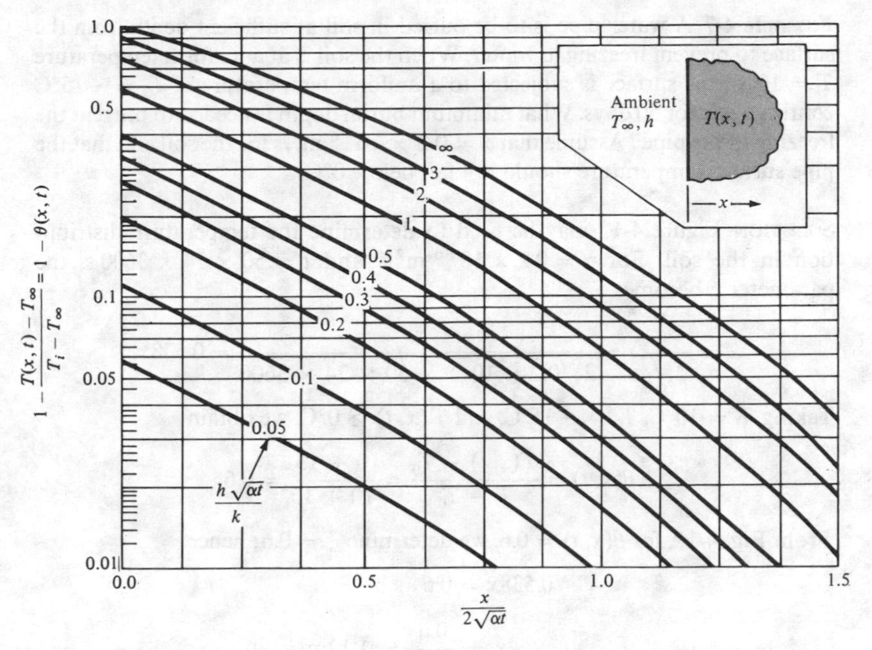

Figure 4-13 Transient temperature $T(x, t)$ in a semi-infinite solid subjected to convection at the boundary surface. (From Ref. 9.)

For $x = 0.05$ m: $\xi = 0.7$, and

$$\frac{T(x, t) - T_0}{T_i - T_0} = \frac{T(x, t) - 125}{25 - 125} = 0.67$$

$$T(x, t) = 58°\text{C}$$

For $x = 0.10$ m: $\xi = 1.4$, and

$$\frac{T(x, t) - 125}{25 - 125} = 0.95$$

$$T(x, t) = 30°\text{C}$$

For $x = 0.15$ m: $\xi = 2.1$, and

$$\frac{T(x, t) - 125}{25 - 125} = 1$$

$$T(x, t) = 25°\text{C}$$

It is apparent from these calculations that the effect of heating has not penetrated to a depth $x = 0.15$ m in 30 min.

Example 4-7 A water pipe is to be buried in soil at sufficient depth from the surface to prevent freezing in winter. When the soil is at a uniform temperature $T_i = 10°C$, the surface is subjected to a uniform temperature of $T_0 = -15°C$ continuously for 50 days. What minimum burial depth is needed to prevent the freezing of the pipe? Assume that $\alpha = 0.2 \times 10^{-6}$ m^2/s for the soil and that the pipe surface temperature should not fall below 0°C.

SOLUTION Figure 4-12 may be used to determine the temperature distribution in the soil. For $\alpha = 0.2 \times 10^{-6}$ m^2/s and $t = 50 \times 24 \times 3600$ s, the parameter ξ becomes

$$\xi = \frac{x}{2\sqrt{\alpha t}} = \frac{x}{2\sqrt{0.2 \times 10^{-6} \times 50 \times 24 \times 3600}} = 0.538x$$

Taking $T_i = 10°C$, $T_0 = -15°C$, and $T(x, t) \geq 0°C$, we obtain

$$\theta(x, t) = \frac{T(x, t) - T_0}{T_i - T_0} = \frac{0 + 15}{10 + 15} = 0.6$$

From Fig. 4-12, for $\theta(x, t) = 0.6$, we determine $\xi = 0.6$; hence

$$0.538x = 0.6$$

$$x = \frac{0.6}{0.538} = 1.12 \text{ m}$$

That is, the pipe should be buried at least to a depth of $x = 1.12$ m.

4-5 PRODUCT SOLUTION—Use of Transient-Temperature Charts

When the temperature gradients are important in not one, but, say, two different directions within the solid, then the problem is one of two-dimensional transient heat conduction in a solid. When there is no internal energy generation in the medium, it is possible to combine the solutions obtained from one-dimensional transient-temperature charts and to construct the solution for a two-dimensional transient heat conduction problem. Such an approach, called the method of *product solution*, is applicable if the solution of a two-dimensional, time-dependent heat conduction problem can be shown to be equivalent to the product of the solutions of two one-dimensional, transient heat conduction problems.

Demonstration of the Concept of Product Solution

The basis of the product solution is better envisioned with the following example.

Consider a rectangular bar of sides $2L_1$ and $2L_2$, confined to the region $-L_1 \leq x \leq L_1$ and $-L_2 \leq y \leq L_2$, as illustrated in Fig. 4-14. Initially the slab

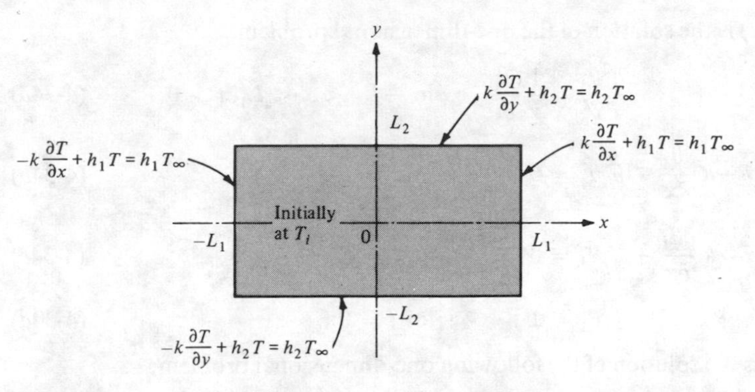

Figure 4-14 Product solution for transient heat conduction in a rectangular bar.

is at a uniform temperature T_i. Suddenly at $t = 0$ all boundary surfaces are subjected to convection to an ambient at a constant temperature T_∞. The mathematical formulation of this heat conduction problem, in terms of the dimensionless temperature

$$\theta(x, y, t) = \frac{T(x, y, t) - T_\infty}{T_i - T_\infty}$$

is given by

$$\frac{\partial^2 \theta}{\partial x^2} + \frac{\partial^2 \theta}{\partial y^2} = \frac{1}{\alpha}\frac{\partial \theta}{\partial t} \qquad \text{in } -L_1 < x < L_1, -L_2 < y < L_2 \text{ for } t > 0 \qquad (4\text{-}33a)$$

$$-k\frac{\partial \theta}{\partial x} + h_1 \theta = 0 \qquad \text{at } x = -L_1 \qquad (4\text{-}33b)$$

$$k\frac{\partial \theta}{\partial x} + h_1 \theta = 0 \qquad \text{at } x = L_1 \qquad (4\text{-}33c)$$

$$-k\frac{\partial \theta}{\partial y} + h_2 \theta = 0 \qquad \text{at } y = -L_2 \qquad (4\text{-}33d)$$

$$k\frac{\partial \theta}{\partial y} + h_2 \theta = 0 \qquad \text{at } y = L_2 \qquad (4\text{-}33e)$$

$$\theta = 1 \qquad \text{for } t = 0 \qquad (4\text{-}33f)$$

It can be shown that the solution of this two-dimensional problem can be expressed as a product of the solutions of two one-dimensional problems $\theta_1(x, t)$ and $\theta_2(y, t)$ in the form [5, p. 54]

$$\theta(x, y, t) = \theta_1(x, t)\theta_2(y, t)$$

where $\theta_1(x, t)$ is the solution of the one-dimensional problem

$$\frac{\partial^2 \theta_1}{\partial x^2} = \frac{1}{\alpha}\frac{\partial \theta_1}{\partial t} \qquad \text{in } -L_1 < x < L_1, t > 0 \tag{4-34a}$$

$$-k\frac{\partial \theta_1}{\partial x} + h_1\theta_1 = 0 \qquad \text{at } x = -L_1 \tag{4-34b}$$

$$k\frac{\partial \theta_1}{\partial x} + h_1\theta_1 = 0 \qquad \text{at } x = L_1 \tag{4-34c}$$

$$\theta_1 = 1 \qquad \text{for } t = 0 \tag{4-34d}$$

and $\theta_2(y, t)$ is the solution of the following one-dimensional problem:

$$\frac{\partial^2 \theta_2}{\partial y^2} = \frac{1}{\alpha}\frac{\partial \theta_2}{\partial t} \qquad \text{in } -L_2 < y < L_2, t > 0 \tag{4-35a}$$

$$-k\frac{\partial \theta_2}{\partial y} + h_2\theta_2 = 0 \qquad \text{at } y = -L_2 \tag{4-35b}$$

$$k\frac{\partial \theta_2}{\partial y} + h_2\theta_2 = 0 \qquad \text{at } y = L_2 \tag{4-35c}$$

$$\theta_2 = 1 \qquad \text{for } t = 0 \tag{4-35d}$$

The validity of the above decomposition can be verified by substituting $\theta = \theta_1\theta_2$ in the original two-dimensional problem [Eqs. (4-33)] and utilizing the above problems, defined by Eqs. (4-34) and (4-35).

Clearly, the above one-dimensional problems for the functions $\theta_1(x, t)$ and $\theta_2(y, t)$ are exactly the same as that whose solution is given by the transient-temperature chart in Fig. 4-5.

From the previous illustration we conclude that the solution of the two-dimensional heat conduction problem defined by Eqs. (4-33) for a rectangular region $-L_1 < x < L_1$, $-L_2 < y < L_2$ can be constructed as the product of the solutions of two one-dimensional, time-dependent problems for slabs, whose solutions are obtained from the chart given in Fig. 4-5.

The concept of the product solution described above—decomposing the solution for a rectangular bar to the product of the solutions of two slab problems—is illustrated in Fig. 4-15*a*.

The basic idea developed here with reference to a rectangular bar can be extended to other configurations. For example, the two-dimensional, transient-temperature distribution $T(r, z, t)$ in a solid cylinder of radius b and finite height $2L$, initially at temperature T_i and subjected to convection at the boundaries, can be expressed in the dimensionless form as

$$\theta(r, z, t) = \frac{T(r, z, t) - T_\infty}{T_i - T_\infty}$$

Then the solution of this problem can be constructed as the product of the solutions

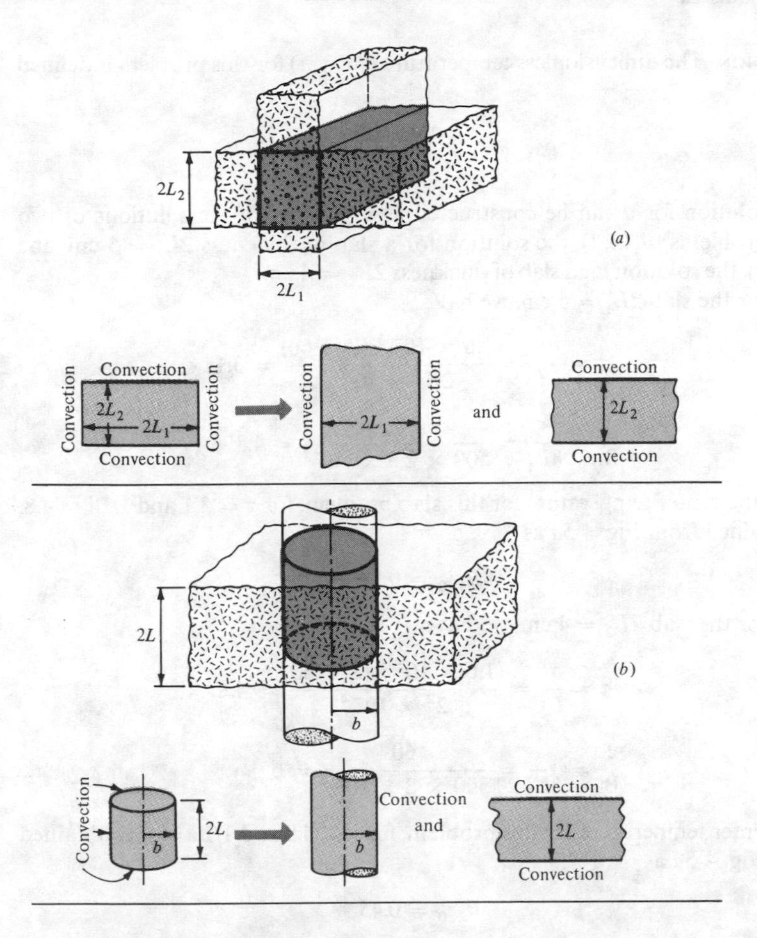

Figure 4-15 Decomposition for product solution.

of two one-dimensional problems, that is, one for the slab problem given in Fig. 4-5 and the other for a long, solid cylinder given in Fig. 4-7. Such a decomposition process is illustrated in Fig. 4-15*b*.

Clearly, there are numerous other multidimensional configurations which permit decomposition into one-dimensional problems, and their solutions can be constructed by the product solution.

Example 4-8 A rectangular iron bar 5 cm by 4 cm [$k = 60$ W/(m · °C), $\alpha = 1.6 \times 10^{-5}$ m^2/s] is initially at a uniform temperature $T_i = 225$°C. Suddenly the surfaces of the bar are subjected to convective cooling with a heat transfer coefficient $h = 500$ W/(m^2 · °C) into an ambient at $T_\infty = 25$°C. Calculate the center temperature T_0 of the bar $t = 2$ min after the start of the cooling.

SOLUTION The dimensionless temperature $\theta(x, y, t)$ for this problem is defined as

$$\theta(x, y, t) = \frac{T(x, y, t) - T_\infty}{T_i - T_\infty}$$

The solution for θ can be constructed as a product of the solutions of two slab problems: $\theta_1(x, t)$, the solution for a slab of thickness $2L_1 = 5$ cm, and $\theta_2(y, t)$, the solution for a slab of thickness $2L_2 = 4$ cm.

For the slab $2L_1 = 5$ cm, we have

$$\tau = \frac{\alpha t}{L_1^2} = \frac{(1.6 \times 10^{-5})(2 \times 60)}{2.5^2 \times 10^{-4}} = 3.1$$

and
$$\frac{1}{\text{Bi}} = \frac{k}{hL_1} = \frac{60}{500 \times 2.5 \times 10^{-2}} = 4.8$$

Then the center temperature for this slab problem, for $\tau = 3.1$ and $1/\text{Bi} = 4.8$, is obtained from Fig. 4-5*a* as

$$\theta_{1,0} = 0.58$$

For the slab $2L_2 = 4$ cm, we have

$$\tau = \frac{\alpha t}{L_2^2} = \frac{(1.6 \times 10^{-5})(2 \times 60)}{2^2 \times 10^{-4}} = 4.8$$

and
$$\frac{1}{\text{Bi}} = \frac{k}{hL_2} = \frac{60}{500 \times 2 \times 10^{-2}} = 6$$

The center temperature for this problem, for $\tau = 4.8$ and $1/\text{Bi} = 6$, is obtained from Fig. 4-5*a* as

$$\theta_{2,0} = 0.45$$

Then the dimensionless center temperature θ_0 for the two-dimensional problem is determined by product solution:

$$\theta_0 = \theta_{1,0} \cdot \theta_{2,0} = 0.58 \times 0.45 = 0.26$$

And the center temperature T_0 for the two-dimensional problem becomes

$$\theta_0 = \frac{T_0 - T_\infty}{T_i - T_\infty} = 0.26$$

$$T_0 = T_\infty + (T_i - T_\infty)(0.26)$$
$$= 25 + 200(0.26) = 77°\text{C}$$

A comparison of this result with that for the slab $2L_1 = 5$ cm thick considered in Example 4-4 shows that the center temperature is lower for the rectangular bar because of increased lateral cooling.

Example 4-9 A short, iron cylinder [$k = 60$ W/(m · °C), $\alpha = 1.6 \times 10^{-5}$ m²/s] of diameter $D = 5$ cm and height $H = 4$ cm is initially at a uniform temperature $T_i = 225$°C. Suddenly the boundary surfaces are exposed to an ambient at $T_\infty = 25$°C with a heat transfer coefficient $h = 500$ W/(m² · °C). Calculate the center temperature T_0 at $t = 2$ min after the start of cooling.

SOLUTION The dimensionless temperature $\theta(r, z, t)$ for this problem is defined as

$$\theta(r, z, t) = \frac{T(r, z, t) - T_\infty}{T_i - T_\infty}$$

The solution for θ can be constructed as a product of the solutions of the following two problems: $\theta_1(r, t)$, the solution for a long cylinder of diameter $D = 5$ cm, and $\theta_2(z, t)$, the solution for a slab of thickness $2L_2 = 4$ cm.

For the cylinder with $D = 5$ cm, we have

$$\tau = \frac{\alpha t}{(D/2)^2} = \frac{(1.6 \times 10^{-5})(2 \times 60)}{2.5^2 \times 10^{-4}} = 3.1$$

and $$\frac{1}{\text{Bi}} = \frac{k}{hD/2} = \frac{60}{500(2.5 \times 10^{-2})} = 4.8$$

The center temperature for this cylinder problem, for $\tau = 3.1$ and $1/\text{Bi} = 4.8$, is obtained from Fig. 4-7*a* as

$$\theta_{1,0} = 0.31$$

For the slab with $2L_2 = H = 4$ cm, we have

$$\tau = \frac{\alpha t}{L_2^2} = \frac{(1.6 \times 10^{-5})(2 \times 60)}{(2 \times 10^{-2})^2} = 4.8$$

and $$\frac{1}{\text{Bi}} = \frac{k}{hL_2} = \frac{60}{500(2 \times 10^{-2})} = 6$$

The center temperature for this slab problem, for $\tau = 4.8$ and $1/\text{Bi} = 6$, is obtained from Fig. 4-5*a* as

$$\theta_{2,0} = 0.45$$

Then the dimensionless center temperature θ_0 for the two-dimensional short cylinder problem is determined by product solution:

$$\theta_0 = \theta_{1,0} \cdot \theta_{2,0} = 0.31(0.45) = 0.14$$

And the center temperature T_0 becomes

$$\theta_0 \equiv \frac{T_0 - T_\infty}{T_i - T_\infty} = 0.14$$

$$T_0 = T_\infty + (T_i - T_\infty)(0.14)$$

$$= 25 + 200(0.14) = 53°\text{C}$$

4-6 TWO-DIMENSIONAL STEADY-STATE HEAT CONDUCTION—Use of Conduction Shape Factors

An approximate, yet very simple approach to determining the steady-state heat conduction in two-dimensional systems uses conduction shape factors. Let T_1 and T_2 be the two characteristic temperature potentials in a two-dimensional, steady-state heat flow system, and let Q be the total heat flow rate across these two potentials. The conduction shape factor S is defined to relate the total heat flow rate Q to the temperature difference $T_1 - T_2$ as

$$\boxed{Q = Sk(T_1 - T_2)} \tag{4-36}$$

The conduction shape factors S have been computed for a variety of geometries and the results tabulated in Refs. 10 and 11. Table 4-1 shows some of these shape factors, and we illustrate their use with the following examples.

Example 4-10 A hot water pipe with diameter $D = 10$ cm and $L = 8$ m long is buried horizontally in the earth [$k = 0.9$ W/(m · °C)] at a depth $z = 0.6$ m from the surface. The pipe is at a uniform temperature $T_1 = 80$°C, and the earth surface is at $T_2 = 10$°C. Calculate the heat loss from the pipe.

SOLUTION The physical situation and the configuration for this problem are similar to those of case 7 in Table 4-1. By setting $L = 8$ m, $z = 0.6$ m, and $R = 0.05$ m, the shape factor S is determined as follows:

$$S = \frac{2\pi L}{\cosh^{-1}(z/R)} = \frac{2\pi(8)}{\cosh^{-1}(0.6/0.05)} = 15.8 \text{ m}$$

The heat loss is calculated from Eq. (4-36) as

$$Q = Sk(T_1 - T_2) = 15.8(0.9)(80 - 10) = 997 \text{ W}$$

Example 4-11 A spherical tank of diameter $D = 0.5$ m containing radioactive materials is buried in the earth [$k = 0.8$ W/(m · °C)] at a depth $z = 1.25$ m from the earth surface to the center of the sphere. The surface of the tank is maintained at a uniform temperature of $T_1 = 100$°C as a result of the radioactive decay while the earth surface is at a uniform temperature $T_2 = 15$°C. Calculate the rate of heat generation in the tank as a result of radioactive disintegration.

SOLUTION The physical situation and the configuration are similar to those in case 5 in Table 4-1. By taking $R = 0.25$ m and $z = 1.25$ m, the shape factor S is found:

$$S = \frac{4\pi R}{1 - R/(2z)} = \frac{4\pi(0.25)}{1 - 0.25/(2 \times 1.25)} = 3.49 \text{ m}$$

Table 4-1 Conduction shape factors; S defined by $Q = Sk(T_1 - T_2)$

	Physical situation		Shape factor S	Remark
1	One-dimensional heat conduction through a slab	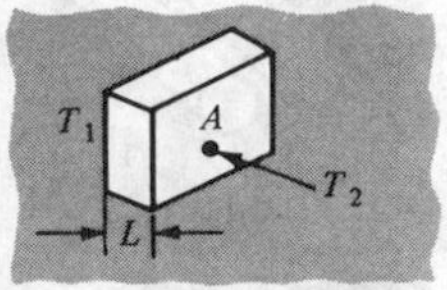	$\dfrac{A}{L}$	See Example 3-1
2	One-dimensional heat conduction through a long, hollow cylinder	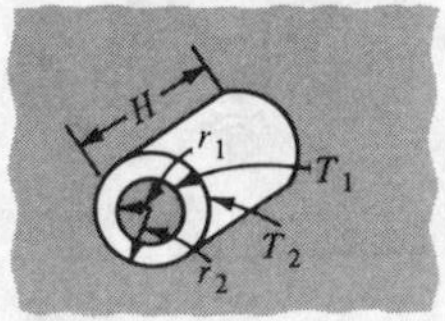	$\dfrac{2\pi H}{\ln(r_2/r_1)}$	See Example 3-6
3	One-dimensional heat conduction through a hollow sphere	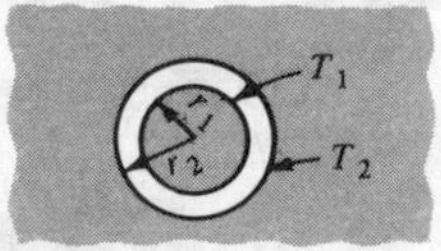	$\dfrac{4\pi r_1 r_2}{r_2 - r_1}$	See Example 3-8
4	An isothermal sphere at T_1 placed in an infinite medium at T_2		$4\pi R$	
5	An isothermal sphere at T_1 placed in a semi-infinite medium having a surface temperature T_2	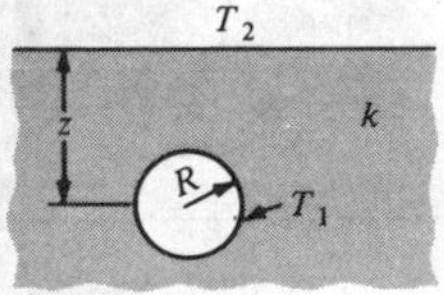	$\dfrac{4\pi R}{1 - R/(2z)}$	

Table 4-1 *(continued)*

	Physical situation	Shape factor S	Remark
6	An isothermal sphere at T_1, placed near the insulated boundary of a semi-infinite medium at T_2	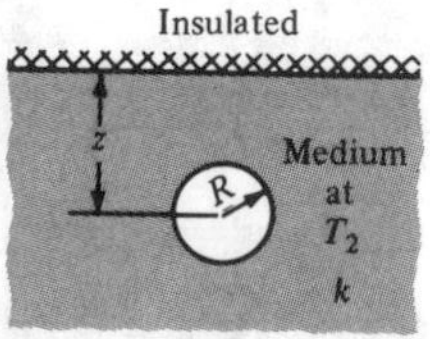 $\dfrac{4\pi R}{1 + R/(2z)}$	
7	Isothermal cylinder of length L, at T_1, placed horizontally in a semi-infinite medium having a surface at T_2	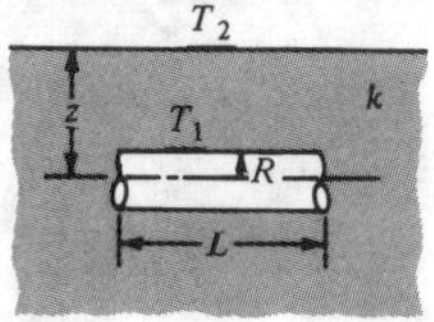 $\dfrac{2\pi L}{\cosh^{-1}(z/R)}$	$L \gg R$
8	Thin, circular disk at T_1 placed horizontally in a semi-infinite medium having a surface at T_2	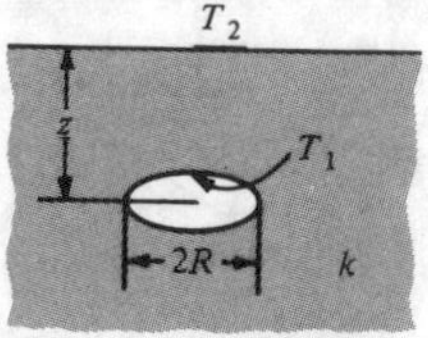 $8R$ More accurate $\dfrac{8.88R}{1 - R/(2.83z)}$	$z \gg 2R$
9	Two parallel, isothermal cylinders of length L at T_1 and T_2 placed in an infinite medium	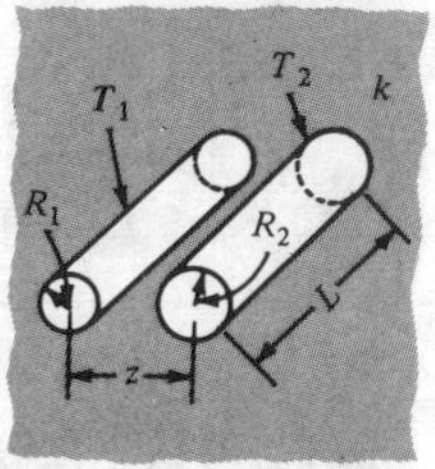 $\dfrac{2\pi L}{\cosh^{-1}[(z^2 - R_1^2 - R_2^2)/(2R_1R_2)]}$	$L \gg R_1, R_2$ $L \gg z$

	Physical system	Schematic	Shape factor	Restrictions
10	Isothermal cylinder of length L and at T_1 placed vertically in a semi-infinite medium having a surface at T_2	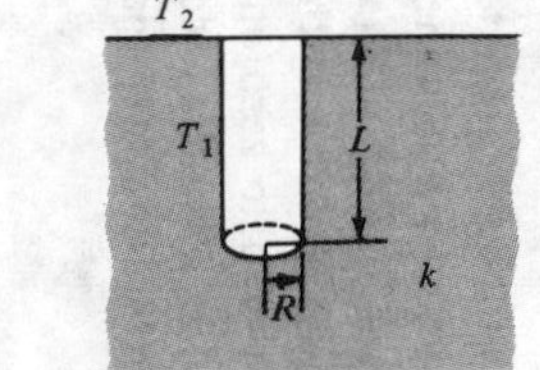	$\dfrac{2\pi L}{\ln (2L/R)}$	$L \gg 2R$
11	Circular hole centered in a square solid of length L	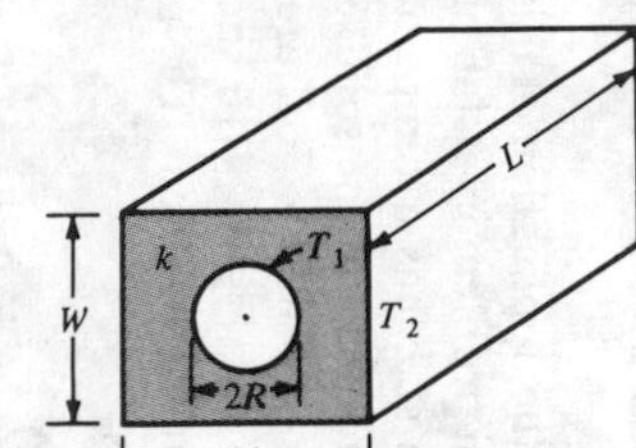	$\dfrac{2\pi L}{\ln (0.54W/R)}$	$L \gg W$
12	Eccentric circular hole in a cylindrical solid of length L	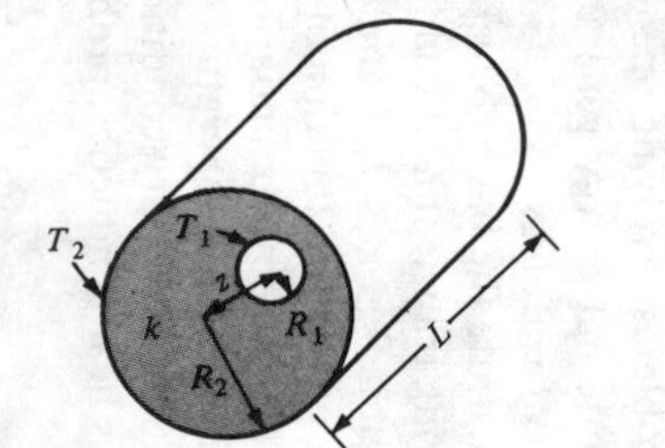	$\dfrac{2\pi L}{\cosh^{-1} [(R_1^2 + R_2^2 - z^2)/(2R_1R_2)]}$	$L \gg R_2$

The heat loss from the sphere is calculated from Eq. (4-35) as

$$Q = Sk(T_1 - T_2) = 3.49(0.8)(100 - 15) = 237.3 \text{ W}$$

4-7 TRANSIENT HEAT CONDUCTION IN A SLAB—Analytic Solution

There are numerous physical situations for which transient-temperature charts are not available. Consider, for example, the slab problem with an initial temperature that varies with the position instead of being uniform within the solid. The transient-temperature charts given in this chapter are not applicable for such situations. Furthermore, it is not practical to construct such charts for all possible spacial variations of the initial temperature distribution. Therefore, it is useful to have analytic solutions for such problems.

The solution of the transient heat conduction problem in its general form is a very complicated matter, and various methods of analysis have been discussed in several texts [1–5] in an advanced treatment of heat conduction. However, simple problems involving one-dimensional transient heat conduction in a slab with no internal energy generation can be solved readily with the method of *separation of variables.*

To illustrate the basic concepts in the application of this method, we consider the following transient heat conduction problem in a slab of thickness L, confined to the region $0 \leq x \leq L$. Initially the slab has a temperature distribution $F(x)$ which is a prescribed function of position within the solid. Suddenly, at $t = 0$, the temperatures at both boundaries at $x = 0$ and $x = L$ are lowered to zero and maintained at that temperature for $t > 0$. Figure 4-16 shows the geometry, coordinates, and boundary conditions for the problem. The mathematical formulation of this heat conduction problem is

$$\frac{\partial^2 T(x, t)}{\partial x^2} = \frac{1}{\alpha}\frac{\partial T(x, t)}{\partial t} \qquad \text{in } 0 < x < L,\ t > 0 \tag{4-37a}$$

subject to the boundary conditions

$$T(x, t) = 0 \qquad \text{at } x = 0,\ t > 0 \tag{4-37b}$$

$$T(x, t) = 0 \qquad \text{at } x = L,\ t > 0 \tag{4-37c}$$

and the initial condition

$$T(x, t) = F(x) \qquad \text{for } t = 0, \text{ in } 0 \leq x \leq L \tag{4-37d}$$

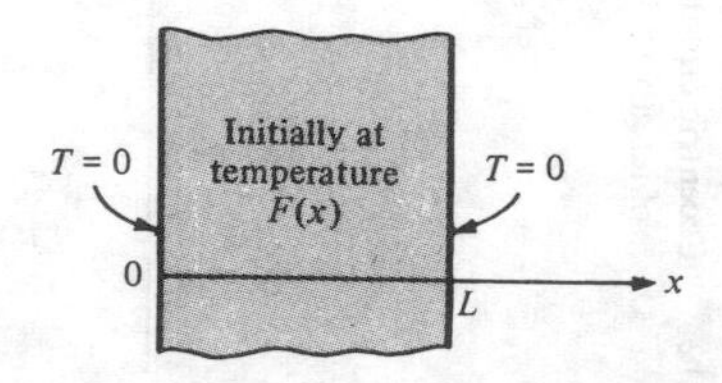

Figure 4-16 Boundary and initial conditions for transient heat conduction in a slab.

Method of Solution

We now focus our attention on the solution of the heat conduction problem given by Eqs. (4-37).

To solve such a problem, we apply the *method of separation of variables*. That is, we assume that the temperature $T(x, t)$ can be represented as a product of two functions in the form

$$T(x, t) = \psi(x)\Gamma(t) \tag{4-38}$$

where $\psi(x)$ is a function of x only and $\Gamma(t)$ is a function of t only.

Our objective is to separate the partial differential equation of heat conduction, Eq. (4-37*a*), into two ordinary differential equations—one for the function $\psi(x)$ and the other for the function $\Gamma(t)$. Then it is easier to solve the resulting ordinary differential equations. Once the functions $\psi(x)$ and $\Gamma(t)$ are known, the solution for $T(x, t)$ is constructed by taking a linear sum of these elementary solutions, because the problem is linear.

To achieve our first objective for separation, we substitute Eq. (4-38) into the differential equation (4-37*a*) and arrange the result in the form

$$\frac{1}{\psi}\frac{d^2\psi}{dx^2} = \frac{1}{\alpha\Gamma}\frac{d\Gamma}{dt} \tag{4-39}$$

where $\psi \equiv \psi(x)$ and $\Gamma \equiv \Gamma(t)$. The left-hand side of this equality is a function of x only, whereas the right-hand side is a function of t only. Such an equality is possible only if both sides are equal to the same constant, say, $-\lambda^2$. Then Eq. (4-39) becomes

$$\frac{1}{\psi}\frac{d^2\psi}{dx^2} = \frac{1}{\alpha\Gamma}\frac{d\Gamma}{dt} = -\lambda^2 \tag{4-40}$$

Here, the negative sign is chosen for λ^2 to ensure that the solution for $\Gamma(t)$ will decay with time. This will become apparent later.

So Eq. (4-40) results in two ordinary differential equations. One is for the function $\Gamma(t)$, and it yields a first-degree, ordinary differential equation in the form

$$\frac{d\Gamma(t)}{dt} + \alpha\lambda^2\Gamma(t) = 0 \tag{4-41}$$

The other is a second-degree, ordinary differential equation for the function $\psi(x)$. The two boundary conditions needed for the solution of this ordinary differential equation are determined by introducing Eq. (4-38) into the boundary conditions (4-37*a*) and (4-37*b*) of the original heat conduction problem. Then the ordinary differential equation and the boundary conditions defining the function $\psi(x)$ are

$$\frac{d^2\psi(x)}{dx^2} + \lambda^2\psi(x) = 0 \qquad \text{in } 0 < x < L \tag{4-42a}$$

$$\psi(x) = 0 \qquad \text{at } x = 0 \tag{4-42b}$$

$$\psi(x) = 0 \qquad \text{at } x = L \tag{4-42c}$$

We now examine the solutions of Eqs. (4-41) and (4-42) to determine the functions $\Gamma(t)$ and $\psi(x)$.

Solution for $\Gamma(t)$

Equation (4-41) is an ordinary differential equation, and its solution is written as

$$\Gamma(t) = e^{-\alpha\lambda^2 t} \tag{4-43}$$

Here we did not include the integration constant, because it will not be needed in this problem. Clearly, because of the choice of a negative sign for λ^2 in Eq. (4-40), the resulting solution for $\Gamma(t)$ is a decaying function of time.

Solution for $\psi(x)$

The problems of the type given by Eqs. (4-42) belong to a special class because both the differential equation and its boundary conditions are homogeneous. Therefore, such problems have solutions only for certain values of the constant λ, and for all other values of λ it has zero solutions. Problems of this type are generally referred to as *eigenvalue problems*, the values of λ for which the problem has solutions are called the *eigenvalues*, and the corresponding solutions $\psi(\lambda, x)$ are called the *eigenfunctions* of the problem.

The solution of Eq. (4-42*a*) is taken as

$$\psi(\lambda, x) = C_1 \sin \lambda x + C_2 \cos \lambda x \tag{4-44}$$

The boundary condition (4-42*b*) requires

$$C_2 = 0 \tag{4-45}$$

Then the solution reduces to

$$\psi(\lambda, x) = C_1 \sin \lambda x \tag{4-46}$$

If this solution satisfies the boundary condition (4-42*c*) for $C_1 \neq 0$, we should have

$$\sin \lambda L = 0 \tag{4-47a}$$

which establishes the permissible values of λ for which the solution exists as

$$\lambda_n L = n\pi$$

or

$$\lambda_n = \frac{n\pi}{L} \qquad n = 1, 2, 3, \ldots \tag{4-47b}$$

Thus we establish the solution of (4-42) and summarize the results: The eigenfunctions are given by

$$\psi(\lambda_n, x) = \sin \lambda_n x \tag{4-48}$$

and the eigenvalues λ_n are the roots of

$$\sin \lambda L = 0 \tag{4-49a}$$

or

$$\lambda_n = \frac{n\pi}{L} \qquad n = 1, 2, 3, \ldots \tag{4-49b}$$

Finally, we present the following property of the functions $\psi(\lambda_n, x)$:

$$\int_0^L \sin \lambda_n x \sin \lambda_m x \, dx = \begin{cases} 0 & \text{for } \lambda_n \neq \lambda_m \\ N & \text{for } \lambda_n = \lambda_m \end{cases} \tag{4-50a}$$

This relation is called the *orthogonality* property of the eigenfunctions. The constant N is called the *normalization integral*, and for this particular case it is evaluated as

$$N = \int_0^L (\sin \lambda_n x)^2 \, dx = \frac{L}{2} \tag{4-50b}$$

The physical significance of the orthogonality property given by Eqs. (4-50) is as follows: If two solutions of the eigenvalue problem in (4-42), one for the eigenvalue λ_m and the other for λ_n, are multiplied and then integrated over the region from $x = 0$ to $x = L$, the resulting integral vanishes if $\lambda_m \neq \lambda_n$ and equals a constant N if $\lambda_m = \lambda_n$. This important property will be needed in developing a solution to the heat conduction problem in (4-37).

Solution for $T(x, t)$

According to the assumption in Eq. (4-38), the product of the separated solutions $\Gamma(t)$ and $\psi(\lambda_n, x)$, given by

$$\Gamma(t)\psi(\lambda_n, x) = e^{-\alpha\lambda_n^2 t} \sin \lambda_n x \tag{4-51}$$

is a solution of the heat conduction equation (4-37*a*). However, there are numerous such solutions, each corresponding to a consecutive value of the eigenvalues λ_n, for $n = 1, 2, 3, \ldots$. Then the complete solution for the temperature $T(x, t)$ is constructed by taking a linear sum of all individual solutions given by Eq. (4-51). That is, each solution (4-51) is multiplied by a constant C_n, and the results are added to yield the solution for $T(x, t)$ in the form

$$T(x, t) = \sum_{n=1}^{\infty} C_n e^{-\alpha\lambda_n^2 t} \sin \lambda_n x \tag{4-52}$$

where the constants C_n are yet to be determined.

The solution (4-52) satisfies the differential equation of heat conduction (4-37*a*) and its two boundary conditions (4-37*b*) and (4-37*c*); but it does not satisfy the initial condition (4-37*d*). Therefore, this initial condition can be utilized to determine the unknown coefficients C_n.

The solution (4-52) is forced to satisfy the initial condition (4-37*d*); that is, in Eq. (4-52) we set $T(x, t) = F(x)$ for $t = 0$ and obtain

$$F(x) = \sum_{n=1}^{\infty} C_n \sin \lambda_n x \qquad \text{in } 0 < x < L \tag{4-53}$$

To determine the coefficients C_n, the orthogonality property of the eigenfunctions, defined by Eq. (4-50*a*), is utilized.

To begin we need to generate on the right-hand side of Eq. (4-53) the integral term given by Eq. (4-50*a*). To achieve this, both sides of Eq. (4-53) are multiplied by $\sin \lambda_m x$, and the resulting expression is integrated with respect to x from $x = 0$ to $x = L$. We obtain

$$\int_0^1 F(x) \sin \lambda_m x \, dx = \sum_{n=1}^{\infty} C_n \int_0^L \sin \lambda_n x \sin \lambda_m x \, dx \tag{4-54}$$

In view of the orthogonality relation (4-50*a*), all the terms of the summation on the right-hand side of Eq. (4-54) vanish, except the term for $\lambda_m = \lambda_n$, and the integral equals N. As a result, the summation drops out, and Eq. (4-54) reduces to

$$\int_0^L F(x) \sin \lambda_n x \, dx = C_n N \tag{4-55a}$$

or

$$C_n = \frac{1}{N} \int_0^L F(x) \sin \lambda_n x \, dx \tag{4-55b}$$

where the subscript m is replaced by n since $m = n$.

The substitution of C_n from Eq. (4-55*b*) into Eq. (4-52) gives the solution of the heat conduction problem (4-37) as

$$T(x, t) = \sum_{n=1}^{\infty} \frac{1}{N} e^{-\alpha \lambda_n^2 t} \sin \lambda_n x \int_0^L F(x') \sin \lambda_n x' \, dx' \tag{4-56a}$$

where

$$\frac{1}{N} = \frac{2}{L} \tag{4-56b}$$

and

$$\lambda_n = \frac{n\pi}{L} \tag{4-56c}$$

Example 4-12 A slab of thickness L confined to the region $0 \le x \le L$ has an initial temperature distribution given by

$$F(x) = T_0 \sin \frac{\pi}{L} x$$

where T_0 is the initial temperature at the center of the plate. Suddenly, at time $t = 0$, both boundary surfaces of the slab are lowered to and maintained at zero temperature for $t > 0$.

Develop an expression for the temperature distribution in the slab.
Develop an expression for the heat flux at the boundary surface $x = 0$.

SOLUTION This problem is exactly the same as that defined by Eqs. (4-37), except the initial temperature distribution is specified as given above. Therefore, the solution (4-56) is applicable, by setting in that equation $F(x) = T_0 \sin[(\pi/L)x]$. We find

$$T(x, t) = T_0 \sum_{n=1}^{\infty} \frac{1}{N} e^{-\alpha\lambda_n^2 t} \sin \lambda_n x \int_0^L \sin \frac{\pi}{L} x' \sin \lambda_n x' \, dx' \tag{4-57a}$$

where

$$\lambda_n = \frac{n\pi}{L} \tag{4-57b}$$

and

$$\frac{1}{N} = \frac{2}{L} \tag{4-57c}$$

To evaluate the integral, we note that according to the definition of λ_n we have

$$\lambda_1 = \frac{\pi}{L}$$

Then the integral term is written as

$$I \equiv \int_0^L \sin \lambda_1 x' \sin \lambda_n x' \, dx'$$

where $n = 1, 2, 3, \ldots$. This integral is similar to that given by Eq. (4-50a). Then the integral vanishes for all values of λ_n except $\lambda_n = \lambda_1$; and for this particular case the integral becomes equal to zero. That is,

$$I = \int_0^L \sin \lambda_1 x' \sin \lambda_n x' \, dx' = \begin{cases} 0 & \text{for } \lambda_n \neq \lambda_1 \\ N & \text{for } \lambda_n = \lambda_1 \end{cases} \tag{4-58}$$

Introducing this result into Eq. (4-57a), we find all the terms of the summation vanish except the term for $n = 1$, and the solution of the problem reduces to

$$T(x, t) = T_0 e^{-\alpha\lambda_1^2 t} \sin \lambda_1 x \tag{4-59a}$$

where

$$\lambda_1 = \frac{\pi}{L} \tag{4-59b}$$

Figure 4-17 illustrates the transient-temperature distribution in the slab as given by Eq. (4-59).

Once the temperature distribution $T(x, t)$ in the slab is known, the heat flux $q(x, t)$, anywhere in the medium, is determined from its definition:

$$q(x, t) = -k \frac{\partial T(x, t)}{\partial x} \tag{4-60}$$

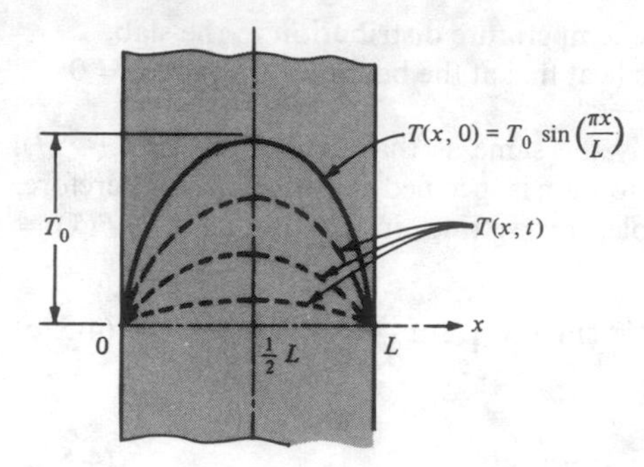

Figure 4-17 Transient-temperature distribution in a slab for an initial temperature $T_0 \sin(\pi x/L)$.

Then the heat flux at the boundary surface $x = 0$ is determined by introducing the solution (4-59) into Eq. (4-60) and evaluating it at $x = 0$. We find

$$q(0, t) = -kT_0 e^{-\alpha\lambda_1^2 t}\lambda_1[\cos \lambda_1 x]_{x=0}$$

$$q(0, t) = -kT_0 \lambda_1 e^{-\alpha\lambda_1^2 t} \tag{4-61a}$$

where

$$\lambda_1 = \frac{\pi}{L} \tag{4-61b}$$

Since k, T_0, and λ_1 are all positive quantities, the right-hand side of Eq. (4-61*a*) is negative. The negative value implies that the heat flow at the boundary surface $x = 0$ is in the negative x direction, or outward. This result is consistent with the physical reality for the problem.

Example 4-13 A slab of thickness L is initially at a uniform temperature T_i. Suddenly at time $t = 0$ the temperatures of both boundary surfaces are lowered to and maintained at T_∞ for all $t > 0$. Develop an expression for the temperature distribution $T(x, t)$ in the slab.

SOLUTION The boundary condition for this problem is not homogeneous; that is, the temperature at the boundaries is not zero. Therefore, a new temperature $\theta(x, t)$ is defined as

$$\theta(x, t) = T(x, t) - T_\infty$$

Then the mathematical formulation of this problem becomes

$$\frac{\partial^2\theta(x, t)}{\partial x^2} = \frac{1}{\alpha}\frac{\partial\theta(x, t)}{\partial t} \qquad \text{in } 0 < x < L, t > 0 \tag{4-62a}$$

$$\theta(x, t) = 0 \qquad \text{at } x = 0 \tag{4-62b}$$

$$\theta(x, t) = 0 \qquad \text{at } x = L \tag{4-62c}$$

$$\theta(x, t) = T_i - T_\infty \equiv \theta_0 \qquad \text{for } t = 0 \tag{4-62d}$$

This problem is now in exactly the same form as (4-37) with

$$F(x) \equiv \theta_0$$

Therefore, the solution (4-56) is applicable and we obtain

$$\theta(x, t) = \theta_0 \sum_{n=1}^{\infty} \frac{1}{N} e^{-\alpha\lambda_n^2 t} \sin \lambda_n x \int_0^L \sin \lambda_n x' \, dx'$$

The integral is evaluated as

$$\begin{aligned} I &\equiv \int_0^L \sin \lambda_n x' \, dx' \\ &= -\frac{1}{\lambda_n}[\cos \lambda_n x']_0^L = \frac{1}{\lambda_n}(1 - \cos \lambda_n L) \\ &= \frac{1}{\lambda_n}(1 - \cos n\pi) = \frac{1}{\lambda_n}[1 - (-1)^n] \\ &= \begin{cases} 0 & \text{for } n = \text{even} \\ \dfrac{2}{\lambda_n} & \text{for } n = \text{odd} \end{cases} \end{aligned}$$

Then $\theta(x, t)$ becomes

$$\theta(x, t) = \theta_0 \sum_{n=1,3,5,\dots}^{\infty} \frac{2}{N\lambda_n} e^{-\alpha_n^2 t} \sin \lambda_n x$$

Noting that

$$\frac{1}{N} = \frac{2}{L} \qquad \frac{1}{\lambda_n} = \frac{L}{n\pi}$$

we see that the solution is written as

$$\theta(x, t) = \frac{4}{\pi}\theta_0 \sum_{n=1,3,5,\dots}^{\infty} \frac{1}{n} e^{-\alpha(n\pi/L)^2 t} \sin \frac{n\pi}{L} x \tag{4-63}$$

which gives the temperature distribution within the slab, in excess of the ambient temperature T_∞, as a function of position and time.

4-8 TRANSIENT HEAT CONDUCTION IN A SLAB—Use of Tabulated Solutions

In the last section, the analytic solution of transient heat conduction in a slab by separation of variables was illustrated for a sample problem given by Eqs. (4-37). In that example, for simplicity, the boundary conditions for both surfaces are taken to be of the first kind. However, in engineering applications, other combinations of boundary conditions are encountered. For the slab problem, there are nine different combinations of boundary conditions, and each such case requires a separate analysis.

Therefore, in this section we develop one general formal solution for the transient heat conduction problem in a slab. Then this solution is used in conjunction with a table from which the eigenfunctions, eigenvalues, and normalization integral appropriate for any given combination of the boundary conditions can be obtained. To develop such a solution, we consider the following heat conduction problem for a slab with convection at both boundaries.

A slab of thickness L, confined to the region $0 < x < L$, initially (i.e., at $t = 0$) has a specified temperature distribution $f(x)$. Suddenly, at time $t = 0$, both boundary surfaces are subjected to convection with a heat transfer coefficient h into an ambient at temperature T_∞. We now solve this problem.

Let $T(x, t)$ be the solution of this problem. We define a new temperature $\theta(x, t)$, measured in excess of the ambient temperature T_∞, as

$$\theta(x, t) = T(x, t) - T_\infty \tag{4-64}$$

This choice of a new temperature is made to obtain homogeneous boundary conditions in the mathematical formulation of the problem. Then the mathematical formulation in terms of the temperature $\theta(x, t)$ is given by

$$\frac{\partial^2 \theta(x, t)}{\partial x^2} = \frac{1}{\alpha} \frac{\partial \theta(x, t)}{\partial t} \qquad \text{in } 0 < x < L, t > 0 \tag{4-65a}$$

$$-k \frac{\partial \theta(x, t)}{\partial x} + h\theta(x, t) = 0 \qquad \text{at } x = 0 \tag{4-65b}$$

$$k \frac{\partial \theta(x, t)}{\partial x} + h\theta(x, t) = 0 \qquad \text{at } x = L \tag{4-65c}$$

$$\theta(x, t) = f(x) - T_\infty \equiv F(x) \qquad \text{for } t = 0 \tag{4-65d}$$

We note that the boundary conditions (4-65b) and (4-65c) are now homogeneous. Furthermore, the heat conduction problem defined by Eqs. (4-65) is sufficiently general to include numerous other combinations of boundary condition as special cases.

By separating the variables as defined by Eq. (4-38), it can be shown that the appropriate eigenvalue problem is given by

$$\frac{d^2 \psi(x)}{dx^2} + \lambda^2 \psi(x) = 0 \qquad \text{in } 0 < x < L \tag{4-66a}$$

$$-k \frac{d\psi(x)}{dx} + h\psi(x) = 0 \qquad \text{at } x = 0 \tag{4-66b}$$

$$k \frac{d\psi(x)}{dx} + h\psi(x) = 0 \qquad \text{at } x = L \tag{4-66c}$$

and the solution for the separated function $\Gamma(t)$ is again an exponential in the form $e^{-\alpha\lambda^2 t}$.

The heat conduction problem (4-65) can be solved by following a procedure described previously, and the solution for $\theta(x, t)$ can be expressed in the form

$$\theta(x, t) = \sum_{n=1}^{\infty} \frac{1}{N} e^{-\alpha\lambda_n^2 t}\psi(\lambda_n, x) \int_0^L F(x')\psi(\lambda_n, x')\, dx' \tag{4-67}$$

where

$$\psi(\lambda_n, x) = \text{eigenfunctions} \tag{4-68a}$$

$$\lambda_n = \text{eigenvalues} \tag{4-68b}$$

$$N = \text{normalization integral} \tag{4-68c}$$

of the eigenvalue problem (4-66).

It is now apparent from the solution (4.67) that the temperature distribution $\theta(x, t)$ within the slab can be determined for any of the nine different combinations of boundary conditions if the three quantities $\psi(\lambda_n, x)$, λ_n, and N are available from the solution of the appropriate eigenvalue problem.

Table 4-2 shows a systematic tabulation of the three quantities defined by Eqs. (4-68) for each of the nine different combinations of boundary conditions. For example, case 1 of this table corresponds to the situation in which the boundary conditions at the surfaces $x = 0$ and $x = L$ are both of the *first kind*. For this case, the eigenfunctions are given by

$$\psi(\lambda_n, x) = \sin \lambda_n x \tag{4-69a}$$

where the eigenvalues λ_n are the roots of $\sin \lambda L = 0$, or

$$\lambda_n = \frac{n\pi}{L} \tag{4-69b}$$

and the normalization integral is given by

$$\frac{1}{N} = \frac{2}{L} \tag{4-69c}$$

These results are indeed the same as those given by Eqs. (4-48), (4-49*b*), and (4-50*b*).

Clearly, all the other cases in this table can be interpreted in a similar manner.

Therefore, the eigenfunctions, the eigenvalues, and the normalization integrals listed in Table 4-2 can be used in conjunction with Eq. (4-67) to obtain solutions for the transient heat conduction in a slab for any of the nine different combinations of boundary conditions.

Example 4-14 A slab of thickness L, confined to a region $0 \le x \le L$, is initially at a uniform temperature T_i. For $t > 0$, the boundary surface at $x = 0$ is kept insulated and the boundary surface at $x = L$ is maintained at constant

Table 4-2 Eigenfunctions $\psi(\lambda, x)$, eigenvalues λ, and the normalization integral N of the eigenvalue problem

$$\frac{d^2\psi(x)}{dx^2} + \lambda^2\psi(x) = 0 \quad \text{in } 0 \le x \le L$$

subject to boundary conditions at $x = 0$ and $x = L$ as shown

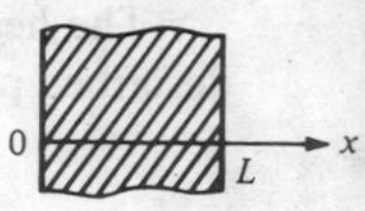

	Boundary condition at: $x = 0$	Boundary condition at: $x = L$	Eigenfunction $\psi(\lambda,x)$	Eigenvalues† λs are positive roots of:	$\frac{1}{N} = \frac{1}{\int_0^L \psi^2(\lambda, x)\,dx}$ †
1	$\psi = 0$	$\psi = 0$	$\sin \lambda x$	$\sin \lambda L = 0$	$\frac{2}{L}$
2	$\psi = 0$	$\frac{d\psi}{dx} = 0$	$\sin \lambda x$	$\cos \lambda L = 0$	$\frac{2}{L}$
3	$\psi = 0$	$k\frac{d\psi}{dx} + h\psi = 0$	$\sin \lambda x$	$\lambda \cot \lambda L = -H$	$\frac{2(\lambda^2 + H^2)}{L(\lambda^2 + H^2) + H}$
4	$\frac{d\psi}{dx} = 0$	$\psi = 0$	$\cos \lambda x$	$\cos \lambda L = 0$	$\frac{2}{L}$
5	$\frac{d\psi}{dx} = 0$	$\frac{d\psi}{dx} = 0$	$\cos \lambda x$	$\sin \lambda L = 0$	$\frac{2}{L}$ for $\lambda \ne 0$; $\frac{1}{L}$ for $\lambda = 0$‡
6	$\frac{d\psi}{dx} = 0$	$k\frac{d\psi}{dx} + h\psi = 0$	$\cos \lambda x$	$\lambda \tan \lambda L = H$	$\frac{2(\lambda^2 + H^2)}{L(\lambda^2 + H^2) + H}$
7	$-k\frac{d\psi}{dx} + h\psi = 0$	$\psi = 0$	$\sin \lambda(L - x)$	$\lambda \cot \lambda L = -H$	$\frac{2(\lambda^2 + H^2)}{L(\lambda^2 + H^2) + H}$
8	$-k\frac{d\psi}{dx} + h\psi = 0$	$\frac{d\psi}{dx} = 0$	$\cos \lambda(L - x)$	$\lambda \tan \lambda L = H$	$\frac{2(\lambda^2 + H^2)}{L(\lambda^2 + H^2) + H}$
9	$-k\frac{d\psi}{dx} + h\psi = 0$	$k\frac{d\psi}{dx} + h\psi = 0$	$\lambda \cos \lambda x + H \sin \lambda x$	$\tan \lambda L = \frac{2\lambda H}{\lambda^2 - H^2}$	$\frac{2}{L(\lambda^2 + H^2) + 2H}$

† H is defined as $H = h/k$.
‡ $\lambda = 0$ is an eigenvalue for this case.

temperature T_∞. Develop an expression for the transient-temperature distribution in the slab.

Solution The boundary condition at $x = L$ also can be made homogeneous by defining a new temperature $\theta(x, t)$ as

$$\theta(x, t) = T(x, t) - T_\infty$$

Then the mathematical formulation of this heat conduction problem becomes

$$\frac{\partial^2\theta(x, t)}{\partial x^2} = \frac{1}{\alpha}\frac{\partial\theta(x, t)}{\partial t} \qquad \text{in } 0 < x < L \tag{4-70a}$$

$$\frac{\partial\theta(x, t)}{\partial x} = 0 \qquad \text{at } x = 0 \tag{4-70b}$$

$$\theta(x, t) = 0 \qquad \text{at } x = L \tag{4-70c}$$

$$\theta(x, t) = T_i - T_\infty \equiv \theta_i \qquad \text{for } t > 0 \tag{4-70d}$$

This problem is a special case of the general problem (4-65). Therefore, its solution can be obtained immediately from Eq. (4-67) by utilizing the results in Table 4-2. The problem (4-70) corresponds to case 4 in this table, because it has boundary condition of the second kind at $x = 0$ and of the first kind at $x = L$. Therefore, we obtain

$$\psi(\lambda_n, x) = \cos \lambda x \tag{4-71a}$$

where the λ_n are the roots of $\cos \lambda L = 0$, or

$$\lambda_n = \frac{(2n-1)\pi}{2L} \tag{4-71b}$$

and the normalization integral is given by

$$\frac{1}{N} = \frac{2}{L} \tag{4-71c}$$

These results are introduced into Eq. (4-67), and the initial condition is taken as $F(x) = \theta_i$. We find

$$\theta(x, t) = \frac{2}{L}\theta_i \sum_{n=1}^{\infty} e^{-\alpha\lambda_n^2 t} \cos \lambda_n x \int_0^L \cos \lambda_n x' \, dx'$$

The integral is evaluated as

$$I \equiv \int_0^L \cos \lambda_n x' \, dx' = \frac{1}{\lambda_n} \sin \lambda_n x \Big|_0^L$$

$$= \frac{1}{\lambda_n} \sin \lambda_n L = \frac{1}{\lambda_n} \sin \left[(2n-1)\frac{\pi}{2}\right]$$

$$= \frac{1}{\lambda_n}(-1)^{n+1}$$

Then the solution becomes

$$\theta(x, t) = \frac{2}{L}\theta_i \sum_{n=1}^{\infty} (-1)^{n+1} e^{-\alpha\lambda_n^2 t} \frac{1}{\lambda_n} \cos \lambda_n x \tag{4-72a}$$

where
$$\lambda_n = \frac{(2n-1)\pi}{2L} \tag{4-72b}$$

Determination of Eigenvalues

In the previous examples we chose problems for which the determination of the eigenvalues λ_n was a simple matter. However, in many other situations the determination of the eigenvalues is not so straightforward.

Consider, for example, case 6 in Table 4-2. The eigenvalues λ_n are given as the roots of the transcendental equation

$$\lambda \tan \lambda L = H \tag{4-73}$$

where $$H \equiv \frac{h}{k}$$

Clearly, it is not possible to obtain an explicit solution for the eigenvalues λ_n for this case. However, the transcendental equation (4-73) can be solved by a numerical scheme, as discussed in Refs. 12 and 13, and the λ_n can be determined. The results also can be presented in tabular form if Eq. (4-73) is rearranged as

$$\beta \tan \beta = c \tag{4-74}$$

where $\beta \equiv \lambda L$ and $c \equiv HL = hL/k$. The first six roots of this transcendental equation are tabulated in Table D-2 from $c = 0$ to $c = \infty$.

Case 4 in Table 4-2 gives the eigenvalues λ_n as the roots of the transcendental equation

$$\lambda \cot \lambda L = -H \tag{4-75}$$

This equation also can be rearranged in the form

$$\beta \cot \beta = -c \tag{4-76}$$

where $\beta \equiv \lambda L$ and $c \equiv HL = hL/k$. The first six roots of this transcendental equation are tabulated in Table D-3.

PROBLEMS

Lumped system analysis

4-1 A hot metal block, initially at a uniform temperature T_i, is suddenly immersed in a well-stirred, cold liquid bath which is maintained at a uniform temperature T_∞. The heat transfer coefficient between the block and the liquid is h W/(m$^2 \cdot$ °C). The metal block has a weight M kg, surface area A m^2, and specific heat c_p J/(kg · °C). Assuming that the lumped system analysis is applicable, develop an expression for the determination of the temperature $T(t)$ of the block as a function of time.

4-2 A solid copper sphere of 10-cm diameter [$\rho = 8954$ kg/m^3, $c_p = 383$ J/(kg · °C), $k = 386$ W/m · °C)], initially at a uniform temperature $T_i = 250$°C, is suddenly immersed in a well-stirred fluid which is maintained at a uniform temperature $T_\infty = 50$°C. The heat transfer coefficient between the sphere and the fluid is $h = 200$ W/(m$^2 \cdot$ °C).

(*a*) Check whether the lumped system analysis is suitable.

(*b*) If it is suitable, determine the temperature of the copper block at $t = 5$, 10, and 20 min after the immersion.

Answer: (*b*) 120°C, 74.5°C, 53°C

4-3 A long, cylindrical iron bar [$\rho = 7800$ kg/m^3, $c_p = 460$ J/(kg · °C), and $k = 60$ W/(m · °C)] of diameter $D = 5$ cm, initially at temperature $T_i = 700$°C, is exposed to a cool airstream at $T_\infty = 100$°C.

The heat transfer coefficient between the airstream and the surface of the iron bar is $h = 80\ \text{W/(m}^2 \cdot {}^\circ\text{C)}$.

(*a*) Check whether the lumped system analysis is suitable.

(*b*) If it is suitable, determine the time required for the temperature of the rod to reach 300°C.

4-4 Using the lumped system analysis, determine the time required for a solid steel ball of diameter $D = 5$ cm [$\rho = 7833\ \text{kg/m}^3$, $c_p = 0.465\ \text{kJ/(kg} \cdot {}^\circ\text{C)}$, and $k = 54\ \text{W/(m} \cdot {}^\circ\text{C)}$] to cool from 600 to 200°C if it is exposed to an airstream at 50°C having a heat transfer coefficient $h = 100\ \text{W/(m}^2 \cdot {}^\circ\text{C)}$.

Answer: 6 min 34 s

4-5 A large aluminum plate [$\rho = 2707\ \text{kg/m}^3$, $c_p = 0.896\ \text{kJ/(kg} \cdot {}^\circ\text{C)}$, and $k = 204\ \text{W/(m} \cdot {}^\circ\text{C)}$] of thickness $L = 0.10$ m, initially at a uniform temperature $T_i = 250$°C, is cooled by exposing it to an airstream at $T_\infty = 40$°C. Using the lumped system analysis, determine the time required to cool the aluminum plate from 250 to 75°C if the heat transfer coefficient between the airstream and the surface is $h = 80\ \text{W/(m}^2 \cdot {}^\circ\text{C)}$.

4-6 A 2-cm-diameter, stainless-steel ball [$\rho = 7865\ \text{kg/m}^3$, $c_p = 0.46\ \text{kJ/(kg} \cdot {}^\circ\text{C)}$, and $k = 61\ \text{W/(m} \cdot {}^\circ\text{C)}$] is uniformly heated to $T_i = 800$°C. It is to be hardened by suddenly dropping it into an oil bath at $T_\infty = 50$°C. If the quenching occurs when the ball reaches 100°C and the heat transfer coefficient between the oil and the sphere is $300\ \text{W/(m}^2 \cdot {}^\circ\text{C)}$, how long should the ball be kept in the oil bath?

If 100 balls are to be quenched per minute, determine the rate of heat removal from the oil bath per minute needed to maintain its temperature at 40°C.

Answer: 1 min 6 s; 3580 kJ/min

4-7 A steel bar [$\rho = 7800\ \text{kg/m}^3$, $c = 0.5\ \text{kJ/(kg} \cdot {}^\circ\text{C)}$, and $k = 50\ \text{W/(m} \cdot {}^\circ\text{C)}$] of diameter $D = 5$ cm is to be annealed by slowly cooling from $T_i = 800$°C to 120°C in an ambient at $T_\infty = 50$°C. If the heat transfer coefficient between the ambient air and the surface of the bar is $h = 45\ \text{W/(m}^2 \cdot {}^\circ\text{C)}$, determine the time required for the annealing process by applying the lumped system analysis.

Answer: 42 min 49 s

4-8 A metal plate of thickness L and surface area A is initially at a uniform temperature T_i. Suddenly one of its surfaces is subjected to a uniform heat flux $q\ \text{W/m}^2$, while the other surface is exposed to a cool airstream at temperature T_∞. The heat transfer coefficient between the airstream and the plate's surface is $h\ \text{W/(m}^2 \cdot {}^\circ\text{C)}$. Assuming that the lumped system analysis is applicable, develop an expression for the temperature $T(t)$ of the plate as a function of time.

4-9 An aluminum plate [$\rho = 2707\ \text{kg/m}^3$, $c_p = 0.896\ \text{kJ/(kg} \cdot {}^\circ\text{C)}$, and $k = 204\ \text{W/(m} \cdot {}^\circ\text{C)}$] of thickness $L = 3$ cm is initially at a uniform temperature $T_i = 50$°C. Suddenly it is subjected to a uniform heat flux $q = 7500\ \text{W/m}^2$ at one of its surfaces and is exposed to a cool airstream at $T_\infty = 30$°C at the other surface. The heat transfer coefficient between the airstream and the surface is $h = 60\ \text{W/(m}^2 \cdot {}^\circ\text{C)}$.

By using the lumped system analysis:

(*a*) Determine the temperature of the plate as a function of time, and plot it against t.

(*b*) Calculate the steady-state temperature of the plate.

4-10 A household electric iron has an aluminum base [$\rho = 2700\ \text{kg/m}^3$, $c_p = 0.896\ \text{kJ/(kg} \cdot {}^\circ\text{C)}$, and $k = 204\ \text{W/(m} \cdot {}^\circ\text{C)}$] which weighs 1.5 kg. The base has an ironing surface of $0.06\ \text{m}^2$ and is heated from the other side with a 500-W heating element. Initially the iron is at the same temperature as the ambient air, $T_\infty = 20$°C. How long will it take for the iron to reach 120°C if the heat transfer coefficient between the iron and the ambient air is $h = 20\ \text{W/(m}^2 \cdot {}^\circ\text{C)}^2$.

4-11 Consider an aluminum cube [$\rho = 2700\ \text{kg/m}^3$, $c_p = 0.896\ \text{kJ/(kg} \cdot {}^\circ\text{C)}$, and $k = 204\ \text{W/(m} \cdot {}^\circ\text{C)}$] that is 5 cm by 5 cm by 5 cm, and initially at $T_i = 20$°C. For $t > 0$, two of the boundary surfaces are kept insulated, two are subjected to uniform heating at a rate $q_0\ \text{W/m}^2$, and the remaining two surfaces dissipate heat by convection into an ambient at $T_\infty = 20$°C with a heat transfer coefficient $h = 50\ \text{W/(m}^2 \cdot {}^\circ\text{C)}$. Assuming the lumped system analysis is applicable, develop an expression for the temperature $T(t)$ of the aluminum block as a function of time. Calculate the equilibrium temperature of the block for $q_0 = 10{,}000\ \text{W/m}^2$.

Answer: 220°C

4-12 Consider a large plate of thickness L, initially at a uniform temperature T_0. Suddenly one of its surfaces is exposed to convection with an ambient at temperature T_1 with a heat transfer coefficient h.

The other surface exchanges heat by radiation with an ambient at temperature T_2 which is regarded as a blackbody. Applying the lumped system analysis, develop an expression for the temperature $T(t)$ in the slab as a function of time.

4-13 A thermocouple junction may be approximated as a sphere of diameter $D = 2$ mm, with $k = 30$ W/(m · °C), $\rho = 8600$ kg/m^3, and $c_p = 0.4$ kJ/(kg · °C). The heat transfer coefficient between the gas stream and the junction is $h = 280$ W/(m^2 · °C). How long will it take for the thermocouple to record 98 percent of the applied temperature difference?

Answer: 8 s

4-14 A thermocouple is to be used to measure the temperature in a gas stream. The junction may be approximated as a sphere having thermal conductivity $k = 25$ W/(m · °C), $\rho = 8400$ kg/m^3, and $c_p =$ 0.4 kJ/(kg · °C). The heat transfer coefficient between the junction and the gas stream is $h = 560$ W/(m^2 · °C). Calculate the diameter of the junction if the thermocouple should measure 95 percent of the applied temperature difference in 3 s.

4-15 A 2-cm-thick copper plate [$\rho = 8954$ kg/m^3, $c_p = 0.3830$ kJ/(kg · °C), and $k = 386$ W/(m · °C)] is initially at $T_0 = 25$°C. For $t > 0$, it is subjected to a heat flux $q = 8000$ W/m^2 at one of its surface and is cooled by convection from the other surface with a heat transfer coefficient $h = 200$ W/(m^2 · °C) into an ambient at $T_\infty = 25$°C. Using the lumped system analysis, develop an expression for the transient temperature $T(t)$ in the plate.

4-16 A 3-cm-diameter aluminum sphere [$k = 204$ W/(m · °C), $\rho = 2700$ kg/m^3, and $c_p = 0.896$ kJ/(kg · °C)] is initially at $T_0 = 175$°C. It is suddenly immersed in a well-stirred fluid at $T_\infty = 25$°C. The temperature of the sphere is lowered to $T(t) = 100$°C in $t = 42$ s. Calculate the heat transfer coefficient.

Answer: $h = 200.1$ W/(m^2 · °C)

Use of transient-temperature charts for the slab, cylinder, and sphere

4-17 A 0.10-m-thick brick wall [$\alpha = 0.5 \times 10^{-6}$ m^2/s, $k = 0.69$ W/(m · °C), and $\rho = 2300$ kg/m^3] is initially at $T_i = 230$°C. The wall is suddenly exposed to a convective environment at $T_\infty = 30$°C with a heat transfer coefficient $h = 60$ W/(m^2 · °C). By using the transient-temperature charts, determine:

(*a*) The center temperature at $\frac{1}{2}$, 2, and 4 h after exposure to the cooler ambient

(*b*) The surface temperature at $\frac{1}{2}$ and 2 h

(*c*) Energy removed from the plate per square meter during $\frac{1}{2}$ h

4-18 A long 8-cm-diameter chrome-steel rod [$\alpha = 1.1 \times 10^{-5}$ m^2/s and $k = 40$ W/(m · °C)] is initially at a uniform temperature $T_i = 225$°C. It is suddenly exposed to a convective environment at $T_\infty = 25$°C with a surface heat transfer coefficient $h = 50$ W/(m^2 · °C). By using the transient-temperature chart, determine (*a*) the center temperature and (*b*) the surface temperature at $t = \frac{1}{10}$ and 1 h after exposure to the cooler ambient.

4-19 A solid iron rod [$\alpha = 2 \times 10^{-5}$ m^2/s and $k = 60$ W/(m · °C)] of diameter $D = 6$ cm, initially at temperature $T_i = 800$°C, is suddenly dropped into an oil bath at $T_\infty = 50$°C. The heat transfer coefficient between the fluid and the surface is $h = 400$ W/(m^2 · °C).

(*a*) Using the transient-temperature charts, determine the centerline temperature 10 min after immersion in the fluid.

(*b*) How long will it take the centerline temperature to reach 100°C?

Answer: (*a*) 54.5°C; (*b*) 5 min 47 s

4-20 An orange of diameter 10 cm is initially at a uniform temperature of 30°C. It is placed in a refrigerator in which the air temperature is 2°C. If the transfer coefficient between the air and the surface of the orange is $h = 50$ W/(m^2 · °C), determine the time required for the center of the orange to reach 10°C. Assume the thermal properties of the orange are the same as those of water at the same temperature [$\alpha = 1.4 \times 10^{-7}$ m^2/s and $k = 0.59$ W/(m · °C)].

Answer: 1 h 32 min

4-21 A marble plate 3 cm thick [$\alpha = 1.3 \times 10^{-6}$ m^2/s and $k = 3$ W/(m · °C)] is initially at a uniform temperature $T_i = 130$°C. The surfaces are suddenly lowered to 30°C. Determine the center-plane temperature 2 min after the lowering of the temperature.

4-22 A long hot dog [$\alpha = 1.6 \times 10^{-7}$ m^2/s and $k = 0.5$ W/(m·°C)] of diameter $D = 2$ cm, initially at a uniform temperature of 7°C, is dropped suddenly into boiling water at $T_\infty = 100$°C. The heat transfer coefficient between the water and the surface is $h = 150$ W/(m^2·°C). The hot dog is considered cooked when its center temperature reaches 80°C. How long will it take the centerline temperature to reach 80°C?

Answer: 8 min 20 s

4-23 A 6-cm-diameter potato, initially at a uniform temperature of 20°C, is suddenly dropped into boiling water at 100°C. The heat transfer coefficient between the water and the surface is $h = 6000$ W/(m^2·°C). The thermophysical properties of potato can be taken the same as those of water [$\alpha = 1.6 \times 10^{-7}$ m^2/s and $k = 0.68$ W/(m·°C)]. Determine the time required for the center temperature of the potato to reach 95°C and the energy transferred to the potato during this time.

Answer: 33 min; 37.8 kJ

4-24 A solid brass sphere [$\alpha = 1.8 \times 10^{-5}$ m^2/s and $k = 61$ W/(m·°C)] of diameter $D = 25$ cm is initially at $T_i = 120$°C. It is cooled with an airstream at $T_\infty = 15$°C. The heat transfer coefficient between the airstream and the surface is $h = 500$ W/(m^2·°C). How long will it take for the center of the sphere to cool to 30°C? What fraction of the initial energy will be removed during this time?

4-25 A steel plate [$\alpha = 1.2 \times 10^{-5}$ and $k = 43$ W/(m·°C)] of thickness $2L = 10$ cm, initially at a uniform temperature of 240°C, is suddenly immersed in an oil bath at $T_\infty = 40$°C. The convection heat transfer coefficient between the fluid and the surface is $h = 600$ W/(m^2·°C). How long will it take for the center plane to cool to 100°C? What fraction of the initial energy is removed during this time?

4-26 A solid aluminum sphere [$\alpha = 8.4 \times 10^{-5}$ m^2/s and $k = 204$ W/(m·°C)] of diameter $D = 10$ cm is initially at $T_i = 250$°C. Suddenly it is immersed in a well-stirred bath at $T_\infty = 80$°C. The heat transfer coefficient between the fluid and the surface is $h = 1000$ W/(m^2·°C). How long will it take for the center of the sphere to cool to 100°C?

Answer: 80.4 s

4-27 Consider a slab of thickness 10 cm, a cylinder of diameter 10 cm, and a sphere of diameter 10 cm, each made of steel [$\alpha = 1.6 \times 10^{-5}$ m^2/s and $k = 61$ W/(m·°C)] and initially at uniform temperature $T_i = 300$°C. Suddenly, they are all immersed into a well-stirred bath at $T_\infty = 50$°C. The heat transfer coefficient between the surface and the fluid is $h = 1000$ W/(m^2·°C). Calculate the time required for the centers of slab, cylinder, and sphere to cool to 80°C.

Answer: 547, 266, and 188 s

Use of transient-temperature charts for a semi-infinite medium

4-28 A thick concrete slab ($\alpha = 7 \times 10^{-7}$ m^2/s) is initially at a uniform temperature $T_i = 60$°C. One of its surfaces is suddenly lowered to 10°C. By treating this as a one-dimensional transient heat conduction problem in a semi-infinite medium, determine the temperatures at depths 5 and 10 cm from the surface 30 min after the surface temperature is lowered.

4-29 A thick stainless-steel slab [$\alpha = 1.6 \times 10^{-5}$ m^2/s and $k = 61$ W/(m·°C)] is initially at a uniform temperature $T_i = 150$°C. Its surface is suddenly lowered to 20°C. By treating this as a one-dimensional transient heat conduction problem in a semi-infinite medium, determine the temperature at a depth 2 cm from the surface and the heat flux at the surface 1 min after the surface temperature is lowered.

Answer: 65.5°C; −144.4 kW/m^2

4-30 A thick copper slab [$\alpha = 1.1 \times 10^{-4}$ m^2/s and $k = 380$ W/(m·°C)] is initially at a uniform temperature 10°C. Suddenly the surface is raised to 100°C. Calculate the heat flux at the surface 5 and 10 min after the raising of the surface temperature. How long will it take the temperature at a depth 5 cm from the surface to reach 90°C?

4-31 A thick bronze [$\alpha = 0.86 \times 10^{-5}$ m^2/s and $k = 26$ W/(m·°C)] is initially at a uniform temperature 250°C. Suddenly the surface is exposed to a coolant 25°C. Assuming that the heat transfer coefficient for convection between the fluid and the surface is 150 W/(m^2·°C), determine the temperature 5 cm from the surface 10 min after the exposure.

Answer: 205°C

4-32 A thick wood wall [$\alpha = 0.82 \times 10^{-7}$ m²/s and $k = 0.15$ W/(m · °C)] is initially at a uniform temperature of 20°C. The wood may ignite at 400°C. If the surface is exposed to hot gases at $T_\infty = 500$°C and the heat transfer coefficient between the gas and the surface is $h = 45$ W/(m² · °C), how long will it take for the surface of the wood to reach 400°C?

4-33 A thick wood wall [$\alpha = 0.8 \times 10^{-7}$ m²/s and $k = 0.15$ W/(m · °C)] is initially at a uniform temperature $T_i = 20$°C. Suddenly the surface is raised to 120°C. Calculate the temperature 2 cm from the surface at 2 and 20 min after the exposure

Answer: 20°C; 36°C

4-34 A thick concrete slab [$\alpha = 7 \times 10^{-7}$ m²/s and $k = 1.37$ W/(m · °C)] is initially at a uniform temperature $T_i = 340$°C. Suddenly its surface is subjected to convective cooling with a heat transfer coefficient $h = 100$ W/(m² · °C) into an ambient at $T_\infty = 40$°C. Calculate the temperature 10 cm from the surface 1 h after the start of cooling.

Product solution

4-35 A rectangular aluminum bar 6 cm by 3 cm [$k = 200$ W/(m · °C), $c_p = 890$ J/(kg · °C), $\rho = 2700$ kg/m³, and $\alpha = 8.4 \times 10^{-5}$ m²/s] is initially at a uniform temperature $T_i = 175$°C. Suddenly the surfaces are subjected to convective cooling with a heat transfer coefficient $h = 250$ W/(m² · °C) into an ambient at $T_\infty = 25$°C as shown in Fig. P4-35. Determine the center temperature T_0 of the bar $t = 1$ min after the start of the cooling.

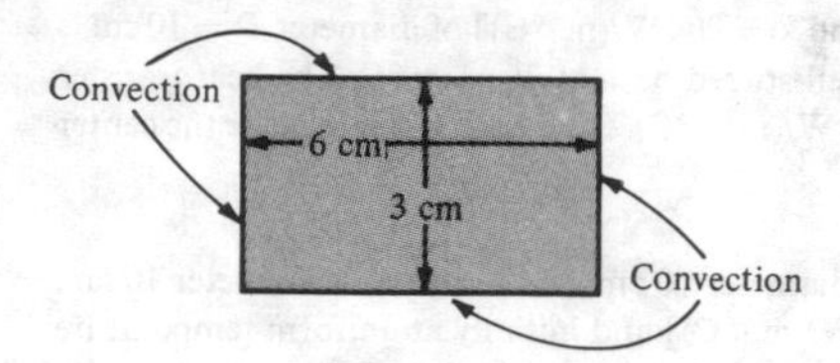

Figure P4-35

Answer: 107.5°C

4-36 A short, cylindrical aluminum bar [$k = 200$ W/(m · °C) and $\alpha = 8.4 \times 10^{-5}$ m²/s] of diameter $D = 6$ cm and height $L = 3$ cm is initially at a uniform temperature $T_i = 175$°C. Suddenly the surfaces are subjected to convective cooling with a heat transfer coefficient $h = 250$ W/(m² · °C) into an ambient at $T_\infty = 25$°C. Calculate the center temperature T_0 of the cylinder $t = 1$ min after the start of the cooling.

Answer: 93.3°C

4-37 A brick column with cross section 10 cm by 10 cm [$k = 0.69$ W/(m · °C) and $\alpha = 0.5 \times 10^{-6}$ m²/s] is initially at a uniform temperature $T_i = 225$°C. Suddenly the surfaces are subjected to convective cooling with a heat transfer coefficient $h = 60$ W/(m² · °C) into an ambient at $T_\infty = 25$°C. Calculate the center temperature T_0 at $t = 1$ h after the start of the cooling. Compare this temperature with that of a plane wall of thickness $L = 10$ cm of the same material and under the same conditions.

4-38 A semi-infinite corner, $0 \le x < \infty$, $0 \le y < \infty$, of concrete cinder stone [$k = 1.3$ W/(m · °C)

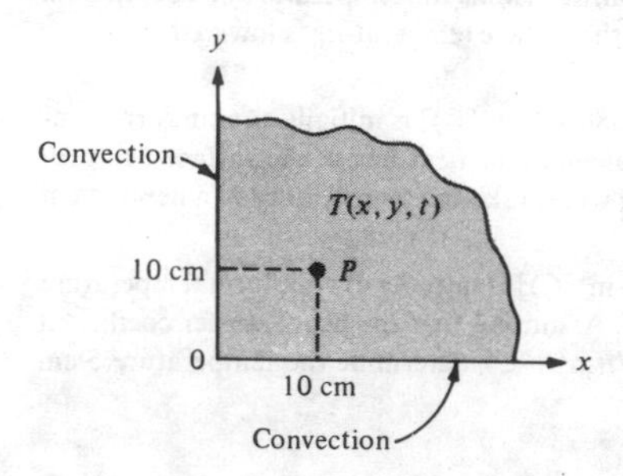

Figure P4-38

and $\alpha = 8 \times 10^{-7}$ m²/s] is initially at a uniform temperature $T_i = 130°C$. Suddenly the surfaces are subjected to convective cooling, with a heat transfer coefficient $h = 100$ W/(m² · °C) into an ambient at $T_\infty = 30°C$. Determine the temperature at a point P at a location $x = 10$ cm, $y = 10$ cm, as illustrated in Fig. P4-38, $t = 2$ h after the start of the cooling.

Answer: 76°C

4-39 A semi-infinite strip, $0 < x < \infty, 0 < y < 10$ cm, of fireclay brick [$k = 1$ W/(m · °C) and $\alpha = 5.4 \times 10^{-7}$ m²/s] is initially at a uniform temperature $T_i = 340°C$. Suddenly all surfaces are subjected to convection, with a heat transfer coefficient $h = 100$ W/(m² · °C) into an ambient at $T_\infty = 40°C$. Calculate the temperature T_0 of a point P located along the midplane at a distance $L = 5$ cm from the surface, as shown in Fig. P4-39, $t = 2$ h after the start of the cooling.

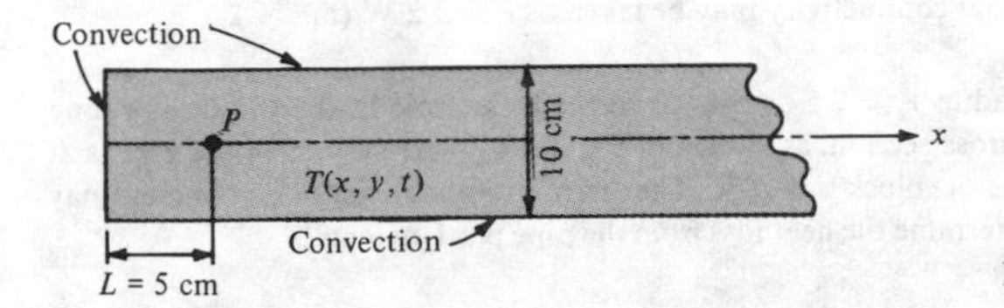

Figure P4-39

Answer: 51.4°C

4-40 A semi-infinite, cylindrical iron bar [$k = 60$ W/(m · °C) and $\alpha = 1.6 \times 10^{-5}$ m²/s] of diameter $D = 5$ cm, confined to the region $0 \le x < \infty$, is initially at a uniform temperature $T_i = 330°C$. Suddenly the surfaces are subjected to convection with a heat transfer coefficient $h = 200$ W/(m² · °C) into an ambient at $T_\infty = 30°C$. Determine the temperature T_0 of a point P located along the axis $L = 3$ cm from the flat surface $t = 2$ min after the start of the cooling. See Fig. P4-40.

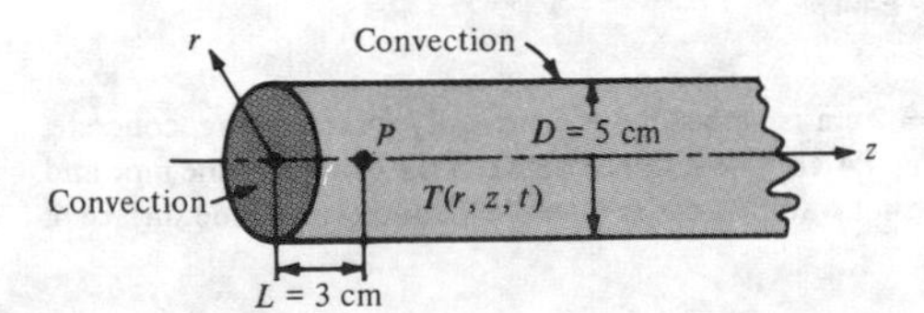

Figure P4-40

Conduction shape factors

4-41 An isothermal sphere of radius $r = 10$ cm is buried in a large body of earth. The sphere is maintained at a uniform temperature of 100°C while the temperature of the earth at large distances from the sphere is 10°C. Calculate the rate of heat loss from the sphere by using the shape factor data if the thermal conductivity of the earth is $k = 1.2$ W/(m · °C).

Answer: 135.8 W

4-42 A pipe carrying a hot fluid at $T_f = 150°C$ is buried in the earth. The distance between the pipe center and the earth's surface is $H = 1$ m. The pipe has an outer radius $r_o = 10$ cm and is covered with a 5-cm-thick fiberglass insulation [$k = 0.04$ W/(m · °C)]. The earth has a thermal conductivity $k = 1.3$ W/(m · °C), and the earth's surface is at 0°C. Determine the heat loss per 10-m length of pipe.

Answer: 777 W

4-43 A spherical tank of radius $r = 1$ m containing a fluid maintained at $T_f = 200°C$ is buried in the earth. The distance between the center of the tank and the earth's surface is $H = 3$ m. The tank is covered with a 5-cm-thick insulation layer of thermal conductivity $k = 0.038$ W/(m · °C), and the thermal conductivity of the earth at this location may be taken as $k = 1.1$ W/(m · °C). If the earth's surface is at 0°C, determine the rate of heat loss from the tank.

4-44 Two parallel pipes are buried in the earth with a spacing of $H = 0.6$ m between their centerlines. One of the pipes has radius $r_1 = 10$ cm, and its surface is maintained at $T_1 = 60°C$; the other pipe has radius $r_2 = 5$ cm, and its surface is maintained at $T_2 = 150°C$. Calculate the heat transfer rate between the two pipes per 1-m length for earth conductivity $k = 1.2$ W/(m · °C).

4-45 A spherical tank of radius $r = 1$ m containing radioactive material is buried in the earth. The distance between the earth's surface and the tank's center is $H = 5$ m. Heat release resulting from radioactive decay in the tank is 700 W. Calculate the steady-state temperature of the tank's surface if the earth's surface is at 10°C. The thermal conductivity of the earth at this location may be taken as $k = 1.0$ W/(m · °C).

Answer: 60°C

4-46 A cylindrical storage tank of radius $r = 0.5$ m and length $L = 2.5$ m is buried in the earth with its axis parallel to the earth's surface. The distance between the earth's surface and the tank axis is $H = 2$ m. If the tank's surface is maintained at 70°C and the earth's surface is at 20°C, determine the rate of heat loss from the tank. The earth's thermal conductivity may be taken as $k = 1.2$ W/(m · °C).

Answer: 457.5 W

4-47 A hot water pipe of outside radius $r_o = 2.5$ cm passes through the hole at the center of a long concrete block 40 cm by 40 cm in cross section, as shown in Fig. P4-47. The surface of the pipe is at 70°C, and the outer surface of the concrete block is at 20°C. The thermal conductivity of the concrete may be taken as $k = 0.76$ W/(m · °C). Determine the heat loss from the pipe per 1-m length.

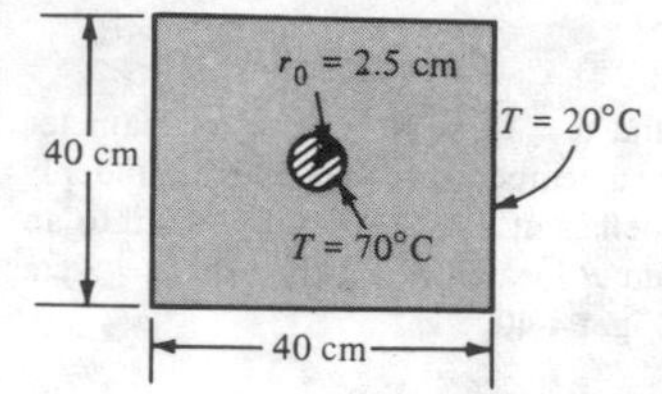

Figure P4-47

4-48 A hot water pipe of outside radius $r_1 = 2$ cm is embedded eccentrically inside a long, concrete, cylindrical block of radius $r_2 = 8$ cm, as in Fig. P4-48. The distance between the centers of the pipe and of the cylinder is $L = 4$ cm. The surface of the hot water pipe is at $T_1 = 70°C$, and the outside surface of

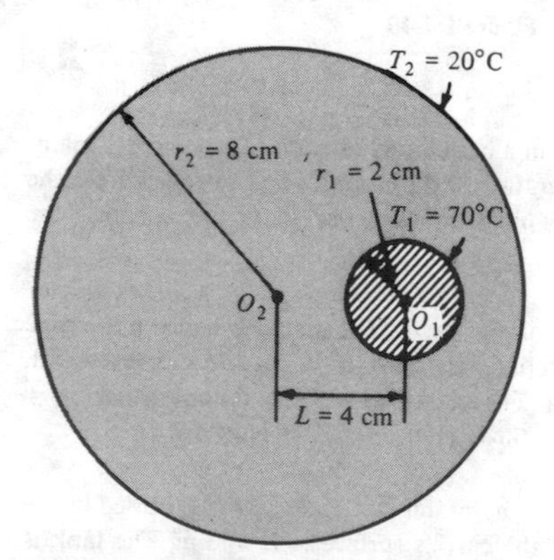

Figure P4-48

the concrete cylinder is maintained at $T_2 = 20°C$. The thermal conductivity of concrete is $k = 0.76$ W/(m · °C). Determine the rate of heat loss from a 1-m length of pipe.

Answer: 129.3 W

4-49 A spherical tank of radius $r = 1$ m containing some radioactive material is totally buried in a large mass of earth of thermal conductivity $k = 1.6$ W/(m° · C). If the surface of the tank is at 150°C and the bulk temperature of the earth is 20°C, determine the rate of energy generated by the radioactive material in the tank that is due to radioactive decay.

Analytic solution of transient heat conduction in a slab

4-50 A slab of thickness L is initially at a uniform temperature T_i. For $t > 0$, the boundary surface at $x = 0$ is kept insulated, and that at $x = L$ is kept at a constant temperature T_0. The thermal properties of the slab are constant, and there is no heat generation within the medium.

(*a*) Develop an expression for the one-dimensional, time-dependent temperature distribution $T(x, t)$ in the slab.

(*b*) Develop an expression for the temperature of the insulated surface.

(*c*) Develop an expression for the heat flux at the boundary surface $x = L$.

4-51 A brick wall confined to the region $-L < x < L$, as illustrated in Fig. P4-51, is initially at a uniform temperature T_i. For $t > 0$, both boundary surfaces are maintained at a uniform temperature T_0.

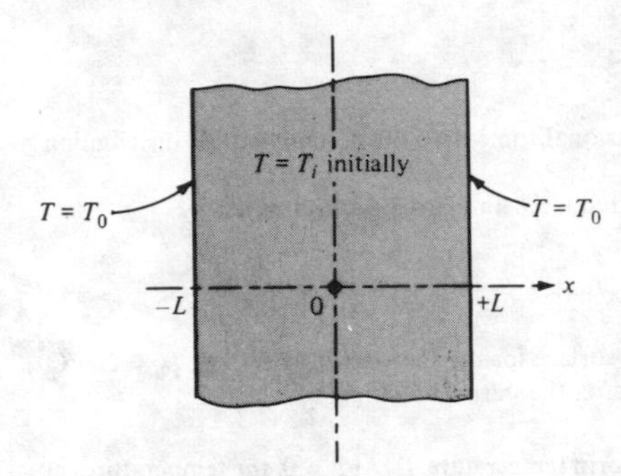

Figure P4-51

(*a*) Develop an expression for the time-dependent temperature distribution $T(x, t)$ in the slab. (*Hint*: Compare this problem with Problem 4.50.)

(*b*) Compute the temperature at the center plane for $L = 5$ cm, $k = 1$ W/(m · °C), $\alpha = 0.5 \times 10^{-6}$ m²/s, $T_i = 230$°C, $T_0 = 30$°C at $t = \frac{1}{10}$, 1, and 3 h after the exposure to cooling.

4-52 Develop an expression for the one-dimensional, time-dependent temperature distribution $T(x, t)$ for a slab in $0 \le x \le L$, which is initially at temperature $T = F(x)$ and for $t > 0$ both boundaries are maintained at zero temperature, as in Fig. P4-52.

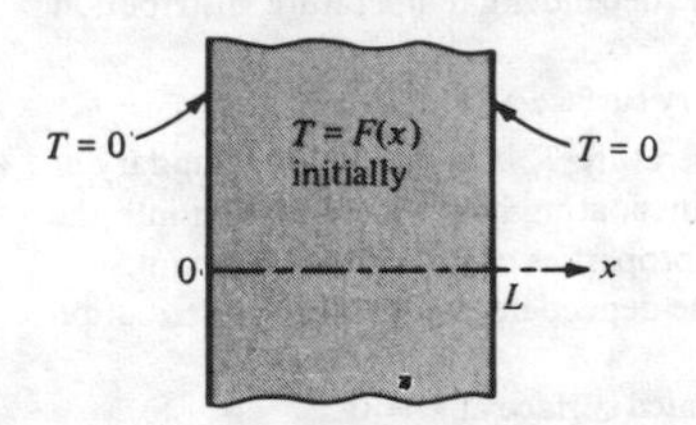

Figure P4-52

(*a*) Calculate the center temperature for the special case of initial distribution given by $F(x) = 100 \sin (\pi x/L)$ by taking $L = 10$ cm, $\alpha = 0.5 \times 10^{-6}$ m²/s, and $k = 1$ W/(m · °C) at $t = \frac{1}{10}$, 1, and 2 h after the exposure to cooling.

(*b*) Calculate the heat fluxes at the boundary surfaces $x = 0$ and $x = 10$ cm at $t = \frac{1}{10}$, 1, and 2 h for the case considered in (*a*).

4-53 A slab of thickness L has an initial temperature distribution $T_i = T_0 \cos [\pi x/(2L)]$. For $t > 0$, the boundary surface at $x = 0$ is kept insulated, and that at $x = L$ is kept at zero temperature.

(*a*) Develop an expression for the one-dimensional, time-dependent temperature distribution $T(x, t)$ in the slab.

(*b*) Calculate the temperature of the insulated boundary at $t = \frac{1}{10}$, 1, and 2 h for a plate of thickness $L = 10$ cm, and thermal diffusivity $\alpha = 10^{-7}$ m^2/s by taking $T_0 = 200$°C.

4-54 A slab in $0 \leq x \leq L$ has an initial temperature distribution $T = F(x)$. For $t > 0$, the boundary at $x = 0$ is kept at zero temperature, and that at $x = L$ is kept insulated, as illustrated in Fig. P4-54.

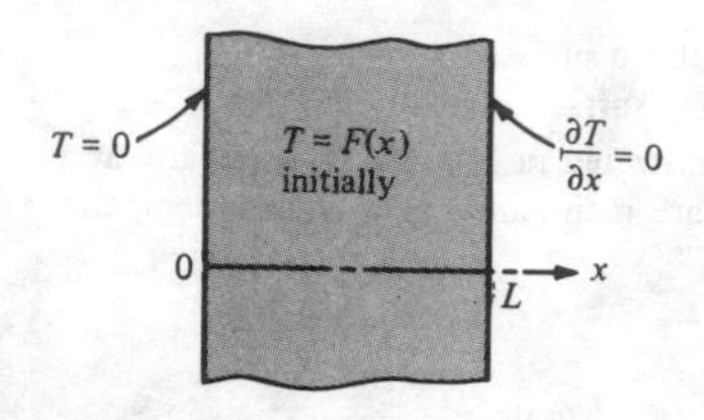

Figure P4-54

(*a*) Develop an expression for the one-dimensional, time-dependent temperature distribution $T(x, t)$ in the slab.

(*b*) Obtain a solution for the special case of initial temperature distribution given by

$$F(x) = T_0 \sin \frac{\pi}{2L} x$$

(*c*) Calculate the temperature of the insulated surface for the case in (*b*) by setting $T_0 = 200$°C, $L = 10$ cm, and $\alpha = 10^{-7}$ m^2/s at $t = \frac{1}{10}$, 1, and 2 h after the start of cooling.

Answer: (c) 198.2, 183, and 167.4°C

4-55 A large plate of thickness L is initially at a uniform temperature T_i. At $t = 0$, the temperatures at both surfaces are suddenly lowered to a temperature T_e.

(*a*) Develop an expression for the one-dimensional, time-dependent temperature distribution $T(x, t)$ in the slab.

(*b*) Calculate the center temperature at $t = \frac{1}{10}$ and 1 h for $L = 20$ cm, $\alpha = 0.5 \times 10^{-6}$ m^2/s, $T_i = 250$°C, and $T_e = 50$°C.

Answer: (*b*) 250 and 211.7°C

4-56 A large plate of thickness L is initially at a prescribed temperature $F(x)$. For $t > 0$, the boundary surface at $x = 0$ is suddenly lowered to zero temperature, and the boundary surface at $x = L$ is kept insulated.

(*a*) Develop an expression for the one-dimensional, time-dependent temperature distribution $T(x, t)$ in the slab.

(*b*) Develop an expression for the heat flux at the boundary surface at $x = 0$.

4-57 A large plate of thickness L is initially at a uniform temperature T_i. For $t > 0$, the boundary at $x = 0$ is kept insulated, and the boundary surface at $x = L$ is dissipating heat by convection into the ambient air at T_∞ with a heat transfer coefficient h. The thermal properties of the slab are constant.

(*a*) Develop an expression for the one-dimensional, time-dependent temperature distribution $T(x, t)$ in the slab.

(*b*) Develop an expression for the temperature of the insulated surface at $x = 0$.

(*c*) Develop an expression for the heat flux at the boundary surface $x = L$.

REFERENCES

1. Carslaw, H. S., and J. C. Jaeger: *Conduction of Heat in Solids*, 2d ed., Oxford University Press, London, 1959.

2. Özişik, M. N.: *Boundary Value Problems of Heat Conduction*, International Textbook, Scranton, Pa., 1968.
3. Arpaci, V. S.: *Conduction Heat Transfer*, Addison-Wesley, Reading, Mass., 1966.
4. Mikhailov, M. D., and M. N. Özişik, *Unified Analysis and Solutions of Heat and Mass Diffusion*, Wiley, New York, 1984.
5. Özişik, M. N.: *Heat Conduction*, Wiley, New York, 1980.
6. Heisler, M. P.: "Temperature Charts for Induction and Constant Temperature Heating," *Trans. ASME*, **69**: 227–236 (1947).
7. Schneider, P. J.: *Temperature Response Charts*, Wiley, New York, 1963.
8. Gröber, H., S. Erk, and U. Grigull: *Fundamentals of Heat Transfer*, McGraw-Hill, New York, 1961.
9. Schneider, P. J.: *Conduction Heat Transfer*, Addison-Wesley, Reading, Mass., 1955.
10. Sunderland, J. E., and K. R. Johnson: "Shape Factors for Heat Conduction through Bodies with Isothermal or Convective Boundary Conditions," *Trans. ASHRAE*, **70**: 237–241 (1964).
11. Hahne, E., and U. Grigul: "Formfaktor und Formwiderstand der Stationären Mehrdimensionalen Wärmeleitung," *Int. J. Heat Mass Transfer*, **18**: 751–767 (1975).
12. James, M. L., G. M. Smith, and J. C. Wolford: *Applied Numerical Methods for Digital Computations with Fortran and CSMP*, 2d ed., IEP, New York, 1977.
13. Groove, W. E.: *Brief Numerical Methods*, Prentice-Hall, Englewood Cliffs, N.J., 1966.

CHAPTER

FIVE

FINITE-DIFFERENCE METHODS FOR SOLVING HEAT CONDUCTION PROBLEMS

The numerical method of solution is used extensively in practical applications to determine the temperature distribution and heat flow in solids having complicated geometries, boundary conditions, and temperature-dependent thermal properties. A commonly used numerical scheme is the *finite-difference* method, described in Refs. 1 to 10. In this approach, the partial differential equation of heat conduction is approximated by a set of algebraic equations for temperature at a number of nodal points over the region. Therefore, the first step in the analysis is the finite-difference representation, or the transformation into a set of algebraic equations, of the differential equation of heat conduction. This matter is discussed in this chapter, first, with emphasis on the finite-difference representation of one-dimensional, steady-state heat conduction problems in rectangular, cylindrical, and spherical coordinates; and, second, by the finite-difference representation of the two-dimensional, steady-state heat conduction problems in rectangular coordinates. Since the method transforms the analysis of the heat conduction problem to the solution of a set of coupled algebraic equations, we briefly discuss the methods of solving simultaneous algebraic equations. Finally, the solution of one-dimensional, time-dependent heat conduction problems by finite differences is described.

In recent years, another numerical scheme called the *finite-element* method, developed originally for the solution of structural problems, has been applied to the solution of heat conduction problems. For problems with complex geometries, the finite-element method offers some advantage over the finite-difference method in the solution of heat conduction problems. The application of this method is described in Refs. 11 to 16. In this approach, either the variational principles or, preferably, the Galerkin method is used to transform the heat conduction problem to a set of algebraic equations; hence some background is needed in the use of these methods as well as the selection of finite elements. Therefore, in the limited space available here, this method is not discussed.

5-1 ONE-DIMENSIONAL, STEADY-STATE HEAT CONDUCTION

When a heat conduction problem is solved exactly by an analytical method, the resulting solution satisfies the governing differential equation at every point in the region as well as at the boundaries. However, when the problem is solved by a numerical method, such as finite differences, the differential equation is satisfied only at a selected number of discrete nodes within the region. Therefore, the starting point in an analysis by the finite-difference method is the finite-difference representation of the heat conduction equation and its boundary conditions.

The finite-difference form of the heat conduction equation can be developed by replacing the partial derivatives of temperature in the heat conduction equation with their equivalent finite-difference forms or by writing an energy balance for a differential volume element. In this book, we develop the finite-difference equations by writing an energy balance. However, it is instructive to use the former approach to illustrate the relation between the heat conduction equation and its finite-difference form. Therefore, we also illustrate the use of this approach in order to provide a better understanding of the finite-difference formulations.

Finite-Difference Forms from the Differential Equation

Consider the following one-dimensional, steady-state heat conduction equation with energy generation

$$\frac{d^2T(x)}{dx^2} + \frac{1}{k}\, g(x) = 0 \qquad \text{in } 0 < x \le L \tag{5-1}$$

The region $0 \le x \le L$ is divided into M equal subregions, each in size

$$\Delta x = \frac{L}{M} \tag{5-2}$$

and there are $M + 1$ nodes for $m = 0$ to $m = M$, as illustrated in Fig. 5-1. Node m corresponds to a location whose coordinate is $x = m\,\Delta x$. Let T_m be the temperature at node m; then the region contains $M + 1$ node temperatures for $m = 0, 1, 2, \ldots, M$, nodes.

The second derivative of temperature $d^2T(x)/dx^2$ at a node m can be represented in finite differences. Consider the locations $(m + \frac{1}{2})$ and $(m - \frac{1}{2})$ as illustrated

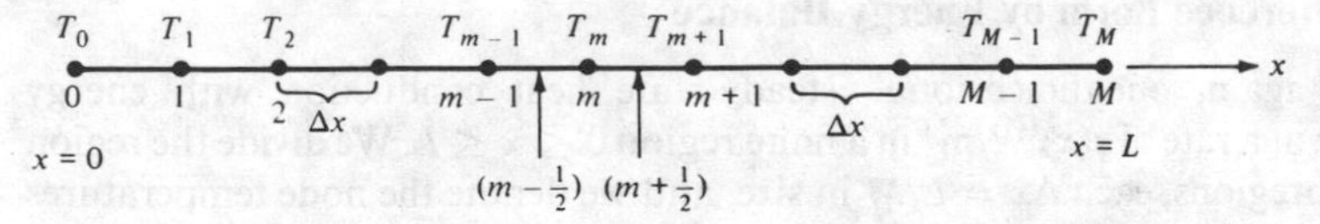

Figure 5-1 Nomenclature for finite-difference representation of derivatives.

in Fig. 5-1. The first derivative of temperature $dT(x)/dx$ at these two locations can be approximated as

$$\left.\frac{dT(x)}{dx}\right|_{m+1/2} \cong \frac{T_{m+1} - T_m}{\Delta x} \tag{5-3a}$$

and

$$\left.\frac{dT(x)}{dx}\right|_{m-1/2} \cong \frac{T_m - T_{m-1}}{\Delta x} \tag{5-3b}$$

Then the second derivative of temperature d^2T/dx^2 at node m can be approximated as

$$\left.\frac{d^2T(x)}{dx^2}\right|_m \cong \frac{dT/dx\,|_{m+1/2} - dT/dx\,|_{m-1/2}}{\Delta x}$$

$$= \frac{T_{m-1} - 2T_m + T_{m+1}}{(\Delta x)^2} \tag{5-4}$$

Equation (5-4) is introduced into the heat conduction equation (5-1):

$$\boxed{(T_{m-1} - 2T_m + T_{m+1}) + \frac{(\Delta x)^2\, g_m}{k} = 0} \tag{5-5}$$

where $g_m \equiv$ energy generation rate at node m
$m = 1, 2, 3, \ldots, M - 1$

Equation (5-5) is called the finite-difference form of the heat conduction equation (5-1), and it is valid at the interior nodes, that is, $m = 1, 2, \ldots, M - 1$, of the region. In this equation the energy generation term g_m, thermal conductivity k, and mesh size Δx are considered known quantities. Then the system (5-5) provides $M - 1$ simultaneous algebraic equations for the $M + 1$ unknown node temperatures T_m, $m = 0, 1, 2, \ldots, M$. Two additional relations are needed to make the number of equations equal to the number of unknown node temperatures T_m. These two additional relationships are obtained from the finite-difference representation of the boundary conditions at nodes $m = 0$ and $m = M$, as is discussed later.

An examination of Eq. (5-5) reveals that the terms inside the parentheses represent the sum of the temperatures of the two neighboring nodes of node m minus twice the temperature of node m.

Finite-Difference Form by Energy Balance

Consider, again, one-dimensional, steady-state heat conduction with energy generation at a rate of $g(x)$ W/m^3 in a finite region $0 \le x \le L$. We divide the region into M subregions, each $\Delta x = L/M$ in size, and we denote the node temperatures by T_m, $m = 0, 1, \ldots, M$.

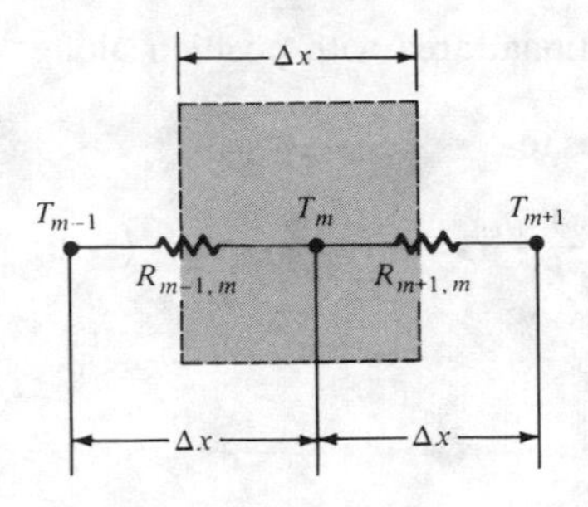

Figure 5-2 Nomenclature for energy balance at a node m.

To develop the finite-difference equation, we consider a differential volume element Δx about node m, as illustrated in Fig. 5-2. The steady-state energy balance equation for this volume element can be stated as

$$\begin{pmatrix}\text{Rate of heat}\\ \text{entering the}\\ \text{volume element}\\ \text{by conduction}\end{pmatrix} + \begin{pmatrix}\text{rate of energy}\\ \text{generation}\\ \text{within the}\\ \text{volume element}\end{pmatrix} = 0 \tag{5-6}$$

Let A be the cross-sectional area of the element normal to the $0x$ axis and k the thermal conductivity of the material. It is assumed that A and k may vary with position along the x axis. By writing the appropriate mathematical expressions, Eq. (5-6) takes the form

$$(Ak)_{m-1,m}\frac{T_{m-1}-T_m}{\Delta x}+(Ak)_{m+1,m}\frac{T_{m+1}-T_m}{\Delta x}+A_m\,\Delta x\,g_m=0 \tag{5-7}$$

By utilizing the thermal resistance concept, this expression can be written more compactly as

$$\boxed{\frac{T_{m-1}-T_m}{R_{m-1,m}}+\frac{T_{m+1}-T_m}{R_{m+1,m}}+\Delta V\,g_m=0} \tag{5-8}$$

where

$$R_{m-1,m}=\left(\frac{\Delta x}{kA}\right)_{m-1,m}=\text{thermal resistance between nodes } m-1 \text{ and } m \tag{5-9a}$$

$$R_{m+1,m}=\left(\frac{\Delta x}{kA}\right)_{m+1,m}=\text{thermal resistance between nodes } m+1 \text{ and } m \tag{5-9b}$$

$$\Delta V=A_m\,\Delta x=\text{volume of the element about node } m \tag{5-9c}$$

$$g_m=\text{energy generation rate at node } m \tag{5-9d}$$

Equation (5-8) is the finite-difference form of the one-dimensional, steady-state heat conduction equation with energy generation in the rectangular coordinate system. This equation is more general than Eq. (5-5), because it allows for the variation of thermal conductivity and cross-sectional area with position along the $0x$ axis.

When k and A are constants, Eq. (5-7) reduces to

$$(T_{m-1} - 2T_m + T_{m+1}) + \frac{(\Delta x)^2 g_m}{k} = 0 \tag{5-10}$$

which is the same as Eq. (5-5).

5-2 BOUNDARY CONDITIONS

The boundary conditions for the heat conduction problem may be a prescribed temperature, prescribed heat flux, or convection boundary condition. There are others, such as the radiation boundary condition. Here we consider only the finite-difference representation for the first three of these boundary conditions.

Prescribed Temperature

Suppose the temperatures are specified at the boundaries $x = 0$ and $x = L$ as

$$T(x)|_{x=0} = f_0 \tag{5-11a}$$

and

$$T(x)|_{x=L} = f_M \tag{5-11b}$$

where f_0 and f_M are two known temperatures. Then the temperatures T_0 and T_M at the nodes $m = 0$ and $m = M$ in Fig. 5-1 are taken as

$$T_0 = f_0 \tag{5-12a}$$

and

$$T_M = f_M \tag{5-12b}$$

which provides the two additional relations needed to make the number of equations equal the number of unknown node temperatures.

Prescribed Heat Flux

Suppose the heat fluxes q_0 and q_M, entering the medium through the boundaries at $x = 0$ and $x = L$, respectively, are prescribed as illustrated in Fig. 5-3. To develop the finite-difference form of these boundary conditions, we need to write energy balance equations for a differential volume element $\Delta x/2$ at nodes $m = 0$ and $m = M$. The energy balance equation for the differential volume element at the boundary may be stated as

$$\begin{pmatrix}\text{Rate of heat}\\ \text{supply through}\\ \text{the boundary}\\ \text{surface}\end{pmatrix} + \begin{pmatrix}\text{rate of heat}\\ \text{entering by}\\ \text{conduction}\end{pmatrix} + \begin{pmatrix}\text{rate of}\\ \text{energy}\\ \text{generation}\end{pmatrix} = 0 \tag{5-13}$$

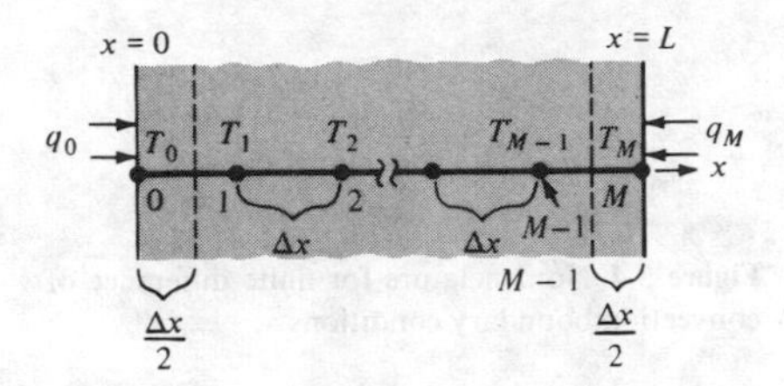

Figure 5-3 Nomenclature for finite difference of prescribed heat flux boundary conditions.

Applying this equation for the volume elements $\Delta x/2$ at the boundaries at $x = 0$ and $x = L$, we obtain respectively,

$$q_0 + k\frac{T_1 - T_0}{\Delta x} + \frac{1}{2}\Delta x\, g_0 = 0 \qquad q_M + k\frac{T_{M-1} - T_M}{\Delta x} + \frac{1}{2}\Delta x\, g_M = 0$$

These equations are rearranged as

$$2T_1 - 2T_0 + \frac{(\Delta x)^2 g_0}{k} + \frac{2\,\Delta x\, q_0}{k} = 0 \qquad \text{for } m = 0 \tag{5-14}$$

$$2T_{M-1} - 2T_M + \frac{(\Delta x)^2 g_M}{k} + \frac{2\,\Delta x\, q_M}{k} = 0 \qquad \text{for } m = M \tag{5-15}$$

Equations (5-14) and (5-15) are the finite-difference representation of the prescribed heat flux boundary conditions for nodes $m = 0$ and $m = M$, respectively.

For the *insulated* or *symmetry* boundary conditions, we have

$$q_0 = q_M = 0 \tag{5-16}$$

Then Eqs. (5-14) and (5-15), respectively, reduce to

$$2T_1 - 2T_0 + \frac{(\Delta x)^2 g_0}{k} = 0 \tag{5-17}$$

$$2T_{M-1} - 2T_M + \frac{(\Delta x)^2 g_M}{k} = 0 \tag{5-18}$$

We note that Eq. (5-17) can be obtained from Eq. (5-10) by setting $m = 0$ and $T_{-1} = T_1$. Here T_{-1} represents the mirror image of T_1 because of symmetry. Similarly, Eq. (5-18) is also obtainable from Eq. (5-10) by setting $m = M$ and

$$T_{M+1} = T_{M-1}$$

Convection

Suppose the boundary surfaces at $x = 0$ and $x = L$ are subjected to convection with a heat transfer coefficient h into an ambient at temperature T_∞, as illustrated in Fig. 5-4. We consider a differential volume element $\Delta x/2$ at nodes $m = 0$ and

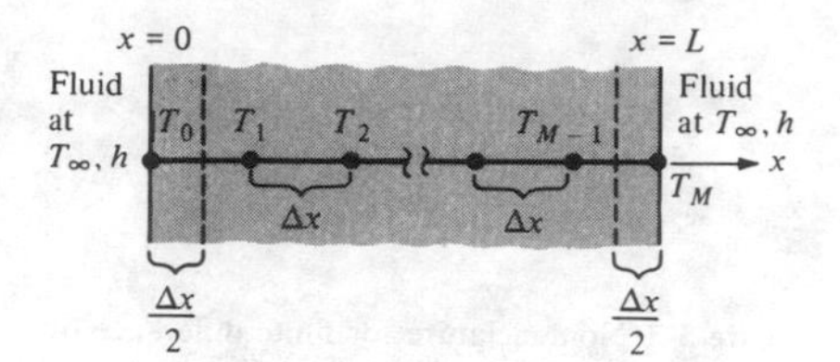

Figure 5-4 Nomenclature for finite difference of convection boundary conditions.

$m = M$. The energy balance equation for the differential volume elements at the boundaries may be stated as

$$\begin{pmatrix}\text{Rate of heat}\\ \text{gain through}\\ \text{the surface}\\ \text{by convection}\end{pmatrix} + \begin{pmatrix}\text{rate of heat}\\ \text{entering by}\\ \text{conduction}\end{pmatrix} + \begin{pmatrix}\text{rate of}\\ \text{energy}\\ \text{generation}\end{pmatrix} = 0 \tag{5-19}$$

Applying this equation for the boundaries at $x = 0$ and $x = L$, we obtain, respectively,

$$h(T_\infty - T_0) + k\,\frac{T_1 - T_0}{\Delta x} + \frac{1}{2}\,\Delta x\, g_0 = 0$$

and

$$h(T_\infty - T_M) + k\,\frac{T_{M-1} - T_M}{\Delta x} + \frac{1}{2}\,\Delta x\, g_M = 0$$

These results are rearranged as

$$2T_1 - \left(2 + \frac{2\,\Delta x\, h}{k}\right)T_0 + \frac{(\Delta x)^2 g_0}{k} + \frac{2\,\Delta x\, h}{k}\,T_\infty = 0 \qquad \text{at } m = 0 \tag{5-20}$$

$$2T_{M-1} - \left(2 + \frac{2\,\Delta x\, h}{k}\right)T_M + \frac{(\Delta x)^2\, g_M}{k} + \frac{2\,\Delta x\, h}{k}\,T_\infty = 0 \qquad \text{at } m = M \tag{5-21}$$

Equations (5-20) and (5-21) are the finite-difference representation of convection boundary conditions at nodes $m = 0$ and $m = M$.

Example 5-1 Consider the steady-state heat conduction in a slab of thickness L, in which energy is generated at a constant rate of g W/m^3. The boundary surface at $x = 0$ is maintained at a constant temperature f_0, while the boundary surface at $x = L$ dissipates heat by convection with a heat transfer coefficient h into an ambient at temperature T_∞. Dividing the region into five equal subregions, write the finite-difference formulation of this heat conduction problem.

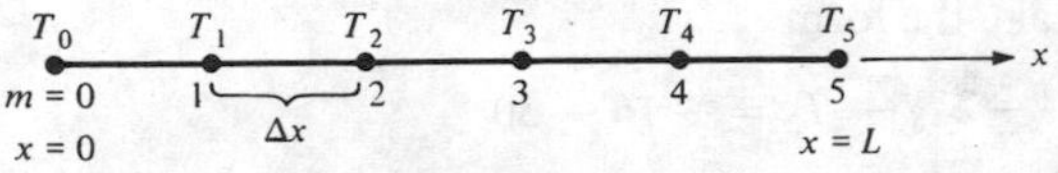

T_0 T_1 T_2 T_3 T_4 T_5 x
$m = 0$ 1 2 3 4 5
$x = 0$ Δx $x = L$

Figure 5-5 Finite-difference nodes for Example 5-1.

Compute the temperature at the nodes by finite differences for $h = 200$ W/(m$^2 \cdot$ °C), $k = 18$ W/(m $\cdot$ °C), $L = 0.01$ m, $T_\infty = 100$°C, $f_0 = 50$°C, and $g = 7.2 \times 10^7$ W/m^3. Compare the numerical solution with the exact result obtained from the analytic solution of the problem.

SOLUTION The mathematical formulation of this heat conduction problem is given by

$$\frac{d^2T(x)}{dx^2} + \frac{1}{k} g = 0 \qquad \text{in } 0 < x < L$$

$$T(x) = f_0 \qquad \text{at } x = 0$$

$$k \frac{dT(x)}{dx} + hT = hT_\infty \qquad \text{at } x = L$$

This problem can readily be solved analytically. However, to illustrate the application of the finite-difference method, the region is divided into five subregions of equal size, as illustrated in Fig. 5-5. There are five unknown node temperatures T_m, $m = 1$ to 5; hence five equations are needed for their determination.

Equation (5-10) is applicable for the four interior nodes $m = 1$ to 4, and Eq. (5-21) is applicable for node $m = 5$ on the convection boundary. These equations are

$$T_{m-1} - 2T_m + T_{m+1} + \frac{(\Delta x)^2 g}{k} = 0 \qquad \text{for } m = 1 \text{ to } 4$$

$$2T_4 - \left(2 + \frac{2 \Delta x\, h}{k}\right) T_5 + \frac{(\Delta x)^2 g}{k} + \frac{2 \Delta x\, h}{k} T_\infty = 0 \qquad \text{for } m = 5$$

The numerical values of various coefficients are

$$\Delta x = \frac{L}{5} = \frac{0.01}{5} = 2 \times 10^{-3} \text{ m}$$

$$\frac{(\Delta x)^2 g}{k} = \frac{(2 \times 10^{-3})^2 (7.2 \times 10^7)}{18} = 16$$

$$\frac{2 \Delta x\, h}{k} = \frac{2(2 \times 10^{-3})(200)}{18} = 0.0444$$

$$\frac{2 \Delta x\, h}{k} T_\infty = (0.0444)(100) = 4.44$$

$$T_0 = f_0 = 50°\text{C}$$

Then these five equations take the form

$$-2T_1 + T_2 = -16 - 50$$

$$T_1 - 2T_2 + T_3 = -16$$

$$T_2 - 2T_3 + T_4 = -16$$

$$T_3 - 2T_4 + T_5 = -16$$

$$2T_4 - 2.044T_5 = -16 - 4.44$$

This system is written in matrix form as

$$\begin{bmatrix} -2 & 1 & 0 & 0 & 0 \\ 1 & -2 & 1 & 0 & 0 \\ 0 & 1 & -2 & 1 & 0 \\ 0 & 0 & 1 & -2 & 1 \\ 0 & 0 & 0 & 2 & -2.044 \end{bmatrix} \begin{bmatrix} T_1 \\ T_2 \\ T_3 \\ T_4 \\ T_5 \end{bmatrix} = \begin{bmatrix} -66 \\ -16 \\ -16 \\ -16 \\ -20.44 \end{bmatrix}$$

Some features of the coefficient matrix should be noted. It is a banded matrix; if the problem involved more than five nodes, the banded form would still be retained. The solution of this matrix equation yields the node temperatures T_m, $m = 1$ to 5. A computer program for the solution of matrix algebraic equations is discussed next.

A Computer Program for Solving Simultaneous Algebraic Equations

We present in Table 5-1 a computer program written in Fortran for solving a system of N simultaneous algebraic equations, applied to the solution of the system in Example 5-1 containing only $N = 5$ equations. This program is sufficiently general to be used for the solution of systems of algebraic equations encountered in this book by changing only a statement. The basic features of this program and its application for the solution of other simultaneous algebraic equations are described in the note below.*

* The program shown in Table 5-1 uses the International Mathematical and Statistical Libraries (IMSL) subroutine [18] LEQT1F for solving N simultaneous algebraic equations, expressed in the matrix form as

$$[A][X] = [B]$$

where $[A]$ is the N-by-N input coefficient matrix, $[B]$ is the N-by-1 right-hand side vector of input data, and $[X]$ is the N-by-1 vector that contains the N unknown node temperatures. This program can be used for solving N simultaneous algebraic equations as follows:

1. If $N \leq 40$: Replace in line 5 the value of 5 assigned to N by the number of equations for the new system; and read in the matrices A and B for the new system.
2. If $N > 40$: In addition to the change made above, replace in lines 2 and 3 the value of 40 by the new value of N; also in line 2 replace the value of 1600 by a value at least equal to N^2.

Table 5-1 Computer program for solving a system of N simultaneous equations applied to the solution of Example 5-1

```
              JOB
 1            IMPLICIT REAL*8 (A-H,O-Z)
 2            REAL*8 A(40,40),B(40,1),WKAREA(1600)
 3            NDIMEN=40
 4            IDGT=0
       C
       C      READ THE MATRICES [A] AND [B] BY ROWS
       C
 5            N=5
 6            DO 50  I=1,N
 7            READ, (A(I,J),J=1,N),B(I,1)
 8     50     CONTINUE
       C
       C      PRINT THE INPUT DATA FOR CHECKING PURPOSES
       C
 9            DO 60 I=1,N
10            WRITE(3,70)(A(I,J),J=1,N),B(I,1)
11     70     FORMAT(IX/TIO,13F9.3)
12     60     CONTINUE
13            CALL LEQT1F(A,1,N,NDIMEN,B,IDGT,WKAREA,IER)
       C
       C      WHEN RETURNED, THE MATRIX [B] CONTAINS THE SOLUTION.
       C      THE SOLUTION IS NOW PRINTED.
       C
14            WRITE(3,80)
15     80     FORMAT(1X///T30,'NODE #', 5X,'TEMPERATURE'/)
16            DO 90 I=1,N
17            WRITE(3,100)I, B(I,1)
18    100     FORMAT(1X,T28,I5,7X,F9.3)
19     90     CONTINUE
20            STOP
21            END
-2.000        1.000      0.000      0.000      0.000     -66.000
 1.000       -2.000      1.000      0.000      0.000     -16.000
 0.000        1.000     -2.000      1.000      0.000     -16.000
 0.000        0.000      1.000     -2.000      1.000     -16.000
 0.000        0.000      0.000      2.000     -2.044     -20.440

                 NODE #      TEMPERATURE
                    1          119.045
                    2          172.090
                    3          209.135
                    4          230.180
                    5          235.225
```

The exact analytic solution of the heat conduction problem in Example 5-1 is straightforward and yields the following expression for the temperature distribution.

$$T(x) = 50 + 5\frac{x}{L} + 200\left(1.9 - \frac{x}{L}\right)\left(\frac{x}{L}\right)$$

We compare the finite-difference solution using five nodes and the exact solution of the problem; the results are very close as shown below.

x/L	T_m	Finite difference	Exact
0.2	T_1	119.05	119.00
0.4	T_2	172.09	172.00
0.6	T_3	209.14	209.00
0.8	T_4	230.18	230.00
1.0	T_5	235.22	235.00

In this particular problem, we have the exact solution to check the accuracy of the numerical solution. When no exact solution is available, the convergence of the numerical solution should be checked by performing further computations using finer subdivisions, until the solutions from subsequent computations satisfy a specified convergence criterion.

Example 5-2 An iron rod $L = 5$ cm long of diameter $D = 2$ cm with thermal conductivity $k = 50\ \mathrm{W/(m \cdot °C)}$ protrudes from a wall and is exposed to an ambient at $T_\infty = 20°\mathrm{C}$ and $h = 100\ \mathrm{W/(m^2 \cdot °C)}$. The base of the rod is at $T_0 = 320°\mathrm{C}$, and its tip is insulated. Assuming one-dimensional steady-state heat flow, calculate the temperature distribution along the rod and the rate of heat loss into the ambient by using finite differences. Compare the finite-difference solution with the exact analytical solution of this problem.

SOLUTION This problem is exactly like the fin problem considered in Chap. 3, and its mathematical formulation is given by [see Eq. (3-54*a*) for $m \equiv N$]

$$\frac{d^2\theta(x)}{dx^2} - N^2\theta(x) = 0 \qquad \text{in } 0 < x < L$$

$$\theta(x) = \theta_0 \qquad \text{at } x = 0$$

$$\frac{d\theta(x)}{dx} = 0 \qquad \text{at } x = L$$

where $\theta(x) = T(x) - T_\infty$

$$\theta_0 = T_0 - T_\infty$$

$$N^2 = \frac{Ph}{Ak} = \frac{\pi D h}{(\pi/4)D^2 k} = \frac{4h}{Dk}$$

The analytic solution of this problem for the temperature distribution $\theta(x)$ and the heat flow rate Q is given in Chap. 3 as [see Eqs. (3-52) and (3-53) for $m \equiv N$]

$$\theta(x) = \theta_0 \frac{\cosh N(L - x)}{\cosh NL}$$

and

$$Q = \theta_0 AkN \tanh NL$$

To solve this problem by finite differences, the finite-difference form of this problem is needed. This can be done either by considering an energy balance for a typical internal node of the rod or by directly writing the finite-difference form of the above mathematical problem. We prefer the latter approach and express the second derivative of temperature in finite differences. The resulting finite-difference equation becomes

$$\frac{\theta_{m-1} - 2\theta_m + \theta_{m+1}}{(\Delta x)^2} - N^2\theta_m = 0$$

For finite-difference calculations, the rod is divided into five subregions of equal size, as illustrated in Fig. 5-6. In this arrangement there are five unknown node temperatures θ_m, $m = 1$ to 5; therefore we need five equations for their determination. The above finite-difference equation is applicable for the four internal nodes

$$\theta_{m-1} - [2 + (N\,\Delta x)^2]\theta_m + \theta_{m+1} = 0 \qquad \text{for } m = 1 \text{ to } 4$$

This equation is also applicable for the node $m = 5$ on the insulated (symmetry) boundary if we set $\theta_{m-1} = \theta_{m+1}$ as discussed previously. We find

$$2\theta_4 - [2 + (N\,\Delta x)^2]\theta_5 = 0 \qquad \text{for } m = 5$$

The numerical values of various coefficients are

$$\Delta x = \frac{L}{5} = \frac{0.05}{5} = 0.01 \text{ m}$$

$$\theta_0 = 320 - 20 = 300°\text{C}$$

$$N^2 = \frac{4h}{Dk} = \frac{4 \times 100}{0.02 \times 50} = 400$$

$$(N\,\Delta x)^2 = 0.04$$

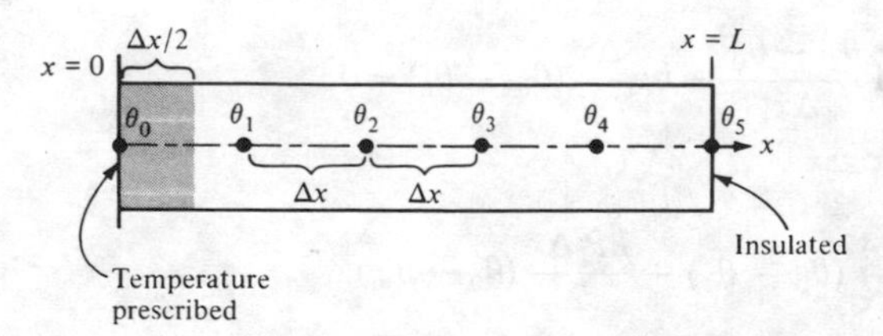

Figure 5-6 Finite-difference nodes for Example 5-2

Then the equations become

$$m = 1: \qquad -2.04\theta_1 + \theta_2 = -300$$

$$m = 2: \qquad \theta_1 - 2.04\theta_2 + \theta_3 = 0$$

$$m = 3: \qquad \theta_2 - 2.04\theta_3 + \theta_4 = 0$$

$$m = 4: \qquad \theta_3 - 2.04\theta_4 + \theta_5 = 0$$

$$m = 5: \qquad 2\theta_4 - 2.04\theta_5 = 0$$

These equations are arranged in matrix form as

$$\begin{bmatrix} -2.04 & 1 & 0 & 0 & 0 \\ 1 & -2.04 & 1 & 0 & 0 \\ 0 & 1 & -2.04 & 1 & 0 \\ 0 & 0 & 1 & -2.04 & 1 \\ 0 & 0 & 0 & 2 & -2.04 \end{bmatrix} \begin{bmatrix} \theta_1 \\ \theta_2 \\ \theta_3 \\ \theta_4 \\ \theta_5 \end{bmatrix} = \begin{bmatrix} -300 \\ 0 \\ 0 \\ 0 \\ 0 \end{bmatrix}$$

We now compare the finite-difference solution with the exact results for the above five nodes:

x/L	θ_m	Exact	Finite difference
0.2	θ_1	260.0	260.1
0.4	θ_2	230.5	230.6
0.6	θ_3	210.2	210.4
0.8	θ_4	198.3	198.6
1.0	θ_5	194.4	194.7

Clearly, the finite-difference solution of this problem, with a five-node analysis considered here, is sufficiently accurate.

To determine the heat flow rate, we consider an energy balance for a differential volume element $\Delta x/2$ at node $m = 0$, as illustrated in Fig. 5-6:

$$\begin{pmatrix} \text{Heat entering} \\ \text{the element} \\ \text{from the base} \end{pmatrix} + \begin{pmatrix} \text{heat entering} \\ \text{the element} \\ \text{by conduction} \end{pmatrix} + \begin{pmatrix} \text{heat entering} \\ \text{the element} \\ \text{by convection} \end{pmatrix} = 0$$

$$Q + kA\,\frac{\theta_1 - \theta_0}{\Delta x} + hp\,\frac{\Delta x}{2}(\theta_\infty - \theta_0) = 0$$

or

$$Q = \frac{kA}{\Delta x}(\theta_0 - \theta_1) + \frac{hP\,\Delta x}{2}(\theta_0 - \theta_\infty)$$

The numerical values of the various coefficients are

$$\theta_0 = T_0 - T_\infty = 300°\text{C}$$

$$\theta_\infty = T_\infty - T_\infty = 0°\text{C}$$

$$\frac{kA}{\Delta x} = \frac{50}{0.01}\left(\frac{\pi}{4} \times 0.02^2\right) = 1.571$$

$$\frac{hP\,\Delta x}{2} = \frac{100}{2}(\pi \times 0.02)(0.01) = 0.0314$$

Then the heat transfer rate is found to be

$$\begin{aligned} Q &= 1.571(300 - 260.1) + 0.0314(300) \\ &= 62.68 + 9.42 = 72.1 \text{ W} \end{aligned}$$

which is very close to the exact result $Q = 71.76$ W determined from the analytic solution of the problem.

5-3 TWO-DIMENSIONAL, STEADY-STATE HEAT CONDUCTION

Consider the two-dimensional, steady-state heat conduction equation with the energy generation term given in the form

$$\frac{\partial^2 T}{\partial x^2} + \frac{\partial^2 T}{\partial y^2} + \frac{g(x, y)}{k} = 0 \qquad \text{in region } R \tag{5-22}$$

subject to a set of boundary conditions. To replace the region by a set of discrete nodes, a rectangular net of mesh size Δx, Δy is constructed over the region, as shown in Fig. 5-7*a*. The symbols m, n are used to denote the location of a nodal point whose coordinates are $x = m\,\Delta x$, $y = n\,\Delta y$. Then the temperature at a node (m, n) is denoted by $T_{m,n}$. By following a procedure similar to that for the derivation of Eq. (5-4), the finite-difference forms of the second derivatives of temperature at a node (m, n) can be determined:

$$\left.\frac{\partial^2 T}{\partial x^2}\right|_{m,n} \cong \frac{\partial T/\partial x|_{m+1/2,n} - \partial T/\partial x|_{m-1/2,n}}{\Delta x} = \frac{T_{m-1,n} + T_{m+1,n} - 2T_{m,n}}{(\Delta x)^2} \tag{5-23a}$$

and similarly,

$$\left.\frac{\partial^2 T}{\partial y^2}\right|_{m,n} \cong \frac{T_{m,n-1} + T_{m,n+1} - 2T_{m,n}}{(\Delta y)^2} \tag{5-23b}$$

Introducing Eqs. (5-23) into Eq. (5-22), we obtain

$$\frac{T_{m-1,n} + T_{m+1,n} - 2T_{m,n}}{(\Delta x)^2} + \frac{T_{m,n-1} + T_{m,n+1} - 2T_{m,n}}{(\Delta y)^2} + \frac{g_{m,n}}{k} = 0 \tag{5-24}$$

In finite-difference analysis, usually a square mesh is used. For a square mesh

$$\Delta x = \Delta y = l$$

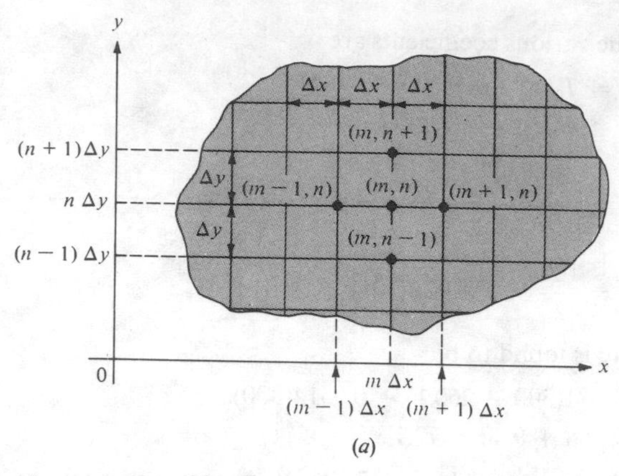

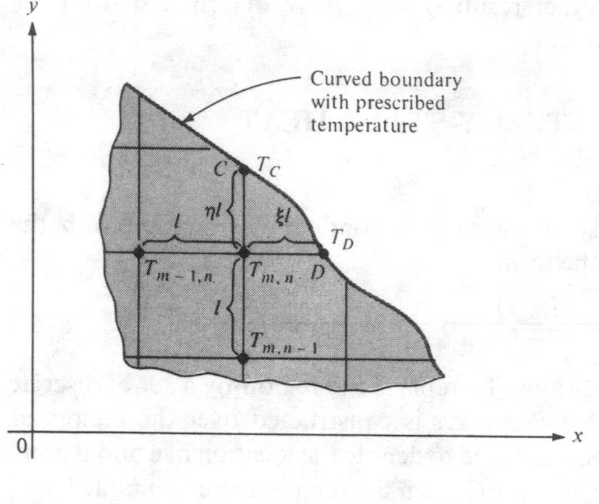

Figure 5-7 Rectangular net of mesh size Δx, Δy and a node (m, n) next to a curved boundary.

Equation (5-24) simplifies to

$$(T_{m-1,n} + T_{m+1,n} + T_{m,n-1} + T_{m,n+1} - 4T_{m,n}) + \frac{g_{m,n} l^2}{k} = 0 \tag{5-25}$$

Equation (5-25) is the *finite-difference form* of the heat conduction equation (5-22) for an interior node (m, n).

It is also possible for an internal node (m, n) to be located next to a curved boundary in such a manner that the curved boundary intersects the two strings from the node (m, n) as illustrated in Fig. 5-7*b*. It is assumed that temperature is prescribed at the curved boundary so that temperatures T_C and T_D are known at points C and D, respectively, where the boundary intersects the strings. If the lengths of the strings from the point m, n to the points C and D are, respectively,

ηl and ξl, where $\eta, \xi \le 1$, then the finite-difference equation for the node m, n is given by [17]

$$2\left[\frac{T_D}{\xi(1+\xi)} + \frac{T_C}{\eta(1+\eta)} + \frac{T_{m-1,n}}{1+\xi} + \frac{T_{m,n-1}}{1+\eta} - \left(\frac{1}{\xi} + \frac{1}{\eta}\right)T_{m,n}\right] + \frac{g_{m,n}l^2}{k} = 0 \tag{5-26}$$

where $\qquad 0 < \xi \le 1 \quad$ and $\quad 0 < \eta \le 1$

We note that for $\xi = \eta = 1$ this equation reduces to Eq. (5-25).

Thermal Resistance Concept

A more general expression for the finite-difference equation, which will allow for the variation of thermal conductivity with position within the medium, can be written in terms of thermal resistances. In formulating the finite-difference equation, we consider a node p and four of its neighboring nodes 1, 2, 3, and 4, as illustrated in Fig. 5-8. The steady-state energy balance equation for the node p contained by the volume element shown shaded in this figure is written as

$$\boxed{\sum_{j=1}^{4} \frac{T_j - T_p}{R_{jp}} + \Delta V_p\, g_p = 0} \tag{5-27}$$

where l = distance between nodes j and p

$R_{jp} = \left(\dfrac{l}{kA}\right)_{jp}$ = thermal resistance between nodes j and p

ΔV_p = volume of the element about node p

g_p = energy generation rate per unit volume at node p

In Eq. (5-27), in general, the subscript p denotes any node in the region, and the subscript j denotes the neighboring nodes around the node p. In Fig. 5-8, there are four neighboring nodes.

Nodes on a Boundary

Finite-difference equations (5-25) or (5-27) are valid for an interior node of the region. When a node is on the boundary, an appropriate finite-difference equation

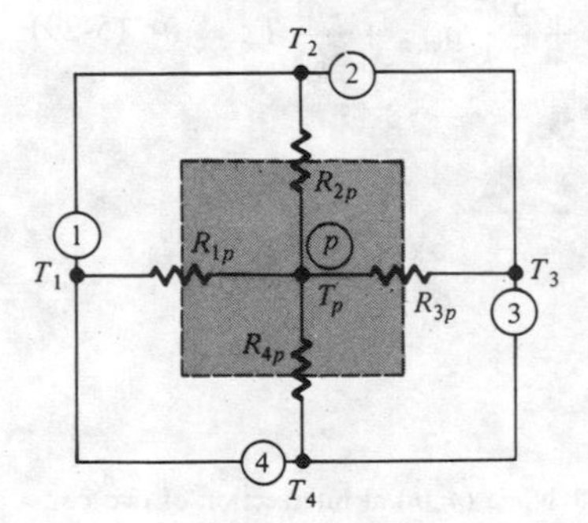

Figure 5-8 Nomenclature for energy balance at a node p.

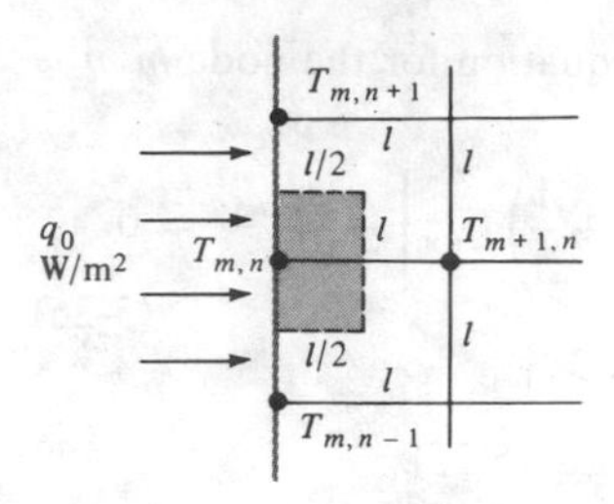

Figure 5-9 Node (m, n) on a boundary subjected to prescribed heat flux.

can be developed by writing an energy balance equation for the node on the boundary, as illustrated below.

1. *Node (m, n) on a prescribed heat flux boundary.* Figure 5-9 shows a node (m, n) on a boundary subjected to a prescribed heat flux q_0, W/m². By writing an energy balance equation for the volume element shown shaded, we obtain

$$q_0 l + k\frac{l}{2}\frac{T_{m,n+1} - T_{m,n}}{l} + kl\frac{T_{m+1,n} - T_{m,n}}{l} + k\frac{l}{2}\frac{T_{m,n-1} - T_{m,n}}{l} + \frac{1}{2}l^2 g_{m,n} = 0$$

After rearrangement,

$$T_{m,n+1} + 2T_{m+1,n} + T_{m,n-1} - 4T_{m,n} + \frac{l^2 g_{m,n}}{k} + \frac{2lq_0}{k} = 0 \qquad (5\text{-}28)$$

2. *Node (m, n) at the intersection of two convection boundaries.* Figure 5-10 shows node (m, n) at the intersection of two convection boundaries. Writing an energy balance equation for the volume element shown shaded, we obtain

$$k\frac{l}{2}\frac{T_{m,n-1} - T_{m,n}}{l} + kl\frac{T_{m-1,n} - T_{m,n}}{l} + kl\frac{T_{m,n+1} - T_{m,n}}{l} + k\frac{l}{2}\frac{T_{m+1,n} - T_{m,n}}{l} + hl(T_\infty - T_{m,n}) + \tfrac{3}{4}l^2 g_{m,n} = 0$$

After rearrangement,

$$T_{m,n-1} + 2T_{m-1,n} + 2T_{m,n+1} + T_{m+1,n} - \left(6 + \frac{2hl}{k}\right)T_{m,n} + \frac{3}{2}\frac{l^2}{k}g_{m,n} + \frac{2hl}{k}T_\infty = 0 \quad (5\text{-}29)$$

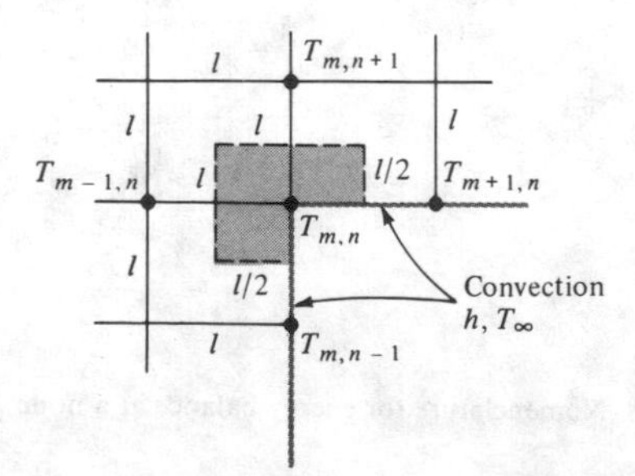

Figure 5-10 Node (m, n) at intersection of two convection boundaries.

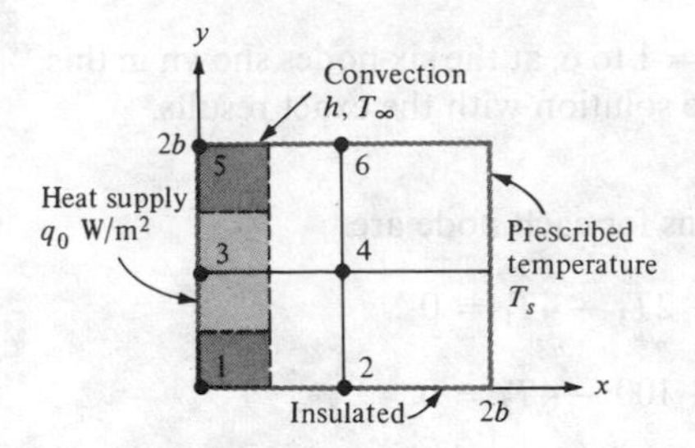

Figure 5-11 Boundary conditions for Example 5-3.

Example 5-3 Consider steady-state heat conduction in a square region of side $2b$, in which energy is generated at a constant rate of g W/m^3. The boundary conditions for the problem are shown in Fig. 5-11. Write the finite-difference equations for nodes 1, 3, and 5 in this figure.

SOLUTION By writing energy balance equations for each of the volume elements shown shaded in this figure, the finite-difference equations for these three nodes are determined:

Node 1:
$$q_0 \frac{b}{2} + k\frac{b}{2}\frac{T_3 - T_1}{b} + k\frac{b}{2}\frac{T_2 - T_1}{b} + \left(\frac{b}{2}\right)^2 g = 0$$

or
$$2T_2 + 2T_3 - 4T_1 + \frac{b^2 g}{k} + \frac{2bq_0}{k} = 0$$

Node 3:
$$q_0 b + k\frac{b}{2}\frac{T_5 - T_3}{b} + kb\frac{T_4 - T_3}{b} + k\frac{b}{2}\frac{T_1 - T_3}{b} + \frac{1}{2}b^2 g = 0$$

or
$$T_1 + 2T_4 + T_5 - 4T_3 + \frac{b^2 g}{k} + \frac{2bq_0}{k} = 0$$

Node 5:
$$q_0 \frac{b}{2} + h\frac{b}{2}(T_\infty - T_5) + k\frac{b}{2}\frac{T_6 - T_5}{b} + k\frac{b}{2}\frac{T_3 - T_5}{b} + \left(\frac{b}{2}\right)^2 g = 0$$

or
$$2T_3 + 2T_6 - \left(4 + \frac{2hb}{k}\right)T_5 + \frac{b^2 g}{k} + \frac{2hb}{k}T_\infty + \frac{2b}{k}q_0 = 0$$

Example 5-4 Consider steady-state heat conduction in a rectangular region $0 \le x \le 3b$, $0 \le y \le 2b$, subjected to the boundary conditions shown in Fig.

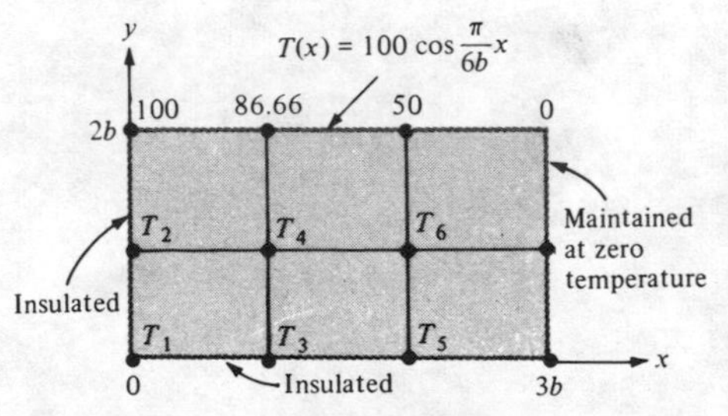

Figure 5-12 The geometry and boundary conditions for Example 5-4.

5-12. Calculate the temperatures T_m, $m = 1$ to 6, at the six nodes shown in this figure, and compare the finite-difference solution with the exact results.

SOLUTION The finite-difference equations for each node are

Node 1: $$2T_2 + 2T_3 - 4T_1 = 0$$

Node 2: $$T_1 + 2T_4 + 100 - 4T_2 = 0$$

Node 3: $$T_1 + 2T_4 + T_5 - 4T_3 = 0$$

Node 4: $$T_2 + T_3 + T_6 + 86.66 - 4T_4 = 0$$

Node 5: $$T_3 + 2T_6 - 4T_5 = 0$$

Node 6: $$T_4 + T_5 + 50 - 4T_6 = 0$$

These equations can be arranged in matrix form as follows:

$$\begin{bmatrix} -4 & 2 & 2 & 0 & 0 & 0 \\ 1 & -4 & 0 & 2 & 0 & 0 \\ 1 & 0 & -4 & 2 & 1 & 0 \\ 0 & 1 & 1 & -4 & 0 & 1 \\ 0 & 0 & 1 & 0 & -4 & 2 \\ 0 & 0 & 0 & 1 & 1 & -4 \end{bmatrix} \begin{bmatrix} T_1 \\ T_2 \\ T_3 \\ T_4 \\ T_5 \\ T_6 \end{bmatrix} = \begin{bmatrix} 0 \\ -100 \\ 0 \\ -86.66 \\ 0 \\ -50 \end{bmatrix}$$

The solution of this matrix equation gives the six node temperatures T_m, $m = 1$ to 6.

The exact analytic solution of this heat conduction problem for the temperature distribution within the region is

$$T(x, y) = 100 \frac{\cosh [\pi y/(6b)]}{\cosh [\pi/(3b)]} \cosh \left(\frac{\pi}{6b} x \right)$$

We now compare the finite-difference solution with the exact results:

T_m	Finite difference	Exact
T_1	63.6	62.4
T_2	72.2	71.1
T_3	55.1	54.0
T_4	62.5	61.7
T_5	31.8	31.2
T_6	36.1	35.6

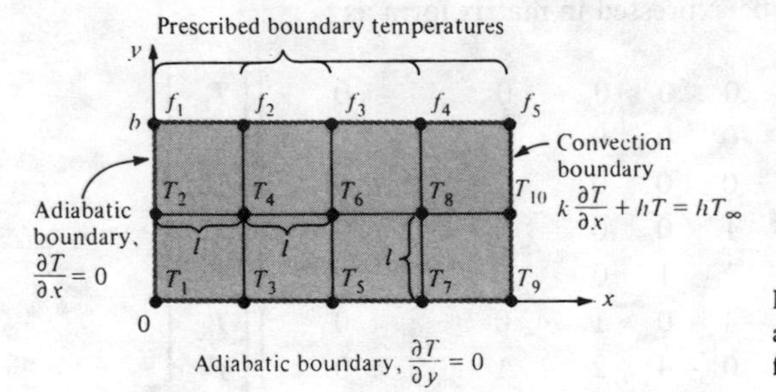

Figure 5-13 Boundary conditions and the finite-difference network for Example 5-5.

The accuracy of the finite-difference solution can be improved if a finer mesh is used.

Example 5-5 Consider a rectangular region subdivided with a square mesh of side l and subjected to the boundary conditions shown in Fig. 5-13. There is energy generation in the medium at a constant rate of g W/m^3. Write the finite-difference equations for the 10 node temperatures T_m, $m = 1$ to 10. Express these equations in matrix form.

SOLUTION The finite-difference equation for the internal nodes is immediately written by utilizing Eq. (5-25). The same equation also can be used for a node on the insulated boundary if proper adjustment is made for the symmetry requirement. An energy balance equation is readily written for a node on the convection boundary. We summarize the resulting 10 equations:

Node number	Finite-difference equation
1	$2T_2 + 2T_3 - 4T_1 + gl^2/k = 0$
2	$T_1 + f_1 + 2T_4 - 4T_2 + gl^2/k = 0$
3	$T_1 + T_5 + 2T_4 - 4T_3 + gl^2/k = 0$
4	$T_2 + T_3 + T_6 + f_2 - 4T_4 + gl^2/k = 0$
5	$T_3 + T_7 + 2T_6 - 4T_5 + gl^2/k = 0$
6	$T_4 + T_5 + T_8 + f_3 - 4T_6 + gl^2/k = 0$
7	$T_5 + T_9 + 2T_8 - 4T_7 + gl^2/k = 0$
8	$T_6 + T_{10} + T_7 + f_4 - 4T_8 + gl^2/k = 0$
9	$2T_7 + 2T_{10} - \left(4 + \frac{2hl}{k}\right)T_9 + \left(\frac{2hlT_\infty}{k} + \frac{gl^2}{k}\right) = 0$
10	$2T_8 + T_9 + f_5 - \left(4 + \frac{2hl}{k}\right)T_{10} + \left(\frac{2hlT_\infty}{k} + \frac{gl^2}{k}\right) = 0$

These 10 equations can be expressed in matrix form as

$$\begin{bmatrix} -4 & 2 & 2 & 0 & 0 & 0 & 0 & 0 & 0 & 0 \\ 1 & -4 & 0 & 2 & 0 & 0 & 0 & 0 & 0 & 0 \\ 1 & 0 & -4 & 2 & 1 & 0 & 0 & 0 & 0 & 0 \\ 0 & 1 & 1 & -4 & 0 & 1 & 0 & 0 & 0 & 0 \\ 0 & 0 & 1 & 0 & -4 & 2 & 1 & 0 & 0 & 0 \\ 0 & 0 & 0 & 1 & 1 & -4 & 0 & 1 & 0 & 0 \\ 0 & 0 & 0 & 0 & 1 & 0 & -4 & 2 & 1 & 0 \\ 0 & 0 & 0 & 0 & 0 & 1 & 1 & -4 & 0 & 1 \\ 0 & 0 & 0 & 0 & 0 & 0 & 2 & 0 & -(4+H) & 2 \\ 0 & 0 & 0 & 0 & 0 & 0 & 0 & 2 & 1 & -(4+H) \end{bmatrix} \begin{bmatrix} T_1 \\ T_2 \\ T_3 \\ T_4 \\ T_5 \\ T_6 \\ T_7 \\ T_8 \\ T_9 \\ T_{10} \end{bmatrix} = \begin{bmatrix} -G \\ -G - f_1 \\ -G \\ -G - f_2 \\ -G \\ -G - f_3 \\ -G \\ -G - f_4 \\ -G - HT_\infty \\ -G - HT_\infty - f_5 \end{bmatrix}$$

where we have defined

$$G \equiv \frac{gl^2}{k} \qquad \text{and} \qquad H \equiv \frac{2hl}{k}$$

Once the numerical values of the parameters G and H are specified, these equations can be solved for the node temperatures.

It is now apparent that in the finite-difference approach, the heat conduction equation is replaced by a set of simultaneous algebraic equations for temperatures at the nodal points of a network constructed over the region. These equations can be solved readily with a high-speed digital computer. The larger the number of nodes, the closer the finite-difference approximation to the exact results; but larger number of nodes requires more equations, hence more computer time and may introduce some roundoff errors. However, in most practical applications, it may not be necessary to choose a large number of nodes. The results obtained with a small number of nodes may be sufficiently accurate in view of the uncertainties in the determination of thermal properties of the material and the heat transfer coefficient at the boundaries.

5-4 METHODS OF SOLVING SIMULTANEOUS ALGEBRAIC EQUATIONS

The analytic solution of simultaneous algebraic equations of order greater than 3 is quite tedious, and it is quite prohibitive for orders such as those encountered in the finite-difference analysis of heat conduction problems. Therefore, various numerical schemes have been developed to solve simultaneous algebraic equations by a digital computer. Table 5-1 shows a computer program calling a computer subroutine that utilizes the *gaussian elimination* method: Here we discuss the basic features of the gaussian elimination method as well as other methods utilizing an iterative scheme and the matrix inversion.

Gaussian Elimination Method

The finite-difference formulation of heat conduction problems, as discussed previously, leads to a system of algebraic equations that form a banded matrix. The gaussian elimination method is very efficient for solving such systems of algebraic equations. In this approach, the matrix is transformed to an upper diagonal form.

To illustrate the basis of the gaussian elimination process, we consider a system of algebraic equations forming a banded matrix as shown in Eqs. (5-30).

$$\begin{bmatrix} a_{11} & a_{12} & a_{13} & 0 & \cdots & 0 & 0 \\ a_{21} & a_{22} & a_{23} & a_{24} & \cdots & 0 & 0 \\ a_{31} & a_{32} & a_{33} & a_{34} & \cdots & \cdots & 0 \\ 0 & a_{42} & a_{43} & a_{44} & \cdots & \cdots & 0 \\ \cdots & \cdots & \cdots & \cdots & \cdots & \cdots & \cdots \\ 0 & 0 & 0 & \cdots & a_{n,n-2} & a_{n,n-1} & a_{m,n} \end{bmatrix} \begin{bmatrix} T_1 \\ T_2 \\ T_3 \\ T_4 \\ \vdots \\ T_n \end{bmatrix} = \begin{bmatrix} C_1 \\ C_2 \\ C_3 \\ C_4 \\ \vdots \\ C_n \end{bmatrix} \tag{5-30}$$

The matrix in this system can be transformed to upper diagonal form. The first equation is used to eliminate the nonzero elements a_{21} and a_{31} in the first column. That is, the first equation is multiplied by a_{21}/a_{11}, and the resulting equation is subtracted from the second equation in order to eliminate a_{21}; then the first equation is utilized in a similar manner to eliminate a_{31}. Next the second equation is used to eliminate a_{32} and a_{42}. The third equation is used to eliminate a_{43}, and so forth. When this process is carried out to the last equation, the resulting system forms an upper diagonal matrix, as shown in Eqs. (5-31).

$$\begin{bmatrix} a_{11}^* & a_{12}^* & a_{13}^* & 0 & \cdots & \cdots & \cdots & 0 \\ 0 & a_{22}^* & a_{23}^* & a_{24}^* & \cdots & \cdots & \cdots & 0 \\ 0 & 0 & a_{33}^* & a_{34}^* & \cdots & \cdots & \cdots & \cdots \\ 0 & 0 & 0 & a_{44}^* & \cdots & \cdots & \cdots & \cdots \\ \cdots & \cdots & \cdots & \cdots & \cdots & \cdots & \cdots & \cdots \\ 0 & \cdots & \cdots & \cdots & 0 & 0 & a_{n-1,n-1}^* & a_{n-1,n}^* \\ 0 & 0 & 0 & \cdots & 0 & 0 & 0 & a_{n,n}^* \end{bmatrix} \begin{bmatrix} T_1 \\ T_2 \\ T_3 \\ T_4 \\ \vdots \\ T_{n-1} \\ T_n \end{bmatrix} = \begin{bmatrix} C_1^* \\ C_2^* \\ C_3^* \\ C_4^* \\ \vdots \\ C_{n-1}^* \\ C_n^* \end{bmatrix} \tag{5-31}$$

Once the upper diagonal form is obtained, the solution follows immediately. That is, the last equation in the system (5-31) immediately gives T_n. Given T_n, the $(n-1)$st equation gives T_{n-1}; and given T_{n-1}, the $(n-2)$d equation gives T_{n-2}; and so forth. The calculations are carried out until T_1 is determined from the first equation.

When the number of equations to be solved is very large, the roundoff errors can accumulate and sometimes cause a deterioration in accuracy. Rearranging the equations so that the coefficients which are largest in magnitude are put on the main diagonal may improve the accuracy.

Various computer subroutines are available for solving simultaneous linear algebraic equations by the gaussian elimination process or by the methods which are some variation of it. The International Mathematical and Statistical Libraries [18] gives numerous such subroutines. For example, LEQT1F and LEQT1B are typical subroutines for solving systems of algebraic equations, while LEQT2F and LEQT2B are the high-accuracy solution versions of these subroutines, respectively.

Gauss-Seidel Iteration

When the number of equations is very large, the matrix is not sparse, and the computer storage is critical, an iterative technique frequently is preferred for the solution of such a system. The Gauss-Seidel iteration (often called the Liebmann iteration) is one of the most efficient procedures for solving large systems of equations. The procedure involves the following steps: (1) Solve each equation for one of the unknowns. (2) Make initial guesses for all unknowns, and compute the unknowns from the equations developed in step 1 (use the most recently computed values for the unknowns in each equation). (3) Repeat the procedure until a specified convergence criterion is satisfied.

To illustrate the application of the Gauss-Seidel iteration process, we consider the following three simultaneous algebraic equations for the three unknowns T_1, T_2, and T_3:

$$a_{11}T_1 + a_{12}T_2 + a_{13}T_3 = f_1 \tag{5-32a}$$

$$a_{21}T_1 + a_{22}T_2 + a_{23}T_3 = f_2 \tag{5-32b}$$

$$a_{31}T_1 + a_{32}T_2 + a_{33}T_3 = f_3 \tag{5-32c}$$

The first equation is solved for T_1, the second for T_2 and the third for T_3:

$$T_1 = \frac{f_1 - a_{12}T_2 - a_{13}T_3}{a_{11}} \tag{5-33a}$$

$$T_2 = \frac{f_2 - a_{21}T_1 - a_{23}T_3}{a_{22}} \tag{5-33b}$$

$$T_3 = \frac{f_3 - a_{31}T_1 - a_{32}T_2}{a_{33}} \tag{5-33c}$$

Let $T_1^{(0)}$, $T_2^{(0)}$, and $T_3^{(0)}$ be the initial guesses for the temperatures. These values are introduced into Eq. (5-33*a*) to obtain a first approximation for T_1 as

$$T_1^{(1)} = \frac{f_1 - a_{12}T_2^{(0)} - a_{13}T_3^{(0)}}{a_{11}} \tag{5-34a}$$

This most recent value of $T_1^{(1)}$ and the initial guesses are introduced into Eq. (5-33*b*) to obtain $T_2^{(1)}$:

$$T_2^{(1)} = \frac{f_2 - a_{21}T_1^{(1)} - a_{23}T_3^{(0)}}{a_{22}} \tag{5-34b}$$

By using the initial guess or the most recent values of temperature in Eq. (5-33*c*), a first approximation $T_3^{(1)}$ is obtained for temperature T_3.

The above procedure is repeated until a specified convergence criterion is satisfied. The convergence criterion may be specified as

$$|T_i^{(n+1)} - T_i^{(n)}| \leq \varepsilon \tag{5-35}$$

which should be satisfied for all T_i.

Another convergence criterion may be specified in the form

$$\left| \frac{T_i^{(n+1)} - T_i^{(n)}}{T_i^{(n+1)}} \right| \leq \varepsilon \tag{5-36}$$

which should be satisfied for all T_i.

The convergence criterion given by Eq. (5-36) is the safest when the magnitudes of the T_i cannot be guessed beforehand, but the testing process requires more computer time than checking the criterion (5-35). If the approximate magnitudes of the T_i are known beforehand, the convergence criterion (5-35) is useful.

Whether the solution of a system of algebraic equations with an iterative scheme converges or diverges, does not depend on the initial guess for the unknowns; it depends on the character of the coefficient matrix. The system of equations in which the diagonal elements are the largest elements in each row is most suitable for the iterative method of solution. For the example given by Eqs. (5-32), the diagonal elements a_{jj} should be the largest elements in their respective rows. Furthermore, the convergence cannot be guaranteed unless for each row i the following criterion is satisfied:

$$|a_{ii}| > \sum_{j=1,\, i \neq j}^{N} |a_{ij}| \tag{5-37}$$

where N is the total number of equations in the system.

We illustrated the application of the iterative method of solution with a specific example given by Eqs. (5-32). We now illustrate its application to the solution of a more general system of finite-difference equations (5-27). We solve Eqs. (5-27) for the temperature T_p

$$T_p = \frac{\Delta V_p g_p + \sum_j (T_j/R_{jp})}{\sum_j (1/R_{jp})} \tag{5-38}$$

where $p = 1, 2, \ldots, N$ and N is the total number of nodes in the region.

The general procedure for iterative method of solving Eq. (5-38) is as follows:

1. An initial guess is made for the node temperature $T_p^{(0)}, p = 1, 2, \ldots, N$.
2. A first approximation for the node temperatures $T_p^{(1)}$ is calculated, by introducing the initial guess values or, if available, the most recent values of the node temperatures on the right-hand side of Eq. (5-38).
3. The procedure is repeated until a convergence criterion, such as that given by Eq. (5-35) or (5-36), is satisfied.

Matrix Inversion

The *matrix inversion* is another method that can be used for solving simultaneous algebraic equations. This method, like the gaussian elimination method, provides an exact solution to the unknowns, instead of approaching the exact results gradually by iteration, as in the Gauss-Seidel iteration. Consider, for example, the matrix equations (5-30) written compactly in the form

$$[A][T] = [B] \tag{5-39}$$

where $[A]$ is the coefficient matrix, $[T]$ is the vector for the unknown temperatures, and $[B]$ is the known coefficient vector whose elements involve contributions due to energy generation and/or the boundary conditions for the problem. The solution of the problem by matrix inversion is given as

$$[T] = [A]^{-1}[B] \tag{5-40}$$

where $[A]^{-1}$ is the inverse of matrix $[A]$. The matrix inversion can be performed readily by using the standard computer subroutine for matrix inversion.

Accuracy of Finite-Difference Solution

In the numerical examples in this chapter, we assessed the accuracy of the finite-difference solution by comparing it with the exact analytic solutions. Therefore, if it is available, the numerical solution should be compared with an analytical solution for a similar problem. We stated that performing the calculations with smaller values of mesh size would improve the accuracy of the finite-difference solution. However, there is a limit to the use of very small mesh size, because an increase in the number of nodes increases the number of machine calculations, which in turn increases the computational roundoff errors. Therefore, one should start the finite-difference calculations with a coarse mesh, gradually refining the mesh size and observing the convergence of the solution before choosing the mesh size to perform the final computations.

Finer mesh size is needed in the regions where the boundary conditions exhibit steep temperature variations and where heat flow rates are to be determined. An accurate determination of the heat flow rate requires the use of finer mesh size.

5-5 CYLINDRICAL AND SPHERICAL SYMMETRY

The formal representation of the finite-difference formulation given by Eq. (5-27) is now applied to develop the specific relations for the finite difference of the one-dimensional, steady-state heat conduction in cylindrical and spherical coordinates.

Cylindrical Symmetry

Consider radial heat conduction in a long, solid cylinder of radius $r = b$ in which energy is generated at a rate of $g(r)$ W/m^3. The region $0 \leq r \leq b$ is divided into M cylindrical subregions, each of thickness

$$\Delta r = \frac{b}{M}$$

as illustrated in Fig. 5-14. The application of Eq. (5-27) for node m gives

$$\frac{T_{m-1} - T_m}{R_{m-1,m}} + \frac{T_{m+1} - T_m}{R_{m+1,m}} + \Delta V_m g_m = 0 \tag{5-41a}$$

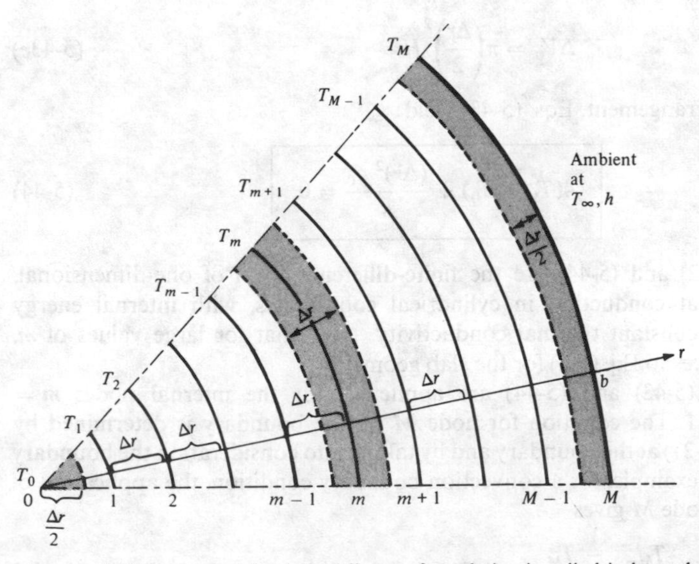

Figure 5-14 Nomenclature for finite-difference formulation in cylindrical or spherical symmetry.

where

$$R_{m-1,m} = \frac{\Delta r}{2\pi(m\,\Delta r - \frac{1}{2}\,\Delta r)Hk} = \frac{1}{2\pi m[1 - 1/(2m)]Hk} \tag{5-41b}$$

$$R_{m+1,m} = \frac{\Delta r}{2\pi(m\,\Delta r + \frac{1}{2}\,\Delta r)Hk} = \frac{1}{2\pi m[1 + 1/(2m)]Hk} \tag{5-41c}$$

$$\Delta V_m = 2\pi m\,\Delta r\,\Delta r\,H \tag{5-41d}$$

H = the length of the cylinder

Equations (5-41) can be rearranged as

$$\left(1 - \frac{1}{2m}\right)T_{m-1} - 2T_m + \left(1 + \frac{1}{2m}\right)T_{m+1} + \frac{(\Delta r)^2\,g_m}{k} = 0 \qquad \text{for } m = 1, 2, \ldots, M-1 \tag{5-42}$$

At the center $r = 0$, Eq. (5-27) yields

$$\frac{T_1 - T_0}{R_{1,0}} + \Delta V_0\,g_0 = 0 \tag{5-43a}$$

where

$$R_{1,0} = \frac{\Delta r}{2\pi(\Delta r/2)\,Hk} = \frac{1}{\pi Hk} \tag{5-43b}$$

$$\Delta V_0 = \pi\left(\frac{\Delta r}{2}\right)^2 H \tag{5-43c}$$

After some rearrangement, Eqs. (5-43) yield

$$4(T_1 - T_0) + \frac{(\Delta r)^2\,g_0}{k} = 0 \tag{5-44}$$

Equations (5-42) and (5-44) are the finite-difference form of one-dimensional, steady-state heat conduction in cylindrical coordinates, with internal energy generation for constant thermal conductivity. Note that for large values of m, Eq. (5-41) reduces to Eq. (5-5) for the slab geometry.

Equations (5-42) and (5-44) are applicable for the internal nodes $m = 0, 1, \ldots, M-1$. The equation for node M on the boundary is determined by applying Eq. (5-27) at the boundary and by taking into consideration the boundary condition. For example, for a convection boundary condition, the application of Eq. (5-27) for node M gives

$$\frac{T_{M-1} - T_M}{R_{M-1,M}} + Ah(T_\infty - T_M) + \Delta V_M\,g_M = 0 \tag{5-45a}$$

where

$$R_{M-1,M} = \frac{\Delta r}{2\pi(M\,\Delta r - \frac{1}{2}\,\Delta r)Hk} = \frac{1}{2\pi M[1 - 1/(2M)]Hk} \tag{5-45b}$$

$$\Delta V_M \cong (2\pi M\,\Delta r)\frac{\Delta r}{2}H \tag{5-45c}$$

$$A = (2\pi M\,\Delta r)H$$

After rearranging the terms in Eq. (5-45), we obtain

$$\boxed{\left(1 - \frac{1}{2M}\right)T_{M-1} - \left[\left(1 - \frac{1}{2M}\right) + \frac{\Delta r\,h}{k}\right]T_M + \frac{\Delta r\,h}{k}\,T_\infty + \frac{(\Delta r)^2 g_M}{2k} = 0} \tag{5-46}$$

which is the finite-difference equation at the boundary node M which is subjected to convection with an ambient at temperature T_∞, with a heat transfer coefficient h. For large values of M, Eq. (5-46) reduces to Eq. (5-21) for the slab geometry.

Example 5-6 A long, cylindrical fuel element of radius $b = 1$ cm and thermal conductivity $k = 25$ W/(m · °C) generates energy at a constant rate of $g = 5 \times 10^8$ W/m^3. The boundary surface at $r = b$ can be assumed to be maintained at 100°C. Assuming one-dimensional, radial heat flow, calculate the radial temperature distribution in the fuel element by using finite differences. Compare the results with the exact solution of the problem.

SOLUTION The region $0 \le r \le b$ is divided into $M = 10$ cylindrical subregions, each of thickness

$$\Delta r = \frac{0.01}{10} = 1 \times 10^{-3} \text{ m}$$

The finite-difference equation for the center node $m = 0$ is obtained from Eq. (5-44):

$$4(T_1 - T_0) + \frac{(\Delta r)^2 g}{k} = 0 \qquad \text{for } m = 0$$

The equations for the remaining internal nodes are obtained from Eq. (5-42) as

$$\left(1 - \frac{1}{2m}\right)T_{m-1} - 2T_m + \left(1 + \frac{1}{2m}\right)T_{m+1} + \frac{(\Delta r)^2 g}{k} = 0 \qquad \text{for } m = 1 \text{ to } 9$$

Since the temperature for the boundary surface has been specified, for node $M = 10$ we have

$$T_{10} = 100°\text{C}$$

Thus the above system provides 10 simultaneous algebraic equations for the 10 unknown temperatures T_m, $m = 0$ to 9.

The energy generation term in these equations is evaluated as

$$\frac{(\Delta r)^2 g}{k} = \frac{(0.001)^2(5 \times 10^8)}{25} = 20$$

The resulting system of equations is given in matrix form as

$$\begin{bmatrix} -4 & 4 & 0 & 0 & 0 & 0 & 0 & 0 & 0 & 0 \\ 0.5 & -2 & 1.5 & 0 & 0 & 0 & 0 & 0 & 0 & 0 \\ 0 & 0.75 & -2 & 1.25 & 0 & 0 & 0 & 0 & 0 & 0 \\ 0 & 0 & 0.833 & -2 & 1.167 & 0 & 0 & 0 & 0 & 0 \\ 0 & 0 & 0 & 0.875 & -2 & 1.125 & 0 & 0 & 0 & 0 \\ 0 & 0 & 0 & 0 & 0.9 & -2 & 1.1 & 0 & 0 & 0 \\ 0 & 0 & 0 & 0 & 0 & 0.917 & -2 & 1.083 & 0 & 0 \\ 0 & 0 & 0 & 0 & 0 & 0 & 0.929 & -2 & 1.071 & 0 \\ 0 & 0 & 0 & 0 & 0 & 0 & 0 & 0.938 & -2 & 1.062 \\ 0 & 0 & 0 & 0 & 0 & 0 & 0 & 0 & 0.944 & -2 \end{bmatrix} \begin{bmatrix} T_0 \\ T_1 \\ T_2 \\ T_3 \\ T_4 \\ T_5 \\ T_6 \\ T_7 \\ T_8 \\ T_9 \end{bmatrix} = \begin{bmatrix} -20 \\ -20 \\ -20 \\ -20 \\ -20 \\ -20 \\ -20 \\ -20 \\ -20 \\ -125.6 \end{bmatrix}$$

This system of equations is readily solved by the gaussian elimination procedure.

The exact solution of this problem is given by

$$T(r) = 100 + 500\left[1 - \left(\frac{r}{b}\right)^2\right]$$

Here we compare the finite-difference solution with the exact results:

T_m	Finite difference, °C	Exact, °C
T_0	600.27	600.00
T_1	595.27	595.00
T_2	580.27	580.00
T_3	555.27	555.00
T_4	520.28	520.00
T_5	475.30	475.00
T_6	420.31	420.00
T_7	355.28	355.00
T_8	280.20	280.00
T_9	195.05	195.00

The finite-difference results obtained with the 10 subdivisions of the region are sufficiently accurate. The finite-difference calculations have also been tried with $M = 5$ subdivisions, but the results were not so accurate.

Spherical Symmetry

We consider radial heat conduction in a sphere of radius $r = b$ in which energy is generated at a rate of $g(r)$ W/m^3. The region $0 \le r \le b$ is divided into M spherical subregions, each of thickness

$$\Delta r = \frac{b}{M}$$

as illustrated in Fig. 5-14. The application of Eq. (5-27) for node m gives

$$\frac{T_{m-1} - T_m}{R_{m-1,m}} + \frac{T_{m+1} - T_m}{R_{m+1,m}} + \Delta v_m \, g_m = 0 \tag{5-47a}$$

where

$$R_{m-1,m} = \frac{\Delta r}{4\pi(m\,\Delta r - \Delta r/2)^2 k} = \frac{1}{4\pi m^2\,\Delta r[1 - 1/(2m)]^2 k} \tag{5-47b}$$

$$R_{m+1,m} = \frac{\Delta r}{4\pi(m\,\Delta r + \Delta r/2)^2 k} = \frac{1}{4\pi m^2\,\Delta r[1 + 1/(2m)]^2 k} \tag{5-47c}$$

$$\Delta v_m \cong 4\pi(m\,\Delta r)^2\,\Delta r \tag{5-47d}$$

Equations (5-47) can be rearranged as

$$\boxed{\left(1 - \frac{1}{2m}\right)^2 (T_{m-1} - T_m) + \left(1 + \frac{1}{2m}\right)^2 (T_{m+1} - T_m) + \frac{(\Delta r)^2\, g_m}{k} = 0} \tag{5-48}$$

The following approximation can be made:

$$\left(1 - \frac{1}{2m}\right)^2 \cong 1 - \frac{1}{m} \qquad \text{and} \qquad \left(1 + \frac{1}{2m}\right)^2 \cong 1 + \frac{1}{m}$$

Then Eq. (5-48) reduces to

$$\boxed{\left(1 - \frac{1}{m}\right) T_{m-1} - 2T_m + \left(1 + \frac{1}{m}\right) T_{m+1} + \frac{(\Delta r)^2\, g_m}{k} = 0} \tag{5-49}$$

This equation is similar to Eq. (5.42) for a cylinder, except $1/2m$ is replaced by $1/m$.

At the center $r = 0$, Eq. (5-27) yields

$$\frac{T_1 - T_0}{R_{1,0}} + \Delta V_0\, g_0 = 0 \tag{5-50a}$$

where

$$R_{1,0} = \frac{\Delta r}{4\pi(\Delta r/2)^2 k} = \frac{1}{\pi\,\Delta r\,k} \tag{5-50b}$$

$$\Delta V_0 = \tfrac{4}{3}\pi\left(\frac{\Delta r}{2}\right)^3 = \tfrac{1}{6}\pi\,(\Delta r)^3 \tag{5-50c}$$

After rearrangement, Eqs. (5-50) reduce to

$$6(T_1 - T_0) + \frac{(\Delta r)^2 g_0}{k} = 0 \tag{5-51}$$

Equations (5-49) and (5-51) are the finite-difference form of one-dimensional, steady-state heat conduction in spherical coordinates, with internal energy generation and constant thermal conductivity. We note that for large values of m, Eq. (5-49) reduces to Eq. (5-5) for the slab geometry.

These equations are applicable for all the internal nodes. The equation for node M on the boundary is determined by applying Eq. (5-27) at the boundary and taking into consideration the boundary condition. For a convection boundary condition at the surface, the application of Eq. (5-27) for node M gives

$$\frac{T_{M-1} - T_M}{R_{M-1,M}} + Ah(T_\infty - T_M) + \Delta V_M\, g_M = 0 \tag{5-52a}$$

where

$$R_{M-1,M} = \frac{\Delta r}{4\pi(M\,\Delta r - \Delta r/2)^2 k} = \frac{1}{4\pi M^2\,\Delta r[1 - 1/(2M)]^2 k} \tag{5-52b}$$

$$A = 4\pi(M\,\Delta r)^2 \tag{5-52c}$$

$$\Delta V_m = 4\pi(M\,\Delta r)^2\,\frac{\Delta r}{2} \tag{5-52d}$$

After rearrangement, Eqs. (5-52) yield

$$\left(1 - \frac{1}{2M}\right)^2 T_{M-1} - \left[\left(1 - \frac{1}{2M}\right)^2 + \frac{h\,\Delta r}{k}\right] T_M + \frac{h\,\Delta r}{k}\, T_\infty + \frac{(\Delta r)^2 g_M}{2k} = 0 \tag{5-53}$$

or this equation can be linearized as

$$\left(1 - \frac{1}{M}\right) T_{M-1} - \left[\left(1 - \frac{1}{M}\right) + \frac{h\,\Delta r}{k}\right] T_M + \frac{h\,\Delta r}{k}\, T_\infty + \frac{(\Delta r)^2\, g_M}{2k} = 0 \tag{5-54}$$

which is applicable for node M at the boundary. This equation is similar to Eq. (5.46), except $1/2M$ is replaced by $1/M$.

Example 5-7 A spherical fuel element of radius $b = 1$ cm and thermal conductivity $k = 25$ W/(m · °C) generates energy at a constant rate of $g = 7.5 \times 10^8$ W/m^3. The boundary surface at $r = b$ is maintained at 100°C. Calculate the radial, steady-state temperature distribution in the sphere by using finite differences, and compare the results with the exact solution.

SOLUTION The region $0 \le r \le b$ is divided into $M = 10$ spherical shells, each of thickness

$$\Delta r = \frac{0.01}{10} = 10^{-3}\ \text{m}$$

The finite-difference equations for various nodes are determined as follows: At the center node $m = 0$, Eq. (5-51) gives

$$6(T_1 - T_0) + \frac{(\Delta r)^2 g_0}{k} = 0$$

For the internal nodes $m = 1$ to 9, Eq. (5-49) is applicable. That is,

$$\left(1 - \frac{1}{m}\right)T_{m-1} - 2T_m + \left(1 + \frac{1}{m}\right)T_{m+1} + \frac{(\Delta r)^2 g_m}{k} = 0$$

for $m = 1$ to 9. Finally, node $M = 10$ on the boundary surface has a specified temperature of 100°C; that is,

$$T_{10} = 100°\text{C}$$

This system provides 10 simultaneous algebraic equations for the unknown node temperatures T_m, $m = 0$ to 9.

The energy generation term is evaluated as

$$\frac{(\Delta r)^2 g}{k} = \frac{(0.001)^2(7.5 \times 10^8)}{25} = 30$$

This system of equations is written in matrix form as

$$\begin{bmatrix} -6 & 6 & 0 & 0 & 0 & 0 & 0 & 0 & 0 & 0 \\ 0 & -2 & 2 & 0 & 0 & 0 & 0 & 0 & 0 & 0 \\ 0 & 0.5 & -2 & 1.5 & 0 & 0 & 0 & 0 & 0 & 0 \\ 0 & 0 & 0.667 & -2 & 1.333 & 0 & 0 & 0 & 0 & 0 \\ 0 & 0 & 0 & 0.75 & -2 & 1.25 & 0 & 0 & 0 & 0 \\ 0 & 0 & 0 & 0 & 0.8 & -2 & 1.2 & 0 & 0 & 0 \\ 0 & 0 & 0 & 0 & 0 & 0.833 & -2 & 1.167 & 0 & 0 \\ 0 & 0 & 0 & 0 & 0 & 0 & 0.857 & -2 & 1.143 & 0 \\ 0 & 0 & 0 & 0 & 0 & 0 & 0 & 0.875 & -2 & 1.125 \\ 0 & 0 & 0 & 0 & 0 & 0 & 0 & 0 & 0.889 & -2 \end{bmatrix} \begin{bmatrix} T_0 \\ T_1 \\ T_2 \\ T_3 \\ T_4 \\ T_5 \\ T_6 \\ T_7 \\ T_8 \\ T_9 \end{bmatrix} = \begin{bmatrix} -30 \\ -30 \\ -30 \\ -30 \\ -30 \\ -30 \\ -30 \\ -30 \\ -30 \\ -141.1 \end{bmatrix}$$

The above system of equations is solved by the gaussian elimination procedure.

The exact solution of this problem is given by

$$T(r) = 100 + 500\left[1 - \left(\frac{r}{b}\right)^2\right]$$

Note that for this problem the magnitude of the generation term was chosen so as to make the temperature distribution in the spherical region exactly the same as that for the cylinder considered in Example 5-6.

We compare the finite-difference solution with the exact results. The agreement is very good.

T_m	Finite difference, °C	Exact, °C
T_0	599.9	600.0
T_1	594.9	595.0
T_2	579.9	580.0
T_3	554.9	555.0
T_4	519.9	520.0
T_5	474.9	475.0
T_6	419.9	420.0
T_7	354.9	355.0
T_8	280.0	280.0
T_9	195.0	195.0

5-6 UNSTEADY HEAT CONDUCTION—EXPLICIT METHOD

Unsteady heat conduction problems can be solved numerically by transforming the partial differential equation of heat conduction to finite-difference equations in both space and time domains.

Various schemes have been developed for the finite-difference representation of the time-dependent heat conduction equation. Here we present the *explicit method*, which is very easy to apply and leads to a set of uncoupled algebraic equations which are very easy to solve.

One-Dimensional Heat Conduction

Consider a one-dimensional, time-dependent heat conduction equation confined to the domain $0 < x < L$:

$$\frac{\partial T(x, t)}{\partial t} = \alpha \frac{\partial^2 T}{\partial x^2} \tag{5-55}$$

To obtain the finite-difference form of this differential equation, the x and t domains are divided into small steps Δx and Δt, as illustrated in Fig. 5-15, so that

$$x = m\,\Delta x \qquad m = 0, 1, 2, \ldots, M \tag{5-56a}$$

$$t = i\,\Delta t \qquad i = 0, 1, 2, \ldots \tag{5-56b}$$

where

$$M = \frac{L}{\Delta x} \tag{5-56c}$$

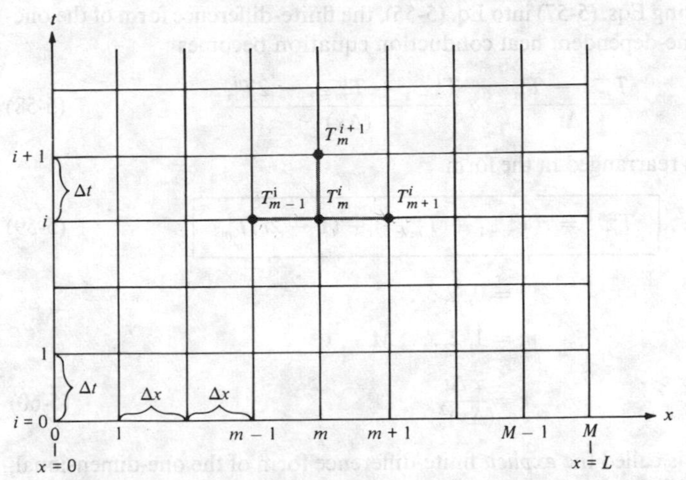

Figure 5-15 Subdivision of the *xt* domain into intervals of Δx and Δt for finite-difference representation of the one-dimensional, time-dependent heat conduction equation.

Then the temperature $T(x, t)$ at a location x and time t is denoted by the symbol T_m^i; that is,

$$T(x, t) = T(m\,\Delta x, i\,\Delta t) \equiv T_m^i \tag{5.56d}$$

By using this formalism, various derivatives appearing in the heat conduction equation (5-55) are represented in finite differences.

The second derivative of temperature with respect to x, at a position $m\,\Delta x$ and at a time $i\,\Delta t$, is represented in the finite-difference form as described previously [see Eq. (5-4)]. We find

$$\left.\frac{\partial^2 T}{\partial x^2}\right|_{m,i} \cong \frac{T_{m-1}^i + T_{m+1}^i - 2T_m^i}{(\Delta x)^2} \tag{5-57a}$$

where T_{m-1}^i and T_{m+1}^i are the two neighboring points of the node T_m^i, and all of which are evaluated at the time $i\,\Delta t$.

The first derivative of temperature with respect to the time variable t at a position $n\,\Delta x$ and at a time $i\,\Delta t$ is represented by

$$\left.\frac{\partial T}{\partial t}\right|_{m,i} \cong \frac{T_m^{i+1} - T_m^i}{\Delta t} \tag{5-57b}$$

where T_m^{i+1} is the temperature at the location $m\,\Delta x$ at the time $(i + 1)\,\Delta t$. The numerator in Eq. (5-57*b*) is a *forward* finite-difference formulation of the change in temperature at node m from $i\,\Delta t$ to $(i + 1)\,\Delta t$.

By introducing Eqs. (5-57) into Eq. (5-55), the finite-difference form of the one-dimensional, time-dependent heat conduction equation becomes

$$\frac{T_m^{i+1} - T_m^i}{\Delta t} = \alpha \frac{T_{m-1}^i + T_{m+1}^i - 2T_m^i}{(\Delta x)^2} \tag{5-58}$$

This equation is rearranged in the form

$$\boxed{T_m^{i+1} = r(T_{m-1}^i + T_{m+1}^i) + (1 - 2r)T_m^i} \tag{5-59}$$

where

$$i = 0, 1, \ldots$$

$$m = 1, 2, \ldots, M - 1$$

$$r = \frac{\alpha\,\Delta t}{(\Delta x)^2} \tag{5-60}$$

Equation (5-59) is called the *explicit* finite-difference form of the one-dimensional, time-dependent heat conduction equation (5-55).

The method is called explicit because the temperature T_m^{i+1} at node m at a time step $i + 1$ is immediately determined from Eq. (5-59) if the temperatures of node m and its neighboring points at the previous time step i are available and the value of the parameter r is specified.

The procedure is very simple. Starting with $i = 0$, which corresponds to the initial condition, we calculate the temperatures at each subsequent step according to Eq. (5-59) and continue the procedure until the specified time. The only disadvantage of this method arises from the restriction on the permissible maximum value of r, which in turn imposes a restriction on the permissible maximum value of the time step Δt. If computations are to be carried out over a long time, a very large number of calculations are needed with very small Δt. We briefly discuss the restriction imposed on the value of r.

Restriction on *r* The value of the parameter r in the explicit finite-difference formula Eq. (5-59) is restricted to

$$\boxed{0 < r \le \tfrac{1}{2}} \tag{5-61a}$$

where

$$r = \frac{\alpha\,\Delta t}{(\Delta x)^2} \tag{5-61b}$$

This restriction implies that for given values of α and Δx, the time step Δt cannot exceed the limit imposed on it by Eqs. (5-61). This is called the *stability criterion.* If this stability criterion is violated, numerical calculations become unstable. Figure 5-16 illustrates what happens to the numerical results if the above stability criterion is violated. This figure shows that the numerical results obtained with the time step satisfying the condition $r = \frac{5}{11} < \frac{1}{2}$ are in good agreement with the exact solution. However, the numerical solution of the same problem with a slightly

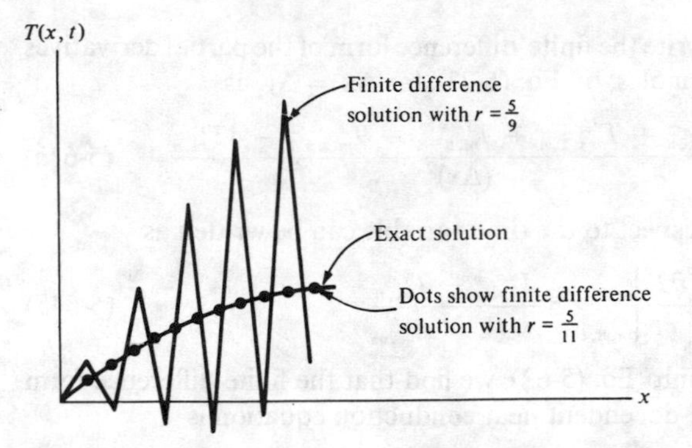

Figure 5-16 Effects of parameter $r = \alpha \, \Delta t/(\Delta x)^2$ on stability of finite-difference solution.

larger time step, which violates the above stability criterion of $r = \frac{5}{9} > \frac{1}{2}$ results in an unstable solution.

There are various mathematical techniques for the determination of the stability criterion associated with the finite-difference representation of the time-dependent heat conduction equation. It is instructive to determine this stability criterion by the following physical argument.

Suppose at any time step i the temperatures T^i_{m-1} and T^i_{m+1} at nodes $m - 1$ and $m + 1$ are equal but less than T^i_m at node m between them. Then if the value of r exceeds $\frac{1}{2}$, the coefficient $1 - 2r$ becomes negative. Then, according to the finite-difference equation (5-59), for $1 - 2r$ negative, the temperature T^{i+1}_m at node m at the next time step should be less than that at the neighboring two nodes. This is not possible thermodynamically, since we assumed that T^i_m was higher than that at the neighboring nodes. Therefore, to obtain meaningful solutions from Eq. (5-59), the coefficient $1 - 2r$ of T^i_m *should not be negative*; that is,

$$1 - 2r \geq 0 \qquad \text{or} \qquad r \leq \tfrac{1}{2}$$

which is the stability criterion given by Eq. (5-61).

Two-Dimensional Heat Conduction

We now generalize the finite-difference procedure to the finite-difference representation of the following two-dimensional, time-dependent heat conduction equation:

$$\frac{\partial T}{\partial t} = \alpha\left(\frac{\partial^2 T}{\partial x^2} + \frac{\partial^2 T}{\partial y^2}\right) \tag{5-62}$$

Suppose this equation is to be solved over a finite region subject to a prescribed temperature at all boundary surfaces.

First, the region is divided into square subregions by a network of mesh size $\Delta x = \Delta y$, and the time domain is divided into small time steps Δt. Then the temperature $T(x, y, t)$ at any location (x, y) and at any time t can be denoted by

$$T(x, y, t) = T(m \, \Delta x, n \, \Delta y, i \, \Delta t) \equiv T^i_{m,n} \tag{5-63}$$

Using this notation, we can write the finite-difference form of the partial derivatives with respect to the space variables, by Eq. (5-23), for $\Delta x = \Delta y$, as

$$\left[\frac{\partial^2 T}{\partial x^2} + \frac{\partial^2 T}{\partial y^2}\right]_{m,n,i} \cong \frac{T^i_{m-1,n} + T^i_{m+1,n} + T^i_{m,n-1} + T^i_{m,n+1} - 4T^i_{m,n}}{(\Delta x)^2} \tag{5-64a}$$

The partial derivative with respect to the time variable can be written as

$$\left[\frac{\partial T}{\partial t}\right]_{m,n,i} \cong \frac{T^{i+1}_{m,n} - T^i_{m,n}}{\Delta t} \tag{5-64b}$$

Introducing Eqs. (5-64) into Eq. (5-62), we find that the finite-difference form of the two-dimensional, time-dependent heat conduction equation is

$$\boxed{T^{i+1}_{m,n} = r[T^i_{m-1,n} + T^i_{m+1,n} + T^i_{m,n-1} + T^i_{m,n+1}] + (1 - 4r)T^i_{m,n}} \tag{5-65}$$

where $$r = \frac{\alpha\,\Delta t}{(\Delta x)^2} \qquad \Delta x = \Delta y \tag{5-66}$$

Equation (5-65) is the explicit finite-difference formulation of the heat conduction equation (5-62). This equation provides a simple expression for calculating the temperature $T^{i+1}_{m,n}$ at a node (m, n) at the time step $i + 1$ if the temperatures at the node (m, n) and at its neighboring nodes are known for the previous time step i. Therefore, by starting with the initial condition $i = 0$, the node temperatures at the subsequent time steps are calculated.

The procedure is simple, but there is a restriction on the permissible maximum value of the parameter r.

Restriction on r By following an argument similar to that described for the one-dimensional equation (5-59), we conclude that to obtain physically meaningful results, the coefficient $1 - 4r$ of $T^i_{m,n}$ in Eq. (5-65) *should not be negative*. This requirement leads to the following stability criterion:

$$\boxed{0 < r \leq \tfrac{1}{4}} \tag{5-67a}$$

where $$r = \frac{\alpha\,\Delta t}{(\Delta x)^2} \tag{5-67b}$$

This condition implies that for given values of α and Δx, the time step Δt cannot exceed the upper limit imposed by the stability criterion given by Eqs. (5-67); otherwise, the numerical calculations become unstable.

Nodes on a Boundary

The finite-difference equations (5-59) and (5-65) are developed for nodes located in the interior of the region, and it is assumed that the temperatures are prescribed on the boundaries.

When the problem involves boundaries subjected to convection or prescribed heat flux or adiabatic conditions, the temperature at the boundaries is not known. Therefore, when nodes are considered on such boundaries, the finite-difference equations are needed for each node on the boundary.

We describe now two different approaches for developing finite-difference equations for the boundary nodes.

Negligible heat capacity If the step size Δx is sufficiently small, the heat capacity associated with the volume element at the boundary node can be neglected. Then the finite-difference equation for the boundary node is developed by considering a steady-state energy balance for the node, which can be stated as

$$\begin{pmatrix} \text{Rate of heat entering} \\ \text{the volume element} \\ \Delta V \text{ from all its} \\ \text{surfaces} \end{pmatrix} + \begin{pmatrix} \text{rate of energy} \\ \text{generated in} \\ \text{the volume element} \\ \Delta V \end{pmatrix} = 0$$

To illustrate the application of this approach, we consider a slab of thickness L, subjected to convection at both its boundary surfaces, as shown in Fig. 5-17, and write the finite-difference equations for nodes $m = 0$ and $m = M$ at the boundary surfaces.

The steady-state energy balance for node $m = 0$ at the boundary surface $x = 0$, at any time step $i + 1$, gives

$$h_0(T_\infty - T_0^{i+1}) + k\,\frac{T_1^{i+1} - T_0^{i+1}}{\Delta x} = 0$$

or

$$\boxed{T_0^{i+1} = \frac{1}{1 + \Delta x\, h_0/k}\left(T_1^{i+1} + \frac{\Delta x\, h_0}{k}\,T_\infty\right)} \tag{5-68}$$

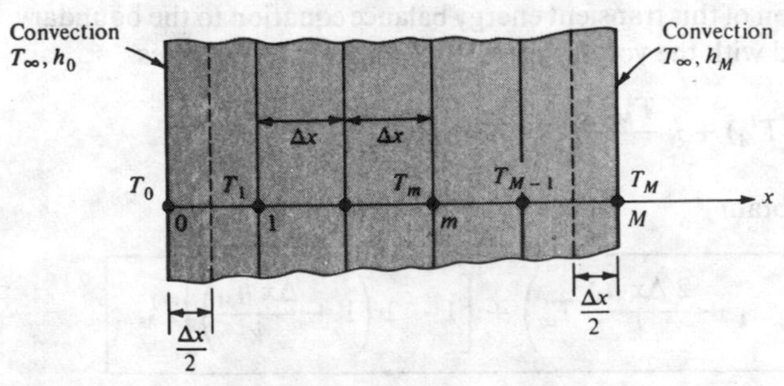

Figure 5-17 Finite-difference equations for nodes on the boundaries.

Similarly, an energy balance for node $m = M$, at the boundary $x = L$, gives

$$k \frac{T_{M-1}^{i+1} - T_M^{i+1}}{\Delta x} + h_M(T_\infty - T_M^{i+1}) = 0$$

or

$$T_M^{i+1} = \frac{1}{1 + \Delta x\, h_M/k}\left(T_{M-1}^{i+1} + \frac{\Delta x\, h_M}{k} T_\infty\right) = 0 \tag{5-69}$$

Equations (5-68) and (5-69) provide the finite-difference equations for the nodes at the boundaries $x = 0$ and $x = L$ at the time step $i + 1$.

Effects of heat capacity If the step size Δx is not sufficiently small, it may be necessary to include the effects of the heat capacitance by considering a time-dependent energy balance for the volume element associated with the node at the boundary. The transient energy balance for a volume element ΔV can be stated as

$$\begin{pmatrix}\text{Rate of heat}\\ \text{entering } \Delta V\\ \text{from all its}\\ \text{surfaces at}\\ \text{time step } i\end{pmatrix} + \begin{pmatrix}\text{rate of energy}\\ \text{generation}\\ \text{in } \Delta V \text{ at}\\ \text{time step } i\end{pmatrix} = \begin{pmatrix}\text{rate of increase}\\ \text{of internal}\\ \text{energy of}\\ \Delta V\end{pmatrix}$$

This energy balance is applied to a volume element of thickness $\Delta x/2$, associated with the boundary node $m = 0$, shown in Fig. 5-17. We obtain

$$h_0(T_\infty - T_0^i) + k\frac{T_1^i - T_0^i}{\Delta x} = \rho c_p \frac{\Delta x}{2} \frac{T_0^{i+1} - T_0^i}{\Delta t}$$

Solving for T_0^{i+1}, we get

$$T_0^{i+1} = r\left(2T_1^i + \frac{2\,\Delta x\, h_0}{k} T_\infty\right) + \left[1 - 2r\left(1 + \frac{\Delta x\, h_0}{k}\right)\right]T_0^i \tag{5-70}$$

Similarly, the application of this transient energy balance equation to the boundary node $m = M$ associated with the volume element of thickness $\Delta x/2$ gives

$$h_M(T_\infty - T_M^i) + k\,\frac{T_{M-1}^i - T_M^I}{\Delta x} = \rho c_p \frac{\Delta x}{2} \frac{T_M^{i+1} - T_M^i}{\Delta t}$$

Solving for T_M^{i+1}, we obtain

$$T_M^{i+1} = r\left(2T_{M-1}^i + \frac{2\,\Delta x\, h_M}{k} T_\infty\right) + \left[1 - 2r\left(1 + \frac{\Delta x\, h_M}{k}\right)\right]T_M^i \tag{5-71}$$

Equations (5-70) and (5-71) are the finite-difference equations for the boundary nodes $m = 0$ and $m = M$, subjected to convection. These equations include the effects of the heat capacity of the volume elements associated with the boundary nodes $m = 0$ and $m = M$.

The effects of Eqs. (5-70) and (5-71) on the stability criteria associated with the solution of the finite-difference equations should be recognized. By following an argument similar to that described for Eq. (5-59) associated with the interior nodes of the slab, we conclude that, to obtain physically meaningful results, the coefficients of T_0^i and T_M^i in these equations should be positive that is,

$$1 - 2r\left(1 + \frac{\Delta x\, h_0}{k}\right) \geq 0 \qquad \text{for } m = 0$$

$$1 - 2r\left(1 + \frac{\Delta x\, h_M}{k}\right) \geq 0 \qquad \text{for } m = M$$

These conditions imply that the following restriction should be imposed on the value of r:

$$\boxed{0 < r \leq \frac{1}{2(1 + \Delta x\, h_0/k)}} \qquad \text{for } m = 0 \tag{5-72}$$

$$\boxed{0 < r \leq \frac{1}{2(1 + \Delta x\, h_M/k)}} \qquad \text{for } m = M \tag{5-73}$$

Clearly, this restriction is more severe than that given by Eq. (5-61) for the interior nodes in the region; only the smaller of the above two different values of r should be considered as the stability criterion for the numerical calculations.

The finite-difference equations developed above for the boundary nodes of the one-dimensional slab problem can be generalized readily for two-dimensional problems by writing a time-dependent energy balance for the node on the boundary. To illustrate, we consider the node 0 at the intersection of two boundaries in a two-dimensional, transient heat conduction problem, as illustrated in Fig. 5-18. The

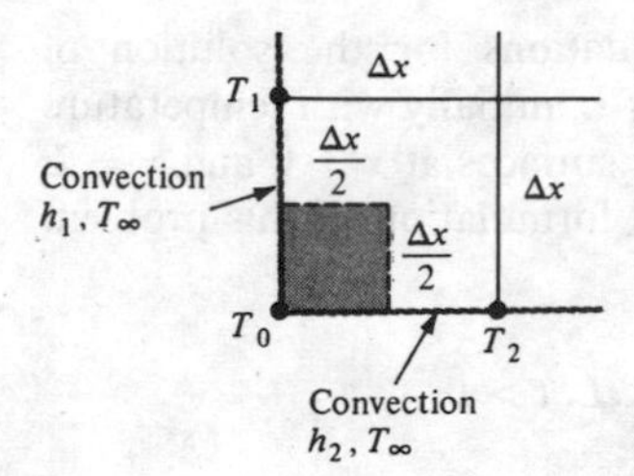

Figure 5-18 Finite-difference equation for a corner node.

transient energy balance equation for this node associated with the shaded volume element is written as

$$h_1 \frac{\Delta x}{2}(T_\infty - T_0^i) + h_2 \frac{\Delta x}{2}(T_\infty - T_0^i) + k\frac{\Delta x}{2}\frac{T_2^i - T_0^i}{\Delta x} + k\frac{\Delta x}{2}\frac{T_1^i - T_0^i}{\Delta x} = \rho c_p \left(\frac{\Delta x}{2}\right)^2 \frac{T_0^{i+1} - T_0^i}{\Delta t}$$

Solving this equation for T_0^{i+1}, we obtain

$$T_0^{i+1} = 2r\left(T_1^i + T_2^i + \frac{h_1 \Delta x}{k} T_\infty + \frac{h_2 \Delta x}{k} T_\infty\right) + \left[1 - 2r\left(2 + \frac{h_1 \Delta x}{k} + \frac{h_2 \Delta x}{k}\right)\right] T_0^i \qquad (5\text{-}74)$$

An examination of this equation reveals that the coefficient of T_0^i *should not be negative* in order to obtain meaningful solutions. Hence we require

$$1 - 2r\left(2 + \frac{h_1 \Delta x}{k} + \frac{h_2 \Delta x}{k}\right) \geq 0 \qquad (5\text{-}75)$$

or the parameter r should satisfy the following stability criterion:

$$\boxed{0 < r \leq \frac{1}{2(2 + h_1 \Delta x/k + h_2 \Delta x/k)}} \qquad (5\text{-}76)$$

This analysis demonstrates that when there are nodal points on a convection boundary and a time-dependent energy balance is used to obtain the finite-difference equations for such nodes, the selection of the stability criterion is not as simple as that for the interior nodes. In such situations, it is usually a good policy to develop all the finite-difference equations for the system and then examine the elements that lie along the main diagonal in the coefficient matrix. *These elements must be greater than or equal to zero for a stable system, and the smallest value of r that satisfies this requirement among the main diagonal elements gives the upper limit for r that satisfies the stability criterion.* This matter will be apparent in Example 5-10.

Example 5-8 Develop the finite-difference equations for the solution of transient heat conduction in a slab of thickness L, initially with temperature distribution $F(x)$, when for $t > 0$ the boundary surfaces at $x = 0$ and $x = L$ are subjected to convection. The mathematical formulation of this problem can be taken as

$$\frac{\partial T}{\partial t} = \alpha \frac{\partial^2 T}{\partial x^2} \qquad \text{in } 0 < x < L,\ t > 0$$

subject to the boundary conditions

$$-k \frac{\partial T}{\partial x} + h_0 T = h_0 T_{\infty 0} \qquad \text{at } x = 0, t > 0$$

$$k \frac{\partial T}{\partial x} + h_M T = h_M T_{\infty M} \qquad \text{at } x = L, t > 0$$

and the initial condition

$$T = F(x) \qquad \text{for } t - 0, \text{ in } 0 \le x \le L$$

SOLUTION The region $0 \le x \le L$ is divided into M equal intervals with the mesh size equal to $\Delta x = L/M$, as illustrated in Fig. 5-17.

The finite-difference equations for the interior nodes $m = 1, 2, \ldots, M - 1$ are immediately obtained from Eqs. (5-59):

$$T_m^{i+1} = r(T_{m-1}^i + T_{m+1}^i) + (1 - 2r)T_m^i \qquad m = 1, 2, \ldots, M - 1 \qquad (a)$$

Now we need two more equations for nodes $m = 0$ and $m = M$ on the boundary surfaces. If we assume that the heat capacitance of the volume element associated with the boundary nodes is negligible, the finite-difference equations for nodes $m = 0$ and $m = M$ are obtained, respectively, from Eqs. (5-68) and (5-69), as

$$T_0^{i+1} = \frac{1}{1 + \Delta x\, h_0/k} \left(T_1^{i+1} + \frac{\Delta x\, h_0}{k} T_{\infty 0} \right) \qquad (b)$$

$$T_M^{i+1} = \frac{1}{1 + \Delta x\, h_M/k} \left(T_{M-1}^{i+1} + \frac{\Delta x\, h_M}{k} T_{\infty M} \right) \qquad (c)$$

Finally, the initial condition for the problem is written as

$$T_m^0 = F(m\, \Delta x) \equiv F_m \qquad m = 0, 1, 2, \ldots, M \text{ for } i = 0 \qquad (d)$$

where

$$r = \frac{\alpha\, \Delta t}{(\Delta x)^2}$$

Equations (*a*) to (*d*) provide the complete system of equations for the solution of this transient heat conduction problem.

The first step in the analysis is to establish the stability criterion. Equations (*a*) for the interior nodes govern the stability criterion, since the steady-state formulation is used for nodes at the boundaries.

All the elements in the main diagonal of the system of Eqs. (*a*) are equal to $1 - 2r$; hence the stability criterion becomes

$$r \le \tfrac{1}{2}$$

Once the stability criterion is established and a suitable value is chosen for r

(for this particular case, it is best to choose $r = \frac{1}{2}$ so that the coefficients of T_m^i will become zero), the calculations are performed in the following order:

1. Calculations are started with the system of Eqs. (*a*) by setting $i = 0$. Then the node temperatures on the right-hand side become the initial condition, and the interior node temperatures T_m^1, $m = 1, 2, \ldots, M - 1$, for the time step $i = 1$ are determined. The temperatures T_0^1 and T_M^1 for time step $i = 1$ at the boundary nodes $m = 0$ and $m = M$ are calculated, respectively, from Eqs. (*b*) and (*c*), because the temperatures T_1^1 and T_{M-1}^1 are now known.
2. Given the node temperatures T_m^1, $m = 0, 1, 2, \ldots, M$, we repeat the procedure in step 1 by setting $i = 2$ in order to calculate the node temperatures T_m^2, $m = 0, 1, \ldots, M$. The node temperatures for the subsequent times are found in a similar manner.

Example 5-9 A marble slab [$k = 2$ W/(m · °C), $\alpha = 1 \times 10^{-6}$ m²/s] that is $L = 2$ cm thick is initially at a uniform temperature $T_i = 200$°C. Suddenly one of its surfaces is lowered to 0°C and is maintained at that temperature, while the other surface is kept insulated. Develop an explicit finite-difference scheme for the determination of the temperature distribution in the slab as a function of position and time as well as the heat flux at the boundary surface.

SOLUTION The mathematical formulation of this heat conduction problem is given by

$$\frac{\partial T}{\partial t} = \alpha \frac{\partial^2 T}{\partial x^2} \qquad \text{in } 0 < x < L,\ t > 0$$

$$\frac{\partial T}{\partial x} = 0 \qquad \text{at } x = 0,\ t > 0$$

$$T = 0 \qquad \text{at } x = L,\ t > 0$$

$$T = 200 \qquad \text{for } t = 0$$

This problem is a special case of that in Example 5-8; hence its finite-difference formulation is immediately obtained from that example by setting $h_0 = 0$, $h_M = \infty$, and $T_{\infty M} = 0$:

$$T_m^{i+1} = rT_{m-1}^i + (1 - 2r)T_m^i + rT_{m+1}^i \qquad m = 1, 2, \ldots, M - 1$$

$$T_0^{i+1} = T_1^{i+1} \qquad \text{for node } m = 0$$

$$T_M^{i+1} = 0 \qquad \text{for node } m = M$$

$$T_m^0 = 200 \qquad \text{for } i = 0,\ m = 0, 1, \ldots, M$$

To solve this problem, we divide the region into $M = 5$ equal parts. Hence,

$$\Delta x = \frac{L}{M} = \frac{2}{5} = 0.4 \text{ cm}$$

The value of the parameter r is taken as $r = \frac{1}{2}$; then the corresponding time step Δt becomes

$$\Delta t = \frac{r(\Delta x)^2}{\alpha} = \frac{1}{2}\frac{(0.4 \times 10^{-2})^2}{10^{-6}} = 8 \text{ s}$$

By setting $r = \frac{1}{2}$, the above set of finite-difference equations becomes

$$T_m^{i+1} = \tfrac{1}{2}(T_{m-1}^i + T_{m+1}^i) \qquad m = 1, 2, 3, 4$$

$$T_0^{i+1} = T_1^{i+1} \qquad m = 0$$

$$T_5^{i+1} = 0 \qquad m = 5$$

$$T_m^0 = 200 \qquad m = 0, 1, 2, 3, 4, 5, i = 0$$

Clearly, the temperature T_m^{i+1} at an interior node m at time step $i + 1$ is equal to the arithmetic average of the two adjacent node temperatures at the previous time step i.

For illustration purposes, the numerical computations are performed by hand calculations, and the results for 10 consecutive time steps are listed in Table 5-2. A programmable calculator or a digital computer can readily be used to perform such calculations.

In Table 5-2, the first row, $i = 0$, is the initial temperature distribution. The second, third, etc. rows are the temperatures at the end of $\Delta t = 8$ s, $2\,\Delta t = 16$ s, and so on.

The heat flux at the boundary surface $x = L$ (that is, $M = 5$) can be determined from

$$q_M^i = k\,\frac{T_4^i - T_5^i}{\Delta x} \qquad \text{W/m}^2$$

For example, the heat flux q_M^i at the end of time step $i = 10$ (i.e., $t = 80$ s) becomes

$$q_M^i = 2\,\frac{48.5 - 0}{0.4 \times 10^{-2}} = 24{,}250 \text{ W/m}^2$$

Table 5-2 Finite-difference calculations for Example 5-8

		$m = 0$	1	2	3	4	5
i	t, s	$x = 0$	0.4	0.8	1.2	1.6	2.0
0	0	200	200	200	200	200	200
1	8	200	200	200	200	200	0
2	16	200	200	200	200	100	0
3	24	200	200	200	150	100	0
4	32	200	200	175	150	75	0
5	40	187.5	187.5	175	125	75	0
6	48	181.2	181.2	156.2	125	62.5	0
7	56	168.7	168.7	153.1	109.4	62.5	0
8	64	160.9	160.9	139.1	107.8	54.7	0
9	72	150.0	150.0	134.4	96.9	53.9	0
10	80	142.2	142.2	123.5	94.2	48.5	0

The exact analytic solution of this problem for the temperature distribution is

$$\frac{T(x, t)}{200} = \frac{4}{\pi} \sum_{n=1}^{\infty} \frac{(-1)^{n+1}}{2n-1} e^{-\alpha \lambda_n^2 t} \cos \lambda_n x$$

where $$\lambda_n = \frac{(2n-1)\pi}{2L}$$

Then the heat flux at the boundary surface $x = L$ becomes

$$q(L, t) = -k \left. \frac{\partial T}{\partial x} \right|_{x=L} = \frac{400k}{L} \sum_{n=1}^{\infty} e^{-\alpha \lambda_n^2 t}$$

The exact value of the heat flux at the boundary surface $x = L$ at $t = 80$ s is

$$q_{\text{exact}} = 24{,}890 \text{ W/m}^2$$

The finite-difference solution obtained by using a rather coarse mesh size compares well with the exact solution.

Example 5-10 Consider two-dimensional, time-dependent heat conduction for the region shown in the accompanying figure. The boundary conditions are

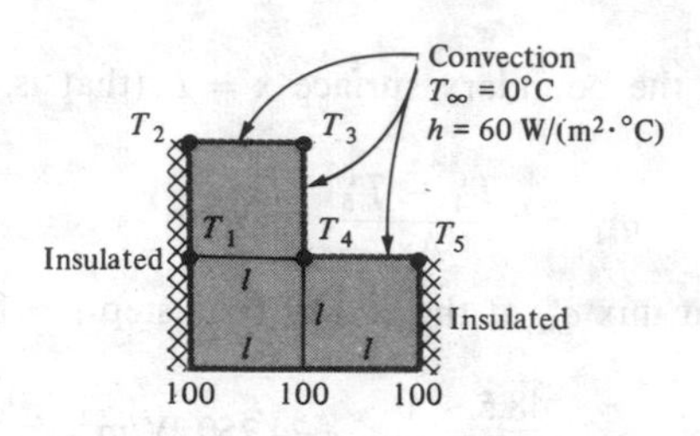

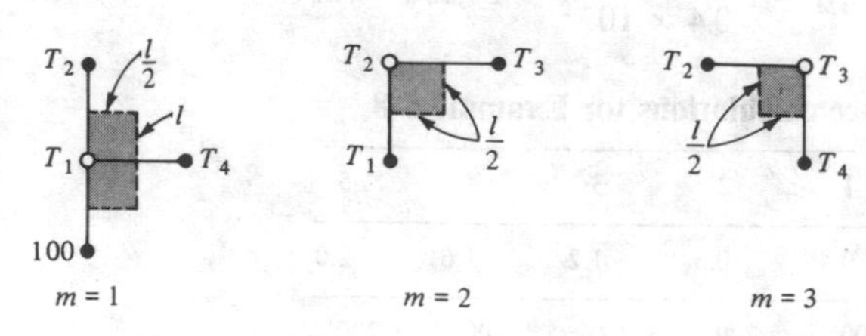

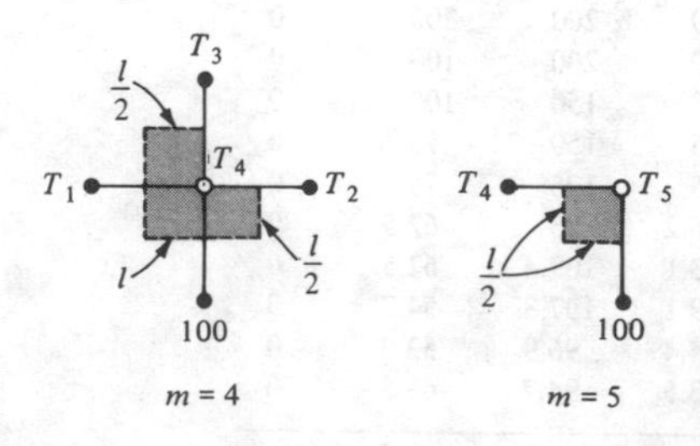

stated in this figure, and a very coarse network of mesh size $\Delta x = \Delta y = l =$ 1 cm is used for finite-difference formulation. Develop the finite-difference equations for each of the five nodes, $m = 1$ to 5, and establish the stability criterion for the resulting system. Take $h = 60\ \text{W/(m}^2 \cdot {}^\circ\text{C)}$, $k = 30\ \text{W/(m} \cdot {}^\circ\text{C)}$, and $T_\infty = 0^\circ\text{C}$.

SOLUTION Since the finite-difference network for this problem is very coarse, there are only a few nodes for the entire region. Therefore, we need to use a transient energy balance to determine the finite-difference equations for all the nodes on the boundaries, because their thermal capacitance cannot be neglected. The accompanying figure also shows the volume element associated with each node used for writing the energy balance equation. The results are summarized here:

Node $m = 1$:

$$k \frac{l}{2} \frac{T_2^i - T_1^i}{l} + kl \frac{T_4^i - T_1^i}{l} + k \frac{l}{2} \frac{100 - T_1^i}{l} = \rho c_p \frac{l^2}{2} \frac{T_1^{i+1} - T_1^i}{\Delta t}$$

Solving for T_1^{i+1}, we obtain

$$T_1^{i+1} = (1 - 4r)T_1^i + rT_2^i + 2rT_4^i + 100r \tag{a}$$

where

$$r = \frac{k}{\rho c_p} \frac{l^2}{\Delta t} = \frac{\alpha l^2}{\Delta t}$$

Node $m = 2$:

$$k \frac{l}{2} \frac{T_1^i - T_2^i}{l} + k \frac{l}{2} \frac{T_3^i - T_2^i}{l} + h \frac{l}{2} (T_\infty - T_2^i) = \rho c_p \frac{l^2}{2} \frac{T_2^{i+1} - T_2^i}{\Delta t}$$

Setting $T_\infty = 0$, $h = 60\ \text{W/(m}^2 \cdot {}^\circ\text{C)}$, $k = 30\ \text{W/(m} \cdot {}^\circ\text{C)}$, and $l = 0.01$ m and then solving for T_2^{i+1}, we obtain

$$T_2^{i+1} = 2rT_1^i + (1 - 4.04r)T_2^i + 2rT_3^i \tag{b}$$

Node $m = 3$:

$$k \frac{l}{2} \frac{T_2^i - T_3^i}{l} + k \frac{l}{2} \frac{T_4^i - T_3^i}{l} + hl(T_\infty - T_3^i) = \rho c_p \frac{l^2}{2} \frac{T_3^{i+1} - T_3^i}{\Delta t}$$

Substituting the appropriate values for T_∞, h, and k and then solving for T_3^{i+1}, we obtain

$$T_3^{i+1} = 2rT_2^i + (1 - 4.08r)T_3^i + 2rT_4^i \tag{c}$$

Node $m = 4$:

$$k \frac{l}{2} \frac{T_3^i - T_4^i}{l} + k \frac{l}{2} \frac{T_1^i - T_4^i}{l} + k \frac{l}{2} \frac{100 - T_4^i}{l} + k \frac{l}{2} \frac{T_5^i - T_4^i}{l}$$
$$+ hl(T_\infty - T_4^i) = \rho c_p \frac{3}{4} l^2 \frac{T_4^{i+1} - T_4^i}{\Delta t}$$

Substituting the appropriate values for T_∞, h, and k and then solving for T_4^{i+1}, we obtain

$$T_4^{i+1} = 1.333rT_1^i + 0.67rT_3^i + (1 - 4.027r)T_4^i + 0.67rT_5^i + 133.3r \quad (d)$$

Node $m = 5$:

$$k\frac{l}{2}\frac{T_4^i - T_5^i}{l} + kl\frac{100 - T_5^i}{l} + hl(T_\infty - T_5^i) = \rho c_p \frac{l^2}{4}\frac{T_5^{i+1} - T_5^i}{\Delta t}$$

Substituting the appropriate values for T_∞, h, and k and then solving for T_5^{i+1}, we get

$$T_5^{i+1} = 2rT_4^i + (1 - 4.04r)T_5^i + 200r \quad (e)$$

Equations (a) to (e) are now expressed in matrix form:

$$\begin{bmatrix} T_1^{i+1} \\ T_2^{i+1} \\ T_3^{i+1} \\ T_4^{i+1} \\ T_5^{i+1} \end{bmatrix} = \begin{bmatrix} 1-4r & r & 0 & 2r & 0 \\ 2r & 1-4.04r & 2r & 0 & 0 \\ 0 & 2r & 1-4.08r & 2r & 0 \\ 1.333r & 0 & 0.67r & 1-4.027r & 0.67r \\ 0 & 0 & 0 & 2r & 1-4.04r \end{bmatrix} \begin{bmatrix} T_1^i \\ T_2^i \\ T_3^i \\ T_4^i \\ T_5^i \end{bmatrix} + \begin{bmatrix} 100r \\ 0 \\ 0 \\ 133.3r \\ 200r \end{bmatrix}$$

An examination of the elements in the main diagonal of the coefficient matrix reveals that the smallest value for r is associated with

$$1 - 4.08r = 0$$

Hence the stability criterion for the system becomes

$$0 < r \leq \frac{1}{4.08}$$

where

$$r = \frac{\alpha l^2}{\Delta t} = \frac{\alpha \times 10^{-4}}{\Delta t}$$

5-7 UNSTEADY HEAT CONDUCTION—IMPLICIT METHOD

The explicit method of the finite-difference scheme discussed earlier has a restriction on the maximum size of the time step because of stability considerations. If computations are to be performed for large periods of time, the number of time steps required for the calculations can become enormous with small Δx. To alleviate such difficulties, other methods of finite-difference schemes that are not restricted by the stability criteria have been developed.

The *implicit* methods of finite-difference schemes, which involve the use of *backward-difference* formulation in time increments, eliminate the restrictions imposed on the size of the time step Δt. Numerous implicit schemes have been reported in the literature. We describe here some of these schemes as applied to one-dimensional, transient heat conduction in a slab.

Fully Implicit Scheme

The finite-difference formulation of the one-dimensional, time-dependent heat conduction equation (5-55) with the fully implicit scheme is given by

$$\frac{T_m^{i+1} - T_m^i}{\Delta t} = \alpha \frac{T_{m-1}^{i+1} - 2T_m^{i+1} + T_{m+1}^{i+1}}{(\Delta x)^2} \tag{5-77}$$

A comparison of this result with that given by Eq. (5-58) obtained by the explicit scheme reveals that the right-hand side of Eq. (5-77) is evaluated at time step $i + 1$, whereas that of Eq. (5-58) is evaluated at time step i.

The implicit schemes are advantageous in that there is no restriction on the size of the time step Δt by the stability considerations. However, to determine the node temperatures T_m^i, a simultaneous solution of all the equations for the nodes at each time step is required.

The basic idea in the use of the implicit scheme has been further developed by numerous investigators with the objective of finding more efficient schemes that are stable but lack restrictions on the size of the time step. We present one such scheme which has been successful in achieving such an objective.

Crank-Nicolson Method

Crank and Nicolson [10] proposed a modified implicit form. In this formulation, the left-hand side of the finite-difference equation (5-77) is retained, but for the right-hand side the arithmetic average of the right-hand sides of the explicit formulation (5-58) and the fully implicit formulation (5-77) are used.

To illustrate the Crank-Nicolson method of finite-difference representation, we consider the following one-dimensional, time-dependent heat conduction problem for a slab subjected to convection boundary conditions:

$$\frac{\partial T}{\partial t} = \alpha \frac{\partial^2 T}{\partial x^2} \quad \text{in } 0 \le x \le L, \text{ for } t > 0 \tag{5-78}$$

subject to the boundary conditions

$$-k\frac{\partial T}{\partial x} + h_1 T = f_1 \quad \text{at } x = 0, \text{ for } t > 0 \tag{5-79a}$$

$$k\frac{\partial T}{\partial x} + h_2 T = f_2 \quad \text{at } x = L, \text{ for } t > 0 \tag{5-79b}$$

$$T = F(x) \quad \text{for } t = 0 \text{ in } 0 \le x \le L \tag{5-79c}$$

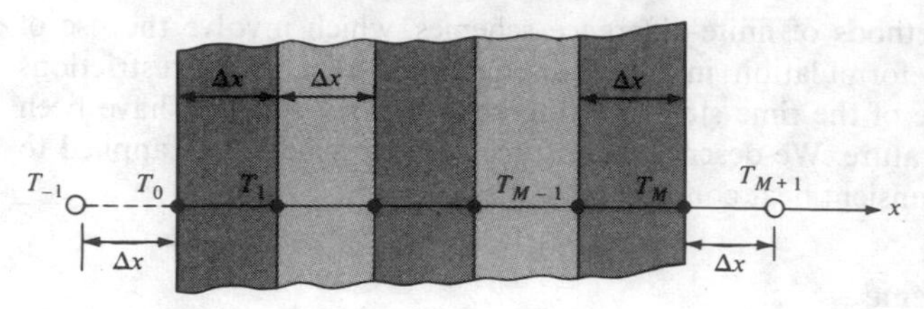

Figure 5-19 Fictitious image nodes T_{-1} and T_{M+1}.

The finite-difference form of the differential equation (5-78) obtained with the Crank-Nicolson method becomes

$$\frac{T_m^{i+1} - T_m^i}{\Delta t} = \frac{\alpha}{2}\left[\frac{T_{m-1}^{i+1} - 2T_m^{i+1} + T_{m+1}^{i+1}}{(\Delta x)^2} + \frac{T_{m-1}^i - 2T_m^i + T_{m+1}^i}{(\Delta x)^2}\right]$$

which is rearranged as

$$\boxed{-rT_{m-1}^{i+1} + (2 + 2r)T_m^{i+1} - rT_{m-1}^{i+1} = rT_{m-1}^i + (2 - 2r)T_m^i + rT_{m+1}^i} \tag{5-80}$$

for $m = 0, 1, 2, \ldots, M$, where

$$r = \frac{\alpha\,\Delta t}{(\Delta x)^2}$$

The finite-difference representations of boundary conditions (5-79*a*) and (5-79*b*) are given, respectively, as

$$-k\,\frac{T_1^i - T_{-1}^i}{2\,\Delta x} + h_1 T_0^i = f_1 \tag{5-81a}$$

$$k\,\frac{T_{M+1}^i - T_{M-1}^i}{2\,\Delta x} + h_2 T_M^i = f_2 \tag{5-81b}$$

and the initial condition (5-79*c*) becomes

$$T_m^0 = F(m\,\Delta x) \qquad m = 0, 1, 2, \ldots, M \tag{5-81c}$$

In Eqs. (5-81*a*) and (5-81*b*), T_{-1}^i and T_{M+1}^i are the fictitious temperatures at the fictitious image nodes $m = -1$ and $m = M + 1$, respectively, as illustrated in Fig. 5-19. These fictitious temperatures are introduced because we use a central-difference approximation in representing the derivatives at the boundaries associated with the convection boundary condition.

We now summarize the results of the above finite-difference formulation:

1. Equations (5-80) provide $M + 1$ simultaneous algebraic equations. The unknowns include $M + 1$ node temperatures T_m^{i+1}, $m = 0, 1, 2, \ldots, M$, and the two fictitious image temperatures T_{-1}^{i+1} and T_{M+1}^{i+1}. Therefore, two more equations are needed.

2. The boundary conditions, Eqs. (5-81), provide the two additional relations that make the number of equations equal to the number of unknowns.

For computational purposes, it is better to solve for T^i_{-1} and T^i_{M+1} (or T^{i+1}_{-1} and T^{i+1}_{M+1}) from Eqs. (5-81*a*) and (5-81*b*) and to introduce them into Eqs. (5-80) in order to eliminate T^{i+1}_{-1} and T^{i+1}_{M+1}. The resulting system provides $M + 1$ simultaneous algebraic equations for the $M + 1$ unknown node temperatures T_m, $m = 0, 1, \ldots, M$. These equations can be expressed in matrix form as

$$
\begin{bmatrix}
2+2r\beta_1 & -2r & 0 & 0 & \cdots & 0 & 0 & 0 \\
-r & 2+2r & -r & 0 & \cdots & 0 & 0 & 0 \\
0 & -r & 2+2r & -r & \cdots & 0 & 0 & 0 \\
\cdots & \cdots & \cdots & \cdots & \cdots & \cdots & \cdots & \cdots \\
0 & 0 & 0 & 0 & \cdots & -r & 2+2r & -r \\
0 & 0 & 0 & 0 & \cdots & 0 & -r & 2+2r\beta_2
\end{bmatrix}
\begin{bmatrix}
T^{i+1}_0 \\ T^{i+1}_1 \\ T^{i+1}_2 \\ \vdots \\ T^{i+1}_N \\ T^{i+1}_N
\end{bmatrix}
$$

$$
= \begin{bmatrix}
2-2r\beta_1 & 2r & 0 & 0 & \cdots & 0 & 0 & 0 \\
r & 2-2r & r & 0 & \cdots & 0 & 0 & 0 \\
0 & r & 2-2r & t & \cdots & 0 & 0 & 0 \\
\cdots & \cdots & \cdots & \cdots & \cdots & \cdots & \cdots & \cdots \\
0 & 0 & 0 & 0 & \cdots & r & 2-2r & r \\
0 & 0 & 0 & 0 & \cdots & 0 & 2r & 2-2r\beta_2
\end{bmatrix}
\times
\begin{bmatrix}
T^i_0 \\ T^i_1 \\ T^i_2 \\ \vdots \\ T^i_{N-1} \\ T^i_N
\end{bmatrix}
+
\begin{bmatrix}
\dfrac{4r\,\Delta x\, f_1}{k} \\ 0 \\ 0 \\ \vdots \\ 0 \\ \dfrac{4r\,\Delta x\, f_2}{k}
\end{bmatrix}
\qquad (5\text{-}82)
$$

where $i = 0, 1, 2, 3, \ldots$, and β_1 and β_2 are defined as

$$\beta_1 = 1 + \frac{h_1\,\Delta x}{k} \qquad \text{and} \qquad \beta_2 = 1 + \frac{h_2\,\Delta x}{k}$$

Once the values of the boundary condition functions f_1 and f_2, the heat transfer coefficients h_1 and h_2, the thermal conductivity k, the space step Δx, and the parameter r are chosen, the system of Eqs. (5-82) is solved for each time step in the following manner:

1. The calculations are started by setting $i = 0$. Then the temperatures T^0_0, T^0_1, $T^0_2, \ldots, T^0_M$ on the right-hand side of Eqs. (5-82) are the initial conditions

and are known. Then $M + 1$ algebraic equations (5-82) are simultaneously solved, and the temperatures $T_0^1, T_1^1, T_2^1, \ldots, T_M^1$ for the nodes $m = 0, 1, \ldots, M$ at the end of the first step are determined.

2. By setting $i = 1$, the temperatures $T_0^1, T_1^1, T_2^1, \ldots, T_M^1$ on the right-hand side of Eqs. (5-82) are known from the calculations in step 1. Then the equations are solved simultaneously, and the temperatures $T_0^2, T_1^2, T_2^2, \ldots, T_M^2$ at the end of the second time step are determined.
3. By setting $i = 2$, the temperatures $T_0^2, T_1^2, \ldots, T_M^2$ on the right-hand side of Eqs. (5-82) become known from step 2. Then a simultaneous solution of the equations give the temperatures $T_0^3, T_1^3, \ldots, T_M^3$ at the end of the third time step.

The temperatures at the following time steps are computed in a similar manner.

We note that the implicit method results in a set of coupled equations to be solved for each time step, whereas with the explicit method equations are uncoupled. Although it is more difficult to solve a system of coupled equations than uncoupled equations, the implicit method has no restriction imposed on the size of the time step Δt, hence has the advantage for situations where larger Δt may be used to proceed with the solution more rapidly. However, Δt should not be chosen so large as to cause inaccuracy in the solutions.

PROBLEMS

One-dimensional, steady-state heat conduction

5-1 Consider the following one-dimensional, steady-state heat conduction problem:

$$\frac{d^2T(x)}{dx^2} + \frac{1}{k}g = 0 \qquad \text{in } 0 < x < L$$

$$\frac{dT(x)}{dx} = 0 \qquad \text{at } x = 0$$

$$k\frac{dT(x)}{dx} + hT(x) = 0 \qquad \text{at } x = L$$

Write the finite-difference formulation of this heat conduction problem by dividing the region $0 < x < L$ into four equal parts.

5-2 In a parallel-plate fuel element for a gas-cooled nuclear reactor, the heat generation in the fuel element has approximately a cosine distribution. The simplest steady-state model for the temperature distribution in the fuel element may be taken as

$$\frac{d^2T(x)}{dx^2} + \frac{1}{k}g_0 \cos\frac{\pi x}{2L} = 0 \qquad \text{in } 0 < x < L$$

$$\frac{dT}{dx} = 0 \qquad \text{at } x = 0$$

$$T = 0 \qquad \text{at } x = L$$

where L is the half-thickness of the fuel element. By dividing the region into four equal parts, calculate the temperature distribution with finite differences for $k = 12$ W/(m · °C), $L = 5 \times 10^{-3}$ m, and $g = 6 \times 10^8$ W/m³. Compare the numerical results with the exact solution.

5-3 Consider the following one-dimensional, steady-state heat conduction problem:

$$\frac{d^2T(x)}{dx^2} + \frac{1}{k} g = 0 \qquad \text{in } 0 < x < L$$

$$-k\frac{dT(x)}{dx} = q_0 \qquad \text{at } x = 0$$

$$T = 0 \qquad \text{at } x = L$$

(*a*) Write the finite-difference formulation of this heat conduction problem by dividing the region $0 < x < L$ into four equal parts.

(*b*) Compute the node temperatures for $k = 12$ W/(m · °C), $L = 0.012$ m, $q_0 = 10^5$ W/m², and $g_0 = 4 \times 10^7$ W/m³.

(*c*) Compare the numerical solution at the nodes with the exact solution.

5-4 Consider an iron rod $L = 10$ cm long of diameter $D = 1$ cm with thermal conductivity $k = 50$ W/(m · °C). One end of the rod is maintained at $T_0 = 200$°C, the other end at 0°C, while it is exposed to convection from its lateral surfaces into ambient air at 0°C, with a heat transfer coefficient $h = 200$ W/(m² · °C). If we assume one-dimensional, steady-state heat flow, the mathematical formulation of this problem is given by

$$\frac{d^2T(x)}{dx^2} - N^2T(x) = 0 \qquad \text{in } 0 < x < L$$

$$T(x) = 200°\text{C} \qquad \text{at } x = 0$$

$$T(x) = 0°\text{C} \qquad \text{at } x = L$$

where

$$N^2 = \frac{Ph}{kA} = \frac{4h}{kD}$$

By dividing the region $0 \leq x \leq L$ into five equal parts, calculate the temperature distribution along the rod, using finite differences. Compare the finite-difference solution with the exact solution.

5-5 Consider a straight fin of rectangular cross section having thermal conductivity $k = 40$ W/(m · °C), length $L = 3.0$ cm, thickness $t = 0.5$ cm, and a large width perpendicular to the plane of Fig. P5-5. The

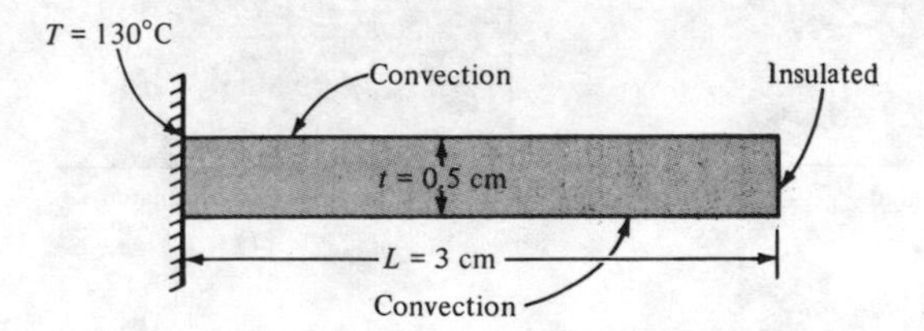

Figure P5-5

fin base is at $T_0 = 130$°C, and the fin tip is regarded insulated (that is, there is negligible heat loss). The fin dissipates heat by convection with a heat transfer coefficient $h = 400$ W/(m² · °C) into a fluid at $T_\infty = 30$°C.

(*a*) By using a one-dimensional mesh of size $\Delta x = 0.3$ cm, calculate by finite differences the temperature distribution along the fin.

(*b*) Estimate the heat transfer rate through the fin per 10-cm width perpendicular to the plane of the paper.

(*c*) Compare the numerical results with those obtained from the analytic solution of the one-dimensional fin equation.

Two-dimensional, steady-state formulations

5-6 By writing an energy balance on a differential volume element, derive the finite-difference form of the heat conduction equation

$$\frac{\partial^2 T}{\partial x^2} + \frac{\partial^2 T}{\partial y^2} + \frac{g}{k} = 0$$

for the nodal point A in each of the accompanying figures for the boundary conditions indicated.

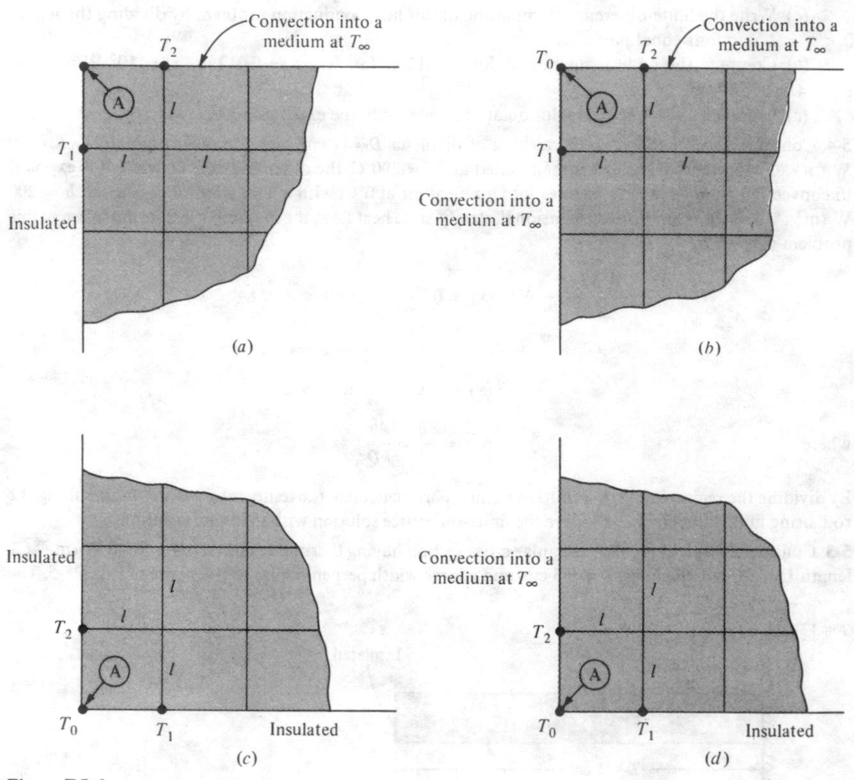

Figure P5-6

5-7 Write the finite-difference formulation of the heat conduction equation

$$\frac{\partial^2 T}{\partial x^2} + \frac{\partial^2 T}{\partial y^2} = 0$$

for the square region of side L by using a mesh size $\Delta x = \Delta y = L/3$ for the boundary conditions shown in Fig. P5-7. Express the resulting equations in matrix form for the unknown node temperatures T_m, $m = 1$ to 4.

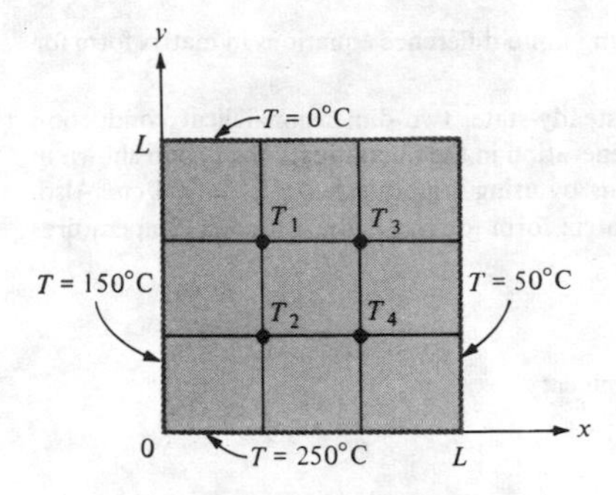

Figure P5-7

5-8 Write the finite-difference formulation of the heat conduction equation

$$\frac{\partial^2 T}{\partial x^2} + \frac{\partial^2 T}{\partial y^2} = 0$$

for the rectangular region $\frac{3}{4}L$ by L by using a mesh size $\Delta x = \Delta y = L/3$ for the boundary conditions shown in Fig. P5-8. Also, express the resulting finite-difference equations in matrix form for the unknown node temperatures T_m, $m = 1$ to 6.

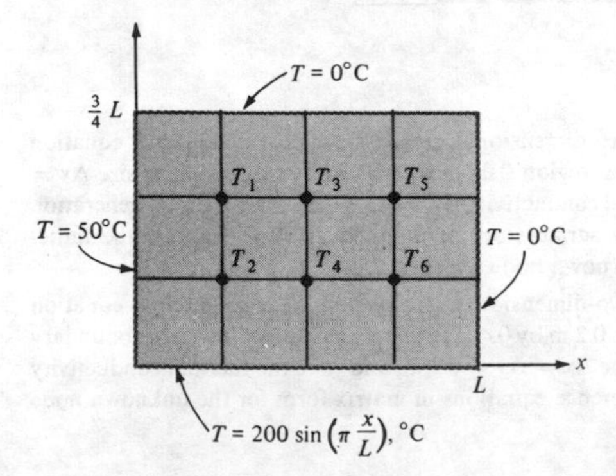

Figure P5-8

5-9 Write the finite-difference formulation for two-dimensional, steady-state heat conduction with no heat generation for a square region of side L by using a mesh size $\Delta x = \Delta y = L/3$ for the boundary

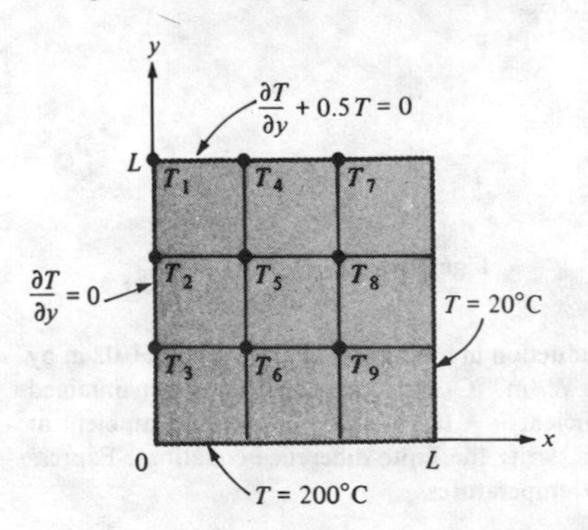

Figure P5-9

conditions shown in Fig. P5-9. Also, express the resulting finite-difference equations in matrix form for the nine unknown node temperatures T_m, $m = 1$ to 9.

5-10 Write the finite-difference formulation of the steady-state, two-dimensional heat conduction equation for constant thermal property and no heat generation in the medium, for the region shown in Fig. P5-10 subject to the specified boundary conditions by using a mesh size $\Delta x = \Delta y = 1$ cm. Also, express the resulting finite-difference equations in matrix form for the unknown node temperatures T_m, $m = 1$ to 7 [$k = 30$ W/(m · °C)].

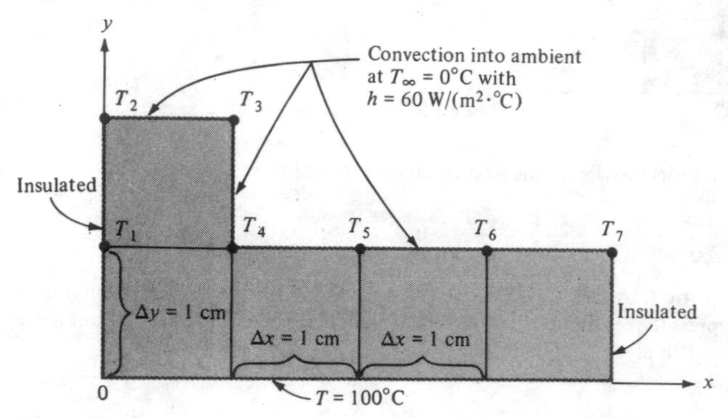

Figure P5-10

5-11 Write the finite-difference form of the two-dimensional, steady-state heat conduction equation with heat generation at a constant rate g_0 for a region 0.03 m by 0.03 m by using a mesh size $\Delta x = \Delta y = 0.01$ m for the following case: The thermal conductivity is $k = 25$ W/(m · °C), the heat generation rate is $g_0 = 10^7$ W/m³, and all the boundary surfaces are maintained at 10°C. Express the finite-difference equations in matrix form for the unknown node temperatures.

5-12 Write the finite-difference form of the two-dimensional, steady-state heat conduction equation with no heat generation for a rectangular region 0.2 m by 0.4 m in cross section, subject to the boundary conditions shown in Fig. P5-12. Use a mesh size $\Delta x = \Delta y = 0.1$ m, and take the thermal conductivity to be $k = 3$ W/(m · °C). Express the finite-difference equations in matrix form for the unknown node temperatures.

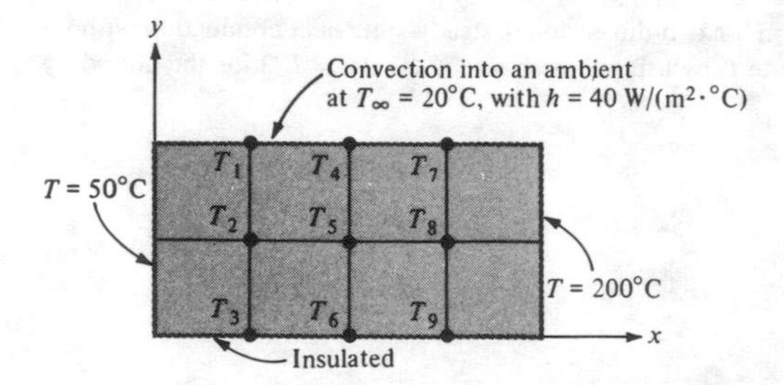

Figure P5-12

5-13 Consider two-dimensional, steady-state heat conduction in a long bar of cross section 0.02 m by 0.05 m. The material has a thermal conductivity $k = 60$ W/(m · °C). One surface of the rod is maintained at 250°C (see Fig. P5-13) while the remaining coefficient $h = 100$ W/(m² · °C) into an ambient at $T_\infty = 30$°C. By using a mesh size $\Delta x = \Delta y = 0.01$ m, write the finite-difference equations. Express these equations in matrix form for the unknown node temperatures.

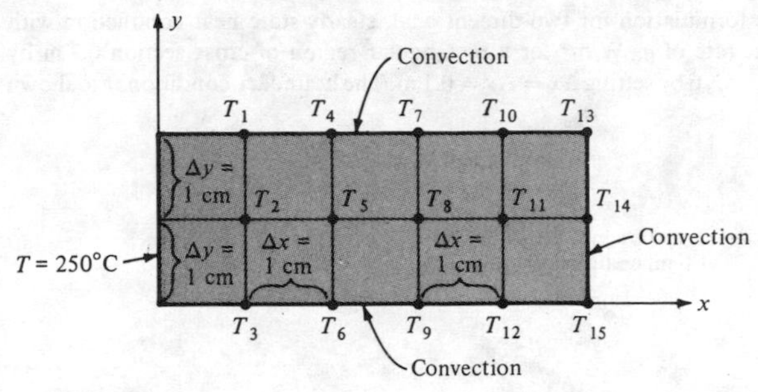

Figure P5-13

5-14 Write the finite-difference form of the heat conduction equation

$$\frac{\partial^2 T}{\partial x^2} + \frac{\partial^2 T}{\partial y^2} + \frac{g}{k} = 0$$

for the nine nodes $m = 1, 2, \ldots, 9$ at which the temperatures T_i are unknown, as shown in Fig. P5-14. The temperatures f_1, f_2, and f_3 are considered specified. Express these equations in matrix form for the unknown node temperatures T_m, $m = 1$ to 9.

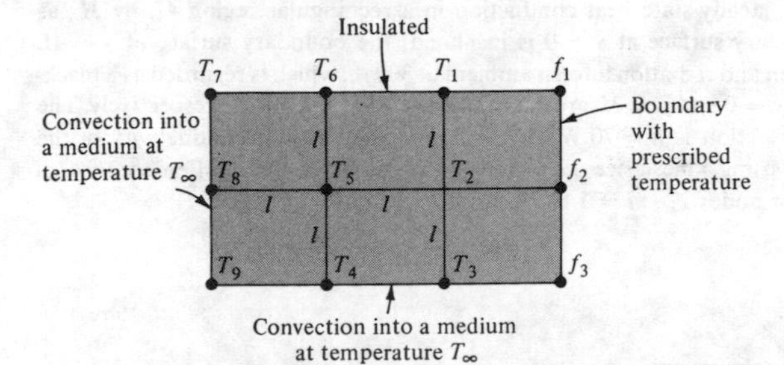

Figure P5-14

5-15 Write the finite-difference formulation of the heat conduction equation

$$\frac{\partial^2 T}{\partial x^2} + \frac{\partial^2 T}{\partial y^2} + \frac{g}{k} = 0$$

for nodes $m = 1, 2, \ldots, 12$ at which the temperatures T_i are unknown, as shown in Fig. P5-15. The temperatures f_1, f_2, and f_3 are considered specified. Express these results in matrix form for the unknown node temperatures T_m, $m = 1$ to 10.

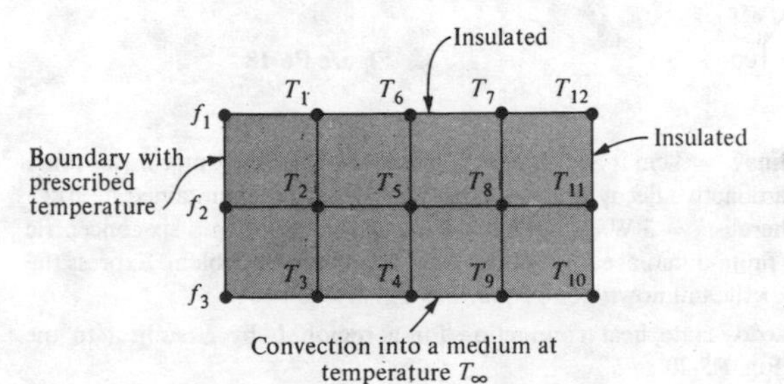

Figure P5-15

5-16 Write the finite-difference formulation for two-dimensional, steady-state heat conduction with energy generation at a constant rate of g_0 W/m³ for a rectangular region of cross section 0.2 m by 0.3 m for the six nodes $m = 1, 2, \ldots, 6$ by setting $\Delta x = \Delta y = 0.1$ m. The boundary conditions are shown in Fig. P5-16.

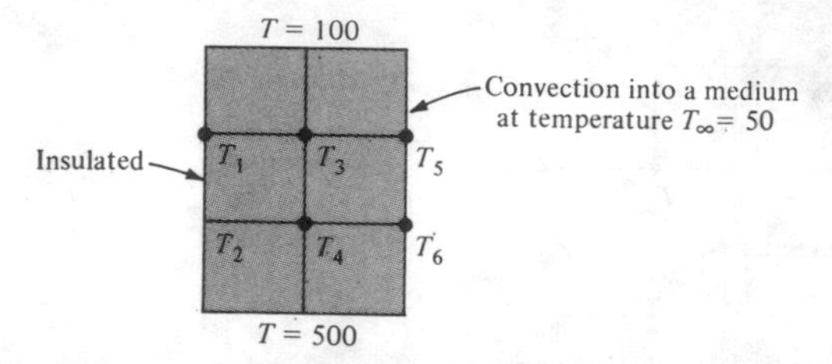

Figure P5-16

5-17 Consider one-dimensional, steady-state heat conduction in a conducting rod of radius $b = 0.5$ cm in which heat is generated at a constant rate of $g_0 = 6 \times 10^8$ W/m³, while the boundary surface at $b = 0.5$ cm is maintained at a uniform temperature $T_s = 25°C$. The thermal conductivity of the conductor is $k = 20$ W/(m · °C). By dividing the region into five concentric cylinders, each with $\Delta r = 0.1$ cm, and considering a unit length $\Delta z = 1$ cm, write the finite-difference form of this heat conduction problem. Express the resulting equations in matrix form for the node temperature T_m, $m = 0$ to 4.

5-18 Consider two-dimensional, steady-state heat conduction in a rectangular region $4L$ by $3L$, as shown in Fig. P5-18. The boundary surface at $x = 0$ is insulated; the boundary surface at $x = 4L$ dissipates heat by both convection and radiation into an ambient at 300 K, which is regarded as a blackbody; and boundary surfaces at $y = 0$ and $y = 3L$ are maintained at 1000 and 500 K, respectively. The heat transfer coefficient for convection is $h = 70$ W/(m² · °C), and the thermal conductivity of the material is $k = 4$ W/(m · °C). By using a mesh size $\Delta x = \Delta y = 0.1$ m, write the finite-difference form of this heat conduction problem for nodes T_m, $m = 1$ to 10.

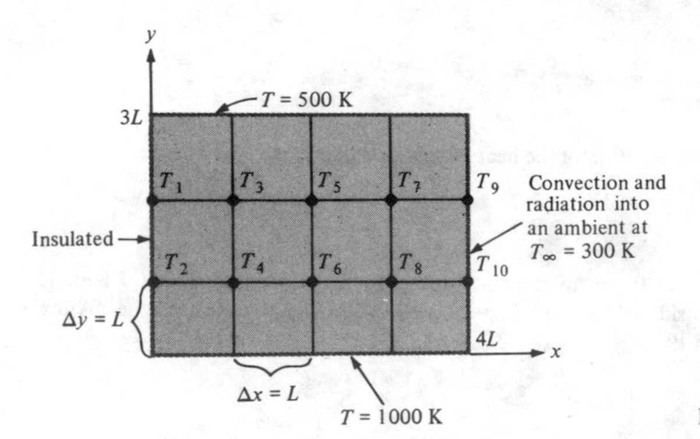

Figure P5-18

5-19 Consider a solid sphere of radius $b = 3$ cm in which heat is generated at a uniform, constant rate of $g_0 = 2 \times 10^6$ W/m³ owing to radioactive decay and the boundary surface is maintained at 20°C. The thermal conductivity of the sphere is $k = 3$ W/(m · °C). By dividing the region into six concentric spheres with $\Delta r = 0.5$ cm, write the finite-difference form of this heat conduction problem. Express the resulting equation in matrix form for the unknown node equations T_m, $m = 0$ to 5.

5-20 Consider two-dimensional, steady-state heat conduction for a region L by L subject to the boundary conditions illustrated in Fig. P5-20.

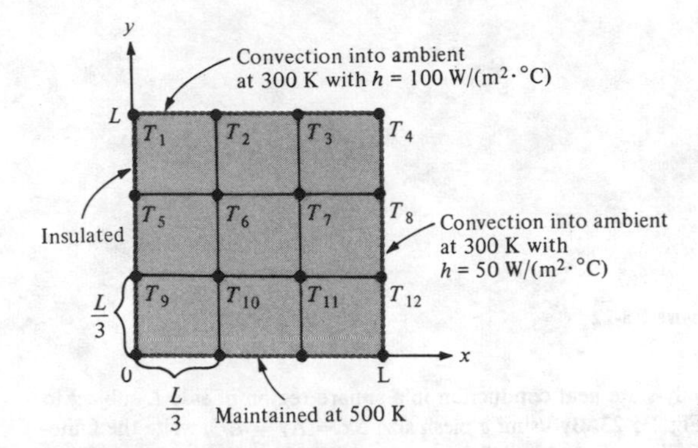

Figure P5-20

1. The boundary at $x = 0$ is insulated.
2. The boundary at $x = L$ exchanges heat by convection with an ambient at $T_\infty = 300$ K with a heat transfer coefficient $h = 50$ W/(m$^2 \cdot$ °C).
3. The boundary at $y = 0$ is kept at 500 K.
4. The boundary at $y = L$ dissipates heat by convection into an ambient at 300 K with a heat transfer coefficient $h = 100$ W/(m$^2 \cdot$ °C).

By using a mesh size $\Delta x = \Delta y = L/3$, write the finite-difference form of this heat conduction problem for nodes T_m, $m = 1$ to 12.

Two-dimensional steady-state solutions

5-21 Consider two-dimensional, steady-state heat conduction in a region L by L subject to the boundary conditions shown in Fig. P5-21. By using a coarse mesh $\Delta x = \Delta y = L/3$, write the finite-difference formulation of this heat conduction problem, and calculate the node temperatures T_m, $m = 1$ to 4. Repeat the analysis by refining the mesh size to $\Delta x = \Delta y = L/6$, and compare the results with those obtained by $\Delta x = \Delta y = L/3$.

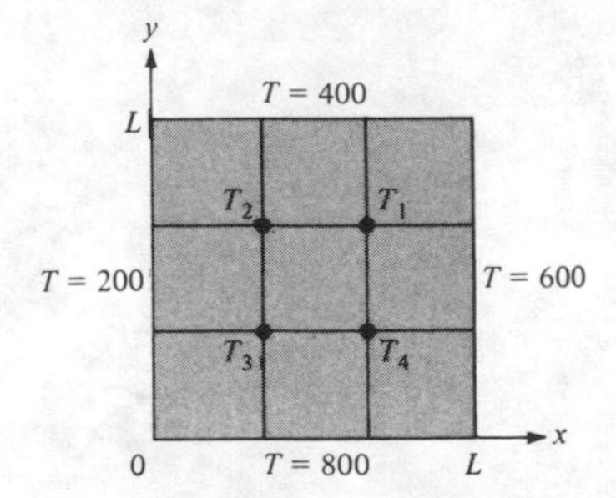

Figure P5-21

5-22 Consider two-dimensional, steady-state heat conduction in a rectangular region of cross section $2L$ by $3L$ subject to the boundary conditions shown in Fig. P5-22. By using a mesh size $\Delta x = \Delta y = L$, write the finite-difference formulation of this heat conduction problem, and calculate the node temperatures T_m, $m = 1$ to 4. Repeat the analysis by refining the mesh size to $\Delta x = \Delta y = L/2$, and compare the node temperatures with those obtained by $\Delta x = \Delta y = L$.

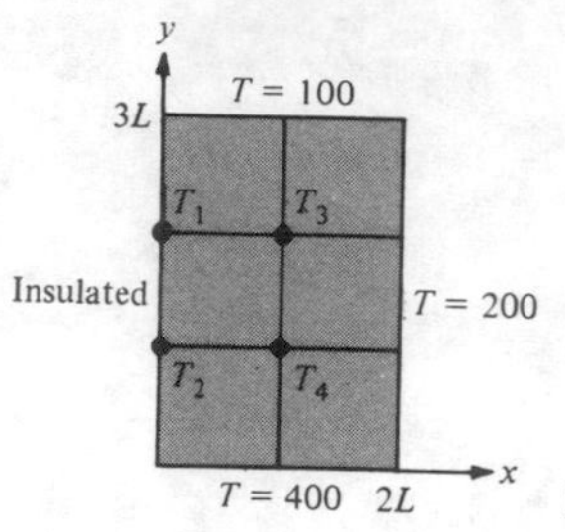

Figure P5-22

5-23 Consider two-dimensional, steady-state heat conduction in a square region of side L subject to the boundary conditions shown in Fig. P5-23. By using a mesh size $\Delta x = \Delta y = L/3$, write the finite-difference formulation of this heat conduction problem. Repeat the analysis by refining the mesh size to $\Delta x = \Delta y = L/6$, and compare the results with those obtained by $\Delta x = \Delta y = L/3$. Calculate the heat transfer rate through the boundary surface at $y = 0$ per 1-m length perpendicular to the plane of the figure for $L = 0.10$ m by taking the thermal conductivity as $k = 20$ W/(m · °C).

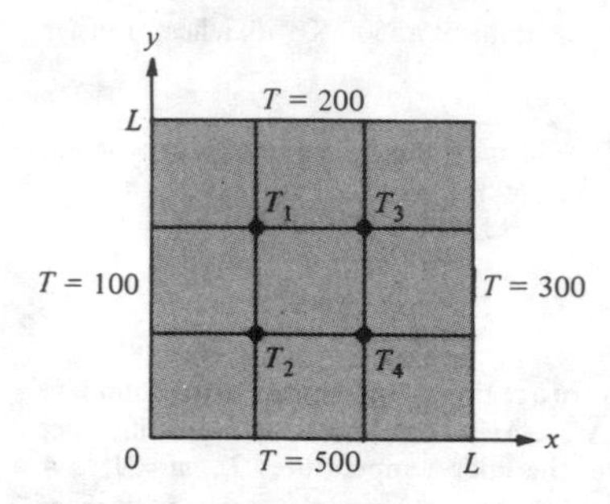

Figure P5-23

5-24 Solve by finite differences the steady-state temperature distribution in a rectangular region subject to the boundary conditions shown in Fig. P5-24 by setting $L = 4$ cm and $T_0 = 200$°C and using a mesh size $\Delta x = \Delta y = 1$ cm.

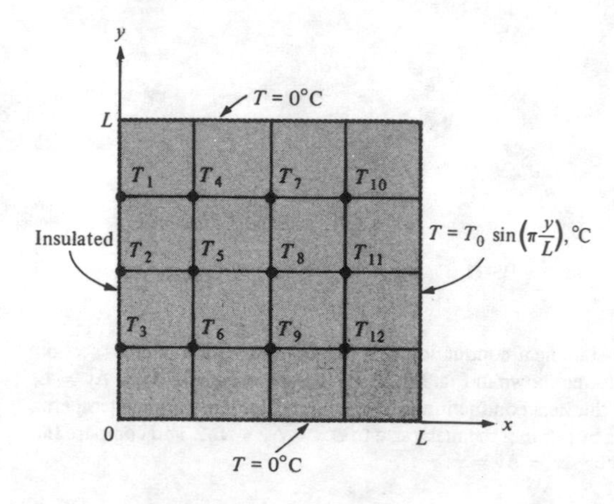

Figure P5-24

5-25 Solve by finite differences the steady-state temperature distribution in the rectangular region subject to the boundary conditions shown in Fig. P5-25 by setting $L = 4$ cm and $T_0 = 100°\text{C}$ and using a mesh size $\Delta x = \Delta y = 1$ cm.

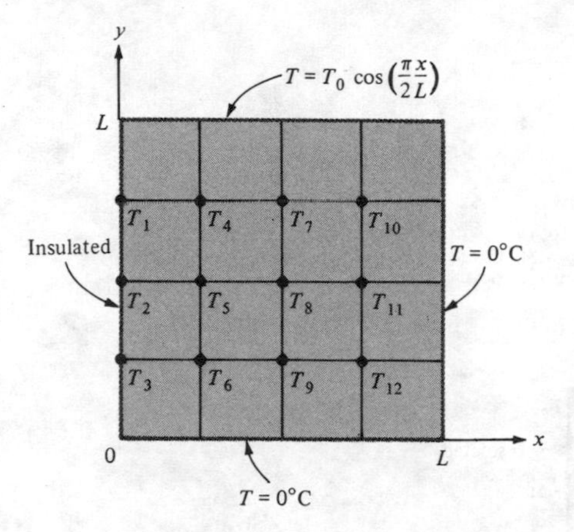

Figure P5-25

5-26 Solve by finite differences the steady-state temperature distribution in a rectangular region subject to the boundary conditions shown in Fig. P5-26 by setting $a = 6$ cm, $b = 4$ cm, and $T_0 = 200°\text{C}$ and using a mesh size $\Delta x = \Delta y = 1$ cm.

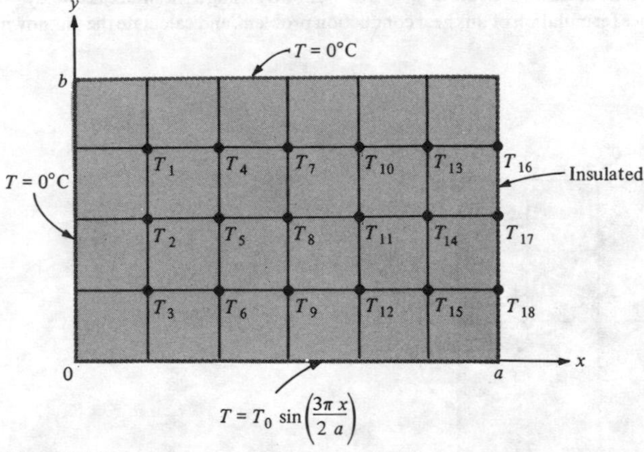

Figure P5-26

5-27 Consider two-dimensional, steady-state heat conduction in a region subject to the boundary conditions shown in Fig. P5-27. The material has a thermal conductivity $k = 60$ W/(m · °C). By using a finite-difference mesh $\Delta x = \Delta y = 0.3$ cm, develop the matrix equation for the unknown node temperatures. By solving these equations, calculate the node temperatures.

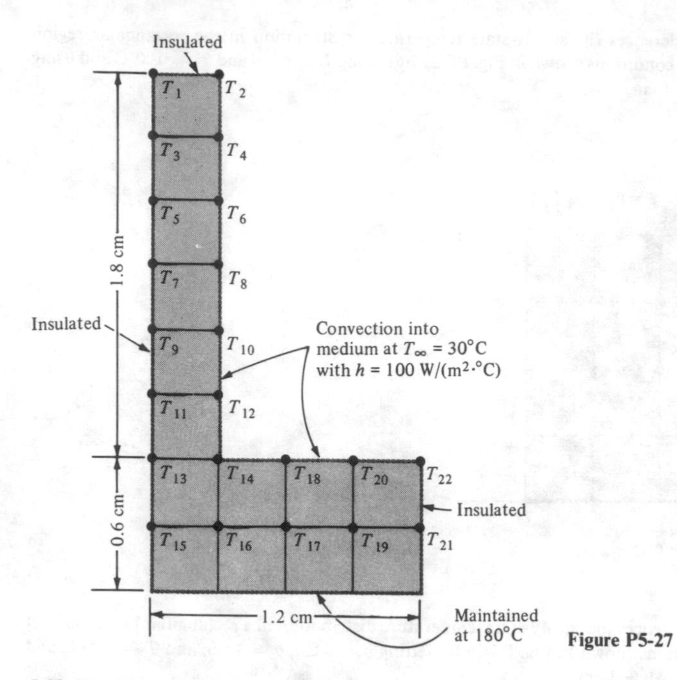

Figure P5-27

5-28 Consider two-dimensional, steady-state heat conduction in a region 5 cm by 5 cm subject to the boundary conditions shown in Fig. P5-28. The thermal conductivity of the material is $k = 60$ W/(m · °C), and there is heat generation in the medium at a rate of $g_0 = 10^7$ W/m^3. By using a mesh size $\Delta x = \Delta y = 1$ cm, develop a finite-difference formulation of this heat conduction problem, and calculate the unknown node temperatures.

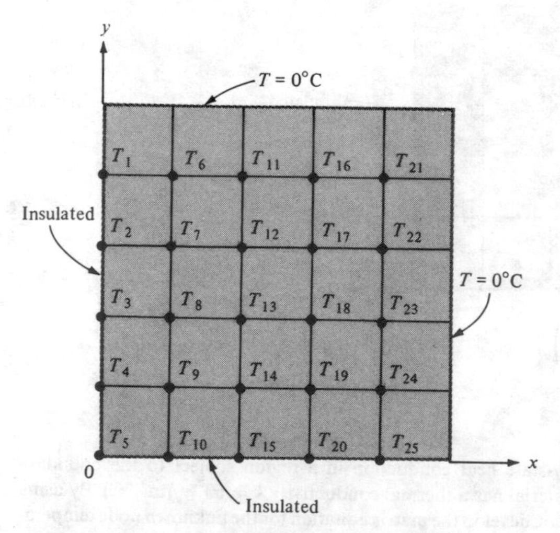

Figure P5-28

5-29 Consider two-dimensional, steady-state heat conduction in a 3 cm by 5 cm rectangular region subject to the boundary conditions shown in Fig. P5-29. By using a mesh size $\Delta x = \Delta y = 1$ cm, develop a finite-difference formulation of this heat conduction problem, and calculate the node temperatures.

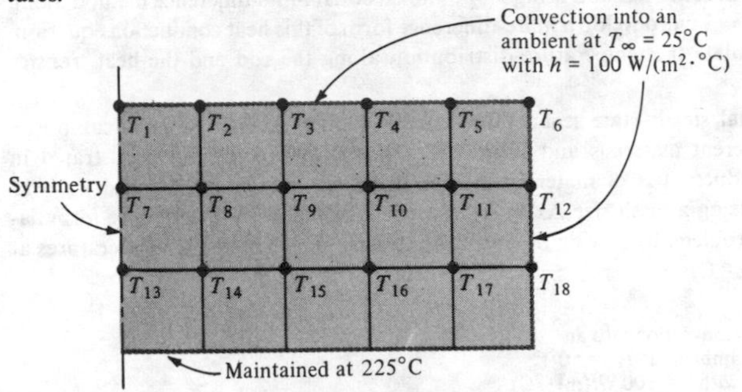

Figure P5-29

5-30 Heat transfer through the walls of a furnace can be represented as a two-dimensional, steady-state heat conduction problem for a geometry illustrated in Fig. P5-30 subject to the boundary conditions indicated. The thermal conductivity of the brick walls is $k = 1.2$ W/(m · °C); the convection heat transfer coefficient and the temperature for the external ambient are $h = 100$ W/(m^2 · °C) and $T_\infty = 30$°C, respectively. By using a mesh size $\Delta x = \Delta y = 5$ cm, develop a finite-difference formulation of this heat conduction problem; by solving the finite-difference equations, calculate the node temperatures.

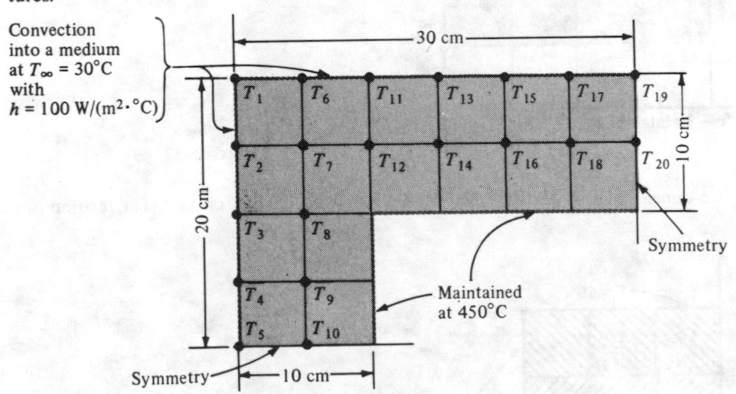

Figure P5-30

5-31 Consider a tapered rod $L = 8$ cm long having its radius r varying in the x direction according to the relation $r = 0.2(10 - x)$ cm, as in Fig. P5-31. The base at $x = 0$ is maintained at $T_0 = 230$°C, and the tip at $x = 8$ cm is assumed to be insulated. The rod dissipates heat by convection with a heat transfer

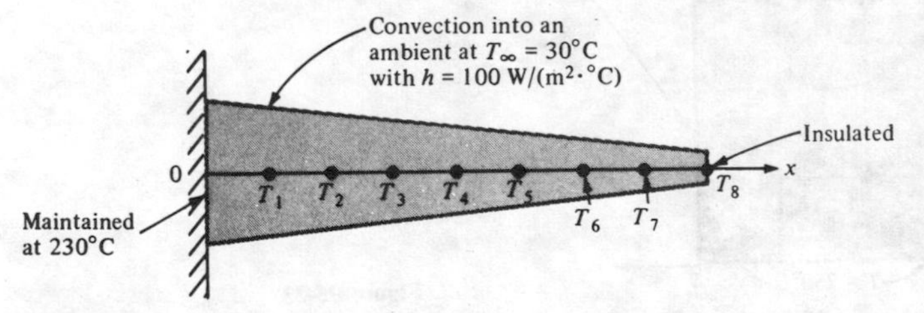

Figure P5-31

coefficient $h = 100$ W/(m$^2 \cdot$ °C) from its lateral surfaces into an ambient at $T_\infty = 30$°C. By approximating this problem as a one-dimensional one and using a one-dimensional finite-difference method along the x axis with a mesh size $\Delta x = 1$ cm, develop a finite-difference form of this heat conduction equation. Solving these equations, calculate the temperature distribution along the rod and the heat transfer rate through the rod base.

5-32 Consider two-dimensional, steady-state heat conduction in a 4 cm by 5 cm rectangular composite region consisting of two different materials and subject to the boundary conditions illustrated in Fig. P5-32. The thermal conductivities of materials A and B are $k_A = 1$ W/(m · °C) and $k_B = 15$ W/(m · °C), respectively. By using a mesh size $\Delta x = \Delta y = 1$ cm, develop a finite-difference formulation of this heat conduction problem. By solving the resulting equations, calculate the temperatures at the nodes. Take $k = 15$ W/(m · °C).

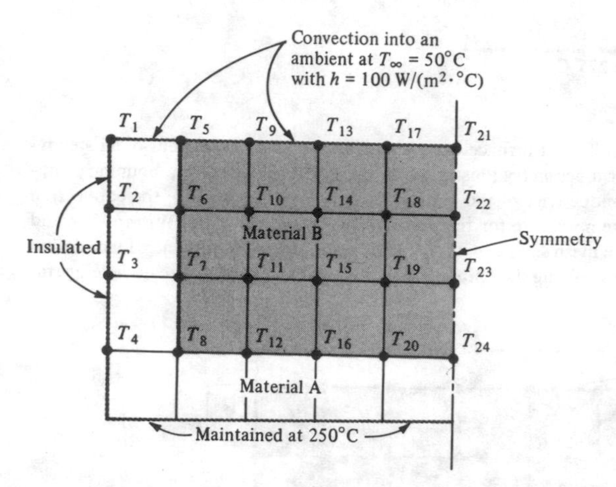

Figure P5-32

5-33 Consider two-dimensional, steady-state heat conduction in a 3 cm by 6 cm rectangular composite

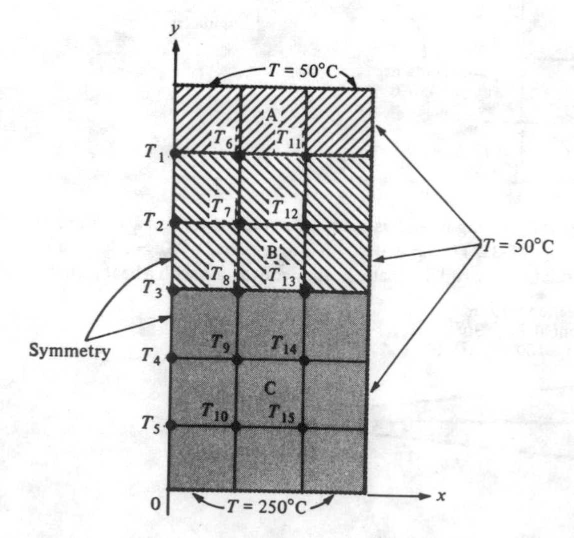

Figure P5-33

region consisting of three different materials A, B, and C, subject to the boundary conditions illustrated in Fig. P5-33. Materials A, B, and C have thermal conductivities $k_A = 1$, $k_B = 2$, and $k_C = 15$ W/(m · °C). By using a mesh size $\Delta x = \Delta y = 1$ cm, develop the finite-difference formulation of this heat conduction problem, and calculate the node temperatures.

5-34 Consider two-dimensional, steady-state heat conduction in a 3 cm by 3 cm composite region consisting of two different materials and subjected to the boundary conditions as shown in Fig. P5-34. Materials A and B have thermal conductivities $k_A = 60$ and $k_B = 20$ W/(m · °C), respectively. By using a mesh $\Delta x = \Delta y = 1$ cm, develop a finite-difference formulation of this heat conduction problem, and calculate the node temperatures.

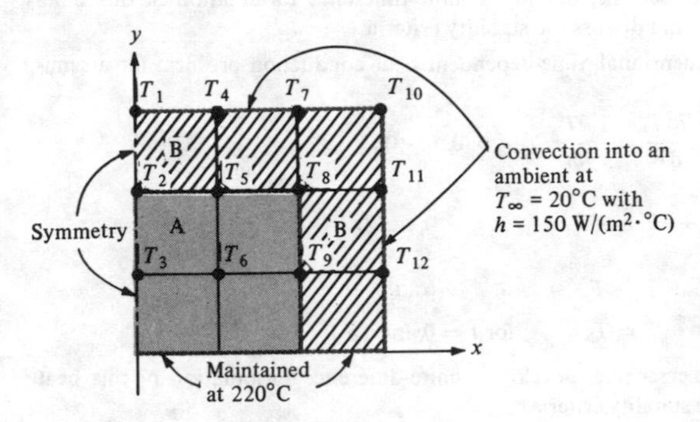

Figure P5-34

Cylindrical and spherical symmetry

5-35 (*a*) Consider one-dimensional, steady-state, radial heat conduction in a rod of radius $b = 2$ cm, thermal conductivity $k = 350$ W/(m · °C), in which energy is generated at a rate of $h = 1.4 \times 10^9$ W/m^3. The boundary surface at $b = 0$ is kept at zero temperature. By dividing the region $0 \leq r \leq b$ into five equal parts, calculate the radial temperature distribution in the rod by finite differences. Compare the finite-difference calculations with the exact solution. (*b*) Repeat the solution by dividing the region $0 \leq r \leq b$ into 10 equal parts, if the accuracy is not good.

5-36 (*a*) Consider one-dimensional, steady-state, radial heat conduction in a solid sphere of radius $b = 2$ cm. thermal conductivity $k = 350$ W/(m · °C), in which energy is generated at a rate of $g = 1.4 \times 10^9$ W/m^3. The boundary surface at $b = 0$ is kept at zero temperature. By dividing the region $0 \leq r \leq b$ into five equal parts, calculate the radial temperature distribution in the sphere by finite differences. Compare the finite-difference calculations with the exact solution. (*b*) Repeat the solution by dividing the region $0 \leq r \leq b$ into 10 equal parts, if the accuracy is not good.

5-37 A 0.5-cm-diameter, long copper rod of thermal conductivity $k = 380$ W/(m · °C) has heat generation at a rate of $g_0 = 10^8$ W/m^3 as a result of the passage of electric current. The rod dissipates heat from its outer surface by convection with a heat transfer coefficient $h = 150$ W/(m^2 · °C) into an ambient at $T_\infty = 25$°C. The axial variation of temperature is neglected. Calculate the radial distribution of steady-state temperature by finite differences by dividing the region into five circular elements, each of radial thickness $\Delta r = 0.05$ cm, and taking $\Delta z = 1$ cm.

5-38 A 0.5-cm-diameter, solid-copper sphere of thermal conductivity $k = 380$ W/(m · °C) has energy generation at a rate of $g_0 = 10^8$ W/m^3. The sphere dissipates heat from its outer surface by convection with a heat transfer coefficient $h = 150$ W/(m^2 · °C) into an ambient at $T_\infty = 25$°C. Calculate the radial distribution of steady-state temperature by finite differences by dividing the region into five elements, each of radial thickness $\Delta r = 0.05$ cm.

Unsteady heat conduction formulations

5-39 Consider the following one-dimensional, time-dependent heat conduction problem for a semi-infinite region $x \geq 0$:

$$\frac{\partial^2 T}{\partial x^2} = \frac{1}{\alpha}\frac{\partial T}{\partial t} \qquad x > 0, t > 0$$

$$T = T_0 \qquad x = 0, t > 0$$

$$T \to T_b \qquad \text{as } x \to \infty, t > 0$$

$$T = T_b \qquad \text{for } t = 0, \text{ in } x \geq 0$$

By using the *explicit* finite-difference scheme, develop a finite-difference formulation of this time-dependent heat conduction problem, and discuss the stability criteria.

5-40 Consider the following one-dimensional, time-dependent heat conduction problem for a semi-infinite region $x \leq 0$:

$$\frac{\partial^2 T}{\partial x^2} = \frac{1}{\alpha}\frac{\partial T}{\partial t} \qquad x > 0, t > 0$$

$$-k\frac{\partial T}{\partial x} + hT = hT_\infty \qquad \text{at } x = 0, t > 0$$

$$T \to T_b \qquad \text{at } x \to \infty, t > 0$$

$$T = T_b \qquad \text{for } t = 0, \text{ in } x \geq 0$$

By using the *explicit* finite-difference scheme, develop a finite-difference formulation of this heat conduction problem, and discuss the stability criteria.

5-41 Consider the following one-dimensional, time-dependent heat conduction problem for a slab of thickness L:

$$\frac{\partial^2 T}{\partial x^2} = \frac{1}{\alpha}\frac{\partial T}{\partial t} \qquad \text{in } 0 < x < L, t > 0$$

$$-k\frac{\partial T}{\partial x} + hT = hT_\infty \qquad \text{at } x = 0, t > 0$$

$$T = T_L \qquad \text{at } x = L, t > 0$$

$$T = 0 \qquad \text{for } t = 0 \text{ in } 0 \leq x \leq L$$

By using the *explicit* finite-difference scheme, develop a finite-difference formulation of this heat conduction problem, and discuss the stability criteria.

5-42 By using the *implicit* finite-difference scheme, develop the finite-difference formulation of Problem 5-41, and discuss the stability criteria.

5-43 By using the *Crank-Nicolson* finite-difference scheme, develop the finite-difference formulation of Problem 5-41, and discuss the stability criteria.

5-44 By using the *explicit* finite-difference scheme, develop the finite-difference formulation of the following heat conduction problem:

$$k\frac{\partial^2 T}{\partial x^2} + g = \rho c_p \frac{\partial T}{\partial t} \qquad \text{in } 0 < x < L$$

$$\frac{\partial T}{\partial x} = 0 \qquad \text{at } x = 0$$

$$k\frac{\partial T}{\partial x} + hT = 0 \qquad \text{at } x = L$$

$$T = T_0 \qquad \text{for } t = 0$$

5-45 By using the *explicit* finite-difference scheme, develop a finite-difference formulation for the following time-dependent heat conduction equation:

$$k\left(\frac{\partial^2 T}{\partial x^2} + \frac{\partial^2 T}{\partial y^2}\right) + g = \rho c_p \frac{\partial T}{\partial t}$$

Discuss the stability criteria.

5-46 Consider one-dimensional, time-dependent heat conduction in a slab of thickness L, which is initially at a uniform temperature T_0. For $t > 0$, the boundary surface at $x = 0$ is kept insulated and the boundary surface at $x = L$ dissipates heat by natural convection and radiation into an ambient which is at a temperature T_∞ K and can be regarded as a blackbody. Heat transfer by radiation and convection at the boundary surface $x = L$ can be calculated from

$$q^r = \sigma\varepsilon A(T^4 - T_\infty^4) \qquad \text{W}$$

$$q^c = C_1 A(T - T_\infty)^{5/4} \qquad \text{W}$$

where ε is the emissivity and $C_1 = 1.92$ is a constant for free convection. By using a mesh size $\Delta x = L/4$ and the explicit scheme, develop the finite-difference formulation of this heat conduction problem.

5-47 Consider two-dimensional, time-dependent heat conduction in a region L by L which is initially at a uniform temperature T_0 and for $t > 0$ is subjected to the boundary conditions shown in Fig. P5-47. By using an *explicit* scheme and a mesh size $\Delta x = \Delta y = L/3$, develop the finite-difference formulation of this time-dependent heat conduction problem.

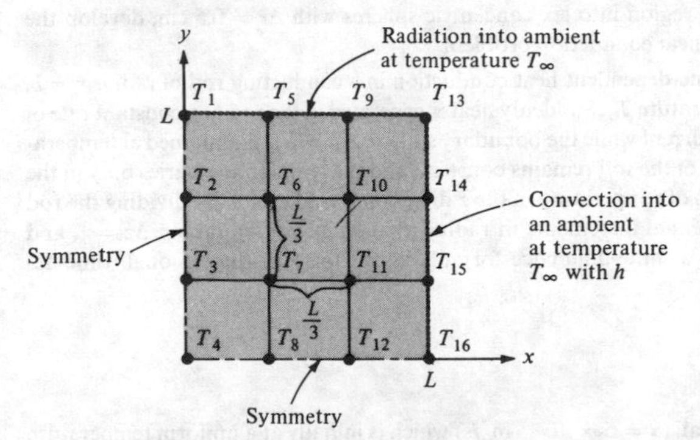

Figure P5-47

5-48 Consider two-dimensional, time-dependent heat conduction in a region 3 cm by 5 cm, which is initially at a uniform temperature $T_0 = 300°C$ and for $t > 0$ is subjected to the boundary conditions shown in Fig. P5-48. By using an explicit scheme and a mesh size $\Delta x = \Delta y = 1$ cm, develop the finite-difference representation of this heat conduction problem.

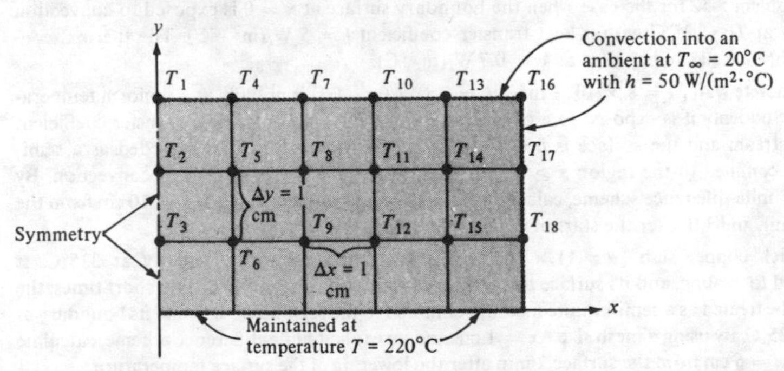

Figure P5-48

5-49 Consider two-dimensional, time-dependent heat conduction in a region which is initially at a uniform temperature T_0 and for $t > 0$ is subjected to the boundary conditions illustrated in Fig. P5-49. By using an explicit scheme and a mesh size $\Delta x = \Delta y = 1$ cm, develop the finite-difference formulation of this heat conduction problem.

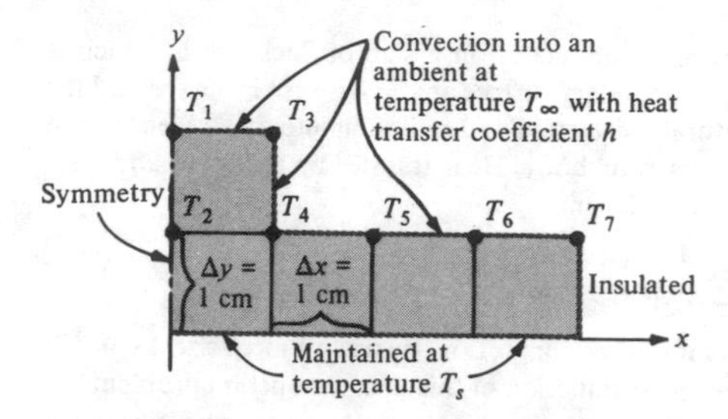

Figure P5-49

5-50 Consider one-dimensional, time-dependent heat conduction in a sphere of radius $b = 3$ cm, in which heat is generated at a rate of g_0 W/m^3. Initially, the sphere is at a uniform temperature T_0, and for $t > 0$ the boundary surface at $r = b$ is maintained at 0°C. By using an explicit method of the finite-difference scheme and dividing the region into six concentric spheres with $\Delta r = 0.5$ cm, develop the finite-difference formulation of this heat conduction problem.

5-51 Consider one-dimensional, time-dependent heat conduction in a conducting rod of radius $r = b$, which is initially at a uniform temperature T_0. Suddenly heat is generated in the rod at a constant rate of g_0 W/m^3 by the passage of electric current while the boundary surface at $r = b$ is maintained at temperature T_0. The thermal conductivity k of the rod remains constant, and the temperature varies only in the radial (r) direction (that is, variation of temperature in the z direction is negligible). By dividing the rod into five concentric cylinders with equal increments in radius (that is, $\Delta r = b/5$), taking $\Delta z = 1$, and using the *explicit* scheme, develop a finite-difference formulation of this one-dimensional, time-dependent heat conduction problem.

Unsteady heat conduction solutions

5-52 A large and very thick brick wall ($\alpha = 5 \times 10^{-7}$ m^2/s) which is initially at a uniform temperature $T_i = 125$°C is suddenly exposed to cooling by maintaining its surface at $x = 0$ at $T_0 = 25$°C. To calculate the temperature transients at depths small in comparison to the thickness, the wall can be regarded as a semi-infinite medium confined to the region $x \geq 0$. By using an explicit scheme and a mesh size $\Delta x =$ 0.3 cm, calculate the temperature $x = 1.2$ cm from the surface $t = 1$ and 5 min after the exposure.

5-53 Repeat Problem 5-52 for the case when the boundary surface at $x = 0$ is exposed to convection into an ambient at $T_\infty = 25$°C with a heat transfer coefficient $h = 5$ W/(m$^2 \cdot$ °C). The thermal conductivity of the brick wall can be taken as $k = 0.7$ W/(m · °C).

5-54 A thick concrete wall [$\alpha = 8 \times 10^{-7}$ m^2/s, $k = 0.8$ W/(m · °C)] is initially at a uniform temperature $T_i = 60$°C. Suddenly it is exposed to a cool airstream at $T_\infty = 5$°C. The heat transfer coefficient between the airstream and the surface is $h = 10$ W/(m$^2 \cdot$ °C). The wall can be regarded as a semi-infinite medium confined to the region $x \geq 0$ with the surface at $x = 0$ subjected to convection. By using an explicit finite-difference scheme, calculate the wall temperature at a depth $x = 10$ cm from the surface $t = 10$ min and 1 h after the start of cooling, by taking $\Delta x = 2$ cm.

5-55 A very thick copper slab [$\alpha = 11 \times 10^{-5}$ m^2/s, $k = 386$ W/(m · °C)], initially at 325°C, is suddenly exposed to cooling, and its surface temperature is lowered to $T_0 = 25$°C. For short times, the copper slab can be treated as a semi-infinite medium confined to the region $x \geq 0$, while its boundary at $x = 0$ is kept at 25°C. By using a mesh size $\Delta x = 1$ cm and an explicit finite-difference scheme, calculate the temperature $x = 6$ cm from the surface 2 min after the lowering of the surface temperature.

5-56 A thick copper plate [$\alpha = 11 \times 10^{-5}$ m^2/s, $k = 386$ W/(m · °C)] is initially at a uniform temperature $T_i = 25$°C. Suddenly one of its surfaces is exposed to a constant heat flux $q_0 = 2 \times 10^5$ W/m^2. For short times the copper plate can be treated as a semi-infinite medium confined to the region $x \geq 0$, subjected to a constant heat flux at the boundary surface $x = 0$. By using a mesh size $\Delta x = 2$ cm and an explicit finite-difference scheme, calculate the temperature $x = 10$ cm from the surface $t = 5$ min after the start of surface heating.

5-57 A very thick, nickel-steel plate [$\alpha = 0.5 \times 10^{-5}$ m^2/s, $k = 20$ W/(m · °C)], initially at a uniform temperature $T_i = 280$°C, is suddenly exposed to a cool airstream at $T_\infty = 30$°C. The heat transfer coefficient between the airstream and the surface is $h = 60$ W/(m^2 · °C). For short times, the plate can be treated as a semi-infinite medium confined to the region $x \geq 0$ with surface at $x = 0$ subjected to convective cooling. By using a mesh size $\Delta x = 1$ cm and an explicit finite-difference scheme, calculate the temperature at a depth $x = 4$ cm at $t = 2$ and 5 min after the start of cooling.

5-58 An 8-cm-thick chrome-steel plate [$\alpha = 1.6 \times 10^{-5}$ m^2/s, $k = 61$ W/(m · °C)], initially at a uniform temperature $T_i = 325$°C, is suddenly exposed to a cool airstream at $T_\infty = 25$°C at both of its surfaces. The heat transfer coefficient between the air and the surface is $h = 400$ W/(m^2 · °C). By using an explicit finite-difference scheme and a mesh size $\Delta x = 1$ cm, determine the center plane temperature $t = 5$ and 15 min after the start of cooling.

5-59 A slab of thickness $L = 12$ cm and thermal diffusivity $\alpha = 2 \times 10^{-5}$ m^2/s has an initial temperature distribution $T_i = 100 \sin (\pi x/L)$. For $t > 0$, both boundaries are kept at zero temperature. By using an explicit finite-difference scheme and a mesh size $\Delta x = 2$ cm, calculate the center temperature $t = 1$, 5, and 15 min after the start of cooling.

5-60 A slab of thickness $L = 6$ cm and thermal diffusivity $\alpha = 8 \times 10^{-5}$ m^2/s has an initial temperature distribution $T_i = 100 \cos [(\pi/2)(x/L)]$. For $t > 0$, the boundary surface at $x = 0$ is kept insulated, and that at $x = 6$ cm is kept at zero temperature. By using an explicit finite-difference scheme and a mesh size $\Delta x = 1$ cm, calculate the temperature of the insulated boundary $t = 1$ and 10 min after the start of cooling.

5-61 A slab of thickness $L = 6$ cm and thermal diffusivity $\alpha = 8 \times 10^{-5}$ m^2/s has an initial temperature distribution $T_i = 100 \sin [(\pi/2)(x/L)]$. For $t > 0$, the boundary surface at $x = 0$ is kept at zero temperature, and that at $x = 6$ cm is kept insulated. By using an explicit finite-difference scheme and a mesh size $\Delta x = 1$ cm, calculate the temperature of the insulated boundary $t = 1$ and 5 min after the start of cooling.

5-62 An aluminum plate ($\alpha = 8 \times 10^{-5}$ m^2/s) $L = 4$ cm thick is initially at a uniform temperature $T_i = 20$°C. Suddenly one of its surfaces is raised to $T_0 = 220$°C while the other surface is kept insulated. By using an explicit finite-difference scheme and a mesh size $\Delta x = 1$ cm, calculate the temperature of the insulated surface at 1 and 3 min after the other surface is exposed to high temperature.

5-63 Repeat Problem 5-56 for a slab of thickness $L = 10$ cm for the case when one of the boundary surfaces is exposed to hot air at $T_\infty = 220$°C with a heat transfer coefficient $h = 50$ W/(m^2 · °C) while the other surface is kept insulated. The thermal conductivity of aluminum is $k = 204$ W/(m · °C).

5-64 A carbon-steel bar of 4 cm by 4 cm cross section [$\alpha = 1 \times 10^{-5}$ m^2/s, $k = 35$ W/(m · °C)] is initially at a uniform temperature $T_i = 425$°C. Suddenly all its surfaces are exposed to cooling by an

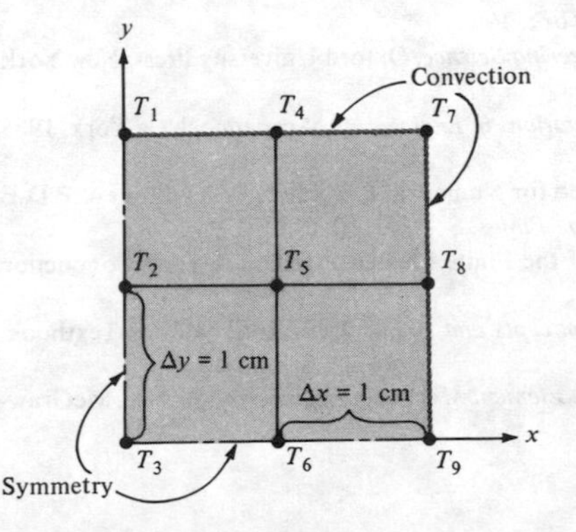

Figure P5-64

airstream at $T_\infty = 25°C$ with a heat transfer coefficient $h = 100\ W/(m^2 \cdot °C)$. By using an *explicit* finite-difference scheme and mesh size $\Delta x = \Delta y = 1$ cm, calculate the center temperature $t = 1$ and 5 min after the start of cooling. Because of symmetry consider only one-quarter of the cross section, as illustrated in Fig. P5-64.

5-65 A chrome-steel bar of 6 cm by 3 cm cross section [$\alpha = 1.6 \times 10^{-5}\ m^2/s$, $k = 60\ W/(m \cdot °C)$] is initially at a uniform temperature $T_i = 20°C$. Suddenly one of its surfaces having 6-cm width is exposed to hot air at $T_\infty = 220°C$ with a heat transfer coefficient $h = 350\ W/(m^2 \cdot °C)$, while the remaining surfaces are kept insulated, as illustrated in Fig. P5-65. By using an explicit finite-difference scheme and mesh size $\Delta x = \Delta y = 1$ cm, calculate how long it will take for the midpoint of the insulated surface of width $L = 6$ cm to reach 120°C.

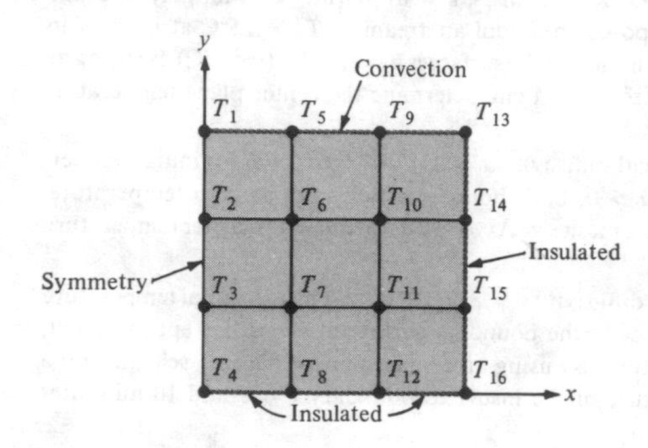

Figure P5-65

REFERENCES

1. Fox, L.: *Numerical Solution of Ordinary and Partial Differential Equations*, Addison-Wesley, Reading, Mass., 1962.
2. Smith, G. D.: *Numerical Solution of Partial Differential Equations with Exercises and Worked Solutions*, Oxford University Press, London, 1965.
3. Richtmeyer, R. D.: *Difference Methods for Initial Value Problems*, Interscience, New York, 1957.
4. Dusinberre, G. M.: *Heat Transfer Calculations by Finite Differences*, International Textbook, Scranton, Pa., 1961.
5. Forsythe, G. E., and W. R. Wasow: *Finite Differences Method for Partial Differential Equations*, Wiley, New York, 1960.
6. Larkin, B. K.: "Some Finite Difference Methods for Problems in Transient Heat Flow," *Chem. Eng. Prog., Symp. Ser.* 59, **61** (1965).
7. Macon, N.: *Numerical Analysis*, Wiley, New York, 1963.
8. Southwell, R. V.: *Relaxation Methods in Engineering Science*, Oxford University Press, New York, 1940.
9. Ames, W. F.: *Nonlinear Partial Differential Equations in Engineering*, Academic, New York, 1965, pp. 365–389.
10. Crank, J., and P. Nicolson: "A Practical Method for Numerical Evaluation of Solutions of P.D.E. of the Heat Conduction Type," *Proc. Cambridge Philos. Soc.*, **43**:50–67 (1947).
11. Nickell, R. E., and E. Wilson: "Application of the Finite Element Method to Heat Conduction Analysis," *Nucl. Eng. Des.*, **4**:276–286 (1966).
12. Oktay, Ural: *Finite Element Method: Basic Concepts and Applications*, International Textbook, Scranton, Pa., 1973.
13. Zienkiewicz, O. C., and I. K. Cheung: *The Finite Element Method in Engineering Science*, McGraw-Hill, New York, 1971.

14. Martin, H. C., and G. F. Carey: *Introduction to Finite Element Analysis*, McGraw-Hill, New York, 1973.
15. Huebner, K. H.: *Finite Element Method for Engineers*, Wiley, New York, 1975.
16. Segerlind, L. J.: *Applied Finite Element Analysis*, Wiley, New York, 1976.
17. Özişik, M. N.: *Heat Conduction*, Wiley, New York, 1980.
18. International Mathematical and Statistical Libraries, 7th ed.: GNB Bldg., 7500 Ballaire Blvd., Houston, TX 77036, 1979.

CHAPTER

SIX

CONVECTION—CONCEPTS AND BASIC RELATIONS

So far we considered heat transfer by conduction in solids in which no motion of the medium was involved. In conduction problems, the convection entered the analysis merely as a boundary condition in the form of a heat transfer coefficient.

Our objective in this and the following chapters on convection is to establish the physical and mathematical basis for the understanding of convective transport and to reveal various heat transfer correlations.

In engineering applications, the pressure drop or the drag force associated with flow inside ducts or over bodies is also of interest. Therefore, appropriate correlations are presented to predict pressure drop or drag force in flow.

The analysis of convection is complicated, because the fluid motion affects the pressure drop, the drag force, and the heat transfer. To determine the drag force or the pressure drop, the velocity field in the immediate vicinity of the surface must be known. To determine the heat transfer by convection, the velocity distribution in the flow also is needed, because the velocity enters the energy equation; the solution of the energy equation yields the temperature distribution in the flow field.

The literature on convection heat transfer is overhelming and ever-growing. In recent years, with the availability of high-speed, large-capacity digital computers, great advances have been made in the analysis of very complicated heat transfer problems in great detail. Nonetheless, a large number of simpler engineering problems can be handled with the use of standard heat transfer correlations. Therefore, we focus our attention on such situations. To achieve this objective, in this chapter we present a coherent view of the subject of convection in order to provide a firm basis for applications. Basic concepts associated with flow over a body, flow inside a duct, and turbulence are discussed. The role of temperature and velocity distribution in flow on heat transfer and drag force is illustrated.

The velocity and temperature distributions in flow are determined from the solution of the equations of motion and energy. Therefore, such equations are presented for the case of two-dimensional, constant-property, incompressible flow in rectangular and cylindrical coordinate systems. The simplification of these equations is illustrated in order to obtain the governing equations for the analysis of simpler heat transfer problems.

Finally, the physical significance of dimensionless parameters is discussed, and the boundary-layer equations are presented.

6-1 FLOW OVER A BODY

When a fluid flows over a body, the velocity and temperature distribution at the immediate vicinity of the surface strongly influence the heat transfer by convection. The *boundary-layer* concept frequently is introduced to model the velocity and temperature fields near the solid surface in order to simplify the analysis of convective heat transfer. So we are concerned with two different kinds of boundary layers, the *velocity boundary layer* and the *thermal boundary layer*.

Velocity Boundary Layer

To illustrate the concept of the velocity boundary layer, we consider the flow of a fluid over a flat plate, as illustrated in Fig. 6-1. The fluid at the *leading edge of the plate* (i.e., at $x = 0$) has a velocity u_∞ which is parallel to the plate surface. As the fluid moves in the x direction along the plate, those fluid particles that make contact with the plate surface assume zero velocity (i.e., no slip at the wall). Therefore, starting from the plate surface there will be a retardation in the x direction component of the velocity $u(x, y) \equiv u$. That is, at the plate surface $y = 0$, the axial velocity component is zero, or $u = 0$. The retardation effect is reduced when the fluid is moving at a location away from the plate surfaces at distances sufficiently far from the plate the retardation effect is considered zero, that is, $u = u_\infty$ for large y. Therefore, at each location x along the plate, one considers a distance $y = \delta(x)$ from the surface of the plate where the axial velocity component u equals 99 percent of the free-stream velocity u_∞, that is, $u = 0.99u_\infty$. The locus of such points where $u = 0.99u_\infty$ is called the *velocity boundary layer* $\delta(x)$. With the boundary-layer concept thus introduced for flow over a flat plate, the flow field can be separated

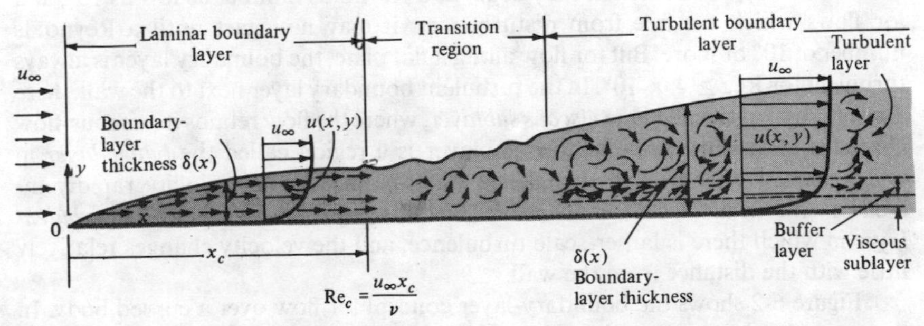

Figure 6-1 Boundary-layer concept for flow along a flat plate.

into two distinct regions. (1) In the *boundary-layer region*, the axial velocity component $u(x, y)$ varies rapidly with the distance y from the plate; hence the velocity gradients and the shear stress are considered large. (2) In the region outside the boundary layer, called the *potential-flow region*, the velocity gradients and shear stresses are negligible.

Referring to the illustration in Fig. 6-1, we now examine the behavior of flow in the boundary layer with the distance x from the leading edge of the plate. The characteristic of the flow is governed by the magnitude of the quantity called the *Reynolds number*. For flow over a flat plate as illustrated in Fig. 6-1, it is defined as

$$\boxed{\mathrm{Re}_x \equiv \frac{u_\infty x}{\nu}} \tag{6-1}$$

where u_∞ = free-stream velocity
x = distance from leading edge
ν = kinematic viscosity of fluid

The boundary layer starts at the leading edge (that is, $x = 0$) of the plate as a *laminar boundary layer*, in which the flow remains orderly and fluid particles move along streamlines. This orderly motion continues along the plate until a critical distance is reached or the Reynolds number attains a critical value. After this critical Reynolds number is attained, the small disturbances in the flow begin to be amplified and fluid fluctuations begin to develop, which characterize the end of the laminar boundary layer and the beginning of transition from the laminar to *turbulent boundary layer*. For flow along a flat plate, the *critical Reynolds number* at which the transition from laminar to turbulent flow takes place is generally taken, for most analytical purposes, as

$$\boxed{\mathrm{Re}_x = \frac{u_\infty x}{\nu} \cong 5 \times 10^5} \tag{6-2}$$

However, this critical value is strongly dependent on the surface roughness and the turbulence level of the free stream. For example, with very large disturbances in the free stream, the transition may begin at a Reynolds number as low as 10^5, and for flows which are free from disturbances, it may not start until a Reynolds number of 10^6 or more. But for flow along a flat plate, the boundary layer is always turbulent for $\mathrm{Re}_x \geq 4 \times 10^6$. In the turbulent boundary layer next to the wall, there is a very thin layer, called the *viscous sublayer*, where the flow retains its viscous-flow character. Adjacent to the viscous sublayer is a region called the *buffer layer* in which there is fine-grained turbulence, and the mean axial velocity rapidly increases with the distance from the wall. The buffer layer is followed by the *turbulent layer* in which there is larger-scale turbulence, and the velocity changes relatively little with the distance from the wall.

Figure 6-2 shows the boundary-layer concept for flow over a curved body. In this case, the x coordinate is measured along the curved surface of the body; by

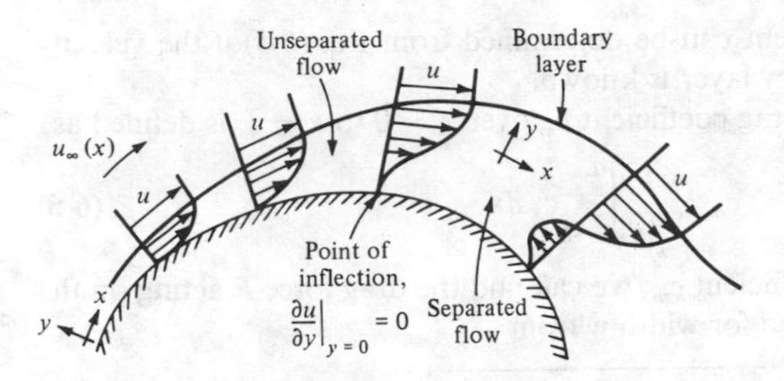

Figure 6-2 Boundary-layer concept for flow along a curved body and the flow separation.

starting from the stagnation point and at each x location, the y coordinate is measured normal to the surface of the body. The free-stream velocity $u_\infty(x)$ is not constant, but varies with distance along the curved surface. The boundary-layer concept discussed above also applies to this particular situation. The boundary-layer thickness $\delta(x)$ increases with the distance x along the surface. However, because of the curvature of the surface, after some distance x, the velocity profile $u(x, y)$ exhibits a point of inflection; that is, $\partial u/\partial y$ becomes zero at the wall surface. Beyond the point of inflection, the flow reversal takes place, and the boundary layer is said to be detached from the wall surface. Beyond the point of flow reversal, the flow patterns are very complicated and the boundary-layer analysis is no longer applicable.

Drag Coefficient and Drag Force

Suppose the velocity profile $u(x, y)$ in the boundary layer is known. The viscous shear stress τ_x acting on the wall at any location x is determined from its definition by

$$\tau_x = \mu \left.\frac{\partial u(x, y)}{\partial y}\right|_{y=0} \tag{6-3}$$

Here, the constant of proportionality μ is the viscosity of the fluid. Thus, knowing the velocity distribution in the boundary layer, one can determine the shear force acting on the wall owing to the flow. The definition of shear stress as given by Eq. (6-3), however, is not practical in engineering applications. In practice, the shear stress or the local drag force τ_x per unit area is related to the *local drag coefficient* c_x by the relation

$$\tau_x = c_x \frac{\rho u_\infty^2}{2} \tag{6-4}$$

where ρ is the density of the fluid and u_∞ is the free-stream velocity. Thus, knowing the drag coefficient, we can calculate the drag force exerted by the fluid flowing over the flat plate. Equating Eqs. (6-3) and (6-4), we obtain

$$c_x = \frac{2\nu}{u_\infty^2} \left.\frac{\partial u(x, y)}{\partial y}\right|_{y=0} \tag{6-5}$$

Thus, the local drag coefficient can be determined from Eq. (6-5) if the velocity profile $u(x, y)$ in the boundary layer is known.

The mean value of the drag coefficient c_m over $x = 0$ to $x = L$ is defined as

$$c_m = \frac{1}{L}\int_{x=0}^{L} c_x \, dx \tag{6-6}$$

Knowing the mean drag coefficient c_m, we can find the drag force F acting on the plate from $x = 0$ to $x = L$ and for width w from

$$\boxed{F = wLc_m \frac{\rho u_\infty^2}{2} \quad \text{N}} \tag{6-7}$$

Example 6-1 The velocity profile $u(x, y)$ for boundary-layer flow over a flat plate is given by

$$\frac{u(x, y)}{u_\infty} = \frac{3}{2}\left[\frac{y}{\delta(x)}\right] - \frac{1}{2}\left[\frac{y}{\delta(x)}\right]^3 \tag{a}$$

where the boundary-layer thickness $\delta(x)$ is

$$\delta(x) = \sqrt{\frac{280}{13}\frac{\nu x}{u_\infty}} \tag{b}$$

Develop an expression for the local drag coefficient c_x.

Develop an expression for the average drag coefficient c_m over a distance $x = L$ from the leading edge of the plate.

SOLUTION The local drag coefficient c_x is related to the velocity profile by Eq. (6-5) as

$$c_x = \frac{2\nu}{u_\infty^2}\left.\frac{\partial u(x, y)}{\partial y}\right|_{y=0}$$

For the velocity profile given above, by Eq. (*a*) we have

$$\left.\frac{\partial u(x, y)}{\partial y}\right|_{y=0} = \frac{3u_\infty}{2\delta(x)}$$

Then

$$c_x = \frac{3\nu}{u_\infty \delta(x)}$$

Introducing the above expression for $\delta(x)$ into this relation, we find

$$c_x = \frac{3\nu}{u_\infty}\sqrt{\frac{13}{280}\frac{u_\infty}{\nu x}} = \sqrt{\frac{117}{280}\frac{\nu}{u_\infty x}}$$

$$= \frac{0.646}{\mathrm{Re}_x^{1/2}} \tag{c}$$

The average c_m over $0 < x < L$ is defined by Eq. (6-6) as

$$c_m = \frac{1}{L} \int_{x=0}^{L} c_x \, dx$$

Introducing Eq. (*c*) into this expression, we obtain

$$c_m = \frac{1}{L} \sqrt{\frac{117}{280} \frac{\nu}{u_\infty}} \int_0^L x^{-1/2} \, dx = 2\sqrt{\frac{117}{280} \frac{\nu}{u_\infty L}} = 2 \frac{0.646}{\mathrm{Re}_x^{1/2}}$$

$$= 2c_x|_{x=L}$$

Thus, the mean value of the drag coefficient over $0 \le x \le L$ is equal to twice the value of the local drag coefficient at $x = L$.

Note that the expression for the drag coefficient developed here is an approximate one, because the velocity profile used in the analysis is approximate.

Example 6-2 The exact expression for the local drag coefficient c_x for laminar flow over a flat plate is given by

$$c_x = \frac{0.664}{\mathrm{Re}_x^{1/2}}$$

Air at atmospheric pressure and at $T_\infty = 300$ K flows with a velocity of $u_\infty = 1.5$ m/s along the plate. Determine the distance from the leading edge of the plate where transition begins from laminar to turbulent flow. Calculate the drag force F acting per 1-m width of the plate over the distance from $x = 0$ to where the transition starts.

SOLUTION The physical properties of atmospheric air at 300 K are

$$\rho = 1.177 \text{ kg/m}^3 \qquad \nu = 0.168 \times 10^{-4} \text{ m}^2\text{/s}$$

The transition is assumed to occur at a distance $x = L$, where $\mathrm{Re}_L = 5 \times 10^5$. Then the distance L is

$$\mathrm{Re}_L = \frac{u_\infty L}{\nu} = \frac{1.5L}{0.168 \times 10^{-4}} = 5 \times 10^5$$

or

$$L = 5.6 \text{ m}$$

The average drag coefficient c_m over the length $x = 0$ to $x = L$ is determined by Eq. (6-6) as

$$c_m = \frac{1}{L} \int_0^L c_x \, dx = \frac{0.664}{L(u_\infty \nu)^{1/2}} \int_0^L x^{1/2} \, dx$$

$$= 2 \frac{0.664}{\mathrm{Re}_L^{1/2}} = 2[c_x]_{x=L}$$

That is, the average value of the drag coefficient over the length $x = 0$ to $x = L$ is twice the value of the local drag coefficient at $x = L$. By introducing the numerical values, c_m becomes

$$c_m = 2\,\frac{0.664}{(5 \times 10^5)^{1/2}} = 1.88 \times 10^{-3}$$

Then the drag force acting on the plate is determined by Eq. (6-7) as

$$F = wLc_m\,\frac{\rho u_m^2}{2} = (1)(5.6)(1.88 \times 10^{-3})\,\frac{1.77 \times 1.5^2}{2}$$

$$= 2.09 \times 10^{-2}\ \text{N}$$

Thermal Boundary Layer

Analogous to the concept of velocity boundary layer, one can envision the development of a *thermal boundary layer* along the flat plate associated with the temperature profile in the fluid. To illustrate the concept, we consider that a fluid at a uniform temperature T_∞ flows along a flat plate maintained at a constant temperature T_w. Let x and y be the coordinate axes along and perpendicular to the plate surface, respectively, as illustrated in Fig. 6-3. We define the dimensionless temperature $\theta(x, y)$ as

$$\theta(x, y) = \frac{T(x, y) - T_w}{T_\infty - T_w} \tag{6-8}$$

where $T(x, y)$ is the local temperature in the fluid. At the wall surface, the fluid temperature is equal to the wall temperature, hence

$$\theta(x, y) = 0 \qquad \text{at } y = 0 \text{ (wall surface)} \tag{6-9a}$$

At distances sufficiently far from the wall, the fluid temperature remains the same as T_∞; then

$$\theta(x, y) \to 1 \qquad \text{as} \qquad y \to \infty \tag{6-9b}$$

Therefore, at each location x along the plate, one envisions a location $y = \delta_t(x)$ in the fluid where $\theta(x, y)$ equals 0.99. The locus of such points where $\theta(x, y) = 0.99$ is called the *thermal boundary layer* $\delta_t(x)$.

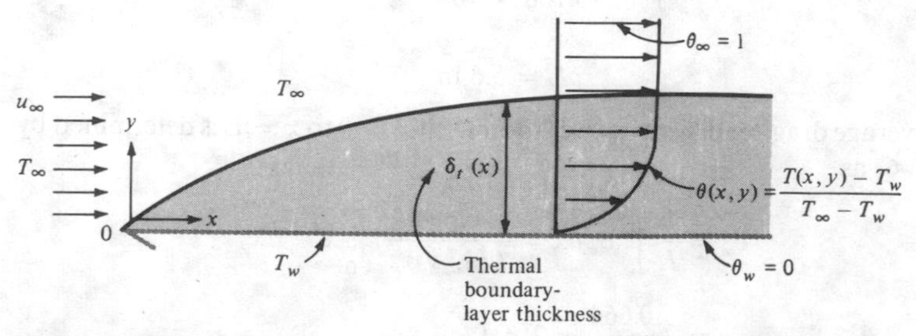

Figure 6-3 Thermal boundary-layer concept for the flow of a hot fluid over a cold wall.

The relative thicknesses of the thermal boundary layer $\delta_t(x)$ and the velocity boundary layer $\delta(x)$ depend on the magnitude of the Prandtl number for the fluid. For fluids having a Prandtl number equal to unity, such as gases, $\delta_t(x) = \delta(x)$. The thermal boundary layer is much thicker than the velocity boundary layer for fluids having $\Pr \ll 1$, such as liquid metals, and is much thinner than the velocity boundary layer for fluids having $\Pr \gg 1$.

Heat Transfer Coefficient

Suppose the temperature distribution $T(x, y)$ in the thermal boundary layer is known. Then the heat flux $q(x)$ from the fluid to the wall is determined from

$$q(x) = k \left. \frac{\partial T(x, y)}{\partial y} \right|_{y=0} \tag{6-10a}$$

where k is the thermal conductivity of the fluid. However, in engineering applications it is not practical to use Eq. (6-10*a*) to calculate the heat transfer rate between the fluid and the wall. In practice, a local heat transfer coefficient $h(x)$ is defined to calculate the heat flux between the fluid and the wall:

$$q(x) = h(x)(T_\infty - T_w) \tag{6-10b}$$

Equating (6-10*a*) and (6-10*b*), we obtain

$$h(x) = k \frac{[\partial T/\partial y]_{y=0}}{T_\infty - T_w} \tag{6-11a}$$

This expression is now written in terms of the dimensionless temperature $\theta(x, y)$ as

$$h(x) = k \left. \frac{\partial \theta(x, y)}{\partial y} \right|_{y=0} \tag{6-11b}$$

Thus, Eqs. (6-11) provide the relation for the determination of the local heat transfer coefficient $h(x)$ from the knowledge of the dimensionless temperature distribution $\theta(x, y)$ in the thermal boundary layer.

The *mean heat transfer coefficient* h_m over the distance $x = 0$ to $x = L$ along the plate surface is determined from

$$\boxed{h_m = \frac{1}{L} \int_0^L h(x)\, dx} \tag{6-12}$$

Knowing the mean heat transfer coefficient h_m, we can find the heat transfer rate Q from the fluid to the wall from $x = 0$ to $x = L$ and for the width w:

$$Q = wLh_m(T_\infty - T_w) \tag{6-13}$$

Example 6-3 An approximate expression for the temperature profile in the thermal layer is given by

$$\theta(x, y) \equiv \frac{T(x, y) - T_w}{T_\infty - T_w} = \frac{3}{2} \frac{y}{\delta_t(x)} - \frac{1}{2} \left[\frac{y}{\delta_t(x)} \right]^3$$

and the thickness of the thermal boundary layer $\delta_t(x)$ is given by

$$\delta_t(x) = 4.53 \frac{x}{\text{Re}_x^{1/2}\ \text{Pr}^{1/3}}$$

where Pr is the Prandtl number. Develop an expression for the local heat transfer coefficient $h(x)$.

SOLUTION The local heat transfer coefficient is related to $\theta(x, y)$ by Eq. (6-11*b*) as

$$h(x) = k \left. \frac{\partial \theta(x, y)}{\partial y} \right|_{y=0}$$

For the temperature profile given above, we have

$$\left. \frac{\partial \theta(x, y)}{\partial y} \right|_{y=0} = \frac{3}{2\delta_t(x)}$$

Then

$$h(x) = \frac{3}{2} \frac{k}{\delta_t(x)}$$

Introducing the above relation for $\delta_t(x)$, we find

$$h(x) = 0.331 \frac{k}{x} \text{Pr}^{1/3}\ \text{Re}_x^{1/2}$$

This expression is arranged in the dimensionless form as

$$\text{Nu}_x \equiv \frac{h(x)x}{k} = 0.331\ \text{Pr}^{1/3}\ \text{Re}_x^{1/2}$$

where Nu_x is called the local *Nusselt number*. We note that the Nusselt number is a function of the Prandtl and Reynolds numbers.

Example 6-4 The exact expression for the local Nusselt number for laminar flow along a flat plate is given by

$$\text{Nu}_x = \frac{h(x)x}{k} = 0.332\ \text{Pr}^{1/3}\ \text{Re}_x^{1/2}$$

Develop a relation for the average heat transfer coefficient $h(x)$ from $x = 0$ to $x = L$.

Atmospheric air at $T_\infty = 400$ K with a velocity $u_\infty = 1.5$ m/s flows over a flat plate $L = 2$ m long maintained at a uniform temperature $T_w = 300$ K.

Calculate the average heat transfer coefficient h_m from $x = 0$ to $x = L = 2$ m. Calculate the heat transfer rate from the airstream to the plate from $x = 0$ to $x = L = 2$ m for $w = 0.5$ m.

SOLUTION The average heat transfer coefficient h_m from $x = 0$ to $x = L$ is determined by Eq. (6-12) as

$$h_m = \frac{1}{L}\int_0^L h(x)\,dx = \frac{1}{L}\,0.332k\,\mathrm{Pr}^{1/3}(u_\infty \nu)^{1/2}\int_0^L \frac{1}{x^{1/2}}\,dx$$

$$= 2\left(0.332k\,\mathrm{Pr}^{1/3}\,\frac{\mathrm{Re}_L^{1/2}}{L}\right) = 2[h(x)]_{x=L}$$

That is, the average heat transfer coefficient h_m from $x = 0$ to $x = L$ is twice the value of the local heat transfer coefficient $h(x)$ evaluated at $x = L$.

The physical properties of atmospheric air evaluated at the arithmetic average of the free-stream and wall temperatures, 350 K, are $\nu = 0.21 \times 10^{-4}\ \mathrm{m^2/s}$, $k = 0.03\ \mathrm{W/(m \cdot {}^\circ C)}$, and $\mathrm{Pr} = 0.697$. Then, at $L = 2\,\mathrm{m}$,

$$\mathrm{Re}_L = \frac{u_\infty L}{\nu} = \frac{1.5 \times 2}{0.21 \times 10^{-4}} = 1.43 \times 10^5$$

$$h_m = 2\left(0.332k\,\mathrm{Pr}^{1/3}\,\frac{\mathrm{Re}_L^{1/2}}{L}\right) = 2\left(0.332 \times 0.03 \times 0.887 \times \frac{378.2}{2}\right)$$

$$= 3.34\ \mathrm{W/(m^2 \cdot {}^\circ C)}$$

Then the heat transfer rate Q is determined by Eq. (6-13) as

$$Q = wLh_m(T_\infty - T_w) = 0.5 \times 2 \times 3.34\,(400 - 300)$$

$$= 334\ \mathrm{W}$$

Relation between c_x and $h(x)$

In Example 6-1 we showed that if the velocity profile and the boundary-layer thickness are known, then an expression can be developed for the local drag coefficient c_x for laminar flow along a flat plate. Similarly, in Example 6-3 we showed that if the temperature profile and the thermal boundary-layer thickness are available, an expression can be developed for the local heat transfer coefficient for laminar flow along a flat plate.

Now we seek a relation between the heat transfer and drag coefficients.

We consider the exact expressions for the local drag coefficient and the local Nusselt number for laminar flow along a flat plate given, respectively, by

$$\boxed{\frac{c_x}{2} = 0.332\,\mathrm{Re}_x^{-1/2}} \qquad (6\text{-}14a)$$

$$\boxed{\mathrm{Nu}_x = 0.332\,\mathrm{Pr}^{1/3}\,\mathrm{Re}_x^{1/2}} \qquad (6\text{-}14b)$$

We define the *local Stanton number* St_x as

$$St_x = \frac{h(x)}{\rho c_p u_\infty}$$

which can be rearranged in the form

$$St_x = \frac{h(x)x/k}{(\nu/\alpha)(u_\infty x/\nu)} = \frac{Nu_x}{Pr\ Re_x}$$

Then the expression (6-14*b*) for the local Nusselt number can be rewritten as

$$St_x = 0.332\ Pr^{-2/3}\ Re_x^{-1/2} \tag{6-14c}$$

From Eqs. (6-14*a*) and (6-14*c*), the following relation is obtained between the Stanton number and the drag coefficient:

$$\boxed{St_x\ Pr^{2/3} = \frac{c_x}{2}} \tag{6-15a}$$

This expression is referred to as the *Reynolds-Colburn analogy* that relates the local drag coefficient c_x to the local Stanton number St_x for laminar flow along a flat plate. Thus, by making frictional drag measurements for laminar flow along a flat plate with no heat transfer involved, the corresponding heat transfer coefficient can be determined by Eq. (6-15*a*). It is much easier to make drag measurements than heat transfer measurements.

Equation (6-15*a*) is also applicable for turbulent flow along a flat plate, but it does not apply to laminar flow inside a tube.

In the case of the average values, Eq. (6-15*a*) is written as

$$\boxed{St_m\ Pr^{2/3} = \frac{c_m}{2}} \tag{6-15b}$$

where St_m and c_m are, respectively, the mean Stanton number and the mean drag coefficient.

Example 6-5 Atmospheric air at 300 K flows with a velocity of $u_\infty = 5$ m/s along a flat plate $L = 1$ m long. The plate has a width $w = 0.5$ m. The total drag force acting on the plate is determined to be $F = 18 \times 10^{-3}$ N. By using the Reynolds-Colburn analogy, estimate the corresponding average heat transfer coefficient h_m for flow of air over the plate.

SOLUTION The total force acting on the plate is related to the mean drag coefficient by Eq. (6-7) as

$$F = wLc_m \frac{\rho u_\infty^2}{2}$$

and the average heat transfer coefficient is related to the average drag coefficient by the Reynolds-Colburn analogy, given by Eq. (6-15*b*), as

$$\frac{h_m}{\rho c_p u_\infty} \text{Pr}^{2/3} = \frac{c_m}{2}$$

The physical properties of atmospheric air at 300 K are taken as $\rho = 1.177\ \text{kg/m}^3$, $c_p = 1.006 \times 10^3\ \text{J/(kg}\cdot{}^\circ\text{C)}$, and $\text{Pr} = 0.708$. Then the mean drag coefficient c_m is determined from the first of these equations as

$$18 \times 10^{-3} = (0.5)(1)c_m \frac{1.177 \times 5^2}{2}$$

or

$$c_m = 2.447 \times 10^{-3}$$

Knowing c_m, we can find the mean heat transfer coefficient h_m from the second equation as

$$\frac{h_m}{(1.177)(1.006 \times 10^3)(5)} (0.708)^{2/3} = \frac{2.447 \times 10^{-3}}{2}$$

or

$$h_m = 9.12\ \text{W/(m}^2\cdot{}^\circ\text{C)}$$

Example 6-6 Atmospheric air at $T_\infty = 400$ K flows with a velocity of $u_\infty = 4$ m/s along a flat plate $L = 1$ m long maintained at a uniform temperature $T_w = 300$ K. The average heat transfer coefficient is determined to be $h_m = 7.75\ \text{W/(m}^2\cdot{}^\circ\text{C)}$. Using the Reynolds-Colburn analogy, estimate the drag force exerted on the plate per 1-m width.

SOLUTION The Reynolds-Colburn analogy is given by Eq. (6-15*b*) as

$$\frac{h_m}{\rho c_p u_\infty} \text{Pr}^{2/3} = \frac{c_m}{2}$$

The physical properties of atmospheric air at the mean film temperature $T_m = (400 + 300)/2 = 350$ K are given by

$$\rho = 0.998\ \text{kg/m}^3 \qquad c_p = 1009\ \text{J/(kg}\cdot{}^\circ\text{C)} \qquad \text{Pr} = 0.697$$

Then the drag coefficient c_m is

$$\frac{7.75}{0.998 \times 1009 \times 4} (0.697)^{2/3} = \frac{c_m}{2}$$

or

$$c_m = 3.03 \times 10^{-3}$$

The drag force F is given by Eq. (6-7) as

$$F = wLc_m \frac{\rho u_\infty^2}{2}$$

$$= (1)(1)(3.03 \times 10^{-3}) \frac{(0.998)(4^2)}{2}$$

$$= 24.2 \times 10^{-3}\ \text{N}$$

6-2 FLOW INSIDE A DUCT

The basic concepts discussed in the last section on the development of velocity and thermal boundary layers for flow along a flat plate also apply to flow at the entrance region of ducts. We illustrate this matter by considering flow inside a circular tube.

Velocity Boundary Layer

Consider the flow inside a circular tube, as illustrated in Fig. 6-4. The fluid has a uniform velocity u_0 at the tube inlet. As the fluid enters the tube, a *velocity boundary layer* starts to develop along the wall surface. The velocity of fluid particles at the wall surface becomes zero, and that at the vicinity of the wall is retarded; as a result, the velocity in the central portion of the tube increases to satisfy the requirement of the continuity of flow. The thickness of the velocity boundary layer $\delta(z)$ continuously grows along the tube surface until it fills the entire tube. The region from the tube inlet to little beyond the hypothetical location where the boundary layer reaches the tube center is called the *hydrodynamic entry region.* In this region the shape of the velocity profile changes in both the axial and radial direction. The region beyond the hydrodynamic entry length is called the *hydrodynamically developed region,* because in this region the velocity profile is invariant with distance along the tube.

If the boundary layer remains laminar until it fills the tube, fully developed laminar flow of parabolic velocity profile prevails in the hydrodynamically developed region. However, if the boundary layer changes to turbulent before its thickness reaches the tube center, fully developed turbulent flow is experienced in the hydrodynamically developed region. When the flow is turbulent, the velocity profile is flatter than the parabolic velocity profile of laminar flow.

For flow inside a circular tube, the Reynolds number, defined as

$$\boxed{\mathrm{Re} \equiv \frac{u_m D}{\nu}} \tag{6-16}$$

is used as a criterion for change from laminar to turbulent flow. In this definition, u_m is the mean flow velocity, D is the tube's inside diameter, and ν is the kinematic

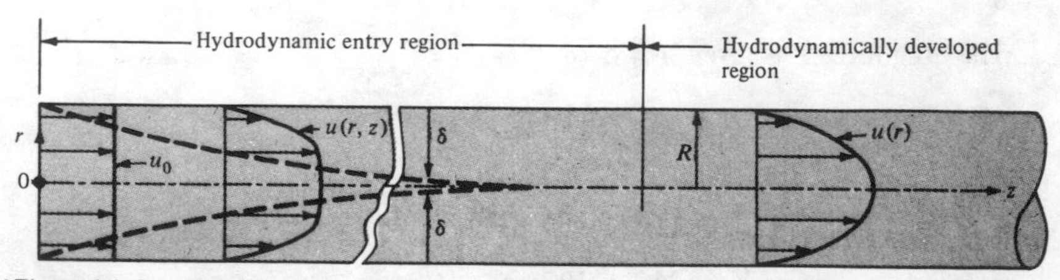

Figure 6-4 Concept of development of velocity boundary layer at entrance region of a circular tube.

viscosity of fluid. For flow inside a circular tube, the turbulent flow is usually observed for

$$\boxed{\text{Re} = \frac{u_m D}{\nu} > 2300} \tag{6-17}$$

However, this critical value is strongly dependent on the surface roughness, the inlet conditions, and the fluctuations in the flow. In general, the transition may occur in the range 2000 < Re < 4000.

Friction Factor and Pressure Drop

In engineering applications, the pressure gradient dP/dz associated with the flow is a quantity of interest, because the pressure drop along a given length of tube can be determined by the integration of dP/dz over the length. To develop an expression defining dP/dz, we consider a force balance over a differential length dz of the tube. By equating the pressure force to the shear force at the wall, we obtain (see Fig. 6-5)

$$(PA)_z - (PA)_{z+\Delta z} = S\,\Delta z\,\tau_w$$

$$\frac{dP}{dz} = -\frac{S}{A}\tau_w = -\frac{\pi D}{(\pi/4)D^2}\tau_w = -\frac{4}{D}\tau_w \tag{6-18a}$$

where A is the cross-sectional area and S the perimeter.

The shear stress τ_w at the wall is related to the velocity gradient by

$$\tau_w = \mu \left.\frac{\partial u}{\partial y}\right|_{\text{wall}} = -\mu \left.\frac{\partial u}{\partial r}\right|_{\text{wall}} \tag{6-18b}$$

since $r = D/2 - y$. Then from Eqs. (6-18a) and (6-18b), we have

$$\frac{dP}{dz} = \frac{4\mu}{D}\left.\frac{\partial u}{\partial r}\right|_{\text{wall}} \tag{6-18c}$$

In engineering applications Eq. (6-18c) is not practical for the determination of dP/dz, because it requires the evaluation of the velocity gradient at the wall. To

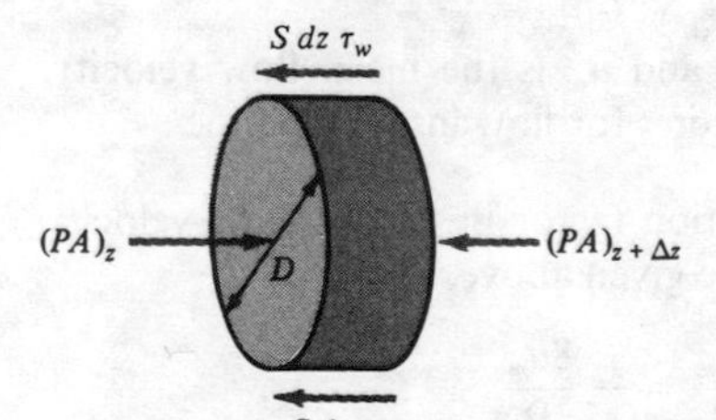

Figure 6-5 A force balance on a differential volume element.

calculate the pressure drop in engineering applications, a *friction factor* f is defined as

$$\frac{dP}{dz} = -f\frac{\rho u_m^2}{2D} \tag{6-18d}$$

where u_m is the mean velocity of flow inside the tube and ρ is the density of the fluid.

By equating (6-18*c*) and (6-18*d*), the following expression is obtained for the friction factor:

$$f = -\frac{8\mu}{\rho u_m^2}\frac{\partial u}{\partial r}\bigg|_{\text{wall}} \tag{6-18e}$$

Thus, given the velocity distribution u for the flow inside the tube, the friction factor f can be determined from Eq. (6-18*e*).

Given the friction factor, the pressure drop $P_1 - P_2 \equiv \Delta P$ over the length $z_2 - z_1 \equiv L$ of the tube is determined by the integration of Eq. (6-18*b*):

$$\int_{P_1}^{P_2} dP = -f\frac{\rho u_m^2}{2D}\int_{z_1}^{z_2} dz$$

or the pressure drop ΔP becomes

$$\boxed{\Delta P = f\frac{L}{D}\frac{\rho u_m^2}{2} \qquad \frac{\text{N}}{\text{m}^2}} \tag{6-19a}$$

If M is the flow rate in cubic meters per second through the pipe, the pumping power required to get the fluid through the pipe against the pressure drop ΔP becomes

$$\text{Pumping power} = \left(M\frac{\text{m}^3}{\text{s}}\right)\left(\Delta P\frac{\text{N}}{\text{m}^2}\right)$$

$$\boxed{\text{Pumping power} = M\,\Delta P \qquad \frac{\text{N}\cdot\text{m}}{\text{s}} \text{ or W}} \tag{6-19b}$$

Example 6-7 The velocity profile for hydrodynamically developed laminar flow inside a circular tube is given by

$$u(r) = 2u_m\left[1 - \left(\frac{r}{R}\right)^2\right]$$

where R is the inside radius of the tube and u_m is the mean flow velocity. Develop an expression for the friction factor f for flow inside the tube.

SOLUTION Equation (6-18*e*) gives the friction factor in terms of the velocity gradient at the wall. For the velocity profile given above, we have

$$\frac{du(r)}{dr}\bigg|_{r=R} = -\frac{4u_m}{R} = -\frac{8u_m}{D}$$

where D is the inside diameter of the tube. By introducing this expression into Eq. (6-18*e*), the friction factor f for hydrodynamically developed laminar flow inside a circular tube is determined as

$$f = \frac{64\mu}{\rho u_m D} = \frac{64}{\text{Re}}$$

where the Reynolds number is defined as

$$\text{Re} = \frac{\rho u_m D}{\mu} = \frac{u_m D}{\nu}$$

Example 6-8 Engine oil ($\rho = 868\ \text{kg/m}^3$, $\nu = 0.75 \times 10^{-4}\ \text{m}^2/\text{s}$) flows with a mean velocity of $u_m = 0.15$ m/s inside a circular tube having an inside diameter $D = 2.5$ cm. Calculate the friction factor and the pressure drop over the length $L = 100$ m of the tube.

SOLUTION First we calculate the Reynolds number in order to establish whether the flow is laminar or turbulent.

$$\text{Re} = \frac{u_m D}{\nu} = \frac{0.15 \times 0.025}{0.75 \times 10^{-4}} = 50$$

The flow is laminar, hence the expression developed in Example 6-7 for the friction factor is applicable. We find

$$f = \frac{64}{\text{Re}} = \frac{64}{50} = 1.28$$

The pressure drop is calculated by Eq. (6-19*a*) as

$$\Delta P = f\,\frac{L}{D}\,\frac{\rho u_m^2}{2}$$

$$= 1.28\left(\frac{100}{0.025}\right)\frac{(868\ \text{kg/m}^2)(0.15^2\ \text{m}^2/\text{s}^2)}{2}$$

$$\cong 50{,}000\ \text{N/m}^2$$

Thermal Boundary Layer

In the case of temperature distribution in flow inside a circular tube, it is more difficult to visualize the development of the thermal boundary layer and the existence of a thermally developed region. However, under certain heating or cooling conditions, such as *constant heat flux* or *uniform temperature* at the tube wall, such a concept is possible.

Consider a laminar flow inside a circular tube subjected to uniform heat flux at the wall. Let r and z be the radial and axial coordinates, respectively. A dimensionless temperature $\theta(r, z)$ is defined as

$$\theta(r, z) = \frac{T(r, z) - T_w(z)}{T_m(z) - T_w(z)} \tag{6-20a}$$

where $T_w(z)$ = tube wall temperature
$T_m(z)$ = bulk mean fluid temperature over cross-sectional area of tube at z
$T(r, z)$ = local fluid temperature

Clearly, $\theta(r, z)$ is zero at the tube wall surface and attains some finite value at the tube center. Then, one envisions the development of a thermal boundary layer along the wall surface. The thickness of the thermal boundary layer $\delta_t(z)$ continuously grows along the tube surface until it fills the entire tube. The region from the tube inlet to the hypothetical location where the thermal boundary-layer thickness reaches the tube center is called the *thermal entry region.* In this region, the shape of the dimensionless temperature profile $\theta(r, z)$ changes in both the axial and radial direction. The region beyond the thermal entry length is called the *thermally developed region,* because in this region the dimensionless temperature profile remains invariant with the distance along the tube, that is,

$$\theta(r) = \frac{T(r, z) - T_w(z)}{T_m(z) - T_w(z)} \tag{6-20b}$$

It is difficult to explain qualitatively why $\theta(r)$ should be independent of the z variable while the temperatures on the right-hand side of Eq. (6-20b) depend on both r and z. However, it can be shown mathematically that for either constant temperature or constant heat flux at the wall the dimensionless temperature $\theta(r)$ depends on only r for sufficiently large values of z.

Heat Transfer Coefficient

In engineering applications involving fluid flow in a tube, the heat transfer rate between the fluid and the tube is of interest. Here we discuss the concept of heat transfer coefficient that is frequently used in engineering applications for the determination of heat transfer between the fluid and the wall surface.

Consider a fluid flowing inside a circular tube of inside radius R. Let $T(r, z)$ be the temperature distribution in the fluid, where r and z are the radial and axial coordinates, respectively. The heat flux from the fluid to the tube wall is determined from its definition

$$q(z) = -k \left. \frac{\partial T(r, z)}{\partial r} \right|_{\text{wall}} \tag{6-21a}$$

where k is the thermal conductivity of fluid.

In engineering applications, it is not practical to use Eq. (6-21a) to determine the heat transfer between the fluid and the tube wall, because it involves the evaluaation of the derivative of temperature at the wall. To avoid this difficulty, *a local heat transfer coefficient* $h(z)$ is defined as

$$q(z) = h(z)[T_m(z) - T_w(z)] \tag{6-21b}$$

where $T_m(z)$ = bulk mean fluid temperature over the tube cross-sectional area at location z

$T_w(z)$ = tube wall temperature at z

Clearly, if the heat transfer coefficient is available, it is a very simple matter to determine the heat flux at the wall for a given temperature difference between the mean fluid and the tube wall. Therefore, the use of heat transfer coefficient is very convenient in engineering applications, and its determination under various flow conditions has been the subject of numerous experimental and analytical investigations. Here we treat the relation between the heat transfer coefficient $h(z)$ and the fluid temperature $T(r, z)$, in order to determine $h(z)$ from $T(r, z)$.

By equating (6-21*a*) and (6-21*b*), we obtain

$$h(z) = -\frac{k}{T_m(z) - T_w(z)} \left.\frac{\partial T(r, z)}{\partial r}\right|_{r=R\,\text{wall}} \tag{6-22a}$$

where $T_m(z)$ and $T_w(z)$, for a circular tube of radius R, are determined from

$$T_m(z) = \frac{\int_0^R u(r)T(r, z)2\pi r\,dr}{\int_0^R u(r)2\pi r\,dr} = \frac{\int_0^R u(r)T(r, z)2\pi r\,dr}{u_m \pi R^2} \tag{6-22b}$$

$$T_w(z) = T(r, z)|_{r=R\,\text{wall}} \tag{6-22c}$$

The mean fluid temperature $T_m(z)$ is a definition based on the thermal energy transport with the bulk motion of the fluid as it passes through the cross section, because the quantity "$\rho c_p u T$" represents energy flux per unit area. For incompressible, constant-property fluid, the ρc_p term is canceled.

Equation (6-22*a*) can be written in terms of the dimensionless temperature $\theta(r, z)$, defined by Eq. (6-20*a*), as

$$\boxed{h(z) = -k \left.\frac{\partial \theta(r, z)}{\partial r}\right|_{r=R\,\text{wall}}} \tag{6-23a}$$

In the thermally developed region, the dimensionless temperature $\theta(r)$ is independent of z. Then Eq. (6-23*a*) reduces to

$$\boxed{h = -k \left.\frac{d\theta(r)}{dr}\right|_{r=R\,\text{wall}}} \tag{6-23b}$$

where $\theta(r)$ is defined by Eq. (6-20*b*). This result implies that in the thermally developed region the heat transfer coefficient does not vary with the distance along the tube; and it is valid for heat transfer under conditions of constant wall heat flux or constant wall temperature.

The definitions given by Eqs. (6-23) can be used to develop relations for the heat transfer coefficient if the dimensionless temperature distribution in the fluid, defined by Eq. (6-20*b*), is available.

Example 6-9 Consider laminar forced convection inside a circular tube of inside radius R and subjected to a uniform heat flux at the tube wall. In the region where the velocity and temperature profiles are fully developed, the dimensionless temperature $\theta(r)$, defined by Eq. (6-20*b*), is given in the form

$$\theta(r) = \frac{T(r, z) - T_w(z)}{T_m(z) - T_w(z)} = \frac{96}{11}\left[\frac{3}{16} + \frac{1}{16}\left(\frac{r}{R}\right)^4 - \frac{1}{4}\left(\frac{r}{R}\right)^2\right]$$

Develop an expression for the heat transfer coefficient.

SOLUTION The heat transfer coefficient can be determined by Eq. (6-23). In the thermally developed region, the dimensionless temperature does not depend on the axial coordinate. Hence by Eq. (6-23*b*) we write

$$h = -k\left.\frac{d\theta(r)}{dr}\right|_{r=R} \tag{a}$$

For the temperature profile given above, we have

$$\left.\frac{d\theta(r)}{dr}\right|_{r=R} = \frac{96}{11}\left(\frac{1}{4R} - \frac{1}{2R}\right) = -\frac{24}{11R} = -\frac{48}{11D} \tag{b}$$

where D is the diameter. Introducing this result in expression (*a*), we obtain the heat transfer coefficient in the thermally developed region of laminar flow inside a circular tube subjected to uniform heat flux at the wall as

$$h = \frac{48}{11}\frac{k}{D}$$

This expression is arranged in the dimensionless form as

$$\boxed{\mathrm{Nu} \equiv \frac{hD}{k} = \frac{48}{11}}$$

where Nu is the Nusselt number.

Example 6-10 Engine oil [$\nu = 0.8 \times 10^{-4}$ m^2/s, $k = 0.14$ W/(m · °C)] flows with a mean velocity of $v_m = 0.2$ m/s inside a 1.25-cm-diameter tube which is electrically heated at the wall at a uniform rate of $q = 2450$ W/m^2. The heat transfer is taking place in the thermally developed region. Calculate the temperature difference between the tube wall surface and the mean flow temperatures.

SOLUTION To ensure that the flow is laminar, we check the R

$$\text{Re} = \frac{u_m D}{\nu} = \frac{0.2 \times 0.0125}{0.8 \times 10^{-4}} = 31.25$$

Thus the flow is laminar.

The heat transfer coefficient for laminar in the thermally deve
subjected to a uniform heat flux at the wall is determined in Exa as

$$h = \frac{48}{11} \frac{k}{D}$$

For this problem, h becomes

$$h = \frac{48}{11} \frac{0.14}{0.0125} = 48.87 \text{ W/(m}^2 \cdot {}^\circ\text{C)}$$

Then the temperature difference ΔT between the wall surface and the mean fluid temperatures is

$$\Delta T = \frac{q}{h} = \frac{2450}{48.87} = 50.1^\circ\text{C}$$

6-3 CONCEPTS ON TURBULENCE

Osborne Reynolds [1], in his classic experiment of injecting dye into the water flowing through a transparent pipe, showed that at low flow rates the flow was streamlined; but as the flow rate was increased, the streamlines became unstable and the laminar motion changed to turbulent flow. The term *turbulent* is used to denote that the motion of the fluid is chaotic in nature and involves crosswise mixing or eddying superimposed on the motion of the mainstream.

The eddying or crosswise mixing in turbulent flow is advantageous in that it assists greatly in improving the heat transfer in flow, but it has the disadvantage of increasing the resistance to flow. The flow patterns in turbulent flow are so complex that they consist of a spectrum of coexisting vortices varying in size from the larger ones down to those of microscopic dimension. For example, in turbulent flow inside a circular tube, the size of the vortices varies from a significant fraction of tube radius in the central region to microscopic sizes near the wall where the velocity approaches zero.

In turbulent flow, the properties such as velocity, temperature, and pressure are subject to fluctuations both with location in the fluid and with time. Therefore, the instantaneous values of these properties can be represented as a sum of a time-averaged mean part and a fluctuating part in the form

$$\begin{aligned} u_i &= u + u' \\ v_i &= v + v' \\ T_i &= T + T' \\ P_i &= P + P' \end{aligned} \tag{6-24}$$

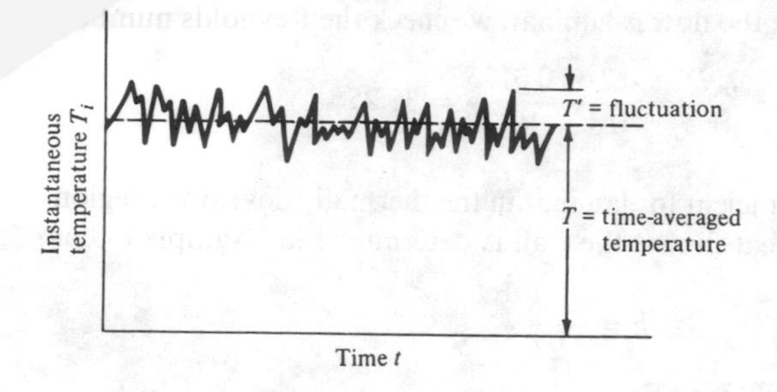

Figure 6-6 Temperature fluctuations with time in turbulent flow.

where u_i, v_i, T_i, and P_i are the *instantaneous* values; u, v, T, and P are the *time-averaged* values; and u', v', T', and P' are the *fluctuations*. For example, if a thermocouple with a sufficiently small time constant is placed in a given location in turbulent flow and the instantaneous value of temperature T_i is recorded as a function of time, temperature may exhibit fluctuations, as illustrated in Fig. 6-6. In turbulent flow, the fluctuations are considered superimposed on local average flow. For example, the time-averaged value T of temperature is defined as

$$T = \frac{1}{\Delta t} \int_{t}^{t+\Delta t} T_i \, dt \tag{6-25}$$

where Δt is a very small time interval which is large enough for recording the turbulent fluctuations but sufficiently small for the temperature to be unaffected by the external disturbances on the system. The averaging process similar to that defined by Eq. (6-25) is applicable for the averaging of velocity, pressure, etc.

The effects of these fluctuations occurring in velocity and temperature in turbulent flow are to increase heat transfer and resistance to flow (drag).

Consider turbulent flow along a flat plate, as illustrated in Fig. 6-1. The y coordinate is measured normal to the plate surface. Let u_i, v_i, and T_i be the instantaneous values of the axial and normal components of velocity and temperature; and let u', v', and T' denote the corresponding fluctuations from the average values u, v, and T, respectively. We now examine the effects of turbulence on the shear stress and heat flux in the fluid.

Let τ be the shear stress in the fluid in the x direction at a location y from the wall surface. This shear stress can be considered to be composed of two parts: *viscous shear stress* resulting from the mean flow velocity u and *turbulent shear stress* resulting from the velocity fluctuations u' and v' from the mean values. With this consideration we write

$$\tau = \tau_{\text{viscous}} + \tau_{\text{turbulent}} \tag{6-26a}$$

where

$$\tau_{\text{viscous}} = \mu \frac{\partial u}{\partial y} \tag{6-26b}$$

$$\tau_{\text{turbulent}} = -\rho \overline{u'v'} \tag{6-26c}$$

Here the viscous shear stress term given by Eq. (6-26*b*) is similar to the one encountered in laminar flow. The physical significance of the turbulent shear stress term given by Eq. (6-26*c*) is not quite apparent, but it can readily be derived by introducing the instantaneous velocities u_i and v_i as defined by Eqs. (6-24) into the x momentum equations and by applying the rules of averaging to the terms involving the cross products of the fluctuations. See Refs. 2 to 7 for a more comprehensive discussion of turbulence.

Now, let q be the heat flux in the fluid in the y direction. This heat flux can be considered as composed of two parts: the *diffusive or conductive heat flux* due to the gradient of the mean temperature T in the y direction and the *turbulent heat flux* resulting from the temperature fluctuations T' and the velocity fluctuations v' in the y direction. With this consideration, we write

$$q = q_{\text{diffusive}} + q_{\text{turbulent}} \tag{6-27a}$$

where

$$q_{\text{diffusive}} = -k\frac{\partial T}{\partial y} \tag{6-27b}$$

$$q_{\text{turbulent}} = \rho c_p \overline{v'T'} \tag{6-27c}$$

Here the diffusive heat flux given by Eq. (6-27*b*) is similar to that encountered in conduction, but the significance of the turbulent heat flux component given by Eq. (6-27*c*) is not quite apparent. It is derived by introducing u_i, v_i, and T_i as defined by Eqs. (6-24) into the energy equation for forced convection and manipulating by utilizing the rules of averaging.

The relations given by Eqs. (6-26*c*) and (6-27*c*) for the turbulent shear stress and the *turbulent heat flux* are not useful for computational purposes unless they are related to the mean quantities u and T. Because of the complexity of turbulence, attempts have been made to develop such relations by semiempirical hypothesis. For example, Prandtl [3] and von Kármán [4] proposed the mixing-length concept that has been used extensively and successively in relating turbulent stress and turbulent heat flux to the gradients of average velocity and average temperature.

Mixing-Length Concept

The basic idea is analogous to the *mean-free-path* concept for molecules in the kinetic theory of gases. The main difference is that in turbulent motion macroscopic lumps are envisioned. That is, for turbulent flow in the x direction along the surface, it is postulated that fluid particles a distance y from the wall surface coalesce into macroscopic lumps and then travel, on the average, a distance l in the direction normal to the main flow while retaining their x direction momentum before they are dispersed. Thus, if the slow-moving lumps enter the fast-moving layer, they act as a drag on it, and the momentum is transferred between layers as a result of transverse mixing. Of course, l is an unknown quantity, and in reality there is no such clearly defined distance. Although the concept lacks generality, it has been found useful in the study of turbulent exchange in most engineering applications.

Prandtl postulated that the velocity fluctuations can be related to $\partial u/\partial y$ by

$$u' \cong l_1 \frac{\partial u}{\partial y} \qquad v' \cong l_2 \frac{\partial u}{\partial y} \tag{6-28a, b}$$

where u' and v' are of opposite sign and l_1 and l_2 are the *mixing lengths for momentum transport.*

A similar approach can be applied to relate T' to $\partial T/\partial y$ as

$$T' \cong l_3 \frac{\partial T}{\partial y} \tag{6-28c}$$

where l_3 is *the mixing length for energy transport.*

We now define

$$l_1 l_2 \equiv l_m^2 \qquad l_2 l_3 = l_h^2 \tag{6-29}$$

In view of these relations, we write

$$\overline{u'v'} = -l_m^2 \left|\frac{\partial u}{\partial y}\right| \frac{\partial u}{\partial y} \equiv -\varepsilon_m \frac{\partial u}{\partial y} \tag{6-30a}$$

$$\overline{v'T'} = -l_h^2 \left|\frac{\partial u}{\partial y}\right| \frac{\partial T}{\partial y} \equiv -\varepsilon_h \frac{\partial T}{\partial y} \tag{6-30b}$$

Introducing Eqs. (6-30) into Eqs. (6-26) and (6-27), we obtain the relations for the shear stress and heat flux in turbulent flow:

$$\tau = \mu \frac{\partial u}{\partial y} + \rho \varepsilon_m \frac{\partial u}{\partial y} \tag{6-31a}$$

$$q = -k \frac{\partial T}{\partial y} - \rho c_p \varepsilon_h \frac{\partial T}{\partial y} \tag{6-31b}$$

where ε_m = eddy diffusivity of momentum
ε_h = eddy diffusivity of heat (or eddy conductivity)

Equations (6-31) can be rearranged as

$$\frac{\tau}{\rho} = (\nu + \varepsilon_m) \frac{\partial u}{\partial y} \tag{6-32a}$$

$$\frac{q}{\rho c_p} = -(\alpha + \varepsilon_h) \frac{\partial T}{\partial y} \tag{6-32b}$$

where ν = kinematic viscosity and α = thermal diffusivity. Equations (6-32) clearly demonstrate the effects of turbulent flow in enhancing both the shear stress (drag) and the heat flux. Depending on the level of turbulence, the turbulent transport properties ε_m and ε_h may be an order of magnitude larger than the diffusive properties ν and α.

The eddy diffusivity of momentum ε_m and the eddy diffusivity of heat ε_h are not necessarily the same. Their ratio is defined as

$$\frac{\varepsilon_m}{\varepsilon_h} = \mathrm{Pr}_t = \text{turbulent Prandtl number} \tag{6-33}$$

This definition is analogous to the definition of the Prandtl number:

$$\frac{\nu}{\alpha} = \mathrm{Pr} = \text{Prandtl number} \tag{6-34}$$

The physical significance of Pr_t and Pr should be distinguished. The Prandtl number is a physical property of the fluid, and it varies from about unity for gases to very large values for ordinary liquids and oils. However, the turbulent Prandtl number is a property of the flow field more than of the fluid. Various models have been developed for the determination of Pr_t. The simplest is due to Reynolds who assumed $\mathrm{Pr}_t = 1$, which implies that heat and momentum transfer in turbulent flow takes place exactly by the same process. The numerical values of Pr_t may well vary between 1 and 2.

Although the mixing-length and the eddy diffusivity concepts have been used extensively and successfully to solve many problems in engineering applications, the very basis of the method and the empirical constants associated with it are not universal. Thus, their scope is very limited with respect to the geometries and flow conditions experimentally investigated. Hence the results cannot be readily extrapolated to other, more complicated flow situations. To overcome these deficiencies, new approaches have been tried in recent years to determine the transport properties for turbulent flow by solving a system of partial differential equations similar in form to the conservation equations; a number of constants associated with such equations are determined experimentally. Although such equations for determining the turbulent flow properties have existed since the middle 1940s and early 1950s, renewed interest in this area has been generated in recent years with the availability of high-speed, large-memory digital computers. Launder and Spalding [8, 9] have summarized the governing equations for the calculation of transport properties for turbulent flow.

Velocity Distribution in Turbulent Flow

Velocity distribution in turbulent flow has been investigated extensively because of its importance in practice, but no fundamental theory is yet available to determine this velocity distribution rigorously by purely theoretical approaches. Therefore, empirical and semiempirical relations are used to correlate the velocity field in turbulent flow.

Nikuradse [10a,b] was an early investigator who presented careful measurement of velocity distribution in turbulent flow through a smooth pipe. Later, experiments were performed by other investigators for turbulent flow along a flat plate [11,12] and inside a pipe [13]. Attempts were then made to develop empirical relations that would fit the velocity distribution in turbulent flow

[5,7,14,15]. Here we discuss the velocity distribution law based on the concept of separating the flow field into three distinct layers as illustrated in Fig. 6-1. That is, (1) a very thin layer immediately adjacent to the wall in which laminar or viscous shear stress is dominant is called the *viscous sublayer*; (2) adjacent to this layer is the *buffer layer* in which viscous and turbulent shear stresses are equally important; and (3) the third layer that follows the buffer layer is called the *turbulent layer* in which turbulent shear stress is dominant. We now examine the velocity distribution laws for each layer for the steady, turbulent flow of an incompressible, constant-property fluid over a *smooth surface.*

In the study of velocity distribution for turbulent flow, the following two dimensionless quantities are introduced:

$$u^+ = \frac{u}{\sqrt{\tau_0/\rho}} = \text{dimensionless velocity}$$

$$y^+ = \frac{y}{\nu}\sqrt{\frac{\tau_0}{\rho}} = \text{dimensionless distance}$$

where ρ is the density, τ_0 is the shear stress at the wall, ν is the kinematic viscosity, and u is the velocity component parallel to the wall surface.

Experiments have shown that the *viscous sublayer* is maintained in the region $y^+ < 5$, where the laminar shear stress is dominant and the turbulent shear stress is virtually zero. Therefore, the shear stress is taken in the form $\tau_0 = \mu\, \partial u/\partial y$. The integration of this expression for constant τ_0 with $u = 0$ for $y = 0$ leads to the following result for the velocity distribution in this layer:

$$u^+ = y^+ \qquad \text{for viscous sublayer, } 0 < y^+ < 5 \tag{6-35}$$

The *buffer layer* is considered to extend from $y^+ = 5$ to $y^+ = 30$, and a logarithmic velocity distribution law in the form $u^+ = A \ln y^+ + B$ is assumed. The constants A and B are determined from the requirement that the velocity u^+ be equal to that of viscous sublayer and of the turbulent layer at $y^+ = 5$ and $y^+ = 30$, respectively. The resulting velocity distribution becomes

$$u^+ = 5.0 \ln y^+ - 3.05 \qquad \text{for buffer layer, } 5 \le y^+ \le 30 \tag{6-36}$$

The region $y^+ > 30$ is considered to be the *turbulent layer* where the laminar shear stress is negligible in comparison to the turbulent shear stress. By utilizing the mixing-length concept and assuming that the mixing length varies linearly with the distance from the wall in the form $l = \kappa y$, it can be shown that the velocity distribution in the turbulent layer has a logarithmic profile in the form

$$u^+ = \frac{1}{\kappa} \ln y^+ + C \tag{6-37}$$

where κ is called the *universal constant*. Experiments have shown that $\kappa = 0.4$, and the constant C has been determined by the correlation of Eq. (6-37) with the measured velocity profile. For turbulent flow inside a smooth pipe, $C = 5.5$.

Then the velocity distribution in the turbulent layer is given as

$$u^+ = 2.5 \ln y^+ + 5.5 \quad \text{for turbulent layer, } y^+ > 30 \tag{6-38}$$

We now summarize the foregoing velocity distribution laws for turbulent flow along a smooth surface:

$$u^+ = y^+ \quad \text{for viscous sublayer, } 0 < y^+ < 5 \tag{6-39a}$$

$$u^+ = 5.0 \ln y^+ - 3.05 \quad \text{for buffer layer, } 5 \leq y^+ \leq 30 \tag{6-39b}$$

$$u^+ = 2.5 \ln y^+ + 5.5 \quad \text{for turbulent layer, } y^+ > 30 \tag{6-39c}$$

where

$$u^+ \equiv \frac{u}{\sqrt{\tau_0/\rho}} \quad \text{and} \quad y^+ \equiv \frac{y}{\nu}\sqrt{\frac{\tau_0}{\rho}} \tag{6-39d}$$

and τ_0 is the total wall shear stress.

Figure 6-7 shows a correlation of the velocity distribution law given by Eqs. (6-39) with Nikuradse's [10a] measured velocity distribution for turbulent flow inside smooth pipes.

Although the velocity distribution law obtained by separating the flow field into three distinct layers appears to be in reasonably good agreement with the experimental data, the transition from a viscous to a turbulent flow regime in reality takes place gradually. Therefore, the representation of velocity distribution by three different curves having discontinuous slopes at locations where they join is not realistic. A more serious inconsistency of the logarithmic velocity distribution law Eq. (6-39c) is that it does not give zero velocity gradient at the tube center. For this reason, the average velocity for flow inside a pipe as determined by using the above equations overestimates the velocity. Despite these shortcomings, the velocity distribution laws given by Eqs. (6-39) have been used extensively in the literature to study the relation between momentum and heat transfer.

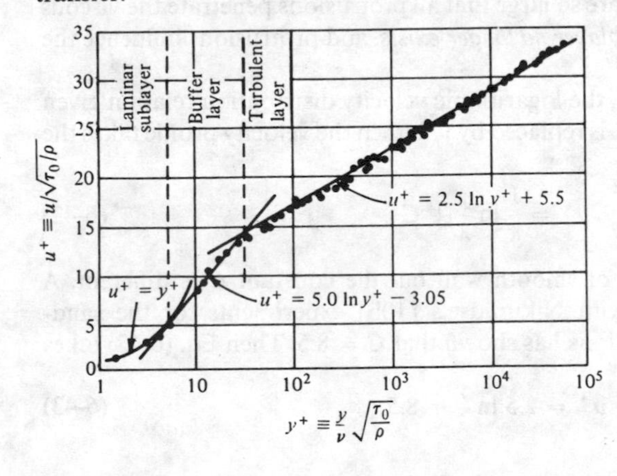

Figure 6-7 Logarithmic velocity distribution law and Nikuradse's [10a] experimental data for turbulent flow inside smooth pipes.

Effects of Surface Roughness on Velocity Distribution

The velocity distribution relations discussed earlier are applicable to turbulent flow over surfaces which are *hydrodynamically smooth.* A surface is considered hydrodynamically smooth if the heights λ of the protrusions are much smaller than the thickness of the viscous sublayer. Surfaces encountered in engineering applications generally are not perfectly smooth, and since for most cases the viscous sublayer is very thin, the protrusions may penetrate it. Varied geometric forms of roughness and the variety of ways that the protrusions may be distributed over the surface make it difficult to analyze the effects of roughness on velocity distribution. Nikuradse [10b] made extensive experiments with turbulent flow inside artificially roughened pipes over a wide range of *relative roughness* λ/D (i.e., protrusion height-to-diameter ratio) from about $\frac{1}{1000}$ to $\frac{1}{30}$. The *sand-grain* roughness used in these experiments has been adopted as a standard for the effects of roughness. These experiments showed that to study the effects of roughness, it is desirable to introduce a *roughness Reynolds number* λ^+ (i.e., a dimensionless protrusion height) defined as

$$\lambda^+ \equiv \frac{\lambda}{\nu}\sqrt{\frac{\tau_0}{\rho}} \tag{6-40}$$

and when λ^+ is less than about 5, the roughness has no effect on the friction due to flow. With this consideration three distinct situations are envisioned for the effects of roughness:

$$\begin{aligned} &\text{Hydrodynamically smooth:} && 0 \le \lambda^+ \le 5 \\ &\text{Transitional:} && 0 \le \lambda^+ \le 70 \\ &\text{Fully rough:} && \lambda^+ > 70 \end{aligned} \tag{6-41}$$

For the hydrodynamically smooth case, the heights of roughness are so small that all protrusions are covered by the viscous sublayer; hence roughness has no effect. For the transitional case, the protrusions are partly outside the viscous sublayer and cause some additional resistance to flow. For the fully rough case, the heights of the protrusions are so large that all protrusions penetrate the viscous sublayer; hence the *viscous sublayer no longer exists,* and protrusions influence the turbulent mixing.

For the *fully rough regime*, the logarithmic velocity distribution relation given by Eq. (6-37) is applicable if y^+ is replaced by y/λ; then the velocity profile takes the form

$$u^+ = \frac{1}{\kappa}\ln\frac{y}{\lambda} + C \tag{6-42}$$

where $\kappa = 0.4$ as in the case of smooth wall but the constant C is different. A correlation of this relation with Nikuradse's [10b] experiments for the sand-roughened wall surface conditions has shown that $C = 8.5$. Then Eq. (6-42) takes the form

$$u^+ = 2.5\ln\frac{y}{\lambda} + 8.5 \tag{6-43}$$

where

$$u^+ \equiv \frac{u}{\sqrt{\tau_0/\rho}}$$

which is called *the logarithmic velocity distribution law for turbulent flow in rough pipes in the fully rough region.*

A significant difference between laminar and turbulent flow is in the fact that in turbulent flow the velocity profile is affected by the surface roughness, whereas in laminar flow the roughness has no effect. As the velocity distribution affects the shear stress at the wall, and hence the friction factor, it is expected that in turbulent flow the friction factor will depend not only on the Reynolds number but also on the surface roughness. This matter is discussed further in the following chapter.

6-4 EQUATIONS OF MOTION

We showed that to determine the drag and the friction coefficients, the velocity distribution in the flow was needed. The velocity distribution can be determined from the solution of the equations of motion subject to appropriate boundary conditions. The problem of velocity distribution in flow for general flow conditions is a very complicated matter, but there are several very simple, yet physically meaningful and practical situations for which the velocity distribution can be determined readily. Our understanding of the physical significance of various parameters in influencing the friction factor and the drag coefficient can be improved if we have some knowledge of the equations governing the fluid motion. With this objective, we present here the equations of continuity and momentum for the two-dimensional, steady motion of constant-property, incompressible newtonian fluid in the two-dimensional, rectangular coordinate system for the x and y variables. We also present the equivalent equations in the cylindrical coordinate system for the (r, z) variables, and we illustrate the simplification of these equations for simpler flow conditions.

Continuity Equation

The continuity equation is essentially the equation for the conservation of mass; it is derived by a mass balance on the fluid entering and leaving a volume element taken in the flow field. Consider a differential volume element $\Delta x\, \Delta y\, 1$ about a point (x, y) in the flow field, as illustrated in Fig. 6-8. The equation for the conservation of mass for two-dimensional steady flow may be stated as

$$\begin{pmatrix}\text{Net rate of mass flow entering}\\ \text{volume element in } x \text{ direction}\end{pmatrix} + \begin{pmatrix}\text{net rate of mass flow entering}\\ \text{volume element in } y \text{ direction}\end{pmatrix} = 0 \quad (6\text{-}44)$$

Let $u \equiv u(x, y)$ and $v \equiv v(x, y)$ be the velocity components in the flow in the x and y directions, respectively. If $M_x \equiv \rho u\, \Delta y\, 1$ is the mass flow rate into the element in

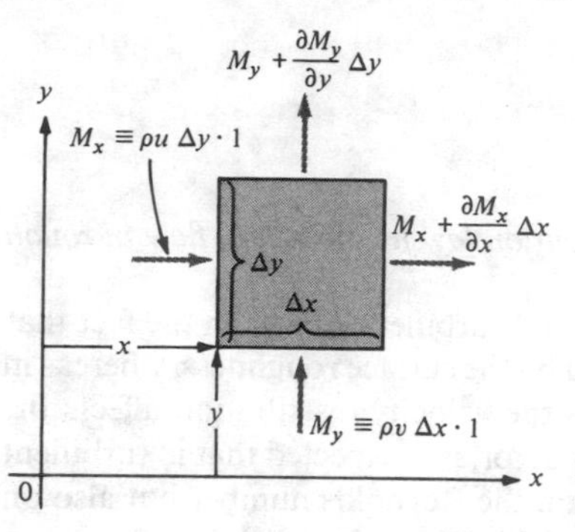

Figure 6-8 Nomenclature for the derivation of the continuity equation.

the x direction through the surface at x, then $M_x + (\partial M_x/\partial x)\,\Delta x$ is the mass flow rate leaving the element in the x direction through the surface at $x + \Delta x$. The net rate of mass flow into the element in the x direction is the difference between the entering and leaving flow rates, given by

$$\begin{pmatrix}\text{Net rate of mass flow entering}\\ \text{element in } x \text{ direction}\end{pmatrix} = -\frac{\partial M_x}{\partial x}\Delta x = -\frac{\partial(\rho u)}{\partial x}\Delta x\,\Delta y\,1 \tag{6-45a}$$

Similarly, for the y direction, we write

$$\begin{pmatrix}\text{Net rate of mass flow entering}\\ \text{element in } y \text{ direction}\end{pmatrix} = -\frac{\partial(\rho v)}{\partial y}\Delta x\,\Delta y\,1 \tag{6-45b}$$

Substituting Eqs. (6-45) into Eq. (6-44), we obtain

$$\frac{\partial(\rho u)}{\partial x} + \frac{\partial(\rho v)}{\partial y} = 0 \tag{6-46a}$$

When density ρ is treated as a constant, Eq. (6-46a) simplifies to

$$\boxed{\frac{\partial u}{\partial x} + \frac{\partial v}{\partial y} = 0} \tag{6-46b}$$

Equation (6-46b) is the *continuity equation* in rectangular coordinates for the steady, two-dimensional flow of an incompressible fluid.

For problems of flow inside a circular tube, the continuity equation in the cylindrical coordinate system is needed. Such an equation can be obtained from Eq. (6-46b) by the standard coordinate transformation.

Let $u \equiv u(r, z)$ and $v \equiv v(r, z)$ denote the velocity components in the axial z and the radial r directions, respectively. Then the *continuity equation* for steady, two-dimensional flow of incompressible fluid in the (r, z) cylindrical coordinates becomes

$$\boxed{\frac{1}{r}\frac{\partial}{\partial r}(rv) + \frac{\partial u}{\partial z} = 0} \tag{6-47}$$

Momentum Equations

The momentum equations are derived from *Newton's second law* of motion, which states that mass times the acceleration in a given direction is equal to the external forces acting on the body in the same direction. The *external forces* acting on a volume element in a flow field are considered to consist of the *body forces* and the *surface forces*. The body forces may result from such effects as the gravitational, electric, and magnetic fields acting on the body of the fluid, and the surface forces result from the stresses acting on the surface of the volume element. With this consideration Newton's second law may be stated for flow in direction i as

$$(\text{Mass})\begin{pmatrix}\text{acceleration in}\\ \text{direction } i\end{pmatrix} = \begin{pmatrix}\text{body forces acting}\\ \text{in direction } i\end{pmatrix} + \begin{pmatrix}\text{surface force}\\ \text{acting in}\\ \text{direction } i\end{pmatrix} \tag{6-48}$$

for a three-dimensional flow, for example, in the rectangular coordinate system $i = x, y$, and z; hence Eq. (6-48) provides three independent momentum equations. In this analysis, we consider two-dimensional, steady, incompressible, constant-property flow in the (x, y) rectangular coordinate system. Therefore, for $i = x$ and y, Eq. (6-48) will provide two momentum equations, one for the x direction and the other for the y direction.

Let $u \equiv u(x, y)$ and $v \equiv v(x, y)$ be the velocity components in the x and y directions, respectively. We consider a volume element $\Delta x\, \Delta y\, 1$ about a point (x, y) in the flow field. Various terms in Eq. (6-48) are determined as follows:

First, if ρ is the density of the fluid, the *mass* term in this equation is given by

$$(\text{Mass}) = (\Delta x\, \Delta y\, 1)\rho \tag{6-49}$$

Second, Eq. (6-48) contains a term called *acceleration*. Commonly, acceleration implies time rate of change of velocity, but for the steady flow considered here, there is an acceleration associated with the convective motion of fluid in other directions, because we have a two-dimensional flow. Consider, for example, motion of the fluid in the x direction. If $u \equiv u(x, y)$ is the velocity component in the x direction, there is an acceleration of the fluid in the x direction associated with the motion of the fluid in other directions, given by*

$$\begin{pmatrix}\text{Acceleration}\\ \text{in } x \text{ direction}\end{pmatrix} = u\frac{\partial u}{\partial x} + v\frac{\partial u}{\partial y} \tag{6-50a}$$

* The derivation of Eq. (6-50*a*) is as follows:

Consider $u \equiv u(x, y)$ for the two-dimensional steady flow. The total derivative of u is

$$du = \frac{\partial u}{\partial x}dx + \frac{\partial u}{\partial y}dy$$

Dividing both sides by dt, we have

$$\frac{du}{dt} = \frac{\partial u}{\partial x}\frac{dx}{dt} + \frac{\partial u}{\partial y}\frac{dy}{dt} = u\frac{\partial u}{\partial x} + v\frac{\partial u}{\partial y}$$

which is the same as Eq. (6-50*a*). Similarly, one obtains Eq. (6-50*b*) by considering the total derivative of $v \equiv v(x, y)$.

Similarly, if $v \equiv v(x, y)$ is the velocity component in the y direction, the acceleration of the fluid in the y direction associated with the motion of the fluid in other directions is given by

$$\begin{pmatrix}\text{Acceleration} \\ \text{in } y \text{ direction}\end{pmatrix} = u\frac{\partial v}{\partial x} + v\frac{\partial v}{\partial y} \tag{6-50b}$$

Third, Eq. (6-48) contains a term called *body forces* acting on the fluid. Let F_x and F_y be the *body forces acting per unit volume* of the fluid in the x and y directions, respectively (that is, ρg denotes the gravitational force acting per unit volume). Then

$$\begin{pmatrix}\text{Body forces acting} \\ \text{in } x \text{ direction}\end{pmatrix} = F_x(\Delta x\ \Delta y\ 1) \tag{6-51a}$$

$$\begin{pmatrix}\text{Body forces acting} \\ \text{in } y \text{ direction}\end{pmatrix} = F_y(\Delta x\ \Delta y\ 1) \tag{6-51b}$$

Fourth, Eq. (6-48) contains a term called the *surface forces* acting on the fluid. The surface forces acting per unit area are called *stresses*. When the stress acts normal to the surface, it is called the *normal stress*: when it acts along the surface, it is called the *shear stress*.

Figure 6-9 shows various stresses acting on the surfaces of a differential volume element. In this figure σ_x and σ_y denote the normal stresses in the x and y directions, respectively. The shear stresses are denoted by τ_{xy} and τ_{yx}, where *the first subscript indicates the axis to which the surface is perpendicular and the second subscript indicates the direction of the shear stress*. Thus, τ_{xy} is the shear stress acting on the surface Δy 1 (i.e., the surface perpendicular to the x axis) at x in the direction y. Then *the net normal surface force acting on the element in the positive x direction* is $(\partial/\partial y)(\sigma_x\ \Delta y\ 1)\ \Delta x$, and *the net shear force acting on the element in the positive x direction* is $(\partial/\partial y)(\tau_{yx}\ \Delta x\ 1)\ \Delta y$. Hence, the net surface forces acting on the element in the positive x direction becomes

$$\begin{pmatrix}\text{Net surface forces acting} \\ \text{in } x \text{ direction}\end{pmatrix} = \left(\frac{\partial \sigma_x}{\partial x} + \frac{\partial \tau_{yx}}{\partial y}\right)(\Delta x\ \Delta y\ 1) \tag{6-52a}$$

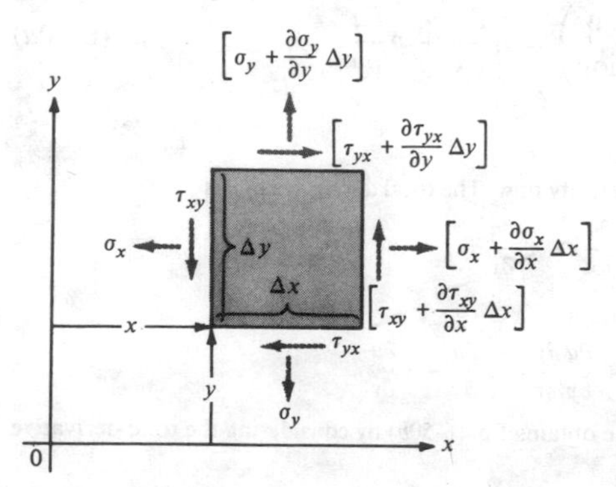

Figure 6-9 Nomenclature for the various stresses acting on the surfaces of the volume element.

Similarly, the net surface force acting in the y direction is

$$\begin{pmatrix}\text{Net surface forces acting}\\ \text{in } y \text{ direction}\end{pmatrix} = \left(\frac{\partial \sigma_y}{\partial y} + \frac{\partial \tau_{xy}}{\partial x}\right)(\Delta x\, \Delta y\ 1) \qquad (6\text{-}52b)$$

When Eqs. (6-49) to (6-52) are introduced into Eq. (6-48) and $\Delta x\, \Delta y$ terms are canceled, the x *momentum* and the y *momentum* equations, respectively, become

$$x \text{ Momentum:} \qquad \rho\left(u\frac{\partial u}{\partial x} + v\frac{\partial u}{\partial y}\right) = F_x + \frac{\partial \sigma_x}{\partial x} + \frac{\partial \tau_{yx}}{\partial y} \qquad (6\text{-}53a)$$

$$y \text{ Momentum:} \qquad \rho\left(u\frac{\partial v}{\partial x} + v\frac{\partial v}{\partial y}\right) = F_y + \frac{\partial \sigma_y}{\partial y} + \frac{\partial \tau_{xy}}{\partial x} \qquad (6\text{-}53b)$$

The final stage in the analysis involves the determination of the expressions for various stresses appearing in these equations. Such relations depend on the type of fluid considered, and a discussion of this matter can be found in several references [16–18]. For the two-dimensional, incompressible, constant-property flow and the newtonian fluid considered here, various stresses are related to the velocity components by [16]

$$\tau_{xy} = \tau_{yx} = \mu\left(\frac{\partial u}{\partial y} + \frac{\partial v}{\partial x}\right) \qquad (6\text{-}54a)$$

$$\sigma_x = -P + 2\mu\frac{\partial u}{\partial x} \qquad (6\text{-}54b)$$

$$\sigma_y = -P + 2\mu\frac{\partial v}{\partial y} \qquad (6\text{-}54c)$$

where P is the pressure and μ is the viscosity in the flow fleld.

When Eqs (6-54) are introduced into Eqs. (6-53), after some manipulation, one obtains

$$x \text{ Momentum:} \quad \rho\left(u\frac{\partial u}{\partial x} + v\frac{\partial u}{\partial y}\right) = F_x - \frac{\partial P}{\partial x} + \mu\left(\frac{\partial^2 u}{\partial x^2} + \frac{\partial^2 u}{\partial y^2}\right) \qquad (6\text{-}55)$$

$$y \text{ Momentum:} \quad \rho\left(u\frac{\partial v}{\partial x} + v\frac{\partial v}{\partial y}\right) = F_y - \frac{\partial P}{\partial y} + \mu\left(\frac{\partial^2 v}{\partial x^2} + \frac{\partial^2 v}{\partial y^2}\right) \qquad (6\text{-}56)$$

where F_x and F_y are the body forces per unit volume acting in the x and y directions, respectively. Equations (6-55*a*) and (6-55*b*) are called, respectively, the x and y momentum equations for the steady, two-dimensional flow of an incompressible, constant-property, newtonian fluid in the rectangular coordinate system.

The physical significance of the various terms in Eqs. (6-55) is as follows: The terms on the left-hand side represent the *inertia forces*, the first term on the right-hand side is the *body force*, the second term is the *pressure force*, and the last term in the parentheses is the *viscous forces* acting on the fluid element.

If the body forces F_x and F_y are known, the continuity equation (6-46*b*) and the two momentum equations (6-55) provide three independent equations for the determination of the three unknown quantities u, v, and P for the steady, two-dimensional flow of an incompressible fluid. The analytical solution of these equations is extremely difficult except for very simple situations.

For problems of flow inside a circular tube, for example, the momentum equations are needed in the cylindrical coordinate system. Let $u \equiv u(r, z)$ and $v \equiv v(r, z)$ be the velocity components in the axial z and the radial r directions, respectively. Then in the two-dimensional, (r, z) cylindrical coordinate system, the equivalent of the momentum equations (6-55) and (6-56) is given by

$$r\text{ Momentum:}\quad \rho\left(v\frac{\partial v}{\partial r} + u\frac{\partial v}{\partial z}\right) = F_r - \frac{\partial P}{\partial r} + \mu\left(\frac{\partial^2 v}{\partial r^2} + \frac{1}{r}\frac{\partial v}{\partial r} - \frac{v}{r^2} + \frac{\partial^2 v}{\partial z^2}\right) \tag{6-57}$$

$$z\text{ Momentum:}\quad \rho\left(v\frac{\partial u}{\partial r} + u\frac{\partial u}{\partial z}\right) = F_z - \frac{\partial P}{\partial z} + \mu\left(\frac{\partial^2 u}{\partial r^2} + \frac{1}{r}\frac{\partial u}{\partial r} + \frac{\partial^2 u}{\partial z^2}\right) \tag{6-58}$$

Here F_r and F_z are, respectively, the body forces acting on the fluid in the r and z directions. For simplicity in the nomenclature we used the symbols u and v to denote the velocity components in the z and r directions, instead of the commonly used notations v_z and v_r, respectively.

Example 6-11 An incompressible, constant-property, newtonian fluid is in steady forced flow between two parallel plates separated by a distance $2L$. The x and y coordinates are chosen along and normal to the main flow, as illustrated in Fig. 6-10. There are no body forces acting on the fluid. Develop the differential equation governing the velocity distribution in the flow in the *hydrodynamically developed* region by simplifying the equations of motion given above.

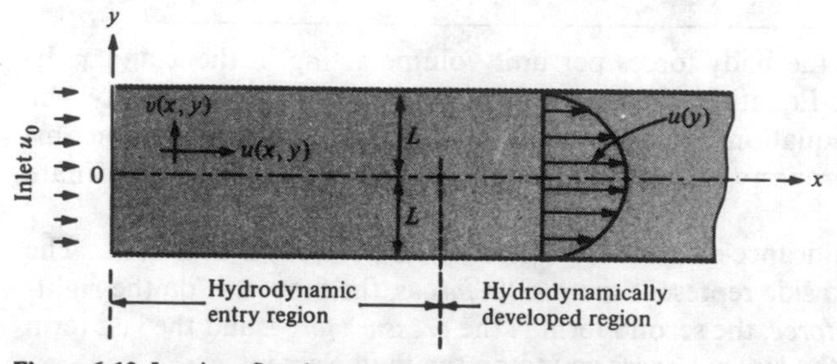

Figure 6-10 Laminar flow between two parallel plates.

SOLUTION In the hydrodynamically developed region, all fluid particles move in the direction parallel to the plates. Therefore, in this region, the velocity component normal to the wall v must be zero:

$$v(x, y) = 0$$

Then the continuity equation (6-46*b*) reduces to

$$\frac{\partial u}{\partial x} = 0$$

which implies that the axial velocity component u does not depend on x, or we have

$$u \equiv u(y)$$

The y momentum equation (6-56), subject to the requirements $v = 0$ and $F_y = 0$, simplifies to

$$\frac{\partial P}{\partial y} = 0$$

which implies that the pressure P does not depend on y, or we have

$$P \equiv P(x)$$

Finally, the x momentum equation (6-55)—subject to the requirements $v = 0$, $u \equiv u(y)$, $P \equiv P(x)$, and $F_x = 0$—reduces to

$$\boxed{\frac{d^2u(y)}{dy^2} = \frac{1}{\mu}\frac{dP(x)}{dx}} \tag{6-59}$$

This is a simple, ordinary differential equation for the axial velocity component in the hydrodynamically developed region, and its solution subject to appropriate boundary conditions gives $u(y)$.

Example 6-12 An incompressible, constant-property, newtonian fluid is in steady forced flow inside a circular tube of radius R. The z and r coordinates are chosen along and normal to the direction of main flow, as illustrated in Fig. 6-11. There are no body forces acting on the fluid. Develop the differential

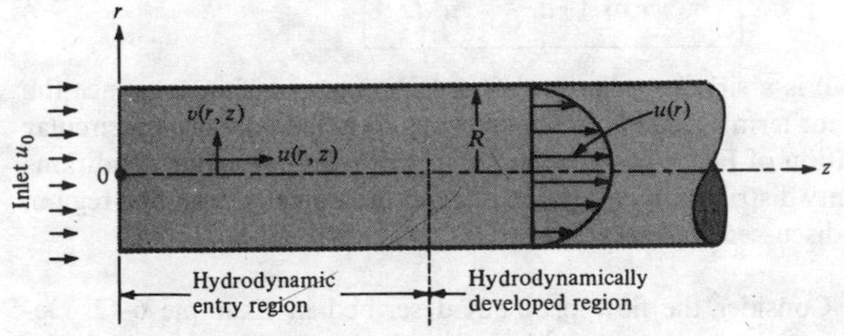

Figure 6-11 Laminar flow inside a circular tube.

equation governing the velocity distribution in the flow in the hydrodynamically developed region by simplifying the equations of motion in cylindrical coordinates.

SOLUTION In the hydrodynamically developed region, all fluid particles move parallel to the z axis. Therefore, in this region, the velocity component v normal to the direction of main flow must be zero:

$$v(r, z) = 0$$

Then the continuity equation (6-47) reduces to

$$\frac{\partial u}{\partial z} = 0$$

which implies that the axial velocity component u does not depend on z, or we have

$$u \equiv u(r)$$

The r momentum equation (6-57), subject to the requirement $v = 0$ and $F_r = 0$, simplifies to

$$\frac{\partial P}{\partial r} = 0$$

which implies that the pressure P does not depend on r, or we have

$$P \equiv P(z)$$

Thus, the r momentum equation is no longer needed.

The z momentum equation (6-58)—subject to the requirement $v = 0$, $u \equiv u(r)$, $P \equiv P(z)$, and $F_z = 0$—reduces to

$$\frac{d^2u}{dr^2} + \frac{1}{r}\frac{du}{dr} = \frac{1}{\mu}\frac{dP}{dz}$$

or

$$\frac{1}{r}\frac{d}{dr}\left(r\frac{du}{dr}\right) = \frac{1}{\mu}\frac{dP}{dz} \tag{6-60}$$

Equation (6-60) is a simple ordinary differential equation for $u(r)$, since the pressure gradient term dP/dz is considered constant for flow inside a circular tube. The solution of Eq. (6-60) subject to appropriate boundary conditions gives the velocity distribution $u(r)$ in the hydrodynamically developed region. This matter is discussed further in Chap. 7.

Example 6-13 Consider the flow problem described in Example 6-12. Develop the differential equation governing the velocity distribution $u(r)$ in the

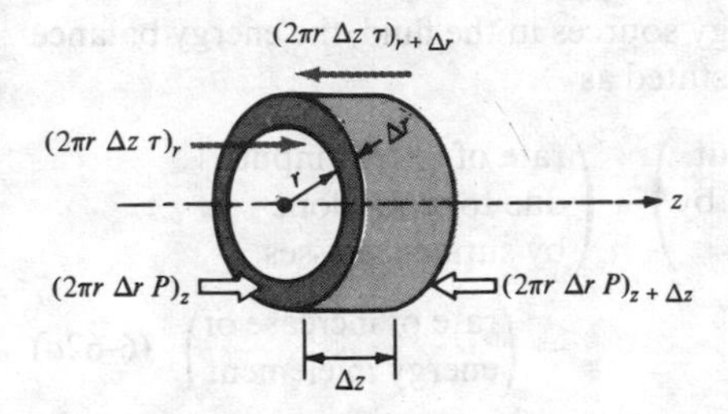

Figure 6-12 Force balance on an annular control volume.

hydrodynamically developed region by writing a force balance equation for an annular control volume.

SOLUTION Figure 6-12 shows an annular control volume of thickness Δr about r and length Δz about z in the fluid. For the hydrodynamically developed flow, the axial velocity component u depends only on r, that is, $u(r, z) \equiv u(r)$.

A force balance on the annular volume element shown in Fig. 6-12 gives

$$(2\pi r\,\Delta r\,P)_z - (2\pi r\,\Delta r\,P)_{z+\Delta z} = (2\pi r\,\Delta z\,\tau)_{r+\Delta r} - (2\pi r\,\Delta z\,\tau)_r$$

$$\frac{P|_{z+\Delta z} - P|_z}{\Delta z} = -\frac{1}{r}\frac{(r\tau)_{r+\Delta r} - (r\tau)_r}{\Delta r}$$

For the limit $(\Delta r, \Delta z) \to 0$, we obtain

$$\frac{dP}{dz} = -\frac{1}{r}\frac{d}{dr}(r\tau)$$

For the newtonian fluid, the shear stress τ is given by [see Eq. (6-18*b*)]

$$\tau = -\mu\frac{du(r)}{dr}$$

Introducing this into the above expression, we obtain the differential equation governing the velocity distribution $u(r)$ as

$$\boxed{\frac{1}{r}\frac{d}{dr}\left(r\frac{du}{dr}\right) = \frac{1}{\mu}\frac{dP}{dz}} \tag{6-61}$$

where we assumed constant μ. We note that this equation is exactly the same as developed in Example 6-12 by the simplification of the more general equations of motion.

6-5 EQUATION OF ENERGY

The temperature distribution in the flow field is governed by the energy equation, which can be derived by writing an energy balance according to the first law of thermodynamics for a differential volume element in the flow field. If radiation is absent and there are no distributed energy sources in the fluid, the energy balance on a differential volume element may be stated as

$$\begin{pmatrix}\text{Rate of energy}\\ \text{input due to}\\ \text{conduction}\end{pmatrix} + \begin{pmatrix}\text{rate of energy input}\\ \text{due to work done by}\\ \text{body forces}\end{pmatrix} + \begin{pmatrix}\text{rate of energy input}\\ \text{due to work done}\\ \text{by surface stresses}\end{pmatrix}$$

$$= \begin{pmatrix}\text{rate of increase of}\\ \text{energy in element}\end{pmatrix} \tag{6-62a}$$

To derive the energy equation, each term in this expression should be evaluated. Here we consider the energy equation in the rectangular coordinate system for steady, two-dimensional (x, y) flow of an incompressible, constant-property, newtonian fluid. Let $\Delta x\, \Delta y\, 1$ be the differential volume element about a point (x, y) in the flow field. Various terms in Eq. (6-62*a*) are evaluated now.

First, the heat addition into the element $\Delta x\, \Delta y\, 1$ by *conduction* occurs in the x and y directions. Referring to the nomenclature shown in Fig. 6-13, we write

$$\begin{pmatrix}\text{Rate of energy}\\ \text{addition by}\\ \text{conduction}\end{pmatrix} = -\left(\frac{\partial Q_x}{\partial x}\Delta x + \frac{\partial Q_y}{\partial y}\Delta y\right) = -\left(\frac{\partial q_x}{\partial x} + \frac{\partial q_y}{\partial y}\right)\Delta x\, \Delta y\, 1$$

$$= k\left(\frac{\partial^2 T}{\partial x^2} + \frac{\partial^2 T}{\partial y^2}\right)\Delta x\, \Delta y\, 1 \tag{6-62b}$$

since

$$q_x = -k\frac{\partial T}{\partial x} \qquad \text{and} \qquad q_y = -k\frac{\partial T}{\partial y}$$

Second, if F_x and F_y are the body forces acting per unit volume of the element and u and y are the velocity components in the x and y directions, respectively, the energy input into the volume element $\Delta x\, \Delta y\, 1$ resulting from the increase in potential energy becomes

$$\begin{pmatrix}\text{Rate of energy input}\\ \text{by body forces}\end{pmatrix} = (uF_x + vF_y)\,\Delta x\, \Delta y\, 1 \tag{6-63}$$

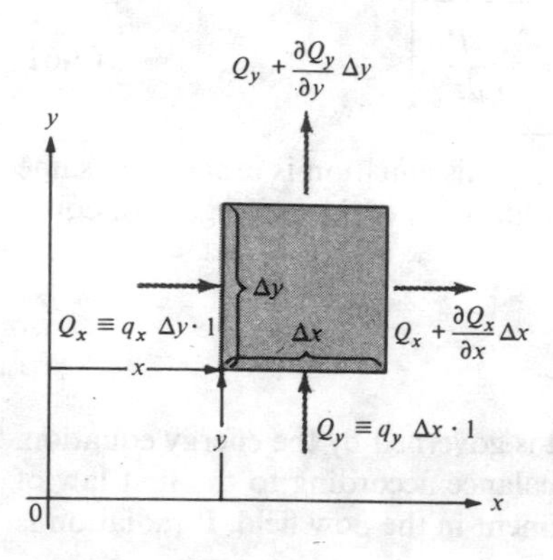

Figure 6-13 Nomenclature for heat addition by conduction.

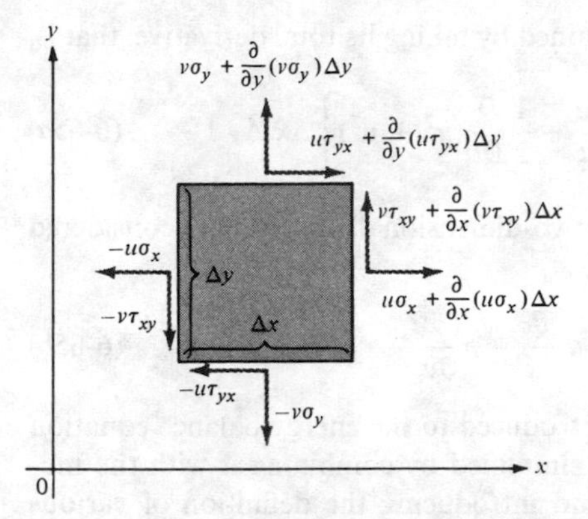

Figure 6-14 Nomenclature for frictional work done by the surface forces.

Third, the rate of energy input to the volume element $\Delta x\ \Delta y\ 1$ due to surface stresses consists of the contributions from the stresses σ_x, σ_y, τ_{yx}, and τ_{xy}. By referring to the illustration and nomenclature in Fig. 6-14, the energy input due to the normal stress σ_x is given by

$$\left\{-u\sigma_x + \left[u\sigma_x + \frac{\partial}{\partial x}(u\sigma_x)\,\Delta x\right]\right\}\Delta y\ 1 = \Delta x\ \Delta y\ 1\frac{\partial}{\partial x}(u\sigma_x)$$

and, due to the normal stress σ_y, is given by

$$\left\{-v\sigma_y + \left[v\sigma_y + \frac{\partial}{\partial y}(v\sigma_y)\,\Delta y\right]\right\}\Delta x\ 1 = \Delta x\ \Delta y\ 1\frac{\partial}{\partial y}(v\sigma_y)$$

Similarly, the energy input due to the stresses τ_{yx} and τ_{xy} are given, respectively, by

$$-u\tau_{yx} + \left[u\tau_{yx} + \frac{\partial}{\partial y}(u\tau_{yx})\,\Delta y\right]\Delta x\ 1 = \Delta x\ \Delta y\ 1\frac{\partial}{\partial y}(u\tau_{yx})$$

$$-v\tau_{xy} + \left[v\tau_{xy} + \frac{\partial}{\partial x}(v\tau_{xy})\,\Delta x\right]\Delta y\ 1 = \Delta x\ \Delta y\ 1\frac{\partial}{\partial x}(v\tau_{xy})$$

The total rate of energy input into the element due to the stresses is obtained by summing the above four quantities:

$$\begin{pmatrix}\text{Rate of energy}\\ \text{input by surface}\\ \text{stresses}\end{pmatrix} = \left[\frac{\partial}{\partial x}(u\sigma_x) + \frac{\partial}{\partial y}(v\sigma_y) + \frac{\partial}{\partial y}(u\tau_{yx}) + \frac{\partial}{\partial x}(v\tau_{xy})\right]\Delta x\ \Delta y\ 1 \tag{6-64}$$

Fourth the energy contained in the volume element is considered to consist of the *specific internal energy e* per unit mass and the *kinetic energy* $\frac{1}{2}(u^2 + v^2)$ per unit mass of the fluid. Then the energy content of the volume element $\Delta x\ \Delta y\ 1$ becomes

$$\rho[e + \tfrac{1}{2}(u^2 + v^2)]\ \Delta x\ \Delta y\ 1$$

The rate of increase of this energy is obtained by taking its total derivative, that is,

$$\begin{pmatrix}\text{Rate of increase}\\ \text{of energy of}\\ \text{element}\end{pmatrix} = \rho\left[\frac{De}{Dt} + \frac{1}{2}\frac{D}{Dt}(u^2 + v^2)\right]\Delta x\,\Delta y\,1 \tag{6-65a}$$

where the *total derivative* D/Dt for the two-dimensional, steady flow considered here is defined as

$$\frac{D}{Dt} \equiv u\frac{\partial}{\partial x} + v\frac{\partial}{\partial y} \tag{6-65b}$$

Finally, Eqs. (6-62*b*) to (6-65) are introduced to the energy balance equation (6-62*a*), and the resulting expression is simplified by combining it with the momentum equations (6-55) and (6-56) and introducing the definition of various stress terms given by Eqs. (6-54). After quite lengthy manipulations, the energy equation in the rectangular coordinate system for steady, two-dimensional (x, y) flow of an incompressible, constant-property, newtonian fluid is determined as

$$\rho c_p\left(u\frac{\partial T}{\partial x} + v\frac{\partial T}{\partial y}\right) = k\left(\frac{\partial^2 T}{\partial x^2} + \frac{\partial^2 T}{\partial y^2}\right) + \mu\Phi \tag{6-66a}$$

where the *viscous-energy-dissipation function* Φ is defined as

$$\Phi \equiv 2\left[\left(\frac{\partial u}{\partial x}\right)^2 + \left(\frac{\partial v}{\partial y}\right)^2\right] + \left(\frac{\partial v}{\partial x} + \frac{\partial u}{\partial y}\right)^2 \tag{6-66b}$$

The physical significance of various terms in Eq. (6-66*a*) is as follows: The left-hand side represents the net energy transfer due to mass transfer; on the right-hand side the terms in parentheses represent conductive heat transfer; and the last term on the right-hand side is the viscous-energy dissipation in the fluid due to internal fluid friction.

For most engineering applications, the flow velocities are moderate; hence the viscous-energy dissipation term can be neglected. Then, the energy equation (6-66) simplifies to

$$\frac{1}{\alpha}\left(u\frac{\partial T}{\partial x} + v\frac{\partial T}{\partial y}\right) = \frac{\partial^2 T}{\partial x^2} + \frac{\partial^2 T}{\partial y^2} \tag{6-67}$$

where $\alpha = k/(\rho c_p)$. For the case of no flow (that is, $u = v = 0$), the energy equation simplifies to the two-dimensional, steady-state heat conduction equation with no heat generation.

For forced convection inside a circular tube, the energy equation is needed in the cylindrical coordinate system. Let $u \equiv u(r, z)$ and $v \equiv v(r, z)$ be the velocity components in the axial z and the radial r directions, respectively. Then in the two-dimensional (r, z) cylindrical coordinate system, the equivalent of the energy equation (6-66) is given by

$$\rho c_p\left(v\frac{\partial T}{\partial r}+u\frac{\partial T}{\partial z}\right)=k\left(\frac{\partial^2 T}{\partial r^2}+\frac{1}{r}\frac{\partial T}{\partial r}+\frac{\partial^2 T}{\partial z^2}\right)+\mu\Phi \tag{6-68a}$$

where the viscous-dissipation function is defined as

$$\Phi\equiv 2\left[\left(\frac{\partial v}{\partial r}\right)^2+\frac{v^2}{r^2}+\left(\frac{\partial u}{\partial z}\right)^2\right]+\left(\frac{\partial u}{\partial r}+\frac{\partial v}{\partial z}\right)^2 \tag{6-68b}$$

The physical significance of various terms in these equations is the same as given above for the rectangular coordinate system.

For moderate flow velocities, the viscous-energy dissipation term Φ is negligible, and Eq. (6-68) reduces to

$$\frac{1}{\alpha}\left(v\frac{\partial T}{\partial r}+u\frac{\partial T}{\partial z}\right)=\frac{\partial^2 T}{\partial r^2}+\frac{1}{r}\frac{\partial T}{\partial r}+\frac{\partial^2 T}{\partial z^2} \tag{6-69}$$

Example 6-14 An incompressible, constant-property fluid is in steady forced flow between two parallel plates. The x and y coordinate axes are chosen along and normal to the plate surface, respectively. By simplifying Eq. (6-66), obtain the energy equation for the region where the flow is hydrodynamically developed.

SOLUTION In the hydrodynamically developed region, $v = 0$ and $u \equiv u(y)$. Then the energy equation (6-66a) reduces to

$$\rho c_p u(y)\frac{\partial T}{\partial x}=k\left(\frac{\partial^2 T}{\partial x^2}+\frac{\partial^2 T}{\partial y^2}\right)+\mu\Phi$$

and the viscous-energy-dissipation function Φ, defined by Eq. (6-66b), simplifies to

$$\Phi\equiv\left[\frac{du(y)}{dy}\right]^2$$

Hence the energy equation becomes

$$\rho c_p u(y)\frac{\partial T}{\partial x}=k\left(\frac{\partial^2 T}{\partial x^2}+\frac{\partial^2 T}{\partial y^2}\right)+\mu\left[\frac{du(y)}{dy}\right]^2 \tag{6-70}$$

which is valid in the hydrodynamically developed region, and $u(y)$ is the fully developed velocity profile.

Example 6-15 An incompressible, constant-property fluid is in steady forced flow inside a circular tube. The coordinate axes z and r are chosen along the tube axis and radially, respectively. The flow velocity is moderate, hence the viscous-energy dissipation in the fluid can be neglected. Give the energy equation valid for the region where the flow is hydrodynamically developed.

SOLUTION The energy equation for flow inside a circular tube with negligible viscous-energy dissipation is given by Eq. (6-69). In the hydrodynamically developed region, we have $v = 0$ and $u \equiv u(r)$. Then the energy equation (6-69) simplifies to

$$\frac{1}{\alpha} u(r) \frac{\partial T}{\partial z} = \frac{\partial^2 T}{\partial r^2} + \frac{1}{r}\frac{\partial T}{\partial r} + \frac{\partial^2 T}{\partial z^2}$$

or

$$\frac{1}{\alpha} u(r) \frac{\partial T}{\partial z} = \frac{1}{r}\frac{\partial}{\partial r}\left(r\frac{\partial T}{\partial r}\right) + \frac{\partial^2 T}{\partial z^2} \tag{6-71a}$$

Example 6-16 Consider the forced convection problem described in Example 6-15 with the additional assumption that the heat conduction in the axial direction is negligible in comparison to that in the radial direction. Develop the energy equation governing the temperature distribution $T(r)$ in the fluid by writing an energy balance equation for an annular control volume.

SOLUTION Figure 6-15 shows an annular control volume of thickness Δr about r and length Δz about z in the fluid. The energy convected out must be balanced by the net heat conducted in radially, because there is no axial conduction or viscous-energy dissipation.

The energy balance equation for the annular volume element shown in Fig. 6-15 can be written as

$$\text{(Net energy convected out)} = \text{(net heat conducted in)}$$

or

$$(2\pi r\,\Delta r)[(\rho c_p u T)_{z+\Delta z} - (\rho c_p u T)_z] = (2\pi r\,\Delta z\, q)_r - (2\pi r\,\Delta z\, q)_{r+\Delta r}$$

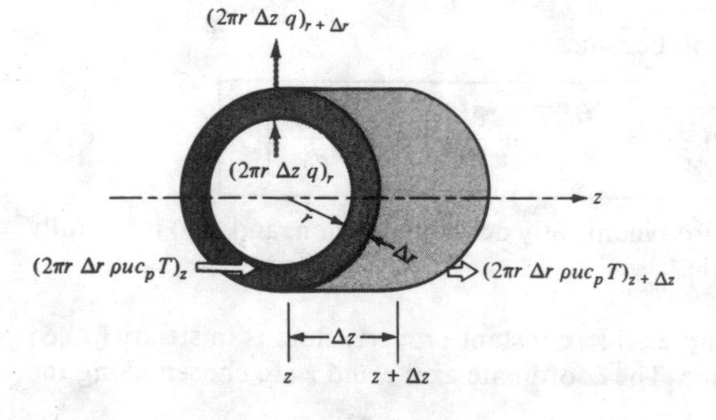

Figure 6-15 An annular control volume for energy balance on fluid in fully developed laminar flow inside a circular tube.

where q is the conduction heat flux in the r direction. This expression is rearranged as

$$\rho c_p u \frac{(T)_{z+\Delta z} - (T)_z}{\Delta z} = -\frac{1}{r}\frac{(rq)_{r+\Delta r} - (rq)_r}{\Delta r}$$

For the limit $(\Delta r, \Delta z) \to 0$, we obtain

$$\rho c_p u \frac{\partial T}{\partial z} = -\frac{1}{r}\frac{\partial}{\partial r}(rq)$$

The conduction heat flux in the radial direction is given by

$$q = -k\frac{\partial T}{\partial r}$$

Then the energy equation takes the form

$$\boxed{\frac{1}{\alpha}u(r)\frac{\partial T}{\partial z} = \frac{1}{r}\frac{\partial}{\partial r}\left(r\frac{\partial T}{\partial r}\right)} \tag{6-71b}$$

where we assumed constant k. We note that this equation is similar to the energy equation (6-71*a*) except the axial conduction term has been neglected.

6-6 DIMENSIONLESS PARAMETERS

In this chapter dimensionless parameters, such as the Reynolds, Prandtl, Nusselt, and Stanton numbers, are introduced and the physical significance of these dimensionless parameters in the interpretation of the conditions associated with fluid flow or heat transfer is discussed.

We consider the Reynolds number based on a characteristic length L, rearranged in the form

$$\text{Re} = \frac{u_\infty L}{\nu} = \frac{u_\infty^2/L}{\nu u_\infty/L^2} = \frac{\text{inertia force}}{\text{viscous force}} \tag{6-72a}$$

Then the Reynolds number represents the ratio of the inertia to viscous force. This result implies that viscous forces are dominant for small Reynolds numbers and inertia forces are dominant for large Reynolds numbers. Recall that the Reynolds number was used as the criterion to determine the change from laminar to turbulent flow. As the Reynolds number is increased, the inertia forces become dominant and small disturbances in the fluid may be amplified to cause the transition from laminar to turbulent flow.

The Prandtl number can be arranged in the form

$$\text{Pr} = \frac{c_p \mu}{k} = \frac{\mu\rho}{k/(\rho c_p)} = \frac{\nu}{\alpha} = \frac{\text{molecular diffusivity of momentum}}{\text{molecular diffusivity of heat}} \tag{6-72b}$$

Thus it represents the relative importance of momentum and energy transport by the diffusion process. Hence for gases with $\Pr \cong 1$, the transfer of momentum and energy by the diffusion process is comparable. For oils, $\Pr \gg 1$, hence the momentum diffusion is much greater than the energy diffusion; but for liquid metals, $\Pr \ll 1$ and the situation is reversed. We recall that in discussing the development of velocity and thermal boundary layers for flow along a flat plate, the relative thickness of velocity and thermal boundary layers depended on the magnitude of the Prandtl number.

Consider the Nusselt number, based on a characteristic length L, rearranged in the form

$$\mathrm{Nu} = \frac{hL}{k} = \frac{h\,\Delta T}{k\,\Delta T/L} \tag{6-73a}$$

where ΔT is the reference temperature difference between the wall surface and fluid temperatures. Then the Nusselt number may be interpreted as the ratio of heat transfer by convection to conduction across the fluid layer of thickness L. Based on this interpretation, the value of the Nusselt number equal to unity implies that there is no convection—the heat transfer is by pure conduction. A larger value of the Nusselt number implies enhanced heat transfer by convection.

The Stanton number can be rearranged as

$$\mathrm{St} = \frac{h}{\rho c_p u_m} = \frac{h\,\Delta T}{\rho c_p u_m\,\Delta T} \tag{6-73b}$$

where ΔT is a reference temperature difference between the wall surface and the fluid. The numerator represents heat flux to the fluid, and the denominator represents the heat transfer capacity of the fluid flow.

The dimensionless parameter, the *Eckert* number, defined as $E \equiv u_\infty^2/(c_p\,\Delta T)$, frequently arises in high-speed, heat transfer problems. The Eckert number can be rearranged as

$$E = \frac{u_\infty^2}{c_p\,\Delta T} = \frac{u_\infty^2/c_p}{\Delta T} = \frac{\text{dynamic temperature due to fluid motion}}{\text{temperature difference}} \tag{6-74}$$

Here $u_\infty^2/(2c_p)$ represents an ideal temperature rise if an ideal gas with a velocity u_∞ is slowed down adiabatically to zero velocity. This definition implies that if the Eckert number is small, the viscous-energy generation effects due to the motion of the fluid can be neglected in comparison with the temperature differences involved in the heat transfer process. We recall that the viscous-energy-dissipation term appeared in the energy equation, and the magnitude of the Eckert number becomes the criterion in deciding whether the viscous-energy-dissipation effects should be considered in the heat transfer analysis.

Our understanding of the physical significance of the dimensionless parameters is further improved if we examine how these dimensionless parameters enter the equations of motion and energy. If the equations of momentum and energy are expressed in the dimensionless form, the dimensionless parameters Re, Pr, and

E appear in these equations. To illustrate this matter we consider the x momentum equation (6-55) by neglecting the body forces:

$$\rho\left(u\frac{\partial u}{\partial x}+v\frac{\partial u}{\partial y}\right)=-\frac{\partial p}{\partial x}+\mu\left(\frac{\partial^2 u}{\partial x^2}+\frac{\partial^2 u}{\partial y^2}\right) \tag{6-75}$$

To nondimensionalize this equation, we select a characteristic length L and a reference velocity u_∞, and we introduce the following dimensionless variables:

$$X=\frac{x}{L} \qquad Y=\frac{y}{L} \qquad P^*=\frac{P}{\rho u_\infty^2} \qquad U=\frac{u}{u_\infty} \qquad V=\frac{v}{u_\infty} \tag{6-76}$$

Here the quantity ρu_∞^2 represents the double of the *dynamic head* (that is, $\frac{1}{2}\rho u_\infty^2$ is called the dynamic head). By introducing these dimensionless variables into the above momentum equation, the dimensionless form of the x momentum equation becomes

$$U\frac{\partial U}{\partial X}+V\frac{\partial U}{\partial Y}=-\frac{\partial P^*}{\partial X}+\frac{1}{\text{Re}}\left(\frac{\partial^2 U}{\partial X^2}+\frac{\partial^2 U}{\partial Y^2}\right) \tag{6-77}$$

The left-hand side of this equation represents the inertial forces, and on the right-hand side the terms in the parentheses represent the viscous forces. Increasing the Reynolds number decreases the relative importance of the viscous forces in the system.

We now consider the energy equation (6-66) written as

$$\rho c_p\left(u\frac{\partial T}{\partial x}+v\frac{\partial T}{\partial y}\right)=k\left(\frac{\partial^2 T}{\partial x^2}+\frac{\partial^2 T}{\partial y^2}\right)+\mu\left[2\left(\frac{\partial u}{\partial x}\right)^2+2\left(\frac{\partial v}{\partial y}\right)^2+\left(\frac{\partial v}{\partial x}+\frac{\partial u}{\partial y}\right)^2\right] \tag{6-78}$$

and we define a dimensionless temperature θ as

$$\theta=\frac{T-T_\infty}{\Delta T} \tag{6-79}$$

where T_∞ is a reference temperature and ΔT is a reference temperature difference. The dimensionless variables (6-76) and (6-79) are introduced to Eq. (6-78). We obtain

$$U\frac{\partial \theta}{\partial X}+V\frac{\partial \theta}{\partial Y}=\frac{1}{\text{Re Pr}}\left(\frac{\partial^2 \theta}{\partial X^2}+\frac{\partial^2 \theta}{\partial Y^2}\right)+\frac{E}{\text{Re}}\left[2\left(\frac{\partial U}{\partial X}\right)^2+2\left(\frac{\partial V}{\partial Y}\right)^2+\left(\frac{\partial V}{\partial X}+\frac{\partial U}{\partial Y}\right)^2\right] \tag{6-80}$$

Here the Eckert number appears as a multiplier to the viscous-energy-dissipation term. Thus, the viscous-energy dissipation can be neglected if the Eckert number is small.

6-7 BOUNDARY-LAYER EQUATIONS

The analysis of convection heat transfer is complicated by the fact that it requires the solution of equations of motion and energy. For example, the two-dimensional, steady-state formulation considered in Sec. 6-6 involves one continuity, two momentum, and one energy equation, which should be solved over the entire flow field. If the flow field is separated into a boundary-layer region where the velocity and temperature gradients are steep and the potential-flow region, significant simplifications can be achieved in the analysis of convection heat transfer. The reason is that the analysis of flow for the potential-flow region is relatively easy and simplifications occur in the momentum and energy equations when they are restricted to only the boundary-layer region.

For the two-dimensional, steady flow, the boundary-layer equations are obtainable from the continuity, momentum, and energy equations given previously, with the simplification of these equations by an order-of-magnitude study. The basic assumption made in the boundary-layer simplification is that the velocity and the thermal boundary-layer thicknesses are small compared with a characteristic dimension of the body. The reader should consult Ref. 16 for a detailed discussion of such an order-of-magnitude study. Here we present only the results of the simplification of the continuity equation (6-54*b*), the momentum equations (6-55) and (6-56), and the energy equation (6-66). If we assume no body forces are acting on the fluid, the resulting *boundary-layer equations* are given by

Continuity:
$$\frac{\partial u}{\partial x} + \frac{\partial v}{\partial y} = 0 \tag{6-81}$$

x Momentum:
$$\rho\left(u\frac{\partial u}{\partial x} + v\frac{\partial u}{\partial y}\right) = -\frac{dP}{dx} + \mu\frac{\partial^2 u}{\partial y^2} \tag{6-82}$$

Energy:
$$\rho c_p\left(u\frac{\partial T}{\partial x} + v\frac{\partial T}{\partial y}\right) = k\frac{\partial^2 T}{\partial y^2} + \mu\left(\frac{\partial u}{\partial y}\right)^2 \tag{6-83}$$

We note that the continuity equation remains unchanged. The y momentum equation is no longer needed, because since the boundary layer is thin, it reduces to $\partial P/\partial y \cong 0$, which merely implies that the pressure across the boundary layer is practically constant. The x momentum equation no longer contains the term $\mu\,\partial^2 v/\partial x^2$. In the energy equation, the axial conduction term $k\,\partial^2 T/\partial x^2$ is dropped, and the viscous-energy-dissipation term is drastically reduced. Clearly, this is a significant simplification of the governing equations.

The momentum equation (6-82) contains the pressure-gradient term dP/dx. If Eq. (6-82) is evaluated at the edge of the velocity boundary layer where $u(x, y)$ becomes equal to the external stream velocity $u_\infty(x)$, then Eq. (6-82) reduces to

$$-\frac{dP}{dx} = \rho u_\infty(x)\frac{du_\infty(x)}{dx} \tag{6-84}$$

This equation relates the pressure-gradient term to the external stream velocity $u_\infty(x)$, which is assumed to be available from the solution of the potential flow problem. Thus dP/dx term in Eq. (6-82) is a known quantity.

In the case of flow over a flat plate, the external flow velocity u_∞ is constant; then

$$\frac{dP}{dx} = 0 \tag{6-85}$$

and the pressure-gradient term in Eq. (6-82) vanishes.

Example 6-17 The drag coefficient for laminar flow along a flat plate is to be determined. Write the governing differential equations needed to solve this problem.

SOLUTION If the velocity distribution $u(x, y)$ in the flow is known, the shear stress at the wall and hence the drag coefficient can be established. The governing equations for the determination of the velocity distribution in laminar flow along a flat plate are obtained from Eqs. (6-81) and (6-82) as

$$\frac{\partial u}{\partial x} + \frac{\partial v}{\partial y} = 0$$

$$u\frac{\partial u}{\partial x} + v\frac{\partial u}{\partial y} = \nu\frac{\partial^2 u}{\partial y^2}$$

These equations should be solved subject to appropriate boundary conditions for the flow.

PROBLEMS

Flow over a body

6-1 Compare the distance from the leading edge of a flat plate at which the transition occurs from laminar to turbulent flow for atmospheric air at 27°C with (*a*) 2, (*b*) 10, and (*c*) 20 m/s. (Assume transition takes place at $\text{Re}_c = 5 \times 10^5$.)

Answer: (*a*) 4.21 m, (*b*) 0.842 m, (*c*) 0.421 m.

6-2 In a low-speed wind tunnel, air flows at 1 atm and 27°C with a velocity of 40 m/s. How long must the length of a flat-plate test section be to produce a transition Reynolds number of 5×10^5?

6-3 Determine the distances from the leading edge of a flat plate at which transition occurs from laminar to turbulent for the flow of air at 77°C with a velocity of $u_\infty = 20$ m/s at pressure of 0.5, 1, and 4 atm. (Assume $\text{Re}_c = 5 \times 10^5$.)

Answer: 0.519 m for the case of 1 atm.

6-4 Assuming the transition from laminar to turbulent flow takes place at a Reynolds number 5×10^5, determine the distance from the leading edge of a flat plate at which transition occurs for the flow of each of the following fluids with a velocity of $u_\infty = 2$ m/s at 40°C: air at atmospheric pressure; hydrogen at atmospheric pressure; water; ethylene glycol, engine oil, and mercury.

6-5 An approximate expression for the velocity profile $u(x, y)$ for laminar boundary-layer flow along a flat plate is given by

$$\frac{u(x, y)}{u_\infty} = 2\frac{y}{\delta(x)} - 2\left[\frac{y}{\delta(x)}\right]^3 + \left[\frac{y}{\delta(x)}\right]^4$$

where the boundary-layer thickness $\delta(x)$ is

$$\frac{\delta(x)}{x} = \frac{5.83}{\mathrm{Re}_x^{1/2}}$$

(*a*) Develop an expression for the local drag coefficient c_x.

(*b*) Develop an expression for the average drag coefficient c_m over a distance $x = L$ from the leading edge of the plate.

(*c*) Determine the drag force F acting on a plate 2 m by 2 m for the flow of air at atmospheric pressure and at $T_\infty = 350$ K with a velocity of $u_\infty = 4$ m/s.

6-6 An approximate expression for the velocity profile $u(x, y)$ for laminar boundary-layer flow along a flat plate is given by

$$\frac{u(x, y)}{u_\infty} = \sin\left[\frac{\pi}{2}\frac{y}{\delta(x)}\right]$$

where the boundary-layer thickness $\delta(x)$ is

$$\frac{\delta(x)}{x} = \frac{4.8}{\mathrm{Re}_x^{1/2}}$$

(*a*) Develop an expression for the local drag coefficient c_x.

(*b*) Develop an expression for the average drag coefficient c_m over a distance $x = L$ from the leading edge of the plate.

6-7 The exact expression for the local drag coefficient c_x for laminar flow over a flat plate is given by

$$c_x = \frac{0.664}{\mathrm{Re}_x^{1/2}}$$

Air at atmospheric pressure and at $T_\infty = 350$ K flows with a velocity of 30 m/s over a flat plate $L = 0.2$ m long. Determine the drag force acting per 1-m width of the plate.

Answer: 0.222 N.

6-8 The local heat transfer coefficient $h(x)$ for laminar boundary-layer flow over a flat plate is given by

$$\frac{xh(x)}{k} = 0.332\ \mathrm{Pr}^{1/3}\ \mathrm{Re}_x^{1/2}$$

Develop an expression for the average heat transfer coefficient h_m over a distance $x = L$ from the leading edge of the plate.

6-9 An approximate expression for temperature profile $\theta(x, y)$ in the thermal boundary layer is given by

$$\theta(x, y) = 2\frac{y}{\delta_t(x)} - \left[\frac{y}{\delta_t(x)}\right]^2$$

and the thermal boundary-layer thickness $\delta_t(x)$ is

$$\frac{\delta_t(x)}{x} = \frac{5.5}{\mathrm{Re}_x^{1/2}\ \mathrm{Pr}^{1/3}}$$

Develop an expression for the local heat transfer coefficient h_x.

6-10 The following expressions are given for the velocity boundary-layer thickness $\delta(x)$ and the thermal boundary-layer thickness $\delta_t(x)$:

$$\frac{\delta(x)}{x} = \frac{4.96}{\text{Re}_x^{1/2}} \quad \text{and} \quad \frac{\delta_t(x)}{x} = \frac{4.53}{\text{Re}_x^{1/2}\,\text{Pr}^{1/3}}$$

Determine the velocity and thermal boundary-layer thickness at $L = 0.8$ m from the leading edge of a flat plate for flow at $T_\infty = 40°\text{C}$ with a velocity of $u_\infty = 1.5$ m/s of (*a*) air at atmospheric pressure, (*b*) water, (*c*) ethylene glycol.

6-11 Atmospheric air at $T_\infty = 375$ K flows with a velocity of $u_\infty = 4$ m/s along a flat plate $L = 1$ m long, maintained at a uniform temperature $T_w = 325$ K. The average heat transfer coefficient is determined to be $h_m = 8\ \text{W/(m}^2\cdot°\text{C)}$. Using the Colburn-Reynolds analogy, estimate the drag force acting on the plate over the width $w = 2$ m.

Answer: 0.05 N

6-12 Atmospheric air at $T_\infty = 350$ K flows with a velocity of $u_\infty = 5$ m/s along a flat plate $L = 1$ m long. The drag force acting on the plate per 1-m width is $F = 4 \times 10^{-2}$ N. By using the Reynolds-Colburn analogy, estimate the corresponding average heat transfer coefficient h_m.

Answer: $10.25\ \text{W/(m}^2\cdot°\text{C)}$

Flow inside a duct

6-13 The velocity profile $u(y)$ for hydrodynamically developed laminar flow between two parallel plates a distance $2L$ apart is given by

$$\frac{u(y)}{u_m} = \frac{3}{2}\left[1 - \left(\frac{y}{L}\right)^2\right]$$

where u_m is the mean flow velocity and the coordinate axis is shown in Fig. P6-13.

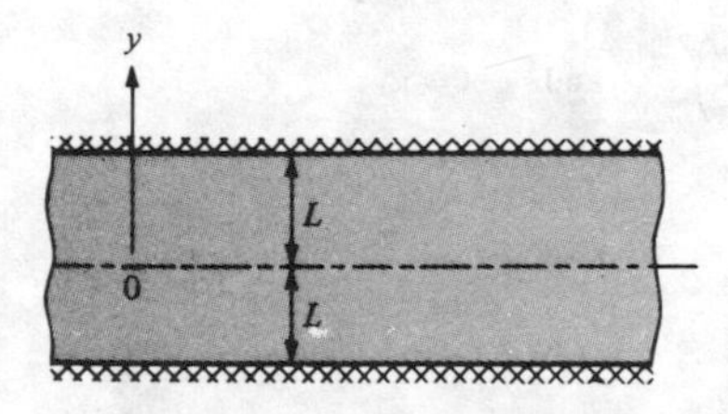

Figure P6-13

(*a*) Develop an expression for the friction factor f.
(*b*) Write the expression for calculating pressure drop ΔP over a length H of the channel.

6-14 The friction factor f for hydrodynamically developed laminar flow inside a circular tube is given by

$$f = \frac{64}{\text{Re}} \quad \text{where } \text{Re} = \frac{u_m D}{\nu}$$

Water at a mean temperature $T_m = 60°\text{C}$ and a mean velocity $u_m = 0.1$ m/s flows inside a 1-cm-ID circular tube. Calculate the pressure drop over $H = 10$ m of the tube. Also determine the pumping power required.

6-15 Repeat Problem 6-14 for the flow of engine oil [$\rho = 868\ \text{kg/m}^3$, $\mu = 65 \times 10^{-3}\ \text{kg/(m}\cdot\text{s)}$]. Compare the results for the water and engine oil; explain the reason for the big difference in pressure drops.

Answer: $4166.4\ \text{N/m}^3$; $0.0654\ \text{N}\cdot\text{m/s}$.

6-16 Engine oil at a mean temperature $T_m = 80°C$ ($\rho = 852$ kg/m^3, $\nu = 0.375 \times 10^{-4}$ m^2/s) flows with a velocity of 0.02 m/s inside a 1-cm-ID circular tube. The flow is hydrodynamically developed, and the friction factor is

$$f = \frac{64}{\text{Re}} \qquad \text{where Re} = \frac{u_m D}{\nu}$$

Calculate the pressure drop over $L = 25$ m of the tube.

6-17 Engine oil [$\nu = 0.8 \times 10^{-4}$ m^2/s, $k = 0.14$ W/($m \cdot$ °C)] is in laminar flow between two parallel plates a distance $2L = 3$ cm apart and subjected to a constant heat flux $q_w = 2500$ W/m^2. The mean heat transfer coefficient for the hydrodynamically and thermally developed region is given by

$$\text{Nu}_m \equiv \frac{h_m 4L}{k} = 8.235$$

Calculate the temperature difference between the wall surface and the mean fluid temperatures.
Answer: 130.2°C

Concepts on turbulence

6-18 What is the magnitude of the critical Reynolds number for transition from laminar to turbulent flow for (*a*) flow inside a circular tube, (*b*) flow over a flat plate?

6-19 Define viscous shear stress and turbulent shear stress.

6-20 Define diffusive heat flux and turbulent heat flux.

6-21 Discuss the *eddy viscosity of momentum* (or *turbulent diffusivity of momentum*) and the *eddy conductivity* (or *turbulent diffusivity of heat*) based on the mixing-length concept suggested by Prandtl.

6-22 Define *turbulent shear stress* and *turbulent heat flux* in terms of eddy viscosity, respectively.

6-23 Define the turbulent Prandtl number.

6-24 Discuss the basic concepts in separating the turbulent boundary layer into three layers.

6-25 Starting with the definition of the total shear stress in turbulent flow given by

$$\tau_0 = \mu \frac{du}{dy} - \overline{\rho u' v'} \qquad \overline{u'v'} = -l^2 \left(\frac{du}{dy}\right)^2 \qquad \text{and} \qquad l = \kappa y$$

derive the logarithmic velocity distribution

$$u^+ = \frac{1}{\kappa} \ln y^+ + C$$

for the fully turbulent flow region.

6-26 Show that the velocity distribution in the viscous sublayer is given by $u^+ = y^+$.

6-27 Write the velocity distribution in the viscous sublayer, buffer layer, and turbulent core in turbulent flow.

6-28 Discuss the effect of surface roughness on the velocity distribution in turbulent flow.

Equations of motion

6-29 By writing a mass balance on a differential volume element, derive the steady, two-dimensional continuity equation in (x and y) the rectangular coordinate system.

6-30 By writing a mass balance on a differential volume element, derive the steady, three-dimensional continuity equation in (x, y, and z) the rectangular coordinate system. Denote the velocity components in the x, y, and z directions by u, v, and w, respectively.

6-31 By writing a mass balance on a differential volume element, derive the steady, two-dimensional continuity equation in (r, z) the cylindrical coordinate system. Denote the velocity components in the r and z directions by v_r and v_z, respectively.

6-32 Consider the x momentum equation for the two-dimensional, incompressible, constant-property fluid in (x, y) the rectangular coordinate system. Describe briefly the physical significance of each term in the equation.

6-33 Consider the z direction momentum equation for the two-dimensional, constant-property fluid in (r, z) the cylindrical coordinate system. Discuss the physical significance of each term in this equation with regard to flow inside a circular tube with v_r and v_z referring to the velocity components in the radial and axial directions, respectively.

6-34 The continuity equation for flow between two parallel plates with the main flow velocity u being in the axial x direction may be simplified to

$$\frac{\partial u}{\partial x} = 0$$

Discuss the assumptions made in simplifying the continuity equation $\partial u/\partial x + \partial v/\partial y = 0$ to the above form. Also discuss the physical significance of the equation $\partial u/\partial x = 0$, regarding the dependence of u on x and y.

6-35 Consider the steady flow of an incompressible, constant-property flow between two parallel plates. Let u denote the main flow velocity in the axial x direction and v the velocity component in the y direction normal to the wall. The flow is called *fully developed* if $v = 0$ everywhere in the flow. The following assumptions are made: (1) The flow is fully developed ($v = 0$). (2) No body forces are acting on the fluid. By utilizing these assumptions, simplify the continuity and x and y momentum equations for flow between two parallel plates. Discuss the physical significance of each term in the simplified equations.

6-36 The continuity equation for flow inside a circular tube with the main flow velocity u being in the axial z direction may be simplified to

$$\frac{\partial u}{\partial z} = 0$$

Discuss the assumptions made in simplifying the continuity equation

$$\frac{1}{r}\frac{\partial}{\partial r}(rv) + \frac{\partial u}{\partial z} = 0$$

to the above form. Also discuss the physical significance of the equation $\partial u/\partial z = 0$, regarding the dependence of u on r and z.

6-37 Consider the steady flow of an incompressible, constant-property flow inside a circular tube. Let v_r and v_z be the velocity components in the radial r and the axial z directions, respectively. The flow is called *fully developed* if $v_r = 0$ everywhere in the flow.

Assuming that the flow is *fully developed* and there are no body forces acting on the fluid, simplify the continuity and r and z direction momentum equations for flow inside a circular tube. Discuss the physical significance of each term in the simplified equation.

6-38 For the fully developed flow in the axial x direction between two parallel plates, the momentum equation can be simplified to

$$\frac{d^2u}{dy^2} = \frac{1}{\mu}\frac{dp}{dx}$$

where u is the velocity component in the x direction. Discuss the assumptions made in simplifying the continuity and momentum equations given in this chapter to this form.

6-39 For the fully developed flow inside a circular tube, the momentum equation can be simplified to

$$\frac{d^2v_z}{dr^2} + \frac{1}{r}\frac{dv_z}{dr} = \frac{1}{\mu}\frac{dp}{dz}$$

where v_z is the axial velocity component. Discuss the assumptions made in simplifying the continuity and momentum equations given in this chapter in the cylindrical coordinate system to this form.

Equation of energy

6-40 The energy equation in the rectangular coordinate system can be used to find the temperature distribution for flow between parallel plates. Consider the steady flow between two parallel plates with u and v being the flow velocities in the axial x and the normal y directions, respectively. Discuss the physical significance of each term in the energy equation for flow between two parallel plates.

6-41 The energy equation in the cylindrical coordinate system can be used to determine the temperature distribution for flow inside a circular tube. Consider the steady flow inside a circular tube with v_z and v_r being the flow velocities in the axial z and the radial r directions, respectively. Discuss the physical significance of each term in the energy equation for flow inside a circular tube.

6-42 Consider steady flow of a fluid between two parallel plates with u and v being the velocity components in the axial x and the normal y directions, respectively. Simplify the energy equation given in rectangular coordinates in this chapter for flow between two parallel plates under the following assumptions.

(*a*) The flow is fully developed.

(*b*) The heat conduction in the x direction is negligible in comparison with heat conduction in the y direction and convection in the x direction.

6-43 Further simplify the energy equation obtained in Problem 6-42 by assuming that the viscous-energy dissipation is negligible. Discuss the physical significance of each term in the simplified equation. What kind of equation results if the flow velocity becomes zero?

6-44 Consider the steady flow of a fluid inside a circular tube with v_z and v_r being the velocity components in the axial z and radial r directions, respectively. Simplify the energy equation in the cylindrical coordinate system given in this chapter for flow inside a circular tube under the following assumptions.

(*a*) The flow is fully developed.

(*b*) The heat conduction in the z direction is negligible in comparison with heat conduction in the r direction and convection in the z direction.

6-45 Further simplify the energy equation obtained in Problem 6-44 by assuming that the viscous-energy dissipation is negligible. Discuss the physical significance of each term in the simplified equation. What kind of equation results in the flow velocity becomes zero? Discuss under which conditions the viscous-energy-dissipation term is negligible.

6-46 Consider steady flow of a fluid between two parallel plates with u and v being the velocity components in the axial x and the normal y directions, respectively. Simplify the energy equation in the rectangular coordinates given in this chapter for flow between two parallel plates under the following assumptions.

(*a*) The flow is fully developed.

(*b*) The temperature gradients in the axial x direction are negligible.

Discuss the physical significance of each term in the resulting energy equation.

6-47 The steady-state energy equation for flow between two parallel plates is taken as

$$\rho c_p u \frac{\partial T}{\partial x} = k \frac{\partial^2 T}{\partial y^2} + \mu \left(\frac{\partial u}{\partial y} \right)^2$$

where u is the flow velocity in the axial x direction. Discuss the assumptions made to simplify the energy equation to this form.

6-48 The steady-state energy equation for flow between two parallel plates is given in the form

$$\rho c_p u \frac{\partial T}{\partial x} = k \frac{\partial^2 T}{\partial y^2}$$

where u is the flow velocity in the axial x direction. Discuss the assumptions made to simplify the energy equation to this form.

6-49 The steady-state energy equation for flow between two parallel plates is given in the form

$$\rho c_p u \frac{\partial T}{\partial x} = k \left(\frac{\partial^2 T}{\partial x^2} + \frac{\partial^2 T}{\partial y^2} \right)$$

where u is the flow velocity in the axial x direction. Discuss the assumptions made to simplify the energy equation to this form.

6-50 The steady-state energy equation for flow inside a circular tube is given in the form

$$\rho c_p v_z \frac{\partial T}{\partial z} = k\left(\frac{\partial^2 T}{\partial r^2} + \frac{1}{r}\frac{\partial T}{\partial r}\right) + \mu\left(\frac{\partial v_z}{\partial r}\right)^2$$

where v_z is the flow velocity in the axial z direction. Discuss the assumptions made to simplify the energy equation to this form. Explain the physical significance of each term in this equation.

Dimensionless parameters

6-51 Consider the x direction momentum equation given by

$$\rho\left(u\frac{\partial u}{\partial x} + v\frac{\partial u}{\partial y}\right) = -\frac{\partial p}{\partial x} + \mu\left(\frac{\partial^2 u}{\partial x^2} + \frac{\partial^2 u}{\partial y^2}\right)$$

By introducing the following dimensionless variables

$$X = \frac{x}{L} \qquad Y = \frac{y}{L} \qquad P = \frac{p}{\rho U_\infty^2} \qquad U = \frac{u}{U_\infty} \qquad V = \frac{v}{U_\infty} \qquad \text{Re} = \frac{\rho U_\infty L}{\mu}$$

where L is a reference length and u_∞ is a reference velocity, express this momentum equation in the dimensionless form.

6-52 Consider the y direction momentum equation given in the form

$$\rho\left(u\frac{\partial v}{\partial x} + v\frac{\partial v}{\partial y}\right) = -\frac{\partial P}{\partial y} + \mu\left(\frac{\partial^2 v}{\partial x^2} + \frac{\partial^2 v}{\partial y^2}\right)$$

Express this equation in the dimensionless form by utilizing the dimensionless variables defined in Problem 6-51.

6-53 Consider the z direction momentum equation in the cylindrical coordinates given by

$$\rho\left(v\frac{\partial u}{\partial r} + u\frac{\partial u}{\partial z}\right) = -\frac{\partial p}{\partial z} + \mu\left[\frac{1}{r}\frac{\partial}{\partial r}\left(r\frac{\partial u}{\partial r}\right) + \frac{\partial^2 u}{\partial z^2}\right]$$

By introducing the following dimensionless variables

$$R = \frac{r}{L} \qquad Z = \frac{z}{L} \qquad P = \frac{p}{\rho U_\infty^2} \qquad V = \frac{v}{U_\infty} \qquad U = \frac{u}{U_\infty} \qquad \text{Re} = \frac{\rho U_\infty L}{\mu}$$

where L is a reference length and u_∞ is a reference velocity, express this momentum equation in the dimensionless form.

6-54 The energy equation for flow between two parallel plates is

$$\rho c_p u(y) \frac{\partial T(x, y)}{\partial x} = k\frac{\partial^2 T(x, y)}{\partial y^2}$$

and the following dimensionless variables are defined:

$$U = \frac{u(y)}{u_0} \qquad Y = \frac{y}{L} \qquad X = \frac{\alpha x}{u_0 L^2} \qquad \theta = \frac{T(x, y) - T_0}{\Delta T}$$

where L, u_0, T_0, and ΔT are, respectively, a reference length (i.e., the spacing between the plates), velocity, temperature, and temperature difference, and α is the thermal diffusivity.

By utilizing these dimensionless variables, express the above energy equation in dimensionless form.

6-55 The energy equation for flow inside a circular tube is

$$\rho c_p u(r) \frac{\partial T(r, z)}{\partial z} = k\frac{1}{r}\frac{\partial}{\partial r}\left(r\frac{\partial T}{\partial r}\right)$$

and the following dimensionless variables are defined:

$$U = \frac{u(r)}{u_0} \qquad R = \frac{r}{L} \qquad Z = \frac{\alpha z}{u_0 L^2} \qquad \theta = \frac{T(r, z) - T_0}{\Delta T}$$

where L, u_0, T_0, and ΔT are, respectively, a reference length, velocity, temperature, and temperature difference, and α is the thermal diffusivity.

By utilizing these dimensionless variables, express the above energy equation in the dimensionless form.

6-56 The energy equation for flow between two parallel plates is

$$\rho c_p u(y) \frac{\partial T(x, y)}{\partial x} = k \frac{\partial^2 T(x, y)}{\partial y^2} + \mu \left[\frac{du(y)}{dy}\right]^2$$

By utilizing the dimensionless variables defined in Problem 6-54, express this energy equation in dimensionless form and show that the following two dimensionless groups will appear in the equation:

$$E = \frac{u_0^2}{c_p \Delta T} = \text{Eckert number}$$

$$\text{Pr} = \frac{c_p \mu}{k} = \frac{\nu}{\alpha} = \text{Prandtl number}$$

Discuss the physical significance of the Eckert number in relation to the relative importance of the viscous-energy-dissipation term in the energy equation.

6-57 The energy equation for flow inside a circular tube is given in the form

$$\rho c_p u(r) \frac{\partial T(r, z)}{\partial z} = k \frac{1}{r} \frac{\partial}{\partial r}\left(r \frac{\partial T}{\partial r}\right) + \mu \left[\frac{du(r)}{dr}\right]^2$$

By utilizing the dimensionless variables defined in Problem 6-55, express this energy equation in dimensionless form and show that the following two dimensionless groups will appear in the equation:

$$E = \frac{u_0^2}{c_p \Delta T} = \text{Eckert number}$$

$$\text{Pr} = \frac{c_p \mu}{k} = \frac{\nu}{\alpha} = \text{Prandtl number}$$

Discuss the physical significance of the Eckert number.

Boundary-layer equations

6-58 Discuss the physical significance of the velocity and thermal boundary layers.

6-59 Discuss the physical significance of the dimensionless groups Prandtl, Reynolds, and Eckert numbers.

6-60 Discuss the approximations and the simplifying assumptions made in obtaining the boundary-layer equations from the general momentum and energy equations.

6-61 Discuss the physical significance of each term in the boundary-layer x momentum equation given in this chapter.

6-62 Discuss the physical significance of each term in the boundary-layer energy equation given in this chapter.

6-63 Explain why the pressure-gradient term $dp(x)/dx$ will vanish in the boundary-layer x momentum equation for flow along a flat plate.

6-64 Consider the boundary-layer energy equation without the viscous-energy dissipation and the

x momentum equation for flow along a flat plate. Explain under which condition these two equations have the same form, so that the heat and momentum transfer analogy applies.

6-65 Consider heat transfer in boundary-layer flow over a flat plate in which heat transfer starts at the leading edge of the plate. Sketch the velocity and thermal boundary layers, and illustrate their relative thicknesses for $\Pr \gg 1$, $\Pr = 1$, and $\Pr \ll 1$.

REFERENCES

1. Reynolds, O.: "On the Experimental Investigation of the Circumstances Which Determine Whether the Motion of Water Shall Be Direct or Sinuous, and the Law of Resistance in Parallel Channels," *Philos. Trans. R. Soc. London, Ser. A*, **174**:935 (1883).
2. Burgess, J. M.: "The Motion of Fluid in the Boundary Layer along a Plane Smooth Surface," *Proc. First Int. Congr. Appl. Mech., Delf*, 1924.
3. Prandtl, L.: *Z. Angew. Math. Mech.*, **5**:136 (1925); *NACA Tech. Memo.* 1231, 1949; *Proc. Second Int. Congr. Appl. Mech., Zurich*, 1926, pp. 62–75.
4. Von Kármán, Th.: *Nach. Ges. Wiss. Göttingen, Math.-Phys. Klasse*, 1930, p. 58. *NACA Tech. Memo.* 611, 1931.
5. Schlichting, H.: *Boundary Layer Theory*, 6th ed., McGraw-Hill, New York, 1968.
6. Hinze, J. O.: *Turbulence*, McGraw-Hill, New York, 1959.
7. Cebeci, T., and A. M. O. Smith: *Analysis of Turbulent Boundary Layers*, Academic, New York, 1974.
8. Launder, B. E., and D. B. Spalding: *Mathematical Models of Turbulence*, Academic, New York, 1972.
9. Launder, B. E., and D. B. Spalding: *Heat and Fluid Flow*, **2**:43–54 (1972).

10a. Nikuradse, J.: *Forsch. Arb. Ing. Wes.*, no. 346, 1932.

10b. Nikuradse, J.: *Forsch. Arb. Ing. Wes.*, no. 361, 1933; also, "Laws of Flow in Rough Pipes" (translation), *NACA Tech. Memo.* 1292, 1950.

11. Wieghardt, K.: Zum Reibungswiderstand rauher Platten, *Kaiser-Wilhelm-Institut fur Stromunsforschung*, Göttingen, UM-6612, 1944.
12. Klebanoff, P. S.: "Characteristics of Turbulence in a Boundary Layer with Zero Pressure Gradient," *NACA Tech. Note* 3178, 1954.
13. Laufer, J.: "The Structure of Turbulence in Fully Developed Pipe Flow," *NACA Tech. Note* 1174, 1954.
14. Coles, D.: "The Law of Wake in the Turbulent Boundary Layers," *J. Fluid Mech.*, **1**:191–226 (1956).
15. Spalding, D. B.: "Heat Transfer to a Turbulent Stream from a Surface with a Step-wise Discontinuity in Wall Temperature," *Conf. Int. Dev. Heat Transfer, ASME*, Boulder, Colo., pt. II, 1961, pp. 439–446.
16. Schlichting, H.: *Boundary Layer Theory*, 6th ed., McGraw-Hill, New York, 1968.
17. Bird, R. B., W. E. Stewart, and E. N. Lightfoot: *Transport Phenomena*, Wiley, New York, 1960.
18. Pai, S. I.: *Viscous Flow Theory*, Van Nostrand, New York, 1956.
19. Eckert, E. R. G., and R. M. Drake: *Analysis of Heat and Mass Transfer*, McGraw-Hill, New York, 1972.
20. Kays, W. M., and M. E. Crowford.: *Convective Heat and Mass Transfer*, McGraw-Hill, New York, 1980.
21. Knudsen, J. G., and D. L. Katz: *Fluid Dynamics and Heat Transfer*, McGraw-Hill, New York, 1958.
22. Jischa, Michael: "Konvektiver Impuls-, Wärme- und Stoffaustausch," Friedr. Vieweg & Sohn, Braunschweig/Wiesbaden, 1982.
23. Welty, J. R., R. E. Wilson, and C. E. Wicks: *Fundamentals of Momentum, Heat and Mass Transfer*, Wiley, New York, 1976.

24. Patankar, S. V., and D. B. Spalding: *Heat and Mass Transfer in Boundary Layers*, Intertext, London, 1970.
25. Rohsenow, W. M., and H. Choi: *Heat, Mass, and Momentum Transfer*, Prentice-Hall, Englewood Cliffs, N.J., 1961.
26. Hinze, J. O.: *Turbulence*, 2d ed., McGraw-Hill, New York, 1975.
27. Cebeci, T., and P. Bradshaw; *Momentum Transfer in Boundary Layers*. Hemisphere Publishing, Washington, 1977.
28. White, F. M.: *Viscous Fluid Flow*, McGraw-Hill, New York, 1974.

CHAPTER

SEVEN

FORCED CONVECTION FOR FLOW INSIDE DUCTS

This chapter is devoted to heat transfer and pressure drop in forced convection inside ducts under both laminar and turbulent flow. Turbulent pipe flow is used widely in various industrial applications, and the available correlations of heat transfer and friction factor are mostly of empirical or semiempirical nature. We present such correlations with particular emphasis on their ranges of validity. Laminar pipe flow is encountered generally in compact heat exchangers, cryogenic coolant systems, the heating or cooling of heavy fluids such as oils, and in many other applications. Numerous analytic expressions are available for the prediction of the friction factor and heat transfer coefficient in laminar tube flow. We present some of these results and discuss their range of validity.

Heat transfer correlations developed for ordinary fluids break down when they are applied to liquid metals, because the Prandtl number for liquid metals is very low. Therefore, heat transfer to liquid metals has been the subject of numerous investigations, but reliable correlations are still rather limited. We discuss some of the available correlations of heat transfer for the flow of liquid metals in tubes.

The experimental determination of pressure drop, and hence the friction factor for flow inside a conduit, is much easier than the determination of the heat transfer coefficient. So when the heat transfer correlation is not available for a certain flow condition inside a conduit, it may be possible to estimate the heat transfer coefficient for turbulent flow from the friction factor data by utilizing the analogies between heat and momentum transfer. We briefly discuss the analogies between heat and momentum transfer in turbulent flow.

In recent years, energy- and material-saving considerations have increased the emphasis on efforts to produce more efficient heat exchange equipment through the augmentation of heat transfer in the flow. The work in this area is progressing rapidly. To familiarize the reader with such developments, we briefly discuss the principles of heat transfer augmentation for flow inside conduits.

7-1 HYDRODYNAMICALLY AND THERMALLY DEVELOPED LAMINAR FLOW

We now consider the determination of the friction factor and the heat transfer coefficient for laminar flow inside conduits in the region where the velocity and temperature profiles are fully developed in the sense discussed in Chap. 6. The velocity and temperature distributions in the flow are needed for the solution of such problems. For geometries such as a circular tube and a parallel-plate channel, the analysis of this problem is very simple. Therefore, our understanding of the physical significance of the parameters governing the velocity and temperature distribution in the flow will be greatly improved if we analyze the problem for simple geometries such as a circular tube and a parallel-plate channel. The friction factor and heat transfer results for flow inside ducts having more complicated geometries are also presented.

Couette Flow

Couette flow provides the simplest model for the analysis of heat transfer for flow between parallel plates. Figure 7-1 illustrates the geometry and the coordinates. The space between the two infinite parallel plates separated by a distance L is filled with a liquid having viscosity μ, density ρ, and thermal conductivity k. The upper plate at $y = L$ moves with a constant velocity u_1 and sets the fluid particles moving in the direction parallel to the plates while the lower plate remains stationary. The lower and upper plates are kept at uniform temperatures T_0 and T_1, respectively. The heat transfer problem characterized with this simple model is important for a journal and its bearing in which one surface is stationary while the other is rotating and the clearance between them is filled with a lubricant oil of high viscosity. When the clearance is small in comparison with the radius of the bearing, the geometry can be considered as two parallel plates. The oil being viscous, the temperature rise in the fluid due to friction (i.e., viscous energy dissipation) may become considerable even at moderate flow velocities. Therefore, the temperature rise in the fluid and the amount of heat transfer through the walls are of interest in engineering applications. In solving this heat transfer problem,

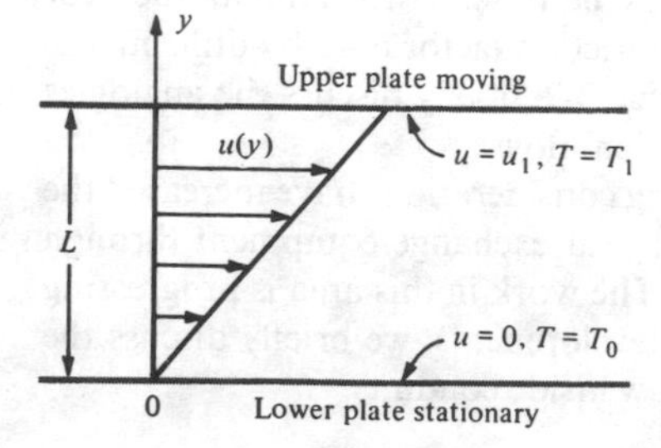

Figure 7-1 Heat transfer in Couette flow.

we determine first the velocity distribution in the flow and then the temperature distribution, since the velocity profile is needed in the energy equation.

The flow is fully developed since all the fluid particles move in the direction parallel to the plates. The simplification of the momentum equations for such a case is discussed in Example 6-11, and the resulting momentum equation is Eq. (6-59). In Couette flow, this equation is further simplified because the fluid motion is set by simple shear flow due to viscosity, and no pressure gradient is involved in the direction of motion. With this consideration, the pressure-gradient term dP/dx is zero, and Eq. (6-59) reduces to

$$\frac{d^2u}{dy^2} = 0 \qquad \text{in } 0 < y < L \tag{7-1}$$

The boundary conditions for this equation are taken as velocity u equal to zero at the surface of the lower plate at $y = 0$ and equal to u_1 at the surface of the upper plate at $y = L$; that is,

$$u = \begin{cases} 0 & \text{at } y = 0 \qquad (7\text{-}2a) \\ u_1 & \text{at } y = L \qquad (7\text{-}2b) \end{cases}$$

The solution of Eq. (7-1) subject to the boundary conditions (7-2) gives the velocity distribution in the Couette flow as

$$u(y) = \frac{y}{L}\, u_1 \tag{7-3}$$

The temperature distribution in the flow is governed by the energy equation. In Example 6-14, the energy equation for laminar flow between parallel plates in the hydrodynamically developed region was obtained by the simplification of the general energy equation, and the result was Eq. (6-70). For this present problem, the temperature varies in the y direction only, that is $T \equiv Y(y)$. Then the derivatives with respect to x vanish, and Eq. (6-70) reduces to

$$\frac{d^2T(y)}{dy^2} = -\frac{\mu}{k}\left(\frac{du}{dy}\right)^2 \tag{7-4}$$

The velocity profile $u(y)$ is needed to determine the temperature distribution from this energy equation. By introducing the velocity profile given by Eq. (7-3) into Eq. (7-4), the energy equation takes the form

$$\frac{d^2T(y)}{dy^2} = -\frac{\mu u_1^2}{kL^2} \qquad \text{in } 0 < y < L \tag{7-5}$$

subject to the boundary conditions

$$T(y) = \begin{cases} T_0 & \text{at } y = 0 \qquad (7\text{-}6a) \\ T_1 & \text{at } y = L \qquad (7\text{-}6b) \end{cases}$$

This system is just like the steady-state heat conduction problem with internal energy generation for a slab; the viscous-energy-dissipation term represents the energy generation. The solution is straightforward and is given as

$$T(y) - T_0 = \frac{y}{L}\left[(T_1 - T_0) + \frac{\mu u_1^2}{2k}\left(1 - \frac{y}{L}\right)\right] \tag{7-7}$$

We now examine the physical significance of this solution and heat flow at the wall for $T_0 \neq T_1$ and $T_0 = T_1$.

First, when the upper and lower plate temperatures are not equal, that is, $T_0 \neq T_1$, both sides of Eq. (7-7) are divided by $(T_1 - T_0)$ to yield

$$\frac{T(y) - T_0}{T_1 - T_0} = \frac{y}{L}\left[1 + \frac{1}{2}\frac{\mu u_1^2}{k(T_1 - T_0)}\left(1 - \frac{y}{L}\right)\right] \tag{7-8}$$

which is written more compactly in the form

$$\frac{T(y) - T_0}{T_1 - T_0} = \frac{y}{L}\left[1 + \frac{1}{2}\,\mathrm{Pr}\,E\left(1 - \frac{y}{L}\right)\right] \tag{7-9}$$

where

$$\mathrm{Pr} = \frac{c_p \mu}{k} = \text{Prandtl number} \tag{7-10a}$$

$$E = \frac{u_1^2}{c_p(T_1 - T_0)} = \text{Eckert number} \tag{7-10b}$$

Figure 7-2 shows a plot of $[T(y) - T_0]/(T_1 - T_0)$ as a function of y/L for several different values of the parameter $\mathrm{Pr} \cdot E$. The case $\mathrm{Pr} \cdot E = 0$ corresponds to no flow condition; hence there is no viscous dissipation in the medium, and the temperature distribution is a straight line, which characterizes pure conduction across the fluid layer. The physical significance of other curves is envisioned better

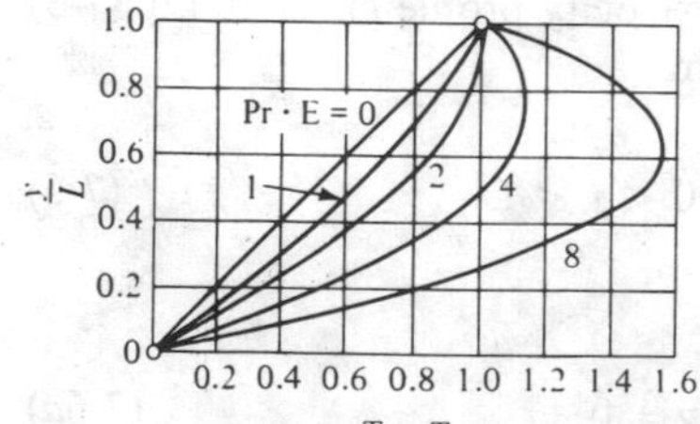

Figure 7-2 Temperature distribution in Couette flow ($T_1 > T_0$).

if heat transfer at the wall, say at the upper wall, is considered. The heat flux at the upper wall is determined from its definition

$$q\bigg|_{y=L} = -k\frac{dT(y)}{dy}\bigg|_{y=L} \tag{7-11}$$

The derivative is obtained from Eq. (7-9):

$$\frac{dT(y)}{dy}\bigg|_{y=L} = \frac{T_1 - T_0}{L}(1 - \tfrac{1}{2}\,\mathrm{Pr}\cdot E) \tag{7-12}$$

Then the heat flux at the upper wall becomes

$$q\bigg|_{y=L} = -k\frac{T_1 - T_0}{L}(1 - \tfrac{1}{2}\,\mathrm{Pr}\cdot E) \tag{7-13}$$

This result implies that the direction of heat flow at the upper wall—that is, whether the heat flow is into the fluid or the wall for $T_1 > T_0$—depends on the magnitude of the parameter $\mathrm{Pr}\cdot E$.

For $\mathrm{Pr}\cdot E > 2$, the right-hand side of Eq. (7-13) is positive; hence the heat flows in the positive y direction, or from the liquid into the wall, even though the upper wall is at a higher temperature than the lower wall.

For $\mathrm{Pr}\cdot E < 2$, the right-hand side of Eq. (7-13) is negative; hence the heat flows is in the negative y direction, or from the upper wall into the fluid.

For $\mathrm{Pr}\cdot E = 2$, the right-hand side of Eq. (7-13) vanishes; hence there is no heat flow at the upper wall.

Second, when $T_0 = T_1$, both plates are at the same temperature, and (7-7) reduces to

$$T(y) - T_0 = \frac{\mu u_1^2}{2k}\frac{y}{L}\left(1 - \frac{y}{L}\right) \tag{7-14}$$

The maximum temperature in the fluid occurs at the midpoint between the plates; by setting $y = L/2$, Eq. (7-14) becomes

$$T_{\max} - T_0 = \frac{\mu u_1^2}{8k} \tag{7-15}$$

By combining Eqs. (7-14) and (7-15) the temperature distribution in the fluid is expressed as

$$\frac{T(y) - T_0}{T_{\max} - T_0} = 4\frac{y}{L}\left(1 - \frac{y}{L}\right) \tag{7-16}$$

Example 7-1 A heavy lubricating oil [$\mu = 0.25$ kg/(m · s), $k = 0.125$ W/(m · °C)] at room temperature flows in the clearance between a journal and its bearing. Assuming both the bearing and the journal are at the same temperature, determine the maximum temperature rise in the fluid for a velocity of $u_1 =$ 6 m/s.

SOLUTION The maximum temperature rise ΔT_{max}, for $T_0 = T_1$, is obtained from Eq. (7-15) as

$$\Delta T_{max} = \frac{\mu u_1^2}{8k}$$

Introducing the numerical values, we obtain

$$\Delta T_{max} = \frac{1}{8}\left(0.25\,\frac{\text{kg}}{\text{m}\cdot\text{s}}\right)\left(6^2\,\frac{\text{m}^2}{\text{s}^2}\right)\left(\frac{1}{0.125}\,\frac{\text{m}\cdot{}^\circ\text{C}}{\text{W}}\right)$$

$$= 9^\circ\text{C}$$

since $1\ \text{kg}\cdot\text{m}^2/\text{s}^2 = 1\ \text{N}\cdot\text{m} = 1\ \text{W}\cdot\text{s}$.

Flow Inside a Circular Tube

The problems of steady-state heat transfer and pressure drop in laminar forced convection inside a circular tube in regions away from the inlet where the velocity and temperature profiles are fully developed are of interest in numerous engineering applications. The friction factor and the heat transfer coefficient for the flow are determined, respectively, from the knowledge of the velocity and temperature distributions in the fluid.

Friction factor Consider an incompressible, constant-property fluid in laminar forced convection inside a circular tube of radius R in the region where the flow is hydrodynamically developed. The friction factor for flow inside a circular tube is related to the velocity gradient at the wall by Eq. (6-18*e*) as

$$f = -\frac{8\mu}{\rho u_m^2}\frac{du}{dr}\bigg|_{r=R} \tag{7-17}$$

The velocity distribution $u(r)$ can be determined from the solutions of the equations of motion. In Example 6-12, it is shown that for hydrodynamically developed laminar flow inside a circular tube, the equations of motion reduce to a simple equation given by Eq. (6-60), which is written in the form

$$\frac{1}{r}\frac{d}{dr}\left(r\frac{du}{dr}\right) = \frac{1}{\mu}\frac{dP}{dz} \qquad \text{in } 0 < r < R \tag{7-18}$$

subject to the boundary conditions

$$\frac{du}{dr} = 0 \qquad \text{at } r = 0 \tag{7-19a}$$

$$u = 0 \qquad \text{at } r = R \tag{7-19b}$$

The first boundary condition is the symmetry of the velocity profile about the tube axis, and the second is that the velocity vanishes at the wall.

For steady fully developed laminar flow inside a circular tube, the pressure gradient dP/dz is constant. Then the solution of Eq. (7-18) for constant dP/dz, subject to the boundary conditions (7-19), gives the fully developed velocity profile $u(r)$ as

$$u(r) = -\left(\frac{1}{4\mu}\frac{dP}{dz}\right)R^2\left[1 - \left(\frac{r}{R}\right)^2\right] \tag{7-20}$$

Here the velocity $u(r)$ is always a positive quantity because for flow in the positive z direction, but the pressure gradient dP/dz is a negative quantity.

The *mean flow velocity* u_m over the tube cross section is determined from the definition as

$$u_m = \frac{1}{\pi R^2}\int_0^R 2\pi r u(r)\,dr = -\frac{R^2}{8\mu}\frac{dP}{dz} \tag{7-21}$$

since $u(r)$ is given by Eq. (7-20).

The physical significance of the mean velocity u_m implies that the flow rate through the tube is determined from

$$\text{Flow rate} = (\text{area cross section})\ u_m = \pi R^2 u_m$$

Now from Eqs. (7-20) and (7-21) we obtain

$$\boxed{\frac{u(r)}{u_m} = 2\left[1 - \left(\frac{r}{R}\right)^2\right]} \tag{7-22}$$

This relation shows that the velocity profile $u(r)/u_m$ in the hydrodynamically developed region is parabolic. The velocity u_0 at the tube axis is obtained from Eq. (7-20) by setting $r = 0$;

$$u_0 = -\frac{R^2}{4\mu}\frac{dP}{dz} \tag{7-23}$$

A comparison of the results given by Eqs. (7-21) and (7-23) shows that the velocity at the tube axis is equal to twice the mean flow velocity;

$$\boxed{u_0 = 2u_m} \tag{7-24}$$

The friction factor f for laminar flow inside a circular tube in the hydrodynamically developed region is determined by obtaining the velocity gradient from Eq. (7-22) as

$$\left.\frac{du(r)}{dr}\right|_{r=R} = -\frac{4u_m}{R} = -\frac{8u_m}{D} \tag{7-25}$$

and introducing this result into Eq. (7-17). We find

$$\boxed{f = \frac{64\mu}{\rho u_m D} = \frac{64}{\text{Re}}} \tag{7-26a}$$

where D is the tube inside radius and

$$\mathrm{Re} = \frac{\rho u_m D}{\mu} = \frac{u_m D}{\nu} \tag{7-26b}$$

is the Reynolds number.

In the literature, the friction factor also has been defined on the basis of *hydraulic radius*. If f_r denotes the friction factor based on the hydraulic radius, it is related to the friction factor f defined by Eq. (7-26*a*) by $f = 4f_r$. That is, Eq. (7-26*a*), on the basis of f_r, would be $f_r = 16/\mathrm{Re}$, where $\mathrm{Re} = \rho u_m D/\mu$. This result is sometimes referred to as the *Hagen-Poiseuille relation* for friction factor in tubes, because Hagen's [1] experimental data were later verified theoretically by Poiseuille [2].

Heat transfer coefficient The heat transfer coefficient for flow inside a circular tube in the thermally developed region is related to the dimensionless temperature gradient at the wall by Eq. (6-23*b*) as

$$h = -k \left. \frac{d\theta(r)}{dr} \right|_{r=R} \tag{7-27}$$

where $\theta(r)$ is defined by Eq. (6-20*b*) as

$$\theta(r) = \frac{T(r, z) - T_w(z)}{T_m(z) - T_w(z)} \tag{7-28}$$

To determine h, the temperature distribution in the flow is needed, and it can be established from the solution of the energy equation.

In the hydrodynamically developed region, the energy equation for laminar flow of an incompressible fluid inside a circular tube with negligible viscous-energy dissipation is discussed in Example 6-15 and given by Eq. (6-71*a*) as

$$\frac{1}{\alpha} u(r) \frac{\partial T}{\partial z} = \frac{1}{r} \frac{\partial}{\partial r}\left(r \frac{\partial T}{\partial r}\right) + \frac{\partial^2 T}{\partial z^2} \tag{7-29}$$

In general, this is a partial differential equation for the determination of temperature distribution in the flow, and its solution is rather involved. However, for forced convection inside a circular tube in the *thermally developed region*, with constant wall temperature or constant wall heat flux, it can be shown [3, 4] that the axial temperature-gradient term in Eq. (7-29) becomes a constant; that is,

$$\frac{\partial T}{\partial z} = \text{constant}$$

Then the partial differential equation (7-29) reduces to an ordinary differential equation for the fully developed temperature profile $T(r)$, because the term $\partial^2 T/\partial z^2$ vanishes for constant $\partial T/\partial z$. We now examine such a heat transfer problem for the constant wall heat flux and constant wall temperature boundary conditions for forced convection inside a circular tube.

Constant wall heat flux In Ref. 3 it has been shown that under constant wall heat flux conditions, the temperature gradient in the direction of flow anywhere in the flow is constant and equal to the axial gradient of the mean temperature of the fluid. That is,

$$\frac{\partial T(r, z)}{\partial z} = \frac{dT_m(z)}{dz} = \text{constant} \tag{7-30}$$

This result implies that under constant wall heat flux, the mean fluid temperature $T_m(z)$, in the thermally developed region, increases linearly with the distance z along the tube.

When Eq. (7-30) is introduced into Eq. (7-29), the term $\partial^2 T/\partial z^2$ vanishes for constant $\partial T/\partial z$, and the following ordinary differential equation is obtained for $T(r)$:

$$\frac{1}{r}\frac{d}{dr}\left(r\frac{dT}{dr}\right) = \frac{1}{\alpha}u(r)\frac{dT_m(z)}{dz} \tag{7-31}$$

This equation is written in terms of the dimensionless temperature $\theta(r)$, defined by Eq. (7-28), as

$$\frac{1}{r}\frac{d}{dr}\left(r\frac{d\theta}{dr}\right) = \frac{1}{\alpha}u(r)\frac{dT_m(z)}{dz}[T_m(z) - T_w(z)]^{-1} \tag{7-32a}$$

where the fully developed velocity profile $u(r)$ is given by Eq. (7-22) as

$$u(r) = 2u_m\left[1 - \left(\frac{r}{R}\right)^2\right] \tag{7-32b}$$

Equations (7-32*a*) and (7-32*b*) are combined and written more compactly as

$$\frac{d}{dr}\left(r\frac{d\theta}{dr}\right) = Ar\left[1 - \left(\frac{r}{R}\right)^2\right] \qquad \text{in } 0 < r < R \tag{7-33a}$$

where the constant A is defined as

$$A = \frac{2u_m}{\alpha[T_m(z) - T_w(z)]}\frac{dT_w(z)}{dz} = \text{constant} \tag{7-33b}$$

The boundary conditions for Eq. (7-33) are taken as

$$\frac{d\theta}{dr} = 0 \qquad \text{at } r = 0 \tag{7-34a}$$

$$\theta = 0 \qquad \text{at } r = R \tag{7-34b}$$

The first boundary condition states that θ has symmetry about the tube axis, and the second results from the definition of θ given by Eq. (7-28) that θ must be zero at the wall.

Equation (7-33a) is similar to the steady-state heat conduction equation in cylindrical coordinates, and it can readily be integrated subject to the boundary conditions in Eqs. (7-34) to give

$$\theta(r) = -AR^2\left[\frac{3}{16} + \frac{1}{16}\left(\frac{r}{R}\right)^4 - \frac{1}{4}\left(\frac{r}{R}\right)^2\right] \tag{7-35}$$

The unknown constant A appearing in this equation can be determined by utilizing the definition of the bulk mean fluid temperature.

According to the definition of the bulk mean fluid temperature given by Eq. (6-22b), we write

$$\theta_m = \frac{\int_0^R u(r)\theta(r)2\pi r\, dr}{u_m \pi R^2} \tag{7-36}$$

where the fully developed velocity profile $u(r)$ is given by Eq. (7-32b), that is,

$$u(r) = 2u_m\left[1 - \left(\frac{r}{R}\right)^2\right] \tag{7-37}$$

Equations (7-35) and (7-37) are introduced into Eq. (7-36), and the integrations are performed. We obtain

$$\theta_m = -\tfrac{11}{96}AR^2 \tag{7-38a}$$

Also, from the definition of $\theta(r)$ given by Eq. (7-28), we write

$$\theta_m = \frac{T_m(z) - T_w(z)}{T_m(z) - T_w(z)} = 1 \tag{7-38b}$$

Equating (7-38a) and (7-38b), we find

$$AR^2 = -\tfrac{96}{11} \tag{7-39}$$

Introducing this result for AR^2 into Eq. (7-35), we obtain

$$\theta(r) = \frac{96}{11}\left[\frac{3}{16} + \frac{1}{16}\left(\frac{r}{R}\right)^4 - \frac{1}{4}\left(\frac{r}{R}\right)^2\right] \tag{7-40}$$

Equation (7-40) is the dimensionless temperature profile for forced convection in a circular tube in the hydrodynamically and thermally developed region under the *constant wall heat flux* boundary condition. We recall that this temperature profile was used in Example 6-9 to determine the heat transfer coefficient.

Given the temperature profile in the fluid, the heat transfer coefficient h is immediately obtained from its definition given by Eq. (7-27):

$$h = \frac{48}{11}\frac{k}{D} \tag{7-41a}$$

or

$$\text{Nu} \equiv \frac{hD}{k} = \frac{48}{11} = 4.364 \tag{7-41b}$$

where D is the tube's inside diameter and Nu is the Nusselt number.

The result given by Eqs. (7-41) represents the *heat transfer coefficient for laminar forced convection inside a circular tube in the hydrodynamically and thermally developed region under constant wall heat flux boundary condition.*

Constant wall temperature The heat transfer problem described above for the hydrodynamically and thermally developed region also can be solved under constant wall temperature boundary condition; but the analysis is more involved and is not presented here. The result is

$$\boxed{\text{Nu} \equiv \frac{hD}{k} = 3.66} \tag{7-42}$$

which represents *the Nusselt number (or the heat transfer coefficient) for laminar forced convection inside a circular tube in the hydrodynamically and thermally developed region under constant wall temperature boundary condition.*

Evaluation of physical properties In the results given by Eqs. (7-41) and (7-42), the thermal conductivity of the fluid k depends on temperature. When the fluid temperature varies along the tube, k may be evaluated at the fluid bulk mean temperature T_b, defined as

$$\boxed{T_b = \tfrac{1}{2}(T_i + T_o)} \tag{7-43}$$

where T_i = bulk fluid temperature at the inlet and T_o = bulk fluid temperature at the outlet.

The logarithmic and arithmetic mean temperature differences The logarithmic mean (LMTD) of the two quantities ΔT_1 and ΔT_2 is defined as

$$\Delta T_{\ln} = \frac{\Delta T_1 - \Delta T_2}{\ln(\Delta T_1/\Delta T_2)} \tag{7-44a}$$

whereas the arithmetic mean (AM) of ΔT_1 and ΔT_2 is defined as

$$\Delta T_{\text{AM}} = \tfrac{1}{2}(\Delta T_1 + \Delta T_2) \tag{7-44b}$$

In Chap. 11 we show that for the heating or cooling of fluid flowing inside a duct, the appropriate mean temperature difference between the fluid and the tube wall is the LMTD, where

$$\Delta T_1 = T_w - T_{\text{in}} = \text{inlet temperature difference}$$

$$\Delta T_2 = T_w - T_{\text{out}} = \text{outlet temperature difference}$$

Note that $\Delta T_{\ln}$ is always less than ΔT_{AM}. If the ratio of ΔT_1 and ΔT_2 is not greater than 50 percent, then $\Delta T_{\ln}$ can be approximated by the arithmetic mean difference within about 1.4 percent.

Example 7-2 Engine oil is pumped with a mean velocity of $u_m = 0.6$ m/s through a bundle of $n = 80$ tubes each of inside diameter $D = 2.5$ cm and length $L = 10$ m. The physical properties of the oil are $\nu = 0.75 \times 10^{-4}$ m^2/s and $\rho = 868$ kg/m^3. Calculate the pressure drop across each tube and the total power required for pumping the oil through 80 tubes to overcome the fluid friction to flow.

SOLUTION The Reynolds number for the flow is

$$\text{Re} = \frac{u_m D}{\nu} = \frac{(0.6)(0.025)}{0.75 \times 10^{-4}} = 200$$

The flow is laminar, hence the friction factor f is computed by Eq. (7-26):

$$f = \frac{64}{\text{Re}} = \frac{64}{200} = 0.32$$

The pressure drop across a tube, according to Eq. (6-19a), becomes

$$\Delta P = f \frac{L}{D} \frac{\rho u_m^2}{2}$$

$$= (0.32)\left(\frac{10}{0.025}\right)\frac{(868 \text{ kg/m}^3)(0.6^2 \text{ m}^2/\text{s}^2)}{2}$$

$$\cong 2 \times 10^4 \text{ N/m}^2$$

The flow rate through the 80 tubes is

$$M = n\left(\frac{\pi}{4} D^2\right)u_m$$

$$= 80 \frac{\pi}{4}(0.025 \text{ m}^2)(0.6 \text{ m/s}) = 2.356 \times 10^{-2} \text{ m}^3/\text{s}$$

The pumping power requirement is determined by (6-19b):

$$W = \Delta P\, M = \left(2 \times 10^4 \frac{\text{N}}{\text{m}^2}\right)\left(2.356 \times 10^{-2} \frac{\text{m}^3}{\text{s}}\right)$$

$$= 471.2 \frac{\text{N} \cdot \text{m}}{\text{s}}$$

$$= \left(471.2 \frac{\text{N} \cdot \text{m}}{\text{s}}\right)\left(\frac{1}{745.7} \frac{\text{hp} \cdot \text{s}}{\text{N} \cdot \text{m}}\right) \cong 0.63 \text{ hp}$$

Example 7-3 Consider the heating of atmospheric air flowing with a velocity of $u_m = 0.5$ m/s inside a thin-walled tube 2.5 cm in diameter in the hydrodynamically and thermally developed region. Heating can be done either by condensing steam on the outer surface of the tube, thus maintaining a uniform surface temperature, or by electric resistance heating, thus maintaining a uniform surface heat flux. Calculate the heat transfer coefficient for both of these heating conditions by assuming air properties can be evaluated at 350 K.

SOLUTION The air properties at 350 K are

$$\nu = 20.76 \times 10^{-6} \text{ m}^2/\text{s} \qquad k = 0.03 \text{ W/(m} \cdot \text{s)}$$

The Reynolds number for the flow is

$$\text{Re} = \frac{u_m D}{\nu} = \frac{(0.5)(0.025)}{20.76 \times 10^{-6}} = 602$$

Hence the flow is laminar. The Nusselt number for laminar flow inside a circular tube in the hydrodynamically and thermally developed region is given by Eqs. (7-41) and (7-42), respectively, for the constant wall heat flux and constant wall temperature boundary conditions. Therefore, the heat transfer coefficients for these two cases are determined as follows:

Heating by condensing steam:

$$h = 3.66 \frac{k}{D} = 3.66 \frac{0.03 \text{ W/(m} \cdot \text{s)}}{0.025 \text{ m}}$$

$$= 4.39 \text{ W/(m}^2 \cdot {}^\circ\text{C)}$$

Electric resistance heating:

$$h = 4.364 \frac{k}{D} = 4.364 \frac{0.03 \text{ W/(m} \cdot \text{s)}}{0.025 \text{ m}}$$

$$= 5.24 \text{ W/(m}^2 \cdot {}^\circ\text{C)}$$

Flow Inside Ducts of Various Cross Sections

The Nusselt number and the friction factor for laminar flow in ducts of various cross sections have been determined in the region where velocity and temperature profiles are fully developed. If the duct cross section for flow is not circular, then the heat transfer and friction factor, for many cases of practical interest, can be based on the *hydraulic diameter* D_h, defined as

$$\boxed{D_h = \frac{4A_c}{P}} \tag{7-45}$$

where A_c = cross-sectional area for flow and P = the wetted perimeter. Then the Nusselt and Reynolds numbers for such cases are

$$\text{Nu} = \frac{hD_h}{k} \tag{7-46a}$$

$$\text{Re} = \frac{u_m D_h}{\nu} \tag{7-46b}$$

The basis for choosing D_h as in Eq. (7-44) is that for a circular tube D_h becomes the tube diameter D, since $A_c = (\pi/4)D^2$ and $P = \pi D$.

The Nusselt number in the hydrodynamically and thermally developed region for laminar flow inside a circular tube was given earlier for two different boundary conditions, namely, the *constant surface temperature* and the *constant wall heat flux*. However, for geometries other than the circular tube and parallel plates, the constant wall heat flux boundary condition can involve two distinct limiting cases.

First, the surface heat flux in the flow direction is uniform while the surface temperature remains uniform around the periphery at a flow cross section. This situation arises in ducts having highly conductive materials resulting in a negligible wall thermal resistance. The reason is that highly conductive material tends to make the temperature around the perimeter at any cross section uniform. Except for a circular tube and a parallel-plate duct, the flow resistance around the duct is not uniform.

Second, the surface heat flux both in the flow direction and around the periphery is uniform. This situation arises in the ducts having walls of very low conducting material.

To distinguish the Nusselt number for the three different cases—the ducts subjected to uniform surface temperature and the ducts subjected to two different cases of uniform heat flux boundary conditions—we adopt the following notation:

Nu_T = Nusselt number for uniform surface temperature
Nu_{H1} = Nusselt number for uniform surface heat flux in the flow direction while the surface temperature remains uniform around the periphery
Nu_{H2} = Nusselt number for uniform surface heat flux both in the flow direction and around the periphery

Shah and London [5] have compiled these three different types of Nusselt number and the quantity $f \cdot Re$ (i.e., the product of the friction factor and the Reynolds number) for laminar flow through ducts of various cross sections in the hydrodynamically and thermally developed region. Their results are listed in Table 7-1. We note that only for flow inside a circular tube and a parallel-plate geometry are Nu_{H1} and Nu_{H2} the same. These results are strictly applicable in the region where the velocity and temperature profiles are fully developed. The determination of hydrodynamic and thermal entry lengths is discussed next.

Example 7-4 Air at atmospheric pressure and with a mean velocity of u_m = 0.5 m/s flows inside thin-walled, square cross-section ducts of sides b = 2.5 cm. The air is heated from the walls of the duct, which are maintained at a uniform temperature by condensing steam on the outside surface. Calculate the friction factor and the heat transfer coefficient in the hydrodynamically and thermally developed region. Air properties can be evaluated at 350 K.

SOLUTION The air properties at 350 K are

$$\nu = 20.76 \times 10^{-6}\ \text{m}^2/\text{s} \qquad k = 0.03\ \text{W/(m}\cdot\text{s)}$$

Table 7-1 Nusselt number and friction factor for hydrodynamically and thermally developed laminar flow in ducts of various cross sections*

Geometry ($L/D_h > 100$)	Nu_T	Nu_{H1}	Nu_{H2}	$f\,Re$
Circle	3.657	4.364	4.364	64.00
Hexagon	3.34	4.002	3.862	60.22
Triangle, 60°, $\frac{2b}{2a} = \frac{\sqrt{3}}{2}$	2.47	3.111	1.892	53.33
Rectangle, $\frac{2b}{2a} = 1$	2.976	3.608	3.091	56.91
Rectangle, $\frac{2b}{2a} = \frac{1}{2}$	3.391	4.123	3.017	62.20
Rectangle, $\frac{2b}{2a} = \frac{1}{4}$	3.66	5.099	4.35	74.8
Rectangle, $\frac{2b}{2a} = \frac{1}{8}$	5.597	6.490	2.904	82.34
Parallel plates, $\frac{2b}{2a} = 0$	7.541	8.235	8.235	96.00
Parallel plates, one side insulated, $\frac{b}{a} = 0$	4.861	5.385	—	96.00

* From Shah and London [5].

The hydraulic diameter of the duct is

$$D_h = \frac{4b^2}{4b} = b = 2.5 \text{ cm}$$

and the Reynolds number becomes

$$Re = \frac{u_m D_h}{\nu} = \frac{(0.5)(0.025)}{20.76 \times 10^{-6}} = 602$$

From Table 7-1, for a square duct we obtain

$$f\,\text{Re} = 56.91$$

$$f = \frac{56.91}{602} = 9.45 \times 10^{-2}$$

and

$$\text{Nu}_T = 2.976$$

$$h = 2.976\,\frac{k}{D_h} = 2.976\,\frac{0.03\ \text{W/(m}\cdot{}^\circ\text{C)}}{0.025\ \text{m}} = 3.57\,\frac{\text{W}}{\text{m}^2\cdot{}^\circ\text{C}}$$

Hydrodynamic and Thermal Entry Lengths

It is of practical interest to know the *hydrodynamic entrance length* L_h and the *thermal entrance length* L_t for flow inside ducts.

The hydrodynamic entrance length L_h is defined, somewhat arbitrarily, as the length required from the duct inlet to achieve a maximum velocity of 99 percent of the corresponding fully developed magnitude.

The thermal entrance length L_t is defined, somewhat arbitrarily, as the length required from the beginning of the heat transfer section to achieve a local Nusselt number Nu_x equal to 1.05 times the corresponding fully developed value.

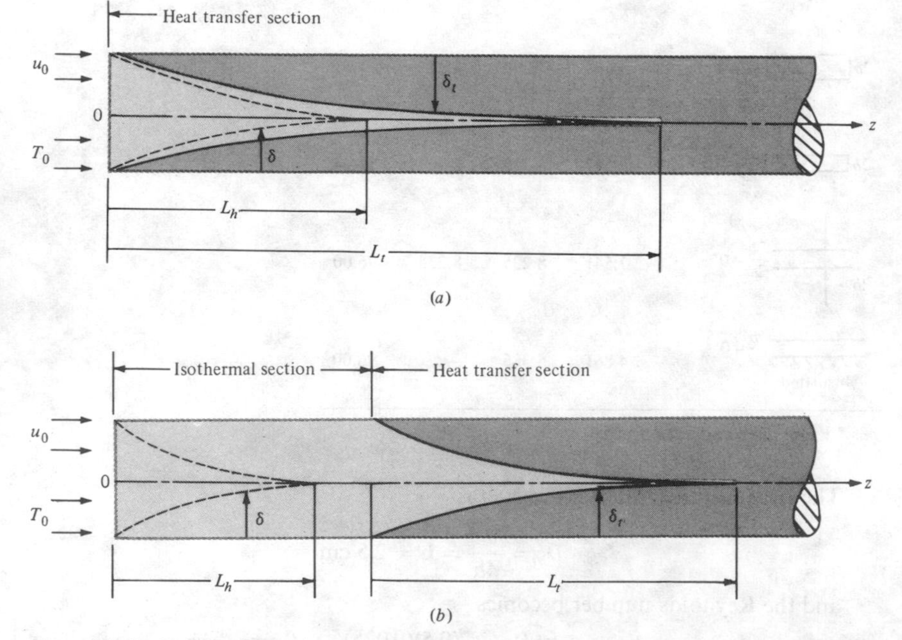

Figure 7-3 Hydrodynamic and thermal entrance lengths: (*a*) heat transfer starts at the duct inlet, (*b*) heat transfer starts after an isothermal section.

If heat transfer to the fluid starts as soon as fluid enters the duct, both the velocity and thermal boundary layers begin to develop immediately, and L_h and L_t are both measured from the tube inlet, as illustrated in Fig. 7-3*a*.

In some situations the heat transfer to the fluid begins after an isothermal calming section, as illustrated in Fig. 7-3*b*. For such a case, L_h is measured from the duct inlet because the velocity boundary layer begins to develop as soon as the fluid enters the duct, but L_t is measured from the location where the heat transfer starts, because the thermal boundary layer begins to develop in the heat transfer section.

The hydrodynamic and thermal entrance lengths for laminar flow inside conduits have been given by various authors [3, 5–8]. We present in Table 7-2 the hydrodynamic entrance length L_h for laminar flow inside conduits of various cross sections based on the definition discussed previously. Included in this table are the thermal entrance lengths for constant wall temperature and constant wall heat flux boundary conditions for thermally developing, hydrodynamically developed flow. In this table, D_h is the hydraulic diameter, and the Reynolds number is based on the hydraulic diameter.

We note from Table 7-2 that for a given geometry, the hydrodynamic entry length L_h depends on the Reynolds number only, whereas the thermal entry length

Table 7-2 Hydrodynamic entrance length L_h and thermal entrance length L_t for laminar flow inside ducts*

Geometry	$\frac{L_h/D_h}{\text{Re}}$	$\frac{L_t/D_h}{\text{Pe}}$ Constant wall temperature	$\frac{L_t/D_h}{\text{Pe}}$ Constant wall heat flux
Circular tube (diameter D)	0.056	0.033	0.043
Parallel plates (spacing $2b$)	0.011	0.008	0.012
Rectangle ($2a \times 2b$):			
$\frac{a}{b} = 0.25$	0.075	0.054	0.042
0.50	0.085	0.049	0.057
1.0	0.09	0.041	0.066

* Based on the results reported in Refs. 5 to 8. The thermal entry lengths are for the hydrodynamically developed, thermally developing flow conditions.

L_t depends on the Péclét number Pe which is equal to the product of the Reynolds and Prandtl numbers. Therefore, for liquids having a Prandtl number of the order of unity, L_h and L_t are of comparable magnitude. For fluids, such as oils, which have a large Prandtl number, $L_t \gg L_h$; and for liquid metals which have small Prandtl number, $L_t \ll L_h$.

The thermal entry lengths given in Table 7-2 are for hydrodynamically developed, thermally developing flow. As we discuss later, in many cases the velocity and temperature profiles develop simultaneously at the entrance region. Such a flow is called the *simultaneously developing flow*. The thermal entry lengths for a simultaneously developing flow also depend separately on the Prandtl number. For example, for simultaneously developing flow inside a circular tube, under constant wall temperature, the thermal entry length L_t is

$$\frac{L_t}{D\,\mathrm{Pe}} = 0.037 \qquad \text{for Pr} = 0.7$$

which should be compared with

$$\frac{L_t}{D\,\mathrm{Pe}} = 0.033 \qquad \text{for Pr} \to \infty$$

which corresponds to that given in Table 7-2 for the hydrodynamically developed, thermally developing flow. Thus L_t increases with decreasing Prandtl number and a weak function of the Prandtl number for Pr > 0.07.

Example 7-5 Determine the hydrodynamic and the thermal entrance lengths in terms of the tube inside diameter D for flow at a mean temperature $T_m = 60°\mathrm{C}$ and Re = 200 inside a circular tube for mercury, air, water, ethylene glycol, and engine oil, under constant wall heat flux boundary condition.

SOLUTION The hydrodynamic entrance length L_h, for laminar flow inside a circular tube, is obtained from Table 7-2 as

$$L_h = 0.056\,\mathrm{Re}\,D$$
$$= (0.056)(200)D \cong 11D$$

Thus, L_h is approximately 11 diameters from the tube inlet for all the fluids considered here.

The thermal entrance length, given heat transfer under the constant wall heat flux boundary condition, is obtained from Table 7-2 as

$$L_t = 0.043\,\mathrm{Re}\,\mathrm{Pr}\,D$$
$$= (0.043)(200)\,\mathrm{Pr}\,D = 8.6\,\mathrm{Pr}\,D$$

Here L_t depends on the Prandtl number, and for the fluids considered in this example it is determined as follows:

Fluid	Pr	$\frac{L_t}{D}$
Mercury	0.02	0.17
Air	0.7	6
Water	3	26
Ethylene glycol	50	430
Engine oil	1050	9030

We note that for flow at Re = 200 the thermal entrance length varies from a fraction of the tube diameter for mercury to about 9000 diameters for engine oil, while the hydrodynamic entrance length is about 11 diameters for all the fluids considered here.

7-2 THERMALLY DEVELOPING, HYDRODYNAMICALLY DEVELOPED LAMINAR FLOW

Consider the flow of a fluid inside a duct, as illustrated in Fig. 7-3*b*, in which there is an isothermal section to allow for the velocity development before the fluid enters the heat transfer zone. As illustrated in Example 7-5, for fluids having a large Prandtl number such as oils, the hydrodynamic entrance length is very small in comparison with the thermal entrance length. In the analysis of heat transfer for such situations, it is reasonable to assume a thermally developing but hydrodynamically developed flow. For flow inside a circular tube, for example, if one neglects the axial heat conduction and viscous-energy dissipation in the fluid, the governing energy equation is obtained from Eq. (6-71*a*) of Example 6-15 [or Eq. (6-71*b*) of Example 6-16] as

$$\frac{1}{\alpha} u(r) \frac{\partial T(r, z)}{\partial z} = \frac{1}{r} \frac{\partial}{\partial r}\left(r \frac{\partial T}{\partial r}\right) \tag{7-47a}$$

and $u(r)$ is the fully developed velocity distribution in flow and is given by Eq. (7-22) as

$$u(r) = 2u_m\left[1 - \left(\frac{r}{R}\right)^2\right] \tag{7-47b}$$

where R is the inside radius of the tube.

The solution of the partial differential equation (7-47) subject to appropriate boundary conditions at the tube inlet and wall surface gives the temperature distribution $T(r, z)$ in the flow. Given the temperature distribution, the local heat transfer coefficient $h(z)$ and the local Nusselt number Nu_z can be determined. Once $h(z)$ and Nu_z are available, their average values h_m and Nu_m from $z = 0$ to $z = L$ along the conduit can be computed readily.

A classic solution of laminar forced convection inside a circular tube subject to uniform wall surface temperature was given by Graetz [9] in 1885 and later quite independently by Nusselt [10] in 1910. The reader should consult Refs. 3 and 11 to 13 for a discussion of the original Graetz problem. A vast amount of literature [14–32] now exists on the extensions of the Graetz problem for boundary conditions other than the uniform surface temperature, geometries other than a circular tube, the effects of energy generation, and viscous dissipation in the fluid. An extensive compilation of literature on the subject of forced convection inside ducts is given in a monograph by Shah and London [5].

Figure 7-4 shows the local and average Nusselt numbers for thermally developing, hydrodynamically developed laminar flow inside a circular tube plotted against the dimensionless parameter $(x/D)/(\mathrm{Re\ Pr})$, where x is the axial distance along the conduit measured from the beginning of the heated section. The inverse of this dimensionless parameter is called the *Graetz number* Gz:

$$(\mathrm{Gz})^{-1} = \frac{x/D}{\mathrm{Re\ Pr}} \tag{7-48}$$

In this figure, the Nusselt numbers are given for both constant wall temperature and constant wall heat flux boundary conditions. We note that the asymptotic values of the Nusselt numbers for the constant wall heat flux and constant wall temperatures are, respectively, 4.364 and 3.66. These values are the same as those given previously by Eqs. (7-41*b*) and (7-42) for the region where the flow is both hydrodynamically and thermally developed.

Figure 7-5 gives the local and average Nusselt numbers for thermally developing, hydrodynamically developed laminar flow between parallel plates plotted against the dimensionless parameter $(x/D_h)/(\mathrm{Re\ Pr})$, where D_h is the hydraulic diameter and x is the distance along the plate measured from the beginning of the heating section in the direction of flow. The Nusselt numbers are given for both constant wall heat flux and constant wall temperatures. The asymptotic values of the Nusselt numbers 8.235 and 7.541 for constant wall heat flux and constant wall temperature, respectively, are the same as those given in Table 7-1 for the hydrodynamically and thermally developed region.

Figure 7-6 shows the mean Nusselt numbers for thermally developing, hydrodynamically developed laminar flow inside a square duct plotted against the dimensionless parameter $(x/D_h)/(\mathrm{Re\ Pr})$. The asypmtotic values 3.608 and 2.976 for the constant wall heat flux and constant wall temperature, respectively, are the same as those Nu_{H1} and Nu_T given in Table 7-1.

Example 7-6 Ethylene glycol at 60°C, with a velocity of $u_m = 4$ cm/s, enters the 6-m-long, heated section of a thin-walled, 2.5-cm-ID tube, after passing through an isothermal calming section. In the heated part, the tube wall is maintained at a uniform temperature $T_w = 100$°C by condensing steam on the outer surface of the tube. Calculate the exit temperature of ethylene glycol.

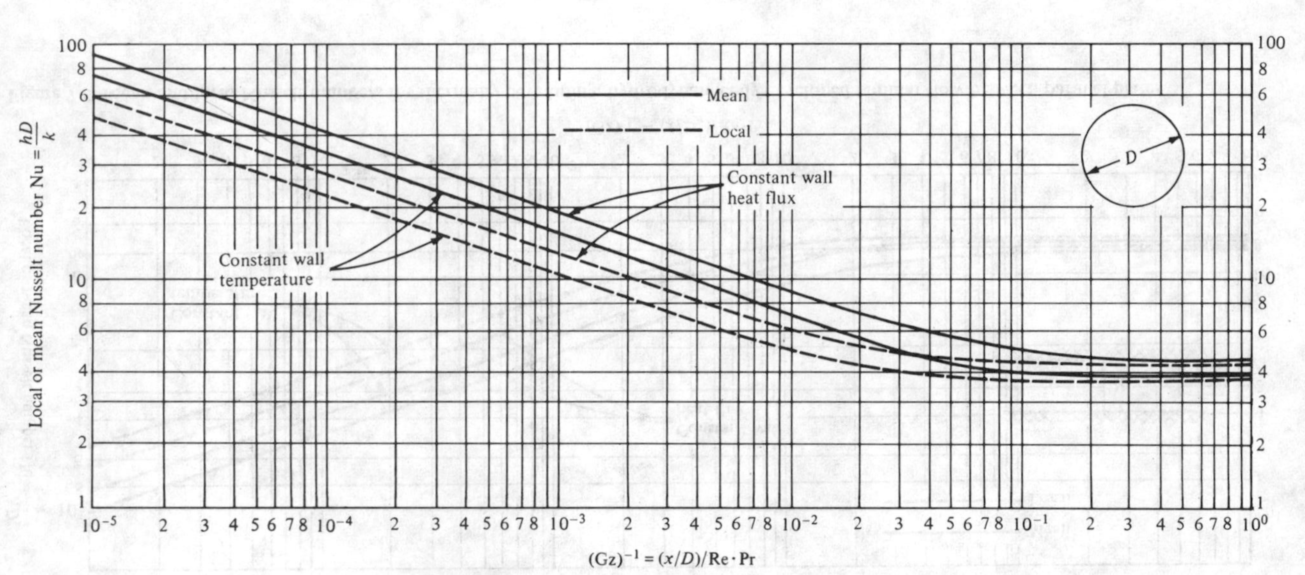

Figure 7-4 Mean and local Nusselt numbers for thermally developing, hydrodynamically developed laminar flow inside a circular tube.

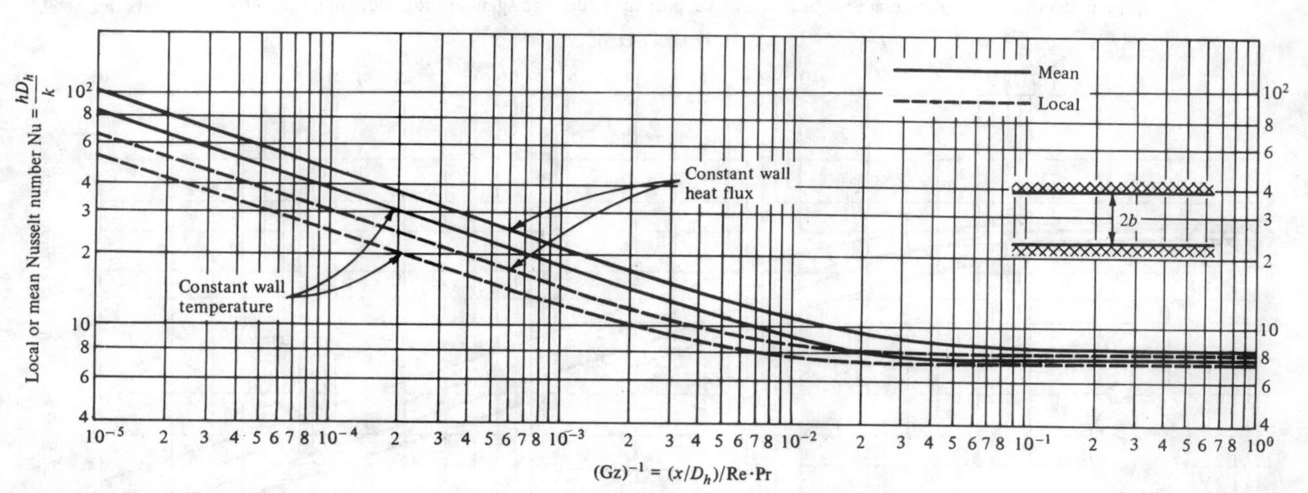

Figure 7-5 Mean and local Nusselt numbers for thermally developing, hydrodynamically developed laminar flow between parallel plates.

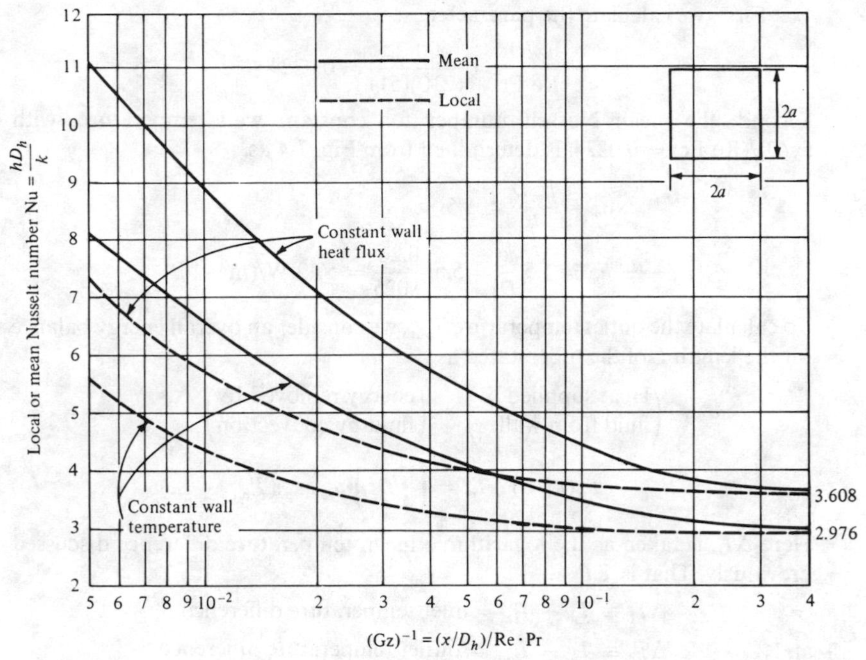

Figure 7-6 Mean and local Nusselt numbers for thermally developing, hydrodynamically developed laminar flow inside a square duct. (*From the data by Chandrupatla and Sastri* [32].)

SOLUTION The Reynolds number should be determined to establish whether the flow is laminar or turbulent. The mean fluid temperature inside the tube cannot be calculated yet because the fluid exit temperature is not known. Therefore, we start by evaluating the physical properties of the fluid at the inlet temperature 60°C. We obtain

$$c_p = 2562 \text{ J/(kg} \cdot {}^\circ\text{C)} \qquad \rho = 1088 \text{ kg/m}^3$$

$$\nu = 4.75 \times 10^{-6} \text{ m}^2/\text{s} \qquad k = 0.26 \text{ W/(m} \cdot {}^\circ\text{C)} \qquad \text{Pr} = 51$$

Then

$$\text{Re} = \frac{u_m D}{\nu} = \frac{(0.04)(0.025)}{4.75 \times 10^{-6}} = 210$$

and the flow is laminar. In the heat transfer section, the flow can be regarded as thermally developing but hydrodynamically developed, because there is an isothermal calming section, and for fluids with high Prandtl number the hydrodynamic entrance length is short compared with the thermodynamic entrance length, as illustrated in Example 7-4. Therefore, Fig. 7-4 can be used to calculate the mean Nusselt number.

First, we calculate the parameter

$$\frac{x/D}{\text{Re Pr}} = \frac{600/2.5}{(210)(51)} = 0.0224$$

Second, the mean Nusselt number for constant wall temperature, with $(x/D)/(\text{Re Pr}) = 0.0224$, is determined from Fig. 7-4 as

$$\text{Nu}_m = \frac{h_m D}{k} \cong 5.5$$

$$h_m = 5.5\frac{k}{D} = 5.5\frac{0.26}{0.025} = 57.2 \text{ W/(m}^2 \cdot {}^\circ\text{C)}$$

To calculate the outlet temperature T_{out}, we consider an overall energy balance for the length L of the tube, stated as

$$\begin{pmatrix}\text{Heat supplied to}\\ \text{fluid from wall}\end{pmatrix} = \begin{pmatrix}\text{energy removed by}\\ \text{fluid by convection}\end{pmatrix}$$

$$h_m(\pi DL)\,\Delta T_m = \left(\frac{\pi}{4}D^2\right)(u_m\rho c_p)(T_{\text{out}} - T_{\text{in}}) \tag{a}$$

Here ΔT_m is taken as the logarithmic mean temperature difference discussed previously. That is, let

$$\Delta T_1 \equiv T_w - T_{\text{in}} = \text{inlet temperature difference}$$

and

$$\Delta T_2 \equiv T_w - T_{\text{out}} = \text{outlet temperature difference}$$

Then the logarithmic mean temperature difference becomes

$$\Delta T_m = \frac{\Delta T_1 - \Delta T_2}{\ln(\Delta T_1/\Delta T_2)} = \frac{T_{\text{out}} - T_{\text{in}}}{\ln[(T_w - T_{\text{in}})/(T_w - T_{\text{out}})]} \tag{b}$$

as ΔT_m is introduced into Eq. (*a*).

$$h_m(\pi DL)\frac{T_{\text{out}} - T_{\text{in}}}{\ln[(T_w - T_{\text{in}})/(T_w - T_{\text{out}})]} = \left(\frac{\pi}{4}D^2\right)(u_m\rho c_p)(T_{\text{out}} - T_{\text{in}})$$

or

$$\ln\frac{T_w - T_{\text{in}}}{T_w - T_{\text{out}}} = \frac{4Lh_m}{Du_m\rho c_p}$$

$$\frac{T_w - T_{\text{in}}}{T_w - T_{\text{out}}} = \exp\left(\frac{4Lh_m}{Du_m\rho c_p}\right) \tag{c}$$

The numerical values are substituted:

$$\frac{100 - 60}{100 - T_{\text{out}}} = \exp\left(\frac{4 \times 6 \times 57.2}{0.025 \times 0.04 \times 1088 \times 2562}\right)$$

$$\frac{40}{100 - T_{\text{out}}} = 1.636$$

$$T_{\text{out}} = 75.6^\circ\text{C}$$

The results can be improved by computing the physical properties at the bulk fluid temperature $(T_{\text{in}} + T_{\text{out}})/2 = (60 + 75.6)/2 = 68^\circ\text{C}$, but the improvement would be very little, that is, $T_{\text{out}} = 75.4^\circ\text{C}$.

7-3 SIMULTANEOUSLY DEVELOPING LAMINAR FLOW

When heat transfer starts as soon as a fluid enters a duct, as illustrated in Fig. 7-3*a*, the velocity and temperature profiles start developing simultaneously. The analysis of temperature distribution in the flow, and hence of heat transfer between the fluid and the walls, for such situations is more involved because the velocity distribution varies in the axial direction as well as normal to it. For example, for simultaneously developing flow inside a circular tube, the velocity distribution in the energy equation (7-47*a*) is a function of both r and z, that is, $u \equiv u(r, z)$.

Heat transfer problems of simultaneously developing flow have been solved mostly by numerical methods for flow inside a circular tube [33–39], a parallel-plate channel [40–49], and a rectangular duct [50, 51].

Figure 7-7 shows the mean Nusselt number for simultaneously developing laminar flow inside a circular tube subject to constant wall temperature. The results are given for the Prandtl numbers 0.7, 2, 5, and ∞ and are plotted against the dimensionless parameter $(x/D)/(\text{Re} \cdot \text{Pr})$. The case for $\text{Pr} = \infty$ corresponds to the thermally developing but hydrodynamically developed flow discussed earlier. Clearly, the Nusselt number for simultaneously developing flow is higher than that for the hydrodynamically developed flow. It is also apparent from this figure that for fluids having a large Prandtl number, the Nusselt number for simultaneously developing flow is very close to that for thermally developing, hydrodynamically developed flow. The asymptotic Nusselt number for all the cases shown in this figure is equal to the fully developed value 3.66.

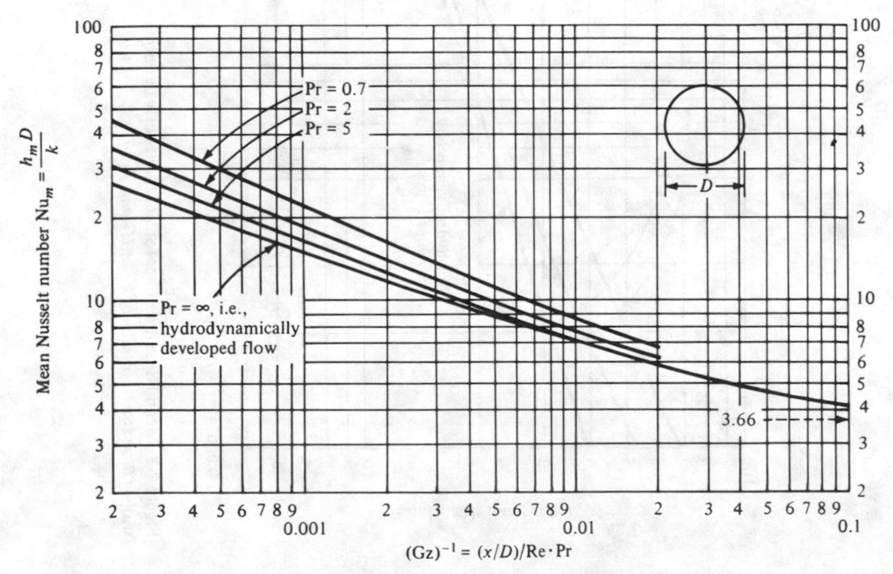

Figure 7-7 Mean Nusselt numbers for simultaneously developing laminar flow inside a circular tube subjected to constant wall temperature. (*Based on the results from Refs. 8 and 35.*)

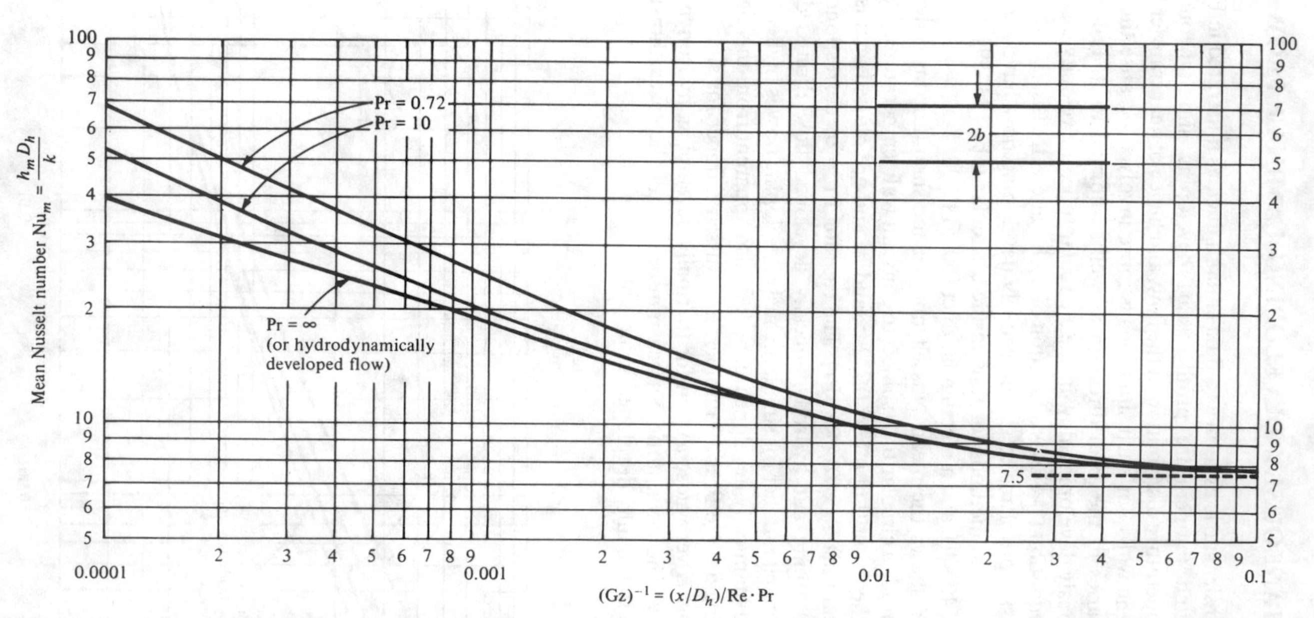

Figure 7-8 Mean Nusselt number for simultaneously developing laminar flow between parallel plates subjected to constant and the same temperature at both walls. (*Based on the results of C. L. Hwang given in Ref. 8.*)

Figure 7-8 shows the mean Nusselt number for simultaneously developing flow inside a parallel-plate channel subjected to a constant and the same temperature at both walls. The results are given for the Prandtl numbers 0.72, 10, and ∞ and are plotted against the dimensionless parameter $(x/D_h)/(\text{Re} \cdot \text{Pr})$. Here, the case for $\text{Pr} = \infty$ corresponds to thermally developing, but hydrodynamically developed flow.

Empirical Correlations

Empirical relations also have been developed for predicting the mean Nusselt number for laminar flow in the entrance region of a circular tube. One such correlation for the mean Nusselt number for laminar flow in a circular tube at constant wall temperature is given by Hausen [52] as

$$\boxed{\text{Nu}_m = 3.66 + \frac{0.0668\,\text{Gz}}{1 + 0.04(\text{Gz})^{2/3}}} \tag{7-49}$$

where
$$\text{Nu}_m = \frac{h_m D}{k} \tag{7-50a}$$

$$\text{Gz} = \frac{\text{Re} \cdot \text{Pr}}{L/D} \tag{7-50b}$$

$$\text{Re} = \frac{u_m D}{\nu} \qquad L = \text{distance from the inlet} \tag{7-50c}$$

This relation is recommended for $\text{Gz} < 100$ (see Table 7-3) and all properties are evaluated at the fluid bulk mean temperature. Clearly, as the length L increases, the Nusselt number approaches the asymptotic value 3.66.

A rather simple empirical correlation has been proposed by Sieder and Tate [53] to predict the mean Nusselt number for laminar flow in a circular tube at constant wall temperature:

$$\boxed{\text{Nu}_m = 1.86(\text{Gz})^{1/3}\left(\frac{\mu_b}{\mu_w}\right)^{0.14}} \tag{7-51}$$

And it is recommended for

$$0.48 < \text{Pr} < 16{,}700 \tag{7-52a}$$

$$0.0044 < \frac{\mu_b}{\mu_w} < 9.75 \tag{7-52b}$$

$$(\text{Gz})^{1/3}\left(\frac{\mu_b}{\mu_w}\right)^{0.14} > 2 \tag{7-52c}$$

All physical properties are evaluated at the fluid bulk mean temperature, except μ_w, which is evaluated at the wall temperature. The Graetz number Gz is defined

Table 7-3 A comparison of empirical and theoretical correlations for the mean Nusselt number for simultaneously developing laminar flow inside a circular tube

$(Gz)^{-1} = \dfrac{x/D}{Re \cdot Pr}$	Nu_m [Hausen], Eq. (7-49)	$Nu_m \left(\dfrac{\mu_w}{\mu_b}\right)^{0.14}$ [Sieder-Tate], Eq. (7-51)	Nu_m from Fig. 7-7		
			Pr = 0.7	Pr = 5	Pr = ∞
0.1	4.22	4.0			4.16
0.02	5.82	6.85	6.8	5.8	5.8
0.01	7.25	8.63	8.7	7.2	7.2
0.001	17.0	18.60	22.2	16.9	15.4
0.0001	30.0	31.80	44.1	30.3	26.7

by Eq. (7-50*b*). The last restriction implies that this equation cannot be used for extremely long tubes, because as $Gz \to 0$ with increasing length, the equation has no provision to yield the correct asymptotic value, as in the case of Hausen equation.

In Table 7-3, we compare the mean Nusselt numbers calculated from Eqs. (7-49) and (7-51) and obtained from Fig. 7-7. The Sieder and Tate equation underestimates Nu_m for $Gz > 10$, which is consistent with the restriction Eq. (7.52*c*). The results given in Table 7-3 show that the theoretical prediction of the Nusselt number for simultaneously developing flow given in Fig. 7-7 is reasonably accurate and includes the effects of the Prandtl number.

Example 7-7 Engine oil is cooled from $T_{in} = 120°C$ to $T_{out} = 80°C$ while it is flowing with a mean velocity of $u_m = 0.04$ m/s through a circular tube of inside diameter 2.5 cm. The tube wall is maintained at a uniform temperature $T_w = 40°C$. Determine the tube length L.

SOLUTION The physical properties of oil at the bulk mean temperature

$$T_m = \frac{T_{in} + T_{out}}{2} = \frac{120 + 80}{2} = 100°C$$

are taken as

$$c_p = 2200 \text{ J/(kg} \cdot °\text{C)} \qquad \rho = 840 \text{ kg/m}^3 \qquad \text{Pr} = 276$$

$$\nu = 0.2 \times 10^{-4} \text{ m}^2/\text{s} \qquad k = 0.137 \text{ W/(m} \cdot °\text{C)}$$

Then the Reynolds number becomes

$$\text{Re} = \frac{u_m D}{\nu} = \frac{(0.04)(0.025)}{0.2 \times 10^{-4}} = 50$$

hence the flow is laminar.

We treat this problem as an entrance region heat transfer problem of simultaneously developing flow. Figure 7-7 or Eq. (7-49) or (7-51) can be used to determine the Nusselt number. However, to perform these calculations, we need the Gratz number

$$\mathrm{Gz} = \frac{\mathrm{Re} \cdot \mathrm{Pr}}{L/D} = \frac{(50)(276)}{L/D} = \frac{13{,}800}{L/D} \tag{a}$$

Here, since the tube length L is unknown, Gz cannot be determined.

Another relation is obtained by writing an overall energy balance for L as

$$\left(\frac{\pi}{4} D^2\right)(\rho u_m c_p)(T_{\text{in}} - T_{\text{out}}) = h_m(\pi D L)\,\Delta T_m \tag{b}$$

We take ΔT_m as the logarithmic means of $T_{\text{in}} - T_w$ and $T_{\text{out}} - T_w$; that is,

$$\Delta T_m = \frac{(T_{\text{in}} - T_w) - (T_{\text{out}} - T_w)}{\ln\left[(T_{\text{in}} - T_w)/(T_{\text{out}} - T_w)\right]} = \frac{T_{\text{in}} - T_{\text{out}}}{\ln\left[(T_{\text{in}} - T_w)/(T_{\text{out}} - T_w)\right]} \tag{c}$$

Equation (b) is rearranged in terms of dimensionless parameters:

$$\mathrm{Re}\ \mathrm{Pr}\left(\tfrac{1}{4}\right)(T_{\text{in}} - T_{\text{out}}) = \mathrm{Nu}_m \left(\frac{L}{D}\right) \Delta T_m \tag{d}$$

The numerical values are

$$(50)(276)\left(\tfrac{1}{4}\right)(120 - 80) = \mathrm{Nu}_m \left(\frac{L}{D}\right) \Delta T_m$$

where

$$\Delta T_m = \frac{120 - 80}{\ln\left[(120 - 40)/(80 - 40)\right]} = 57.71$$

Or, solving for Nu_m, we obtain

$$\mathrm{Nu}_m = \frac{2391.3}{L/D} \tag{e}$$

Equations (a) and (e) can be used in conjunction with the appropriate correlation for the Nusselt number, and the two unknowns L/D and Nu_m can be found.

We use the Sieder and Tate equations (7-51) for this purpose. Introducing Eqs. (a) and (e) into Eq. (7-51), we obtain

$$\frac{2391.3}{L/D} = 1.86\left(\frac{13{,}800}{L/D}\right)^{1/3}\left(\frac{0.17}{0.21}\right)^{0.14}$$

where $\mu = 0.17$ and $\mu_w = 0.21$ are the viscosities evaluated at the fluid bulk mean and the wall temperatures, respectively.

Solving for L/D, we obtain

$$\frac{L}{D} = 410.2 \qquad \text{or} \qquad L = (410.2)(0.025) = 10.3\ \text{m}$$

An iterative solution is needed if Fig. 7-7 is used.

7-4 TURBULENT FLOW INSIDE DUCTS

Turbulent flow is important in engineering applications because it is involved in the vast majority of fluid flow and heat transfer problems encountered in engineering practice. Here we show how to determine the friction factor and heat transfer in turbulent flow inside conduits.

Friction Factor and Pressure Drop

Consider fully developed turbulent flow at a mean velocity of u_m through a circular tube of inside diameter D. The pressure drop ΔP over the length L of the tube can be determined according to the relation given by Eq. (6-19*a*):

$$\boxed{\Delta P = f \frac{L}{D} \frac{\rho u_m^2}{2} \qquad \frac{\text{N}}{\text{m}^2}} \tag{7-53}$$

where f is the *friction factor* for turbulent flow. The friction factor for laminar flow inside a circular tube can be found by a purely theoretical approach and has been shown to be $f = 64/\text{Re}$. In the case of turbulent flow, however, some empiricism is introduced in its derivation because a semiempirical velocity profile is used in the analysis. It is instructive to describe here the development of the relation defining the friction factor in turbulent flow.

A force balance over the length L of a circular tube, shown in Fig. 7-9, is written as

$$\pi D L \tau_0 = \Delta p \frac{\pi D^2}{4}$$

or

$$\Delta p = \frac{4L}{D} \tau_0 \tag{7-54}$$

where τ_0 is the total shear stress at the wall. Combining Eqs. (7-53) and (7-54), we obtain

$$f = \frac{8\tau_0}{\rho u_m^2} = \frac{8}{(u_m/\sqrt{\tau_0/\rho})^2} = \frac{8}{(u_m^+)^2} \tag{7-55}$$

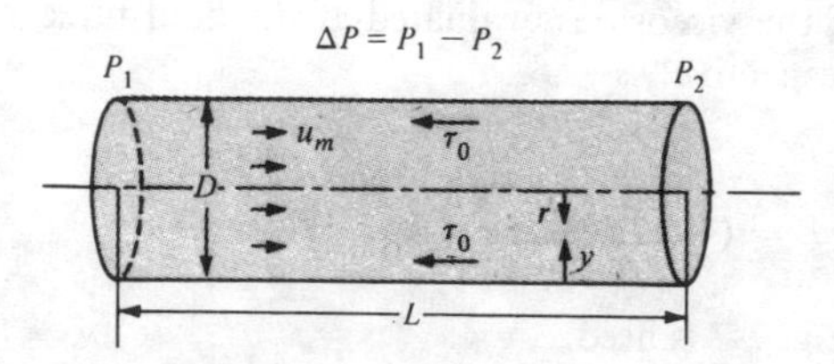

Figure 7-9 A force balance over length L of a tube.

where

$$u_m^+ = \frac{u_m}{\sqrt{\tau_0/\rho}} = \text{dimensionless mean velocity}$$

We assume that the velocity distribution for turbulent flow is given by Eq. (6-39*c*), that is,

$$u^+ = 2.5 \ln y^+ + 5.5 \tag{7-56}$$

where

$$y^+ = \frac{y}{\nu}\sqrt{\frac{\tau_0}{\rho}} = \text{dimensionless distance measured from tube wall}$$

The average value of u^+ over the cross section of the flow through the tube is determined as

$$u_m^+ = 2.5 \ln\left(\frac{R}{\nu}\sqrt{\frac{\tau_0}{\rho}}\right) + 1.75 \tag{7-57}$$

where R is the tube radius. The term $\sqrt{\tau_0/\rho}$ appearing in this expression is obtained from Eq. (7-55) as

$$\sqrt{\frac{\tau_0}{\rho}} = u_m\sqrt{\frac{f}{8}} \tag{7-58}$$

Equation (7-58) is introduced into Eq. (7-57):

$$u_m^+ = 2.25 \ln\left[\left(\frac{Du_m}{\nu}\right)\left(\frac{1}{4\sqrt{2}}\right)\sqrt{f}\right] + 1.75$$

$$= 2.5 \ln(\text{Re}\sqrt{f}) - 2.5 \ln\left(\frac{1}{4\sqrt{2}}\right) + 1.75$$

$$= 5.756 \log(\text{Re}\sqrt{f}) - 2.582 \tag{7-59}$$

Equation (7-59) is substituted into Eq. (7-55). The following expression results for the friction factor f:

$$\frac{1}{\sqrt{f}} = 2.035 \log(\text{Re}\sqrt{f}) - 0.91 \tag{7-60}$$

where $\text{Re} \equiv u_m D/\nu$ = Reynolds number. Equation (7-60) is derived by utilizing a semiempirical velocity profile. A comparison with experiments suggests that Eq. (7-60) should be modified as follows:

$$\boxed{\frac{1}{\sqrt{f}} = 2.0 \log(\text{Re}\sqrt{f}) - 0.8} \tag{7-61a}$$

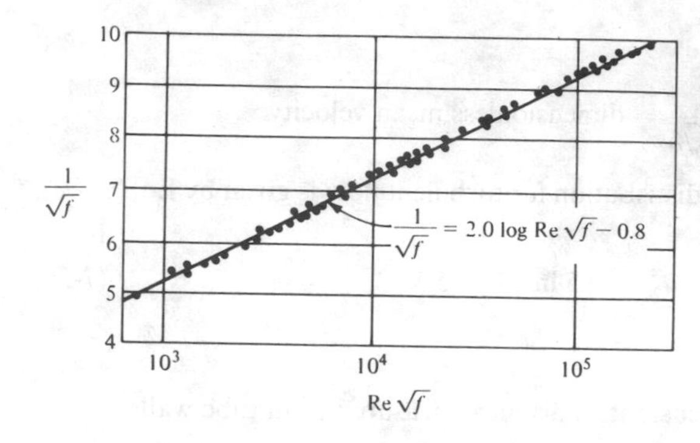

Figure 7-10 Friction law for turbulent flow inside smooth pipes and experimental data of various investigators. (*From Schlichting* [56].)

This relation agrees well with the experiments and is used for the determination of the friction factor for turbulent flow inside smooth pipes. Figure 7-10 shows a comparison of Eq. (7-61*a*) with the experiments of various investigators; here Nikuradse's [54] experiments cover a range of Reynolds numbers up to 3.4×10^6.

The implicit equation (7-61*a*) is closely approximated by the following explicit expression:

$$f = (1.82 \log \mathrm{Re} - 1.64)^{-2} \tag{7-61b}$$

Nikuradse [55] also made extensive experiments with turbulent flow inside artificially roughened pipes over a wide range of *relative roughness* λ/D (i.e., protrusion height-to-diameter ratio) from about $\frac{1}{1000}$ to $\frac{1}{30}$. The *sand-grain* roughness used in these experiments has been adopted as a standard for the effects of roughness. A friction factor correlation also has been developed for turbulent flow inside rough pipes based on experiments performed with rough pipes.

Figure 7-11 shows the friction factor chart, originally presented by Moody [57], for turbulent flow inside smooth and rough pipes. The curve for the smooth pipe is based on Eq. (7-61). Included on this figure is the friction factor $f = 64/\mathrm{Re}$ for laminar flow inside circular pipes. It is apparent that for laminar flow the surface roughness has no effect on the friction factor; for turbulent flow, however, the friction factor is a minimum for a smooth pipe. The laminar flow is confined to the region $\mathrm{Re} < 2000$. The transitional turbulence occurs in the region $2000 < \mathrm{Re} < 10{,}000$. The fully turbulent flow occurs in the region $\mathrm{Re} > 10^4$.

For smooth pipes, simpler but approximate analytic expressions have been given for the friction factor in the form

$$f = 0.316\,\mathrm{Re}^{-0.25} \qquad \text{for } \mathrm{Re} < 2 \times 10^4 \tag{7-62}$$

$$f = 0.184\,\mathrm{Re}^{-0.2} \qquad \text{for } 2 \times 10^4 < \mathrm{Re} < 3 \times 10^5 \tag{7-63}$$

These results apply to hydrodynamically developed turbulent flow. The hydrodynamic development for turbulent flow occurs for x/D much shorter than that for laminar flow. For example, hydrodynamically developed flow conditions occur for x/D greater than about 10 to 20.

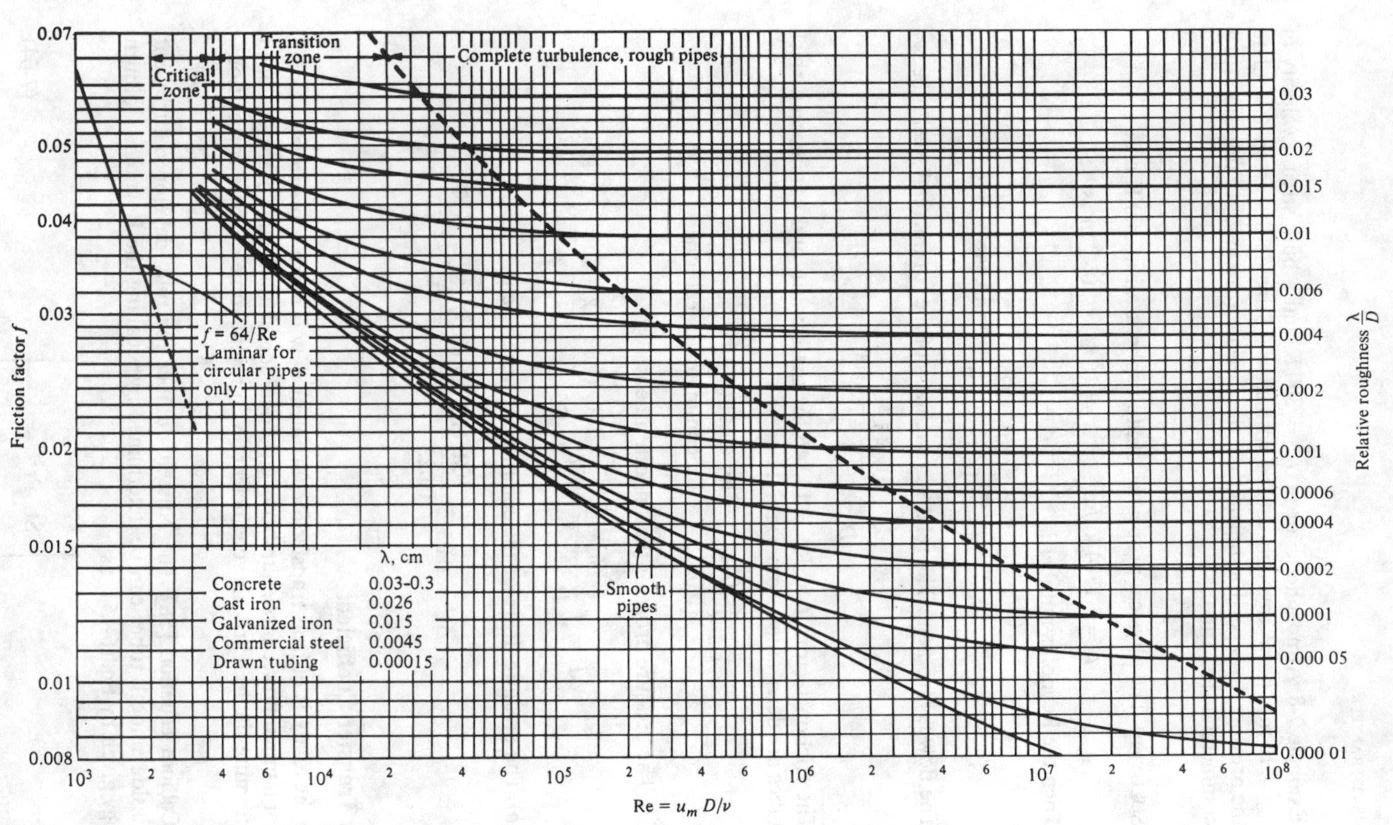

Figure 7-11 Friction factor for use in the relation $\Delta P = f(L/D)\,(\rho U_m^2/2)$ for pressure drop for flow inside circular pipes. (*From Moody* [57].)

Example 7-8 Atmospheric air at $T_m = 300$ K and a bulk stream velocity of $u_m = 10$ m/s flows through a tube with $D = 2.5$ inside diameter. Calculate the pressure drop per 100-m length of the tube for (*a*) a smooth tube and (*b*) a commercial steel tube.

SOLUTION The physical properties of atmospheric air at 300 K are

$$\rho = 1.1774 \text{ kg/m}^3 \qquad \nu = 16.84 \times 10^{-6} \text{ m}^2\text{/s}$$

Then the Reynolds number is

$$\text{Re} = \frac{u_m D}{\nu} = \frac{(10)(0.025)}{16.84 \times 10^{-6}} = 14{,}846$$

The flow is turbulent. The relative roughness of the commercial steel tube is

$$\frac{\lambda}{D} = \frac{0.0045}{2.5} = 0.0018$$

The friction factors f at Re = 14,846 for the smooth and commercial steel tubes are determined from Fig. 7-11 as

$$f = \begin{cases} 0.028 & \text{for smooth tube} \\ 0.0315 & \text{for commercial steel tube} \end{cases}$$

(*a*) The pressure drop for the smooth tube is

$$\Delta P = f \frac{L}{D} \frac{\rho u_m^2}{2} = 0.028 \frac{100}{0.025} \frac{(1.1774)(10^2)}{2} = 6.59 \frac{\text{kN}}{\text{m}^2}$$

(*b*) The pressure drop for the commercial steel tube is

$$\Delta P = 0.0315 \frac{100}{0.025} \frac{(1.1774)(10^2)}{2} = 7.42 \frac{\text{kN}}{\text{m}^2}$$

Heat Transfer Coefficient

Since the analysis of heat transfer for turbulent flow is much more involved than that for laminar flow, a large number of empirical correlations have been developed to determine the heat transfer coefficient. We present some of these correlations.

The Colburn equation [58] The Nusselt number for fully developed turbulent flow inside smooth tubes can be determined by recalling the Reynolds-Colburn analogy given by Eq. (6-15*a*) as

$$\boxed{\text{St}_x \, \text{Pr}^{2/3} = \frac{c_x}{2}} \tag{7-64}$$

and the definitions of c_x and f, given by Eqs. (6-4) and (7-55), respectively, as

$$\frac{c_x}{2} = \frac{\tau_x}{\rho u_\infty^2} \tag{7-65a}$$

$$\frac{f}{8} = \frac{\tau_0}{\rho u_m^2} \tag{7-65b}$$

From Eqs. (7-65*a*) and (7-65*b*) we conclude that $f/8$ for flow inside a circular tube is analogous to $c_x/2$ for flow along a flat plate.

With this consideration, the analogue of Eq. (7-64) can be written for turbulent flow inside a circular tube as

$$\boxed{\text{St Pr}^{2/3} = \frac{f}{8}} \tag{7-66}$$

We substitute the friction factor from Eq. (7-63) into Eq. (7-66):

$$\text{St Pr}^{2/3} = 0.023 \text{ Re}^{-0.2} \tag{7-67a}$$

where

$$\text{St} \equiv \frac{\text{Nu}}{\text{Re Pr}} \tag{7-67b}$$

Equations (7-67) are rearranged to yield the Colburn equation for turbulent flow inside a smooth tube:

$$\boxed{\text{Nu} = 0.023 \text{ Re}^{0.8} \text{ Pr}^{1/3}} \tag{7-68}$$

where $\text{Nu} = hD/k$, $\text{Re} = u_m D/\nu$, and $\text{Pr} = \nu/\alpha$. Equation (7-68) is applicable for:

$$0.7 < \text{Pr} < 160 \qquad \text{Re} > 10{,}000$$

$$\frac{L}{D} > 60 \qquad \text{smooth pipes}$$

and small to moderate temperature differences. Fluid properties are evaluated at the bulk mean temperature T_b.

Dittus-Boelter [59] equation A slightly different form of Eq. (7-68) is given by Dittus and Boelter:

$$\boxed{\text{Nu} = 0.023 \text{ Re}^{0.8} \text{ Pr}^{n}} \tag{7-69}$$

where $n = 0.4$ for heating ($T_w > T_b$) and $n = 0.3$ for cooling ($T_w < T_b$) of the fluid. The range of applicability is the same as for the Colburn equation.

Sieder and Tate [53] equation For situations involving a large property variation, the Sieder and Tate equation is recommended:

$$\boxed{\mathrm{Nu} = 0.027\ \mathrm{Re}^{0.8}\ \mathrm{Pr}^{1/3}\left(\frac{\mu_b}{\mu_w}\right)^{0.14}} \tag{7-70}$$

This equation is applicable for

$$0.7 < \mathrm{Pr} < 16{,}700 \qquad \mathrm{Re} > 10{,}000$$

$$\frac{L}{D} > 60 \qquad \text{smooth pipes}$$

All properties are evaluated at the bulk mean temperature T_b, except μ_w which is evaluated at the wall temperature.

Petukhov [61] equation The previous relations are relatively simple, but they give maximum errors of ± 25 percent in the range of $0.67 < \mathrm{Pr} < 100$ and apply to turbulent flow in smooth ducts. A more accurate correlation, which is also applicable for rough ducts, has been developed by Petukhov and coworkers at the Moscow Institute for High Temperature:

$$\mathrm{Nu} = \frac{\mathrm{Re}\ \mathrm{Pr}}{X}\left(\frac{f}{8}\right)\left(\frac{\mu_b}{\mu_w}\right)^n \tag{7-71a}$$

where

$$X = 1.07 + 12.7(\mathrm{Pr}^{2/3} - 1)\left(\frac{f}{8}\right)^{1/2} \tag{7-71b}$$

and

$$n = \begin{cases} 0.11 & \text{heating with uniform } T_w\ (T_w > T_b) \\ 0.25 & \text{cooling with uniform } T_w\ (T_w < T_b) \\ 0 & \text{uniform wall heat flux or gases} \end{cases}$$

Equations (7-71) are applicable for fully developed turbulent flow in the range

$$10^4 < \mathrm{Re} < 5 \times 10^6$$

$$0.5 < \mathrm{Pr} < 200 \qquad \text{with 5 to 6 percent error}$$

$$0.5 < \mathrm{Pr} < 2000 \qquad \text{with 10 percent error}$$

$$0.08 < \frac{\mu_w}{\mu_b} < 40$$

We note that $\mu_w/\mu_b < 1$ when a liquid is heated and $\mu_w/\mu_b > 1$ when the liquid is cooled. All physical properties, except μ_w, are evaluated at the bulk temperature.

The friction factor f in Eqs. (7-71) can be evaluated by Eq. (7-61b) for smooth tubes or obtained from the Moody chart (Fig. 7-11) for both smooth and rough tubes.

We have given four different equations for the determination of the Nusselt number for turbulent flow inside circular tubes. The Petukhov equation (7-71) is the most up-to-date correlation that is applicable for both smooth and rough tubes, and it appears to correlate with the experimental data very well over a wide range of parameters. Therefore, it should be preferred to the other correlations.

Nusselt [62] equation The previous relations apply to the region $L/D > 60$. Nusselt studied the experimental data for L/D from 10 to 100 and concluded that h, in this region, is approximately proportional to $(D/L)^{1/8}$. Hence, he replaced Eq. (7-70) by

$$\mathrm{Nu} = 0.036\ \mathrm{Re}^{0.8}\ \mathrm{Pr}^{1/3}\left(\frac{D}{L}\right)^{0.055} \qquad \text{for } 10 < \frac{L}{D} < 400 \tag{7-72}$$

where L is the length measured from the beginning of the heat transfer section, and the fluid properties are evaluated at the bulk mean fluid temperature.

Notter and Sleicher [63] equation The Nusselt number is determined theoretically from the solution of the energy equation by using an appropriate velocity profile for turbulent flow. The resulting Nusselt number for the hydrodynamically and thermally developed region was expressed in the form

$$\mathrm{Nu} = 5 + 0.016\ \mathrm{Re}^{a}\ \mathrm{Pr}^{b} \tag{7-73}$$

where

$$a = 0.88 - \frac{0.24}{4 + \mathrm{Pr}} \qquad \text{and} \qquad b = 0.33 + 0.5e^{-0.6\,\mathrm{Pr}}$$

which is applicable for

$$0.1 < \mathrm{Pr} < 10^4$$

$$10^4 < \mathrm{Re} < 10^6$$

$$\frac{L}{D} > 25$$

Equation (7-73) correlates well with experimental data and provides a more accurate representation of the effect of the Prandtl number. It may be preferred to Eq. (7-72).

Noncircular Ducts

So far, discussion of the friction factor and heat transfer coefficient for turbulent flow has been restricted to flow inside circular tubes. Numerous engineering applications involve turbulent forced convection inside ducts of noncircular cross section. The friction factor for a circular tube given by the Moody chart

(Fig. 7-11) applies to turbulent flow inside noncircular ducts if the tube diameter D is replaced by the hydraulic diameter of the noncircular duct, defined by Eq. (7-44), as

$$D_h = \frac{4A_c}{P} \tag{7-74}$$

where A_c is the cross-sectional area for flow and P is the wetted perimeter.

For noncircular ducts, the turbulent flow also occurs for Re > 2300, where the Reynolds number is based on the hydraulic diameter.

With noncircular ducts, the heat transfer coefficient varies around the perimeter and approaches zero near the sharp corners. Therefore, for certain situations difficulties may arise in applying the circular-tube results to a noncircular duct by using the hydraulic diameter concept. Irvine [64] discusses some problems of heat transfer in noncircular ducts.

Effects of Surface Roughness

The heat transfer coefficient for turbulent flow in rough-walled tubes is higher than that for smooth-walled tubes because roughness disturbs the viscous sublayer. The increased heat transfer due to roughness is achieved at the expense of increased friction to fluid flow. The correlation of heat transfer for turbulent flow in rough-walled tubes is very sparse in the literature. The Petukhov equation (7-71) can be recommended for predicting the heat transfer coefficient in hydrodynamically and thermally developed turbulent flow in rough pipes, because the friction factor f can be obtained from the Moody chart (Fig. 7-11) once the relative roughness of the pipe is known.

Effects of Property Variation

When heat transfer takes place to or from a fluid flowing inside a duct, the temperature varies over the flow cross section of the duct. For most *liquids*, although the specific heat and thermal conductivity are rather insensitive to temperature, the viscosity decreases significantly with temperature. For *gases*, the viscosity and thermal conductivity increase approximately by 0.8 power of the temperature. Therefore, the property variation affects both the heat transfer coefficient and the friction factor.

To compensate for the effects of nonisothermal conditions in the fluid, the Sieder and Tate equation (7-70) and the Petukhov equation (7-71) included a viscosity correction term in the form $(\mu_b/\mu_w)^n$.

The variation of the transport properties with temperature changes both the velocity and temperature profiles within the fluid, and an accurate determination of the effects of such changes on the friction factor and the heat transfer coefficient is a very complicated matter. Here we discuss some of the recommended simple correlations, used in engineering applications, in order to adjust the analytic or experimental correlations for constant-property conditions for the effects of property variation.

For *liquids*, the variation of viscosity is responsible for the property effects. Therefore, viscosity corrections of the following power-law form are found to be sufficiently good approximations:

$$\frac{\text{Nu}}{\text{Nu}_{\text{iso}}} = \left(\frac{\mu_b}{\mu_w}\right)^n \tag{7-75a}$$

$$\frac{f}{f_{\text{iso}}} = \left(\frac{\mu_b}{\mu_w}\right)^k \tag{7-75b}$$

where μ_b = viscosity evaluated at bulk mean temperature

μ_w = viscosity evaluated at wall temperature

Nu_{iso}, Nu = Nusselt number under isothermal and nonisothermal conditions, respectively

f_{iso}, f = friction factor under isothermal and nonisothermal conditions, respectively

In the case of *gases*, the viscosity, thermal conductivity, and density depend on the absolute temperature. Therefore, temperature corrections of the following form are found to be adequate for most practical applications:

$$\frac{\text{Nu}}{\text{Nu}_{\text{iso}}} = \left(\frac{T_b}{T_w}\right)^n \tag{7-76a}$$

$$\frac{f}{f_{\text{iso}}} = \left(\frac{T_b}{T_w}\right)^k \tag{7-76b}$$

where T_b and T_w are the absolute bulk mean and wall temperatures, respectively.

A number of experimental investigations and variable-property analyses have been reported in the literature to determine the values of the exponents n and k appearing in Eqs. (7-75) and (7-76). We present in Table 7-4 recommended values of these exponents. Thus, by using the corrections given by Eqs. (7-75) and (7-76), the Nusselt number and the friction factor for ideal isothermal conditions can be adjusted for the effects of property variations, if no viscosity correction is included in the equation.

Table 7-4 The exponents *n* and *m* associated with Eqs. (7-75) and (7-76)

Type of flow	Fluid	T_w = constant condition	n	k	Reference
Laminar	Liquid	Cooling or heating	0.14		[53]
	Gas	Cooling or heating	0	−1	[3]
Turbulent	Liquid	Cooling	0.25		[61]
	Liquid	Heating	0.11		[61]
	Liquid	Cooling or heating		−0.25	[3]
	Gas	Cooling	0	0.1	[3]
	Gas	Heating	0.5	0.1	[3]

Example 7-9 Water flows with a mean velocity of $u_m = 2$ m/s inside a circular pipe of inside diameter $D = 5$ cm. The pipe is of commercial steel, and its wall is maintained at a uniform temperature $T_w = 100°C$ by condensing steam on its outer surface. At a location where the fluid is hydrodynamically and thermally developed, the bulk mean temperature of water is $T_b = 60°C$. Calculate the heat transfer coefficient h by using the Petukhov equation (7-71).

SOLUTION Various properties for water at $T_b = 60°C$ are taken as

$$\rho = 985 \text{ kg/m}^3 \qquad \mu_b = 4.71 \times 10^{-4} \text{ kg/(m} \cdot \text{s)}$$

$$k = 0.651 \text{ W/(m} \cdot °\text{C)} \qquad \text{Pr} = 3.02$$

and the viscosity at the tube wall temperature $T_w = 100°C$ is

$$\mu_w = 2.82 \times 10^{-4} \text{ kg/(m} \cdot \text{s)}$$

Then

$$\text{Re} = \frac{\rho u_m D}{\mu} = \frac{(985)(2)(0.05)}{4.71 \times 10^{-4}} = 2.04 \times 10^5$$

For the heating problem, we set $n = 0.11$, and the viscosity ratio becomes

$$\left(\frac{\mu_b}{\mu_w}\right)^{0.11} = \left(\frac{4.71}{2.82}\right)^{0.11} = 1.06$$

The relative roughness of the tube wall for commercial steel is

$$\frac{\lambda}{D} = \frac{0.0045}{5} = 0.0009$$

The friction factor f is determined from Fig. 7-11 as

$$f = 0.0205$$

We now apply the Petukhov equation (7-71):

$$\text{Nu} = \frac{(2.04 \times 10^5)(3.02)}{X}\left(\frac{0.0205}{8}\right)(1.06)$$

where $$X = 1.07 + 12.7(3.02^{2/3} - 1)\left(\frac{0.0205}{8}\right)^{1/2}$$

Then $$\text{Nu} = 945.28$$

and $$h = \text{Nu}\,\frac{k}{D} = 945.28\,\frac{0.651}{0.05} = 12{,}307 \text{ W/(m}^2 \cdot °\text{C)}$$

Example 7-10 Solve the problem considered in Example 7-9 for a smooth pipe by using the following correlations:

(*a*) The Notter and Sleicher equation (7-73).

(*b*) The Petukhov equation (7-71).
(*c*) The Sieder and Tate equation (7-70).
(*d*) The Dittus and Boelter equation (7-69).

SOLUTION The physical properties at $T_b = 60°\text{C}$ are taken as

$$k = 0.651\ \text{W/(m}\cdot°\text{C)} \qquad \text{Pr} = 3.02 \qquad \text{Re} = 2.04 \times 10^5$$

$$\mu_b = 4.71 \times 10^{-4}\ \text{kg/(m}\cdot\text{s)} \qquad \mu_w = 2.82 \times 10^{-4}\ \text{kg/(m}\cdot\text{s)}$$

The friction factor for smooth pipe at $\text{Re} = 2.04 \times 10^5$ is obtained from Fig. 7-11 as

$$f = 0.0152$$

(*a*) The Notter and Sleicher Eq. (7-73) gives

$$a = 0.88 - \frac{0.24}{4 + \text{Pr}} = 0.88 - \frac{0.24}{4 + 3.02} = 0.846$$

$$b = 0.33 + 0.5e^{-0.6\,\text{Pr}} = 0.412$$

$$\text{Nu} = 5 + 0.016(2.04 \times 10^5)^{0.846}(3.02)^{0.412}$$

$$= 788$$

$$h = 788\,\frac{0.651}{0.05} = 10{,}267\ \text{W/(m}^2\cdot°\text{C)}$$

(*b*) The Petukhov equation (7-71) gives

$$\text{Nu} = \frac{(2.04 \times 10^5)(3.02)}{X}\left(\frac{0.0152}{8}\right)\left(\frac{4.71}{2.82}\right)^{0.11}$$

where $$X = 1.07 + 12.7(3.02^{2/3} - 1)\left(\frac{0.0152}{8}\right)^{1/2}$$

Then $$\text{Nu} = 741.65$$

$$h = 741.65\,\frac{0.651}{0.05} = 9656\ \text{W/(m}^2\cdot°\text{C)}$$

(*c*) The Sieder and Tate equation (7-70) gives

$$\text{Nu} = 0.027(2.04 \times 10^5)^{0.8}(3.02)^{1/3}\left(\frac{4.71}{2.82}\right)^{0.14}$$

Then $$\text{Nu} = 704$$

$$h = 704\,\frac{0.651}{0.05} = 9166\ \text{W/(m}^2\cdot°\text{C)}$$

(*d*) The Dittus and Boelter equation (7-69) gives

$$\mathrm{Nu} = 0.023(2.04 \times 10^5)^{0.8}(3.02)^{0.4}$$
$$= 633$$

$$h = 633\,\frac{0.651}{0.05} = 8242\ \mathrm{W/(m^2 \cdot {}^\circ C)}$$

7-5 HEAT TRANSFER TO LIQUID METALS

The liquid metals are characterized by their very low Prandtl number, varying from about 0.02 to 0.003. Therefore, the heat transfer correlations in previous sections do not apply to liquid metals, because their range of validity does not extend to such low values of the Prandtl number.

Lithium, sodium, potassium, bismuth, and sodium-potassium are among the common low-melting metals which are suitable for heat transfer purposes as liquid metals. There has been interest in liquid-metal heat transfer in engineering applications, because large amounts of heat can be transferred at high temperatures with a relatively low temperature difference between the fluid and the tube wall surface. The high heat transfer rates result from the high thermal conductivity of liquid metals compared with that of ordinary liquids and gases. Therefore, they are particularly attractive as heat transfer media in nuclear reactors and many other high-temperature, high-heat-flux applications. The major difficulty in their use lies in handling them. They are corrosive, and some may cause violent reactions when they come into contact with water or air.

As discussed in Chap. 6 and schematically illustrated in Fig. 6-4, when $\mathrm{Pr} \ll 1$, as in liquid metals, the thermal boundary layer is much thicker than the velocity boundary layer. This implies that the temperature profile, and hence the heat transfer for liquid metals, is not influenced by the velocity sublayer or viscosity. So in such cases one expects rather weak dependence of heat transfer on the Prandtl number. Thus most empirical correlations of liquid-metal heat transfer have been made by plotting the Nusselt number against the Péclét number, $\mathrm{Pe} = \mathrm{Re} \cdot \mathrm{Pr}$. This situation, discussed earlier with reference to flow along a flat plate, also applies to flow inside a circular tube, as illustrated in Fig. 7-12. There the Nusselt numbers from various sources for the heating of liquid metals in long tubes subjected to constant wall heat flux are compiled by Lubarsky and Kaufman [65] and plotted against the Péclét numbers. The data appear to correlate reasonably well, but there is also the scatter of the data. The reason has been attributed to difficulties inherent in liquid-metal experiments, namely, having to deal with high temperatures and very small temperature differences. The nonwetting of some liquid metals on solid surfaces also has been considered as a possible explanation why some measured values of the Nusselt number are lower than the theoretical predictions. The reader should consult Refs. 66 and 67 for extensive data on liquid-metal heat transfer and the heat transfer characteristics of liquid metals.

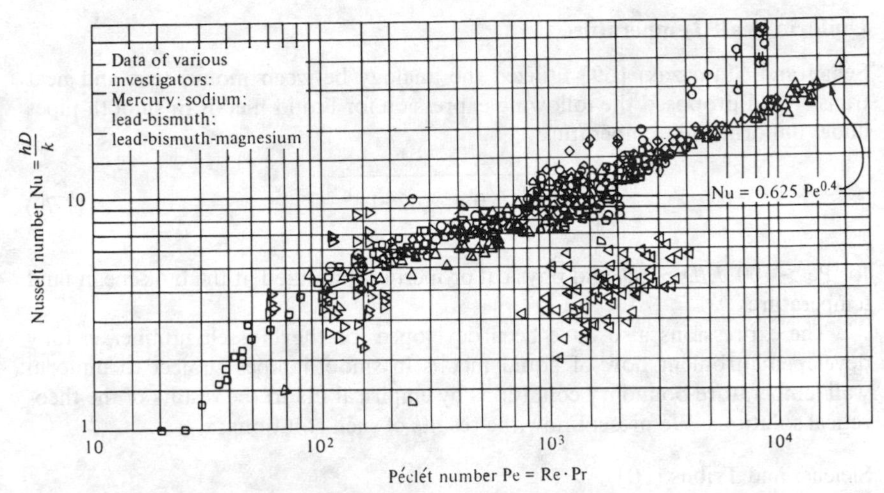

Figure 7-12 Measured Nusselt number for the heating of liquid metals in long, round tubes with constant wall heat flux. (*From Ref. 65.*)

We summarize now some empirical and theoretical correlations for heat transfer to liquid metals in fully developed turbulent flow inside a circular tube under uniform wall heat flux and uniform wall temperature boundary conditions.

Uniform Wall Heat Flux

Lubarsky and Kaufman [65] proposed the following empirical relation for calculating the Nusselt number in fully developed turbulent flow of liquid metals in smooth pipes:

$$\boxed{\mathrm{Nu} = 0.625\,\mathrm{Pe}^{0.4}} \tag{7-77}$$

where

$$\text{Péclét number} \equiv \mathrm{Pe} = \mathrm{Re}\,\mathrm{Pr}$$

for $10^2 < \mathrm{Pe} < 10^4$, $L/D > 60$, and properties evaluated at the bulk mean fluid temperature.

Skupinski, Tortel, and Vautrey [68], based on heat transfer experiments with sodium-potassium mixtures, recommended the following expression for liquid metals in fully developed turbulent flow in smooth pipes:

$$\boxed{\mathrm{Nu} = 4.82 + 0.0185\,\mathrm{Pe}^{0.827}} \tag{7-78}$$

for $3.6 \times 10^3 < \mathrm{Re} < 9.05 \times 10^5$, $10^2 < \mathrm{Pe} < 10^4$, and $L/D > 60$. The physical properties are evaluated at the bulk mean fluid temperature.

Equation (7-77) predicts the Nusselt number lower than Eq. (7-78); therefore it is on the conservative side.

Uniform Wall Temperature

Seban and Shimazaki [69] utilized the analogy between momentum and heat transfer and proposed the following expression for liquid metals in smooth pipes under uniform wall temperature:

$$\boxed{\text{Nu} = 5.0 + 0.025\,\text{Pe}^{0.8}} \tag{7-79}$$

for Pe > 100, $L/D > 60$, and physical properties evaluated at the bulk mean fluid temperature.

The expressions also have been developed for the Nusselt number in fully developed turbulent flow of liquid metals in smooth pipes subject to uniform wall temperature boundary conditions by empirical fits to the results of the theoretical solutions. We present now the results of such solutions:

Sleicher and Tribus [70]:

$$\boxed{\text{Nu} = 4.8 + 0.015\,\text{Pe}^{0.91}\,\text{Pr}^{0.30} \qquad \text{for Pr} < 0.05} \tag{7-80}$$

Azer and Chao [71]:

$$\boxed{\text{Nu} = 5.0 + 0.05\,\text{Pe}^{0.77}\,\text{Pr}^{0.25} \qquad \text{for Pr} < 0.1,\ \text{Pe} < 15{,}000} \tag{7-81}$$

Notter and Sleicher [63]:

$$\boxed{\text{Nu} = 4.8 + 0.0156\,\text{Pe}^{0.85}\,\text{Pr}^{0.08} \qquad \text{for } 0.004 < \text{Pr} < 0.1,\ \text{Re} < 500{,}000} \tag{7-82}$$

The physical properties in calculating Nu, Pe, and Pr in these expressions are evaluated at the bulk mean fluid temperature; the expressions are applicable for $L/D > 60$.

Figure 7-13 shows a comparison of the Nusselt number under uniform tube wall temperature determined from Eqs. (7-79) to (7-82) with the experimental data of Sleicher, Awad, and Notter [72] for heat transfer to NaK in turbulent flow inside a circular tube with uniform wall temperature. It appears that Eq. (7-82) agrees well with the experimental data.

Thermal Entry Region

The previous relations for liquid metals in turbulent flow are applicable in the fully developed region. Sleicher, Awad, and Notter [72] examined the heat transfer calculations of Notter and Sleicher [63] in the thermal entry region for both uniform wall heat flux and uniform wall temperature. They noted that the local

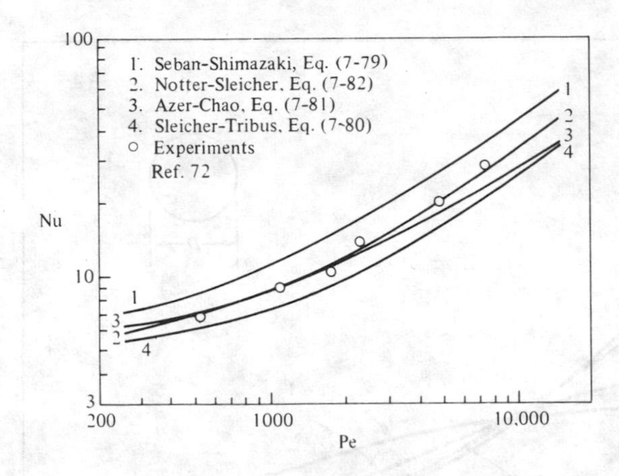

Figure 7-13 Comparison of Nusselt number from various correlations and the experimental results for heat transfer to NaK in turbulent flow inside a circular tube with uniform wall temperature. (*From Sleicher et al.* [72].)

Nusselt number for the thermal entrance region can be correlated within 20 percent with

$$\boxed{\mathrm{Nu}_x = \mathrm{Nu}\left(1 + \frac{2}{x/D}\right) \qquad \text{for } \frac{x}{D} > 4} \tag{7-83}$$

where

$$\mathrm{Nu} = 6.3 + 0.0167\,\mathrm{Pe}^{0.85}\,\mathrm{Pr}^{0.08} \qquad \text{for uniform wall heat flux} \tag{7-84}$$

$$\mathrm{Nu} = 4.8 + 0.0156\,\mathrm{Pe}^{0.85}\,\mathrm{Pr}^{0.08} \qquad \text{for uniform wall temperature} \tag{7-85}$$

and applies in the range $0.004 < \mathrm{Pr} < 0.1$.

Effects of Axial Heat Conduction in Liquid Metals

Liquid metals have very high thermal conductivity. Therefore, in the thermal entrance region where the temperature gradient in the axial direction is high, the heat conduction in the fluid in the axial direction may become important. In general, the effect of axial conduction in the fluid is negligible for $\mathrm{Pe} > 50$; this condition implies that the axial heat conduction in liquid metals may become important for laminar flow.

Heat transfer in laminar forced convection in circular tubes with axial heat conduction in the fluid has been investigated theoretically [73–80]. Figure 7-14 shows the effects of axial heat conduction in the fluid on the local Nusselt number Nu_x at the thermal entrance region for laminar flow inside a circular tube subject to uniform wall heat flux in the downstream region (i.e., $x \geq 0$) while the upstream region (i.e., $x < 0$) is insulated. The axial heat conduction becomes important for $\mathrm{Pe} < 50$, and its effect is to reduce the local Nusselt number in the thermal entrance region; but the length of the thermal entrance region is very short (i.e., about 1 diameter).

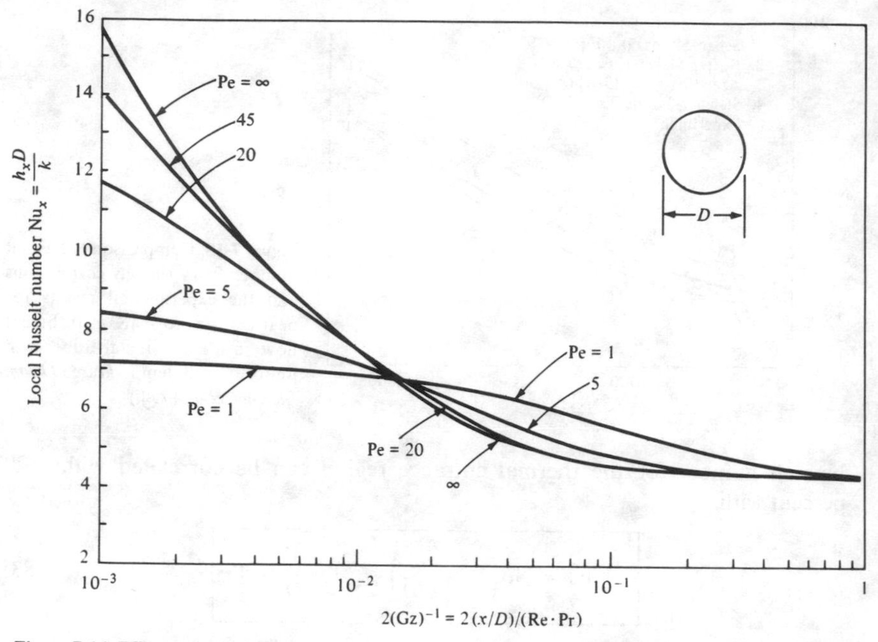

Figure 7-14 Effects of axial heat conduction in fluid on the local Nusselt number for laminar flow in a circular tube with uniform wall heat flux. (*From Vick and Özişik* [80].)

Example 7-11 Liquid NaK (56 percent Na) flows with a mean velocity of $u_m = 3$ m/s inside a smooth tube of inside diameter $D = 2.5$ cm and is heated by the tube wall maintained at a uniform temperature $T_w = 120°\text{C}$. Determine the heat transfer coefficient at a location where the bulk mean fluid temperature is $T_b = 95°\text{C}$ and the flow is fully developed by using Eqs. (7-79), (7-80), and (7-82). Compare the results.

SOLUTION The physical properties of NaK (56 percent Na) are taken as

$$\rho = 887 \text{ kg/m}^3 \qquad \mu = 0.58 \times 10^{-3} \text{ kg/(m} \cdot \text{s)}$$

$$k = 25.6 \text{ W/(m} \cdot °\text{C)} \qquad \text{Pr} = 0.026$$

Then

$$\text{Re} = \frac{\rho u_m D}{\mu} = \frac{(887)(3)(0.025)}{0.58 \times 10^{-3}} = 115{,}000$$

and

$$\text{Pe} = \text{Re Pr} = (115{,}000)(0.026) = 2990$$

Hence Eqs. (7-79), (7-80), and (7-82) are applicable.

Using Eq. (7-79), we find

$$\begin{aligned}\mathrm{Nu} &= 5.0 + 0.025\ \mathrm{Pe}^{0.8} \\ &= 5.0 + 0.025(2990)^{0.8} = 20.1\end{aligned}$$

$$h = \mathrm{Nu}\frac{k}{D} = 20.1\frac{25.6}{0.025} = 20{,}582\ \mathrm{W/(m^2 \cdot {}^\circ C)}$$

Using Eq. (7-80) yields

$$\begin{aligned}\mathrm{Nu} &= 4.8 + 0.015\ \mathrm{Pe}^{0.91}\ \mathrm{Pr}^{0.30} \\ &= 4.8 + 0.015(2990)^{0.91}(0.026)^{0.30} = 12.1\end{aligned}$$

$$h = \mathrm{Nu}\frac{k}{D} = 12.1\frac{25.6}{0.025} = 12{,}877\ \mathrm{W/(m^2 \cdot {}^\circ C)}$$

Using Eq. (7-82), we find

$$\begin{aligned}\mathrm{Nu} &= 4.8 + 0.0156\ \mathrm{Pe}^{0.85}\ \mathrm{Pr}^{0.08} \\ &= 4.8 + 0.0156(2990)^{0.85}(0.026)^{0.08} = 15.3\end{aligned}$$

$$h = \mathrm{Nu}\frac{k}{D} = 15.3\frac{25.6}{0.025} = 15{,}667\ \mathrm{W/(m^2 \cdot {}^\circ C)}$$

Equation (7-82) gives the heat transfer coefficient somewhere between those obtained from Eqs. (7-79) and (7-80). A comparison of these three equations with the experimental results for NaK, given in Fig. 7-13, shows that the result from Eq. (7-82) is closer to the experimental data. So, of the three results calculated, the one from Eq. (7-82) is preferred.

Example 7-12 By using the heat transfer coefficient for the fully developed flow determined as case 3 in Example 7-11, calculate the heat transfer coefficient for the thermal entry region at locations 5 and 10 diameters from the inlet.

SOLUTION Equation (7-83) can be used to determine the thermal entrance h_x:

$$h_x = \left(1 + \frac{2}{x/D}\right)h$$

where h for the fully developed region is determined in case 3, of Example 7-11, as

$$h = 15{,}667\ \mathrm{W/(m^2 \cdot {}^\circ C)}$$

Then

$$\text{At } \frac{x}{D} = 5: \qquad h_x = (1 + \tfrac{2}{5})(15{,}667) = 21{,}934 \text{ W/(m}^2 \cdot {}^\circ\text{C)}$$

$$\text{At } \frac{x}{D} = 10: \qquad h_x = (1 + \tfrac{2}{10})(15{,}667) = 18{,}800 \text{ W/(m}^2 \cdot {}^\circ\text{C)}$$

7-6 ANALOGIES BETWEEN HEAT AND MOMENTUM TRANSFER IN TURBULENT FLOW

In Chap. 6, based on the analysis of laminar flow along a flat plate, we developed a relation between the local heat transfer coefficient h_x and the local drag coefficient c_x, given by Eq. (6-15*a*) as

$$\boxed{\text{St}_x \cdot \text{Pr}^{2/3} = \frac{c_x}{2}} \tag{7-86a}$$

where

$$\text{St}_x = \frac{\text{Nu}_x \, \text{Re}_x}{\text{Pr}} = \frac{h_x}{\rho c_p u_\infty} \tag{7-86b}$$

For turbulent flow inside a circular tube a similar expression is given by Eq. (7-66):

$$\boxed{\text{St} \cdot \text{Pr}^{2/3} = \frac{f}{8}} \tag{7-87a}$$

where

$$\text{St} = \frac{\text{Nu} \cdot \text{Re}}{\text{Pr}} = \frac{h}{\rho c_p u_m} \tag{7-87b}$$

Equations (7-86) and (7-87) are the relations between heat and momentum transfer for turbulent flow along a flat plate and a circular tube, respectively. Given the drag coefficient or the friction factor, the heat transfer coefficient can be found from these equations. It is instructive to examine how equivalent expressions can be developed for turbulent flow along a flat plate or inside a circular tube, by making use of the analogies between heat and momentum transfer. Here we consider only the flow inside a circular tube.

The first and simplest such relation for turbulent flow was developed by Reynolds and is now known as the *Reynolds analogy* for momentum and heat transfer. More refined analogies were developed later by Prandtl [81], von Kármán [82], Martinelli [83], Deissler [60, 84], and many others. The development of all these analogies is based on the solution, with various degrees of approximation, of the relations for the shear stress and the heat flow in turbulent flow given by

Eqs. (6-32). If we assume fully developed turbulent flow, hence $u \equiv u(y)$ and $T \equiv T(y)$, Eqs. (6-32) become

$$\frac{\tau}{\rho} = (\nu + \varepsilon_m)\frac{du}{dy} \tag{7-88}$$

$$\frac{q}{\rho c_p} = -(\alpha + \varepsilon_h)\frac{dT}{dy} \tag{7-89}$$

where y is the distance measured from the tube wall.

To provide some insight into the nature of the approximations involved in the development of such analogies, we now derive the Reynolds analogy.

Reynolds Analogy

Reynolds assumed that the entire flow field consisted of a *single zone* of highly turbulent region. That is, he neglected the presence of the viscous sublayer and the buffer layer. In such a turbulent core, the molecular diffusivities of heat α and of momentum ν are negligible in comparison with turbulent diffusivities. That is,

$$\nu \ll \varepsilon_m \quad \text{and} \quad \alpha \ll \varepsilon_h \tag{7-90a}$$

In addition, he assumed that the turbulent diffusivities are equal:

$$\varepsilon_m = \varepsilon_h \equiv \varepsilon \tag{7-90b}$$

With the assumptions in Eqs. (7-90), the expressions given by Eqs. (7-88) and (7-89) simplify, respectively, to

$$\frac{\tau}{\rho} = \varepsilon\frac{du}{dy} \tag{7-91}$$

$$\frac{q}{\rho c_p} = -\varepsilon\frac{dT}{dy} \tag{7-92}$$

By combining these two equations, we find

$$dT = -\frac{q}{\tau c_p}du \tag{7-93}$$

The integration of Eq. (7-93) from the wall conditions $T = T_w$, $u = 0$ to the mean bulk stream conditions $T = T_m$, $u = u_m$ with the assumption that q/τ remains constant results in

$$\int_{T_w}^{T_m} dT = -\frac{q}{\tau c_p}\int_0^{u_m} du$$

or

$$T_w - T_m = \frac{q u_m}{\tau c_p} \tag{7-94}$$

Now, the heat transfer coefficient h and the friction factor f for flow inside a tube are defined as

$$q = h(T_w - T_m) \tag{7-95a}$$

$$\tau = f\frac{\rho u_m^2}{8} \qquad \text{[see Eq. (7-55)]} \tag{7-95b}$$

The substitution of Eqs. (7-95) into (7-94) yields

$$\boxed{\text{St} \equiv \frac{h}{\rho c_p u_m} = \frac{f}{8}} \tag{7-96}$$

This result is known as the Reynolds analogy for momentum and heat transfer in fully developed turbulent flow in a pipe. It is valid for $\text{Pr} \cong 1$.

Prandtl Analogy

Prandtl assumed that the flow field consisted of *two layers*, a viscous sublayer where the molecular diffusivities are dominant, that is,

$$\varepsilon_m \ll \nu \qquad \text{and} \qquad \varepsilon_h \ll \alpha \tag{7-97}$$

and a turbulent zone where the turbulent diffusivities are dominant, that is

$$\nu \ll \varepsilon_m \qquad \alpha \ll \varepsilon_h \qquad \text{and} \qquad \varepsilon_m = \varepsilon_h \equiv \varepsilon \tag{7-98}$$

These assumptions are utilized to simplify Eqs. (7-88) and (7-89) for each layer, the equations are integrated, and the definitions of the friction factor and the heat transfer coefficient are introduced. The following result is obtained:

$$\boxed{\text{St} \equiv \frac{h}{\rho c_p u_m} = \frac{f}{8}\frac{1}{1 + 5\sqrt{f/8}\,(\text{Pr} - 1)}} \tag{7-99}$$

This relationship is known as the Prandtl analogy for momentum and heat transfer for fully developed turbulent flow in a pipe. We note that for $\text{Pr} = 1$ the Prandtl analogy reduces to the Reynolds analogy.

Von Kármán Analogy

Von Kármán extended Prandtl's analogy by separating the flow field into *three distinct layers*: a viscous sublayer, a buffer layer, and a turbulent core. He made assumptions about the relative magnitudes of the molecular and turbulent diffusivities of heat and momentum in the viscous sublayer and the turbulent core similar to those made by Prandtl, but in addition he included the effects of the buffer layer by assuming that the molecular and eddy diffusivities in this layer were of the same order of magnitude. The following result is obtained:

$$\text{St} \equiv \frac{h}{\rho c_p u_m} = \frac{f}{8} \frac{1}{1 + 5\sqrt{f/8}\{(\text{Pr} - 1) + \ln[(5\,\text{Pr} + 1)/6]\}} \tag{7-100}$$

which is known as the von Kármán analogy for momentum and heat transfer for fully developed turbulent flow in a pipe. We note that for $\text{Pr} = 1$ this result also reduces to the Reynolds analogy. This relation appears to be good for a Prandtl number up to about 30.

The previous relations are given in terms of the Stanton number. They also can be expressed in terms of the Nusselt number by noting that

$$\text{St} = \frac{\text{Nu}}{\text{Re} \cdot \text{Pr}} \tag{7-101}$$

where $\text{Nu} = hD/k$ and $\text{Re} = u_m D/\nu$. The effect of the property variation can be included by utilizing Eq. (7-75), that is,

$$\text{Nu} = \text{Nu}_{\text{iso}} \left(\frac{\mu_b}{\mu_w} \right)^n \tag{7-102}$$

With these considerations, we write *the von Kármán analogy in the alternative form as*

$$\text{Nu} = \frac{\text{Re} \cdot \text{Pr} \cdot f/8}{Z} \left(\frac{\mu_b}{\mu_w} \right)^n \tag{7-103a}$$

where

$$Z = 1 + 5\left[(\text{Pr} - 1) + \ln\left(\frac{5\,\text{Pr} + 1}{6} \right) \right] \left(\frac{f}{8} \right)^{1/2} \tag{7-103b}$$

and the exponent n may be determined from Table 7-4.

Note that the von Kármán analogy, expressed in the alternative form Eq. (7-103), resembles the Petukhov equation (7-71) except for the definition of the Z term in the denominator. However, Eq. (7-71) should be preferred to predict the heat transfer coefficient, because the denominator has been adjusted to closely correlate with the experimental data over a wide range of the parameters. For example, for $\text{Pr} = 1$, in the von Kármán analogy $Z = 1$ and the corresponding term in the Petukhov equation becomes $X = 1.07$, which is about a 7 percent difference.

Example 7-13 Solve the problem considered in Example 7-9 by using the alternative form of the von Kármán analogy, Eqs. (7-103). Compare the resulting heat transfer coefficient h with that obtained by the Petukhov equation (7-71).

SOLUTION The physical properties for water from Example 7-9 are:

$$k = 0.651\ \text{W/(m}^2 \cdot {}^\circ\text{C)} \qquad \text{Re} = 2.04 \times 10^5 \qquad \text{Pr} = 3.02$$

$$\mu_b = 4.71 \times 10^{-4}\ \text{kg/(m} \cdot \text{s)} \qquad \mu_w = 2.82 \times 10^{-4}\ \text{kg/s}$$

and the friction factor is

$$f = 0.0205$$

For the heating problem we set $n = 0.11$. We now apply Eq. (7-103):

$$\text{Nu} = \frac{\text{Re} \cdot \text{Pr}}{Z}\left(\frac{f}{8}\right)\left(\frac{\mu_b}{\mu_w}\right)^n$$

$$= \frac{(2.04 \times 10^5)(3.02)}{Z}\left(\frac{0.0205}{8}\right)\left(\frac{4.71}{2.82}\right)^{0.11} = \frac{1763.42}{Z}$$

where

$$Z = 1 + 5\left[(\text{Pr} - 1) + \ln\left(\frac{5\ \text{Pr} + 1}{6}\right)\right]\left(\frac{f}{8}\right)^{1/2}$$

$$= 1 + 5\left[(3.02 - 1) + \ln\left(\frac{5 \times 3.02 + 1}{6}\right)\right]\left(\frac{0.0205}{8}\right)^{1/2} = 1.761$$

Then $\text{Nu} = 950.27$

and $h = 950.27\,\dfrac{0.651}{0.05} = 12{,}372\ \text{W/(m}^2 \cdot {}^\circ\text{C)}$

This result is very close to 12,307 given by the Petukhov equation (7-71) in Example 7-9. However, the deviation may increase at larger Prandtl numbers.

7-7 HEAT TRANSFER AUGMENTATION

In recent years, energy and material saving considerations have prompted an expansion of the efforts aimed at producing more efficient heat exchange equipment through the augmentation of heat transfer. The potentials of heat transfer augmentation in engineering applications are numerous. For example, the heat exchanger for a projected ocean thermal energy conversion (OTEC) plant [85] requires a heat transfer surface area on the order of 10,000 $\text{m}^2/\text{MW(e)}$. Clearly, an increase in the efficiency of the heat exchanger through augmentation may result in considerable savings in the material need.

Desalinization is another application in which large heat transfer surfaces are required and possibilities exist for the use of augmented systems [86, 87]. Numerous other examples can be cited for the application of heat transfer augmentation.

A vast amount of literature exists on this subject, and a comprehensive survey on heat transfer augmentation is available [88–90]. Here we briefly discuss principal augmentation techniques for single-phase forced flow in ducts, and we cite the pertinent references that present data on heat transfer and friction factor characteristics of various augmentation systems.

Augmentation Techniques

We recall from our discussion of the analogies between momentum and heat transfer that increasing the friction factor increases the heat transfer coefficient. The Moody chart (Fig. 7-11), for example, shows that in turbulent flow increasing the relative roughness of the surface increases the friction factor. This chart is based on the random sand-grain type of surface roughness. Other types of surface roughness have been produced, and their friction factors and heat transfer characteristics have been tested for possible use in heat transfer augmentation. We discuss some of these enhancement techniques.

Roughened surfaces Surface roughnesses can be produced by the machining of the surface as well as by casting, forming, and welding processes. Clearly, an infinite number of geometric configurations are possible, each having its own heat transfer and pressure-drop characteristics. Experimental techniques generally are used to determine the heat transfer coefficient and friction factor for flow, since no unified analysis is yet available for predicting them by purely theoretical means. Bergles and Jensen [91] and Webb, Eckert, and Goldstein [92, 93] presented heat transfer and friction factor data for various surface roughnesses.

Extended surfaces The use of fins on the outer surface of tubes to enhance heat transfer is well known. Internally finned tubes have been used also to enhance heat transfer to fluids flowing inside tubes. Heat transfer and friction factor correlations have been presented for internally finned tubes under laminar flow [94] and turbulent flow [95–97] conditions. A theoretical approach has been applied to predict the effects of internal fins on the turbulent flow of air [98].

Enhancement devices Enhancement devices such as twisted tapes have been employed in the form of inserts into the tubes, to promote increased heat transfer for the laminar and turbulent flow of viscous fluids [99, 100].

Coiled tubes Coiled tubes can serve as a heat transfer enhancement device because the secondary flow produced by the curvature causes an increase in the heat transfer coefficient. Effects of curvature in enhancing heat transfer in curved pipes have been reported [101–103].

Optimization Methods

The increase in heat transfer with augmentation is accompanied by an increase in the friction factor. For example, in some situations the heat transfer coefficients are increased at most about 4 times while the friction factors are increased as much as 50 times or more [91]. An increased friction factor implies an increased power for pumping the fluid. So the results of augmentation, when it is applied to produce more efficient heat exchange equipment, should be weighed against the increased power requirement for pumping the fluid.

Studies have been conducted to develop methodologies for assessing heat transfer, pumping power, and surface area for a system resulting from enhancement versus those for the unenhanced system [104–108]. For a given enhancement technique, if the heat transfer and the friction factor data are available as a function of the Reynolds number, it may be possible to optimize the system to reduce the heat transfer surface, to obtain increased heat transfer capacity, or to reduce the power required for pumping the fluid.

No unified correlation of heat transfer and friction factor characteristics of enhancement techniques is yet available. Each specific enhancement method has its own correlation. For this reason, we do not present here various correlations for specific cases; instead, pertinent references have been cited. However, we give now an example, to illustrate a typical data set on heat transfer augmentation and its use in the optimization processes as a means to reduce the heat transfer surface area for a heat exchanger. In an actual optimization study, numerous other questions arise in connection with the area reduction resulting from heat transfer enhancement. For example, the manufacturing procedure, manufacturing cost, material, and many other factors associated with the augmentation technique used should be considered before a final decision is made.

Example 7-14 Heat transfer augmentation is applied on the inner surface of a circular tube subjected to uniform wall temperature by condensing vapor on the outer surface. Figure 7-15 shows heat transfer and friction factor data for different types of enhancement on the inner surface of the tube for experi-

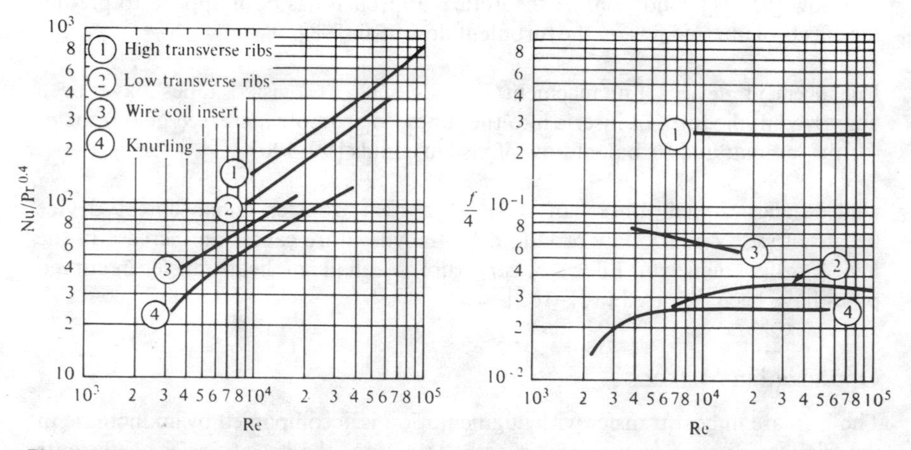

Figure 7-15 Heat transfer and friction factor data for water flowing in internally roughened tubes. (*From Ref. 91.*)

ments performed with water flow. Examine the effectiveness of each enhancement method in reducing the heat transfer surface area while the total heat transfer rate, the pumping power requirement, and the tube length remain the same as for the unaugmented, smooth tubes.

SOLUTION This example is based on the optimization study [91] performed in connection with the inside augmentation of a proposed shell-and-tube heat exchanger for an ocean thermal energy conversion power plant, described in Ref. 17. The total heat transfer rate Q is considered to remain the same for both the augmented and the reference (unaugmented) systems.

As a first approximation, we write

$$Q = A_0 h_0 \, \Delta T = A_a h_a \, \Delta T$$

where the subscripts 0 and a refer, respectively, to the reference and the augmented cases. Here we assume that the temperature difference ΔT does not change for the augmented and the reference systems. And we use h instead of the overall heat transfer coefficient u because our analysis is an approximate one. The area ratio is obtained from the above heat transfer relation and rearranged as

$$\frac{A_a}{A_0} = \frac{h_0}{h_a} = \frac{\text{Nu}_0/\text{Pr}^{0.4}}{\text{Nu}_a/\text{Pr}^{0.4}} = \frac{0.023 \, \text{Re}_0^{0.8}}{\text{Nu}_a/\text{Pr}^{0.4}} \tag{a}$$

where the correlation $\text{Nu}_0 = 0.023 \, \text{Re}^{0.8} \, \text{Pr}^{0.4}$ is used for the reference case.

The pumping power requirement W for both systems is the same and can be expressed as

$$W = cf_0 A_0 u_0^3 = cf_a A_a u_a^3$$

since the constant c is defined as

$$W = (\text{flow rate})(\text{pressure drop})$$

$$= \left(nu \frac{\pi}{4} D^2 \right) \left(f \frac{L}{D} \frac{\rho u^2}{2} \right) \equiv cfAu^3$$

where A is the surface area, u is the flow velocity, n is the number of tubes in the bundle, and f is the friction factor.

The area ratio is now obtained from the above pumping power relationship as

$$\frac{A_a}{A_0} = \frac{f_0 u_0^3}{f_a u_a^3} = \frac{f_0 \, \text{Re}_0^3}{f_a \, \text{Re}_a^3} = \frac{0.184 \, \text{Re}_0^{2.8}}{f_a \, \text{Re}_a^3} \tag{b}$$

where the correlation $f_0 = 0.184 \, \text{Re}_0^{-0.2}$ is used for the smooth tube.

From Eqs. (a) and (b) we obtain

$$\frac{0.023 \, \text{Re}_0^{0.8}}{\text{Nu}_a/\text{Pr}^{0.4}} = \frac{0.184 \, \text{Re}_0^{2.8}}{f_a \, \text{Re}_a^3}$$

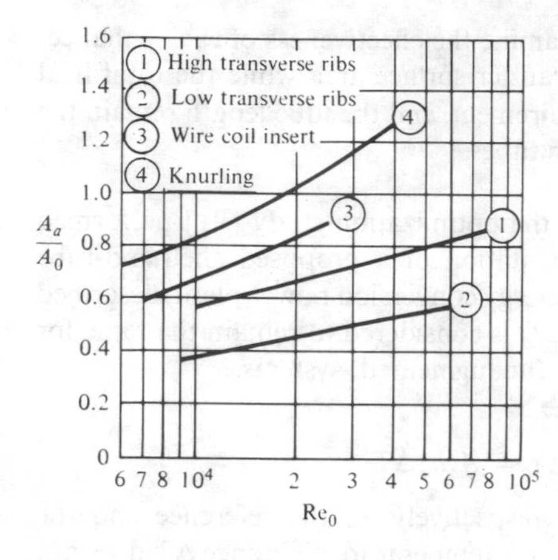

Figure 7-16 Ratio of the heat transfer surface area of augmented heat exchanger to the reference one. (*Derived from data given in Fig. 7-15.*)

or

$$\mathrm{Re}_0 = 0.3535\left(\frac{f_a \mathrm{Re}_a^3}{\mathrm{Nu}_a/\mathrm{Pr}^{0.4}}\right)^{1/2} \tag{c}$$

Now the calculation procedure is as follows:

1. Choose Re_a.
2. Determine from Fig. 7-15 the corresponding values of $\mathrm{Nu}_a/\mathrm{Pr}^{0.4}$ and f_a; then calculate from Eq. (*c*) the corresponding value of Re_0.
3. Given Re_0, calculate the area ratio A_a/A_0 from Eq. (*a*).

Figure 7-16 shows the area ratio A_a/A_0 plotted against Re_0 for each of the various enhancement techniques. The results suggest that the enhancement with the low transverse ribs is most efficient, and such tubes are also acceptable with manufacturing considerations.

For more accurate results, the calculations should be performed with variable temperature differences and by using the overall heat transfer coefficient instead of the h values. The reduction in the heat transfer surface area obtained with such calculations will be less than that shown in Fig. 7-16. However, the area reduction—even with the inclusion of outside resistance for condensation and the tube resistance—may amount to about 30 percent. This is a considerable savings when large surface areas are involved.

7-8 SUMMARY OF CORRELATIONS

We summarize in Table 7-5 the correlations of the heat transfer coefficient and the friction factor for forced convection inside ducts presented in this chapter.

Table 7-5 Summary of correlations for forced convection inside ducts*

Equation or table number	Correlation	Flow regime	Remarks
(7-26)	$f = \dfrac{64}{\mathrm{Re}}$	Laminar	Fully developed flow in a circular tube
(7-41)	$\mathrm{Nu} = 48/11$	Laminar	Fully developed flow in a circular tube, constant wall heat flux
(7-42)	$\mathrm{Nu} = 3.66$	Laminar	Fully developed flow in a circular tube, constant wall temperature
Table 7-1	Nu and f	Laminar	Fully developed flow in ducts of various cross section
Table 7-2	L_h and L_t	Laminar	Hydrodynamic and thermal entry lengths for ducts of various cross sections
(7-49)	$\mathrm{Nu}_m = 3.66 + \dfrac{0.0668\ \mathrm{Gz}}{1 + 0.04(\mathrm{Gz})^{2/3}}$	Laminar	Entrance region, circular tube, constant wall temperature
(7-51)	$\mathrm{Nu}_m = 1.86(\mathrm{Gz})^{1/3}\left(\dfrac{\mu_b}{\mu_w}\right)^{0.14}$	Laminar	Entrance region, circular tube, constant wall temperature, $0.48 < \mathrm{Pr} < 16{,}700$ and $(\mathrm{Gz})^{1/3}\left(\dfrac{\mu_b}{\mu_w}\right)^{0.14} > 2$
(7-61b)	$f = (1.82 \log \mathrm{Re} - 1.64)^{-2}$	Turbulent	Smooth tubes, $\mathrm{Re} > 10^4$
(7-62)	$f = 0.316\ \mathrm{Re}^{-0.25}$	Turbulent	Smooth tubes, $\mathrm{Re} < 2 \times 10^4$
(7-63)	$f = 0.184\ \mathrm{Re}^{-0.2}$	Turbulent	Smooth tubes, $2 \times 10^4 < \mathrm{Re} < 3 \times 10^5$
(7-68)	$\mathrm{Nu} = 0.023\ \mathrm{Re}^{0.8}\ \mathrm{Pr}^{1/3}$	Turbulent	$0.7 < \mathrm{Pr} < 160$; $\mathrm{Re} > 10{,}000$; $L/D > 60$; smooth pipes
(7-69)	$\mathrm{Nu} = 0.023\ \mathrm{Re}^{0.8}\ \mathrm{Pr}^n$ $n = 0.4$ for heating $n = 0.3$ for cooling	Turbulent	$0.7 < \mathrm{Pr} < 160$; $\mathrm{Re} > 10{,}000$; $L/D > 60$; smooth pipes
(7-70)	$\mathrm{Nu} = 0.027\ \mathrm{Re}^{0.8}\ \mathrm{Pr}^{1/3}\left(\dfrac{\mu_b}{\mu_w}\right)^{0.14}$	Turbulent	$0.7 < \mathrm{Pr} < 16{,}700$; $\mathrm{Re} > 10{,}000$; $L/D > 60$; smooth pipes
(7-71)	$\mathrm{Nu} = \dfrac{\mathrm{Re} \cdot \mathrm{Pr}}{X}\left(\dfrac{f}{8}\right)\left(\dfrac{\mu_b}{\mu_w}\right)^n$ where $X = 1.07 + 12.7(\mathrm{Pr}^{2/3} - 1)\left(\dfrac{f}{8}\right)^{1/2}$ $n = 0.11$ heating with uniform T_w $n = 0.2$ cooling with uniform T_w $n = 0$ uniform wall heat flux or gases	Turbulent	Smooth or rough pipes $10^4 < \mathrm{Re} < 5 \times 10^6$ $0.5 < \mathrm{Pr} < 200$ with 5 to 6% error $0.5 < \mathrm{Pr} < 2000$ with 10% error Properties, except μ_w, are evaluated at bulk mean temperature
(7-72)	$\mathrm{Nu} = 0.036\ \mathrm{Re}^{0.8}\ \mathrm{Pr}^{1/3}\left(\dfrac{D}{L}\right)^{0.055}$	Turbulent	$10 < \dfrac{L}{D} < 400$

Table 7-5 (*Continued*)

Equation or table number	Correlation	Flow regime	Remarks
(7-73)	$\mathrm{Nu} = 5 + 0.016\,\mathrm{Re}^a\,\mathrm{Pr}^b$ where $a = 0.88 - \dfrac{0.24}{4 + \mathrm{Pr}}$, $b = 0.33 + 0.5e^{-0.6\mathrm{Pr}}$	Turbulent	$0.1 < \mathrm{Pr} < 10^4$; $10^4 < \mathrm{Re} < 10^6$; $\dfrac{L}{D} > 25$
(7-75*a*, *b*)	$\mathrm{Nu} = \mathrm{Nu}_{\mathrm{iso}}\left(\dfrac{\mu_b}{\mu_w}\right)^n, \quad f = f_{\mathrm{iso}}\left(\dfrac{\mu_b}{\mu_w}\right)^k$	Liquids laminar or turbulent	See Table 7-4 for recommended values of n, k
(7-76*a*, *b*)	$\mathrm{Nu} = \mathrm{Nu}_{\mathrm{iso}}\left(\dfrac{T_b}{T_w}\right)^n, \quad f = f_{\mathrm{iso}}\left(\dfrac{T_b}{T_w}\right)^k$	Gases laminar or turbulent	See Table 7-4 for recommended values of n, k
(7-77)	$\mathrm{Nu} = 0.625\,\mathrm{Pe}^{0.4}$	Turbulent	Liquid metals; uniform wall heat flux; $L/D > 60$; $10^2 < \mathrm{Pe} < 10^4$
(7-78)	$\mathrm{Nu} = 4.82 + 0.0185\,\mathrm{Pe}^{0.827}$	Turbulent	Liquid metals; uniform wall heat flux; $L/D > 60$; $3.6 \times 10^3 < \mathrm{Re} < 9.05 \times 10^5$; $10^2 < \mathrm{Pe} < 10^4$
(7-79)	$\mathrm{Nu} = 5.0 + 0.025\,\mathrm{Pe}^{0.8}$	Turbulent	Liquid metals; uniform wall temperature; $L/D > 60$; $\mathrm{Pe} > 100$
(7-80)	$\mathrm{Nu} = 4.8 + 0.015\,\mathrm{Pe}^{0.91}\,\mathrm{Pr}^{0.30}$	Turbulent	Liquid metals; uniform wall temperature; $L/D > 60$; $\mathrm{Pr} < 0.05$
(7-81)	$\mathrm{Nu} = 5.0 + 0.05\,\mathrm{Pe}^{0.77}\,\mathrm{Pr}^{0.25}$	Turbulent	Liquid metals; uniform wall temperature; $L/D > 60$; $\mathrm{Pr} < 0.1$; $\mathrm{Pe} > 15{,}000$
(7-82)	$\mathrm{Nu} = 4.8 + 0.0156\,\mathrm{Pe}^{0.85}\,\mathrm{Pr}^{0.08}$	Turbulent	Liquid metals; uniform wall temperature; $L/D > 60$; $\mathrm{Re} < 500{,}000$; $0.004 < \mathrm{Pr} < 0.1$
(7-83)	$\mathrm{Nu}_x = \mathrm{Nu}\left(1 + \dfrac{2}{x/D}\right)$ for $x/D > 4$ where Nu is determined from: Eq. (7-84) for uniform wall heat flux; Eq. (7-85) for uniform wall temperature	Turbulent	Liquid metals; entrance region; $0.004 < \mathrm{Pr} < 0.1$

* Unless otherwise stated, fluid properties are evaluated at the bulk mean fluid temperature $T_b = (T_{\mathrm{in}} + T_{\mathrm{out}})/2$. The film temperature T_f is defined as $T_f = (T_w + T_h)/2$. In the viscosity ratio μ_b/μ_w, μ_b is the viscosity at bulk mean fluid temperature and μ_w at the wall temperature. Also $\mathrm{Re} = u_m D_h/\nu$, $\mathrm{Gz} = (\mathrm{Re}\cdot\mathrm{Pr})/(L/D)$, and $\mathrm{Pe} = \mathrm{Re}\,\mathrm{Pr}$.

PROBLEMS

Couette flow

7-1 A lubricating oil of viscosity μ and thermal conductivity k fills the clearance L between two rotating cylinders, which can be regarded as two parallel plates in motion for the purpose of the analysis. Let u_0 and u_1 be the velocities of the inner and outer cylinders, respectively. Develop a relation for the velocity distribution in the oil layer for (*a*) the inner and outer cylinders rotating in the same direction but $u_1 > u_0$ and (*b*) the inner and outer cylinders rotating in opposite directions. Also develop a relation for the shear stress in the fluid resulting from the rotation.

7-2 A lubricating oil of viscosity μ and thermal conductivity k is contained in the clearance L between the journal and the bearing, which can be regarded as two parallel plates. Let u_1 be the velocity of the upper plate while the lower plate is considered stationary. Heat is generated in the oil layer during rotation as a result of viscous-energy dissipation. Develop expressions for the temperature distribution in the fluid for the following cases:

(*a*) The lower plate is maintained at a temperature T_0 and the upper plate at a temperature T_1, with $T_1 > T_0$.

(*b*) The lower plate is kept at a temperature T_0 while the upper plate is insulated. Also develop an expression for the temperature of the insulated surface.

7-3 Develop an expression for the maximum temperature rise in a lubricating oil contained between a journal and its bearing if the velocity of the rotating surface is u_1 m/s and the journal and the bearing are both maintained at the same temperature T_0. Calculate the maximum temperature rise for a rotation velocity of $u_1 = 10$ m/s and $T_0 = 20°C$. The viscosity of the oil is given as $\mu = 0.21$ kg/(m · s) at 40°C, $\mu = 0.45$ at 30°C, and $\mu = 0.8$ at 20°C. The viscosity can be taken at the mean fluid temperature. The thermal conductivity of oil is $k = 0.14$ W/(m · °C).

7-4 Consider Couette flow with heat transfer between two parallel plates separated by a distance L. The spacing between the plates is filled with a fluid of viscosity μ and thermal conductivity k. The lower plate is stationary, and the upper plate is moving with a velocity u_1 and maintained at a temperature T_1, with $T_1 > T_0$ where T_0 is the temperature of the lower plate. Establish the criteria under which there will be no heat transfer in the upper plate. Also develop an expression for heat transfer rate at the lower plate.

7-5 Two large parallel plates separated by a distance $L = 0.2$ cm are maintained at $T = 20°C$. The upper plate is moving with a speed $u_1 = 100$ m/s, while the lower plate is stationary. In one case the fluid between the plates is water; in the other case, it is air. For each fluid determine the following: (*a*) the maximum temperature rise in the fluid, (*b*) the shear stress in the fluid, and (*c*) the heat flux at the plate surface.

Answer: (*a*) 0.93 and 2.1°C; (*b*) 0.957 and 50.33 N/m^2; (*c*) 47.85 and 2516 W/m^2.

7-6 Consider Couette flow between two parallel plates with the lower plate at $y = 0$ stationary and the upper plate at $y = L$ moving with a velocity $u = u_1$. Develop a solution for the velocity distribution $u(y)$ for the case $dP/dx = C \neq 0$.

Hydrodynamically and thermally developed flow inside circular tubes

7-7 Determine the friction factor and the pressure drop for fully developed laminar flow of ethylene glycol at 40°C [$\mu = 0.96 \times 10^{-2}$ kg/(m · s), $\rho = 1101$ kg/m^3] through a 5-cm-diameter, 50-m-long tube at a rate of 0.1 kg/s.

Answers: 0.242; 286 N/m^2.

7-8 Engine oil at 40°C [$\mu = 0.21$ kg/(m · s), $\rho = 875$ kg/m^3] flows inside a 2.5-cm-diameter, 50-m-long tube with a mean velocity of 1 m/s. Determine the pressure drop for flow through the tube.

Answer: 538 kN/m^2.

7-9 Determine the friction factor and the pressure drop for fully developed laminar flow of water at 21°C [$\mu = 9.8 \times 10^{-4}$ kg/(m · s), $\rho = 997.4$ kg/m^3] through a 2.5-cm-diameter, 100-m-long tube at a rate of 0.015 kg/s. What are the mean and the maximum velocities of flow inside the tube?

7-10 Determine the friction factor, the pressure drop, the mean, and the maximum flow velocities for the fully developed laminar flow of mercury at 20°C [$\mu = 1.55 \times 10^{-3}$ kg/(m · s), $\rho = 13{,}579$ kg/m^3] through a 2.5-cm-diameter, 50-m-long tube at a rate of 0.05 kg/s.

Answer: $f = 0.039$; $\Delta P = 29.8$ N/m^2; $u_m = 0.0075$ m/s; $u_{max} = 0.015$ m/s.

7-11 Compare the mean velocities, friction factors, and pressure drops for fully developed laminar flow of Freon at 20°C [$\mu = 2.63 \times 10^{-4}$ kg/(m · s), $\rho = 1330$ kg/m^3], ethylene glycol [$\mu = 2.24 \times 10^{-2}$ kg/(m · s), $\rho = 1117$ kg/m^3], and engine oil [$\mu = 0.8$ kg/(m · s), $\rho = 888$ kg/m^3] through a 2.5-cm-diameter, 100-m-long tube at a rate of 0.01 kg/s.

7-12 Consider hydrodynamically fully developed laminar flow inside a circular tube. Let z be the coordinate measured axially in the direction of flow and r the radial coordinate. By writing a force balance equation for the pressure and the viscous shear forces acting on an annular element of radius r, radial thickness Δr, and axial thickness Δz in the fluid, show that the z momentum equation is

$$\frac{1}{r}\frac{d}{dr}\left[r\frac{du(r)}{dr}\right] = \frac{1}{\mu}\frac{dP}{dz}$$

where μ is the viscosity, dP/dz is the pressure gradient, and $u(r)$ is the axial flow velocity.

7-13 By solving the z momentum equation

$$\frac{1}{r}\frac{d}{dr}\left[r\frac{du(r)}{dr}\right] = \frac{1}{\mu}\frac{dP}{dz}$$

for constant μ and dP/dz, develop an expression for the velocity distribution $u(r)$ for fully developed laminar flow inside a circular tube of radius R. Then utilize this velocity distribution to:

(*a*) Develop a relation for the mean flow velocity u_m.

(*b*) Develop an expression for the friction factor f, defined as

$$f = -\frac{dP/dz}{\frac{1}{2}\rho u_m^2/D}$$

where ρ is the density of fluid and D is the inside diameter of the tube.

(*c*) Determine the ratio u_0/u_m, where u_0 is the centerline velocity and u_m is the mean flow velocity.

7-14 By solving the z momentum equation for constant μ and dP/dz, develop an expression for the velocity distribution in fully developed laminar flow inside a circular annulus of inside radius r_1 and outside radius r_2.

Hydrodynamically and thermally developed flow inside noncircular ducts

7-15 Determine the friction factor for flow of ethylene glycol at 20°C ($\nu = 19.2 \times 10^{-6}$ m^2/s) through a parallel-plate channel having a spacing of $L = 5$ cm between the plates with a velocity of $u_m = 0.25$ m/s. Assume the flow is laminar and fully developed.

Answer: 0.0737

7-16 Consider the fully developed laminar flow of a viscous fluid through a parallel-plate channel of spacing L and a circular tube of diameter D. Determine the ratio L/D for which the friction factors for flow between a parallel-plate channel and circular tube are equal.

7-17 Compare the friction factors associated with the fully developed laminar flow of engine oil at 40°C ($\nu = 0.84 \times 10^{-4}$ m^2/s) through a parallel-plate channel of spacing $L = 1.5$ cm and a circular tube of diameter $D = 2$ cm with a velocity of 0.5 m/s.

7-18 Consider hydrodynamically fully developed laminar flow between two parallel plates. Let x be the coordinate measured axially in the direction of flow and y be the coordinates normal to the walls. By writing a force balance equation for the pressure and the viscous shear forces acting on a differential volume element of sides Δx and Δy and length unity in the fluid, show that the x momentum equation is

$$\frac{d^2u(y)}{dy^2} = \frac{1}{\mu}\frac{dP}{dx}$$

where μ is the viscosity, dP/dx is the pressure gradient in the flow, and $u(y)$ is the axial flow velocity.

7-19 Consider fully developed laminar flow between two parallel plates separated by a distance $2L$. Let the $0x$ axis be taken along the center plane in the direction of flow and the $0y$ axis be normal to the walls. The x momentum equation is given by

$$\frac{d^2u(y)}{dy^2} = \frac{1}{\mu}\frac{dP(x)}{dx}$$

(*a*) By solving this momentum equation for constant μ and $dP(x)/dx$, develop a relation for the fully developed velocity profile $u(y)$.

(*b*) Develop a relation for the mean velocity u_m.

(*c*) If the friction factor is defined by

$$f = -\frac{dP/dx}{\frac{1}{2}\rho u_m^2/D_h}$$

where ρ is the density and $D_h = 4L$ the hydraulic diameter, develop a relation for the friction factor f.

(*d*) Determine the ratio u_0/u_m, where u_0 is the flow velocity at the midplane of the channel and u_m is the mean flow velocity.

7-20 Consider the flow of water at a rate of 0.015 kg/s through a square duct 2 cm by 2 cm whose walls are maintained at a uniform temperature 100°C. Assume that the flow is hydrodynamically and thermally developed. Determine the duct length required to heat the water from 30 to 70°C.

Answer: $L \cong 7$ m

7-21 Consider the flow of water at a rate of 0.01 kg/s through an equilateral triangular duct of sides 2 cm and whose walls are kept at a uniform temperature 100°C. Assume that the flow is hydrodynamically and thermally developed. Determine the duct length required to heat the water from 20 to 70°C.

Answer: 5 m

7-22 Engine oil at 50°C with a flow rate of 10^{-3} kg/s enters a 1-m-long, equilateral triangular duct of sides 0.5 cm whose walls are maintained at a uniform temperature 120°C. Assuming hydrodynamically and thermally fully developed flow, determine the outlet temperature of the oil.

7-23 Oil at 50°C enters a 4-m-long conduit whose cross section can be approximated as two parallel plates with a spacing of 1 cm. The flow rate of oil is 5 kg/(m^2 · s). The walls are subjected to a uniform heating at rate of 1000 W/m^2. Assuming hydrodynamically and thermally fully developed flow, determine the average heat transfer coefficient, the temperature rise of the oil for flow through the conduit, and the wall temperature. [Properties of oil may be taken at the anticipated mean temperature of 80°C as $\mu = 0.032$ kg/(m · s), $c_p = 2130$ W · s/(kg · °C), $k = 0.14$ W/(m · °C), $\rho = 850$ kg/m^3, and Pr = 487.]

7-24 Consider hydrodynamically developed laminar flow of velocity $u(y)$ in the x direction between two parallel plates. Let y be the coordinate axis normal to the plates. Consider a differential volume element $\Delta x\ \Delta y$ 1. By writing an energy balance equation for heat conducted in the y direction and heat convected in the x direction for this volume element, show that the energy equation is

$$\frac{1}{\alpha}u(y)\frac{\partial T(x, y)}{\partial x} = \frac{\partial^2 T(x, y)}{\partial y^2}$$

where α is the thermal diffusivity of the fluid, $T(x, y)$ is the temperature, and $u(y)$ is the laminar velocity profile.

7-25 Water at 20°C with a flow rate of 0.015 kg/s enters a 2.5-cm-ID tube which is maintained at a uniform temperature of 90°C.

(*a*) Determine the thermal entry length.

(*b*) Assuming hydrodynamically and thermally fully developed flow, determine the heat transfer coefficient and the tube length required to heat the water to 70°C.

(*c*) Determine the heat transfer rate to the water.

(*d*) Determine the friction factor and the pressure drop.

Answer: (*a*) 4.4 m; (*b*) 92.7 W/(m^2 · °C) and 10.8 m; (*c*) 3.14 kW; (*d*) 10.4 N/m^2.

7-26 Determine the friction factor f for the fully developed laminar flow of engine oil at 60°C with a flow rate of 0.1 kg/s through (*a*) a circular tube of 1-cm diameter, (*b*) a square duct 1 cm by 1 cm, (*c*) a rectangular duct 0.5 cm by 2 cm, and (*d*) an equilateral-triangle duct of sides 1 cm and having sharp corners.

7-27 Determine the pressure drops over the length $L = 10$ m for each of the cases considered in Prob. 7-26.

Hydrodynamic and thermal entry lengths

7-28 Determine the *hydrodynamic entry lengths* for flow at 60°C and at a rate of 0.015 kg/s of water, ethylene glycol, and engine oil through a circular tube of inside diameter 2.5 cm.

Answer: 2.27, 0.207, 0.015 m

7-29 Determine the *thermal entry lengths* for laminar flow at 60°C at a rate of 0.015 kg/s of water, ethylene glycol, and engine oil through a circular tube of inside diameter 2.5 cm and subjected to uniform wall temperature.

Answer: 4.04, 6.2, 9.1 m

7-30 Determine the thermal entry lengths for hydrodynamically developed laminar flow at 40°C with a flow rate of 50 kg/(m^2 · s) of water, glycerin, ethylene glycol, and engine oil through a parallel-plate channel having a spacing of 1 cm and subjected to uniform wall temperature.

7-31 Repeat Prob. 7-30 for uniform wall heat flux.

7-32 Repeat Probs. 7-30 and 7-31 for a flow rate of 20 kg/(m^2 · s).

7-33 Compare the thermal entry lengths for hydrodynamically developed laminar flow of the same fluid at a rate of Q kg/s through the following ducts subjected to uniform wall temperature: (*a*) a circular duct of diameter D, (*b*) a square duct of side D, and (*c*) a rectangular duct of sides $\frac{1}{2}D$ and $2D$.

7-34 Determine the hydrodynamic entry length, thermal entry length, and heat transfer coefficient for fully developed flow for engine oil at 60°C flowing at 0.01 kg/s through a square duct 1 cm by 1 cm in cross section and subjected to a uniform wall temperature. [Physical properties of engine oil at 60°C may be taken as $\rho = 864$ kg/m^3, $c_p = 2047$ W · s/(kg · °C), $k = 0.140$ W/(m · °C), $\mu = 0.0725$ kg/(m · s), Pr = 1050.]

7-35 Repeat Prob. 7-34 for a flow through a rectangular duct 0.5 cm by 1 cm in cross section and subjected to uniform wall temperature.

7-36 Glycerin at 20°C enters a square duct 1 cm by 1 cm in cross section with a flow rate of 0.01 kg/s. The walls of the duct are subjected to uniform heat flux everywhere (that is, Nu_{H2} condition of Table 7-1).

(*a*) Determine the hydrodynamic entry length.

(*b*) Determine the thermal entry length.

(*c*) Determine the heat transfer coefficient for the region where the velocity and temperature profiles are fully developed.

Answer: (*a*) 0.0006 m; (*b*) 5.6 m; (*c*) 103 W/(m^2 · °C).

7-37 Repeat Prob. 7-36 for a rectangular duct having a cross section 0.5 cm by 2 cm.

7-38 Engine oil at 20°C enters a 0.3-cm-diameter, 15-m-long tube with a flow rate of 0.01 kg/s. The tube wall is maintained at 100°C by condensing steam outside the tube surface.

(*a*) Determine the thermal entry length.

(*b*) Determine the hydrodynamic entry length.

(*c*) Assuming hydrodynamically and thermally fully developed flow, determine the heat transfer coefficient and the outlet temperature of the fluid.

7-39 Glycerin at 20°C with a flow rate of 0.01 kg/s enters a 1-cm-ID tube which is maintained at a uniform temperature of 80°C.

(*a*) Determine the thermal entry length.

(*b*) Assuming hydrodynamically and thermally fully developed flow, determine the heat transfer coefficient and the tube length required to heat the glycerin to 50°C.

Answer: (*a*) 3.5 m; (*b*) 5.2 m.

7-40 Determine the hydrodynamic and thermal entry lengths for fully developed laminar flow of engine oil at 60°C with a flow rate of 0.1 kg/s through (*a*) a circular tube of 1-cm diameter, (*b*) a square duct 1 cm by 1 cm, and (*c*) a rectangular duct 0.5 cm by 2 cm.

7-41 Repeat Prob. 7-40 for the case when the ducts are subjected to uniform wall heat flux.

Thermally developing, hydrodynamically developed laminar flow

7-42 By writing an energy balance for heat conducted radially and energy convected axially in the z direction on an annular element of radius r, radial thickness Δr, and axial thickness Δz in the fluid, show that the energy equation associated with the hydrodynamically developed flow in a circular tube is given by

$$\frac{1}{\alpha} u(r) \frac{\partial T(r, z)}{\partial z} = \frac{1}{r} \frac{\partial}{\partial r}\left(r \frac{\partial T}{\partial r}\right)$$

where $u(r)$ is the velocity distribution in flow, α is the thermal diffusivity, and $T(r, z)$ is the temperature.

7-43 Water at 15°C with a flow rate of 0.01 kg/s enters a 2.5-cm-diameter, 3-m-long tube maintained at a uniform temperature of 100°C by condensing steam on the outer surface of the tube. Assume the flow is hydrodynamically developed. Determine the thermal entry length, the average heat transfer coefficient, and the outlet temperature of water.

Answer: $L_t = 3$ m; $h_m = 95$ W/(m^2 · °C); $T_2 = 50.3$°C.

7-44 Water at 30°C with a flow rate of 0.01 kg/s enters a 2.5-cm-diameter tube which is maintained at a uniform temperature of 100°C. Assuming hydrodynamically developed flow, determine the tube length to heat the water to 65°C.

7-45 Engine oil at 30°C with a flow rate of 0.2 kg/s enters a 1-cm-diameter tube which is maintained at a uniform temperature of 120°C. Assuming hydrodynamically developed flow, determine the tube length to heat the oil to 70°C.

7-46 Engine oil at 40°C with a flow rate of 0.3 kg/s enters a 2.5-cm-diameter, 40-m-long tube maintained at a uniform temperature of 100°C. The flow is hydrodynamically developed. Determine (*a*) the average heat transfer coefficient, (*b*) the outlet temperature of the oil, (*c*) the total heat transfer rate to the oil, and (*d*) the thermal entry length.

7-47 Water at an average temperature of 60°C and with a flow rate of 0.02 kg/s flows through a square duct 2.5 cm by 2.5 cm in cross section which is maintained at a uniform temperature.

(*a*) Assuming a fully developed velocity profile, determine the average heat transfer coefficients over $L = 1$, 2, and 3 m from the inlet.

(*b*) Determine the average heat transfer coefficient for the hydrodynamically and thermally developed region.

Answer: (*a*) 185, 146, and 128 W/(m^2 · °C); (*b*) 77.5 W/(m^2 · °C).

7-48 Repeat Prob. 7-47 for the case when the walls are subjected to uniform heat flux.

7-49 Water at 30°C with a flow rate of 0.02 kg/s enters a square duct 2.5 cm by 2.5 cm whose walls are maintained at a uniform temperature 100°C. Assuming hydrodynamically developed flow, determine the length of the duct needed to heat the water to 70°C.

7-50 Engine oil at 20°C with a flow rate of 0.2 kg/s enters a 30-m-long, square duct 1 cm by 1 cm in cross section whose walls are maintained at a uniform temperature of 120°C. Assuming hydrodynamically developed flow, determine the outlet temperature of the oil.

Answer: 45°C

7-51 Engine oil at 20°C with a flow rate of 0.1 kg/s enters a square duct 1 cm by 1 cm whose walls are maintained at a uniform temperature of 100°C. Assuming hydrodynamically developed flow, determine the duct length required to heat the oil to 60°C.

Answer: 29 m

7-52 Water at 20°C with a flow rate of 0.02 kg/s enters a square duct 2.5 cm by 2.5 cm, which is maintained at a uniform temperature of 90°C. Assuming hydrodynamically developed flow, determine the thermal entry length. Calculate the duct length required to heat the water to 70°C.

7-53 Engine oil at 30°C with a flow rate of 0.002 kg/s enters a 4-m-long, rectangular duct of sides 0.4 cm by 0.8 cm whose walls are maintained at a uniform temperature of 100°C. Assuming hydrodynamically developed flow, determine the outlet temperature of the oil.

7-54 Water at a mean temperature of 60°C with a flow rate of 0.01 kg/s flows through a 1-m-long, rectangular duct maintained at a uniform temperature. Assuming hydrodynamically developed flow, determine the local heat transfer coefficient 1 m from the inlet. Also calculate the average heat transfer coefficient over the 1-m length of duct having a square cross section 2 cm by 2 cm.

7-55 Repeat Prob. 7-54 for the flow of engine oil at a mean temperature of 60°C.

Simultaneously developing laminar flow

7-56 Repeat Prob. 7-46 for the case when both the velocity and the temperature profiles are simultaneously developing.

7-57 Repeat Prob. 7-43 for the case when both the velocity and the temperature profiles are developing simultaneously.

7-58 Water at an average temperature of 60°C and with a flow rate of 0.015 kg/s flows through a 2.5-cm-diameter tube which is maintained at a uniform temperature.

(*a*) Assuming a fully developed velocity profile, determine the average heat transfer coefficients 1, 2, and 3 m from the tube inlet.

(*b*) Assuming simultaneously developing velocity and temperature profiles, determine the average heat transfer coefficients over a distance 1, 2, and 3 m from the tube inlet.

(*c*) Determine the heat transfer coefficient for the case of hydrodynamically and thermally fully developed flow.

Answer: (*a*) 195, 156, and 143 $W/(m^2 \cdot °C)$; (*b*) 203, 161, and 146 $W/(m^2 \cdot °C)$; (*c*) 95.3 $W/(m^2 \cdot °C)$.

7-59 Water at an average temperature of 60°C with a flow rate of 0.015 kg/s flows through a 2.5-cm-diameter tube which is maintained at a uniform temperature.

(*a*) Assuming a fully developed velocity profile, determine the local heat transfer coefficients 1, 2, and 3 m from the inlet.

(*b*) Assuming simultaneously developing velocity and temperature, determine the mean heat transfer coefficients over a distance 1, 2, and 3 m from the inlet.

7-60 Repeat Prob. 7-59*a* for the case when the tube wall is subjected to uniform wall heat flux.

Friction factor and pressure drop in turbulent flow

7-61 Oil having a kinematic viscosity $\nu = 1 \times 10^{-4}$ m^2/s and density $\rho = 890$ kg/m^3 flows through a 0.75-m-diameter commercial-steel pipe with a velocity of 1.5 m/s. The pumping stations are 100 km apart. Determine the pressure drop, hence the pumping power, required for pumping to overcome the friction losses for flow between the pumping stations.

Answer: 3200 hp

7-62 Water at 20°C flows with a velocity of $u = 1.5$ m/s inside a 20-cm-diameter clean cast-iron tube.

(*a*) Determine the friction factor.

(*b*) Determine the pressure drop over $L = 1500$ m of the pipe.

(*c*) Determine the pumping power required to pump the water over $L = 1500$ m of the pipe.

Answer: (*a*) 0.0215; (*b*) 181.5 kN/m^2; (*c*) 11.4 hp.

7-63 After a few years of service, field tests indicate that the roughness of the pipe in Prob. 7-62 increased by 15 percent. What increase in the friction factor, pressure drop, and pumping power is required to pump the water over the same distance with the same velocity?

7-64 Consider hydrodynamically fully developed turbulent flow of water in a circular tube of inside diameter D and length L with a mass flow rate M kg/s. Compare the pressure drops over the length L of the tube if the tube diameter is reduced from D to $D/2$ while the mass flow rate remains the same.

7-65 Water at 20°C is to be pumped through a 15-cm-diameter commercial-steel pipe over $L = 2000$ m. The available head (or the pressure difference) is 450,000 N/m^2. Determine the velocity of the water through the pipe.

Answer: 2.15 m/s

7-66 Determine the friction factor, pressure drop, and pumping power required for the flow of water at 20°C at a rate of 6 kg/s through a 5-cm-diameter, 200-m-long commercial-steel pipe.

7-67 Compare the friction factor, pressure drop, and pumping power required for the flow of air at 20°C at $Q = 0.1$ kg/s through a 2.5-cm-diameter, 100-m-long tube at 1, 2, and 3 atm.

7-68 Compare the friction factors and the pressure drops associated with the flow of water at a mean temperature of 60°C with a mass flow rate of 2 kg/s through 10-m-long, smooth tubes having inside diameters $D = 1.25$, 2.5, and 5 cm.

7-69 Compare the friction factor, pressure drop, and pumping power required for the flow of air, hydrogen, and helium at 80°C at $M = 0.05$ kg/s through a 2.5-cm-diameter, 100-m-long tube at 1 atm pressure.

7-70 Compare the friction factor and pressure drop for the flow of water and engine oil at 40°C at 5 kg/s through a 5-cm-diameter 60-m-long commercial-steel pipe.

7-71 Determine the friction factor, pressure drop, and pumping power required for the flow of water at 40°C at 0.5 kg/s through a square duct of cross section 2 cm by 2 cm and $L = 12$ m long. The surface of the duct is assumed smooth.

7-72 Determine the friction factor, pressure drop, and pumping power required for the flow of water at 40°C at 0.5 kg/s through an equilateral-triangular cross-sectional duct of sides 2 cm and length $L = 5$ m. The surface is assumed smooth.

7-73 Compare the friction factor and the pressure drop for the flow of water at 20°C at a rate of 1 kg/s through a 2.5-cm-diameter, 5-m-long tube made of (*a*) drawn tubing, (*b*) commercial steel, (*c*) galvanized iron, and (*d*) cast iron.

Heat transfer in turbulent flow

7-74 Water at a mean temperature of 60°C flows inside a 5-cm-ID tube with a velocity $u = 3$ m/s. Determine the heat transfer coefficient for the fully developed turbulent flow.

Answer: 10,816 W/(m^2 · °C)

7-75 Water at 20°C with a mass flow rate of 5 kg/s enters a 5-cm-ID, 10-m-long tube whose surface is maintained at a uniform temperature of 80°C. Calculate the outlet temperature of water.

7.76 Water at a mean temperature of 40°C flows inside a 2.5-cm-diameter, 10-m-long tube with a velocity of 6 m/s. The tube wall is maintained at a uniform temperature of 100°C by condensing steam. Determine the heat transfer rate to the water. Assume an inlet temperature of 20°C.

7-77 A fluid at a mean temperature of 40°C flows with a velocity of 10 m/s inside a 5-cm-diameter tube. Assuming fully developed turbulent flow, determine the average heat transfer coefficient for the flow of (*a*) helium at 1 atm, (*b*) air at 1 atm, (*c*) water, and (*d*) glycerin.

7-78 Water at 30°C with a mass flow rate of 2 kg/s enters a 2.5-cm-ID tube whose wall is maintained at a uniform temperature of 90°C. Calculate the tube length required to heat the water to 70°C.

7-79 Air at 27°C with a flow rate of $M = 0.01$ kg/s enters a rectangular tube 0.6 cm by 1.0 cm in cross section and 2 m long. The duct wall is subjected to a uniform heat flux of $q = 5$ kW/m^2. Determine the outlet temperature of the air and the duct surface temperature at the tube outlet. (Assume hydrodynamically and thermally developed flow.)

7-80 Air at atmospheric pressure and 27°C enters a 12-m-long, 1.5-cm-ID tube with a mass flow rate of 0.1 kg/s. The tube surface is maintained at a uniform temperature of 90°C. Calculate the average heat transfer coefficient and the rate of heat transfer to the air.

Answer: 842 W/(m^2 · °C); 6.276 kW

7-81 Consider hydrodynamically and thermally developed turbulent flow of water with a mass flow rate of M kg/s inside a circular tube of diameter D. The Dittus-Boelter equation can be used to determine the heat transfer coefficient. If the tube's inside diameter is changed from D to $D/2$ while the mass flow rate M remains the same, determine the resulting change in the heat transfer coefficient.

7-82 Air at 1 atm and 27°C mean temperature flows with a velocity of 20 m/s through a 5-cm-diameter, 2-m-long tube. Flow is hydrodynamically developed and thermally developing. Determine the average heat transfer coefficient over the entire length of the tube.

7-83 The pressurized water at a mean temperature of 300°C flows with a velocity of 15 m/s between parallel-plate fuel elements in a nuclear reactor. Assuming fully developed turbulent flow and a spacing of 0.25 cm between the plates, determine the average heat transfer coefficient.

7-84 Cooling water at 20°C and with a velocity of 2 m/s enters a condenser tube and leaves the tube at 30°C. The inside diameter of the tube is $D = 1.7$ cm. Assuming a fully developed turbulent flow, determine the average heat transfer coefficient.

Heat transfer to liquid metals in turbulent flow

7-85 Mercury at 20°C with a mass flow rate of 5 kg/s enters a 5-cm-ID tube whose surface is maintained at 100°C. Calculate the required tube length to raise the temperature of the mercury to 80°C.

Answer: 2.77 m

7-86 Mercury at an average temperature of 100°C flows with a velocity of 0.5 m/s inside a 2.5-cm-diameter tube. The flow is hydrodynamically developed but thermally developing. Determine the entry region heat transfer coefficient under both uniform wall heat flux and uniform wall temperature conditions over $x/D = 20$, 40, and 60.

7-87 Repeat Prob. 7-85 for the hydrodynamically and the thermally developed region for the flow of NaK (56 percent Na, 44 percent K) at a mean temperature of 93°C.

7-88 Repeat Prob. 7-86 for the flow of liquid NaK (56 percent Na, 44 percent K) at a mean temperature of 204°C.

7-89 Mercury at a temperature of 100°C and with a velocity of 1 m/s enters a 1.25-cm-diameter tube which is maintained at a uniform temperature of 250°C. Determine the length of the tube required to raise the temperature of mercury to 200°C.

7-90 Liquid sodium at 180°C with a mass flow rate of 3 kg/s enters a 2.5-cm-ID tube whose wall is maintained at a uniform temperature of 240°C. Calculate the tube length required to heat sodium to 230°C.

Answer: 2.18 m

7-91 Liquid bismuth at 500°C and with a velocity of $u = 1$ m/s enters a 1.25-cm-diameter tube which is maintained at a uniform temperature of 600°C. Determine the length of the tube required to raise the temperature of the liquid bismuth to 580°C.

Analogies between heat and momentum transfer

7-92 Explain the basic assumption s made in the Reynolds, Prandtl, and von Kármán analogies between heat and momentum transfer in turbulent flow. Discuss their range of applicability to determine the heat transfer coefficient given the friction factor.

7-93 Using the Reynolds analogy, develop an expression for the heat transfer coefficient for turbulent flow over a flat plate in terms of the drag coefficient.

7-94 Water at a mean temperature of 60°C flows inside a 5-cm-diameter tube with a velocity of 4 m/s. By obtaining the corresponding friction factor from the friction factor chart, determine the heat transfer coefficients by using both the Reynolds and Prandtl analogies. Compare the heat transfer coefficient thus obtained with that determined from the Dittus-Boelter equation.

7-95 By using the von Kármán analogy, determine the heat transfer coefficient for hydrodynamically and thermally developed flow of water at a mean temperature of 40°C at a rate of 15 kg/s through a 10-cm-diameter, galvanized iron pipe by utilizing the friction factor data.

7-96 For turbulent flow inside a smooth pipe, the friction factor f is approximately represented by

$$f = 0.184\ \mathrm{Re}^{-0.2} \qquad \text{for } 20{,}000 < \mathrm{Re} < 300{,}000$$

Colburn, by making use of Reynolds analogy, proposed the following correlation:

$$\mathrm{St} \cdot \mathrm{Pr}^{2/3} = \frac{f}{8}$$

By utilizing these relations, develop a correlation for the Nusselt number and compare it with the Dittus-Boelter equation.

7-97 The measured pressure drop for the flow of water at 20°C with a velocity of $u = 3$ m/s through a 2.5-cm-diameter, 5-m-long tube is 1.8×10^4 N/m². Using the Colburn analogy, determine the heat transfer coefficient.

Answer: 8560 W/(m² · °C)

7-98 Compare the Nusselt numbers for the flow of water at a mean temperature of 20°C inside a 5-cm-ID smooth pipe at Re = 40,000 calculated by using (*a*) the Dittus-Boelter equation, (*b*) the Prandtl analogy, and (*c*) the Colburn analogy.

REFERENCES

1. Hagen, G.: Über die Bewegung des Wassers in engen zylindrischen Röhren, *Pogg. Ann.*, **46**:423 (1839).
2. Poiseuille, J.: Recherches experimentelles sur le mouvement des Liquids dans les tubes de tres petits diamètres, *C. R.*, **11**:961 (1840).
3. Kays, W. M., and M. E. Crowford: *Convective Heat and Mass Transfer*, McGraw-Hill, New York, 1980.
4. Jischa, M.: *Konvektiver Impuls-, Wärme- und Stoffaustausch*, Friedr. Vieweg & Sohn, Braunschweig/Wiesbaden, 1982.
5. Shah, R. K., and A. L. London: *Laminar Flow: Forced Convection in Ducts*, Academic, New York, 1978.
6. Hornbeck, R. W.: "Laminar Flow in the Entrance Region of a Pipe," *Appl. Sci. Res.*, Sec. A, **13**:224–232 (1964).
7. Wiginton, C. L., and C. Dolton: "Incompressible Laminar Flow in the Entrance Region of a Rectangular Duct," *J. Appl. Mech.*, **37**:854–856 (1970).
8. Shah, R. K.: "Thermal Entry Length Solutions for a Circular Tube and Parallel Plates," *Proc. Nat. Heat. Mass Transfer Conf.*, 3d, Indian Inst. Technol., Bombay, vol. 1, Paper no. HMT-11-75 (1975).
9. Graetz, L.: "Uber die Wärmeleitungsfähigkeit von Flüssigkeiten (On the thermal conductivity of liquids). Part 1." *Ann. Phys. Chem.*, **18**:79–94 (1883); Part 2. *Ann. Phys. Chem.*, **25**:337–357 (1885).
10. Nusselt, W.: "Die Abhängigkeit der Wärmeübergangszahl von der Rohrlänge (The Dependence of the Heat-Transfer Coefficient on the Tube Length)," *VDI Z.*, **54**:1154–1158 (1910).
11. Jakob, M.: *Heat Transfer*, vol. 1, Wiley, New York, 1949.
12. Knudsen, J. G., and D. L. Katz: *Fluid Dynamics and Heat Transfer*, McGraw-Hill, New York, 1958.
13. Eckert, E. R. G., and R. M. Drake, Jr.: *Analysis of Heat and Mass Transfer*, McGraw-Hill, New York, 1972.
14. Sellars, S. R., M. Tribus, and J. S. Klein: "Heat Transfer to Laminar Flow in a Round Tube or Flat Plate—The Graetz Problem Extended," *Trans. ASME*, **78**:441–448 (1956).
15. Norris, R. H., and D. D. Streid: "Laminar-Flow Heat-Transfer Coefficient for Ducts," *Trans. ASME*, **62**:525–533 (1940).
16. Reynolds, W. C.: "Heat Transfer to Fully Deveoped Laminar Flow in a Circular Tube with Arbitrary Circumferential Heat Flux," *J. Heat Transfer*, **82**:108–112 (1960).

17. Brown, G. M.: "Heat or Mass Transfer in a Fluid in Laminar Flow in a Circular or Flat Conduit," *AIChE J.*, **6**: 179–183 (1960).
18. Tao, L. N.: "On Some Laminar Force-Convection Problems." *J. Heat Transfer*, **83**: 466–472 (1961).
19. Madejski, J.: "Temperature Distribution in Channel Flow with Friction," *Int. J. Heat Mass Transfer*, **6**: 49–51 (1963).
20. Chen, J. C.: "Laminar Heat Transfer in Tube with Nonlinear Radiant Heat-Flux Boundary Condition," *Int. J. Heat Mass Transfer*, **9**: 433–440 (1966).
21. Hasegawa, S., and Y. Fujita: "Nusselt Numbers for Fully Developed Flow in a Tube with Exponentially Varying Heat Flux," *Mem. Fac. Eng., Kyushu Univ.*, **27** (1): 77–80 (1967).
22. Chandler, R. D., J. N. Panaia, R. B. Stevens, and G. E. Zinsmeister: "The Solution of Steady State Convection Problems by the Fixed Random Walk Method," *J. Heat Transfer*, **90**: 361–363 (1968).
23. Gräber, H.: "Heat Transfer in Smooth Tubes, between Parallel Plates, in Annuli and Tube Bundles with Exponential Heat Flux Distributions in Forced Laminar or Turbulent Flow (in German), "*Int. J. Heat Mass Transfer*, **13**: 1645–1703 (1970).
24. Tay, A. O., and G. De Vahl Davis: "Application of the Finite Element Method to Convection Heat Transfer between Parallel Planes," *Int. J. Heat Mass Transfer*, **14**: 1057–1069 (1971).
25. McKillop, A. A., J. C. Harper, and H. J. Bader: "Transfer in Entrance–Region Flow with External Resistance," *Int. J. Heat Mass Transfer*, **14**: 863–866 (1971).
26. Swearingen, T. W., and D. M. McEligot: "Internal Laminar Heat Transfer with Gas-Property Variation," *J. Heat Transfer*, **93C**: 432–440 (1971).
27. Kadaner, Y. S., Y. P. Rassadkin, and E. L. Spektor: "Heat Transfer during Laminar Fluid Flow in a Pipe with Radiative Heat Removal," *J. Eng. Phys. (USSR)*, **20**: 20–24 (1971); also in *Heat Transfer—Sov. Res.*, **3** (5): 182–188 (1971).
28. Ou, J. W., and K. C. Cheng: "Effects of Pressure Work and Viscous Dissipation on Graetz Problem for Gas Flows in Parallel-Plate Channels," *Waerme- Stoffuebertrag.*, **6**: 191–198 (1973).
29. Ou, J. W., and K. C. Cheng: "Viscous Dissipation Effects on Thermal Entrance Region Heat Transfer in Pipes with Uniform Wall Heat Flux," *Appl. Sci. Res.*, **28**: 289–301 (1973).
30. Ou, J. W., and K. C. Cheng: "Viscous Dissipation Effects on Thermal Entrance Heat Transfer in Laminar and Turbulent Pipe Flows with Uniform Wall Temperature," *Am. Inst. Aeronaut. Astron. Pap.*, **74-743** or *Am. Soc. Mech. Eng., Pap.*, **74-HT-50** (1974).
31. Hong, S. W., and A. E. Bergles: "Laminar Flow Heat Transfer in the Entrance Region of Semi-circular Tubes with Uniform Heat Flux," *Int. J. Heat Mass Transfer*, **19**: 123–124 (1976).
32. Chandrupatla, A. R., and V. M. K. Sastri: "Laminar Forced Convection Heat Transfer of a Non-Newtonian Fluid in a Square Duct," *Int. J. Heat Mass Transfer*, **20**: 1315–1324 (1977).
33. Kays, W. M.: "Numerical Solutions for Laminar Flow Heat Transfer in Circular Tubes," *Trans. ASME*, **77**: 1265–1274 (1955).
34. Ulrichson, D. L., and R. A. Schmitz: "Laminar-Flow Heat Transfer in the Entrance Region of Circular Tubes," *Int. J. Heat Mass Transfer*, **8**: 253–258 (1965).
35. Hornbeck, R. W.: "An All-Numerical Method for Heat Transfer in the Inlet of a Tube," *ASME* paper No. 65-WA/HT-36 (1965).
36. Bender, F., "Wärmeübergang bei ausgebildeter und nicht ausgebildeter laminarer Rohrströmung mit temperatureabhängigen Stoffwerten," *Waerme- Stoffuebertrag*, **1**: 159–168 (1968).
37. Manohar, R. "Analysis of Laminar-Flow Heat Transfer in the Entrance Region of Circular Tubes," *Int. J. Heat Mass Transfer*, **12**: 15–22 (1969).
38. Bankston, C. A., and D. M. McEligot: "Turbulent and Laminar Heat Transfer to Gases with Varying Properties in the Entry Region of Circular Ducts," *Int. J. Heat Mass Transfer*, **13**: 319–344 (1970).
39. Hwang, G. J., and J.-P. Sheu: "Effect of Radial Velocity Component on Laminar Forced Convection in Entrance Region of a Circular Tube," *Int. J. Heat Mass Transfer*, **17**: 372–375 (1974).
40. Sparrow, E. M.: "Analysis of Laminar Forced-Convection Heat Transfer in Entrance Region of Flat Rectangular Ducts," *NACA Tech. Notes*, **TN 3331** (1955).

41. Siegel, R., and E. M. Sparrow: "Simultaneous Development of Velocity and Temperature Distributions in a Flat Duct with Uniform Wall Heating." *AIChE J.*, **5**:73–75 (1959).
42. Slezkin, N. A.: "On the Development of the Flow of a Viscous Heat-Conducting Gas in a Pipe," *J. Appl. Math. Mech.*, **23**:473–489 (1959).
43. Stephan, K.: "Wärmeübergang und druckabfall bei nicht ausgebildeter Laminarströmung in Rohren und in ebenen Spalten," *Chem.-Ing.-Tech.*, **31**:773–778 (1959).
44. Bodoia, J. R., and J. F. Osterle, "Finite Difference Analysis of Plane Poiseuille and Couette Flow Developments," *Appl. Sci. Res., Sec. A*, **10**:265–276 (1961).
45. Hwang, C. L., and L. T. Fan: "Finite Difference Analysis of Forced-Convection Heat Transfer in Entrance Region of a Flat Rectangular Duct," *Appl. Sci. Res., Sec. A*, **13**:401–422 (1964).
46. Miller, J. A., and D. D. Lundberg: "Laminar Convective Heat Transfer in the Entrance Region Bounded by Parallel Flat Plates at Constant Temperature," *Am. Soc. Mech. Eng., Pap.* **67-HT-48** (1967); condensed from TR No. 54, U.S. Naval Postgrad. Sch., Monterey, Calif., 1965.
47. Mercer, W. E., W. M. Pearce, and J. E. Hitchcock: "Laminar Forced Convection in the Entrance Region between Parallel Flat Plates," *J. Heat Transfer*, **89**:251–257 (1967).
48. Kiya, M., S. Fukusako, and M. Arie: "Effect of Non-uniform Inlet Velocity Profile on the Development of a Laminar Flow between Parallel Plates," *Bull. JSME*, **15**(81):324–336 (1972).
49. Bhatti, M. S., and C. W. Savery: "Heat Transfer in the Entrance Region of a Straight Channel: Laminar Flow with Uniform Wall Heat Flux," *Am. Soc. Mech. Eng., Pap.* **76-HT-20** (1976); also in a condensed form in *J. Heat Transfer*, **99**:142–144 (1977).
50. Montgomery, S. R., and P. Wibulswas: "Laminar Flow Heat Transfer for Simultaneously Developing Velocity and Temperature Profiles in Ducts of Rectangular Cross Section," *Appl. Sci. Res.*, **18**:247–259 (1967).
51. Wibulswas, P.: "Laminar-Flow Heat-Transfer in Non-Circular Ducts," Ph.D. thesis. London University, London, 1966.
52. Hausen, H.: "Darstellung des Wärmeüberganges in Rohren durch verallgemeinerte Potenzbeziehungen, *VDI Z*, **4**:91 (1943).
53. Sieder, E. N., and G. E. Tate: "Heat Transfer and Pressure Drop of Liquids in Tubes," *Ind. Eng. Chem.*, **28**:1429–1435 (1936).
54. Nikruadse, J.: *Forsch. Arb. Ing. Wes.* no. 346 (1932).
55. Nikuradse, J.: *Forsch. Arb. Ing. Wes.* no. 361 (1933); also, "Laws of Flow in Rough Pipes" (trans.), *NACA Tech. Memo.* 1292, 1950.
56. Schlichting, H.: *Boundary Layer Theory*, 6th ed., McGraw-Hill, New York, 1968.
57. Moody, L. F.: "Friction Factor for Pipe Flow," *Trans. ASME*, **66**:671–684 (1944).
58. Colburn, A. P.: "A Method of Correlating Forced Convection Heat Transfer Data and a Comparison with Liquid Frictions," *Trans. AIChE*, **29**:174–210 (1933).
59. Dittus, F. W., and L. M. K. Boelter: *Univ. Calif., Berkeley, Publ. Eng.*, **2**:443 (1930).
60. Deissler, R. G.: "Investigation of Turbulent Flow and Heat Transfer in Smooth Tubes Including the Effects of Variable Properties," *Trans. ASME*, **73**:101 (1951).
61. Petukhov, B. S.: "Heat Transfer and Friction in Turbulent Pipe Flow with Variable Physical Properties," in J. P. Hartnett and T. F. Irvine (eds.), *Advances in Heat Transfer*, Academic, New York, 1970, pp. 504–564.
62. Nusselt, W.: "Der Warmeaustausch Zwischen Wand und Wasser im Rohr," *Forsch. Geb. Ingenieurwes*, **2**:309 (1931).
63. Notter, R. H., and C. A. Sleicher: "A Solution to the Turbulent Graetz Problem. III. Fully Developed and Entry Heat Transfer Rates," *Chem. Eng. Sci.*, **27**:2073–2093 (1972).
64. Irvine, T. R.: "Noncircular Convective Heat Transfer," in W. Ible (ed.), *Modern Developments in Heat Transfer*, Academic, New York, 1963.
65. Lubarsky, B., and S. J. Kaufman: "Review of Experimental Investigation of Liquid Metal Heat Transfer," *NACA Tech. Note* 3336, 1955.
66. Lyon, R. D. (ed.): *Liquid Metals Handbook*, 3d ed., U.S. Atomic Energy Commission and Department of the Navy, Washington, 1952.
67. Stein, R.: "Liquid Metal Heat Transfer," *Adv. Heat Transfer*, **3**:(1966).

68. Skupinski, E. S., J. Tortel, and L. Vautrey: "Determination des coefficients de convection d'un alliage sodium-potassium dans un tube circulaire," *Int. J. Heat Mass Transfer*, **8**:937 (1965).
69. Seban, R. A., and T. T. Shimazaki: "Heat Transfer to Fluid Flowing Turbulently in a Smooth Pipe with Walls of Constant Temperature," *Trans. ASME*, **73**:803–808 (1951).
70. Sleicher, C. A., Jr., and M. Tribus: "Heat Transfer in a Pipe with Turbulent Flow and Arbitrary Wall-Temperature Distribution," *Trans. ASME*, **79**:789–797 (1957).
71. Azer, N. Z., and B. T. Chao: "Turbulent Heat Transfer in Liquid Metals—Fully Developed Pipe Flow with Constant Wall Temperature," *Int. Heat Mass Transfer*, **3**:77–83 (1961).
72. Sleicher, C. A., A. S. Awad, and R. H. Notter: "Temperature and Eddy Diffusivity Profiles in NaK," *Int. J. Heat Mass Transfer*, **16**:1565–1575 (1973).
73. Hennecke, D. K.: "Heat Transfer by Hagen-Poiseuille Flow in the Thermal Development Region with Axial Conduction," *Wärme-Stoffubetragung Bd.*, **1**:177–184 (1968).
74. Hsu, Chia-Jung: "Theoretical Solutions for Low-Péclét-Number Thermal-Entry-Region Heat Transfer in Laminar Flow through Concentric Annuli," *Int. J. Heat Mass Transfer*, **13**:1907–1924 (1970).
75. Hsu, Chia-Jung: "An Exact Analysis of Low-Péclét-Number Thermal-Entry-Region Heat Transfer in Transversely Nonuniform Velocity Fields," *AIChE J.*, **17** (3):732–740 (May 1971).
76. Verhoff, F. H., and D. P. Fisher.: "A Numerical Solution of the Graetz Problem with Axial Conduction," *ASME J. Heat Transfer*, **95**:132–134 (1973).
77. Campo, A., and J. C. Auguste.: "Axial Conduction in Laminar Pipe Flows with Nonlinear Wall Heat Fluxes," *Int. J. Heat Mass Transfer*, **21**:1597–1607 (1978).
78. Vick, B., M. N. Özişik, and Y. Bayazitoğlu: "A Method of Analysis of Low Péclét Number Thermal Entry Region Problems with Axial Conduction," *Letters in Heat Mass Transfer*, **7**:235–248 (1980).
79. Vick, B., and M. N. Özişik: "An Exact Analysis of Low Péclét Number Heat Transfer in Laminar Flow with Axial Conduction," *Letters in Heat Mass Transfer*, **8**:1–10 (1981).
80. Vick, B., and M. N. Özişik: "Effects of Axial Conduction and Convective Boundary Conditions in Slug Flow inside a Circular Tube," *J. Heat Transfer*, **103**:436–440, 1981.
81. Prandtl, L.: *Z. Phys.*, **11**:1072 (1910).
82. Von Kármán, Th.: "The Analogy between Fluid Friction and Heat Transfer," *Trans. ASME*, **61**:705–711 (1939).
83. Martinelli, R. C.: "Heat Transfer in Molten Metals," *Trans. ASME*, **69**:947–959 (1947).
84. Deissler, R. G.: "Analysis of Turbulent Heat Transfer, Mass Transfer and Friction in Smooth Tubes at High Prandtl and Schmidt Numbers," *NACA Tech. Note* 3145, 1954.
85. Trimble, L. C. et al.: "Ocean Thermal Energy Conversion System Study Report," *Proc. Third Workshop on Ocean Thermal Energy Conversion*, APL/JHU SR 75-2, August 1975, pp. 3–21.
86. Alexander, L. G., and H. W. Hoffman: "Performance Characteristics of Corrugated Tubes for Vertical Tube Evaporators," *ASME* paper no. 71-HT-30, 1971.
87. Thomas, D. G.: "Prospects for Further Improvements in Enhanced Heat Transfer Surfaces," *Desalination*, **12**:189–215 (1973).
88. Shah, R. K.: "Perforated Heat Exchanger Surfaces. Part 2—Heat Transfer and Fluid Flow Characteristics," *ASME* paper no. 75-WA/HT-9, 1975.
89. Bergles, A. E., R. L. Webb, G. H. Junkhan, and M. K. Jensen: "Bibliography on Augmentation of Convective Heat and Mass Transfer," HTL-19, ISU-ERI-Ames-79206, Iowa State University, 1979.
90. Bergles, A. E., and R. L. Webb: "Bibliography on Augmentation of Convective Heat and Mass Transfer—Part 1 to Part 6," *Previews of Heat and Mass Transfer*, **4** (2):61–73 (1978) to **6** (3):242–313 (1980).
91. Bergles, A. E., and M. L. Jensen: "Enhanced Single-Phase Heat Transfer for OTEC Systems," *Proc. Fourth Annual Conference on Ocean Thermal Energy Conversion*, New Orleans, March 1977, pp. VI-41–VI-54.
92. Webb, R. L., E. R. G. Eckert, and R. J. Goldstein: "Heat Transfer and Friction in Tubes with Repeated Rib Roughness," *Int. J. Heat Mass Transfer*, **14**:601–618 (1971).

93. Webb, R. L., E. R. G. Eckert, and R. J. Goldstein: "Generalized Heat Transfer and Friction Correlations for Tubes with Repeated-Rib Roughness," *Int. J. Heat Mass Transfer*, **15**:180–184 (1972).
94. Watkinson, A. P., D. C. Miletti, and G. R. Kubanek: "Heat Transfer and Pressure Drop of Internally Finned Tubes in Laminar Oil Flow," *ASME* paper no. 75-HT-41, 1975.
95. Carnavos, T. C.: "Cooling Air in Turbulent Flow with Multipassage Internally Finned Tubes," *ASME* paper no. 78-WA/HT-52, 1978.
96. Carnavos, T. C.: "Heat Transfer Performance of Internally Finned Tubes in Turbulent Flow," in *Advances in Enhanced Heat Transfer*, *ASME*, 1979, pp. 61–67.
97. Carnavos, T. C.: "Cooling Air in Turbulent Flow with Internally Finned Tubes," *Heat Transfer Engineering*, **1** (2):41–46 (1979).
98. Patankar, S. V., M. Ivanovic, and E. M. Sparrow: "Analysis of Turbulent Flow and Heat Transfer in Internally Finned Tubes and Annuli," *J. Heat Transfer*, **101**:29–37 (1979).
99. Marner, W. J., and A. E. Bergles: "Augmentation of Tubeside Laminar Flow Heat Transfer by Means of Twisted-Tape Inserts, Static-Mixer Inserts, and Internally Finned Tubes," *Heat Transfer 1978*, vol. 2, Hemisphere, Washington, 1978, pp. 583–588.
100. Hong, S. W., and A. E. Bergles: "Augmentation of Laminar Flow Heat Transfer by Means of Twisted-Tape Inserts," *J. Heat Transfer*, **98**:251–256 (1976).
101. Özişik, M. N., and H. Topakoglu: "Heat Transfer for Laminar Flow in a Curved Pipe," *J. Heat Transfer*, **90**:313–381 (1968).
102. Schmidt, E. F.: "Waermeuebergang and Druckverlust in Rohrschlangen," *Chem.-Ing.-Tech.*, **39**:781–789 (1967).
103. Abul-Hamayel, M. A., and K. J. Bell: "Heat Transfer in Helically-Coiled Tubes with Laminar Flow," *ASME* paper no. 79-WA/HT-11, 1979.
104. Webb, R. L., and E. R. G. Eckert: "Application of Rough Surfaces to Heat Exchanger Design," *Int. J. Heat Mass Transfer*, **15**:1647–1658 (1972).
105. Bergles, A. E., A. R. Blumenkrantz, and J. Taborek: "Performance Evaluation Criteria for Enhanced Heat Transfer Surfaces," AIChE Preprint 9 for *13th National Heat Transfer Conf.*, Denver, Colo., August 1972.
106. Bergles, A. E., R. L. Bunn, and G. H. Junkhan: "Extended Performance Evaluation Criteria for Enhanced Heat Transfer Surfaces," *Letters in Heat and Mass Transfer*, **1**:113–120 (1974).
107. Webb, R. L., and M. J. Scott: "A Parametric Analysis of the Performance of Internally Finned Tubes for Heat Exchanger Applications," *J. Heat Transfer*, **102**:38–43 (1980).
108. Webb, R. L., and J. T. Hong: "Water-Side Enhancement for OTEC Shell and Tube Evaporators," Paper III 13/4, presented at *7th Ocean Energy Conference*, June 1980.

CHAPTER

EIGHT

FORCED CONVECTION FOR FLOW OVER BODIES

In Chap. 7 we discussed the determination of the friction factor and heat transfer coefficient for flow inside ducts. There are numerous important engineering applications in which heat transfer and pressure drop for flow over bodies such as a flat plate, a sphere, a circular tube, or a tube bundle are needed. This chapter is devoted to the discussion and presentation of various correlations of heat transfer and drag coefficients for flow in such situations. To provide better insight into the mechanism of heat transfer and pressure drop for flow over bodies, a simple approximate method of analysis is discussed for the determination of drag and heat transfer coefficients for laminar flow along a flat plate before the exact results are presented. Recommended correlations are given for the drag and heat transfer coefficients for flow over bodies of various geometries for both laminar and turbulent flow conditions. Heat transfer at high speeds also is discussed briefly.

8-1 DRAG COEFFICIENT FOR FLOW OVER A FLAT PLATE

As discussed in Chap. 6, flow over a flat plate remains laminar until the critical Reynolds number (that is, $\mathrm{Re}_c \cong 5 \times 10^5$) is reached, and then the transition into turbulent begins. The correlations for the drag coefficient in the laminar and turbulent flow regimes are different. Therefore, the determination of the drag coefficients for these two regimes is considered separately.

Laminar Boundary Layer

Consider two-dimensional, steady flow of an incompressible, constant-property fluid along a flat plate, as illustrated in Fig. 8-1.

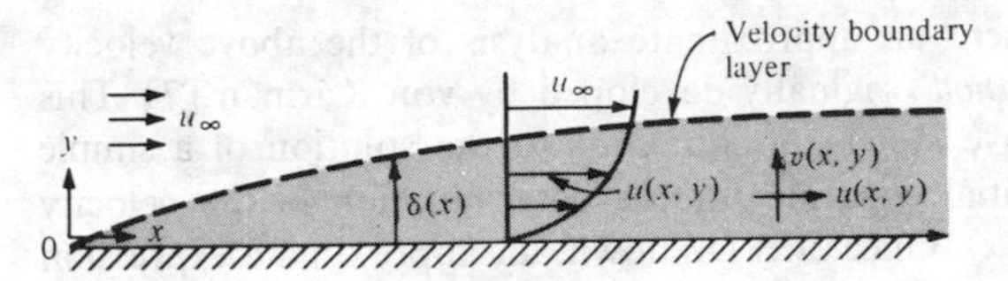

Figure 8-1 Coördinates for the velocity problem for forced laminar flow along a flat plate.

The x axis is chosen along the plate with the origin $x = 0$ at the leading edge and the y axis perpendicular to the plate surface. Let $u(x, y)$ and $v(x, y)$ be the velocity components in the x and y directions, respectively, u_∞ the free-stream velocity, and $\delta(x)$ the thickness of the velocity boundary layer. The velocity components $u(x, y)$ and $v(x, y)$ satisfy the continuity and momentum equations for a boundary layer given in Chap. 6 [see Eqs. (6-81) and (6-82)]:

$$\text{Continuity:} \qquad \frac{\partial u}{\partial x} + \frac{\partial v}{\partial y} = 0 \tag{8-1}$$

$$x \text{ Momentum:} \qquad u\frac{\partial u}{\partial x} + v\frac{\partial u}{\partial y} = \nu\frac{\partial^2 u}{\partial y^2} \tag{8-2}$$

The pressure term dP/dx in the momentum equation vanishes for flow along a flat plate, as discussed in Chap. 6, hence is not included in Eq. (8-2).

The boundary conditions for these equations are

$$u = 0 \qquad v = 0 \qquad \text{at } y = 0 \tag{8-3a}$$

$$u \to u_\infty \qquad \text{at } y = \delta(x) \tag{8-3b}$$

The boundary conditions (8-3*a*) state that the velocity components are zero at the wall surface (i.e., wall surface is impermeable to flow), and the boundary condition (8-3*b*) implies that the axial velocity component is almost equal to the external flow velocity u_∞ at the edge of the velocity boundary layer at $y = \delta(x)$.

Equations (8-1) to (8-3) give the complete mathematical formulation of the velocity problem for laminar flow along a flat plate. If this system of equations is solved and the axial velocity component $u(x, y)$ is determined, the local drag coefficient c_x can be determined readily from its definition, given by Eq. (6-5), that is,

$$c_x = \frac{2\nu}{u_\infty^2}\left.\frac{\partial u(x, y)}{\partial y}\right|_{y=0} \tag{8-4}$$

The first exact analysis of the above velocity problem is due to Blasius [1], who transformed the system to a third-order ordinary differential equation and solved it by a series expansion method. Later, the resulting ordinary differential equation was solved by Howarth [2] more accurately by using a numerical technique. A discussion of the exact solution of this velocity problem can be found in several sources [3–6].

In this section we present an approximate analysis of the above velocity problem by the *integral method* originally developed by von Kármán [7]. This method transforms the above system of equations to the solution of a simple first-order ordinary differential equation for the determination of the velocity boundary-layer thickness $\delta(x)$. Once $\delta(x)$ is known, an approximate expression becomes available for the velocity profile, hence the drag coefficient is determined by Eq. (8-4). It is instructive to analyze the velocity problem by the integral method and develop an approximate expression for the drag coefficient for laminar flow along a flat plate before the exact results are presented.

The basic steps in the analysis with this method are as follows:

1. The x momentum equation (8-2) is integrated with respect to y over the boundary-layer thickness $\delta(x)$, and the velocity component $v(x, y)$ appearing in this equation is eliminated by means of the continuity equation (8-1). The resulting equation, called the *momentum integral equation*, is given (see Note 1 at the end of this chapter for the derivation).

$$\frac{d}{dx}\left[\int_0^{\delta} u(u_\infty - u)\,dy\right] = \nu \frac{\partial u}{\partial y}\bigg|_{y=0} \qquad \text{in } 0 \le y \le \delta \tag{8-5}$$

where $\delta \equiv \delta(x)$ and $u \equiv u(x, y)$. So far the analysis and Eq. (8-5) are exact, but this equation cannot be solved because it involves two unknowns, namely, $\delta(x)$ and $u(x, y)$. Therefore, an additional relationship is needed between δ and u.

2. At this stage an approximation is introduced to the analysis. To obtain a relationship between δ and u, an assumption is made regarding the functional form of the velocity profile $u(x, y)$ within the boundary layer $0 \le y \le \delta(x)$. A polynomial form is generally chosen. A second-, a third-, or a fourth-degree polynomial approximation generally is chosen for this purpose. Experience has shown that there is no significant improvement in the accuracy of the solution by choosing a polynomial greater than the fourth degree.

Suppose a cubic approximation is chosen for $u(x, y)$ in the form

$$u(x, y) = a_0 + a_1 y + a_2 y^2 + a_3 y^3 \qquad \text{in } 0 \le y \le \delta(x) \tag{8-6}$$

where the coefficients a_i are a function of x. The four conditions needed to determine these four coefficients a_0, a_1, a_2, and a_3 are taken as

$$u\bigg|_{y=0} = 0 \qquad u\bigg|_{y=\delta} = u_\infty \qquad \frac{\partial u}{\partial y}\bigg|_{y=\delta} = 0 \qquad \text{and} \qquad \frac{\partial^2 u}{\partial y^2}\bigg|_{y=0} = 0 \tag{8-7}$$

Clearly, the first two relations are the boundary conditions for the problem, the third one results from the boundary-layer concept, and the last one is the derived condition obtained by evaluating the x momentum equation (8-2) at $y = 0$, where $u = v = 0$. The application of these four conditions given by Eqs. (8-7) to Eq. (8-6) results in a velocity profile in the form

$$\frac{u(x, y)}{u_\infty} = \frac{3}{2}\left(\frac{y}{\delta}\right) - \frac{1}{2}\left(\frac{y}{\delta}\right)^3 \tag{8-8}$$

3. The velocity profile given by Eq. (8-8) is introduced to the momentum integral equation (8-5), that is,

$$u_\infty^2 \frac{d}{dx}\left\{\int_0^\delta \left[\frac{3}{2}\frac{y}{\delta} - \frac{1}{2}\left(\frac{y}{\delta}\right)^3\right]\left[1 - \frac{3}{2}\frac{y}{\delta} + \frac{1}{2}\left(\frac{y}{\delta}\right)^3\right] dy\right\} = \nu u_\infty \frac{3}{2\delta} \tag{8-9}$$

where the right-hand side of this equation results from the relation $\partial u/\partial y|_{y=0} = 3u_\infty/(2\delta)$. When the integration with respect to y is performed, Eq. (8-9) becomes

$$\frac{d}{dx}\left[\frac{39}{280}\delta(x)\right] = \frac{3\nu}{2u_\infty \delta(x)}$$

or

$$\delta\, d\delta = \frac{140}{13}\frac{\nu}{u_\infty} dx \tag{8-10}$$

Thus, we transformed the velocity problem given by Eqs. (8-1) to (8-3) to the solution of a simple ordinary differential equation (8-10) for the boundary-layer thickness $\delta(x)$. The physical situation illustrated in Fig. 8-1 implies that the boundary-layer thickness is zero at $x = 0$. Therefore, the boundary condition for Eq. (8-10) is taken as

$$\delta(x) = 0 \qquad \text{at } x = 0 \tag{8-11}$$

The integration of Eq. (8-10) subject to the boundary condition (8-11) yields

$$\delta^2(x) = \frac{280}{13}\frac{\nu x}{u_\infty} \tag{8-12}$$

Then the boundary-layer thickness becomes

$$\delta(x) = \sqrt{\frac{280}{13}\frac{\nu x}{u_\infty}} \tag{8-13}$$

which can be rearranged in the dimensionless form as

$$\frac{\delta(x)}{x} = \sqrt{\frac{280}{13\,\mathrm{Re}_x}} = \frac{4.64}{\mathrm{Re}_x^{1/2}} \tag{8-14}$$

where the local Reynolds number Re_x is defined by

$$\mathrm{Re}_x = \frac{u_\infty x}{\nu} \tag{8-15}$$

4. The velocity gradient at the wall, for the cubic velocity profile given by Eq. (8-8), becomes

$$\left.\frac{\partial u}{\partial y}\right|_{y=0} = \frac{3u_\infty}{2\delta(x)} \tag{8-16}$$

Introducing $\delta(x)$ from Eq. (8-13) to Eq. (8-16), we obtain

$$\left.\frac{\partial u}{\partial y}\right|_{y=0} = \frac{3u_\infty}{2}\sqrt{\frac{13}{280}\frac{u_\infty}{\nu x}} \tag{8-17}$$

The substitution of Eq. (8-17) into Eq. (8-4) yields the local drag coefficient c_x:

$$c_x = 3\sqrt{\frac{13}{280}\frac{\nu}{u_\infty x}} = \frac{0.646}{\text{Re}_x^{1/2}} \tag{8-18}$$

The simple method of analysis just described is approximate because of the approximation introduced in step 2. We summarize in Table 8-1 the boundary-layer thickness $\delta(x)$ and the local drag coefficient c_x obtained by the integral method using second-, third- and fourth-degree polynomial approximations, together with the exact solution for the problem. Note that the result obtained above by a cubic polynomial approximation is within 2.7 percent of the exact value of the local drag coefficient given by

$$\boxed{c_x = \frac{0.664}{\text{Re}_x^{1/2}} \qquad \text{(exact)}} \tag{8-19}$$

which is valid for laminar flow, that is, Re $\leq 5 \times 10^5$. We note that the local drag coefficient $c_x \sim x^{-1/2}$, The average drag coefficient c_m, over the length $x = 0$ to $x = L$, can be determined from the definition [see Eq. (6-6)]:

$$c_m = \frac{1}{L}\int_{x=0}^{L} c_x\, dx = 2c_x\Big|_{x=L} \tag{8-20}$$

Table 8-1 Comparison of exact and approximate solutions for boundary-layer thickness and the local drag coefficient for laminar flow along a flat plate

Velocity profile	$\dfrac{\delta(x)}{x}$	c_x
Exact	$\dfrac{4.96}{\text{Re}_x^{1/2}}$	$\dfrac{0.664}{\text{Re}_x^{1/2}}$
Approximate: Second-degree polynomial	$\dfrac{5.5}{\text{Re}_x^{1/2}}$	$\dfrac{0.727}{\text{Re}_x^{1/2}}$
Approximate: Cubic polynomial	$\dfrac{4.64}{\text{Re}_x^{1/2}}$	$\dfrac{0.646}{\text{Re}_x^{1/2}}$
Approximate: Fourth-degree polynomial	$\dfrac{5.83}{\text{Re}_x^{1/2}}$	$\dfrac{0.686}{\text{Re}_x^{1/2}}$

We find

$$c_m = \frac{1.328}{\mathrm{Re}_L^{1/2}} \qquad \text{(exact)} \tag{8-21}$$

and the boundary-layer thickness is given by

$$\frac{\delta(x)}{x} = \frac{4.96}{\mathrm{Re}_x^{1/2}} \qquad \text{(exact)} \tag{8-22}$$

The results given by Eqs. (8-21) and (8-22) are valid for laminar flow, (that is, $\mathrm{Re} \leq 5 \times 10^5$).

Once the average drag coefficient c_m is known, the drag force F acting on the plate over the length $x = 0$ to $x = L$ and for a width w is determined from [see Eq. (6-7)]

$$F = wLc_m \frac{\rho u_\infty^2}{2} \qquad \mathrm{N} \tag{8-23}$$

Example 8-1 Atmospheric air at 300 K at a velocity of 1 m/s flows over a flat plate. Calculate the boundary-layer thickness $\delta(x)$ and the local drag coefficient c_x at $x = 0.75$ m from the leading edge of the plate. What is the drag force F acting on the plate over the length $x = 0$ to $x = 0.75$ m and width $w = 0.5$ m of the plate?

Solution The physical properties of atmospheric air at $T_\infty = 300$ K are

$$\rho = 1.177 \ \mathrm{kg/m^3} \qquad \nu = 0.168 \times 10^{-4} \ \mathrm{m^2/s}$$

The Reynolds number at $x = 0.75$ m is

$$\mathrm{Re}_x = \frac{u_\infty x}{\nu} = \frac{(1)(0.75)}{0.168 \times 10^{-4}} = 44{,}640$$

The boundary-layer thickness $\delta(x)$ and the local drag coefficient c_x at $x =$ 0.75 m are

$$\delta(x) = \frac{4.96x}{\mathrm{Re}_x^{1/2}} = \frac{(4.96)(0.75)}{(44{,}640)^{1/2}} = 0.0176 \ \mathrm{m}$$

and

$$c_x = \frac{0.664}{\mathrm{Re}_x^{1/2}} = 3.14 \times 10^{-3}$$

The mean drag coefficient over the length $x = 0$ to $x = 0.75$ m is

$$c_m = 2c_x \bigg|_{x=0.75} = 6.28 \times 10^{-3}$$

The drag force acting on the plate becomes

$$F = wLc_m \frac{\rho u_\infty^2}{2}$$

$$= (0.5 \text{ m})(0.75 \text{ m})(6.28 \times 10^{-3}) \frac{(1.177 \text{ kg/m}^3)(1^2 \text{ m}^2/\text{s}^2)}{2}$$

$$= 1.39 \times 10^{-3} \text{ N}$$

since

$$1 \text{ N} = 1 \text{ kg} \cdot \text{m/s}^2$$

Example 8-2 Calculate the boundary-layer thickness $\delta(x)$ at $x = 0.5$ m from the leading edge of a flat plate for flow at $T_\infty = 20°\text{C}$, $u_\infty = 1$ m/s, and atmospheric pressure of hydrogen, ethylene glycol, air, and water.

SOLUTION The boundary-layer thickness is given by Eq. (8-22) as

$$\delta(x) = \frac{4.96x}{\text{Re}_x^{1/2}} \qquad \text{valid for} \qquad \text{Re}_x \leq 5 \times 10^5$$

The results are listed here for $x = 0.5$ m, $u_\infty = 1$ m/s, $T_\infty = 20°\text{C}$, and atmospheric pressure:

Fluid:	Hydrogen	Ethylene glycol	Air	Water
ν, m²/s:	1×10^{-4}	0.19×10^{-4}	0.15×10^{-4}	1×10^{-6}
Re_x:	0.5×10^4	2.63×10^4	3.33×10^4	5×10^5
$\delta(x)$, m:	0.035	0.0153	0.0136	0.0035

Turbulent Boundary Layer

In the study of turbulent flow over bodies, the *local drag coefficient* c_x and the *thickness of the turbulent boundary layer* $\delta(x)$ are quantities of practical interest. The drag coefficient c_x cannot be determined by purely theoretical means as was the case for laminar boundary-layer flow over a flat plate. Also, an experimentally determined velocity profile is used for the determination of the boundary-layer thickness. Here we present the relations for the determination of c_x and $\delta(x)$ for turbulent flow over a flat plate.

Local drag coefficient Schlichting [3] examined the experimental data and recommended the following correlation for local drag coefficient c_x for turbulent flow over a flat plate:

$$\boxed{c_x = 0.0592\ \mathrm{Re}_x^{-0.2}} \tag{8-24}$$

valid for $5 \times 10^5 < \mathrm{Re}_x < 10^7$.

At higher Reynolds numbers, the following correlation is recommended by Schultz-Grunow [8]:

$$\boxed{c_x = 0.370(\log \mathrm{Re}_x)^{-2.584}} \tag{8-25}$$

valid for $10^7 < \mathrm{Re}_x < 10^9$.

We now consider a boundary-layer flow along a flat plate such that the flow is *laminar* over the region $0 \le x \le x_c$ and *turbulent* over the region $x_c < x \le L$. The local drag coefficients for each of the two regions are taken as

$$c_x^l = 0.664\ \mathrm{Re}_x^{-0.5} \qquad \text{in } 0 \le x \le x_c \text{ (laminar)}$$

$$c_x^t = 0.0592\ \mathrm{Re}_x^{-0.2} \qquad \text{in } x_c < x \le L \text{ (turbulent)}$$

The average drag coefficient c_m over the entire region $0 \le x \le L$ is

$$c_m = \frac{1}{L}\left(\int_0^{x_c} c_x^l\, dx + \int_{x_c}^{L} c_x^t\, dx\right)$$

$$= \frac{1}{L}\left[0.664\left(\frac{u_\infty}{\nu}\right)^{-0.5}\int_0^{x_c} x^{-0.5}\, dx + 0.0592\left(\frac{u_\infty}{\nu}\right)^{-0.2}\int_{x_c}^{L} x^{-0.2}\, dx\right]$$

After performing the integration, we obtain

$$c_m = 0.074\ \mathrm{Re}_L^{-0.2} - \frac{0.074\ \mathrm{Re}_c^{0.8} - 1.328\ \mathrm{Re}_c^{0.5}}{\mathrm{Re}_L} \tag{8-25a}$$

valid for $\mathrm{Re}_c \le \mathrm{Re}_L < 10^7$, where $\mathrm{Re}_L = u_\infty L/\nu$ and Re_c is the critical Reynolds number for transition to turbulent flow.

Clearly, the mean drag coefficient c_m, over the region where the flow is partly laminar and partly turbulent, depends on the value of the critical Reynolds number Re_c. Therefore, for ready reference, Eq. (8-25*a*) can be written more compactly in the form

$$\boxed{c_m = 0.074\ \mathrm{Re}_L^{-0.2} - \frac{B}{\mathrm{Re}_L} \qquad \text{for } \mathrm{Re}_c < \mathrm{Re}_L < 10^7} \tag{8-25b}$$

valid over the laminar and turbulent regions. The value of B depends on the choice of the value of Re_c and is given as

$$B = \begin{cases} 700 & \text{for } \text{Re}_c = 2 \times 10^5 \\ 1050 & \text{for } \text{Re}_c = 3 \times 10^5 \\ 1740 & \text{for } \text{Re}_c = 5 \times 10^5 \\ 3340 & \text{for } \text{Re}_c = 1 \times 10^6 \end{cases}$$

We note that $B = 0$ corresponds to the case of turbulent flow starting from $x = 0$.

Boundary-layer thickness The momentum integral equation (8-5) can be used to determine the turbulent boundary-layer thickness because the laminar sublayer is very thin. Calculations have been performed by assuming that the velocity profile u appearing on the left-hand side of this equation can be represented with a power-law velocity distribution determined experimentally in the form $u/u_\infty = (y/\delta)^{1/7}$, where δ is the boundary-layer thickness and u_∞ is the free-stream velocity. This velocity profile, however, cannot be used on the right-hand side of Eq. (8-5), because its derivative becomes infinite at $y = 0$. To alleviate this difficulty, an empirical relation for the wall shear stress, given in the form

$$\tau_w = \mu \frac{\partial u}{\partial y}\bigg|_{y=0} = 0.0296 \left(\frac{\nu}{u_\infty x} \right)^{1/5} \rho u_\infty^2$$

can be used for the right-hand side of this equation. The results of such analysis yield the following expressions for the thickness of turbulent boundary layer over a flat plate:

1. The boundary layer is completely turbulent, starting from the leading edge $y = 0$ of the plate. The boundary-layer thickness becomes

$$\boxed{\frac{\delta(x)}{x} = 0.381\, \text{Re}_x^{-1/5}} \tag{8-26a}$$

2. The boundary-layer thickness is laminar up to the location where $\text{Re}_c = 5 \times 10^5$, and then it becomes fully turbulent. For such a case, $\delta(x)$ is given by

$$\boxed{\frac{\delta(x)}{x} = 0.381\, \text{Re}_x^{-1/5} - 10{,}256\, \text{Re}_x^{-1}} \tag{8-26b}$$

valid for $5 \times 10^5 < \text{Re}_x < 10^7$.

Example 8-3 Water at $T_\infty = 20°\text{C}$ and a velocity of $u_\infty = 1.5$ m/s flows over a flat plate $L = 1.2$ m long. Calculate the mean drag coefficient c_m and the drag force F acting over the length $L = 1.2$ m and width $w = 1$ m of the plate. Assume the transition from laminar to turbulent flow occurs at $\text{Re}_c = 5 \times 10^5$.

SOLUTION The physical properties of water at $T_\infty = 20°C$ are

$$\rho = 998 \text{ kg/m}^3 \qquad \nu = 1 \times 10^{-6} \text{ m}^2\text{/s}$$

The Reynolds number at $L = 1.2$ m is

$$\text{Re}_L = \frac{u_\infty L}{\nu} = \frac{(1.5)(1.2)}{10^{-6}} = 1.8 \times 10^6$$

Thus the flow regime includes both laminar and turbulent flow; c_m is obtained from Eq. (8-25*b*) for $\text{Re}_c = 5 \times 10^5$ as

$$c_m = 0.074 \, \text{Re}_L^{-0.2} - \frac{1740}{\text{Re}_L} = 3.18 \times 10^{-3}$$

The drag force F is determined by Eq. (8-23)

$$F = wLc_m \frac{\rho u_\infty^2}{2}$$

$$= (1)(1.2)(3.18 \times 10^{-3}) \frac{(998)(1.5^2)}{2} = 4.28 \text{ N}$$

8-2 HEAT TRANSFER COEFFICIENT FOR FLOW OVER A FLAT PLATE

We now consider heat transfer to or from a fluid flowing over a flat plate. Suppose heat transfer starts at the leading edge of the plate. As discussed in Chap. 6, the velocity and thermal boundary layers start to develop simultaneously, and their relative thicknesses depend on the magnitude of the Prandtl number. If the temperature distribution $T(x, y)$ in the boundary layer is known, the local heat transfer coefficient $h(x)$ can be determined from its definition, given in Eq. (6-11*a*) as

$$h(x) = k \frac{[\partial T/\partial y]_{y=0}}{T_\infty - T_w} \tag{8-27}$$

where T_∞ and T_w are the fluid free-stream and the wall temperatures, respectively.

Here, first we present an approximate analysis of determining the temperature distribution in the thermal boundary layer, hence the heat transfer coefficient for the special case of $\text{Pr} \ll 1$, that is, liquid metals. The reason for considering the liquid metals first is the simplicity of the analysis for this particular case; yet, it will help improve our understanding of the role of the thermal boundary layer in heat transfer. The case of $\text{Pr} \cong 1$ (gases), which involves a more elaborate analysis, is considered afterward.

Liquid Metals in Laminar Flow

The Prandtl number is very low for liquid metals; therefore, the thermal boundary layer is much thicker than the velocity boundary layer (that is, $\delta_t \gg \delta$). Figure 8-2

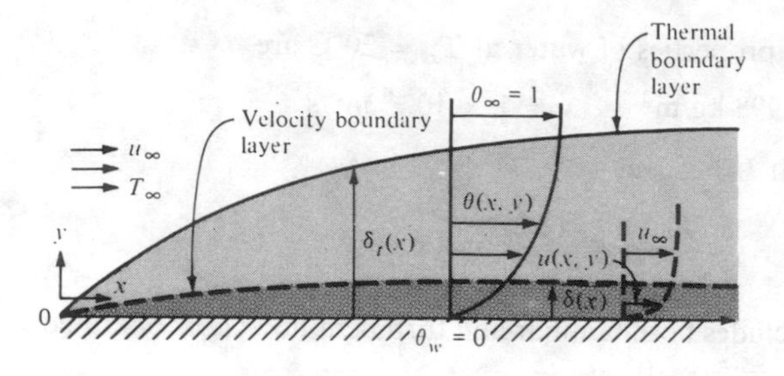

Figure 8-2 Velocity and thermal boundary layers for liquid-metal heat transfer, Pr $\ll$ 1.

illustrates the velocity and the thermal boundary layers when both start to develop from the leading edge of the flat plate. Let T_∞ and u_∞ be the temperature and the velocity of the fluid, respectively, outside the boundary layers; T_w is the surface temperature of the plate. We assume incompressible, constant-property fluid in two-dimensional, steady flow with negligible viscous dissipation. The energy equation governing the temperature distribution $T(x, y)$ in the thermal boundary layer is obtained from Eq. (6-83) by neglecting the viscous-dissipation term:

$$u\frac{\partial T}{\partial x} + v\frac{\partial T}{\partial y} = \alpha\frac{\partial^2 T}{\partial y^2} \tag{8-28}$$

For convenience in the analysis, we define a dimensionless temperature $\theta(x, y)$ as

$$\theta(x, y) = \frac{T(x, y) - T_w}{T_\infty - T_w} \tag{8-29}$$

where $\theta(x, y)$ varies from zero at the wall surface to unity at the edge of the thermal boundary layer. Then the energy equation is written in terms of $\theta(x, y)$ as

$$u\frac{\partial \theta}{\partial x} + v\frac{\partial \theta}{\partial y} = \alpha\frac{\partial^2 \theta}{\partial y^2} \qquad \text{for } x > 0 \tag{8-30}$$

and the boundary conditions are taken as

$$\theta = 0 \qquad \text{at } y = 0 \tag{8-31a}$$

$$\theta = 1 \qquad \text{at } y = \delta_t(x) \tag{8-31b}$$

where Eqs. (8-31*a*) and (8-31*b*) are, respectively, the statements of temperatures equal to T_w at the wall surface and T_∞ at the edge of the thermal boundary-layer thickness $\delta_t(x)$.

The exact analysis of this temperature problem is quite involved, because the velocity components u and v should be determined from the velocity problem given by Eqs. (8-1) to (8-3), before the energy equation (8-30) can be solved.

However, an approximate solution of this problem with the *integral method* is a relatively easy matter. The basic steps are as follows:

1. The energy equation (8-30) is integrated with respect to y over the thermal boundary layer, and the velocity component $v(x, y)$ is eliminated by means of

the continuity equation (8-1). The resulting equation, called the *energy integral equation*, is given by (see Note 2 at the end of this chapter for the derivation)

$$\frac{d}{dx}\left[\int_0^{\delta_t} u(1-\theta)\,dy\right] = \alpha \left.\frac{\partial\theta}{\partial y}\right|_{y=0} \qquad \text{in } 0 \le y \le \delta_t \tag{8-32}$$

where $\delta_t \equiv \delta_t(x)$, $u \equiv u(x, y)$, and $\theta \equiv \theta(x, y)$. So far, the analysis and Eq. (8-32) are exact, but this equation cannot be solved because it involves three unknowns, namely, $\delta_t(x)$, $u(x, y)$, and $\theta(x, y)$. Therefore, additional relationships are needed.

2. Approximations are introduced at this stage to develop simple analytic expressions for $u(x, y)$ and $\theta(x, y)$ consistent with the physical reality. Since the velocity boundary layer is very thin, the flow velocity over the large portion of the thermal boundary layer is uniform and equal to u_∞, as illustrated in Fig. 8-2. Therefore, as a first approximation, the velocity profile is taken as

$$u(x, y) = u_\infty = \text{constant} \tag{8-33}$$

The temperature profile $\theta(x, y)$ can be represented with a polynomial approximation within the thermal boundary layer. Suppose a cubic approximation is chosen for $\theta(x, y)$ in the form

$$\theta(x, y) = c_0 + c_1(x)y + c_2(x)y^2 + c_3(x)y^3 \qquad \text{in } 0 \le y \le \delta_t(x) \tag{8-34}$$

and the four conditions needed to determine the four coefficients are taken as

$$\theta = 0 \qquad \text{at } y = 0 \tag{8-35a}$$

$$\theta = 1 \qquad \text{at } y = \delta_t \tag{8-35b}$$

$$\frac{\partial\theta}{\partial y} = 0 \qquad \text{at } y = \delta_t \tag{8-35c}$$

$$\frac{\partial^2\theta}{\partial y^2} = 0 \qquad \text{at } y = 0 \tag{8-35d}$$

We note that the first two conditions are the boundary conditions for the problem given by Eqs. (8-31), the third condition is based on the definition of the thermal boundary layer, and the last condition is obtained by evaluating the energy equation (8-30) at $y = 0$ and noting that $u = v = 0$ at the wall surface. The application of conditions (8-35) to Eq. (8-34) gives the temperature profile in the form

$$\theta(x, y) = \frac{3}{2}\left(\frac{y}{\delta_t}\right) - \frac{1}{2}\left(\frac{y}{\delta_t}\right)^3 \tag{8-36}$$

3. The velocity and temperature profiles given by Eqs. (8-33) and (8-36) are introduced to the *energy integral equation* (8-32). We obtain

$$\frac{d}{dx}\left\{\int_0^{\delta_t} u_\infty\left[1 - \frac{3}{2}\frac{y}{\delta_t} + \frac{1}{2}\left(\frac{y}{\delta_t}\right)^3\right]dy\right\} = \alpha\frac{3}{2\delta_t} \tag{8-37}$$

where the right-hand side of this equation has resulted from the relation $[\partial\theta/\partial y]_{y=0} = 3/(2\delta_t)$. When the integration with respect to y is performed, Eq. (8-37) reduces to the following ordinary differential equation for the thermal boundary-layer thickness δ_t:

$$u_\infty \frac{3}{8}\frac{d\delta_t}{dx} = \frac{3\alpha}{2\delta_t}$$

or

$$\delta_t \, d\delta_t = \frac{4\alpha}{u_\infty} dx \tag{8-38}$$

The integration of Eq. (8-38) with the condition $\delta_t = 0$ for $x = 0$ gives the thermal boundary-layer thickness as

$$\delta_t^2 = \frac{8\alpha}{u_\infty} x \tag{8-39a}$$

or

$$\delta_t = \sqrt{\frac{8\alpha x}{u_\infty}} \tag{8-39b}$$

4. The temperature gradient at the wall for the cubic temperature profile, Eq. (8-36), becomes

$$\left.\frac{\partial\theta}{\partial y}\right|_{y=0} = \frac{3}{2\delta_t} \tag{8-40}$$

and the heat transfer coefficient defined by Eq. (8-27) is written in terms of $\theta(x, y)$ as

$$h(x) = k \left.\frac{\partial\theta}{\partial y}\right|_{y=0} \tag{8-41}$$

From Eqs. (8-40) and (8-41), we have

$$h(x) = \frac{3}{2}\frac{k}{\delta_t} \tag{8-42}$$

By introducing δ_t from Eq. (8-39b) to Eq. (8-42), the local heat transfer coefficient $h(x)$ is determined as

$$h(x) = \frac{3k}{2\sqrt{8}}\sqrt{\frac{u_\infty}{\alpha x}} = \frac{3}{2\sqrt{8}}\frac{k}{x}\sqrt{\frac{u_\infty x}{\nu}\frac{\nu}{\alpha}} = \frac{3}{2\sqrt{8}}\frac{k}{x}\sqrt{\text{Re}_x \, \text{Pr}} \tag{8-43}$$

The local Nusselt number Nu_x for the laminar flow of liquid metals over a flat plate maintained at a uniform temperature becomes

$$\text{Nu}_x = \frac{h(x)x}{k} = \frac{3}{2\sqrt{8}}\sqrt{\text{Re}_x \, \text{Pr}} = \boxed{0.530 \, \text{Pe}_x^{1/2}} \tag{8-44}$$

where $\mathrm{Re}_x = \dfrac{u_\infty x}{\nu}$ = local Reynolds number

$\mathrm{Pr} = \dfrac{\nu}{\alpha}$ = Prandtl number

$\mathrm{Pe}_x = \mathrm{Re}_x \, \mathrm{Pr} = \dfrac{u_\infty x}{\alpha}$ = local Péclét number

The solution given by Eq. (8-44) is obtained by an approximate analysis. This result should be compared with Pohlhausen's [9] exact solution of this heat transfer problem for the limiting case of Pr → 0, given by

$$\boxed{\mathrm{Nu}_x = 0.564\, \mathrm{Pe}_x^{1/2} \quad \text{(exact) for } \mathrm{Pr} \to 0} \tag{8-45}$$

This equation was derived under the assumption Pr → 0; in practice, this assumption implies liquid metals (that is, $\mathrm{Pr} < 0.05$). The approximate solution given by Eq. (8-44), is reasonably close to the above exact result.

At the beginning of this analysis we stated that for liquid metals the velocity boundary layer is much smaller than the thermal boundary layer. To check the validity of this statement, the velocity boundary-layer thickness $\delta(x)$ given by Eq. (8-13) is divided by the thermal boundary-layer thickness $\delta_t(x)$ given by Eq. (8-39*b*). We obtain

$$\frac{\delta(x)}{\delta_t(x)} = \sqrt{\frac{280}{13}\frac{\nu x}{u_\infty}}\sqrt{\frac{u_\infty}{8\alpha x}} = \sqrt{2.692\,\mathrm{Pr}}$$

For liquid metals with $\mathrm{Pr} \cong 0.01$, we find

$$\frac{\delta(x)}{\delta_t(x)} = 0.164 \tag{8-46}$$

which shows that for liquid metals $\delta(x) \ll \delta_t(x)$.

Ordinary Fluids in Laminar Flow

We now examine the determination of the heat transfer coefficient for laminar flow of ordinary fluids having $\mathrm{Pr} > 1$ over a flat plate maintained at a uniform temperature. It is assumed that a fluid at a temperature T_∞ flows with a velocity u_∞ over a flat plate. The x axis is chosen along the plate in the direction of flow with the origin $x = 0$ at the leading edge, and the y axis is perpendicular to the plate in the outward direction. The plate is maintained at a temperature T_∞ in the region $0 \le x \le x_0$ and at a uniform temperature T_w in the region $x > x_0$. That is, heat transfer between the plate and the fluid does not start until the location $x = x_0$. Figure 8-3 illustrates the velocity and thermal boundary layers for the physical situation just described. We note that the velocity boundary layer is thicker than

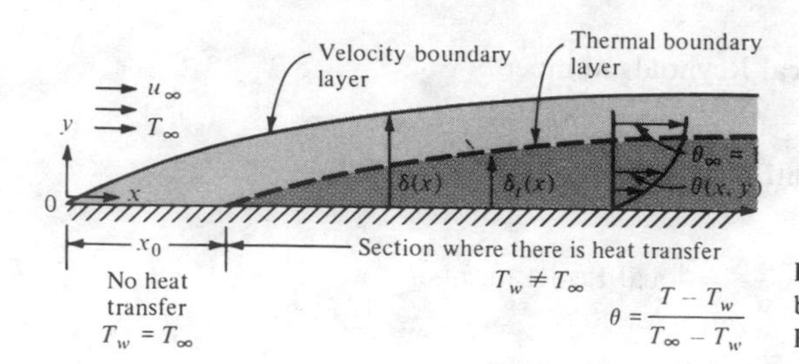

Figure 8-3 Velocity and thermal boundary layers for a fluid having Pr > 1.

the thermal boundary layer because Pr > 1; and $\delta(x)$ starts to develop at the leading edge of the plate, while $\delta_t(x)$ starts to develop at $x = x_0$ where the heat transfer section begins. Again we assume incompressible, constant-property fluid in two-dimensional, steady laminar flow with negligible viscous dissipation. The boundary-layer energy equation is taken as

$$u \frac{\partial \theta}{\partial x} + v \frac{\partial \theta}{\partial y} = \alpha \frac{\partial^2 \theta}{\partial y^2} \qquad \text{for } x > x_0 \tag{8-47}$$

and the boundary conditions are taken as

$$\theta = 0 \qquad \text{at } y = 0 \tag{8-48a}$$

$$\theta = 1 \qquad \text{at } y = \delta_t(x) \tag{8-48b}$$

where θ is defined by Eq. (8-29).

Since the exact analysis of this temperature problem is rather involved, again we consider its solution by the *integral method*:

1. The energy equation (8-47) is integrated with respect to y over the thermal boundary layer, and the velocity component $v(x, y)$ is eliminated by means of the continuity equation (8-1). The *energy integral equation* is determined as

$$\frac{d}{dx}\left[\int_0^{\delta_t} u(1 - \theta)\, dy\right] = \alpha \left.\frac{\partial \theta}{\partial y}\right|_{y=0} \qquad \text{in } 0 \le y \le \delta_t \tag{8-49}$$

which is the same as Eq. (8-32). This equation cannot be solved because it involves three unknowns, namely, $\delta_t(x)$, $u(x, y)$, and $\theta(x, y)$. Therefore, additional relationships are needed.

2. Approximations are introduced to develop analytic expressions for $u(x, y)$ and $\theta(x, y)$. For the velocity profile $u(x, y)$, we choose a cubic polynomial approximation as given by Eq. (8-8) and take it in the form

$$\frac{u(x, y)}{u_\infty} = \frac{3}{2}\left(\frac{y}{\delta}\right) - \frac{1}{2}\left(\frac{y}{\delta}\right)^3 \tag{8-50}$$

For the temperature profile $\theta(x, y)$, we also choose a cubic profile and immediately obtain the resulting expression from Eq. (8-36) as

$$\theta(x, y) = \frac{3}{2}\left(\frac{y}{\delta_t}\right) - \frac{1}{2}\left(\frac{y}{\delta_t}\right)^3 \tag{8-51}$$

3. The velocity and temperature profiles given by Eqs. (8-50) and (8-51) are introduced to the *energy integral equation* (8-49). We obtain

$$\frac{d}{dx}\left\{u_\infty \int_0^{\delta_t}\left[\frac{3}{2}\frac{y}{\delta} - \frac{1}{2}\left(\frac{y}{\delta}\right)^3\right]\left[1 - \frac{3}{2}\frac{y}{\delta_t} + \frac{1}{2}\left(\frac{y}{\delta_t}\right)^3\right] dy\right\} = \frac{3\alpha}{2\delta_t} \tag{8-52a}$$

$$\frac{d}{dx}\left[\int_0^{\delta_t}\left(\frac{3}{2\delta}y - \frac{9}{4\delta\delta_t}y^2 + \frac{3}{4\delta\delta_t^3}y^4 - \frac{1}{2\delta^3}y^3 + \frac{3}{4\delta^3\delta_t}y^4 - \frac{1}{4\delta^3\delta_t^3}y^6\right)dy\right] = \frac{3\alpha}{2\delta_t u_\infty} \tag{8-52b}$$

The integration with respect to y is performed:

$$\frac{d}{dx}\left(\frac{3}{4}\frac{\delta_t^2}{\delta} - \frac{3}{4}\frac{\delta_t^2}{\delta} + \frac{3}{20}\frac{\delta_t^2}{\delta} - \frac{1}{8}\frac{\delta_t^4}{\delta^3} + \frac{3}{20}\frac{\delta_t^4}{\delta^3} - \frac{1}{28}\frac{\delta_t^4}{\delta^3}\right) = \frac{3\alpha}{2\delta_t u_\infty} \tag{8-53}$$

A new variable $\Delta(x)$ is now defined as the ratio of the thermal boundary-layer thickness to the velocity boundary-layer thickness:

$$\Delta(x) = \frac{\delta_t(x)}{\delta(x)} \tag{8-54}$$

Then Eq. (8-53) becomes

$$\frac{d}{dx}\left[\delta\left(\tfrac{3}{20}\Delta^2 - \tfrac{3}{280}\Delta^4\right)\right] = \frac{3\alpha}{2\delta\Delta u_\infty} \tag{8-55}$$

We now consider the situation in which the thermal boundary-layer thickness is smaller than the velocity boundary-layer thickness δ, as illustrated in Fig. 8-3, for $\Pr > 1$. Then $\Delta < 1$, and in Eq. (8-55) the term $\frac{3}{280}\Delta^4$ can be neglected in comparison to $\frac{3}{20}\Delta^2$. Equation (8-55) simplifies to

$$\delta\Delta \frac{d}{dx}(\delta\Delta^2) = \frac{10\alpha}{u_\infty} \tag{8-56}$$

Differentiation with respect to x is performed as

$$2\delta^2\Delta^2 \frac{d\Delta}{dx} + \Delta^3\delta\frac{d\delta}{dx} = \frac{10\alpha}{u_\infty}$$

or

$$\tfrac{2}{3}\delta^2\frac{d\Delta^3}{dx} + \Delta^3\delta\frac{d\delta}{dx} = \frac{10\alpha}{u_\infty} \tag{8-57}$$

since

$$\Delta^2 \frac{d\Delta}{dx} = \frac{1}{3}\frac{d\Delta^3}{dx}$$

The velocity boundary-layer thickness δ was previously determined as [see Eq. (8-12)]:

$$\delta^2 = \frac{280}{13}\frac{\nu x}{u_\infty} \tag{8-58a}$$

and by differentiating we obtain

$$\delta \frac{d\delta}{dx} = \frac{140}{13}\frac{\nu}{u_\infty} \tag{8-58b}$$

The substitution of Eqs. (8-58) into Eq. (8-57) yields

$$x\frac{d\Delta^3}{dx} + \tfrac{3}{4}\Delta^3 = \frac{39}{56}\frac{\alpha}{\nu} \tag{8-59}$$

This is an ordinary differential equation of the first order in Δ^3, and its general solution is written as (see Note 3 at the end of this chapter for derivation)

$$\Delta^3(x) = Cx^{-3/4} + \frac{13}{14}\frac{\alpha}{\nu} \tag{8-60}$$

The integration constant C is determined by the application of the boundary condition $\delta_t = 0$ for $x = x_0$, which is equivalent to

$$\Delta(x) = 0 \qquad \text{for } x = x_0 \tag{8-61}$$

We find

$$\Delta^3(x) = \tfrac{13}{14}\text{Pr}^{-1}\left[1 - \left(\frac{x_0}{x}\right)^{3/4}\right] \tag{8-62}$$

where

$$\text{Pr} = \frac{\nu}{\alpha} = \text{Prandtl number}$$

If it is assumed that the heat transfer to the fluid starts at the leading edge of the plate, we set $x_0 \to 0$ and Eq. (8-62) simplifies to

$$\Delta(x) = \frac{\delta_t(x)}{\delta(x)} = (\tfrac{13}{14})^{1/3}\,\text{Pr}^{-1/3} = 0.976\,\text{Pr}^{-1/3} \tag{8-63}$$

This relation shows that the ratio of the thermal to velocity boundary-layer thickness for laminar flow along a flat plate is inversely proportional to the cube root of the Prandtl number.

The substitution of $\delta(x)$ from Eq. (8-58a) into Eq. (8-63) gives the thermal boundary-layer thickness as

$$\delta_t(x) = 4.53 \frac{x}{\mathrm{Re}_x^{1/2}\,\mathrm{Pr}^{1/3}} \tag{8-64}$$

where

$$\mathrm{Re}_x = \frac{u_\infty x}{\nu}$$

4. For the cubic polynomial approximation considered here for $\theta(x, y)$, the local heat transfer coefficient $h(x)$ has been related to the thermal boundary-layer thickness $\delta_t(x)$ previously, by Eq. (8-42), as

$$h(x) = \frac{3}{2}\frac{k}{\delta_t(x)} \tag{8-65}$$

By introducing $\delta_t(x)$ from Eq. (8-64) into Eq. (8-65), the local Nusselt number Nu_x is found:

$$\mathrm{Nu}_x = \frac{h(x)x}{k} = 0.331\,\mathrm{Pr}^{1/3}\,\mathrm{Re}_x^{1/2} \qquad \text{for } \mathrm{Re}_x < 5 \times 10^5 \tag{8-66}$$

This approximate solution is remarkably close to the exact solution of this problem given by Pohlhausen [9] as

$$\boxed{\mathrm{Nu}_x = 0.332\,\mathrm{Pr}^{1/3}\,\mathrm{Re}_x^{1/2} \qquad \text{(exact) for } \mathrm{Re}_x < 5 \times 10^5} \tag{8-67}$$

Note that the heat transfer relation given by Eq. (8-66) was derived by an approximate analysis on the assumption $\delta_t < \delta$ or $\mathrm{Pr} > 1$. However, its comparison with the exact results shows that it is valid in the range of $0.6 < \mathrm{Pr} < 10$, which covers most gases and liquids.

For very large values of the Prandtl number, Pohlhausen's exact calculations show that the local Nusselt number Nu_x is given by

$$\boxed{\mathrm{Nu}_x = 0.339\,\mathrm{Pr}^{1/3}\,\mathrm{Re}_x^{1/2} \qquad \text{(exact) for } \mathrm{Pr} \to \infty \text{ and } \mathrm{Re}_x < 5 \times 10^5} \tag{8-68}$$

To calculate the heat transfer coefficient from the above relations, it is recommended that the fluid properties be evaluated at the *arithmetic mean* of the *wall temperature* T_w *and the external-flow temperature* T_∞, that is, at $T_f = \frac{1}{2}(T_w + T_\infty)$, which is called the *film temperature.*

In engineering applications, an average heat transfer coefficient h_m over the length of the plate from $x = 0$ to $x = L$ is defined as

$$h_m = \frac{1}{L}\int_0^L h(x)\,dx \tag{8-69}$$

Noting that $h_x \sim x^{-1/2}$, we find the average heat transfer coefficient for laminar flow along a flat plate, over the length $x = 0$ to $x = L$, as

$$h_m = 2h(x)\Big|_{x=L} \tag{8-70}$$

Then the average Nusselt numbers for laminar flow along a flat plate are given by

$$\text{Nu}_m = 0.664\ \text{Pr}^{1/3}\ \text{Re}_L^{1/2} \qquad \text{(exact)}\ 0.6 < \text{Pr} < 10 \tag{8-71a}$$

$$\text{Nu}_m = 0.678\ \text{Pr}^{1/3}\ \text{Re}_L^{1/2} \qquad \text{(exact)}\ \text{Pr} \to \infty \tag{8-71b}$$

where

$$\text{Nu}_m = \frac{h_m L}{k} \qquad \text{Re}_L = \frac{u_\infty L}{\nu}$$

and the properties are evaluated at the *film temperature*. Equation (8-71*b*), derived for the limiting case of $\text{Pr} \to \infty$, is applicable for fluids having large Prandtl number, such as oils.

Turbulent Flow

A transition takes place from laminar to turbulent flow in the range of Reynolds numbers from 2×10^5 to 5×10^5 for flow over a flat plate. Heat transfer correlations can be developed for turbulent flow over a flat plate by utilizing the relationship between the heat transfer and the drag coefficients given by Eq. (6-15*a*):

$$\text{St}_x\ \text{Pr}^{2/3} = \tfrac{1}{2}c_x \tag{8-72}$$

For example, if c_x is obtained from Eq. (8-24), we find

$$\text{St}_x\ \text{Pr}^{2/3} = 0.0296\ \text{Re}_x^{-0.2} \qquad \text{for } 5 \times 10^5 < \text{Re}_x < 10^7 \tag{8-73a}$$

or c_x is obtained from Eq. (8-25) to yield

$$\text{St}_x\ \text{Pr}^{2/3} = 0.185(\log \text{Re}_x)^{-2.584} \qquad \text{for } 10^7 < \text{Re}_x < 10^9 \tag{8-73b}$$

and all properties are evaluated at the *film temperature*.

More recently, Whitaker [10] used the experimental data of Zukauskas and Ambrazyavichyus [11] and modified the Colburn [12] expression Eq. (7-68) to develop the following correlation for turbulent boundary layer along a flat plate:

$$\text{Nu}_x = 0.029\ \text{Re}_x^{0.8}\ \text{Pr}^{0.43} \tag{8-74}$$

valid for $\text{Re}_x > 2 \times 10^5$ to 5×10^5; and all properties are evaluated at the *film temperature*.

In practical applications, the average heat transfer coefficient h_m over the distance $0 \le x \le L$ of the plate is of interest. When the flow is turbulent, it is always preceded by a laminar boundary layer in which the equation governing the heat transfer is different from that for turbulent flow. Therefore, the averaging must be performed over both regions as now described.

Let the flow be *laminar* over the region $0 \le x \le c$ and *turbulent* over the region $c < x \le L$. The local heat transfer coefficients for these two regions are obtained from Eqs. (8-67) and (8-74), respectively, as

$$h_x^l = 0.332\left(\frac{k}{x}\right)\left(\frac{u_\infty x}{\nu}\right)^{1/2} \text{Pr}^{1/3} \qquad \text{in } 0 \le x \le c \text{ (laminar)}$$

$$h_x^t = 0.029\left(\frac{k}{x}\right)\left(\frac{u_\infty x}{\nu}\right)^{0.8} \text{Pr}^{0.43} \qquad \text{in } c < x \le L \text{ (turbulent)}$$

The average heat transfer coefficient h_m over the region $0 \le x \le L$ is defined as

$$h_m = \frac{1}{L}\left(\int_0^c h_x^l \, dx + \int_c^L h^t \, dx\right)$$

$$h_m = \frac{1}{L}\left[0.332k\left(\frac{u_\infty}{\nu}\right)^{0.5} \text{Pr}^{1/3} \int_0^c x^{-0.5} \, dx + 0.029k\left(\frac{u_\infty}{\nu}\right)^{0.8} \text{Pr}^{0.43} \int_c^L x^{-0.2} \, dx\right] \tag{8-75a}$$

and the average Nusselt number Nu_m over the region $0 \le x \le L$ as

$$\text{Nu}_m = \frac{h_m L}{k} \tag{8-75b}$$

After the integrations are performed, the average Nusselt number over the *laminar and turbulent flow regions* is

$$\text{Nu}_m = 0.036\,\text{Pr}^{0.43}(\text{Re}_L^{0.8} - \text{Re}_c^{0.8}) + 0.664\,\text{Pr}^{1/3}\,\text{Re}_c^{0.5} \tag{8-76}$$

valid for $\text{Re}_L > \text{Re}_c$, where $\text{Re}_L = u_\infty L/\nu$ and Re_c = critical Reynolds number for transition.

Clearly, Nu_m, as given by Eq. (8-76), depends on the value of the critical Reynolds number for transition from laminar to turbulent flow. The free-stream turbulence level affects the transition. When high-intensity turbulence is generated in the free stream, the transition to turbulent flow takes place at a lower critical Reynolds number. However, if care is taken to eliminate free-stream turbulence, the transition to turbulent flow is delayed.

For a critical Reynolds number $\text{Re}_c = 2 \times 10^5$, Eq. (8-76) yields

$$\text{Nu}_m = 0.036\,\text{Pr}^{0.43}(\text{Re}_L^{0.8} - 17{,}400) + 297\,\text{Pr}^{1/3} \tag{8-77}$$

The last term on the right-hand side can be approximated as

$$297\ \mathrm{Pr}^{1/3} \cong 297\ \mathrm{Pr}^{0.43}$$

and the viscosity correction can be introduced by multiplying the right-hand side of the resulting expression by $(\mu_\infty/\mu_w)^{0.25}$. Then the following equation is obtained:

$$\mathrm{Nu}_m = 0.036\ \mathrm{Pr}^{0.43}(\mathrm{Re}_L^{0.8} - 9200)\left(\frac{\mu_\infty}{\mu_w}\right)^{0.25} \tag{8-78}$$

All physical properties are evaluated at the free-stream temperature except μ_w, which is evaluated at the wall temperature. For gases, the viscosity correction is neglected, and for such a case the physical properties are evaluated at the film temperature.

Equation (8-78) gives the average Nusselt number over the laminar and turbulent boundary layers over a flat plate for $\mathrm{Re}_L > 2 \times 10^5$. It has been proposed by Whitaker [10] and used to correlate the experimental data of several investigators [11, 13, 14] for air, water, and oil covering the following ranges:

$$2 \times 10^5 < \mathrm{Re}_L < 5.5 \times 10^6$$

$$0.70 < \mathrm{Pr} < 380$$

$$0.26 < \frac{\mu_\infty}{\mu_w} < 3.5$$

Equation (8-78) correlates the experimental data reasonably well when the free-stream turbulence is small. If high-level turbulence is present in the free stream, Eq. (8-78) without the constant 9200 correlates the data reasonably well.

Example 8-4 Atmospheric air at $T_\infty = 275$ K and a free-stream velocity $u_\infty = 20$ m/s flows over a flat plate $L = 1.5$ m long that is maintained at a uniform temperature $T_w = 325$ K.

(*a*) Calculate the average heat transfer coefficient h_m over the region where the boundary layer is laminar.

(*b*) Find the average heat transfer coefficient over the entire length $L = 1.5$ m of the plate.

(*c*) Calculate the total heat transfer rate Q from the plate to the air over the length $L = 1.5$ m and width $w = 1$ m.

Assume transition occurs at $\mathrm{Re}_c = 2 \times 10^5$.

SOLUTION The physical properties of atmospheric air are taken as follows at $(T_w + T_\infty)/2 = 300$ K:

$$k = 0.026\ \mathrm{W/(m \cdot {}^\circ C)} \qquad \mathrm{Pr} = 0.708$$

$$\nu = 16.8 \times 10^{-6}\ \mathrm{m^2/s}$$

$$\mu_\infty = 1.98 \times 10^{-5}\ \mathrm{kg/(m \cdot s)}$$

For $Re_c = 2 \times 10^5$, the location x_c where the transition occurs is determined as

$$x_c = \frac{\nu\, Re_c}{u_\infty} = \frac{(16.8 \times 10^{-6})(2 \times 10^5)}{20}$$

$$= 0.168 \text{ m}$$

(*a*) The average heat transfer coefficient for the laminar boundary layer, if we neglect the viscosity correction, is determined by Eq. (8-71*a*):

$$h_m = 0.664\left(\frac{k}{x_c}\right) Pr^{1/3}\, Re_c^{1/2}$$

$$= 0.664\left(\frac{0.026}{0.168}\right)(0.708)^{1/3}(2 \times 10^5)^{1/2}$$

$$= 41.0 \text{ W/(m}^2 \cdot \text{s)} \qquad \text{for } 0 < x < 0.168 \text{ m}$$

(*b*) The Reynolds number at $L = 1.5$ m is

$$Re_L = \frac{u_\infty L}{\nu} = \frac{(20)(1.5)}{16.8 \times 10^{-6}} = 1.79 \times 10^6$$

Then the average heat transfer coefficient over $L = 1.5$ m, if we neglect the viscosity correction, is determined by Eq. (8-78) as

$$h_m = 0.036\left(\frac{k}{L}\right) Pr^{0.43}(Re_L^{0.8} - 9200)$$

$$= 0.036\left(\frac{0.026}{1.5}\right)(0.708)^{0.43}[(1.79 \times 10^6)^{0.8} - 9200]$$

$$= 49.1 \text{ W/(m}^2 \cdot {}^\circ\text{C)}$$

(*c*) The heat transfer rate is

$$Q = wLh_m(T_w - T_\infty)$$

$$= (1)(1.5)(49.1)(325 - 275) = 3683 \text{ W}$$

8-3 FLOW ACROSS A SINGLE CIRCULAR CYLINDER

Flow across a single circular cylinder frequently is encountered in practice, but the determination of the drag and heat transfer coefficients is a very complicated matter because of the complexity of the flow patterns around the cylinder. Figure 8-4 illustrates with sketches the flow characteristics around a circular cylinder; clearly they depend on the Reynolds number, defined as

$$Re = \frac{u_\infty D}{\nu} \tag{8-79}$$

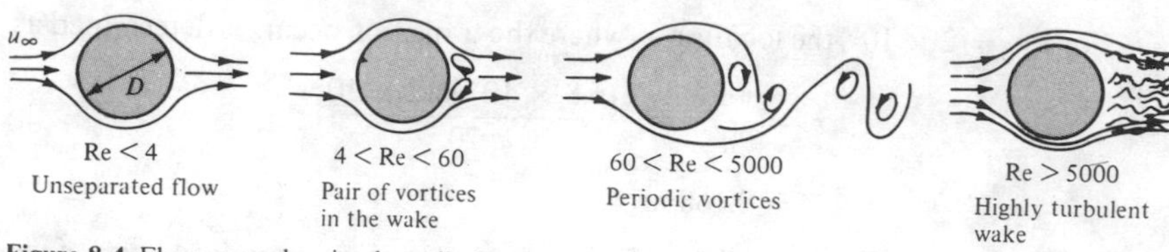

Figure 8-4 Flow around a circular cylinder at various Reynolds numbers.

where D is the cylinder diameter and u_∞ is the free-stream velocity. For a Reynolds number less than about 4, the flow remains unseparated and the velocity field can be analyzed by the solution of equations of motion [15]. For Reynolds numbers above about 4, the vortices start in the wake region and the analysis of velocity and temperature distribution around the cylinder for Re > 4 becomes very complicated [16].

Drag Coefficient

Consider flow at a velocity u_∞ across a circular cylinder of diameter D, and let F be the drag force acting on the length L of the cylinder. A drag coefficient c_D is defined as

$$\boxed{\frac{F}{LD} = c_D \frac{\rho u_\infty^2}{2}} \tag{8-80}$$

Here LD represents the area normal to the flow. The drag coefficient c_D defined by Eq. (8-80) is the average value of the local drag coefficient over the circumference of the cylinder. Thus, given c_D, the drag force F acting over the length L of the cylinder can be calculated according to Eq. (8-80).

Figure 8-5 shows the drag coefficient c_D for flow across a single cylinder. The physical significance of the variation of c_D with the Reynolds number is better envisioned if we examine the results in Fig. 8-5 in relation to the sketches in Fig. 8-4. For Re < 4, the drag is caused by viscous forces only since the boundary layer remains attached to the cylinder. In the region 4 < Re < 5000, vortices are formed in the wake; therefore, the drag is partly due to the viscous forces and partly due to the wake formation, that is, the low pressure caused by the flow separation. In the region $5 \times 10^3 < \text{Re} < 3.5 \times 10^5$, the drag is caused predominantly by the highly turbulent eddies in the wake. The sudden reduction in drag at $\text{Re} \cong 3.5 \times 10^5$ is caused because the boundary layer changes to turbulent, thus causing the point of flow separation to move toward the rear of the cylinder, which in turn reduces the size of the wake, hence the drag.

Heat Transfer Coefficient

Figure 8-6 shows McAdams' [17] correlation of the average heat transfer coefficient h_m for the cooling or heating of air flowing across a single circular cylinder. The properties are to be evaluated at $(T_\infty + T_w)/2$. This correlation for gases does

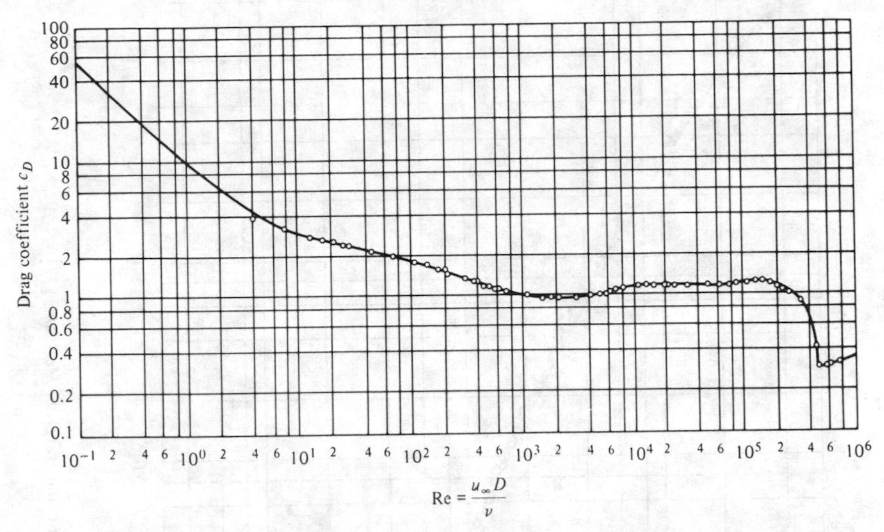

Figure 8-5 Drag coefficient for flow across a single circular cylinder. (*From Schlichting* [3].)

not show explicitly the dependence of the results on the Prandtl number, because gases have a Prandtl number of about unity. Therefore, more elaborate correlations have been developed by several investigators [18–20] in order to include the effects of the Prandtl number and hence extend the applicability of the results to fluids other than gases.

Whitaker [18] correlated the average heat transfer coefficient h_m for the flow of gases or liquids across a single cylinder by

$$\mathrm{Nu}_m \equiv \frac{h_m D}{k} = (0.4\ \mathrm{Re}^{0.5} + 0.06\ \mathrm{Re}^{2/3})\ \mathrm{Pr}^{0.4}\left(\frac{\mu_\infty}{\mu_w}\right)^{0.25} \tag{8-81}$$

which agrees with the experimental data [21–26] within ± 25 percent in the range of variables

$$40 < \mathrm{Re} < 10^5$$

$$0.67 < \mathrm{Pr} < 300$$

$$0.25 < \frac{\mu_\infty}{\mu_w} < 5.2$$

where the physical properties are evaluated at the free-stream temperature except for μ_w, which is evaluated at the wall temperature. For gases, the viscosity correction is neglected, and for such a case properties are evaluated at the film temperature. We note that Eq. (8-81) involves two different functional dependences of the Nusselt number on the Reynolds number. The functional dependence $\mathrm{Re}^{0.5}$ characterizes the contribution from the undetached laminar boundary region, and

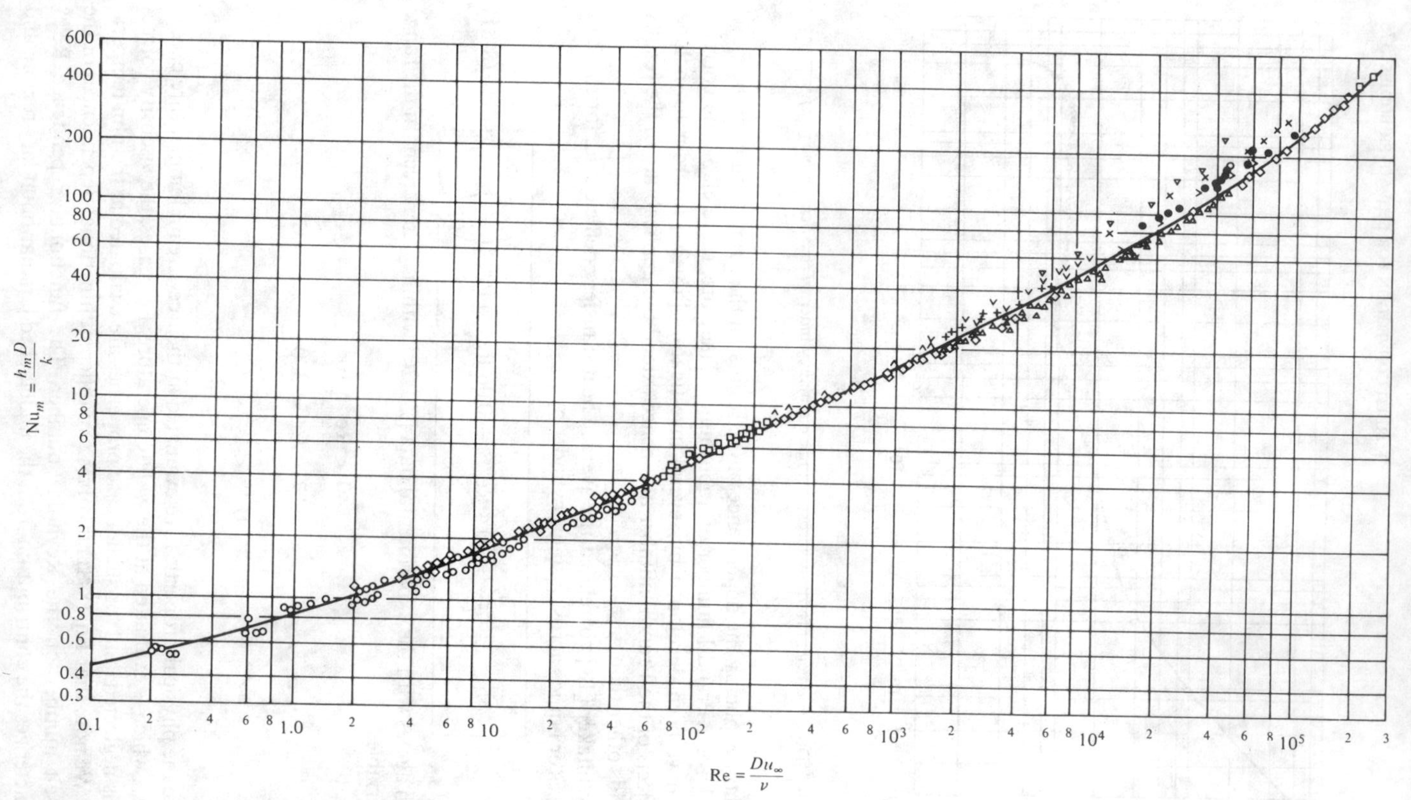

Figure 8-6 Average Nusselt number for heating or cooling of air flowing across a single circular cylinder. (*From McAdams* [17].)

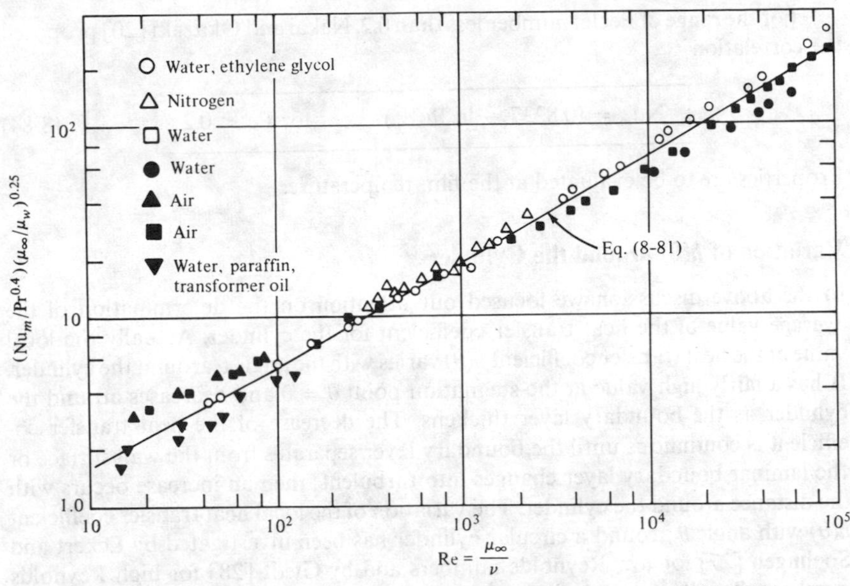

Figure 8-7 Nusselt number for flow across a single circular cylinder. (*From Whitaker* [*18*].)

$\mathrm{Re}^{2/3}$ characterizes the contribution from the wake region around the cylinder. Figure 8-7 shows the correlation of Eq. (8-81) with the experimental data of various investigators [21–26] for different fluids.

A more elaborate but more general correlation is given by Churchill and Bernstein [19] for the average heat transfer coefficient h_m for flow across a single cylinder as

$$\mathrm{Nu}_m = 0.3 + \frac{0.62\,\mathrm{Re}^{1/2}\,\mathrm{Pr}^{1/3}}{[1 + (0.4/\mathrm{Pr})^{2/3}]^{1/4}} \left[1 + \left(\frac{\mathrm{Re}}{282{,}000}\right)^{5/8}\right]^{4/5} \tag{8-82}$$

which is applicable for $10^2 < \mathrm{Re} < 10^7$ and $\mathrm{Pe} = \mathrm{Re} \cdot \mathrm{Pr} > 0.2$.

Equation (8-82) underpredicts most data by about 20 percent in the range $20{,}000 < \mathrm{Re} < 400{,}000$. Therefore, for this particular range of Reynolds number, the following modified form of Eq. (8-82) is recommended:

$$\mathrm{Nu}_m = 0.3 + \frac{0.62\,\mathrm{Re}^{1/2}\,\mathrm{Pr}^{1/3}}{[1 + (0.4/\mathrm{Pr})^{2/3}]^{1/4}} \left[1 + \left(\frac{\mathrm{Re}}{282{,}000}\right)^{1/2}\right] \tag{8-83}$$

for $20{,}000 < \mathrm{Re} < 400{,}000$.

In Eqs. (8-82) and (8-83), all properties are evaluated at the film temperature. Equations (8-82) and (8-83) were developed by correlating the experimental data of many investigators, and the fluids included air, water, and liquid sodium with both constant wall temperature and constant wall heat flux.

For the range of Péclét number less than 0.2, Nakai and Okazaki [20] proposed the correlation

$$\text{Nu}_m = (0.8237 - \ln \text{Pe}^{1/2})^{-1} \qquad \text{for Pe} < 0.2 \tag{8-84}$$

Properties are to be evaluated at the film temperature.

Variation of $h(\theta)$ around the Cylinder

In the above discussion we focused our attention on the determination of the average value of the heat transfer coefficient for the cylinder. Actually the local value of the heat transfer coefficient $h(\theta)$ varies with the angle θ around the cylinder. It has a fairly high value at the stagnation point $\theta = 0$ and decreases around the cylinder as the boundary layer thickens. The decrease of the heat transfer coefficient is continuous until the boundary layer separates from the wall surface or the laminar boundary layer changes into turbulent; then an increase occurs with the distance around the cylinder. The variation of the local heat transfer coefficient $h(\theta)$ with angle θ around a circular cylinder has been investigated by Eckert and Soehngen [27] for low Reynolds numbers and by Giedt [28] for high Reynolds numbers. To illustrate the complicated heat transfer mechanism around the cylinder, we present in Fig. 8-8 Giedt's data. At high Reynolds numbers, the heat transfer coefficient exhibits two minima around the cylinder. For example, in the curve for Re = 140,000, the first minimum occurs at the transition from laminar to turbulent boundary layer at an angle $\theta \cong 80°$; the second minimum occurs at $\theta \cong 130°$, where the flow separation takes place.

Example 8-5 Atmospheric air at $T_\infty = 250$ K and a free-stream velocity $u_\infty = 30$ m/s flows across a circular cylinder of diameter $D = 2.5$ cm. The surface of the cylinder is maintained at a uniform temperature $T_w = 350$ K.
(*a*) Calculate the average heat transfer coefficient h_m.
(*b*) Determine the heat transfer rate Q per 1-m length of the cylinder.
(*c*) Find the average drag coefficient c_D.
(*d*) Calculate the drag force F acting per 1-m length of the cylinder.

SOLUTION Equation (8-81) is used to calculate the heat transfer coefficient. The physical properties of air at the film temperature $T_f = 300$ K are

$k = 0.0262$ W/(m · °C) $\qquad \nu = 16.84 \times 10^{-6}$ m²/s

Pr = 0.708 $\qquad \mu = 1.983 \times 10^{-5}$ kg/(m · s) $\qquad \rho = 1.177$ kg/m³

The Reynolds number becomes

$$\text{Re} = \frac{u_\infty D}{\nu} = \frac{(30)(0.025)}{16.84 \times 10^{-6}} = 44{,}537$$

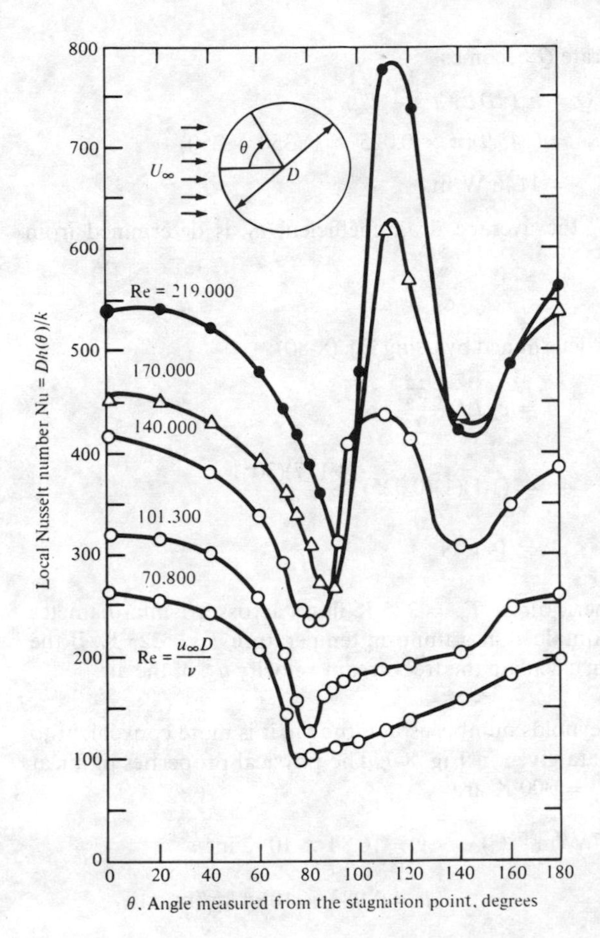

Figure 8-8 Variation of the local heat transfer coefficient $h(\theta)$ around a circular cylinder for flow of air. (*From Giedt* [*28*].)

(*a*) Equation (8-81) is applied to calculate h_m:

$$\text{Nu}_m = (0.4\ \text{Re}^{0.5} + 0.06\ \text{Re}^{2/3})\ \text{Pr}^{0.4}$$

$$= [0.4(44{,}537)^{0.5} + 0.06(44{,}537)^{2/3}](0.708)^{0.4}$$

$$= 139.2$$

$$h_m = \frac{k}{D}\ \text{Nu}_m = \frac{0.0262}{0.025}(139.2) = 145.9\ \text{W/(m}^2 \cdot {}^\circ\text{C)}$$

If Eq. (8-83) were used, the heat transfer coefficient would be $h_m = 143.7$ W/(m$^2 \cdot$ °C); the two results are very close.

(*b*) The heat transfer rate Q becomes

$$Q = h_m(\pi DL)(T_w - T_\infty)$$
$$= (145.9)(\pi \times 0.025 \times 1)(350 - 250)$$
$$= 1146 \text{ W/m}$$

(*c*) With Re = 44,537, the average drag coefficient c_D is determined from Fig. 8-5;

$$c_D \cong 1.1$$

(*d*) The drag force F is determined by using Eq. (8-80):

$$F = c_D LD \frac{\rho u_\infty^2}{2}$$
$$= (1.1)(1)(0.025)\frac{(1.177)(30^2)}{2}$$
$$= 14.6 \text{ N}$$

Example 8-6 Atmospheric air at $T_\infty = 275$ K flows across a 1-mm-diameter electric wire that is maintained at a uniform temperature $T_w = 325$ K. If the wire dissipates 70 W/m, calculate the free-stream velocity u_∞ of the air.

SOLUTION Since the Reynolds number is not known, it is more convenient to use the heat transfer data given in Fig. 8-7. The physical properties of air at the film temperature $T_f = 300$ K are

$$k = 0.0262 \text{ W/(m} \cdot {}^\circ\text{C)} \qquad \nu = 16.84 \times 10^{-6} \text{ m}^2\text{/s}$$
$$\text{Pr} = 0.708 \qquad \mu = 1.983 \times 10^{-5} \text{ kg/(m} \cdot \text{s)}$$

The heat transfer coefficient h_m is

$$h_m = \frac{Q}{\pi DL\,\Delta T} = \frac{70}{\pi \times 10^{-3} \times 1 \times 50} = 445.6 \text{ W/(m}^2 \cdot {}^\circ\text{C)}$$

Then

$$\text{Nu}_m = \frac{h_m D}{k} = \frac{445.6 \times 10^{-3}}{0.0262} = 17$$
$$\text{Pr}^{0.4} = (0.708)^{0.4} = 0.871$$

or

$$\frac{\text{Nu}_m}{\text{Pr}^{0.4}} = \frac{17}{0.871} = 19.5$$

From Fig. 8-7 we obtain

$$\text{Re} \cong 10^3$$

or

$$u_\infty = \frac{\nu}{D} 10^3 = \frac{16.84 \times 10^{-6}}{10^{-3}} (10^3) = 16.84 \text{ m/s}$$

8-4 FLOW ACROSS A SINGLE NONCIRCULAR CYLINDER

The results of experiments for the average heat transfer coefficient h_m for the flow of gases across a single, noncircular, long cylinder of various geometries have been correlated by Jakob [29] with the following simple relationship:

$$\text{Nu}_m = \frac{h_m D_e}{k} = c\left(\frac{u_\infty D_e}{\nu}\right)^n \tag{8-85}$$

Table 8-2 Constants *c* and *n* of Eq. (8-85)

Flow direction and geometry	$\text{Re} = \frac{u_\infty D_e}{\nu}$	n	c
u_∞ → D_e	5,000–100,000	0.588	0.222
u_∞ → D_e	2,500–15,000	0.612	0.224
u_∞ → D_e	2,500–7500	0.624	0.261
u_∞ → D_e	5,000–100,000	0.638	0.138
u_∞ → D_e	5,000–19,500	0.638	0.144
u_∞ → D_e	5,000–100,000	0.675	0.092
u_∞ → D_e	2,500–8,000	0.699	0.160
u_∞ → D_e	4,000–15,000	0.731	0.205
u_∞ → D_e	19,500–100,000	0.782	0.035
u_∞ → D_e	3,000–15,000	0.804	0.085

Source: Jakob [29].

where the constant c, the exponent n, and the characteristic dimension D_e for various geometries are presented in Table 8-2. The physical properties of the fluid are evaluated at the arithmetic mean of the free-stream and the wall temperatures.

8-5 FLOW ACROSS A SINGLE SPHERE

The flow characteristics across a single sphere are somewhat similar to those shown in Fig. 8-4 for a single cylinder. Therefore, the dependence of the drag and heat transfer coefficients on the Reynolds number for a sphere is expected to be of the same form as that for a single cylinder.

Drag Coefficient

If F is the total drag force due to flow across a single sphere, the average drag coefficient c_D is defined by the relation

$$\frac{F}{A} = c_D \frac{\rho u_\infty^2}{2} \tag{8-86}$$

where A is the *frontal area* (that is, $A = \pi D^2/4$) and u_∞ is the free-stream velocity. We note that F/A is the drag force per unit frontal area of the sphere. Figure 8-9 shows the average drag coefficient c_D for flow across a single sphere. A comparison of the drag coefficient curves in Figs. 8-5 and 8-9 for a single cylinder and

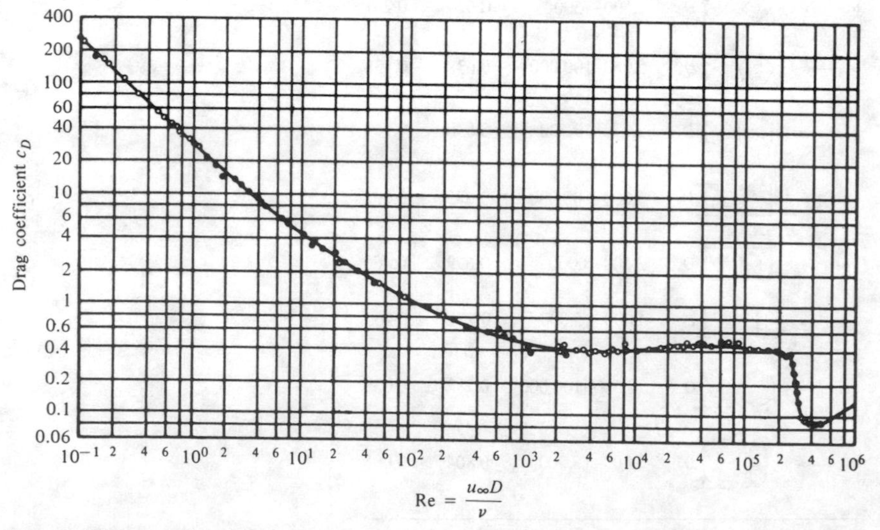

Figure 8-9 Drag coefficient for flow over a single sphere. (*From Schlichting* [3].)

sphere, respectively, reveals that the two curves have similar general characteristics.

Heat Transfer Coefficient

For the flow of gases across a single sphere, McAdams [17] recommends the simple correlation

$$\mathrm{Nu}_m = \frac{h_m D}{k} = 0.37\ \mathrm{Re}^{0.6} \qquad \text{for } 17 < \mathrm{Re} < 70{,}000 \tag{8-87}$$

where h_m is the mean heat transfer coefficient over the entire surface of the sphere. The properties are evaluated at $(T_\infty + T_w)/2$.

A more general correlation for the flow of gases and liquids across a single sphere is presented by Whitaker [18] in the form

$$\mathrm{Nu}_m = 2 + (0.4\ \mathrm{Re}^{0.5} + 0.06\ \mathrm{Re}^{2/3})\ \mathrm{Pr}^{0.4}\left(\frac{\mu_\infty}{\mu_w}\right)^{0.25} \tag{8-88}$$

which is valid over the ranges

$$3.5 < \mathrm{Re} < 8 \times 10^4$$

$$0.7 < \mathrm{Pr} < 380$$

$$1 < \frac{\mu_\infty}{\mu_w} < 3.2$$

and the physical properties are evaluated at the free-stream temperature, except μ_w, which is evaluated at the wall temperature. For gases the viscosity correction is neglected, but the physical properties are evaluated at the film temperature.

Equation (8-88) for a sphere and Eq. (8-81) for a cylinder have the same functional dependence of the Nusselt number on the Reynolds number, except for the constant 2 in Eq. (8-88). As $\mathrm{Re} \to 0$ (i.e., no flow), Eq. (8-88) assumes a limiting value $\mathrm{Nu} = 2$, which represents the steady-state heat conduction from a sphere at a uniform temperature into the surrounding infinite medium.

Figure 8-10 shows a correlation of Eq. (8-88) with the experimental data of Refs. 30 to 32 for air, water, and oil. Equation (8-88) represents the data reasonably well.

Example 8-7 Atmospheric air at $T_\infty = 250$ K and a free-stream velocity $u_\infty = 30$ m/s flows across a sphere of diameter $D = 2.5$ cm. The surface of the sphere is maintained at a uniform temperature $T_w = 350$ K by electric heating.

(*a*) Calculate the average heat transfer coefficient h_m.

(*b*) Find the heat transfer rate Q from the sphere.

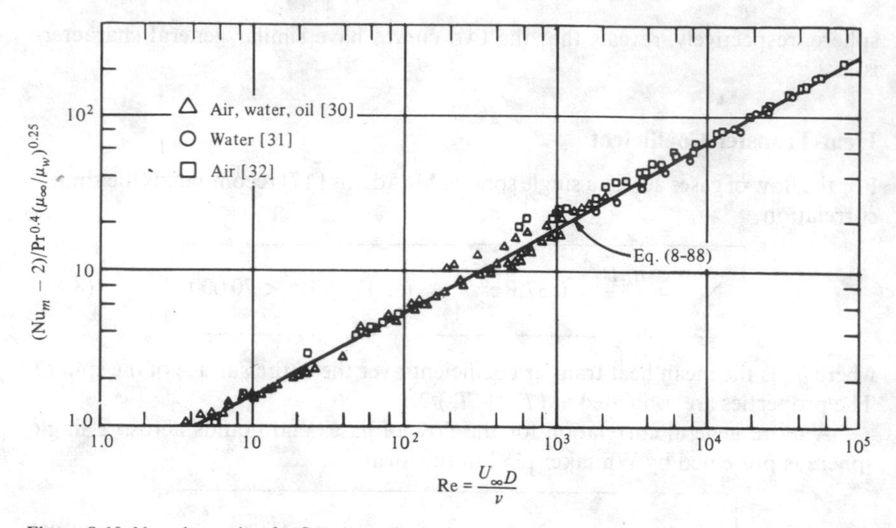

Figure 8-10 Nusselt number for flow across a single sphere. (*From Whitaker [18].*)

(*c*) Find the average drag coefficient c_D.
(*d*) Determine the drag force F acting on the sphere.

SOLUTION Equation (8-88) can be used to calculate the average heat transfer coefficient h_m. The physical properties of air are the same as in Example 8-5. Therefore,

$$\mathrm{Re} = \frac{u_\infty D}{\nu} = \frac{(30)(0.025)}{16.84 \times 10^{-6}} = 44{,}537$$

(*a*) Equation (8-88) is the same as Eq. (8-81), used in Example 8-5, except for the constant of 2. Then Nu_m is obtained immediately from Example 8-5 as

$$\mathrm{Nu}_m = 139.2 + 2 = 141.2$$

and

$$h_m = \frac{k}{D}\,141.2 = \frac{0.0262}{0.025}(141.2)$$

$$= 148 \text{ W/(m}^2 \cdot {}^\circ\text{C)}$$

(*b*) The heat transfer rate becomes

$$\begin{aligned} Q &= h_m(\pi D^2)(T_w - T_\infty) \\ &= (148)(\pi \times 0.025^2)(350 - 250) \\ &= 29 \text{ W} \end{aligned}$$

(*c*) The mean drag coefficient c_D for Re = 44,537 is determined from Fig. 8-9 as

$$c_D = 0.45$$

(*d*) The drag force F is determined from Eq. (8-86) as

$$F = c_D(\tfrac{1}{4}\pi D^2)\frac{\rho u_\infty^2}{2}$$

$$= (0.45)(\tfrac{1}{4}\pi)(0.025^2)\frac{(1.177)(30^2)}{2} = 0.117 \text{ N}$$

8-6 FLOW ACROSS TUBE BUNDLES

Heat transfer and pressure drop characteristics of tube bundles have numerous applications in the design of heat exchangers and industrial heat transfer equipment. For example, a common type of heat exchanger consists of a tube bundle with one fluid passing through the tubes and the other passing across the tubes. Frequently used tube bundle arrangements include the *in-line* and the *staggered* arrangements, illustrated in Fig. 8-11*a* and *b* respectively. The tube bundle geometry is characterized by the *transverse pitch* S_T and the *longitudinal pitch* S_L between the tube centers; the *diagonal pitch* S_D between the centers of the tubes in the diagonal row sometimes is used for the staggered arrangement. To define the Reynolds number for flow through the tube bank, the flow velocity is based on the *minimum free-flow area* available for flow, whether the minimum area occurs between the

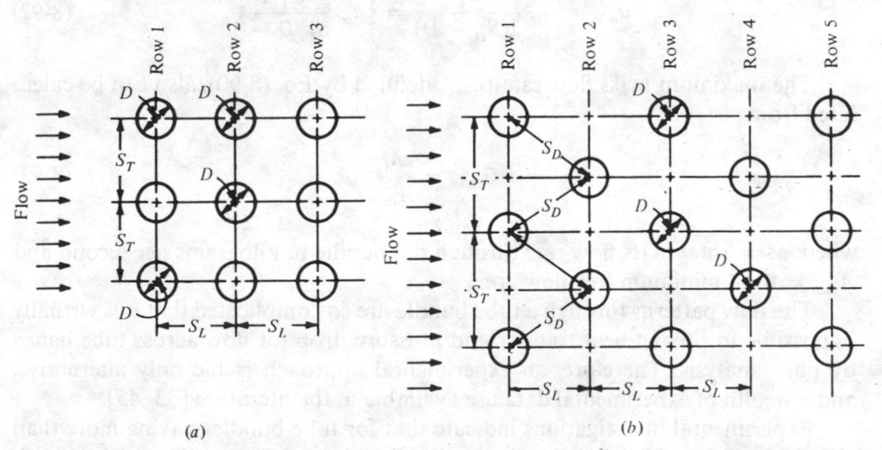

Figure 8-11 Definitions of longitudinal, transverse, and diagonal pitches for in-line and staggered tube bundle arrangements. (*a*) In-line arrangement, (*b*) staggered arrangement.

tubes in a transverse row or in a diagonal row. Then the Reynolds number for flow across a tube bank is defined as

$$\mathrm{Re} = \frac{DG_{max}}{\mu} \tag{8-89}$$

where

$$G_{max} = \rho u_{max} = \text{maximum mass flow velocity} \tag{8-90}$$

is the mass flow rate per unit area where the flow velocity is maximum, and D is the outside diameter of the tube, ρ is the density, and u_{max} the maximum velocity based on the minimum free-flow area available for fluid flow. If u_∞ is the flow velocity measured at a point in the heat exchanger before the fluid enters the tube bank (or the flow velocity based on flow inside the heat exchanger shell without the tubes), then the maximum flow velocity u_{max} for the *in-line arrangement* shown in Fig. 8-11*a* is determined from

$$u_{max} = u_\infty \frac{S_T}{S_T - D} = u_\infty \frac{S_T/D}{S_T/D - 1} \tag{8-91}$$

where S_T is the transverse pitch and D is the outer diameter of the tube. Clearly, for the in-line arrangement, $S_T - D$ is the minimum free-flow area between the adjacent tubes in a transverse row per unit length of the tube.

For the *staggered arrangement* shown in Fig. 8-11*b* the minimum free-flow area may occur between adjacent tubes either in a transverse row or in a diagonal row. In the former case, u_{max} is determined as given above; in the latter case, it is determined from

$$u_{max} = u_\infty \frac{S_T}{2(S_D - D)} = \tfrac{1}{2} u_\infty \frac{S_T/D}{S_D/D - 1} \tag{8-92}$$

The maximum mass flow rate G_{max}, defined by Eq. (8-90), also can be calculated from

$$G_{max} = \frac{M}{A_{min}} \tag{8-93}$$

where M = total mass flow rate through the bundle in kilograms per second and A_{min} = total minimum free-flow area.

The flow patterns through a tube bundle are so complicated that it is virtually impossible to predict heat transfer and pressure drop for flow across tube banks by pure analysis. Therefore, an experimental approach is the only alternative, and a wealth of experimental data are available in the literature [33–45].

Experimental investigations indicate that for tube bundles having more than about N = 10 to 20 rows of tubes in the direction of flow, and the tube length large compared to the tube diameter, the entrance, exit, and edge effects are negligible. For such cases, the Nusselt number for flow across the bundle depends on the following parameters:

$$\text{Re} \qquad \text{Pr} \qquad \frac{S_L}{D} \qquad \frac{S_T}{D}$$

and the geometric arrangement of the tubes, namely, whether the tubes are aligned or staggered.

We now present the heat transfer and pressure drop correlations for flow across the tube bundles.

Heat Transfer Correlations

Grimison [34] correlated heat transfer data for air reported by several investigators for both in-line and staggered tube arrangements, for tube bundles having 10 or more transverse rows in the direction of flow with an expression in the form

$$\frac{h_m D}{k} = c_0 \left(\frac{D G_{\max}}{\mu} \right)^n \tag{8-94}$$

for air in the range $2000 < \text{Re} < 40{,}000$.

This expression has been generalized to fluids other than air by including the Prandtl number effect in the form

$$\boxed{\frac{h_m D}{k} = 1.13 c_0 \,\text{Re}^n \,\text{Pr}^{1/3}} \tag{8-95a}$$

for $2000 < \text{Re} < 40{,}000$, $\text{Pr} > 0.7$, and $N \geq 10$. Here Re is defined as

$$\text{Re} = \frac{D G_{\max}}{\mu} \tag{8-95b}$$

The values of constant c_0 and the exponent n are listed in Table 8-3. All physical properties in Eq. (8-95a) are evaluated at the mean film temperature.

Kays, London, and Lo [39] examined experimentally the effects of the row number on the heat transfer coefficient for a variety of tube arrangements. For tube bundles having less than $N = 10$ transverse rows in the direction of flow, there was some reduction in the heat transfer coefficient. Based on the results of their experiments, the heat transfer coefficient h_N for $N < 10$ could be determined by utilizing the following relation:

$$\boxed{h_N = c_1 h_{N \geq 10} \qquad \text{for } 1 \leq N \leq 10} \tag{8-96}$$

Table 8-4 lists the values of the correction factor c_1 for both in-line and staggered tube arrangements, with N varying from 1 to 9. The results depend only slightly on the Reynolds number.

Table 8-3 Constants c_0 and n of Eq. (8-95)

Arrangement	$\frac{S_L}{D}$	$\frac{S_T}{D}$ = 1.25 c_0	1.25 n	1.50 c_0	1.50 n	2.0 c_0	2.0 n	3.0 c_0	3.0 n
Staggered	0.6	—	—	—	—	—	—	0.213	0.636
	0.9	—	—	—	—	0.446	0.571	0.401	0.581
	1.0	—	—	0.497	0.588	—	—	—	—
	1.125	—	—	—	—	0.478	0.565	0.518	0.560
	1.250	0.518	0.556	0.505	0.554	0.519	0.556	0.522	0.562
	1.50	0.451	0.568	0.460	0.562	0.452	0.568	0.488	0.568
	2.0	0.404	0.572	0.416	0.568	0.482	0.556	0.449	0.570
	3.0	0.310	0.592	0.356	0.580	0.440	0.562	0.421	0.574
In-line	1.25	0.348	0.592	0.275	0.608	0.100	0.704	0.0633	0.752
	1.50	0.367	0.586	0.250	0.620	0.101	0.702	0.0678	0.744
	2.0	0.418	0.570	0.299	0.602	0.229	0.632	0.198	0.648
	3.0	0.290	0.601	0.357	0.584	0.374	0.581	0.286	0.608

Source: Grimison [34].

More recently, Zukauskas [47] reviewed the work of various investigators and proposed the following correlation for the heat transfer coefficient for flow across tube bundles:

$$\boxed{\frac{h_m D}{k} = c_2 \, \mathrm{Re}^m \, \mathrm{Pr}^{0.36} \left(\frac{\mathrm{Pr}}{\mathrm{Pr}_w} \right)^n} \tag{8-97}$$

where Pr_w is the Prandtl number evaluated at the wall temperature, and

$$n = \begin{cases} 0 & \text{for gases} \\ \frac{1}{4} & \text{for liquids} \end{cases}$$

which is valid for $0.7 < \mathrm{Pr} < 500$ and $N \geq 20$. For liquids, the physical properties are evaluated at the bulk mean temperature, since the viscosity correction term is included through the Prandtl number ratio. For gases, the properties are evaluated at the film temperature and the viscosity correction term $(\mathrm{Pr}/\mathrm{Pr}_w)^n$ is omitted.

Table 8-4 Correction factor c_1 for Eq. (8-96)

N	1	2	3	4	5	6	7	8	9
In-line	0.64	0.80	0.87	0.90	0.92	0.94	0.96	0.98	0.99
Staggered	0.68	0.75	0.83	0.89	0.92	0.95	0.97	0.98	0.99

Source: Kays, London, and Lo [39].

Table 8-5 Constant c_2 and exponent m of Eq. (8-97)

Geometry	Re	c_2	m	Remarks
In-line	10 to 10^2	0.8	0.40	
	10^2 to 10^3	Large and moderate longitudinal pitch, can be regarded as a single tube		
	10^3 to 2×10^5	0.27	0.63	
	2×10^5 to 10^6	0.21	0.84	
Staggered	10 to 10^2	0.9	0.40	
	10^2 to 10^3	About 20 percent higher than that for single tube		
	10^3 to 2×10^5	$0.35\left(\frac{S_T}{S_L}\right)^{0.2}$	0.60	$\frac{S_T}{S_L} < 2$
	10^3 to 2×10^5	0.40	0.60	$\frac{S_T}{S_L} > 2$
	2×10^5 to 10^6	0.022	0.84	

Source: Zukauskas [47].

The coefficient c_2 and the exponent m were determined by correlating the experimental data for air, water, and oil reported by numerous investigators. Table 8-5 lists the recommended values of c_2 and m of Eq. (8-97).

Equation (8-97) correlates the experimental data very well for tube bundles having $N = 20$ or more rows in the direction of flow. For bundles having less than $N = 20$ rows, the Nusselt number can be found from

$$\boxed{\mathrm{Nu}_N = c_3 \, \mathrm{Nu}_{N \geq 20}} \tag{8-98}$$

where the correction factor c_3 is given in Fig. 8-12 for both in-line and staggered tube arrangements.

In the previous discussion we considered two correlations, given by Eqs. (8-95) and (8-97), for the heat transfer coefficient for flow across tube bundles. The latter equation is a more recent correlation which has been developed as a result of comparison with the experimental data over a wide range of flow rates,

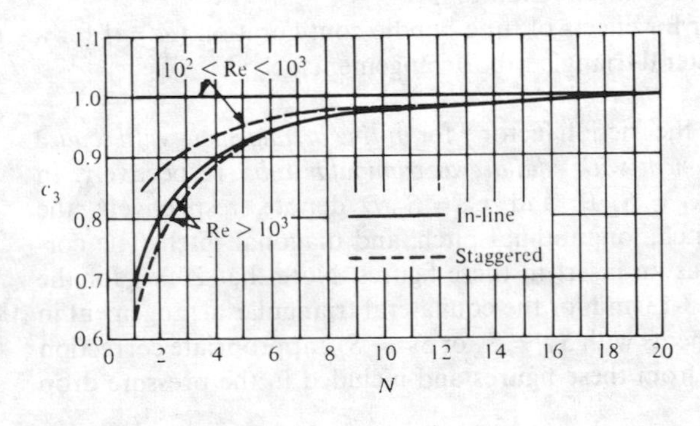

Figure 8-12 Correction factor c_3 of Eq. (8-98). (*From Zukauskas* [47].)

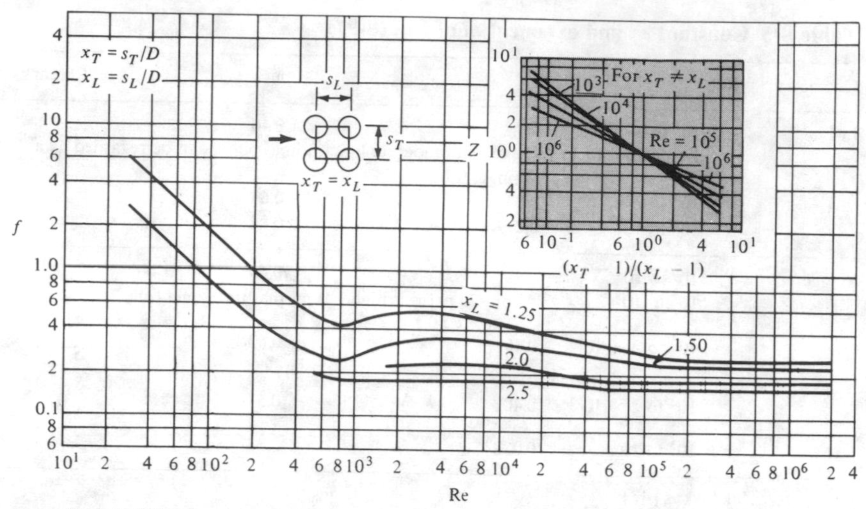

Figure 8-13 Friction factor f and the correction factor Z for use in Eq. (8-99) for in-line tube arrangement. (*From Ref. 47.*)

Prandtl numbers, and tube arrangements. The agreement with the experimental data was shown to be very good; therefore, it should be preferred.

Pressure Drop Correlations

Zukauskas [47] correlated the pressure drop due to fluid friction for flow across tube bundles by

$$\Delta P = f \frac{NG_{\max}^2}{2\rho} Z \tag{8-99}$$

where f = friction factor

$G = \rho u_{\max}$ = maximum mass flow velocity, kg/(m$^2 \cdot$ s)

N = number of tube rows in direction of flow

Z = correction factor for effects of tube bundle configuration ($Z = 1$ for a square or equilateral triangle tube arrangements)

Figures 8-13 and 8-14 show the friction factor f for *in-line arrangement with square tube* and *staggered arrangement with equilateral triangular tube*, respectively. In these figures $x_T = S_T/D$, $x_L = S_L/D$, and $x_D = S_D/D$ denote, respectively, the dimensionless transverse pitch, longitudinal pitch, and diagonal pitch. The correction factor Z is plotted as an insert in these figures. Note that $Z = 1$ for the square arrangement in Fig. 8-13 and for the equilateral triangular arrangement in Fig. 8-14. For tube arrangements with $S_T \neq S_L$ or $S_T \neq S_D$, appropriate correction factors should be obtained from these figures and included in the pressure drop expression, Eq. (8-99).

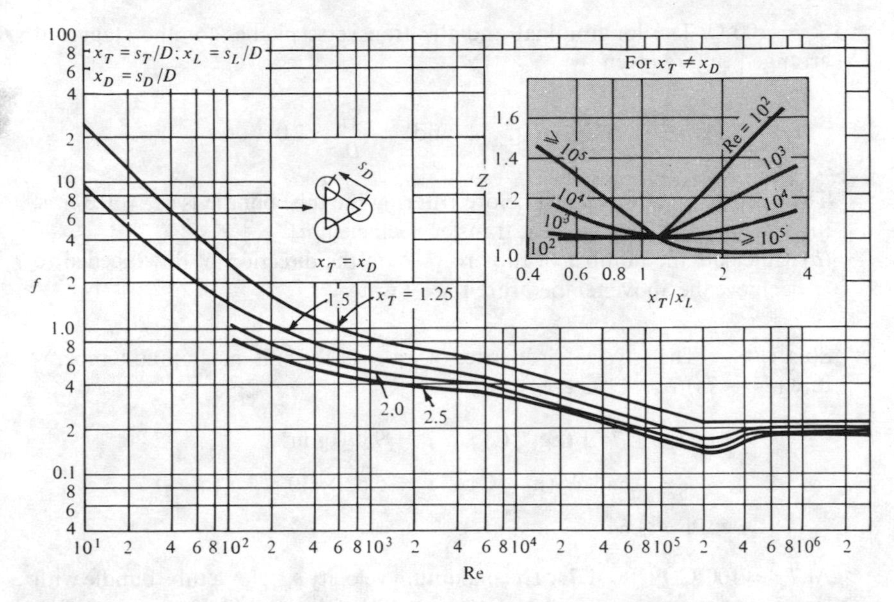

Figure 8-14 Friction factor f and the correction factor Z for use in Eq. (8-99) for staggered tube arrangement. (*From Ref. 47.*)

Liquid Metals

The previous heat transfer correlations are not applicable to fluids having very low Prandtl number, such as liquid metals. Hoe, Dropkin, and Dwyer [48] and Richards, Dwyer, and Dropkin [49] reported experimental data on heat transfer rates for mercury flowing across staggered tube banks. In these experiments the mercury flow was across a tube bank 60 to 70 rows deep and consisting of $\frac{1}{2}$-in tubes arranged in equilateral-triangle array having a pitch-to-diameter ratio of 1.375. The average heat transfer coefficient for these experiments were correlated by

$$\boxed{\mathrm{Nu}_m = \frac{h_m D}{k} = 4.03 + 0.228\,(\mathrm{Re\,Pr})^{0.67}} \tag{8-100}$$

where $\mathrm{Re} = DG_{\max}/\mu$ for $20{,}000 < \mathrm{Re} < 80{,}000$. The physical properties are evaluated at the arithmetic average of the bulk fluid and wall surface temperatures.

Kalish and Dwyer [50] give heat transfer data for NaK flowing through tube bundles.

Example 8-8 Water at $T_1 = 24°\mathrm{C}$ is to be heated to $T_2 = 74°\mathrm{C}$ by passing it through a tube bundle in *staggered* tube arrangement. Tubes have an outside diameter $D = 2.5$ cm and are maintained at a uniform surface temperature

$T_w = 100°C$. The longitudinal and the transverse pitches for the staggered arrangement are given by

$$\frac{S_L}{D} = 1.5 \qquad \text{and} \qquad \frac{S_T}{D} = 2.0$$

The velocity u_∞ of water just before entering the tube bundle is $u_\infty = 0.3$ m/s.
(*a*) Calculate the average heat transfer coefficient h_m.
(*b*) Calculate the number of tube rows N in the direction of flow needed to achieve the above temperature rise of water.

SOLUTION The physical properties of water at the bulk mean fluid temperature $(24 + 74)/2 = 49°C$ are

$$c_P = 4174 \text{ J/(kg} \cdot \text{°C)} \qquad \rho = 989 \text{ kg/m}^3$$

$$k = 0.644 \text{ W/(m} \cdot \text{°C)} \qquad \mu = 5.62 \times 10^{-4} \text{ kg/(m} \cdot \text{s)}$$

$$\text{Pr} = 3.64$$

At $T_w = 100°C$, $\text{Pr}_w = 1.74$. The maximum velocity u_{max} for a tube bundle with staggered tube arrangement is determined by either Eq. (8-92) or Eq. (8-91), depending on which gives the greater value for u_{max}. For the particular geometry, Eq. (8-91) gives greater u_{max}; hence, from Eq. (8-91)

$$u_{max} = u_\infty \frac{S_T/D}{S_T/D - 1} = 0.3 \frac{2}{2 - 1} = 0.6 \text{ m/s}$$

The maximum mass-flow velocity G_{max} becomes

$$G_{max} = \rho u_{max} = (989)(0.6) = 593.4 \text{ kg/(m}^2 \cdot \text{s)}$$

Then the Reynolds number is

$$\text{Re} = \frac{DG_{max}}{\mu} = \frac{(0.025)(593.4)}{5.62 \times 10^{-4}} = 2.64 \times 10^4$$

(*a*) Equation (8-97) now can be used to determine the heat transfer coefficient h_m with the assumption that $N \geq 20$. For the staggered tube arrangement, with $S_T/D = 2.0$, $S_L/D = 1.5$, or $S_T/S_L = 2.0/1.5 < 2$, and $\text{Re} = 2.64 \times 10^4$, the constant c_2 and the exponent m of Eq. (8-97) are obtained from Table 8-5:

$$c_2 = 0.35\left(\frac{S_T}{S_L}\right)^{0.2} = 0.35\left(\frac{2.0}{1.5}\right)^{0.2} = 0.371$$

$$m = 0.6$$

Then Eq. (8-97) becomes

$$\frac{h_m D}{k} = 0.371\ \text{Re}^{0.6}\ \text{Pr}^{0.36}\left(\frac{\text{Pr}}{\text{Pr}_w}\right)^{1/4}$$

$$\frac{h_m(0.025)}{0.644} = 0.371(26{,}400)^{0.6}(3.64)^{0.36}\left(\frac{3.64}{1.74}\right)^{1/4}$$

$$h_m = 8230\ \text{W/(m}^2 \cdot {}^\circ\text{C)}$$

[If Eq. (8-95*a*) were used, the resulting heat transfer coefficient would be 6570 W/(m^2 · °C), which is within 20 percent.]

(*b*) The accompanying figure illustrates the definition of various quantities that we now use to write an energy balance to determine the number of transverse rows N needed:

$$\begin{pmatrix}\text{Heat transferred from}\\ \text{tube surface to water}\end{pmatrix} = \begin{pmatrix}\text{heat carried away}\\ \text{by water}\end{pmatrix}$$

$$A_s h_m \Delta T_m = M c_P (T_2 - T_1) \qquad (a)$$

where $M = A_\infty u_\infty \rho$ = total mass flow rate, kg/s
A_∞ = free-flow area just before entry to tube bundle
= LmS_T
A_s = total heat transfer surface
= (surface area per tube)(number of rows)(number of tubes per row)
= $(\pi DL)(N)(m)$
m = number of tubes per row
N = number of rows
L = tube length
S_T = transverse pitch
ΔT_m = difference between fluid and wall surface temperatures
T_1, T_2 = inlet and outlet temperatures of water, respectively

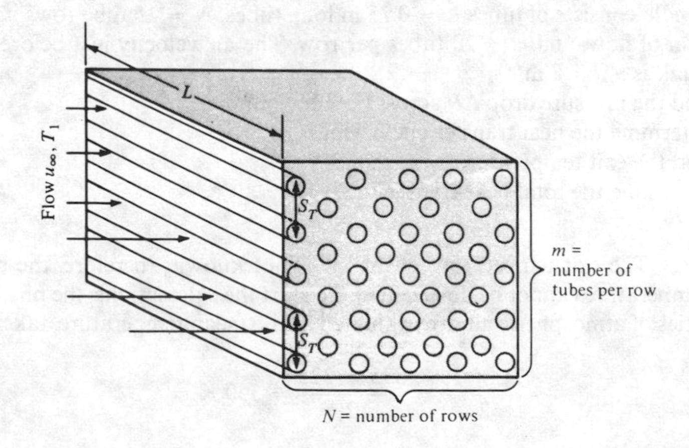

The logarithmic mean temperature difference is used to determine ΔT_m:

$$\Delta T_m = \frac{(T_w - T_1) - (T_w - T_2)}{\ln [(T_w - T_1)/(T_w - T_2)]} = \frac{T_2 - T_1}{\ln [(T_w - T_1)/(T_w - T_2)]}$$

Now, A_∞, A_S, and ΔT_m, as defined above, are introduced into the energy balance equation (*a*):

$$(\pi DLNm)h_m \frac{T_2 - T_1}{\ln [(T_w - T_1)/(T_w - T_2)]} = (LmS_T)(u_\infty \rho c_p)(T_2 - T_1)$$

Solving for N yields

$$N = \frac{1}{\pi}\frac{S_T}{D}\frac{u_\infty \rho c_p}{h_m} \ln \frac{T_w - T_1}{T_w - T_2}$$

$$= \frac{1}{\pi}(2)\frac{(0.3)(989)(4174)}{8230} \ln \frac{100 - 24}{100 - 74}$$

$$= 102.7 \cong 103$$

Thus, 103 rows are needed.

Example 8-9 Air at atmospheric pressure and temperature $T_1 = 325$ K flows through a tube bundle in *in-line* tube arrangement, as shown in Fig. 8-11*a*. Tubes have an outside diameter $D = 1.9$ cm and are maintained at a uniform temperature $T_w = 375$ K. The longitudinal and transverse pitches for the bundle are given by

$$\frac{S_T}{D} = \frac{S_L}{D} = 2$$

The bundle consists of tubes $L = 0.75$ m long tubes, $N = 15$ tube rows in the direction of flow, and $m = 20$ tubes per row. The air velocity just before the tube bank is $u_\infty = 8$ m/s.
(*a*) Find the pressure drop ΔP across the tube bundle.
(*b*) Determine the heat transfer coefficient h_m.
(*c*) Find the exit temperature T_2 of the air.
(*d*) Determine the total heat transfer rate Q.

SOLUTION The exit temperature of the air is not known; therefore, the bulk fluid temperature cannot be determined. To start the calculations, the physical properties of atmospheric air are evaluated at the mean temperature, taken as

$$\frac{T_w + T_1}{2} = \frac{375 + 325}{2} = 350 \text{ K}$$

Then

$$c_P = 1009 \text{ J/(kg} \cdot \text{°C)} \qquad \rho = 0.998 \text{ kg/m}^3$$
$$k = 0.03 \text{ W/(m} \cdot \text{°C)} \qquad \mu = 2.075 \times 10^{-5} \text{ kg/(m} \cdot \text{s)}$$
$$\text{Pr} = 0.697$$

The maximum flow velocity u_{max} is determined by Eq. (8-91):

$$u_{max} = u_\infty \frac{S_T/D}{S_T/D - 1} = 8 \frac{2}{2 - 1} = 16 \text{ m/s}$$

The maximum mass flow velocity G_{max} becomes

$$G_{max} = \rho u_{max} = 0.998 \times 16 = 15.97 \text{ kg/(m}^2 \cdot \text{s)}$$

Then

$$\text{Re} = \frac{DG_{max}}{\mu} = \frac{(0.019)(15.97)}{2.075 \times 10^{-5}} = 14{,}623$$

(*a*) Equation (8-99) is used now to determine the pressure drop:

$$\Delta P = f \frac{NG_{max}^2}{2\rho} Z$$

The friction factor f is obtained from Fig. 8-13, for Re = 14,623 and $X_L = S_L/D = 2$, as

$$f = 0.22$$

The correction factor is $Z = 1$, since $X_T = X_L$. Then

$$\Delta P = 0.22 \frac{(15)(15.97^2)}{2 \times 0.998} = 422 \text{ N/m}^2$$

(*b*) The heat transfer coefficient h_m is determined by using Eq. (8-97). For the in-line arrangement with Re = 14,623, the coefficients c_2 and m are obtained from Table 8-5:

$$c_2 = 0.27 \qquad m = 0.63$$

and for gas we set $n = 0$. Then Eq. (8-97) becomes

$$\frac{h_m D}{k} = 0.27 \text{ Re}^{0.63} \text{ Pr}^{0.36}$$

$$\frac{h_m(0.019)}{0.03} = 0.27(14{,}623)^{0.63}(0.697)^{0.36}$$

$$h_m = 157.5 \text{ W/(m}^2 \cdot \text{°C)}$$

(*c*) To calculate the exit temperature T_2, we consider an energy balance

$$\begin{pmatrix}\text{Heat transferred from}\\ \text{tube surface to air}\end{pmatrix} = \begin{pmatrix}\text{heat carried}\\ \text{away by air}\end{pmatrix}$$

$$Q \equiv A_s h_m \, \Delta T_m = Mc_p(T_2 - T_1) \tag{a}$$

where $M = A_\infty u_\infty \rho$ = total mass flow rate, kg/s
A_∞ = free-flow area just before tube bundle
$= LmS_T$
A_s = total heat transfer surface
= (surface area per tube)(number of rows)(number of tubes per row)
$= (\pi DL)(N)(m)$
ΔT_m = difference between wall surface and air temperatures
T_1, T_2 = inlet and outlet temperatures of air, respectively

The logarithmic mean temperature difference is used to determine ΔT_m:

$$\Delta T_m = \frac{T_2 - T_1}{\ln [(T_w - T_1)/(T_w - T_2)]}$$

Now A_∞, A_s, and ΔT_m are introduced into the energy balance equation (*a*):

$$(\pi DLNm)h_m \frac{T_2 - T_1}{\ln [(T_w - T_1)/(T_w - T_2)]} = (LmS_T)(u_\infty \rho c_P)(T_2 - T_1)$$

Solving for the logarithmic term yields

$$\ln \frac{T_w - T_1}{T_w - T_2} = \pi N \frac{D}{S_T} \frac{h_m}{u_\infty \rho c_p}$$

$$\ln \frac{375 - 325}{375 - T_2} = (\pi)(15) \frac{1}{2} \frac{157.5}{(8)(0.998)(1009)}$$

$$T_2 = 343.5 \text{ K}$$

(*d*) Given the exit temperature T_2, the heat transfer rate is calculated from Eq. (*a*):

$$\begin{aligned} Q &= Mc_p(T_2 - T_1) \\ &= M(1009)(343.5 - 325) \\ &= 18{,}666.5M \end{aligned}$$

where

$$\begin{aligned} M &= A_\infty u_\infty \rho \\ &= (LmS_T)u_\infty \rho \\ &= (0.75)(20)(2 \times 0.019)(8)(0.998) \\ &= 4.551 \text{ kg/s} \end{aligned}$$

Then

$$Q = (18{,}666.5)(4.551) = 84.95 \text{ kW}$$

Calculations can be repeated by evaluating the air properties at

$$\frac{T_1 + T_2}{2} = \frac{325 + 343.5}{2} = 334.3 \text{ K}$$

but the improvement would be negligible.

8-7 HEAT TRANSFER IN HIGH-SPEED FLOW OVER A FLAT PLATE

At flow velocities approaching or exceeding the velocity of sound encountered in applications such as high-speed aircraft, missiles, and reentry vehicles, the effects of compressibility, viscous dissipation, and property variation with temperature become important. The general analysis of such problems is very involved. However, the heat transfer rate in high-speed flow along a flat plate at a uniform temperature T_w can be predicted, for most practical purposes, by using the *low-speed heat transfer coefficient* h_x for flow along a flat plate with the temperature difference $T_w - T_{aw}$. That is, at high-speed flow along a flat plate, the local heat flux q_x at the wall can be computed from

$$q_x = h_x(T_w - T_{aw}) \tag{8-101}$$

where T_w = wall temperature
T_{aw} = adiabatic wall temperature
h_x = local, low-speed heat transfer coefficient

Note that in Eq. (8-101) the adiabatic wall temperature T_{aw} replaces the fluid free-stream temperature T_∞ commonly used in low-speed heat transfer.

To use Eq. (8-101) for high-speed heat transfer calculations, the local heat transfer coefficient h_x is obtained from the low-speed heat transfer correlations for flow along a flat plate presented earlier. For example, for *laminar boundary-layer flow*, h_x is obtained from the Pohlhausen equation (8-67), that is

$$\text{Nu}_x \equiv \frac{h_x x}{k} = 0.332 \text{ Pr}^{1/3} \text{ Re}_x^{1/2} \qquad \text{for Re}_x < 5 \times 10^5 \tag{8-102}$$

For *turbulent boundary-layer flow*, it can be obtained from Eqs. (8-73*a*), (8-73*b*), and (8-74); that is,

$$\frac{h_x}{\rho u_\infty c_p} = 0.0296 \text{ Re}_x^{-0.2} \text{ Pr}^{-2/3} \qquad \text{for } 5 \times 10^5 < \text{Re}_x < 10^7 \tag{8-103}$$

$$\frac{h_x}{\rho u_\infty c_p} = 0.185(\log \text{Re}_x)^{-2.584} \text{Pr}^{-2/3} \qquad \text{for } 10^7 < \text{Re}_x < 10^9 \tag{8-104}$$

$$\frac{h_x x}{k} = 0.029 \text{ Re}_x^{0.8} \text{ Pr}^{0.43} \qquad \text{for Re}_x > 2 \times 10^5 \text{ to } 5 \times 10^5, \tag{8-105}$$

The *adiabatic wall temperature* T_{aw} is determined from

$$T_{aw} = T_\infty + r \frac{u_\infty^2}{2c_p} \tag{8-106}$$

where T_∞ and u_∞ are the free-stream temperature and velocity, respectively. The parameter r is called the *recovery factor*, and it is related to the Prandtl number as follows: For the laminar boundary layer:

$$r \cong \text{Pr}^{1/2} \qquad 0.6 < \text{Pr} < 15\,[3, 51] \tag{8-107}$$

$$r \cong 1.9\,\text{Pr}^{1/3} \qquad \text{for Pr} \to \infty\ [52] \tag{8-108}$$

For the turbulent boundary layer [53]:

$$r \cong \text{Pr}^{1/3} \tag{8-109}$$

The physical significance of the recovery factor is better envisioned by considering an ideal gas at a temperature T_∞ with a velocity u_∞ that is slowed down adiabatically to zero velocity. The conversion of kinetic energy in the gas to internal energy will result in a gas temperature T_0 given by

$$T_0 = T_\infty + \frac{u_\infty^2}{2c_p} \tag{8-110}$$

where T_0 is called the *stagnation temperature*. A comparison of Eqs. (8-106) and (8-110) reveals that for $r = 1$, T_{aw} is equivalent to the stagnation temperature T_0. Note that $r = 1$ for $\text{Pr} = 1$; thus for a gas ($\text{Pr} \cong 1$) flowing at high speed along a flat plate, the adiabatic wall temperature is the same as the stagnation temperature. For fluids with $\text{Pr} > 1$, the recovery factor r is greater than unity, and the adiabatic wall temperature exceeds the stagnation temperature. For a fluid with $\text{Pr} < 1$, the recovery factor r is less than unity, and the adiabatic wall temperature is less than the stagnation temperature.

In high-speed flow, temperature gradients in the boundary layer are generally large; hence the properties of the fluid vary significantly with temperature. Although the exact analysis of such problems is very involved, the effects of variation of properties may be approximately included in the heat transfer relations given above by Eqs. (8-103) to (8-105) if the properties of the fluid are evaluated at the following reference temperature [54]:

$$T_r = T_\infty + 0.5(T_w - T_\infty) + 0.22(T_{aw} - T_\infty) \tag{8-111}$$

Readers should consult Refs. 55 to 58 for further details of high-speed heat transfer.

Example 8-10 Air at a pressure $P = \frac{1}{30}$ atm, temperature $T_\infty = 250$ K, and velocity $u_\infty = 600$ m/s flows over a flat plate $L = 0.80$ m long and $w = 0.30$ m wide. Calculate the amount of cooling needed to maintain the plate surface at a uniform temperature $T_w = 300$ K.

SOLUTION This is a high-speed flow problem, and if it involves both laminar and turbulent regions, the heat transfer analysis for each region should be considered separately because the recovery factor, and hence the adiabatic wall temperatures, is different for each flow regime.

Laminar Flow Region

To start the calculations, we evaluate Pr and c_p at 300 K:

$$\text{Pr} = 0.708 \qquad c_p = 1006 \text{ J/(kg} \cdot {}^\circ\text{C)}$$

Then the recovery factor r is determined by Eq. (8-107):

$$r = \text{Pr}^{1/2} = (0.708)^{1/2} = 0.841$$

and the adiabatic wall temperature T_{aw} is by Eq. (8-106),

$$T_{aw} = T_\infty + r\frac{u_\infty^2}{2c_p} = 250 + 0.841\frac{600^2}{2 \times 1006} = 400.5 \text{ K}$$

The reference temperature T_r is determined by Eq. (8-111):

$$\begin{aligned} T_r &= T_\infty + 0.5(T_w - T_\infty) + 0.22(T_{aw} - T_\infty) \\ &= 250 + 0.5(300 - 250) + 0.22(400.5 - 250) \\ &= 308.1 \text{ K} \end{aligned}$$

The physical properties of air at $T_r = 308$ K and at $P = \frac{1}{30}$ atm are taken as

$$\rho = \frac{1.1475}{30} = 0.0383 \text{ kg/m}^3 \qquad k = 0.0269 \text{ W/(m} \cdot {}^\circ\text{C)}$$

$$\mu = 1.998 \times 10^{-5} \text{ kg/(m} \cdot \text{s)} \qquad \text{Pr} = 0.710$$

$$c_p = 1007 \text{ J/(kg} \cdot {}^\circ\text{C)}$$

Note that the Prandtl number and c_p at T_r are sufficiently close to the values used above to compute the recovery factor; therefore, there is not need to repeat the calculations.

Assuming that the transition takes place at a critical Reynolds number $\text{Re}_c = 5 \times 10^5$, we can find the distance x_c:

$$x_c = \frac{\mu}{\rho u_\infty}\text{Re}_c = \frac{1.998 \times 10^{-5}}{0.0383 \times 600}(5 \times 10^5) = 0.43 \text{ m}$$

The average heat transfer coefficient for laminar boundary-layer flow over $0 \le x \le 0.43$ m is determined by Eq. (8-71*a*):

$$\begin{aligned} h_m &= \frac{k}{x_c} 0.664 \text{ Re}_c^{1/2} \text{ Pr}^{1/3} \\ &= \frac{0.0269}{0.43}(0.664)(5 \times 10^5)^{1/2}(0.710)^{1/3} = 26.2 \text{ W/(m}^2 \cdot {}^\circ\text{C)} \end{aligned}$$

The heat transfer rate Q^l over the laminar flow region is determined by Eq. (8-101):

$$\begin{aligned} Q^l &= wx_c h_m(T_w - T_{aw}) \\ &= 0.3(0.43)(26.2)(300 - 400.5) \\ &= -340 \text{ W} \end{aligned}$$

Turbulent Flow Region

To start the calculations, we evaluate Pr and c_p again at 300 K:

$$\text{Pr} = 0.708 \qquad c_p = 1006 \text{ J/(kg} \cdot {}^\circ\text{C)}$$

Then the recovery factor r is determined by Eq. (8-109):

$$r = \text{Pr}^{1/3} = (0.708)^{1/3} = 0.891$$

And T_{aw} is found by Eq. (8-106):

$$T_{aw} = T_\infty + r\frac{u_\infty^2}{2c_p} = 250 + 0.891\frac{600^2}{2(1006)} = 409.4 \text{ K}$$

The reference temperature T_r is determined by Eq. (8-111):

$$\begin{aligned} T_r &= T_\infty + 0.5(T_w - T_\infty) + 0.22(T_{aw} - T_\infty) \\ &= 250 + 0.5(300 - 250) + 0.22(409.4 - 250) \\ &= 310 \text{ K} \end{aligned}$$

The physical properties of air at 310 K can be taken the same as those given above for the laminar flow region.

The local heat transfer coefficient for the turbulent flow region can be determined from Eq. (8-103);

$$\begin{aligned} h_x &= 0.0296\rho u_\infty c_p \text{Pr}^{-2/3} \text{Re}_x^{-0.2} \\ &= (0.0296)(0.0383)(600)(1007)(0.71)^{-2/3}\left(\frac{0.0383 \times 600}{1.998 \times 10^{-5}}\right)^{-0.2} x^{-0.2} \\ &= 52.8x^{-0.2} \end{aligned}$$

The average value of h_x over the region $0.43 < x < 0.8$ is

$$h_m = \frac{1}{0.8 - 0.43}\int_{0.43}^{0.8} h_x\,dx = 58.4 \text{ W/(m}^2 \cdot {}^\circ\text{C)}$$

The heat transfer rate Q^t over the turbulent flow region is

$$\begin{aligned} Q_t &= h_m w(L - x_c)(T_w - T_{aw}) \\ &= 58.4(0.3)(0.8 - 0.43)(300 - 409.4) \\ &= -709.2 \text{ W} \end{aligned}$$

Table 8-6 Summary of correlations for forced convection over bodies*

Equation number	Correlation	Flow regime	Remarks
(8-19)	$c_x = \dfrac{0.664}{\mathrm{Re}_x^{1/2}}$ (local)	Laminar	Flow over a flat plate for $\mathrm{Re}_x \le 5 \times 10^5$
(8-21)	$c_m = \dfrac{1.328}{\mathrm{Re}_L^{1/2}}$ (average)	Laminar	Flow over a flat plate for $\mathrm{Re}_L \le 5 \times 10^5$
(8-22)	$\dfrac{\delta(x)}{x} = \dfrac{4.96}{\mathrm{Re}_x^{1/2}}$	Laminar	Flow over a flat plate for $\mathrm{Re}_x \le 5 \times 10^5$
(8-24)	$c_x = 0.0592\ \mathrm{Re}_x^{-0.2}$ (local)	Turbulent	Flow over a flat plate for $5 \times 10^5 < \mathrm{Re}_x < 10^7$
(8-25)	$c_x = 0.370(\log \mathrm{Re}_x)^{-2.584}$ (local)	Turbulent	Flow over a flat plate for $10^7 < \mathrm{Re}_x < 10^9$
(8-25*b*)	$c_m = 0.074\ \mathrm{Re}_L^{-0.2} - \dfrac{B}{\mathrm{Re}_L}$ (average)	Laminar and turbulent	Flow over a flat plate for $\mathrm{Re}_c < \mathrm{Re}_L < 10^7$
(8-26*b*)	$\dfrac{\delta(x)}{x} = 0.381\ \mathrm{Re}_x^{-1/5} - 10{,}256\ \mathrm{Re}_x^{-1}$	Laminar and turbulent	Flow over a flat plate for $5 \times 10^5 < \mathrm{Re}_x < 10^7$
(8-45)	$\mathrm{Nu}_x = 0.564\ \mathrm{Pe}_x^{1/2}$ (local)	Laminar	Flow over a flat plate, constant wall temperatures, $\mathrm{Pr} \ll 1$
(8-67)	$\mathrm{Nu}_x = 0.332\ \mathrm{Pr}^{1/3}\ \mathrm{Re}_x^{1/2}$ (local)	Laminar	Flow over a flat plate, constant wall temperature,
(8-73*a*)	$\mathrm{St}_x\ \mathrm{Pr}^{2/3} = 0.0296\ \mathrm{Re}_x^{-0.2}$ (local)	Turbulent	Flow over a flat plate for $5 \times 10^5 < \mathrm{Re}_x < 10^7$ $\mathrm{Re}_x \le 5 \times 10^5$; $0.6 < \mathrm{Pr} < 10$
(8-68)	$\mathrm{Nu}_x = 0.339\ \mathrm{Pr}^{1/3}\ \mathrm{Re}_x^{1/2}$ (local)	Laminar	Flow over a flat plate, constant wall temperature, $\mathrm{Re}_x \le 5 \times 10^5$, $\mathrm{Pr} \gg 1$
(8-71*a*)	$\mathrm{Nu}_m = 0.664\ \mathrm{Pr}^{1/3}\ \mathrm{Re}_L^{1/2}$ (average)	Laminar	Flow over a flat plate, constant wall temperature, $\mathrm{Re}_L \le 5 \times 10^5$; $0.6 < \mathrm{Pr} < 10$

* Fluid properties are evaluated at the film temperature $T_f = (T_w + T_\infty)/2$ if no viscosity correction is included.

Table 8-6 *(Continued)*

Equation number	Correlation	Flow regime	Remarks
(8-71*b*)	$Nu_m = 0.678\ Pr^{1/3}\ Re_L^{1/2}$ (average)	Laminar	Flow over a flat plate, constant wall temperature, $Re_L \leq 5 \times 10^5$, $Pr \gg 1$
(8-73*a*)	$St_x\, Pr^{2/3} = 0.0296\ Re_x^{-0.2}$ (local)	Turbulent	Flow over a flat plate for $5 \times 10^5 > Re_x > 10^7$
(8-73*b*)	$St_x\, Pr^{2/3} = 0.185(\log Re_x)^{-2.584}$ (local)	Turbulent	Flow over a flat plate for $10^7 < Re_x < 10^9$
(8-74)	$Nu_x = 0.029\ Re_x^{0.8}\ Pr^{0.43}$ (local)	Turbulent	Flow over a flat plate for $Re_x > 2 \times 10^5$ to 5×10^5
(8-78)	$Nu_m = 0.036\ Pr^{0.43} \cdot (Re_L^{0.8} - 9200)\left(\frac{\mu_\infty}{\mu_w}\right)^{0.25}$ (average)	Laminar and turbulent	Flow over a flat plate, small free-stream turbulence, for $2 \times 10^5 < Re_L < 5.5 \times 10^6$ $0.70 < Pr < 380$ $0.26 < \frac{\mu_\infty}{\mu_w} < 3.5$
(8-81)	$Nu_m = (0.4\ Re^{0.5} + 0.06\ Re^{2/3}) \cdot Pr^{0.4} \cdot \left(\frac{\mu_\infty}{\mu_w}\right)^{0.25}$		Flow across a single cylinder, for $40 < Re < 10^5$ $0.67 < Pr < 300$ $0.25 < \frac{\mu_\infty}{\mu_w} < 5.2$
(8-82)	$Nu_m = 0.3 + \frac{0.62\ Re^{1/2}\ Pr^{1/3}}{[1 + (0.4/Pr)^{2/3}]^{1/4}} \cdot \left[1 + \left(\frac{Re}{282{,}000}\right)^{5/8}\right]^{4/5}$		Flow across a single cylinder, for $10^2 < Re < 10^7$ and $Pe = Re\ Pr > 0.2$
(8-84)	$Nu_m = (0.8237 - \ln Pe^{1/2})^{-1}$		Flow across a single cylinder for $Pe < 0.2$
(8-85)	$Nu_m = c\left(\frac{u_\infty D_e}{\nu}\right)^n$ constant c and n are given in Table 8-2		Flow across a single noncircular cylinder

(8-87)	$\mathrm{Nu}_m = 0.37\ \mathrm{Re}^{0.6}$	Flow of gases across a single sphere for $17 < \mathrm{Re} < 70{,}000$
(8-88)	$\mathrm{Nu}_m = 2 + (0.4\ \mathrm{Re}^{0.5} + 0.06\ \mathrm{Re}^{2/3}) \cdot \mathrm{Pr}^{0.4}\left(\dfrac{\mu_\infty}{\mu_w}\right)^{0.25}$	Flow across a single sphere for $3.5 < \mathrm{Re} < 8 \times 10^4$, $0.7 < \mathrm{Pr} < 380$, $1 < \dfrac{\mu_\infty}{\mu_w} < 3.2$
(8-95a)	$\dfrac{h_m D}{k} = 1.13 c_0\ \mathrm{Re}^n\ \mathrm{Pr}^{1/3}$ where $\mathrm{Re} = DG_{\max}/\mu$; coefficients c_0 and n given in Table 8-3	Flow across tube bundles having 10 or more transverse rows in direction of flow and for $2000 < \mathrm{Re} < 40{,}000$, $\mathrm{Pr} > 0.7$, $N \ge 10$
(8-96)	$h_N = c_1 h_{N \ge 10}$ Correction factor c_1 is given in Table 8-4.	To correct h_m given by Eq. (8-95a) for tube bundles having $1 \le N < 10$
(8-97)	$\dfrac{h_m D}{k} = c_2\ \mathrm{Re}^m\ \mathrm{Pr}^{0.36}\left(\dfrac{\mathrm{Pr}_b}{\mathrm{Pr}_w}\right)^n$ $n = \begin{cases} 0 & \text{for gases} \\ \frac{1}{4} & \text{for liquids} \end{cases}$ c_2 and m given in Table 8-5	Flow across tube bundles for $0.7 < \mathrm{Pr} < 500$ and $N \ge 20$
(8-98)	$\mathrm{Nu}_N = c_3\ \mathrm{Nu}_{N \ge 20}$ Correction factor c_3 is given in Fig. 8-12.	To correct Nu_m given by Eq. (8-97) for tube bundles having $1 \le N \le 20$
(8-99)	$\Delta P = f\ \dfrac{N G_{\max}^2}{2\rho} Z$ f and Z are given in Figs. 8-13 and 8-14.	Pressure drop for flow across tube bundles
(8-100)	$\mathrm{Nu}_m = 4.03 + 0.228\ (\mathrm{Re\ Pr})^{0.67}$	Mercury flow across a staggered, equilateral tube arrangement; $20{,}000 < \mathrm{Re} < 80{,}000$

The total amount of cooling is determined by adding the heat transfers for the laminar and turbulent regions:

$$\text{Total cooling} = 340 + 709 = 1049 \text{ W}$$

8-8 SUMMARY OF CORRELATIONS

In Table 8-6 we summarize the correlations of heat transfer and drag coefficients for forced convection over bodies presented in this chapter.

PROBLEMS

Drag force for laminar flow over a flat plate

8-1 Atmospheric air at 27°C flows along a flat plate with a velocity of $u_\infty = 8$ m/s. The critical Reynolds number for transition from laminar to turbulent flow is $\text{Re}_c = 5 \times 10^5$.

(*a*) At what distance from the leading edge of the plate does transition occur?

(*b*) Determine the velocity boundary-layer thickness and the local drag coefficient at the location where the transition occurs.

(*c*) Determine the average drag coefficient over the distance where the flow is laminar.

Answer: (*a*) 0.98 m; (*b*) 7 mm, 0.00094; (*c*) 0.00188

8-2 Atmospheric air at 77°C flows over a flat plate with a velocity of $u_\infty = 10$ m/s. Plot the variation of the thickness of the velocity boundary layer $\delta(x)$ and the local drag coefficient as a function of the distance along the plate up to the location where the transition takes place from laminar to turbulent flow at $\text{Re} = 5 \times 10^5$.

8-3 Determine the thickness of the velocity boundary layer $\delta(x)$ and the local shear stress τ_x at $x = 2$ m from the leading edge of a flat plate for the boundary-layer flow of air and hydrogen at atmospheric pressure and 80°C with a velocity of $u_\infty = 2$ m/s.

8-4 Determine the thickness of the velocity boundary layer $\delta(x)$ and the local shear stress τ_x at $x = 0.5$ m from the leading edge of a flat plate for the boundary-layer flow of engine oil and ethylene glycol at 80°C with a velocity of $u_\infty = 2$ m/s.

8-5 Determine the thickness of the velocity boundary layer $\delta(x)$, the local drag coefficient c_x, and the local shear stress τ_x at a distance $x = 0.5$ m from the leading edge of a flat plate for the boundary-layer flow of air at $T_\infty = 77$°C with a velocity of $u_\infty = 2$ m/s at (*a*) 0.5, (*b*) 1, and (*c*) 2 atm.

8-6 Consider a flat plate of length $L = 1$ m in the x direction and width $w = 3$ m in the y direction. Air at $T_\infty = 27$°C and atmospheric pressure flows along this plate in the x direction with a velocity of $u_\infty = 2$ m/s. Calculate the total drag force exerted on the plate. What would the drag force be if the air flow were in the y direction?

Answer: 0.0263 N, 0.0152 N

8-7 Determine the drag force exerted on a 2-m-long flat plate per 1-m width for the flow of the following fluids at atmospheric pressure and 350 K with a velocity of $u_\infty = 5$ m/s: (*a*) air, (*b*) hydrogen, and (*c*) helium.

8-8 Show that the second-degree polynomial representation of the velocity profile for flow along a flat plate subject to the conditions

$$u\Big|_{y=0} = 0 \qquad u\Big|_{y=\delta} = u_\infty \qquad \text{and} \qquad \frac{\partial u}{\partial y}\Big|_{y=\delta} = 0$$

is given by

$$\frac{u}{u_\infty} = 2\frac{y}{\delta} - \left(\frac{y}{\delta}\right)^2$$

8-9 Show that the fourth-degree polynomial representation of the velocity profile for flow along a flat plate is given by

$$\frac{u}{u_\infty} = 2\frac{y}{\delta} - 2\left(\frac{y}{\delta}\right)^3 + \left(\frac{y}{\delta}\right)^4$$

8-10 Using the momentum integral equation given by

$$\frac{d}{dx}\left[\int_0^{\delta(x)} (u_\infty - u)u\, dy\right] = \nu \left.\frac{\partial u}{\partial y}\right|_{y=0}$$

and a velocity profile represented by a sinusoidal expression in the form

$$\frac{u(x, y)}{u_\infty} = \sin\left(\frac{\pi}{2}\frac{y}{\delta}\right)$$

for laminar boundary-layer flow along a flat plate, derive expressions for the boundary-layer thickness $\delta(x)$, the local drag coefficient c_x, and the average drag coefficient c_L over the length $0 \le x \le L$.

8-11 Repeat Prob. 8-10, using a linear velocity profile given in the form

$$\frac{u(x, y)}{u_\infty} = \frac{y}{\delta}$$

8-12 Repeat Prob. 8-10, using a second-degree polynomial representation of the velocity profile given in the form

$$\frac{u(x, y)}{u_\infty} = 2\frac{y}{\delta} - \left(\frac{y}{\delta}\right)^2$$

Drag force for turbulent flow over a flat plate

8-13 Atmospheric air at 27°C flows with a free-stream velocity of $u_\infty = 10$ m/s along a flat plate $L = 4$ m long. Compute the drag coefficient at 2 and 4 m from the leading edge. Assuming an all-turbulent boundary layer, determine the drag force exerted per 1-m width of the plate.

Answer: 3.56×10^{-3}, 3.1×10^{-3}; 0.91 N

8-14 A fluid at 80°C flows with a free-stream velocity of $u_\infty = 8$ m/s along a flat plate $L = 5$ m long. Compute the local drag coefficient and the shear stress at the trailing edge (that is, $L = 5$ m) of the plate for (*a*) air, (*b*) CO_2, (*c*) water, and (*d*) ethylene glycol.

8-15 Air at 27°C flows with a free-stream velocity of $u_\infty = 40$ m/s along a flat plate $L = 2$ m long. Calculate the boundary-layer thickness at the end of the plate for air at (*a*) $\frac{1}{2}$, (*b*) 1, and (*c*) 2 atm.

Answer: (*a*) 3.18, (*b*) 3.07, (*c*) 2.82 cm

8-16 In a low-speed wind tunnel, air at 1 atm and 300 K flows over a flat plate with a free-stream velocity of 50 m/s. Determine the length of the plate to study boundary layers at Reynolds numbers up to 2×10^7. What is the boundary-layer thickness at the trailing edge of the plate?

8-17 Air at a pressure $P_\infty = 0.6$ atm and temperature $T_\infty = -15$°C flows with a free-stream velocity $u_\infty = 120$ m/s over the wing of an airplane. The wing is $L = 2$ m long in the direction of flow and can be regarded as a flat plate. Determine the local drag coefficient and the shear stress at the trailing edge of the wing. What is the drag force per meter width of the wing?

Heat transfer for laminar flow over a flat plate

8-18 Determine the velocity and the thermal boundary-layer thicknesses at $L = 0.5$ m from the leading edge of a plate at 50°C for flow at $T_\infty = 30$°C, velocity $u_\infty = 0.5$ m/s of (*a*) air at atmospheric pressure, (*b*) water, and (*c*) engine oil. Compare the ratio δ_t/δ of thermal to velocity boundary-layer thickness.

8-19 Determine the velocity and the thermal boundary-layer thicknesses $L = 0.5$ m from the leading edge of a flat plate at 74°C for flow at $T_\infty = 80$°C and atmospheric pressure with a velocity of $u_\infty = 3$ m/s of air, hydrogen, and helium, respectively. Compare the ratio δ_t/δ.

Answer: $\delta = 9.2, 24.1, 25.8$ mm; $\delta_t = 9.5, 24.8, 26.4$ mm

8-20 Engine oil at 40°C flows with a velocity of $u_\infty = 1$ m/s over a 2-m-long flat plate whose surface is maintained at a uniform temperature of 80°C. Determine the average heat transfer coefficient over the 2-m length of the plate.

Answer: 74.4 W/(m^2 · °C)

8-21 Air at atmospheric pressure and 40°C flows with a velocity of $u_\infty = 5$ m/s over a 2-m-long flat plate whose surface is kept at a uniform temperature of 120°C. Determine the average heat transfer coefficient over the 2-m length of the plate. Also find the rate of heat transfer between the plate and the air per 1-m width of the plate.

8-22 Air at atmospheric pressure and $T_\infty = 54$°C flows with a velocity of $u_\infty = 10$ m/s over a 1-m-long flat plate maintained at $T_w = 200$°C. Calculate the average drag and heat transfer coefficients over the 1-m length of the plate. Determine the rate of heat transfer between the plate and air per meter width of the plate.

8-23 Mercury at $T_\infty = 80$°C flows with a velocity of $u_\infty = 0.1$ m/s over a flat plate maintained at $T_w = 120$°C. Assuming that the transition from laminar to turbulent flow takes place at $\text{Re}_c = 5 \times 10^5$, determine the average heat transfer coefficient over the length of the plate where the flow is laminar.

Answer: 2300 W/(m^2 · °C)

8-24 The following information is given:

$\text{Pr} \gg 1$ for oils

$\text{Pr} \cong 0.7$ for gases

$\text{Pr} \ll 1$ for liquid metals

Sketch the velocity and thermal boundary-layer thicknesses with distance from the leading edge of a flat plate for laminar flow of each of the above fluids.

8-25 The local heat transfer coefficient h_x for laminar flow along a flat plate varies as $x^{-1/2}$, with x being the distance measured from the leading edge of the plate. Develop an expression for the average value of the heat transfer coefficient h_m over the distance L from the leading edge of the plate.

8-26 Determine the ratio of the average Nusselt number Nu_m over the length L of a flat plate to the local value of the Nusselt number Nu_L at the location L if the local heat transfer coefficient h_x varies as $x^{-1/2}$, where x is the distance from the leading edge of the plate.

8-27 Atmospheric air at $T_\infty = 24$°C flows with a velocity of $u_\infty = 4$ m/s along a flat plate $L = 2$ m long that is maintained at a uniform temperature of 130°C.

(*a*) Determine the thermal boundary-layer thickness δ_t and the local heat transfer coefficient h at the trailing edge (that is, $L = 2$ m) of the plate.

(*b*) Find the average heat transfer coefficient over the entire length of the plate.

(*c*) Calculate the heat transfer rate from the plate to the air per meter width of the plate.

8-28 Consider the flow of air, hydrogen, and helium at atmospheric pressure and 77°C with a velocity of $u_\infty = 4$ m/s along a flat plate $L = 2$ m long. Determine the thicknesses of the velocity and the thermal boundary layers and the value of the local heat transfer coefficient 1 m from the leading edge of the plate.

8-29 Determine the thickness of the thermal boundary layer δ_t and the local heat transfer coefficient h_x a distance $L = 1$ m from the leading edge of a flat plate for the flow of air at 77°C with velocity of $u_\infty = 4$ m/s at pressures 0.5, 1.0, and 2 atm.

8-30 Air at atmospheric pressure at a mean temperature of 77°C flows over a flat plate $L = 2$ m long with a velocity of $u_m = 4$ m/s. Plot the local and average heat transfer coefficients as a function of the distance from the leading edge of the plate.

8-31 Ethylene glycol at a mean temperature of 80°C flows over a 0.5-m-long flat plate with a velocity of $u_\infty = 2$ m/s. Calculate the average heat transfer coefficient over the entire length of the plate.

Answer: 640 W/(m^2 · °C).

8-32 Helium at atmospheric pressure and 20°C flows with a velocity of 10 m/s over a flat plate $L = 2$ m long that is maintained at a uniform temperature of 140°C. Calculate the rate of heat loss from the plate per meter width of the plate. What is the drag force exerted on the plate?

8-33 Atmospheric air at 20°C flows with a velocity of 2 m/s over the 3 m by 3 m surface of a wall which absorbs solar energy flux at a rate of 500 W/m^2 and dissipates heat by convection into the airstream. Assuming that the other surface of the wall has negligible heat loss, determine the average temperature of the wall under equilibrium conditions.

Answer: 180°C

8-34 Atmospheric air at 27°C flows with a speed of 4 m/s over a 0.5 m by 0.5 m flat plate which is uniformly heated with an electric heater at a rate of 2000 W/m^2. Calculate the average temperature of the plate.

8-35 Consider the laminar boundary-layer flow of a liquid metal with velocity u_∞ and temperature T_∞ along a flat plate kept at a uniform temperature T_w. Derive the expressions for the thermal boundary-layer thickness $\delta_t(x)$ and the local Nusselt number $\text{Nu}_x \equiv hx/k$ by using a linear profile for the temperature distribution given in the form

$$\frac{T(x, y) - T_w}{T_\infty - T_w} = \frac{y}{\delta_t(x)}$$

8-36 Repeat Prob. 8-35, using a second-degree polynomial representation for the temperature profile.

8-37 Consider laminar boundary-layer flow of a fluid having a Prandtl number $\text{Pr} \cong 1$ with a velocity u_∞ and themperature T_∞ along a flat plate kept at a uniform temperature T_w. Derive the expressions for the thermal boundary-layer thickness $\delta_t(x)$ and the local Nusselt number $\text{Nu}_x \equiv hx/k$ by using a linear velocity profile for the velocity distribution and a second-degree polynomial representation for the temperature distribution. Compare this result with those derived in this chapter by using cubic velocity and temperature profiles.

8-38 Consider the laminar boundary-layer flow of a liquid metal with velocity u_∞ and temperature T_∞ over a flat plate maintained at a uniform temperature T_∞. Taking the temperature profile in the form

$$\theta(x, y) = \frac{T(x, y) - T_w}{T_\infty - T_w} = \sin\left(\frac{\pi}{2}\frac{y}{\delta_t}\right)$$

where T_w is the wall temperature and δ_t is the thermal boundary-layer thickness, develop an expression for the local heat transfer coefficient h_x.

Heat transfer for turbulent flow over a flat plate

8-39 Air at atmospheric pressure and 24°C flows with a velocity of $u_\infty = 10$ m/s along a flat plate $L = 4$ m which is maintained at a uniform temperature of 130°C. Assume $\text{Re}_c = 2 \times 10^5$.

(*a*) Calculate the local heat transfer coefficient at $x = 2$, 3, and 4 m from the leading edge of the plate.

(*b*) Find the average heat transfer coefficient over $L = 4$ m.

(*c*) Determine the heat transfer rate from the plate to the air per meter width of the plate.

Answer: (*a*) 22.8, 21.1, 19.9 W/(m^2 · °C); (*b*) 23.3 W/(m^2 · °C); (*c*) 9880 W/m

8-40 The local heat transfer coefficient for turbulent flow along a flat plate can be determined from

$$\text{St}_x\,\text{Pr}^{2/3} = 0.0296\,\text{Re}_x^{-1/5} \qquad \text{for } 5 \times 10^5 < \text{Re}_x < 10^7$$

Assuming that this relation is valid from the leading edge of a flat plate, develop an expression for the average value of the heat transfer coefficient over the length L of a flat plate.

8-41 A fluid at $T_\infty = 40$°C flows with a velocity $u_\infty = 8$ m/s along a flat plate $L = 3$ m long which is maintained at a uniform temperature of 100°C. Calculate the local heat transfer coefficient at the end of the plate and the average heat transfer coefficient over the entire length of the plate for (*a*) air at atmospheric pressure and (*b*) ethylene glycol. Assume $\text{Re}_c = 2 \times 10^5$.

8-42 Helium at 1 atm, $u_\infty = 30$ m/s, and 300 K flows over a flat plate $L = 5$ m long and $w = 1$ m wide which is maintained at a uniform temperature of 600 K. Calculate the average heat transfer coefficient and the total heat transfer rate. (Assume $Re_c = 2 \times 10^5$.)

8-43 Air at $T_\infty = 24°C$ flows along a flat plate $L = 4$ m which is maintained at a uniform temperature of 130°C. Calculate the average heat transfer coefficient over the entire length of the plate and the heat transfer rate per meter width of the plate for (*a*) $u_\infty = 5$, (*b*) $u_\infty = 10$, and (*c*) $u_\infty = 20$ m/s. (Assume $Re_c = 2 \times 10^5$.)

8-44 A highly conducting thin wall $L = 2$ m long separates the hot and cold airstreams flowing on both sides parallel to the plate surface. The hot stream is at the atmospheric pressure and has a temperature $T_h = 250°C$ and a velocity $u_h = 50$ m/s. The cold airstream is also at atmospheric pressure and has temperature $T_c = 50°C$ and velocity $u_c = 15$ m/s. Calculate the average heat transfer coefficients for both airstreams and the total heat transfer rate between the streams per meter width of the separating plate. (Assume the wall at the arithmetic mean of T_h and T_c for the calculation of physical properties, and take $Re_c = 2 \times 10^5$.)

Flow across a single cylinder

8-45 A fluid at 27°C flows with a velocity of 10 m/s across a 5-cm-OD tube whose surface is kept at a uniform temperature of 120°C. Determine the average heat transfer coefficients and the heat transfer rates per meter length of the tube for (*a*) air at atmospheric pressure, (*b*) water, (*c*) ethylene glycol.

Answer: (*a*) 58.3, (*b*) 31,800, (*c*) 10,628 W/(m^2 · °C)

8-46 Consider the flow of air at 27°C with a velocity of $u_\infty = 20$ m/s across a single cylinder of outside diameter $D = 2.5$ cm. The surface of the cylinder is maintained at a uniform temperature of $T_s = 120°C$. Determine the average heat transfer coefficient and the heat transfer rate from the tube to the air per meter length of the tube for (*a*) 1, (*b*) 2, and (*c*) 4 atm.

8-47 A fluid at 80°C flows with a free-stream velocity of 10 m/s across a tube with $D = 5$ cm OD. Determine average drag coefficient and the drag force per meter length of the tube for (*a*) air at 1 atm, (*b*) CO_2 at 1 atm, (*c*) water, (*d*) ethylene glycol, and (*e*) engine oil.

8-48 Consider the flow of air at 77°C and with a mass flow rate of $\rho u_\infty = 25$ kg/(m^2 · s) across a single cylinder of outside diameter $D = 2.5$ cm. Determine the drag force exerted per meter length of the tube for air flowing at (*a*) 1, (*b*) 2, and (*c*) 4 atm.

8-49 Engine oil at 20°C flows with a velocity of 1 m/s across a 2.5-cm-diameter tube which is maintained at a uniform temperature $T_s = 100°C$. Determine the average heat transfer coefficient and the heat transfer rate between the tube surface and the oil per meter length of the tube.

Answer: 3206 W/(m^2 · °C), 20.14 kW/m

8-50 The thermal insulation is removed from the 1-m-length section of a steam pipe of outside diameter $D = 25$ cm and carrying high-pressure, high-temperature steam at 180°C. Air at −5°C is flowing across the exposed section with a velocity of $u_\infty = 6$ m/s. Determine the average heat transfer coefficient and the rate of heat loss from the 1-m-length exposed section of the tube into the air.

8-51 Atmospheric air at 27°C flows with a velocity of $u_\infty = 10$ m/s across a single duct whose surface is kept at a uniform temperature $T_w = 100°C$. Determine the average heat transfer coefficient and the heat transfer rate per meter length of the duct having the following dimension and configuration:

(*a*) A circular tube with $D = 1.25$ cm OD

(*b*) A square duct of 1.2 cm by 1.2 cm cross section oriented such that one of its lateral surfaces is perpendicular to the direction of flow

8-52 Water at $T_\infty = 20°C$ with a free-stream velocity $u_\infty = 1.5$ m/s flows across a single circular tube of outside diameter $D = 2.5$ cm. The tube surface is maintained at a uniform temperature $T_w = 80°C$. Calculate the average heat transfer coefficient h_m and the heat transfer rate per meter length of the tube.

8-53 Atmospheric air at $T_\infty = 300$ K and with a free-stream velocity of $u_\infty = 30$ m/s flows across a single cylinder of outside diameter $D = 2.5$ cm. The cylinder's surface is at a uniform temperature of $T_w = 400$ K. Calculate the mean heat transfer coefficient h_m and the heat transfer rate per meter length of the cylinder.

Flow across a single sphere

8-54 Consider the flow of a fluid at 80°C and with a mass flow rate of $\rho u_\infty = 50\ \text{kg/(m}^2 \cdot \text{s)}$ across a single sphere of diameter $D = 5$ cm. Compute the average drag coefficient and the drag force exerted on the sphere for (*a*) air at 1 atm, (*b*) CO_2 at 1 atm, (*c*) water, (*d*) ethylene glycol, and (*e*) engine oil.

8-55 A fluid at 80°C flows with a free-stream velocity of 10 m/s across a 5-cm-diameter single sphere. Compute the average drag coefficient and the total drag force exerted on the sphere for (*a*) air at 1 atm, (*b*) CO_2 at 1 atm, (*c*) water, (*d*) ethylene glycol, and (*e*) engine oil.

8-56 A fluid at 40°C flows with a velocity of 2 m/s across a 2.5-cm-diameter sphere. The surface of the sphere is maintained at a uniform temperature of 100°C. Compute the average heat transfer coefficient and the rate of heat transfer from the sphere to the fluid for (*a*) air at 1 atm, (*b*) CO_2 at 1 atm, (*c*) water, (*d*) ethylene glycol, and (*e*) engine oil.

8-57 Atmospheric air at 20°C flows with a free-stream velocity $u_\infty = 5$ m/s over a 2-m-diameter spherical tank which is maintained at 80°C. Compute the average heat transfer coefficient and the heat transfer rate from the sphere to the air.

Answer: 63.2 W/(m² · °C), 29.8 W

8-58 Water at 20°C flows with a free-stream velocity of 1 m/s over a 2.5-cm-diameter sphere whose surface is maintained at a uniform temperature of 140°C. Determine the average heat transfer coefficient and the rate of heat loss from the sphere to the air.

Flow across tube bundles

8-59 Air at atmospheric pressure and 27°C flows over a tube bank consisting of D = 1-cm-diameter tubes 10 rows deep. The flow velocity before the air enters the tube bundle is 1 m/s. Determine the average heat transfer coefficient for the following two cases:

(*a*) Tubes are in equilateral-triangular arrangement with $S_T/D = S_D/D = 1.25$.

(*b*) Tubes are in square arrangement with $S_T/D = S_L/D = 1.25$.

Answer: (*a*) 106, (*b*) 101 W/(m² · °C)

8-60 Air at atmospheric pressure and 27°C flows over a tube bank consisting of D = 1-cm-diameter tubes 10 rows deep. The flow velocity before the air enters the tube bank is $u_\infty = 1.5$ m/s. Determine the average friction factor and the pressure drop for the following configurations:

(*a*) Tubes are in equilateral-triangular arrangement with $S_T/D = S_D/D = 1.25$.

(*b*) Tubes are in a square arrangement with $S_T/D = S_L/D = 1.25$.

Answer: (*a*) $f = 0.6$, $\Delta P = 198.7\ \text{N/m}^2$; (*b*) $f = 0.53$, $\Delta P = 175.5\ \text{N/m}^2$

8-61 Air at 227°C and 1.5 atm pressure flows over a tube bank consisting of 1.25-cm-OD tubes 10 rows deep in the direction of flow and forming a stack 40 tubes high. Tubes are in *in-line arrangement* with $S_L/D = S_T/D = 2$. The air velocity before entering the tube bank is $u_\infty = 4$ m/s. Determine the friction factor and the pressure drop for flows across this tube bank.

8-62 Repeat Prob. 8-61 for the *staggered arrangement* with $S_L/D = S_T/D = 2$.

8-63 Air at 27°C flows over a tube bank consisting of 2.5-cm-OD, 2-m-long tubes in in-line arrangement with $S_L/D = 1.5$ and $S_T/D = 2$. The tube bank is 10 rows deep, forming a stack 20 tubes high. The air velocity before entering the tube bank is $u_\infty = 10$ m/s. Tubes are maintained at a uniform temperature of 100°C. Determine the average friction factor and the pressure drop for flow across the tube bank.

Answer: 0.3; 410 N/m²

8-64 Air at 100°C and 5 atm pressure flows over a tube bank consisting of 2.5-cm-OD tubes 40 rows deep in the direction of flow; G_{max} for the flow is 20 kg/(m² · s). Determine the average heat transfer coefficient for a staggered arrangement with $S_L/D = 1.5$ and $S_T/D = 2.0$.

8-65 Hot flue gases at 375°C flow across a tube bank consisting of 1.25-cm-OD tubes which are maintained at a uniform temperature of 30°C by flowing water through the tubes. The tube bundle is 10 rows deep in the direction of flow and contains 40 tubes in each row. The tubes are L = 1 m long and have an in-line arrangement with $S_L/D = S_T/D = 2$. The velocity of flue gases before entering the tube matrix is $u_\infty = 7$ m/s. Determine the average heat transfer coefficient and the total heat transfer rate in the tube matrix. (Treat flue gases as air.)

Answer: 155.3 W/(m^2 · °C); 96.9 kW

8-66 Air at atmospheric pressure and 20°C enters a tube bank consisting of 1.25-cm-OD tubes 14 rows deep in the direction of flow and containing 20 tubes in each row. The tubes are $L = 2$ m long and have an in-line arrangement with $S_L/D = S_T/D = 1.5$. The velocity of air before entering the tube bank is $u_\infty = 6$ m/s. Determine the average heat transfer coefficient and the total heat transfer rate for the tube matrix. Assume a wall temperature $T_w = 134$°C.

8-67 Mercury at 200°C flows over a tube bank consisting of 1.25-cm-OD tubes arranged in equilateral-triangular arrangement with a pitch-to-diameter ratio of 1.375. There are 60 rows in the direction of flow, and the flow velocity just before the fluid enters the tube bank is $u_\infty = 0.07$ m/s. Determine the friction factor and the pressure drop for flow across this tube bank.

Answer: 0.25; 6.5 kN/m^2

8-68 Mercury at 250°C flows over a tube bank consisting of 1.25-cm-OD. 1-m-long tubes in equilateral-triangular arrangement with a pitch-to-diameter ratio of 1.375. The matrix has 60 rows in the direction of flow and 30 tubes in each row. The velocity of mercury before entering the matrix is $u_\infty =$ 0.05 m/s, and the tubes are maintained at a uniform temperature of 160°C. Determine the average heat transfer coefficient and the total heat transfer rate through the matrix.

8-69 Liquid sodium at 425°C flows across a tube bank consisting of 1.25-cm-OD tubes 50 rows deep in the direction of flow and having an equilateral-triangular arrangement with a pitch-to-diameter ratio of 1.5. The flow velocity before the liquid enters the tube bank is 0.3 m/s, and the tubes are at a uniform temperature of 200°C. Determine the average heat transfer coefficient.

Answer: 62.84 kW/(m^2 · °C)

High-speed flow over a flat plate

8-70 Air at $\frac{1}{25}$ atm and 300 K flows over a flat plate with a free-stream velocity of 600 m/s. The plate surface is to be maintained at a uniform temperature of 340 K. Determine the amount of cooling needed per meter width of the plate over the region where the flow is laminar. (Assume the critical Reynolds number 5×10^5.)

8-71 Air at $\frac{1}{30}$ atm and 275 K flows over a flat plate with a free-stream velocity of 700 m/s. The surface of the plate is to be maintained at 325 K. Determine the amount of cooling needed per meter width of the plate over the region where the flow is laminar. (Assume $Re_c = 5 \times 10^5$.)

Answer: 1.923 kW/m

8-72 Air at $\frac{1}{20}$ atm and at 250 K flows over a 0.25-m-long flat plate with a free-stream velocity of 800 m/s. How much cooling is needed per meter width of the plate to keep the plate at 300 K over the portion where the flow is laminar?

8-73 Repeat Prob. 8-72 for air pressures of $\frac{1}{15}$ and $\frac{1}{25}$ atm.

8-74 In a wind-tunnel test, air at $\frac{1}{25}$ atm and 250 K flows with a free-stream velocity of $u_\infty = 900$ m/s over a flat plate $L = 1.0$ m long. If the plate surface is to be maintained at a uniform temperature of 300 K, determine the amount of cooling needed when the plate width is $w = 0.5$ m.

8-75 Air at $\frac{1}{20}$ atm and 275 K flows with a free-stream velocity of $u_\infty = 700$ m/s over a flat plate $L =$ 1.2 m long. If the surface of the plate is to be maintained at 325 K, determine the amount of cooling needed per meter width of the plate.

Answer: 14.384 kW/m

8-76 Repeat Prob. 8-75 for air pressures of $\frac{1}{15}$ and $\frac{1}{25}$ atm.

REFERENCES

1. Blasius, H.: "Grenzschleten in Flussigkeiten mit kleiner Reibung," *Z. Angew. Math. Phys.*, **56**:1 (1908).

2. Howarth, L.: "On the Solution of the Laminar Boundary Layer Equations," *Proc. R. Soc.* (*London*), **A164**: 546 (1938).
3. Schlichting, H.: *Boundary Layer Theory*, 7th ed., McGraw-Hill, New York, 1979.
4. Rosenhead, L.: *Laminar Boundary Layers*, Clarendon Press, Oxford, 1963.
5. White, F. M.: *Viscous Fluid Flows*, McGraw-Hill, New York, 1974.
6. Özişik, M. N.: *Basic Heat Transfer*, 1st ed., McGraw-Hill, New York, 1977, chap. 8.
7. Von Kármán, T.: "Über laminare und turbulente Reibung," *Z. Angew. Math. Mech.*, **1**: 233 (1912); also (trans.) *NACA Tech. Memo.* 1092, 1946.
8. Schultz-Grunow, F.: "Neues Widerstandsgesetz für glatte Platten," *Luftfahrtforschung*, **17**: 239 (1940); also *NACA Tech. Memo.* 986, 1941.
9. Pohlhausen, E.: *Z. Angew. Math. Mech.*, **1**: 115 (1921).
10. Whitaker, S.: *Elementary Heat Transfer Analysis*, Pergamon, New York, 1976.
11. Zukauskas, A. A., and A. B. Ambrazyavichyus: "Heat Transfer of a Plate in a Liquid Flow," *Int. J. Heat Mass Transfer*, **3**: 305 (1961).
12. Colburn, A. P.: "A Method of Correlating Forced Convection Heat Transfer Data with Fluid Friction," *Trans. AIChE*, **29**: 174 (1933).
13. Edwards, A., and B. N. Furber, "The Influence of Free-Stream Turbulence on Heat Transfer by Convection from an Isolated Region of a Plane Surface in Parallel Air Flow," *Proc. Inst. Mech. Engrs.* (*London*), **170**: 941 (1956).
14. Parmelee, G. V., and R. G. Huebscher: "Heat Transfer by Forced Convection along a Smooth Flat Surface," *Heating, Piping, and Air Conditioning*, **19**(8): 115 (Aug. 1947).
15. Thom, A.: "The Flow Past Circular Cylinders at Low Speeds," *Proc. Roy. Soc.*, **A141**: 651 (1933).
16. Acrivos, A., D. D. Snowden, A. S. Grove, and E. E. Petersen: "The Steady Separated Flow past a Circular Cylinder at Large Reynolds Numbers," *J. Fluid Mech.*, **21**: 737 (1965).
17. McAdams, W. H.: *Heat Transmission*, 3d ed., McGraw-Hill, New York, 1954.
18. Whitaker, S.: "Forced Convection Heat Transfer Calculations for Flow in Pipes, past Flat Plates, Single Cylinders, and for Flow in Packed Beds and Tube Bundles," *AIChE J.*, **18**: 361–371 (1972).
19. Churchill, S. W., and M. Bernstein: "A Correlating Equation for Forced Convection from Gases and Liquids to a Circular Cylinder in Cross Flow," *J. Heat Transfer*, **99**: 300–306 (1977).
20. Nakai, S., and T. Okazaki: "Heat Transfer from a Horizontal Circular Wire at Small Reynolds and Grashof Numbers—1. Pure Convection," *Int. J. Heat Mass Transfer*, **18**: 387 (1975).
21. King, L. V.: "On the Convection of Heat from Small Cylinders in a Stream of Fluid," *Phil. Trans. Roy. Soc.* (*London*), **214**: 373 (1914).
22. Davis, A. H.: "Convective Cooling of Wires in Streams of Viscous Liquids," *Phil. Mag.*, **47**: 1057 (1924).
23. Hilpert, Von R.: "Warmeabgabe von Geheizten Drahten und Rohren im Luftstrom," *Forsch. Gebiete Ingenieurw.*, **4**: 215 (1933).
24. Piret, E. L., W. James, and M. Stacy: "Heat Transmission from Fine Wires to Water," *I&EC*, **39**:1098 (1947).
25. Perkins, H. C., and G. Leppert: "Forced Convection Heat Transfer from Uniformly Heated Cylinder," *J. Heat Transfer, Trans. ASME, Series C* **84**:257 (1962).
26. Fand R. M.: "Heat Transfer by Forced Convection from a Cylinder to Water in Crossflow," *Int. J. Heat Mass Transfer*, **8**: 995 (1965).
27. Eckert, E. R. G., and E. Soehngen: "Distributions of Heat-Transfer Coefficients around Circular Cylinders in Cross Flow at Reynolds Numbers from 20 to 500," *Trans. ASME*, **74**: 343–347 (1952).
28. Giedt, W. H.: "Investigation of Variation of Point Unit-Heat-Transfer Coefficient around a Cylinder Normal to an Air Stream," *Trans. ASME*, **71**: 375–381 (1949).
29. Jakob, Max: *Heat Transfer*, vol. 1, Wiley, New York, 1949.
30. Kramers, H.: "Heat Transfer from Spheres to a Flowing Media," *Physica*, **12**: 61 (1946).
31. Vliet, G. C., and G. Leppert: "Forced Convection Heat Transfer from an Isothermal Sphere to Water," *Trans. ASME*, **83C**: 163 (1961).
32. Yuge, T.: "Experiments on Heat Transfer from Spheres Including Combined Natural and Forced Convection," *Trans. ASME*, **82C**: 214 (1960).

33. Pierson, O. L.: "Experimental Investigation of the Influence of Tube Arrangement on Convection Heat Transfer and Flow Resistance in Cross Flow of Gases Over Tube Banks," *Trans. ASME*, **59**:563–572 (1937).
34. Grimison, E. D.: "Correlation and Utilization of New Data on Flow Resistance and Heat Transfer for Cross Flow of Gases over Tube Banks," *Trans. ASME*, **59**:583–594 (1937).
35. Jakob, M.: "Heat Transfer and Flow Resistance in Cross-Flow of Gases over Tube Banks," *Trans. ASME*, **60**:384–368 (1938).
36. Bergelin, O. P., A. P. Colburn, and H. L. Hull: *Heat Transfer and Pressure Drop During Viscous Flow across Unbaffled Tube Banks*, Bulletin no. 2, University of Delaware Engineering Experiment Station, June 1950.
37. Bergelin, O. P., G. A. Brown, and S. C. Doberstein: "Heat Transfer and Fluid Friction during Flow across Banks of Tubes," *Trans. ASME*, **74**:953–960 (1952).
38. Kays, W. M., and R. K. Lo: "Basic Heat Transfer and Fluid Friction Data for Gas Flow Normal to Banks of Staggered Tubes—Use of Transient Technique," *Stanford Univ. Dep. Mech. Eng., Tech. Rep.* 15, Navy Contract NG-ONR-251, T.O. 6, 1952.
39. Kays, W. M., A. L. London, and R. K. Lo: "Heat Transfer and Friction Characteristics for Gas Flow Normal to Tube Banks—Use of a Transient Test Technique," *Trans. ASME*, **76**:387 (1954).
40. Zukauskas, A. A., R. V. Ulinskas, and C. S. J. Sipavicius: "Heat Transfer and Pressure Drop in Cross-Flow of Viscous Fluids Over Tube Bundles at Low Re," *Heat Transfer-Soviet Research*, **10**: 90–101 (1978).
41. Hoe, R. J., D. Dropkin, and O. E. Dwyer: "Heat Transfer Rates to Cross Flowing Mercury in a Staggered Tube Bank, I," *Trans. ASME*, **79**:899–908 (1957).
42. Kays, W. M., and A. L. London: *Compact Heat Exchanger Design*, McGraw-Hill, New York, 1958.
43. Bergelin, O. P., M. D. Leighton, W. L. Lafferty, and R. L. Pigford: *Heat Transfer and Pressure Drop during Viscous and Turbulent Flow across Baffled and Unbaffled Tube Banks*, Bulletin no. 4, University of Delaware Engineering Experiment Station, April, 1958.
44. Fairchild, H. N., and C. P. Welch: "Convection Heat Transfer and Pressure Drop of Air-Flowing across In-Line Tube Banks at Close Back Spacings," Paper no. 61-WA-250, presented at ASME annual winter meeting, 1961.
45. Masliyah, J. H.: "Viscous Flow Across Banks of Circular and Elliptical Cylinders: Momentum and Heat Transfer," *Can. J. Chem. Engr.*, **61**:550–555 (1973).
46. (a) Jakob, Max: *Heat Transfer*, vol. 1, Wiley, New York, 1949.
    (b) Jakob, Max: "Heat Transfer and Flow Resistance in Cross Flow of Gases over Tube Banks," *Trans. ASME*, **60**:384–386 (1938).
47. Zukauskas, A.: "Heat Transfer from Tubes in Cross Flow," *Adv. Heat Transfer*, **8**:93–160 (1972).
48. Hoe, R. J, D. Dropkin, and O. E. Dwyer: "Heat Transfer Rates to Cross Flowing Mercury in a Staggered Tube Bank, I," *Trans. ASME*, **79**:899–908 (1957).
49. Richards, C. L., O. E. Dwyer, and D. Dropkin: "Heat Transfer Rates to Cross Flowing Mercury in a Staggered Tube Bank, II," *ASME-AIChE Heat Transfer Conf. Paper* 57-HT-11, 1957.
50. Kalish, S., and O. E. Dwyer: "Heat Transfer to NaK Flowing through Unbaffled Rod Bundles," *Int. J. Heat Mass Transfer*, **10**:1533–1558 (1967).
51. Eckert, E., and O. Drewitz: *Forsch. Geb. Ingenieurwes.*, **7**:116 (1940); (trans.) *NACA Tech. Memo.* 1045, 1943.
52. Meksyn, D.: "Plate Thermometer," *Z. Angew. Math. Phys.*, **11**:63–68 (1960).
53. Kaye, J.: "Survey of Friction Coefficients, Recovery Factors, and Heat Transfer Coefficients for Supersonic Flow," *J. Appl. Sci.*, **21**:117–119 (1954).
54. Eckert, E. R. G.: "Engineering Relations for Heat Transfer and Friction in High-Velocity Laminar and Turbulent Boundary Layer Flow over Surface with Constant Pressure and Temperature," *Trans. ASME*, **78**:1273–1284 (1956).
55. Eckert, E. R. G.: "Engineering Relations for Friction and Heat Transfer to Surfaces in High-Velocity Flow," *J. Appl. Sci.*, **22**:585–587 (1955).
56. Lin, C. C. (ed.): "Turbulent Flows and Heat Transfer," in *High Speed Aerodynamics and Jet Propulsion*, vol. 5, Princeton University Press, Princeton, N.J., 1954.
57. Eckert, E. R. G.: "Survey of Boundary Layer Heat Transfer at High Velocities and High Temperatures," *WADC Tech. Rep.* 59–624, 1960.

NOTES

1. *The derivation of the momentum integral equation* (8-5). The momentum equation (8-2) is integrated with respect to the y variable over the boundary-layer thickness $\delta(x)$; we obtain

$$\int_0^{\delta(x)} u \frac{\partial u}{\partial x} dy + \int_0^{\delta(x)} v \frac{\partial u}{\partial y} dy = \nu\left(\frac{\partial u}{\partial y}\bigg|_{y=\delta} - \frac{\partial u}{\partial y}\bigg|_{y=0}\right) = -\nu \frac{\partial u}{\partial y}\bigg|_{y=0} \tag{1}$$

since $\partial u/\partial y|_{y=\delta} = 0$ by the boundary-layer concept. The velocity component v appearing in Eq. (1) is eliminated by making use of the continuity equation (8-1). The second integral on the left-hand side is evaluated by parts as

$$\int_0^{\delta} v \frac{\partial u}{\partial y} dy = uv\bigg|_0^{\delta} - \int_0^{\delta} u \frac{\partial v}{\partial y} dy = u_\infty v\bigg|_{\delta} - \int_0^{\delta} u \frac{\partial v}{\partial y} dy \tag{2}$$

since $u = u_\infty$ at $y = \delta$ and $u = 0$ at $y = 0$. The terms on the right-hand side of this relation are now determined in the following manner: $\partial v/\partial y$ is immediately obtained from the continuity equation (8-1) as

$$\frac{\partial v}{\partial y} = -\frac{\partial u}{\partial x} \tag{3}$$

and $v|_\delta$ is determined by integrating Eq. (3) from $y = 0$ to δ as

$$v\bigg|_0^{\delta} = -\int_0^{\delta} \frac{\partial u}{\partial x} dy$$

or

$$v\bigg|_{\delta} = -\int_0^{\delta} \frac{\partial u}{\partial x} dy \tag{4}$$

since $v|_{y=0}$ 0. The substitution of Eqs. (3) and (4) into Eq. (2) yields

$$\int_0^{\delta} v \frac{\partial u}{\partial y} dy = -u_\infty \int_0^{\delta} \frac{\partial u}{\partial x} dy + \int_0^{\delta} u \frac{\partial u}{\partial x} \tag{5}$$

When this result is substituted into Eq. (1), we find

$$\int_0^{\delta} 2u \frac{\partial u}{\partial x} dy - u_\infty \int_0^{\delta} \frac{\partial u}{\partial x} dy = -\nu \frac{\partial u}{\partial y}\bigg|_{y=0}$$

or

$$\int_0^{\delta} \frac{\partial u^2}{\partial x} dy - \int_0^{\delta} \frac{\partial(uu_\infty)}{\partial x} dy = -\nu \frac{\partial u}{\partial y}\bigg|_{y=0}$$

or

$$\frac{d}{dx}\left[\int_0^{\delta} u(u_\infty - u)\, dy\right] = \nu \frac{\partial u}{\partial y}\bigg|_{y=0} \tag{6}$$

which is the momentum integral equation given by Eq. (8-5).

2. *The derivation of the energy integral equation* (8-32). The energy equation (8-30) is integrated with respect to y over a distance H that exceeds the thicknesses of both boundary layers

$$\int_0^H u\frac{\partial\theta}{\partial x}\,dy + \int_0^H v\frac{\partial\theta}{\partial y}\,dy = \alpha\left(\frac{\partial\theta}{\partial y}\bigg|_{y=H} - \frac{\partial\theta}{\partial y}\bigg|_{y=0}\right) = -\alpha\frac{\partial\theta}{\partial y}\bigg|_{y=0} \tag{1}$$

since $\partial\theta/\partial y|_{y=H} = 0$ by the definition of the boundary layer. The velocity component v appearing in Eq. (1) is eliminated by making use of the continuity equation (8-1).

The second integral on the left-hand side of this equation is evaluated by parts as

$$\int_0^H v\frac{\partial\theta}{\partial y}\,dy = v\theta\bigg|_0^H - \int_0^H \theta\frac{\partial v}{\partial y}\,dy = v\bigg|_{y=H} - \int_0^H \theta\frac{\partial v}{\partial y}\,dy \tag{2}$$

since $v|_{y=0} = 0$ and $\theta|_{y=H} = 1$. The terms $v|_{y=H}$ and $\partial v/\partial y$ appearing in Eq. (2) are obtained from the continuity equation (8-1) as

$$\frac{\partial v}{\partial y} = -\frac{\partial u}{\partial x} \qquad \text{and} \qquad v\bigg|_{y=H} = -\int_0^H \frac{\partial u}{\partial x}\,dy \tag{3}$$

The substitution of Eqs. (3) into Eq. (2) yields

$$\int_0^H v\frac{\partial\theta}{\partial y}\,dy = -\int_0^H \frac{\partial u}{\partial x}\,dy + \int_0^H \theta\frac{\partial u}{\partial x}\,dy \tag{4}$$

Substituting Eq. (4) into Eq. (1), we obtain

$$\int_0^H \left(u\frac{\partial\theta}{\partial x} + \theta\frac{\partial u}{\partial x} - \frac{\partial u}{\partial x}\right)dy = -\alpha\frac{\partial\theta}{\partial y}\bigg|_{y=0}$$

or

$$\int_0^H \left[\frac{\partial(u\theta)}{\partial x} - \frac{\partial u}{\partial x}\right]dy = -\alpha\frac{\partial\theta}{\partial y}\bigg|_{y=0}$$

or

$$\frac{d}{dx}\left[\int_0^{H=\delta_t} u(1-\theta)\,dy\right] = \alpha\frac{\partial\theta}{\partial y}\bigg|_{y=0} \tag{5}$$

which is the energy integral equation given by Eq. (8-32).

3. *The derivation of equation* (8-60). The differential equation (8-59) is of the following form:

$$x\frac{dy}{dx} + Ay = B \tag{1}$$

where A and B are constants. A particular solution Y_p of this equation is given as

$$Y_p = \frac{B}{A} \tag{2}$$

and the homogeneous solution Y_H that satisfies the homogeneous part of this equation is given as

$$Y_H = x^{-A} \tag{3}$$

Then the complete solution becomes

$$Y = Cx^{-A} + \frac{B}{A} \tag{4}$$

where C is the integration constant. This solution is of the same form as that given by Eq. (8-60).

CHAPTER

NINE

FREE CONVECTION

In Chaps. 7 and 8, we considered heat transfer in forced convection, in which the fluid motion was imposed externally by a fan, a blower, or a pump. Also in some situations convective motion is set up within the fluid without a forced velocity. Consider, for example, a hot plate placed vertically in a body of fluid at rest, which is at a uniform temperature lower than that of the plate. Heat transfer will take place first by pure conduction, and a temperature gradient will be established in the fluid. The temperature variation within the fluid will generate a density gradient which, in a gravitational field, will give rise, in turn, to a convective motion as a result of buoyancy forces. The fluid motion set up as a result of the buoyancy force is called *free convection*, or *natural convection*.

The flow velocity in free convection is much smaller than that encountered in forced convection; therefore, heat transfer by free convection is much smaller than that by forced convection. Figure 9-1*a* illustrates the development of the velocity field in front of a hot vertical plate owing to the *buoyancy force*. The heated fluid in front of the hot plate rises, entraining fluid from the quiescent outer region. Figure 9-1*b* shows a cold vertical plate in a hot fluid. In this case, the direction of motion is reversed; namely, the fluid in front of the cold plate moves vertically down, again entraining fluid from the quiescent outer region. In both cases a velocity boundary layer is developed, with the peak velocity occurring somewhere within the boundary layer. The velocity is zero at both the plate surface and the edge of the boundary layer. Furthermore, in the regions near the leading edge of the plate, the boundary-layer development is *laminar*, but, at a certain distance from the leading edge of the plate, transition to a *turbulent boundary layer* begins. Eventually, a fully developed turbulent boundary layer is established.

We now consider a fluid contained in a cavity or between two parallel plates arranged horizontally as illustrated in Fig. 9-2*a*. Suppose the lower plate is maintained at a temperature higher than that of the upper plate (that is, $T_1 > T_2$).

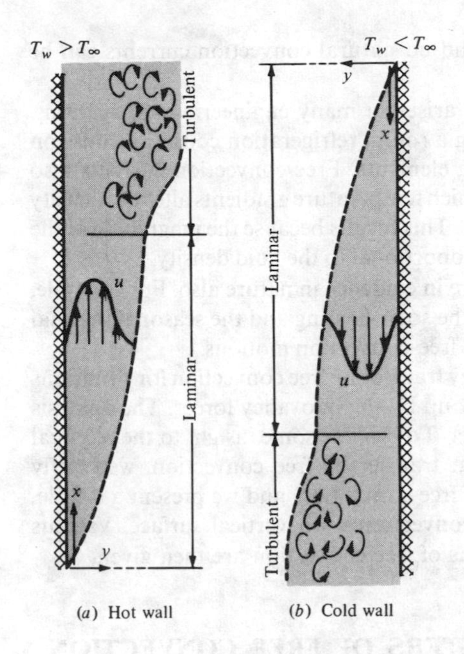

Figure 9-1 The laminar and turbulent velocity boundary layer for free convection on a vertical plate.

A temperature gradient will be established in the vertical direction. The layer will be top-heavy, since the density of the cold fluid at the top is higher than that of the hot fluid at the bottom. If the temperature difference is increased beyond a certain critical value, the viscous forces within the fluid can no longer sustain the buoyancy forces, and a convective motion is set up, giving rise to circulation patterns. Thus again we have natural convection in the fluid.

Suppose, in Fig. 9-2*b*, the lower plate is cold and the upper plate is hot (that is, $T_1 < T_2$). In such a case, the density of the top layer is less than that of the bottom

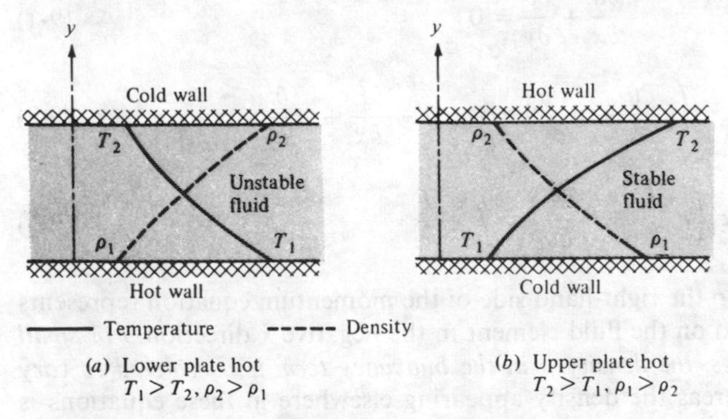

Figure 9-2 Fluid contained between two horizontal plates.

layer; then the fluid is always stable, and no natural convection currents can be set up.

Energy transfer by free convection arises in many engineering applications, such as a hot steam radiator for heating a room, refrigeration coils, transmission lines, electric transformers, and heating elements. Free-convection currents also can be set up at high-speed rotation in which temperature gradients alter the density in the presence of centrifugal body forces. This results because the magnitude of the body force due to centrifugal effects is proportional to the fluid density.

Buoyancy-induced flows in water are in evidence in nature also. For example, short-term circulations of water due to the solar heating and the seasonal thermal inversion of lakes are buoyancy-induced free-convection motions.

In this chapter we examine the energy transfer by free convection for situations in which the convection currents are set up by the buoyancy forces. The analysis of free convection is a complicated matter. To provide some insight to the physical significance of the factors affecting heat transfer by free convection, we briefly discuss the dimensionless parameters of free convection, and we present a simple, approximate analysis of laminar free convection on a vertical surface. Various empirical and semiempirical correlations of free convection are then given.

9-1 DIMENSIONLESS PARAMETERS OF FREE CONVECTION

To develop the principal dimensionless parameters of free convection, we consider free convection on a vertical plate, as illustrated in Fig. 9-1. For simplicity in the analysis, we assume the boundary-layer flow is steady and laminar; the viscous-energy dissipation term in the energy equation can be neglected because of the small flow velocities associated with free convection. Then the governing continuity, momentum, and energy equations are obtained from the boundary-layer equations (6-81), (6-82), and (6-83), respectively, by introducing the appropriate buoyancy term into the momentum equation:

Continuity:
$$\frac{\partial u}{\partial x} + \frac{\partial v}{\partial y} = 0 \tag{9-1}$$

x Momentum:
$$\rho\left(u\frac{\partial u}{\partial x} + v\frac{\partial u}{\partial y}\right) = -\rho g - \frac{\partial P}{\partial x} + \mu\frac{\partial^2 u}{\partial y^2} \tag{9-2}$$

Energy:
$$\rho c_p\left(u\frac{\partial T}{\partial x} + v\frac{\partial T}{\partial y}\right) = k\frac{\partial^2 T}{\partial y^2} \tag{9-3}$$

Here the term $-\rho g$ on the right-hand side of the momentum equation represents the body force exerted on the fluid element in the negative x direction. *For small temperature differences, the density ρ in the buoyancy term is considered to vary with temperature* whereas the density appearing elsewhere in these equations is considered constant.

To determine the pressure gradient term $\partial P/\partial x$, the x momentum equation (9-2) is evaluated at the edge of the velocity boundary layer, where $u \to 0$ and $\rho \to \rho_\infty$. We obtain

$$\frac{\partial P}{\partial x} = -\rho_\infty g \tag{9-4}$$

where ρ_∞ denotes the fluid density outside the boundary layer. Then the term $-\rho g - \partial P/\partial x$ appearing in the momentum equation (9-2) becomes

$$-\rho g - \frac{\partial P}{\partial x} = (\rho_\infty - \rho)g \tag{9-5}$$

If β is the *volumetric coefficient of thermal expansion* of the fluid, the change of density with temperature is related to β by

$$-\frac{1}{\rho}\left(\frac{\partial \rho}{\partial T}\right)_P = \beta \tag{9-6}$$

By expressing the derivative term in this relation with finite difference, Eq. (9-6) is approximated by

$$\Delta\rho = -\beta\rho\,\Delta T$$

or (9-7)

$$\rho_\infty - \rho = -\beta\rho(T_\infty - T)$$

Then Eq. (9-5) becomes

$$-\rho g - \frac{\partial P}{\partial x} = -\beta\rho(T_\infty - T)g \tag{9-8}$$

Equation (9-8) is substituted into the momentum equation (9-2). We summarize the resulting equations for free convection on a vertical plate:

$$\frac{\partial u}{\partial x} + \frac{\partial v}{\partial y} = 0 \tag{9-9}$$

$$u\frac{\partial u}{\partial x} + v\frac{\partial u}{\partial y} = g\beta(T - T_\infty) + \nu\frac{\partial^2 u}{\partial y^2} \tag{9-10}$$

$$u\frac{\partial T}{\partial x} + v\frac{\partial T}{\partial y} = \alpha\frac{\partial^2 T}{\partial y^2} \tag{9-11}$$

If the fluid is considered to be an ideal gas, we have

$$\rho = \frac{P}{\mathcal{R}T} \tag{9-12}$$

Then the coefficient of expansion β in Eq. (9-7) becomes

$$\beta = \frac{(\rho_\infty/\rho) - 1}{T - T_\infty} = \frac{(T/T_\infty) - 1}{T - T_\infty} = \frac{1}{T_\infty} \tag{9-13}$$

For liquids, the value of β can be obtained from the property tables in App. B-2.

To determine the dimensionless parameters that govern heat transfer in free convection, we need to nondimensionalize the above governing equations. The following dimensionless parameters are defined:

$$X = \frac{x}{L} \qquad Y = \frac{y}{L} \qquad U = \frac{u}{U_0} \qquad V = \frac{v}{U_0} \qquad \theta = \frac{T - T_\infty}{T_w - T_\infty} \tag{9-14}$$

Here L is a characteristic length, U_0 is a reference velocity, T_w is the wall surface temperature, and T_∞ is the fluid temperature at a far distance from the hot plate. When these new variables are introduced into Eqs. (9-9) to (9-11), the resulting nondimensional equations become

$$\frac{\partial U}{\partial X} + \frac{\partial V}{\partial Y} = 0 \tag{9-15}$$

$$U\frac{\partial U}{\partial X} + V\frac{\partial U}{\partial Y} = \frac{g\beta(T_w - T_\infty)L}{U_0^2}\theta + \frac{1}{\text{Re}}\frac{\partial^2 U}{\partial Y^2} \tag{9-16}$$

$$U\frac{\partial \theta}{\partial X} + V\frac{\partial \theta}{\partial Y} = \frac{1}{\text{Re Pr}}\frac{\partial^2 \theta}{\partial Y^2} \tag{9-17}$$

Here the Reynolds and Prandtl numbers are defined as

$$\text{Re} = \frac{U_0 L}{\nu} \qquad \text{Pr} = \frac{\nu}{\alpha} \tag{9-18}$$

The dimensionless group in the momentum equation can be rearranged as

$$\frac{g\beta(T_w - T_\infty)L}{U_0^2} = \frac{g\beta L^3(T_w - T_\infty)/\nu^2}{(LU_0/\nu)^2} \equiv \frac{\text{Gr}}{\text{Re}^2} \tag{9-19}$$

where the *Grashof number* Gr is defined as

$$\boxed{\text{Gr} = \frac{g\beta L^3(T_w - T_\infty)}{\nu^2}} \tag{9-20}$$

The Grashof number represents the ratio of the buoyancy force to the viscous force acting on the fluid. We recall that in forced convection, the Reynolds number represents the ratio of the inertial to viscous forces acting on the fluid. Therefore, the Grashof number in free convection plays the same role as the Reynolds number in forced convection. For example, in forced convection the transition from laminar to turbulent flow is governed by the critical value of the Reynolds number.

Similarly, in free convection, the transition from laminar to turbulent flow is governed by the critical value of the Grashof number.

Equations (9-15) to (9-17) imply that when the effects of free and forced convection are of comparable magnitude, the Nusselt number depends on Re, Pr, and Gr:

$$\mathrm{Nu} = f(\mathrm{Re}, \mathrm{Gr}, \mathrm{Pr}) \tag{9-21}$$

The parameter $\mathrm{Gr}/\mathrm{Re}^2$, defined by Eq. (9-19), is a measure of the relative importance of free convection in relation to forced convection. When $\mathrm{Gr}/\mathrm{Re}^2 \cong 1$, free and forced convection are of the same order of magnitude, hence both must be considered.

If $(\mathrm{Gr}/\mathrm{Re}^2) \ll 1$, flow is primarily by forced convection. If $(\mathrm{Gr}/\mathrm{Re}^2) \gg 1$, free convection becomes dominant and the Nusselt number depends on Gr and Pr only:

$$\mathrm{Nu} = f(\mathrm{Gr}, \mathrm{Pr}) \tag{9-22}$$

In free convection flow velocities are produced by the buoyancy forces only, hence there are no externally induced flow velocities. As a result, the Nusselt number does not depend on the Reynolds number.

For gases, $\mathrm{Pr} \cong 1$; hence the Nusselt number for free convection is a function of the Grashof number only:

$$\mathrm{Nu} = f(\mathrm{Gr}) \qquad \text{for gases} \tag{9-23}$$

Sometimes another dimensionless parameter, called the *Rayleigh number* (Ra), which is defined as

$$\boxed{\mathrm{Ra} = \mathrm{Gr}\,\mathrm{Pr} = \frac{g\beta L^3(T_w - T_\infty)}{\nu^2}\,\mathrm{Pr} = \frac{g\beta L^3(T_w - T_\infty)}{\nu\alpha}} \tag{9-24}$$

is used instead of the Grashof number to correlate heat transfer in free convection. For such cases, Eq. (9-22) takes the form

$$\mathrm{Nu} = f(\mathrm{Ra}, \mathrm{Pr}) \tag{9-25}$$

Example 9-1 Examine the dimension of each term in the momentum equation (9-2).

SOLUTION We write Eq. (9-2) and show below the dimension of each term:

$$\underbrace{u\frac{\partial u}{\partial x} + v\frac{\partial u}{\partial y}}_{\frac{\mathrm{m}}{\mathrm{s}^2}} = \underbrace{-g}_{\frac{\mathrm{m}}{\mathrm{s}^2}} - \underbrace{\frac{1}{\rho}\frac{\partial P}{\partial x}}_{\underbrace{\frac{\mathrm{m}^3}{\mathrm{kg}}\cdot\frac{\mathrm{N}}{\mathrm{m}^2\cdot\mathrm{m}}}_{\frac{\mathrm{N}}{\mathrm{kg}}}} + \underbrace{\nu\frac{\partial^2 u}{\partial y^2}}_{\underbrace{\frac{\mathrm{m}^2}{\mathrm{s}}\cdot\frac{\mathrm{m}}{\mathrm{s}\cdot\mathrm{m}^2}}_{\frac{\mathrm{m}}{\mathrm{s}^2}}}$$

Here,

$$\frac{\text{m}}{\text{s}^2} = \frac{\text{kg} \cdot \text{m}}{\text{kg} \cdot \text{s}^2} = \frac{\text{N}}{\text{kg}}$$

Thus, each term in the momentum equation has the dimension of newtons per kilogram.

9-2 AN APPROXIMATE ANALYSIS OF LAMINAR FREE CONVECTION ON A VERTICAL PLATE

Heat transfer by free convection on a vertical or an inclined plate has been the subject of numerous investigations [1–26]. To provide better insight into heat transfer by free convection, we consider here the simplest situation involving a vertical plate under isothermal conditions (i.e., subjected to uniform surface temperature) placed in a large body of fluid at rest, as illustrated in Fig. 9-1 or 9-3.

Let T_w and T_∞ be, respectively, the temperature of the wall surface and the bulk temperature of the fluid. The fluid moves upward along the plate for $T_w > T_\infty$

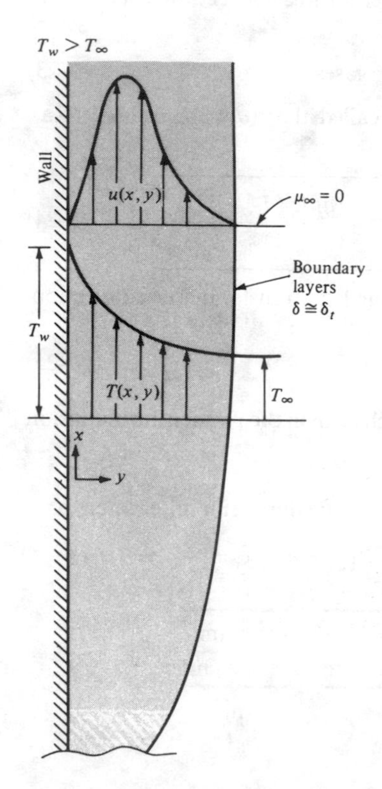

Figure 9-3 Temperature and velocity profiles for free convection on a hot vertical plate.

and flows downward for $T_w < T_\infty$, as illustrated in Fig. 9-1. The following analysis is applicable for both cases.

Mathematical Formulation of Problem

The buoyancy-induced flow problem, as illustrated in Fig. 9-1, is considered a boundary-layer type of flow, and the governing equations of motion and energy are obtained from Eqs. (9-9) to (9-11):

$$\frac{\partial u}{\partial x} + \frac{\partial v}{\partial y} = 0 \tag{9-26}$$

$$u\frac{\partial u}{\partial x} + v\frac{\partial u}{\partial y} = g\beta(T - T_\infty) + \nu\frac{\partial^2 u}{\partial y^2} \tag{9-27}$$

$$u\frac{\partial T}{\partial x} + v\frac{\partial T}{\partial y} = \alpha\frac{\partial^2 T}{\partial y^2} \tag{9-28}$$

The physical boundary conditions for the problem include zero velocity components at the wall surface, no axial velocity outside the boundary layer, temperatures T_w at the wall and T_∞ outside the boundary layer. These boundary conditions are stated as follows:

$$\begin{aligned} &u = 0 \qquad v = 0 \qquad T = T_w && \text{at } y = 0 \text{ (wall)} \\ &u = 0 \qquad T = T_\infty && \text{as } y \to \infty \text{ (or outside boundary layer)} \end{aligned} \tag{9-29}$$

The heat transfer problem posed here yields itself to exact solution, but the analysis is more involved. Here we describe a simple, approximate method of analysis by using the *integral method*, considered in Chap. 8 for the solution of heat transfer in laminar forced convection over a flat plate.

For simplicity, we assume the Prandtl number for the fluid is close to unity, hence the thicknesses of the velocity and the thermal boundary layers are almost equal or $\delta \cong \delta_t$. The basic steps in the analysis are described.

The Momentum and Energy Integral Equations

The first step in the analysis by the *integral method* is the development of the momentum and the energy integral equations. The *momentum integral equation* is developed by integrating the momentum equation (9-27) over the velocity boundary-layer thickness δ, utilizing the continuity equation to eliminate the velocity component v, and making use of the boundary conditions for the velocity component u. If we omit the details of such a development, the resulting momentum integral equation can be expressed in the form

$$\frac{d}{dx}\left(\int_0^\delta u^2\,dy\right) = -\nu\left.\frac{\partial u}{\partial y}\right|_{y=0} + g\beta\int_0^\delta (T - T_\infty)\,dy \tag{9-30}$$

The *energy integral equation* is developed by integrating the energy equation (9-28) over the boundary layer $\delta_t = \delta$, utilizing the continuity equation to eliminate the velocity component v, and employing the boundary conditions imposed on the temperature. Omitting the details of this development, we find the energy integral equation to be

$$\frac{d}{dx}\left[\int_0^\delta u(T - T_\infty)\, dy\right] = -\alpha \left.\frac{\partial T}{\partial y}\right|_{y=0} \tag{9-31}$$

As expected, Eqs. (9-30) and (9-31) are coupled because u and T appear in both; hence they should be solved simultaneously.

The next step in the analysis is the determination of suitable profiles for the distribution of velocity and temperature in the boundary layers.

Choice of Velocity and Temperature Profiles

A suitable choice should be made for the velocity and temperature profiles, consistent with the physical reality. For free convection on a vertical plate, say for $T_w > T_\infty$, the physical nature of the problem gives rise to the velocity and temperature profiles as illustrated in Fig. 9-3. Polynomial representations can be used to approximate these profiles.

The temperature profile is represented by a second-degree polynomial in the form

$$T(x, y) = a_0 + b_0 y + c_0 y^2 \tag{9-32}$$

The following three conditions can be applied to determine the three coefficients a_0, b_0, and c_0 in terms of the boundary-layer thickness:

$$T = T_w \quad \text{at } y = 0 \tag{9-33}$$

$$T = T_\infty \quad \text{at } y = \delta \tag{9-34}$$

$$\frac{\partial T}{\partial y} = 0 \quad \text{at } y = \delta \tag{9-35}$$

The resulting temperature profile becomes

$$\frac{T(x, y) - T_\infty}{T_w - T_\infty} = \left(1 - \frac{y}{\delta}\right)^2 \tag{9-36}$$

Here we assume $\delta \cong \delta_t$.

The determination of a suitable velocity profile, however, is more involved. An examination of the velocity profile in Fig. 9-3 implies that the velocity component $u(x, y)$ is zero both at the wall surface and at the edge of the velocity boundary layer, but has a peak inside the boundary layer. To represent such a situation, we choose a cubic polynomial in the form

$$u(x, y) = u_0^*(a_1 + b_1 y + c_1 y^2 + d_1 y^3) \tag{9-37}$$

where the coefficients a_1, b_1, c_1, d_1 and a reference velocity u_0^* are considered to be functions of x and are yet to be determined. Four conditions are needed on the velocity to determine the four coefficients. Three of these conditions are taken as

$$u = 0 \qquad \text{at } y = 0 \tag{9-38}$$

$$u = 0 \qquad \text{at } y = \delta \tag{9-39}$$

$$\frac{\partial u}{\partial y} = 0 \qquad \text{at } y = \delta \tag{9-40}$$

where δ is the edge of the velocity boundary layer. A fourth condition is obtained by evaluating the momentum equation (9-27) at $y = 0$ and noting that $u = v = 0$ and $T = T_w$ at $y = 0$. We find

$$\frac{\partial^2 u}{\partial y^2} = -\frac{g\beta}{\nu}(T_w - T_\infty) \tag{9-41}$$

With the application of the conditions given by Eqs. (9-38) to (9-41), the velocity profile, Eq. (9-37), becomes

$$u(x, y) = \left[u_0^* \frac{\beta\delta^2 g(T_w - T_\infty)}{4\nu} \right] \frac{y}{\delta} \left(1 - \frac{y}{\delta}\right)^2 \tag{9-42}$$

which can be written more compactly as

$$u(x, y) = u_0 \frac{y}{\delta} \left(1 - \frac{y}{\delta}\right)^2 \tag{9-43}$$

where $u_0 \equiv u_0(x) = u_0^* \beta\delta^2 g(T_w - T_\infty)/(4\nu)$ is an arbitrary function of x with the dimension of velocity.

By differentiating Eq. (9-43) with respect to y, it can be shown that the maximum velocity $u(x, y)$ occurs at a distance $y = \delta/3$.

Solution of Equations

The temperature profile equation (9-36) and the velocity profile equation (9-43) are introduced into the momentum integral equation (9-30) and the energy integral equation (9-31), and the indicated operations are performed. The momentum and the energy integral equations, respectively, become

$$\frac{1}{105}\frac{d}{dx}(u_0^2\delta) = \frac{1}{3} g\beta(T_w - T_\infty)\delta - \nu \frac{u_0}{\delta} \tag{9-44}$$

$$\frac{1}{30}(T_w - T_\infty)\frac{d}{dx}(u_0\delta) = 2\alpha \frac{T_w - T_\infty}{\delta} \tag{9-45}$$

To solve these equations, we assume that $u_0(x)$ and $\delta(x)$ depend on x in the form

$$u_0(x) = c_1 x^m \qquad \text{and} \qquad \delta(x) = c_2 x^n \tag{9-46a, b}$$

where c_1, c_2, m, and n are constants. Equations (9-46) are substituted into Eqs. (9-44) and (9-45) to yield

$$\frac{(2m+n)c_1^2 c_2}{105} x^{2m+n-1} = \frac{1}{3} g\beta(T_w - T_\infty)c_2 x^n - \frac{\nu c_1}{c_2} x^{m-n} \tag{9-47}$$

$$\frac{(m+n)c_1 c_2}{30} x^{m+n-1} = \frac{2\alpha}{c_2} x^{-n} \tag{9-48}$$

To make this system of equations invariant of x, we equate the exponents of x on both sides of Eqs. (9-47) and (9-48), and obtain

$$2m + n - 1 = n = m - n \tag{9-49}$$

$$m + n - 1 = -n \tag{9-50}$$

which yields

$$m = \tfrac{1}{2} \quad \text{and} \quad n = \tfrac{1}{4} \tag{9-51}$$

When these values of m and n are introduced into Eqs. (9-47) and (9-48), the x variable cancels. A simultaneous solution of the resulting equations gives c_1 and c_2 as

$$c_1 = 5.17\nu\left(\frac{20}{21} + \frac{\alpha}{\nu}\right)^{-1/2}\left[\frac{g\beta(T_w - T_\infty)}{\nu^2}\right]^{1/2} \tag{9-52}$$

$$c_2 = 3.93\left(\frac{20}{21} + \frac{\nu}{\alpha}\right)^{1/4}\left[\frac{g\beta(T_w - T_\infty)}{\nu^2}\right]^{-1/4}\left(\frac{\nu}{\alpha}\right)^{-1/2} \tag{9-53}$$

These values of n and c_2 are introduced into Eq. (9-46b); the boundary-layer thickness $\delta(x)$ becomes

$$\delta(x) = 3.93(0.952 + \mathrm{Pr})^{1/4}\left[\frac{g\beta(T_w - T_\infty)}{\nu^2}\right]^{-1/4} \mathrm{Pr}^{-1/2} x^{1/4} \tag{9-54}$$

or

$$\frac{\delta(x)}{x} = 3.93\ \mathrm{Pr}^{-1/2}(0.952 + \mathrm{Pr})^{1/4}\ \mathrm{Gr}_x^{-1/4} \tag{9-55}$$

where the local Grashof number Gr_x is defined as

$$\boxed{\mathrm{Gr}_x = \frac{g\beta(T_w - T_\infty)x^3}{\nu^2}} \tag{9-56}$$

Given $\delta(x)$, the temperature profile $T(x, y)$ can be determined from Eq. (9-36), and the local Nusselt number can be obtained as described below.

Local Nusselt Number

The local Nusselt number Nu_x is related to the temperature gradient of the fluid at the wall surface by

$$\mathrm{Nu}_x \equiv \frac{h_x x}{k} = \frac{q_w}{T_w - T_\infty} \frac{x}{k} = \frac{-k(\partial T/\partial y)|_{y=0}}{T_w - T_\infty} \frac{x}{k} = -\frac{x}{T_w - T_\infty} \left.\frac{\partial T}{\partial y}\right|_{y=0} \tag{9.57}$$

Here the temperature gradient is determined from Eq. (9-36):

$$\left.\frac{\partial T}{\partial y}\right|_{y=0} = -\frac{2(T_w - T_\infty)}{\delta} \tag{9-58}$$

Introducing Eq. (9-58) into Eq. (9-57), the local Nusselt number is related to the boundary-layer thickness.

$$\mathrm{Nu}_x = 2\frac{x}{\delta} \tag{9-59}$$

From Eqs. (9-55) and (9-59), the local Nusselt number for laminar free convection on a vertical wall subjected to uniform surface temperature is

$$\mathrm{Nu}_x \equiv \frac{h_x x}{k} = 0.508\ \mathrm{Pr}^{1/2}(0.952 + \mathrm{Pr})^{-1/4}\ \mathrm{Gr}_x^{1/4} \tag{9-60}$$

This result can be expressed in the alternative form as

$$\mathrm{Nu}_x = 0.508\ \mathrm{Ra}_x^{1/4}\left(\frac{\mathrm{Pr}}{0.952 + \mathrm{Pr}}\right)^{1/4} \tag{9-61}$$

where the local Grashof and Rayleigh numbers, respectively, are defined as

$$\mathrm{Gr}_x = \frac{g\beta(T_w - T_\infty)x^3}{\nu^2} \tag{9-62}$$

$$\mathrm{Ra}_x = \mathrm{Gr}_x\mathrm{Pr} = \frac{g\beta(T_w - T_\infty)x^3}{\nu\alpha} \tag{9-63}$$

The results developed above for $T_w > T_\infty$ also apply for $T_\infty > T_w$. For the latter case, the temperature difference should be replaced by $T_\infty - T_w$.

Note that functional forms of Eqs. (9-60) and (9-61) are, respectively, $\mathrm{Nu}_x = f(\mathrm{Gr}_x, \mathrm{Pr})$ and $\mathrm{Nu}_x = f(\mathrm{Ra}_x, \mathrm{Pr})$. These forms are consistent with those given by Eqs. (9-22) and (9-25), which were developed from the examination of the dimensionless form of the governing equations.

Mean Nusselt Number

In engineering applications, the mean Nusselt number Nu_m over a distance $x = 0$ to $x = L$ along the plate is also of interest. To determine Nu_m, we consider the average heat transfer coefficient h_m, defined as

$$h_m = \frac{1}{L}\int_0^L h_x\, dx \tag{9-64a}$$

According to Eq. (9-61), h_x is proportional to $x^{-1/4}$; then h_m becomes

$$h_m = \tfrac{4}{3}[h_x]_{x=L} \tag{9-64b}$$

The average Nusselt number Nu_m is

$$Nu_m \equiv \frac{Lh_m}{k} = \frac{L}{k}\frac{4}{3}[h_x]_{x=L} = \frac{4}{3}\left[\frac{xh_x}{k}\right]_{x=L}$$

or

$$\boxed{Nu_m = \tfrac{4}{3}[Nu_x]_{x=L}} \tag{9-64c}$$

Note that the mean Nusselt number given by Eq. (9-64c) is developed for the case when h_x is proportional to $x^{-1/4}$.

Suppose the local heat transfer coefficient h_x depends on x in the form

$$h_x = cx^{-n} \tag{9-65a}$$

Then the mean heat transfer coefficients h_m over $x = 0$ to $x = L$ becomes

$$h_m = \frac{c}{L}\int_0^L x^{-n}\,dx = \frac{1}{1-n}cL^{-n} = \frac{1}{1-n}[h_x]_{x=L} \tag{9-65b}$$

and the average Nusselt number becomes

$$\boxed{Nu_m = \frac{1}{1-n}[Nu_x]_{x=L}} \tag{9-65c}$$

Clearly, for $n = \frac{1}{4}$ Eq. (9-65c) reduces to Eq. (9-64c). The relation given by Eq. (9-64c) can be used now to determine the corresponding average value of the local Nusselt number given by Eq. (9-61). Introducing Eq. (9-61) into Eq. (9-64c), we obtain

$$Nu_m = 0.677\,Ra_L^{1/4}\left(\frac{Pr}{0.952 + Pr}\right)^{1/4} \tag{9-66}$$

where

$$Nu_m = \frac{h_m L}{k} \tag{9-67a}$$

and

$$Ra_L = Gr_L Pr = \frac{g\beta(T_w - T_\infty)L^3}{\nu\alpha} \tag{9-67b}$$

The results for Nu_x and Nu_m given by Eqs. (9-61) and (9-66) are applicable for laminar free convection on a vertical plate maintained at a uniform temperature. The fluid properties are evaluated at the film temperature $T_f = (T_w + T_\infty)/2$.

Under the present circumstances, the transition from laminar to turbulent flow has been observed [4, 15, 16] to take place in the range

$$10^8 < Ra_x < 10^9$$

Therefore, the above laminar-flow solution is restricted to the range $Ra_L < 10^9$.

Table 9-1 Exact solutions of the mean Nusselt number for laminar free convection on a vertical plate*

Pr	$\dfrac{Nu_m}{(G_L\ Pr)^{1/4}}$
0.003	0.182
0.008	0.228
0.01	0.242
0.02	0.280
0.03	0.305
0.72	0.516
0.73	0.518
1	0.535
2	0.568
10	0.620
100	0.653
1000	0.665
∞	0.670

* Valid for $Gr_L\ Pr < 10^9$.
Source: Schlichting [14].

Results of Exact Analysis

The previous analysis, even though approximate, helps to envision the implications of heat transfer in free convection and the physical significance of various dimensionless parameters.

The exact analysis of free convection on a vertical plate subject to uniform wall temperature has been performed over a wide range of Prandtl numbers by various investigators. Schlichting [14] compiled the mean Nusselt number obtained by several investigators [1, 3, 6, 13]. We present in Table 9-1 the resulting expressions for the mean Nusselt number; in this table, Nu_m and Gr_L are defined as follows:

$$Nu_m \equiv \frac{h_m L}{k} \qquad Gr_L \equiv \frac{g\beta(T_w - T_\infty)L^3}{\nu^2} \tag{9-68}$$

and they are applicable for $Gr_L\, Pr < 10^9$.

Example 9-2 A square plate 0.4 by 0.4 m, maintained at a uniform temperature of $T_w = 400$ K, is suspended vertically in quiescent atmospheric air at $T_\infty = 300$ K.

(*a*) Determine the boundary-layer thickness $\delta(x)$ at the trailing edge of the plate (at $x = 0.4$ m).

(*b*) Calculate the average heat transfer coefficient h over the entire length of the plate by using the results of the exact analysis listed in Table 9-1.

SOLUTION The physical properties of air at atmospheric pressure and $T_f = (300 + 400)/2 = 350$ K are

$$\nu = 20.75 \times 10^{-6}\ \text{m}^2/\text{s} \qquad \text{Pr} = 0.697$$

$$k = 0.03\ \text{W}/(\text{m} \cdot {}^\circ\text{C}) \qquad \beta = \frac{1}{T_f} = 2.86 \times 10^{-3}\ \text{K}^{-1}$$

The Grashof number at $L = 0.4$ m becomes

$$\text{Gr}_{L=0.4} = \frac{g\beta(T_w - T_\infty)L^3}{\nu^2}$$

$$= \frac{(9.8)(2.86 \times 10^{-3})(400 - 300)(0.4)^3}{(20.76 \times 10^{-6})^2}$$

$$= 4.16 \times 10^8$$

Thus the boundary layer is laminar, and the approximate analysis is applicable.
(*a*) The boundary-layer thickness $\delta(x)$ is determined from Eq. (9-55):

$$\delta|_{L=0.4} = 3.93\ \text{Pr}^{-1/2}(0.952 + \text{Pr})^{1/4}\ \text{Gr}_L^{-1/4}\ L$$

$$= \frac{3.93}{(0.693)^{1/2}}(0.952 + 0.693)^{1/4}\frac{0.4}{(4.16 \times 10^8)^{1/4}}$$

$$= 1.5 \times 10^{-2}\ \text{m} = 1.5\ \text{cm}$$

(*b*) The average Nusselt number is determined from Table 9-1:

$$\text{Nu}_m = \frac{hL}{k} = 0.518(\text{Gr}_L\ \text{Pr})^{1/4}$$

$$h = 0.518(4.16 \times 10^8 \times 0.697)^{1/4}\frac{0.03}{0.4}$$

$$= 5.07\ \text{W}/(\text{m}^2 \cdot {}^\circ\text{C})$$

9-3 CORRELATIONS OF FREE CONVECTION ON A VERTICAL PLATE

The analysis of free convection is complicated, and experimental data are often desirable in order to develop reliable heat transfer relations. Here we present some of the recommended empirical correlations of free convection on a vertical wall in laminar and turbulent flow, for both uniform wall temperature (i.e., isothermal surface) and the uniform wall heat flux boundary conditions.

Table 9-2 Constant *c* and the exponent *n* of Eq. (9-69)

Type of flow	Range of Gr_L Pr	c	n	Reference(s)
Laminar	10^4 to 10^9	0.59	$\frac{1}{4}$	[8]
Turbulent	10^9 to 10^{13}	0.10	$\frac{1}{3}$	[9, 17]

Uniform Wall Temperature

McAdams [8] correlated the average Nusselt number with an expression in the form

$$\text{Nu}_m = c(\text{Gr}_L \cdot \text{Pr})^n = c\,\text{Ra}_L^n \tag{9-69}$$

where L is the height of the vertical plate and Gr_L and Nu_m are defined by Eq. (9-68). The recommended values of the constant c and the exponent n are listed in Table 9-2. The physical properties are evaluated at

$$T_f = \tfrac{1}{2}(T_w + T_\infty)$$

An examination of the values of the exponent n given in this table reveals that in turbulent flow h_m is independent of the plate height L, since $\text{Gr}_L \sim L^3$ and $h_m \sim (1/L)(\text{Gr}_L)^{1/3}$.

More recently, Churchill and Chu [25] proposed two equations for correlating free convection on a vertical plate under isothermal surface conditions. One expression, which applies to only laminar flow and holds for all values of the Prandtl number, is given by

$$\text{Nu}_m = 0.68 + \frac{0.67\,\text{Ra}_L^{1/4}}{[1 + (0.492/\text{Pr})^{9/16}]^{4/9}} \qquad \text{for } 10^{-1} < \text{Ra}_L < 10^9 \tag{9-70}$$

The other expression, which applies to both laminar and turbulent flow, is given by

$$\text{Nu}_m^{1/2} = 0.825 + \frac{0.387\,\text{Ra}_L^{1/6}}{[1 + (0.492/\text{Pr})^{9/16}]^{8/27}} \qquad \text{for } 10^{-1} < \text{Ra}_L < 10^{12} \tag{9-71}$$

In Eqs. (9-70) and (9-71), the Rayleigh number is defined by Eq. (9-67*b*). The physical properties are evaluated at $T_f = (T_w + T_\infty)/2$.

Although Eq. (9-71) is applicable over both laminar and turbulent flow regimes, Eq. (9-70) yields slightly more accurate results for laminar flow.

Example 9-3 A square plate 0.4 m by 0.4 m, maintained at a uniform temperature $T_w = 400$ K, is suspended vertically in quiescent atmospheric air at $T_\infty = 300$ K. Calculate the average heat transfer coefficient h_m over the entire length of the plate by using both correlations given by Eqs. (9-69) and (9-70). Compare these results with that obtained in Example 9-2.

SOLUTION The Grashof number at $L = 0.4$ m is found in Example 9-2:

$$\text{Gr}|_{L=0.4} = 4.16 \times 10^8$$

We also have $k = 0.03$ W/(m · °C) and Pr = 0.697. Equation (9-69) gives

$$\text{Nu}_m = \frac{hL}{k} = 0.59(\text{Gr}_L \, \text{Pr})^{1/4}$$

so

$$h = 0.59 \frac{(0.03)(4.16 \times 10^8 \times 0.697)^{1/4}}{0.4} = 5.77 \text{ W/(m}^2 \cdot \text{°C)}$$

Equation (9-70) gives

$$\text{Nu}_m = \frac{hL}{k} = 0.68 + \frac{0.67(4.16 \times 10^8 \times 0.697)^{1/4}}{[1 + (0.492/0.697)^{9/16}]^{4/9}}$$

so

$$h = \frac{k}{L}(72.61) = \frac{0.03(72.61)}{0.4} = 5.45 \text{ W/(m}^2 \cdot \text{°C)}$$

Note that the results obtained from Eqs. (9-69) and (9-70) are close. The result obtained from the purely theoretical solution in Example 9-2, $h = 5.07$ W/(m^2 · °C), underestimates the heat transfer coefficient.

Example 9-4 A vertical plate $L = 5$ m high and $w = 1.5$ m wide has one of its surfaces insulated; the other surface, maintained at a uniform temperature $T_w = 400$ K, is exposed to quiescent atmospheric air at $T_\infty = 300$ K. Calculate the total rate of heat loss from the plate.

SOLUTION The physical properties of atmospheric air at $T_f = (300 + 400)/2 = 350$ K are

$$\nu = 20.75 \times 10^{-6} \text{ m}^2\text{/s} \qquad \text{Pr} = 0.697$$

$$k = 0.03 \text{ W/(m} \cdot \text{°C)} \qquad \beta = \frac{1}{T_f} = 2.86 \times 10^{-3} \, K^{-1}$$

The Grashof number for $L = 5$ m is

$$\text{Gr}\Big|_{L=5m} = \frac{g\beta(T_w - T_\infty)L^3}{\nu^2}$$

$$= \frac{(9.8)(2.86 \times 10^{-3})(400 - 300)(5^3)}{(20.75 \times 10^{-6})^2} = 8.137 \times 10^{11}$$

Since the Grashof number (hence the Rayleigh number) is beyond the range of Eq. (9-70), we use Eq. (9-71) to calculate Nu_m or the average heat transfer coefficient h:

$$h = \frac{k}{L}\left\{0.825 + \frac{0.387(\mathrm{Gr}_L\,\mathrm{Pr})^{1/6}}{[1 + (0.492/\mathrm{Pr})^{9/16}]^{8/27}}\right\}^2$$

$$= \frac{0.03}{5}\left\{0.825 + \frac{0.387(8.137 \times 10^{11} \times 0.697)^{1/6}}{[1 + (0.492/0.697)^{9/16}]^{8/27}}\right\}^2$$

$$= 5.51\ \mathrm{W/(m^2 \cdot {}^\circ C)}$$

The total heat transfer rate becomes

$$Q = hA(T_w - T_\infty)$$

$$= (5.51)(5 \times 1.5)(400 - 300)$$

$$= 4.133\ \mathrm{kW}$$

Uniform Wall Heat Flux

Free convection on a vertical plate subject to *uniform heat flux* at the wall surface has been investigated by Sparrow and Gregg [11], Vliet and Liu [19], and Vliet [20]. Based on the experimental data of Ref. 20 for air and water and of Ref. 19 for water, the following correlations are proposed for the local Nusselt number under uniform wall heat flux:

$$\boxed{\mathrm{Nu}_x = 0.60(\mathrm{Gr}_x^*\,\mathrm{Pr})^{1/5} \qquad \text{for } 10^5 < \mathrm{Gr}_x^*\,\mathrm{Pr} < 10^{11} \text{ (laminar)}} \tag{9-72}$$

$$\boxed{\mathrm{Nu}_x = 0.568(\mathrm{Gr}_x^*\,\mathrm{Pr})^{0.22} \qquad \text{for } 2 \times 10^{13} < \mathrm{Gr}_x^*\,\mathrm{Pr} < 10^{16} \text{ (turbulent)}} \tag{9-73}$$

where the *modified Grashof number* Gr_x^* is defined as

$$\mathrm{Gr}_x^* = \mathrm{Gr}_x\,\mathrm{Nu}_x = \frac{g\beta(T_w - T_\infty)x^3}{\nu^2}\,\frac{q_w x}{T_w - T_\infty} = \frac{g\beta q_w x^4}{k\nu^2} \tag{9-74a}$$

and the local Nusselt number as

$$\mathrm{Nu}_x = \frac{xh_x}{k} \tag{9-74b}$$

And q_w is the constant wall heat flux.

To determine the average Nusselt number Nu_m, we need to establish the dependence of h_x on x in the form given by Eq. (9-65a). From Eqs. (9-72) and (9-73) we have, respectively,

$$h_x \sim \frac{1}{x}(\text{Gr}_x^*)^{0.2} \sim \frac{1}{x}(x^4)^{0.2} \sim x^{-0.2} \tag{9-74c}$$

$$h_x \sim \frac{1}{x}(\text{Gr}_x^*)^{0.22} \sim \frac{1}{x}(x^4)^{0.22} \sim x^{-0.12} \tag{9-74d}$$

Then Eq. (9-65c) is used to determine the average Nusselt number. Therefore, the average Nusselt numbers for Eqs. (9-72) and (9-73) are, respectively,

$$\text{Nu}_m = \frac{1}{1-0.2}[\text{Nu}_x]_{x=L} = 1.25[\text{Nu}_x]_{x=L} \qquad \text{for } 10^5 < \text{Gr}_x^* \text{Pr} < 10^{11} \tag{9-74e}$$

$$\text{Nu}_m = \frac{1}{1-0.12}[\text{Nu}_x]_{x=L} = 1.136[\text{Nu}_x]_{x=L} \qquad \text{for } 2 \times 10^{13} < \text{Gr}_x^* \text{Pr} < 10^{16} \tag{9-74f}$$

All physical properties are evaluated at the film temperature.

When the wall heat flux q_w is prescribed instead of the wall temperature T_w, the temperature difference $T_w - T_\infty$, and hence the film temperature, is unknown at the onset of the problem. In such situations, an initial guess is made for the film temperature, and the calculations are performed. If the guess and the calculated values are different enough to affect the results, the calculations are repeated with the new value of the film temperature.

In Refs. 19 and 20 the transition from laminar to turbulent regime begins in the range $3 \times 10^{12} < \text{Gr}_x^* \text{Pr} < 4 \times 10^{13}$ and ends in the range $2 \times 10^{13} < \text{Gr}_x^* \text{Pr} < 10^{14}$. Therefore, the *fully turbulent regime* is considered to occur at $\text{Gr}^* \text{Pr} = 10^{14}$, but this value may be as low as 2×10^{13}.

Churchill and Chu [25] also correlated some of the available experimental data under uniform wall heat flux conditions with Eq. (9-70), which was developed strictly for the uniform wall surface temperature. The two problems, because of the boundary conditions (i.e., one is uniform heat flux, the other is uniform wall temperature), are different from a mathematical point of view. Therefore, rigorously speaking, the heat transfer results are not expected to be the same. However, they found that Eq. (9-70) correlated the data well; hence they concluded that Eq. (9-70) also can be used to correlate free convection on a vertical wall subjected to uniform wall heat flux. Similarly, they propose Eq. (9-71) for use with the uniform wall heat flux condition.

If Eq. (9-70) is to be used for uniform wall heat flux condition, it is desirable to express the right-hand side in terms of the modified Grashof number Gr^*. This can be done by noting that

$$\text{Ra}_L = \text{Gr}_L \text{Pr} \qquad \text{and} \qquad \text{Gr}_L^* = \text{Gr}_L \text{Nu}_m$$

Then

$$\text{Ra}_L = \frac{\text{Gr}_L^* \text{ Pr}}{\text{Nu}_m}$$

Introducing this result for Ra_L on the right-hand side of Eq. (9-70) and rearranging, we obtain

$$\boxed{\text{Nu}_m^{1/4}(\text{Nu}_m - 0.68) = \frac{0.67(\text{Gr}_L^* \text{ Pr})^{1/4}}{[1 + (0.492/\text{Pr})^{9/16}]^{4/9}}} \tag{9-75}$$

for laminar free convection on a vertical plate subjected to uniform wall heat flux.

Example 9-5 A thin vertical panel $L = 3$ m high and $w = 1.5$ m wide is thermally insulated on one side and exposed to a solar radiation flux of $q_s = 750 \text{ W/m}^2$ on the other side. The exposed surface has an absorptivity of $\alpha_s = 0.8$ for solar radiation. Assuming that the energy absorbed by the plate is dissipated by free convection into the surrounding quiescent air at atmospheric pressure and $T_\infty = 300$ K, calculate the surface temperature of the panel.

SOLUTION The problem is one of constant wall heat flux, hence the surface temperature is not known and T_f cannot be determined to evaluate the physical properties of air. To determine a film temperature to evaluate properties, we choose an *approximate* value for h in free convection, namely, $h = 6 \text{ W/(m}^2 \cdot {}^\circ\text{C)}$. The solar energy absorbed by the wall is

$$q_w = \alpha_s q_s = 0.8 \times 750 = 600 \text{ W/m}^2$$

Then a first approximation to the wall temperature T_w is

$$h(T_w - T_\infty) = q_w$$

$$T_w = T_\infty + \frac{q_w}{h} = 300 + \frac{600}{6} = 400 \text{ K}$$

Hence, $$T_f = \tfrac{1}{2}(T_w + T_\infty) = \tfrac{1}{2}(400 + 300) = 350 \text{ K}$$

The physical properties of atmospheric air at $T_f = 350$ K are taken as

$$\nu = 2.076 \times 10^{-5} \text{ m}^2\text{/s} \qquad k = 0.03 \text{ W/(m} \cdot {}^\circ\text{C)}$$

$$\text{Pr} = 0.697 \qquad \beta = \frac{1}{T_f} = 2.86 \times 10^{-3} \text{ K}^{-1}$$

Then the *modified Grashof number* at $x = 3$ m is determined by Eq. (9-74*a*):

$$\text{Gr}_x^* = \frac{g\beta q_w x^4}{k\nu^2} = \frac{(9.8)(2.86 \times 10^{-3})(600)(3^4)}{(0.03)(2.076 \times 10^{-5})^2}$$

$$= 1.05 \times 10^{14} \qquad \text{at } x = 3 \text{ m}$$

Therefore, Eq. (9-73) can be used to determined the local h_x at $x = 3$ m:

$$h_x = \frac{k}{x}(0.568)(\text{Gr}_x^* \text{ Pr})^{0.22}$$

$$= \frac{0.03}{3}(0.568)(1.05 \times 10^{14} \times 0.697)^{0.22}$$

$$= 6.38 \text{ W/(m}^2 \cdot {}^\circ\text{C)} \qquad \text{at } x = 3 \text{ m}$$

The average heat transfer coefficient over $0 < x < 3$ m is determined from Eq. (9-74*f*):

$$h_m = 1.136[h_x]_{x=3\text{ m}} = 7.25 \text{ W/(m}^2 \cdot {}^\circ\text{C)}$$

Given h_m, the wall surface temperature T_w is

$$T_w = T_\infty + \frac{q_w}{h_m} = 300 + \frac{600}{7.25} = 382.8 \text{ K}$$

To improve the result, a new film temperature T_f is calculated:

$$T_f = \tfrac{1}{2}(300 + 382.8) \cong 341 \text{ K}$$

Then these calculations are repeated by evaluating the physical properties of air at this new film temperature.

However, another iteration is not warranted, since the film temperature $T_f = 350$ K used in the above calculations is sufficiently close to $T_f = 341$ K and the resulting change in the physical properties will be negligible.

9-4 FREE CONVECTION ON A HORIZONTAL PLATE

The average Nusselt number for free convection on a horizontal plate depends on whether the surface is facing up or down and whether the plate surface is warmer or cooler than the surrounding fluid. Again we consider the cases for uniform wall temperature and uniform wall heat flux separately.

Uniform Wall Temperature

The mean Nusselt number for free convection on a horizontal plate is correlated by McAdams [8] with an expression in the form

$$\boxed{\text{Nu}_m = c(\text{Gr} \cdot \text{Pr})^n} \qquad (9\text{-}76a)$$

where

$$\text{Nu}_m = \frac{h_m L}{k} \qquad \text{Gr}_L \equiv \frac{g\beta(T_w - T_\infty)L^3}{\nu^2} \qquad (9\text{-}76b)$$

Table 9-3 Constant c and exponent n of Eq. (9-76a) for free convection on a horizontal plate at uniform temperature

Orientation of plate	Range of Gr_L Pr	c	n	Flow regime
Hot surface facing up or cold surface facing down	10^5 to 2×10^7	0.54	$\frac{1}{4}$	Laminar
	2×10^7 to 3×10^{10}	0.14	$\frac{1}{3}$	Turbulent
Hot surface facing down or cold surface facing up	3×10^5 to 3×10^{10}	0.27	$\frac{1}{4}$	Laminar

Source: McAdams [8].

The coefficient c and the exponent n are listed in Table 9-3. The characteristic length L of the plate can be taken as the length of a side for a square, the arithmetic mean of the two dimensions for a rectangular surface, and $0.9D$ for a circular disk of diameter D, as suggested by McAdams [8]. We note that for turbulent flow Nu_m is independent of the characteristic length. For the case of a hot surface facing down or a cold surface facing up, the turbulent flow regime is not reached even at $Gr_L \, Pr \cong 3 \times 10^{10}$.

Recent correlations [24, 27] suggest that improved accuracy may be obtained if the characteristic length L for the plate is defined as

$$L \equiv \frac{A}{P} \tag{9-77}$$

where A is the surface area of the plate and P is the perimeter, which encompasses the area. Equation (9-77) is similar to the definition of the equivalent diameter except for a factor of 4.

Uniform Wall Heat Flux

Free convection under uniform wall heat flux has been studied extensively in Ref. 23 for an electrically heated plate in vertical, horizontal, and inclined positions. We present here only the results for the horizontal plate.

For the horizontal plate with the *heated surface facing upward*:

$$Nu_m = 0.13(Gr_L \, Pr)^{1/3} \qquad \text{for } Gr_L \, Pr < 2 \times 10^8 \tag{9-78}$$

$$Nu_m = 0.16(Gr_L \, Pr)^{1/3} \qquad \text{for } 5 \times 10^8 < Gr_L \, Pr < 10^{11} \tag{9-79}$$

Here, the $\frac{1}{3}$ power suggests that the heat transfer coefficient does not change locally or depend on the plate length. These two results have been developed by using 30- and 5-cm-long heated plates, respectively. Equation (9-79) developed with the smaller plate shows higher heat transfer than Eq. (9-78) developed with the larger plate. It appears that the edge effect becomes negligibly small with the larger plate.

For the horizontal plate with the *heated surface facing downward*:

$$\mathrm{Nu}_m = 0.58(\mathrm{Gr}_L \,\mathrm{Pr})^{1/5} \qquad \text{for } 10^6 < \mathrm{Gr}_L \,\mathrm{Pr} < 10^{11} \tag{9-80}$$

Here the $\frac{1}{5}$ power suggests that the flow regime remains laminar. The physical properties in Eqs. (9-78) to (9-80) are to be evaluated at a mean temperature, defined as

$$T_m = T_w - 0.25(T_w - T_\infty) \tag{9-81}$$

and the thermal expansion coefficient β at $(T_w + T_\infty)/2$. In these expressions, the Grashof number is defined as

$$\mathrm{Gr}_L = \frac{g\beta(T_w - T_\infty)L^3}{\nu^2} \tag{9-82a}$$

and the mean Nusselt number over the length L as

$$\mathrm{Nu}_m = \frac{h_m L}{k} = \frac{q_w L}{(T_w - T_\infty)k} \tag{9-82b}$$

Example 9-6 Consider a square plate 0.5 m by 0.5 m with one surface insulated and the other surface maintained at a uniform temperature of $T_w =$ 385 K which is placed in quiescent air at atmospheric pressure and $T_\infty = 315$ K. Calculate the average heat transfer coefficient for free convection for the following three orientations of the hot surface:

(*a*) The plate is horizontal, and the hot surface faces up.
(*b*) The plate is vertical.
(*c*) The plate is horizontal, and the hot surface faces down.

SOLUTION The physical properties of atmospheric air at $T_f = \frac{1}{2}(385 + 315) =$ 350 K are taken as

$$\nu = 2.076 \times 10^{-5}\ \mathrm{m^2/s} \qquad \mathrm{Pr} = 0.697$$

$$k = 0.03\ \mathrm{W/(m \cdot {}^\circ C)} \qquad \beta = \frac{1}{T_f} = 2.86 \times 10^{-3}\ \mathrm{K^{-1}}$$

Then the Grashof number for $L = 0.5$ m becomes

$$\mathrm{Gr}_L = \frac{g\beta(T_w - T_\infty)L^3}{\nu^2}$$

$$= \frac{(9.8)(2.86 \times 10^{-3})(385 - 315)(0.5)^3}{(2.076 \times 10^{-5})^2}$$

$$= 5.7 \times 10^8$$

(*a*) For the horizontal plate with the hot surface up, the average Nusselt number is determined from Eq. (9-76*a*) and Table 9-3. For the turbulent flow condition, we obtain

$$\mathrm{Nu}_m = \frac{h_m L}{k} = 0.14(\mathrm{Gr}_L \, \mathrm{Pr})^{1/3}$$

$$h_m = \frac{k}{L}(0.14)(\mathrm{Gr}_L \, \mathrm{Pr})^{1/3}$$

$$= \frac{0.03}{0.5}(0.14)(5.7 \times 10^8 \times 0.697)^{1/3}$$

$$= 6.18 \ \mathrm{W/(m^2 \cdot {}^\circ C)}$$

(*b*) For the vertical plate; the average Nusselt number is determined from Eq. (9-69) and Table 9-2. For the laminar flow condition we have

$$\mathrm{Nu}_m = \frac{h_m L}{k} = 0.59(\mathrm{Gr}_L \, \mathrm{Pr})^{1/4}$$

$$h_m = \frac{0.03}{0.5}(0.59)(5.7 \times 10^8 \times 0.697)^{1/4}$$

$$= 5.0 \ \mathrm{W/(m^2 \cdot {}^\circ C)}$$

(*c*) For the horizontal plate with the hot surface facing down, the average Nusselt number is determined from Eq. (9-76*a*) and Table 9-3:

$$\mathrm{Nu}_m = \frac{h_m L}{k} = 0.27(\mathrm{Gr}_L \, \mathrm{Pr})^{1/4}$$

$$h_m = \frac{0.03}{0.5}(0.27)(5.7 \times 10^8 \times 0.697)^{1/4}$$

$$= 2.29 \ \mathrm{W/(m^2 \cdot {}^\circ C)}$$

These results show that of the three orientations considered with a square plate, the highest and lowest values of the free-convection heat transfer coefficient occur with the horizontal position, the hot surface facing upward and downward, respectively.

9-5 FREE CONVECTION ON AN INCLINED PLATE

Rich [7] pointed out that the heat transfer coefficient for free convection on an inclined plate can be predicted by the vertical plate formulas if the gravitational term in the Grashof number is adjusted to accommodate the effect of the inclination. Since then, free convection on inclined surfaces has been studied by several investigators [20, 23, 25, 26]. The orientation of the inclined surface, whether the

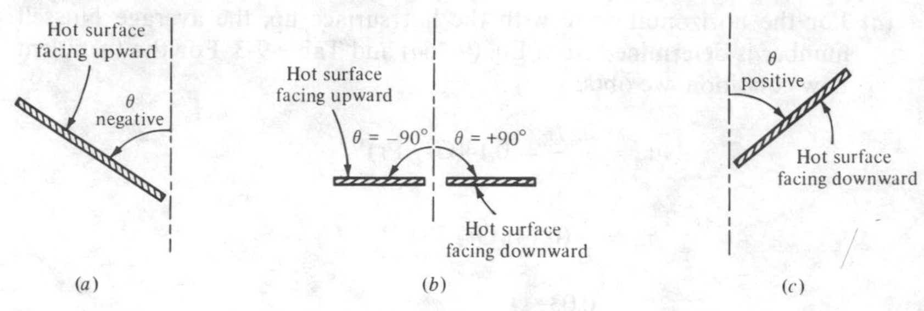

Figure 9-4 The concept of positive and negative inclination angles from the vertical to define the orientation of the hot surface.

surface is facing upward or downward, is also a factor that affects the Nusselt number. To make a distinction in the orientation of the surface, by following Fujii and Imura [23], we designate the sign of the angle θ that the surface makes with the vertical as follows:

1. The angle θ is considered *negative* if the hot surface is *facing up*, as illustrated in Fig. 9-4*a*.
2. The angle θ is considered *positive* if the hot surface is *facing down*, as illustrated in Fig. 9-4*c*.

Figure 9-4*b* illustrates the limiting cases of $\theta \to -90°$, the horizontal plate with hot surface facing upward, and $\theta \to +90°$, the horizontal plate with hot surface facing downward.

Uniform Wall Heat Flux

Here we present the heat transfer correlations based on the extensive experimental investigations of Fujii and Imura [23] for free convection from an inclined plate subjected to approximately uniform wall flux to water.

For an inclined plate with the *heated surface facing downward*:

$$\boxed{\mathrm{Nu}_m = 0.56(\mathrm{Gr}_L \,\mathrm{Pr} \cos\theta)^{1/4} \qquad \text{for } +\theta < 88°,\ 10^5 < \mathrm{Gr}_L \,\mathrm{Pr} < 10^{11}} \tag{9-83}$$

For the plate slightly inclined with the horizontal (that is, $88° < \theta < 90°$) and the heated surface facing downward, Eq. (9-80) is applicable. The $\frac{1}{4}$ power in Eq. (9-83) implies that the flow is always in the laminar regime.

For the inclined plate with the *heated surface facing upward*, the heat transfer correlation has been developed with the following considerations [23]. It is assumed that Eq. (9-83) is applicable in the laminar flow regime of $\mathrm{Gr}_L \,\mathrm{Pr} < \mathrm{Gr}_c \,\mathrm{Pr}$, where Gr_c is the critical Grashof number at which the transition from laminar to turbulent flow takes place. In the turbulent regime, it is assumed that Eq. (9-78) or (9-79) is applicable if $\mathrm{Gr}_L \,\mathrm{Pr}$ is replaced by $\mathrm{Gr}_L \,\mathrm{Pr} \cos\theta$. With this consideration,

Table 9-4 Transition Grashof number (see Fig. 9-4*a* for the definition of θ)

θ, degrees	Gr_c
−15	5×10^9
−30	10^9
−60	10^8
−75	10^6

From the experimental data of Fujii and Imura [23].

two expressions can be developed, one involving the coefficient 0.13 based on the results of experiments with the 30-cm test plate and the other involving the coefficient 0.16 based on a 5-cm test plate. Here we present the average of these two results and give the correlation of free convection on an inclined plate with the *heated surface facing upward* as

$$\boxed{Nu_m = 0.145[(Gr_L\ Pr)^{1/3} - (Gr_c\ Pr)^{1/3}] + 0.56(Gr_c\ Pr\cos\theta)^{1/4}} \tag{9-84}$$

for $Gr_L\ Pr < 10^{11}$, $Gr_L > Gr_c$, and $-15° < \theta < -75°$. Here, the value of the transition Grashof number Gr_c depends on the angle of inclination θ, as listed in Table 9-4.

In Eqs. (9-83) and (9-84), all physical properties are evaluated at the mean temperature

$$T_m = T_w - 0.25(T_w - T_\infty)$$

and β is evaluated at $T_\infty + 0.25(T_w - T_\infty)$.

Example 9-7 A square plate $\frac{1}{2}$ m by $\frac{1}{2}$ m is thermally insulated on one side and subjected to a solar radiation flux $q = 600\ W/m^2$ on the other side which is considered a black surface. The plate makes an angle $\theta = -60°$ with the vertical, as illustrated in Fig. 9-4*a*, so that the hot surface is facing upward. The heated surface dissipates heat by free convection into quiescent atmospheric air at $T_\infty = 300$ K. Calculate the equilibrium temperature of the plate.

SOLUTION This is a problem of constant wall heat flux. Therefore, the surface temperature is not known, and the mean temperature at which physical properties are to be evaluated cannot be determined. To start the calculations, we choose an *approximate* value for h. From our previous experience we take $h = 6\ W/(m^2 \cdot °C)$. Then a first *approximation* to the wall temperature T_w is

$$h(T_w - T_\infty) = q$$

$$T_w = T_\infty + \frac{q}{h} = 300 + \frac{600}{6} = 400\ K$$

We plan to use Eq. (9-84) for which the physical properties are evaluated at a mean temperature

$$T_m = T_w - 0.25(T_w - T_\infty)$$
$$= 400 - 0.25(400 - 300) = 375 \text{ K}$$

and β is evaluated at

$$T_\infty + 0.25(T_w - T_\infty) = 300 + 0.25(400 - 300) = 325 \text{ K}$$

Then the physical properties of air are taken as

$$\nu = 2.33 \times 10^{-5} \text{ m}^2/\text{s} \qquad k = 0.032 \text{ W/(m} \cdot {}^\circ\text{C)}$$
$$\text{Pr} = 0.693 \qquad \beta = \tfrac{1}{325} = 3.08 \times 10^{-3}$$

The transition Grashof number for $\theta = -60°$ is obtained from Table 9-4 as

$$\text{Gr}_c = 10^8$$

The Grashof number Gr_L is

$$\text{Gr}_L = \frac{g\beta(T_w - T_\infty)L^3}{\nu^2}$$
$$= \frac{(9.8)(3.08 \times 10^{-3})(400 - 300)(0.5)^3}{(2.33 \times 10^{-5})^2}$$
$$= 6.95 \times 10^8$$

Equation (9-84) is given as

$$\text{Nu}_m = \frac{hL}{k} = 0.145[(\text{Gr}_L \text{ Pr})^{1/3} - (\text{Gr}_c \text{ Pr})^{1/3}] + 0.56(\text{Gr}_c \text{ Pr} \cos\theta)^{1/4}$$

or

$$h = \frac{0.032}{0.5}(0.145)[(6.95 \times 10^8 \times 0.693)^{1/3} - (10^8 \times 0.693)^{1/3}]$$
$$+ \frac{0.032}{0.5}(0.56)(10^8 \times 0.693 \times 0.5)^{1/4}$$
$$= 6.21 \text{ W/(m}^2 \cdot {}^\circ\text{C)}$$

Given h, the wall surface temperature T_w can be determined from

$$T_w = T_\infty + \frac{q}{h} = 300 + \frac{600}{6.21} = 396.6 \text{ K}$$

Another iteration is not warranted, because the mean temperature used in the above calculations is sufficiently close to the mean temperature obtained by taking $T_w = 396.6$ K.

9-6 FREE CONVECTION ON A LONG CYLINDER

We now examine free convection on a long cylinder for both vertical and horizontal cylinders.

Vertical Cylinder

The average Nusselt number for free convection on a vertical cylinder is the same as that for a vertical plate if the thickness of the thermal boundary layer is much smaller than the cylinder radius, namely, if the curvature effects are negligible. Therefore, for an isothermal vertical cylinder, the average Nusselt number can be found from vertical plate relation, Eq. (9-69), with the coefficients c and n given in Table 9-2. For such cases, the length L in the definition of the Grashof or the Rayleigh number represents the height of the cylinder.

For fluids having a Prandtl number 0.7 and higher, a vertical cylinder may be treated as a vertical flat plate when

$$\frac{L/D}{(\mathrm{Gr}_L)^{1/4}} < 0.025$$

where D is the diameter of the cylinder. The physical significance of this criterion and the error associated with it are better envisioned by referring to Fig. 9-5*b*

In the case of vertical, slender, circular cylinders, the above criterion is not satisfied; hence a vertical cylinder can no longer be treated as a vertical plate. This matter was studied by Sparrow and Gregg [29], Minkowycz and Sparrow [30], and Cebeci [31]. Figure 9-5*a* shows a plot of the ratio of the *local Nusselt number* for a vertical cylinder to that for a flat plate as a function of the parameter $\xi = (2\sqrt{2}/\mathrm{Gr}_x^{1/4})(x/R)$ for several different values of the Prandtl number. Here R is the radius of the cylinder, $\mathrm{Nu}_x = hx/k$ is the local number, and $\mathrm{Gr}_x = g\beta(T_w - T_\infty)x^3/\nu^2$ is the local Grashof number. Similarly, Fig. 9-5*b* shows a plot of the ratio of the *average Nusselt number* for a vertical cylinder to that for a vertical plate. Note that the deviation increases when the Grashof number or the Prandtl number decreases.

For the case of a vertical cylinder subjected to uniform wall heat flux, the local Nusselt number may be determined from Eqs. (9-72) and (9-73).

Horizontal Cylinder

Churchill and Chu [32] correlated the existing data for the average Nusselt number for free convection on an *isothermal* horizontal cylinder with an equation of the same form as Eq. (9-71) and proposed the following correlation:

$$\mathrm{Nu}_m^{1/2} = 0.60 + \frac{0.387\,\mathrm{Ra}_D^{1/6}}{[1 + (0.559/\mathrm{Pr})^{9/16}]^{8/27}} \qquad \text{for } 10^{-4} < \mathrm{Ra}_D < 10^{12} \qquad (9\text{-}85)$$

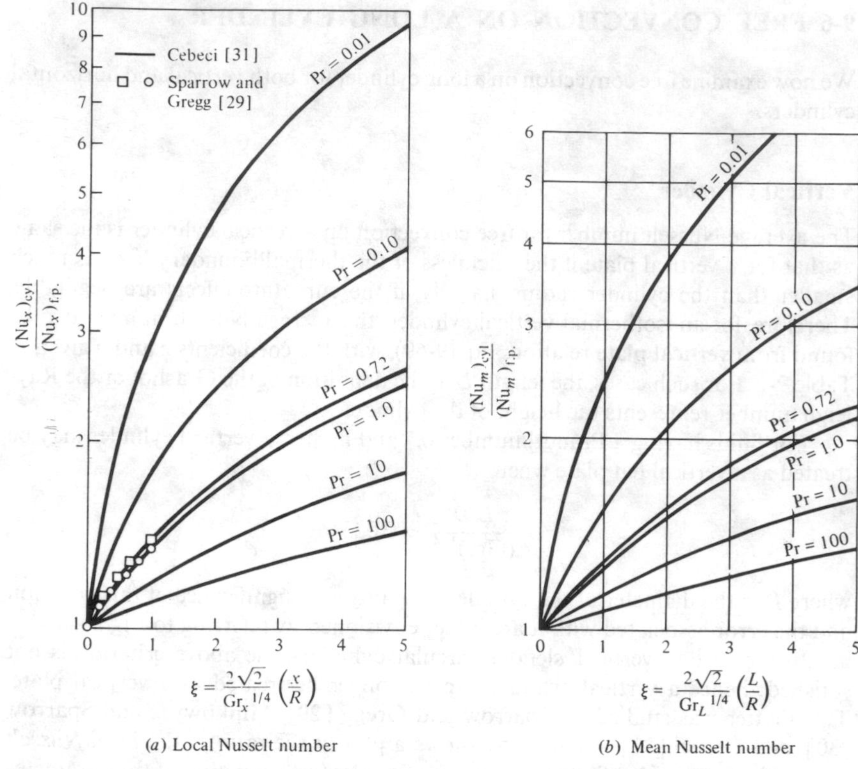

Figure 9-5 The ratio of the Nusselt number for a vertical plate to that for a vertical cylinder (*From Cebeci [31]*.)

where Nu_m and Ra_D are based on the cylinder diameter, namely,

$$Nu_m = \frac{hD}{k} \tag{9-86a}$$

$$Ra_D = Gr_D \, Pr = \frac{g\beta(T_w - T_\infty)D^3}{\nu^2} \, Pr \tag{9-86b}$$

Figure 9-6 shows a comparison of Eq. (9-85) correlating the experimental data for heat transfer with air, water, and mass transfer. The correlation appears to provide a smooth transition from the laminar to the turbulent regime, despite the fact that the actual transition is known to occur in discrete steps.

The experimental data by Parsons and Mulligan [22] with 0.127-mm-diameter wire correlated well with Eq. (9-85) for Rayleigh numbers above 10^{-3}, but Eq. (9-85) was overpredicting the Nusselt number for Rayleigh numbers below about 10^{-4}. A similar trend is also apparent from the correlation shown in Fig. 9-6. Equation (9-85) overpredicts the Nusselt number for $Ra_D < 10^{-4}$ and so

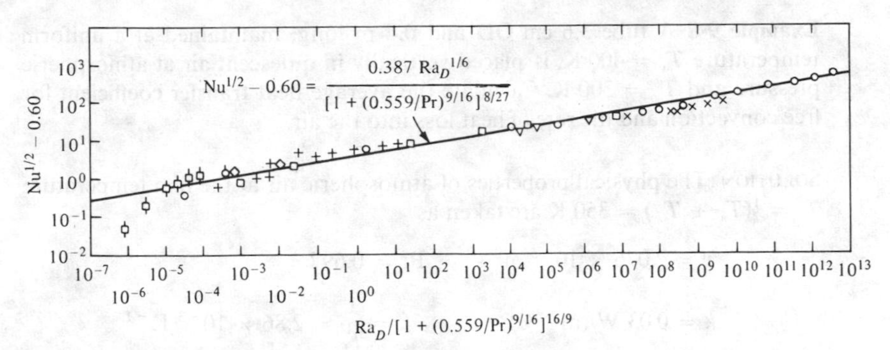

Figure 9-6 Comparison of Eq. (9-85) with the experimental data (Ref. 32).

should not be used for Rayleigh numbers below about 10^{-4}. The range $\mathrm{Ra}_D < 10^{-4}$ has numerous important applications in free convection from small-diameter wires, especially for heat transfer in microelectronic circuits.

Recently, Farouk and Güçeri [34*b*] numerically solved free convection from an isothermal horizontal cylinder for the turbulent flow regime. Their results closely agreed with the empirical correlation given by Kuehn and Goldstein [36]. The correlation given by Eq. (9-85) was little above the theoretical predictions in the range of Rayleigh number $10^7 < \mathrm{Ra} < 10^{11}$.

Morgan [33] presented a simple correlation for free convection from a horizontal isothermal cylinder, covering the range $10^{-10} < \mathrm{Ra}_D < 10^{12}$. It is given in the form

$$\mathrm{Nu}_m = \frac{hD}{k} = c\,\mathrm{Ra}_D^n \tag{9-87}$$

where the constant c and the exponent n are listed in Table 9-5.

For example, for $\mathrm{Ra}_D = 10^{-5}$ and $\mathrm{Pr} = 1$, the Nusselt numbers obtained from Eqs. (9-85) and (9-87) are, respectively, 0.420 and 0.346. The latter is closer to the experimental data.

Table 9-5 Constant *c* and exponent *n* of Eq. (9-87) for free convection on a horizontal cylinder

Ra_D	c	n
10^{-10}–10^{-2}	0.675	0.058
10^{-2}– 10^2	1.02	0.148
10^2– 10^4	0.850	0.188
10^4– 10^7	0.480	0.250
10^7– 10^{12}	0.125	0.333

Source: Morgan [33].

Example 9-8 A tube 3.6 cm OD and 0.4 m long, maintained at a uniform temperature $T_w = 400$ K, is placed vertically in quiescent air at atmospheric pressure and $T_\infty = 300$ K. Calculate the average heat transfer coefficient for free convection and the rate of heat loss into the air.

SOLUTION The physical properties of atmospheric air at the film temperature $T_f = \frac{1}{2}(T_w + T_\infty) = 350$ K are taken as

$$\nu = 2.076 \times 10^{-5}\ \text{m}^2/\text{s} \qquad \text{Pr} = 0.697$$

$$k = 0.03\ \text{W/(m}\cdot{}^\circ\text{C)} \qquad \beta = \frac{1}{T_f} = 2.86 \times 10^{-3}\ \text{K}^{-1}$$

Then the Grashof number for $L = 0.4$ m becomes

$$\text{Gr}\Big|_{L=0.4} = \frac{g\beta(T_w - T_\infty)L^3}{\nu^2} = 4.16 \times 10^8$$

The tube diameter is small compared with the tube length. Therefore, the problem may not be treated as a vertical plate; we need to use the correlation given in Fig. 9-5*b* to establish the effect of slenderness on the Nusselt number.

The parameter ξ of Fig. 9-5*b* is determined for $x = L = 0.4$ m:

$$\xi = \frac{2\sqrt{2}}{\text{Gr}^{1/4}}\left(\frac{L}{R}\right) = \frac{2\sqrt{2}}{(4.16 \times 10^8)^{1/4}}\left(\frac{40}{1.8}\right)$$
$$= 0.44$$

Then, from Fig. 9-5*b*, for $\xi = 0.44$ and Pr $= 0.697$, we obtain

$$\frac{(\text{Nu}_m)_{\text{cyl}}}{(\text{Nu}_m)_{\text{fp}}} = \frac{(h_m)_{\text{cyl}}}{(h_m)_{\text{fp}}} = 1.1$$

Thus, the mean heat transfer coefficient for the cylindrical tube considered here is 1.1 times that for a vertical plate. In Example 9-3, we calculated the mean heat transfer coefficient for a vertical plate under the same conditions and found that

$$(h_m)_{\text{fp}} = 5.77\ \text{W/(m}^2\cdot{}^\circ\text{C)}$$

Then the mean heat transfer coefficient for the vertical, slender cylinder considered in this example becomes

$$(h_m)_{\text{cyl}} = 5.77 \times 1.1 = 6.4\ \text{W/(m}^2\cdot{}^\circ\text{C)}$$

The heat transfer rate Q becomes

$$Q = \pi D L (h_m)_{\text{cyl}}(T_w - T_\infty)$$
$$= \pi(0.036)(0.4)(6.4)(400 - 300)$$
$$= 29\ \text{W}$$

Example 9-9 Compare the average Nusselt number for free convection from an isothermal horizontal cylinder for air as obtained from the correlations given by Eqs. (9-85) and (9-87) at Rayleigh numbers $\mathrm{Ra}_D = 10^3, 10^5$, and 10^{10}.

SOLUTION We calculate Nu_m as obtained from the simple expression given by Eq. (9-87) with the coefficients c and n obtained from Table 9-5:

$\mathrm{Ra}_D = 10^3$: $\quad \mathrm{Nu}_m = 0.850(10^3)^{0.188} = 3.15$

$\mathrm{Ra}_D = 10^5$: $\quad \mathrm{Nu}_m = 0.480(10^5)^{0.250} = 8.54$

$\mathrm{Ra}_D = 10^{10}$: $\quad \mathrm{Nu}_m = 0.125(10^{10})^{0.333} = 269.1$

We now calculate these results from Eq. (9-85), by taking the Prandtl number for air as $\mathrm{Pr} = 0.70$, and obtain

$\mathrm{Ra}_D = 10^3$: $\quad \mathrm{Nu}_m = 2.61$

$\mathrm{Ra}_D = 10^5$: $\quad \mathrm{Nu}_m = 7.76$

$\mathrm{Ra}_D = 10^{10}$: $\quad \mathrm{Nu}_m = 240.1$

A comparison of these two sets of results reveals that the more elaborate correlation given by Eq. (9-85) gives a Nusselt value about 17 to 10 percent lower than that predicted by the simpler correlation given by Eq. (9-87).

9-7 FREE CONVECTION ON A SPHERE

The average Nusselt number for free convection on a single isothermal sphere, for fluids having Prandtl number close to unity, has been correlated by Yuge [35] with the following expression:

$$\mathrm{Nu}_m = \frac{hD}{k} = 2 + 0.43\,\mathrm{Ra}_D^{1/4} \tag{9-88a}$$

for $1 < \mathrm{Ra}_D < 10^5$ and $\mathrm{Pr} \cong 1$.

Amato and Tien [37], based on experimental data for free convection on a single isothermal sphere in water, proposed the following correlation:

$$\mathrm{Nu}_m = 2 + 0.50\,\mathrm{Ra}_D^{1/4} \tag{9-88b}$$

for $3 \times 10^5 < \mathrm{Ra}_D < 8 \times 10^8$ and $10 \le \mathrm{Nu}_m \le 90$. Here the Rayleigh number, based on the sphere diameters, is defined as

$$\mathrm{Ra}_D = \mathrm{Gr}_D\,\mathrm{Pr} = \frac{g\beta D^3 (T_w - T_\infty)}{\nu^2}\,\mathrm{Pr} \tag{9-88c}$$

Properties are evaluated at the film temperature. Note that as $\mathrm{Ra}_D \to 0$, Eqs. (9-88a) and (9-88b) reduce to $\mathrm{Nu}_m \to 2$, which is the limiting value for heat conduction from an isothermal sphere in an infinite medium.

9-8 SIMPLIFIED EQUATIONS FOR AIR

We present in Table 9-6 simplified expressions for rapid but *approximate estimation* of the average heat transfer coefficient from isothermal surfaces in air at *atmospheric pressure and moderate temperatures.* For more accurate results, previously given, more exact expressions should be used. The correlations given in this table apply to air from about 20 to 800°C as well as to CO, CO_2, O_2, N_2, and flue gases. The Grashof number is defined as

$$\mathrm{Gr}_L \equiv \frac{g\beta\,\Delta T L^3}{\nu^2}$$

where L is the characteristic dimension of the body and ΔT is the temperature difference between the surface and the ambient air, that is, $\Delta T = T_w - T_\infty$.

The correlations given in Table 9-6 for air at atmospheric pressure can be extended to higher or lower pressures by multiplying by the following factors:

$$P^{1/2} \quad \text{for laminar regime}$$

$$P^{2/3} \quad \text{for turbulent regime}$$

where P is the pressure in atmospheres.

We reiterate that the expressions given in Table 9-6 are merely approximations; care must be exercised in their use.

Example 9-10 Calculate the free-convection heat transfer coefficients on a square plate 0.5 m by 0.5 m considered in Example 9-6 by using the simplified relations given in Table 9-6.

Table 9-6 Simplified equations for free convection to air at atmospheric pressure and moderate temperatures

Geometry	Characteristic dimension, L	Type of flow	Range of Gr_L Pr	Heat transfer coefficient h_m, W/(m² · °C)
Vertical plates and cylinders	Height	Laminar	10^4 to 10^9	$h_m = 1.42(\Delta T/L)^{1/4}$
		Turbulent	10^9 to 10^{13}	$h_m = 1.31\,\Delta T^{1/3}$
Horizontal cylinders	Outside diameter	Laminar	10^4 to 10^9	$h_m = 1.32(\Delta T/D)^{1/4}$
		Turbulent	10^9 to 10^{12}	$h_m = 1.24\,\Delta T^{1/3}$
Horizontal plates				
(*a*) Upper surface hot or lower surface cold	As defined in the text	Laminar	10^5 to 2×10^7	$h_m = 1.32(\Delta T/L)^{1/4}$
		Turbulent	2×10^7 to 3×10^{10}	$h_m = 1.52\,\Delta T^{1/3}$
(*b*) Lower surface hot or upper surface cold	As defined in the text	Laminar	3×10^5 to 3×10^{10}	$h_m = 0.59(\Delta T/L)^{1/4}$

From McAdams [8]

SOLUTION In Example 9-6, the Grashof number was $Gr = 5.7 \times 10^8$, the temperature difference was $\Delta T = 385 - 315 = 70°C$, and the characteristic dimension of the plate was $L = 0.5$ m. We now use the simplified expressions given in Table 9-6 to determine the free-convection heat transfer coefficients and compare the results with those obtained in Example 9-6.

For the horizontal plate with the hot surface facing up, from Table 9-6 we have

$$h_m = 1.52\,\Delta T^{1/3} = 1.52 \times 70^{1/3} = 6.26\ \text{W/(m}^2 \cdot {}^\circ\text{C)}$$

which is within 1.5 percent of the result obtained in Example 9-6 by using Eq. (9-76*a*).

For the vertical plate, from Table 9-6 we have

$$h_m = 1.42\left(\frac{\Delta T}{L}\right)^{1/4} = 1.42\left(\frac{70}{0.5}\right)^{1/4} = 4.88\ \text{W/(m}^2 \cdot {}^\circ\text{C)}$$

which is within 2.3 percent of the result obtained in Example 9-6 by using Eq. (9-69).

For the horizontal plate with the hot surface facing down, from Table 9-6 we have

$$h_m = 0.59\left(\frac{\Delta T}{L}\right)^{1/4} = 0.59\left(\frac{70}{0.5}\right)^{1/4} = 2.03\ \text{W/(m}^2 \cdot {}^\circ\text{C)}$$

which is about 11 percent less than that obtained in Example 9-6 by using Eq. (9-76*a*).

9-9 MECHANISM OF FREE CONVECTION IN ENCLOSED SPACES

Heat transfer by free convection in enclosed spaces has numerous engineering applications. For example, free convection in wall cavities, between window glazing, in the annulus between concentric cylinders or spheres, and in flat-plate solar collectors are typical examples. Before we discuss the correlations of heat transfer in Sec. 9-10, it is instructive to provide some insight into the physical nature of the problem and give a qualitative discussion of the onset of free convection in enclosed spaces.

Consider a fluid contained in a horizontal enclosed space, as illustrated in Fig. 9-7*a*. Let the lower plate be maintained at a uniform temperature T_h which is higher than the temperature T_c of the upper colder plate. Then there is a heat flow through the fluid layer in the upward direction, and a temperature profile, decreasing upward, is established. The denser, cold fluid layers lie above the lighter, warm layers. For sufficiently small values of the temperature difference $T_h - T_c$ between the plates, the viscous forces overcome the buoyancy forces and the fluid

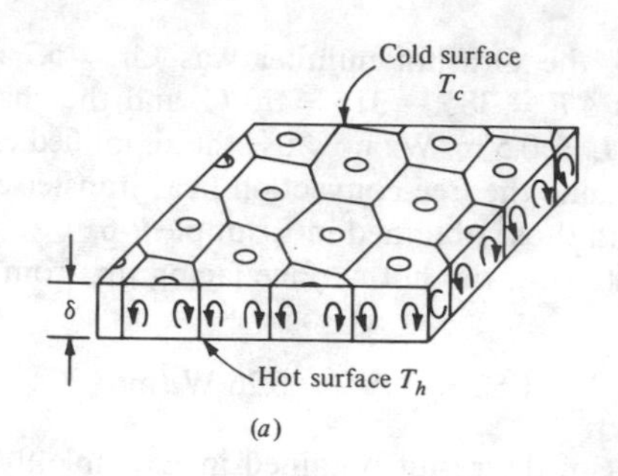

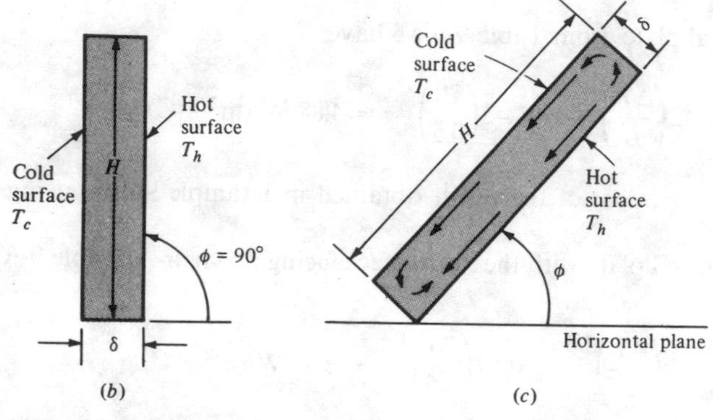

Figure 9-7 Free convection in enclosed spaces: (*a*) Horizontal layer with Bénard cells, (*b*) vertical layer, (*c*) inclined layer.

remains motionless; as a result, heat transfer across the fluid layer is by pure *conduction*, and a Nusselt number can be defined as

$$k\,\frac{T_h - T_c}{\delta} \equiv h(T_h - T_c)$$

or

$$\mathrm{Nu}_\delta \equiv \frac{h\delta}{k} = 1$$

where δ is the thickness of the fluid layer.

Suppose the temperature difference $T_h - T_c$ is increased beyond a value such that the buoyancy forces overcome the viscous forces. Then some kind of convective motion takes place in the fluid. Theoretical and experimental investigations have verified that for a horizontal enclosure if $T_h - T_c$ is increased beyond a value corresponding to the *critical Rayleigh number*

$$\mathrm{Ra} = \mathrm{Gr}_\delta\,\mathrm{Pr} = \frac{g\beta(T_h - T_c)\delta^3}{\nu^2}\,\mathrm{Pr} = 1708 \tag{9-89}$$

then the horizontal fluid layer becomes unstable, giving rise to the formation of flow patterns in the form of hexagonal cells as illustrated in Fig. 9-7*a*. These cells are called *Bénard cells*, after Bénard who observed this phenomenon first in 1900.

If the temperature difference $T_h - T_c$ is further increased beyond a value corresponding to about Ra = 50,000, then turbulent free convection sets in, destroying the regular cellular flow patterns.

It is apparent from the previous discussion that the heating of a horizontal layer of fluid from below is associated with three distinct heat transfer regimes: the *conduction*, the *cellular convection*, and the *turbulent convection* regimes. A knowledge of the Nusselt number for such situations is of interest in engineering applications in order to predict heat transfer across the fluid layer.

We now consider a fluid layer contained between two vertical, isothermal plates maintained at two different temperatures, as illustrated in Fig. 9-7*b*. Let T_h and T_c be the temperatures of the hot and cold surfaces, respectively.

For small temperature differences $T_h - T_c$, the energy is transmitted across the fluid layer by pure conduction, because viscous forces overcome the buoyancy forces and the fluid remain stationary. For such a situation, the Nusselt number is unity:

$$\mathrm{Nu}_\delta \equiv \frac{h\delta}{k} = 1$$

where δ is the spacing between the plates.

For larger temperature differences $T_h - T_c$, a moderate circulatory flow starts, and cells are formed. In these cells, the circulatory flow is symmetric with respect to the cell center, and the velocity distribution is invariant with height over the fluid layer. The heat transfer through the central portion of the layer is by conduction, while there is a complex flow pattern at the ends.

Further increase of the temperature difference $T_h - T_c$ beyond about $\mathrm{Ra} > 10^4$ gives rise to a semiboundary-layer type of flow. Namely, fluid moves upward as a boundary-layer flow along the hot wall and downward along the cold wall. The central layer between the plates, however, remains stationary as a result of a balance between the buoyancy and viscous forces. In this regime, the heat is transferred primarily by convection in the boundary-layer region and by conduction through the stationary, central region of the layer.

Further increase of the temperature difference beyond about $\mathrm{Ra} > 10^5$ gives rise to the vertical row of vortices with horizontal axes. The number of vortices increases with increasing Grashof number.

Finally, at much larger Grashof numbers turbulence sets in.

In the case of vertical layers, the aspect ratio H/δ is another factor which governs the end effects and the transition from one regime to another.

For an inclined layer, as illustrated in Fig. 9-7*c*, with the cold plate above the hot plate, free-convection flow patterns set in when a critical Grashof number is exceeded, but the mechanism of heat transfer in various flow regimes is much more complicated than that of free convection in horizontal or vertical layers.

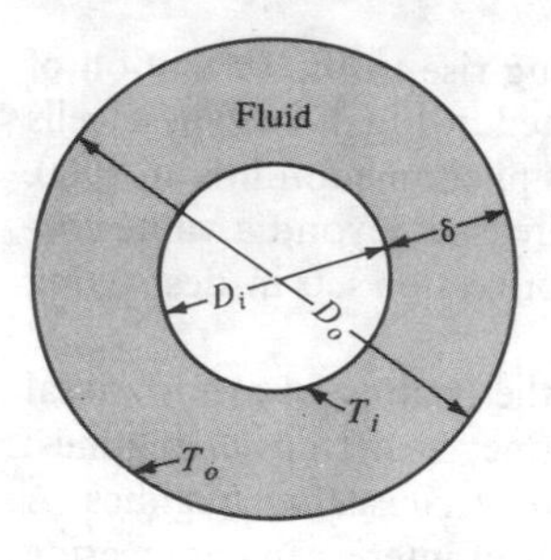

Figure 9-8 Free convection in horizontal cylindrical annulus or spherical annulus.

Therefore, one expects that the heat transfer correlations for free convection in inclined layers to be more complicated than those for vertical and horizontal layers.

In the case of fluid contained in the annulus between two horizontal cylinders or two concentric spheres, as illustrated in Fig. 9-8, free-convection flow patterns start when the temperature difference $T_i - T_o$ between the surfaces is increased, hence a critical Grashof number is exceeded.

The preceding qualitative discussion of free convection in enclosed spaces is intended to illustrate the complexity of the mechanism of flow, hence of heat transfer and the physical significance of various heat transfer regimes. Next we present some of the correlations of heat transfer in free convection in enclosed spaces.

9-10 CORRELATIONS OF FREE CONVECTION IN ENCLOSED SPACES

The determination of the onset of free convection and the heat transfer coefficient associated with free convection has been the subject of numerous investigations [38–53]. Despite the vast amount of experimental and analytic studies of free convection in enclosed spaces, the heat transfer correlations covering all ranges of parameters are still not available.

Here we consider some specific situations and present some recommended correlations, but such results should not be used beyond the ranges of parameters for which they are tested against the experimental data.

Simple Empirical Correlations

As discussed previously, the mechanism of heat transfer in free convection in enclosed spaces is so complicated that the correlation of heat transfer with simple expressions is not expected to be very accurate. However, simple correlations are useful for a quick estimate of the approximate magnitude of the free-convection heat transfer coefficient. Therefore, before presenting more accurate but more elaborate expressions, we consider a very simple correlation of the free-convection heat transfer coefficient for enclosures given in the form

$$\boxed{\text{Nu}_\delta = c(\text{Ra}_\delta)^n \left(\frac{H}{\delta}\right)^m} \tag{9-90a}$$

where $\text{Ra}_\delta = \text{Gr}_\delta \, \text{Pr} = \dfrac{g\beta(T_h - T_c)\delta^3}{\nu^2} \text{Pr}$

δ = thickness of fluid layers; in an annulus, it is $(D_o - D_i)/2$

H = height of fluid layer

Table 9-7 lists the values of the constants c, n, and m for vertical and horizontal enclosures formed by two parallel plates, a horizontal cylindrical annulus, and a spherical annulus. Once the Nusselt number is computed from Eq. (9-90), the mean heat transfer coefficient h_m is determined from its definition

$$h_m = \text{Nu}_\delta \frac{k}{\delta} \tag{9-90b}$$

Given h_m, the total heat transfer rate Q across the fluid layer is

$$Q = A_m h_m (T_h - T_c) \tag{9-90c}$$

where the mean area A_m depends on the geometry:

Plane layer: $A_m = A_w$ wall area

Cylindrical annulus: $A_m = \dfrac{A_o - A_i}{\ln (A_o/A_i)}$

Spherical annulus: $A_m = \sqrt{A_o A_i}$

The vertical cylindrical annulus can be treated as a plane layer for determining the constants c, n, and m in Eq. (9-90), but the logarithmic mean area as defined above is used for calculating A_m.

More Accurate Empirical Correlations for Free Convection in Enclosures

We now present some of the more accurate, and consequently more elaborate, correlations of free convection in vertical, horizontal, and inclined enclosed layers; a horizontal cylindrical annulus; and a spherical annulus.

Vertical layer Consider a vertical layer of fluid contained between two parallel plates of height H and separated by a distance δ, as illustrated in Fig. 9-7*b*. The plates are maintained at uniform temperatures T_h and T_c. Let $a = H/\delta$ be the *aspect ratio*. Heat transfer across such a gap by free convection has been experimentally investigated [53] for air, covering the range of aspect ratios from $a = 5$ to 110 and the range of Rayleigh numbers from $\text{Ra}_\delta = 10^2$ to 2×10^7. Based on the

Table 9-7 Coefficients c, n, and m of Eq. (9-90)

Geometry	Type of heating	Type of fluid	Range of Pr	Range of Ra_δ	$\frac{H}{\delta}$	c	n	m	Reference(s)
Vertical layer or vertical cylindrical annulus	Isothermal	Gas		<2000	—	1	0	0	[40, 41, 43]
		Gas	0.5–2	2×10^3–2×10^5	11–42	0.197	$\frac{1}{4}$	$-\frac{1}{9}$	
		Gas	0.5–2	2×10^5–10^7	11–42	0.073	$\frac{1}{3}$	$-\frac{1}{9}$	
	Isothermal or uniform heat flux	Liquid	1–2×10^4	10^4–10^7	10–40	$0.42\ Pr^{0.012}$	$\frac{1}{4}$	-0.3	[15, 39]
		Liquid	1–20	10^6–10^9	1–40	0.046	$\frac{1}{3}$	0	
Horizontal layer		Gas		<1700	—	1	0	0	[40, 41, 43, 44, 48]
		Gas	0.5–2	1.7×10^3–7×10^3	—	0.059	0.4	0	
	Isothermal, lower plate hot	Gas	0.5–2	7×10^3–3.2×10^5	—	0.212	$\frac{1}{4}$	0	
		Gas	0.5–2	$>3.2 \times 10^5$	—	0.061	$\frac{1}{3}$	0	
		Liquid		<1700	—	1	0	0	
		Liquid	1–5000	1.7×10^3–6×10^3	—	0.012	0.6	0	[41, 45, 48, 49]
		Liquid	1–5000	6×10^3–3.7×10^4	—	0.375	0.2	0	
		Liquid	1–20	3.7×10^4–10^8	—	0.13	0.3	0	
		Liquid	1–20	$>10^8$	—	0.057	$\frac{1}{3}$	0	
Horizontal cylindrical annulus	Isothermal	Gas or liquid	1–5000	6×10^3–10^6	—	0.11	0.29	0	[56]
			1–5000	10^6–10^8	—	0.40	0.20	0	
Spherical annulus	Isothermal	Gas or liquid	0.7–4000	10^2–10^9	—	0.228	0.226	0	[57]

results of these experiments, the following correlation equation is proposed for free convection in vertical layers:

$$\boxed{\text{Nu}_{90^\circ} = [\text{Nu}_1, \text{Nu}_2, \text{Nu}_3]_{\max}} \tag{9-91}$$

which implies that one should select the maximum of the three Nusselts numbers Nu_1, Nu_2, and Nu_3, which are defined as

$$\text{Nu}_1 = 0.0605\,\text{Ra}^{1/3} \tag{9-92a}$$

$$\text{Nu}_2 = \left\{1 + \left[\frac{0.104\,\text{Ra}^{0.293}}{1 + (6310/\text{Ra})^{1.36}}\right]^3\right\}^{1/3} \tag{9-92b}$$

$$\text{Nu}_3 = 0.242\left(\frac{\text{Ra}}{a}\right)^{0.272} \tag{9-92c}$$

where

$$a = \frac{H}{\delta} = \text{aspect ratio}$$

and

$$\text{Ra} = \text{Gr}_\delta\,\text{Pr} = \frac{g\beta(T_h - T_c)\delta^3}{\nu^2}\,\text{Pr}$$

The Nusselt number can be defined as

$$\text{Nu}_{90^\circ} = \frac{q\delta}{k(T_h - T_c)} \qquad \text{since } q = h(T_h - T_c)$$

and all physical properties are evaluated at $(T_h + T_c)/2$. The experimental data for air agree with Eq. (9-91) to within 9 percent. Therefore, knowing the Nusselt number, the heat flux q is determined.

Example 9-11 Air at atmospheric pressure is contained in a vertical enclosure consisting of two square parallel plates 0.6 m by 0.6 m separated by a distance $\delta = 0.05$ m. The temperatures of the hot and cold plates are, respectively, $T_h = 380$ and $T_c = 320$ K. Calculate the rate of heat transfer by free convection across the airspace by using the heat transfer correlation given by Eqs. (9-91) and (9-92). Compare the Nusselt number determined from this more accurate correlation with that obtainable from the approximate correlation given by Eq. (9-90).

SOLUTION The physical properties of air at $T_m = (380 + 320)/2 = 350$ K are

$$\nu = 2.076 \times 10^{-5}\ \text{m}^2/\text{s} \qquad \alpha = 2.983 \times 10^{-5}\ \text{m}^2/\text{s}$$

$$\beta = \tfrac{1}{350} = 2.86 \times 10^{-3}\ \text{K}^{-1} \qquad k = 0.03\ \text{W}/(\text{m}\cdot{}^\circ\text{C})$$

Noting that $\Pr = \nu/\alpha$, we know the Rayleigh number is written as

$$\mathrm{Ra} = \frac{g\beta(T_h - T_c)\delta^3}{\nu\alpha}$$

which becomes

$$\mathrm{Ra} = \frac{(9.8)(2.86 \times 10^{-3})(380 - 320)(0.05)^3}{(2.07 \times 10^{-5})(2.983 \times 10^{-5})}$$

$$= 3.4 \times 10^5$$

The aspect ratio is

$$a = \frac{H}{\delta} = \frac{0.60}{0.05} = 12$$

The correlations given by Eqs. (9-91) and (9-92) are used now to calculate the Nusselt number. From Eqs. (9-92) we have

$$\mathrm{Nu}_1 = 0.0605\,\mathrm{Ra}^{1/3} = 0.0605(3.4 \times 10^5)^{1/3} = 4.22$$

$$\mathrm{Nu}_2 = \left\{1 + \left[\frac{0.104\,\mathrm{Ra}^{0.293}}{1 + (6310/\mathrm{Ra})^{1.36}}\right]^3\right\}^{1/3} = 4.36$$

$$\mathrm{Nu}_3 = 0.242\left(\frac{\mathrm{Ra}}{a}\right)^{0.272} = 3.93$$

Hence the largest of these three values is taken as the Nusselt number for the problem.

$$\mathrm{Nu}_{90^\circ} = \frac{q\delta}{k(T_h - T_c)} = 4.36$$

or

$$q = \frac{4.36k(T_h - T_c)}{\delta}$$

$$= \frac{(4.36)(0.03)(380 - 320)}{0.05} = 157\ \mathrm{W/m^2}$$

The total heat transfer rate across the gap is thus

$$Q = q \cdot \text{area} = (157)(0.6 \times 0.6) = 56.5\ \mathrm{W}$$

We now calculate the Nusselt number for this problem by using the approximate correlation given by Eq. (9-90). We obtain the constants from Table 9-7 and write

$$\mathrm{Nu}_{90^\circ} = 0.073(\mathrm{Ra})^{1/3}\left(\frac{H}{\delta}\right)^{-1/9}$$

$$= 0.073(3.4 \times 10^5)^{1/3}(12^{-1/9}) = 3.86$$

which is about 12 percent below the value predicted by Eq. (9-91).

Inclined layer ($90° < \phi \leq 60°$) Figure 9-7*c* shows a fluid contained between two inclined parallel plates, separated by a distance δ and maintained at uniform temperatures T_h and T_c. The plates make an angle ϕ with the horizontal, and the cold plate is above the hot plate. We consider here the inclinations $90° < \phi \leq 60°$, since the inclinations below $\phi = 60°$ require a different type of correlation.

The experimental data with air for a layer making $\phi = 60°$ with the horizontal have been correlated with the following expression [53]:

$$\text{Nu}_{60°} = [\text{Nu}_1, \text{Nu}_2]_{\max} \tag{9-93}$$

where one should select the maximum of Nu_1 and Nu_2, defined as

$$\text{Nu}_1 = \left[1 + \left(\frac{0.093\,\text{Ra}^{0.314}}{1 + G}\right)^7\right]^{1/7} \tag{9-94a}$$

$$\text{Nu}_2 = \left(0.104 + \frac{0.175}{a}\right)\text{Ra}^{0.283} \tag{9-94b}$$

and

$$G = \frac{0.5}{[1 + (\text{Ra}/3160)^{20.6}]^{0.1}} \tag{9-94c}$$

and the Nusselt number is defined as

$$\text{Nu}_{60°} = \frac{q\delta}{k(T_h - T_c)}$$

To determine the Nusselt number for inclinations in the range $60° \leq \phi \leq 90°$, a straight-line interpolation is used between Eqs. (9-91) and (9-93), namely,

$$\text{Nu}_\phi = \frac{(90° - \phi°)\text{Nu}_{60°} + (\phi° - 60°)\text{Nu}_{90°}}{30°} \quad \text{for } 60° \leq \phi \leq 90° \tag{9-95}$$

The experimental data with air agree with Eq. (9-95) to within 6.5 percent.

Inclined layer ($0° \leq \phi \leq 60°$) Based on the experimental investigations with air, the following correlating equation has been proposed [42] for free convection in inclined layers lying in the range $0° \leq \phi \leq 60°$:

$$\boxed{\text{Nu}_\phi = 1 + 1.44\left[1 - \frac{1708}{\text{Ra}\cos\phi}\right]^*\left[1 - \frac{1708\,(\sin 1.8\phi)^{1.6}}{\text{Ra}\cos\phi}\right] + \left[\left(\frac{\text{Ra}\cos\phi}{5830}\right)^{1/3} - 1\right]^* \quad \text{for } 0° \leq \phi \leq 60°} \tag{9-96a}$$

where

$$\text{Nu}_\phi = \frac{q\delta}{k(T_h - T_c)}$$

$$q = \text{heat flux across layer, W/m}^2$$

and the notation [ ]* is used to denote that if the quantity in the bracket is negative, it should be set equal to zero.

Equation (9-96*a*) is recommended for $0° \le \phi < 60°$ and for high aspect ratios in the range of $0 < \text{Ra} < 10^5$.

The experimental data with air agree with Eq. (9-96*a*) to within 5 percent.

Horizontal layer ($\phi = 0°$) Equation (9-96) includes the horizontal layer as a special case. By setting $\phi = 0°$ in Eq. (9-96*a*), the Nusselt number for the horizontal layer is

$$\text{Nu}_{\phi=0°} = 1 + 1.44\left[1 - \frac{1708}{\text{Ra}}\right]^* + \left[\left(\frac{\text{Ra}}{5830}\right)^{1/3} - 1\right]^* \tag{9-96b}$$

Note that for Ra < 1708, Eq. (9-96*b*) reduces to

$$\text{Nu}_{\phi=0} = 1 \qquad \text{for Ra} < 1708 \tag{9-96c}$$

which is consistent with our previous discussion that the free-convection flow sets up at a critical Rayleigh number 1708, and for Ra < 1708 the conduction regime prevails. [See Eq. (9-89).]

Horizontal cylindrical annulus Theoretical and experimental work on free convection on a horizontal cylindrical annulus and a spherical annulus is rather limited. A comprehensive discussion of the pertinent literature can be found in Refs. 54 and 55.

Consider a fluid contained in a long, horizontal cylindrical annulus, having a gap spacing $\delta = \frac{1}{2}(D_o - D_i)$, where D_o and D_i are, respectively, the diameters of the outer and inner cylinders, as illustrated in Fig. 9-8.

Raithby and Hollands [54] proposed the following correlation for the heat transfer rate Q for the length H of a cylindrical annulus:

$$Q = \frac{2\pi k_{\text{eff}} H}{\ln (D_o/D_i)}(T_i - T_o) \qquad \text{W} \tag{9-97}$$

where

$$\frac{k_{\text{eff}}}{k} = 0.386\left(\frac{\text{Pr}}{0.861 + \text{Pr}}\right)^{1/4}(\text{Ra}^*_{\text{cyl}})^{1/4} \tag{9-98a}$$

$$(\text{Ra}^*_{\text{cyl}})^{1/4} = \frac{\ln (D_o/D_i)}{\delta^{3/4}(D_i^{-3/5} + D_o^{-3/5})^{5/4}}\,\text{Ra}_\delta^{1/4} \tag{9-98b}$$

$$H = \text{length of cylindrical annulus}$$

$$\text{Ra}_\delta = \frac{g\beta(T_i - T_o)\delta^3}{\nu^2}\,\text{Pr} \tag{9-98c}$$

$$\delta = \tfrac{1}{2}(D_o - D_i) \tag{9-98d}$$

This correlation is applicable over the range $10^2 < \text{Ra}^*_{\text{cyl}} < 10^7$.

Spherical annulus We now consider a fluid contained between two concentric spheres of inner and outer diameters D_i and D_o, respectively, as illustrated in Fig. 9-8. The surfaces are maintained at uniform temperatures T_i and T_o. Raithby and Hollands [54] proposed the following correlation for the total heat transfer rate from the spheres:

$$\boxed{Q = k_{\text{eff}} \frac{\pi D_i D_o}{\delta} (T_i - T_o)} \qquad \text{W} \tag{9-99}$$

where

$$\frac{k_{\text{eff}}}{k} = 0.74 \left(\frac{\text{Pr}}{0.861 + \text{Pr}} \right)^{1/4} (\text{Ra}^*_{\text{sph}})^{1/4} \tag{9-100a}$$

and

$$(\text{Ra}^*_{\text{sph}})^{1/4} = \frac{\delta^{1/4}}{D_i D_o} \frac{\text{Ra}_\delta^{1/4}}{(D_i^{-7/5} + D_o^{-7/5})^{5/4}} \tag{9-100b}$$

where Ra_δ and δ are defined by Eqs. (9-98*c*) and (9-98*d*), respectively. The correlation equation (9-99) is applicable over the range $10^2 < \text{Ra}^*_{\text{sph}} < 10^4$.

9-11 COMBINED FREE AND FORCED CONVECTION

Earlier we considered free convection resulting from unstable density gradients caused by the temperature gradients within the fluid. In Chaps. 7 and 8 we completely neglected the free convection and considered forced convection in which the fluid motion was imposed externally by a fan, a blower, or a pump.

We know from our examination of the dimensionless form of the governing equations that the relative magnitude of the dimensionless parameter Gr/Re^2 governs the relative importance of free convection in relation to forced convection. We recall from Eq. (9-19) that this dimensionless group can be expressed in terms of the physical parameters as

$$\frac{\text{Gr}}{\text{Re}^2} = \frac{g\beta(T_w - T_\infty)L}{U_0^2} \tag{9-101}$$

which represents the ratio of the buoyancy forces to inertial forces. When this ratio is of the order of unity, that is, $\text{Gr} \cong \text{Re}^2$, the free and forced convection are of comparable magnitude; hence they should be analyzed simultaneously. Therefore, this parameter can be used as a criterion to establish the dominant regions of free and forced convection:

$$\frac{\text{Gr}}{\text{Re}^2} \gg 1: \quad \text{Free convection dominates} \tag{9-102a}$$

$$\frac{\text{Gr}}{\text{Re}^2} \cong 1: \quad \text{Free and forced convection are of comparable magnitude} \tag{9-102b}$$

$$\frac{\text{Gr}}{\text{Re}^2} \ll 1: \quad \text{Forced convection dominates} \tag{9-102c}$$

In this section we examine the effects of free convection on forced convection.

Vertical and Horizontal Plates

Combined free and forced convection has been studied by various investigators for flow over vertical [58, 59] and horizontal [60–67] plates.

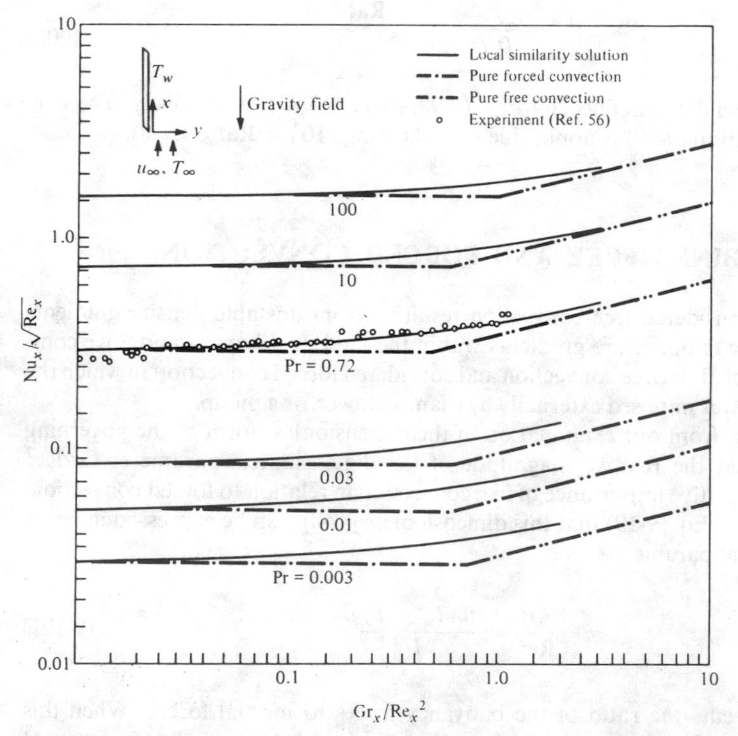

Figure 9-9 Local Nusselt number for combined free and forced convection from an isothermal vertical plate. (*From Lloyd and Sparrow* [58].)

Table 9-8 Threshold values of Gr_x/Re_x^2 for 5 percent deviation of the Nusselt number resulting from neglecting free convection in forced flow along a vertical plate for $T_w > T_\infty$ when u_∞ is upward

Pr	100	10	0.72	0.03–0.003
Gr_x/Re_x^2	0.24	0.13	0.08	0.056–0.05

Based on data from Lloyd and Sparrow [58].

Figure 9-9 illustrates the role of the dimensionless group Gr_x/Re_x^2 on the effects of free convection on forced convection on a vertical, heated (that is, $T_w > T_\infty$), isothermal flat plate subjected to upward forced flow with a free-stream velocity u_∞. The results also apply for a cold plate (that is, $T_w < T_\infty$) when the forced flow velocity u_∞ is downward. In such cases, the effect of free convection is to enhance the total heat transfer.

In Fig. 9-9, the solid lines represent the theoretical predictions of the local Nusselt number when heat transfer by combined free and forced convection is considered. The horizontal and the slanted chain dotted lines are for the cases of pure forced and free convection, respectively. For larger values of Gr_x/Re_x^2, the solid lines asymptotically join the lines for pure free convection. For smaller values of Gr_x/Re_x^2, the solid lines join the lines for pure forced convection. Included on this figure are the experimental data for Pr = 0.72. The agreement between the experiment and the analysis is very good. The theoretical results indicate that the effects of free convection on forced convection become pronounced at low Prandtl numbers. The results shown in Fig. 9-9 can be utilized to establish a threshold for the effects of free convection on force convection. Table 9-8 gives the threshold values of Gr_x/Re_x^2 for 5 percent deviation of the local Nusselt number caused by neglecting free-convection effects in forced flow over a hot, isothermal vertical plate. For example, if the forced convection correlation

$$Nu_x = 0.332\, Re_x^{1/2}\, Pr^{1/3} \tag{9-103}$$

is used for forced convection of air (that is, Pr = 0.72), it underestimates the local Nusselt number by 5 percent at the threshold value $Gr_x/Re_x^2 = 0.08$.

In the case of forced convection over a *horizontal*, isothermal flat plate, Mori's [60] results show that for air the effect of free convection on the local heat transfer coefficient on either side of the plate is less than about 10 percent if

$$\frac{Gr_x}{Re_x^{2.5}} \leq 0.08 \tag{9-104}$$

Flow across a Horizontal Cylinder and Sphere

A number of investigators have reported on combined free and forced convection in external flow over a sphere [68] and a horizontal cylinder [69–71]. A review of the pertinent literature is given in Ref. 70. It is shown that the results depend strongly on the orientation of the flow velocity relative to the gravitational field.

Flow inside Tubes

Combined free and forced convection for laminar flow inside circular tubes has been the subject of numerous investigations [72–82]. The results depend on a number of parameters, including the Reynolds and Prandtl numbers, Rayleigh numbers, and the orientation of the forced flow relative to the gravitational field.

In general, for large Reynolds numbers and small Rayleigh numbers, the effects of free convection on forced convection are negligible.

Metais and Eckert [72] discuss various regimes that characterize the relative importance of free and forced convection for flow inside a horizontal tube. The free and forced convections are of comparable magnitude when

$$\frac{\mathrm{Gr}_D}{\mathrm{Re}_D^2} \cong 1 \tag{9-105}$$

where Gr_D and Re_D are the Grashof and Reynolds numbers, respectively, based on the tube diameter. For combined free and forced convection in the laminar flow regime inside a circular tube, Brown and Gauvin [73] recommend the following correlation for the Nusselt number:

$$\boxed{\mathrm{Nu}_D = 1.75[\mathrm{Gz} + 0.012(\mathrm{Gz}\,\mathrm{Gr}_D^{1/3})^{4/3}]^{1/3}\left(\frac{\mu_b}{\mu_w}\right)^{0.14}} \tag{9-106}$$

where Gz is the Graetz number, defined as

$$\mathrm{Gz} = \mathrm{Re}_D\,\mathrm{Pr}\left(\frac{D}{L}\right) \tag{9-107}$$

and Gr_D and Re_D are the Grashof and Reynolds numbers, respectively, based on the tube's inside diameter. The temperature difference in the Grashof number is taken as $\Delta T = T_w - T_b$, where T_w and T_b are the tube wall and fluid bulk temperatures, respectively.

9-12 SUMMARY OF CORRELATIONS FOR HEAT TRANSFER IN FREE CONVECTION

In Table 9-9 we summarize correlations for heat transfer in free convection.

Table 9-9 Summary of free convection correlations

Equation number	Geometry	Correlation†	Remarks
(9-69)	Vertical isothermal plate	$\mathrm{Nu}_m = c(\mathrm{Gr}_L\,\mathrm{Pr})^n$	c and n from Table 9-2
(9-70)	Vertical isothermal plate (L)	$\mathrm{Nu}_m = 0.68 + \dfrac{0.67\,\mathrm{Ra}_L^{1/4}}{[1 + (0.492/\mathrm{Pr})^{9/16}]^{4/9}}$	$10^{-1} < \mathrm{Ra}_L < 10^9$
(9-71)	Vertical isothermal plate	$\mathrm{Nu}_m^{1/2} = 0.825 + \dfrac{0.387\,\mathrm{Ra}_L^{1/6}}{[1 + (0.492/\mathrm{Pr})^{9/16}]^{8/27}}$	$10^{-1} < \mathrm{Ra}_L < 10^{12}$
(9-72)	Vertical plate, uniform wall heat flux	$\mathrm{Nu}_x = 0.60(\mathrm{Gr}_x^*\,\mathrm{Pr})^{1/5}$	$10^5 < \mathrm{Gr}_x^*\,\mathrm{Pr} < 10^{11}$
(9-73)		$\mathrm{Nu}_x = 0.568(\mathrm{Gr}_x^*\,\mathrm{Pr})^{0.22}$	$2 \times 10^{13} < \mathrm{Gr}_x^*\,\mathrm{Pr} < 10^{16}$
		where $\mathrm{Gr}_x^* = \dfrac{g\beta q_w x^4}{k\nu^2}$	
(9-76)	Hot (upper surface) or Hot (lower surface); Horizontal isothermal plate	$\mathrm{Nu}_m = c(\mathrm{Gr}_L\,\mathrm{Pr})^n$ where L is characteristic dimension	c and n from Table 9-3
	Horizontal plate, uniform wall heat flux	Heated surface facing upward:	
(9-78)		$\mathrm{Nu}_m = 0.13(\mathrm{Gr}_L\,\mathrm{Pr})^{1/3}$	$\mathrm{Gr}_L\,\mathrm{Pr} < 2 \times 10^8$
(9-79)	Hot (upper surface) or Hot (lower surface)	$\mathrm{Nu}_m = 0.16(\mathrm{Gr}_L\,\mathrm{Pr})^{1/3}$	$5 \times 10^8 < \mathrm{Gr}_L\,\mathrm{Pr} < 10^{11}$
		Heated surface facing downward:	
(9-80)		$\mathrm{Nu}_m = 0.58(\mathrm{Gr}_L\,\mathrm{Pr})^{1/5}$	$10^6 < \mathrm{Gr}_L\,\mathrm{Pr} < 10^{11}$

Table 9-9 (*Continued*)

Equation number	Geometry	Correlation†	Remarks
(9-83)	Inclined plate, uniform heat flux, heated surface facing downward 	$\mathrm{Nu}_m = 0.56(\mathrm{Gr}_L \,\mathrm{Pr} \cos\theta)^{1/4}$ $\theta < 88°$	$10^5 < \mathrm{Gr}_L \,\mathrm{Pr} < 10^{11}$
(9-84)	Inclined plate, uniform heat flux, heated surface facing upward 	$\mathrm{Nu}_m = 0.145[(\mathrm{Gr}_L \,\mathrm{Pr})^{1/3} - (\mathrm{Gr}_c \,\mathrm{Pr})^{1/3}] + 0.56(\mathrm{Gr}_c \,\mathrm{Pr} \cos\theta)^{1/4}$ for $-15° < \theta < -75°$ Use Eq. (9-83) for $\mathrm{Gr}_L < \mathrm{Gr}_c$	$\mathrm{Gr}_L \,\mathrm{Pr} < 10^{11}$ $\mathrm{Gr}_L > \mathrm{Gr}_c$ and Gr_c from Table 9-4
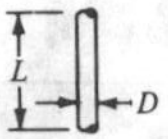	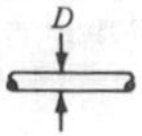 Vertical isothermal cylinder	Can be treated as an isothermal vertical plate. For a slender cylinder, use the chart in Fig. 9-5.	
(9-85)	Horizontal isothermal cylinder	$\mathrm{Nu}_m^{1/2} = 0.60 + \dfrac{0.387 \,\mathrm{Ra}_D^{1/6}}{[1 + (0.559/\mathrm{Pr})^{9/16}]^{8/27}}$	$10^{-4} < \mathrm{Ra}_D < 10^{12}$
(9-87)		$\mathrm{Nu}_m = c \,\mathrm{Ra}_D^n$	c and n from Table 9-5
(9-88*a*)	Isothermal sphere	$\mathrm{Nu}_m = 2 + 0.43 \,\mathrm{Ra}_D^{1/4}$	$1 < \mathrm{Ra}_D < 10^5$; $\mathrm{Pr} \cong 1$
(9-88*b*)	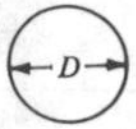	$\mathrm{Nu}_m = 2 + 0.50 \,\mathrm{Ra}_D^{1/4}$	$3 \times 10^5 < \mathrm{Ra}_D < 8 \times 10^8$ $10 \le \mathrm{Nu}_m \le 90$

Equation	Geometry	Correlation
(9-90a)	Vertical or horizontal plane layer; vertical annulus; horizontal cylindrical annulus; spherical annulus	$\mathrm{Nu}_\delta = c(\mathrm{Ra}_\delta)^n\left(\frac{H}{\delta}\right)^m$ constants c, n, and m from Table 9-7
(9-91)	Vertical isothermal parallel plate enclosure	$\mathrm{Nu}_{90^\circ} = \frac{qL}{k(T_h - T_c)} = [\mathrm{Nu}_1, \mathrm{Nu}_2, \mathrm{Nu}_3]_{\max}$ where
(9-92a)		$\mathrm{Nu}_1 = 0.0605\ \mathrm{Ra}^{1/3}$
(9-92b)	$a \equiv \frac{H}{L}$	$\mathrm{Nu}_2 = \left\{1 + \left[\frac{0.104\ \mathrm{Ra}^{0.293}}{1 + (6310/\mathrm{Ra})^{1.36}}\right]^3\right\}^{1/3}$
(9-92c)		$\mathrm{Nu}_3 = 0.242\left(\frac{\mathrm{Ra}}{a}\right)^{0.272}$
(9-93)	Inclined isothermal parallel plate enclosure	$\phi = 60^\circ$: $\mathrm{Nu}_{60^\circ} = [\mathrm{Nu}_1, \mathrm{Nu}_2]_{\max}$ where
(9-94a)		$\mathrm{Nu}_1 = \left[1 + \left(\frac{0.093\ \mathrm{Ra}^{0.314}}{1 + G}\right)^7\right]^{1/7}$
(9-94b)	$a = \frac{H}{L}$	$\mathrm{Nu}_2 = \left(0.104 + \frac{0.175}{a}\right)\mathrm{Ra}^{0.283}$
(9-94c)		$G = \frac{0.5}{[1 + (\mathrm{Ra}/3160)^{20.6}]^{0.1}}$

Table 9-9 *(Continued)*

Equation number	Geometry	Correlation†	Remarks
	Inclined, isothermal parallel-plate enclosure	*$60° < \phi \le 90°$:*	
(9-95)		$\mathrm{Nu}_\phi = \dfrac{(90° - \phi°)\mathrm{Nu}_{60°} + (\phi° - 60°)\mathrm{Nu}_{90°}}{30°}$	
		$0° \le \phi < 60°$:	
(9-96*a*)		$\mathrm{Nu}_\phi = 1 + 1.44\left[1 - \dfrac{1708}{\mathrm{Ra}\cos\phi}\right]^* \times \left[1 - \dfrac{1708(\sin 1.8\phi)^{1.6}}{\mathrm{Ra}\cos\phi}\right] + \left[\left(\dfrac{\mathrm{Ra}\cos\phi}{5830}\right)^{1/3} - 1\right]$	
		where [ ]* implies that if the quantity in the bracket is negative, it should be set equal to zero.	
	Horizontal cylindrical annulus or spherical annulus.	*Horizontal cylindrical annulus:*	
(9-97)		$Q = \dfrac{2\pi k_{\mathrm{eff}} H}{\ln(D_o/D_i)}(T_i - T_o)$ W	$10^2 < \mathrm{Ra}^*_{\mathrm{cyl}} < 10^7$
		k_{eff} is defined by Eq. (9-98).	
		Spherical annulus:	
(9-99)		$Q = k_{\mathrm{eff}}\dfrac{\pi D_i D_o}{\delta}(T_i - T_o)$ W	$10^2 < \mathrm{Ra}^*_{\mathrm{sph}} < 10^4$
		k_{eff} is defined by Eq. (9-100) and δ by Eq. (9-98*d*)	

† Physical properties are evaluated at the film temperature unless otherwise stated.

PROBLEMS

Boundary-layer concept

9-1 A large, vertical plate at a uniform temperature of $T_w = 120°C$ is exposed to atmospheric air at $T_\infty = 24°C$. It is assumed that the transition from laminar to turbulent boundary layer takes place at a critical Grashof number 4×10^8.

(*a*) Determine the location from the lower edge of the plate where the transition from laminar to turbulent flow takes place.

(*b*) Determine the thickness of the velocity boundary layer at that location.

Answer: (a) 0.392 m; (*b*) 1.48 cm

9-2 A large, vertical plate at a uniform temperature $T_w = 125°C$ is suspended in a large body of fluid at $T_\infty = 29°C$. The transition from laminar to turbulent boundary-layer flow takes place at a critical Grashof number 4×10^8. Determine the location from the lower edge of the plate where the transition occurs and the thickness of the velocity boundary layer at that location for (*a*) hydrogen, (*b*) air, and (*c*) carbon dioxide at atmospheric pressure.

9-3 Two plates each at a uniform temperature of $T_w = 80°C$ are immersed vertically and parallel to each other into a large tank containing water at $T_\infty = 20°C$. The plates are 10 cm high. Determine the minimum spacing between the plates to prevent the touching of the velocity boundary layers.

Answer: 0.446 cm

9-4 By utilizing the expression for the velocity component $u(x, y)$ given by

$$u(x, y) = u_0 \frac{y}{\delta}\left(1 - \frac{y}{\delta}\right)^2$$

in this chapter, show that the maximum velocity occurs at a distance $y = \frac{1}{3}\delta$ from the surface of the plate, where δ is the thickness of the laminar boundary layer at location y.

9-5 The local heat transfer coefficient h_x for laminar free convection from a vertical plate can be expressed as

$$h_x = cx^{-1/4}$$

where x is the distance from the leading edge of the plate and c is a constant which depends on fluid properties. Develop an expression for the average value of the heat transfer coefficient h_m over a distance L from the leading edge of the plate.

9-6 A vertical plate, maintained at a uniform temperature of $T_w = 90°C$, is exposed to queiscent atmospheric air at $T_\infty = 25°C$.

(*a*) Compute the thickness of the velocity boundary layer at locations $L = 15$ and 30 cm from the bottom of the plate.

(*b*) Plot the velocity profiles within the boundary layer at these two locations.

Flat plate

9-7 A vertical plate 0.3 m high and 1 m wide, maintained at a uniform temperature 124°C, is exposed to quiescent atmospheric air at 30°C.

(*a*) Calculate the average heat transfer coefficient for free convection.

(*b*) Find the total heat transfer rate from both surfaces of the plate by free convection into the air.

Answer: (*a*) 5.38 W/(m² · °C); (*b*) 303.4 W

9-8 Calculate the heat transfer rates by free convection from a 0.3-m-high vertical plate maintained at a uniform temperature $T_w = 80°C$ to an ambient at $T_\infty = 24°C$ containing air at 1.0, 2.0, and 3.0 atm.

9-9 Determine the heat transfer rates by free convection from a plate 0.3 m by 0.3 m whose one surface is insulated and whose other surface is maintained at $T_w = 100°C$ and exposed to quiescent atmospheric air at $T_\infty = 30°C$ for the following conditions:

(*a*) The plate is vertical.

(*b*) The plate is horizontal with the heated surface facing up.
(*c*) The plate is horizontal with the heated surface facing down.
Answer: (*a*) 36.2, (*b*) 39.8, (*c*) 16.5 W

9-10 Consider a rectangular plate 0.15 m by 0.3 m maintained at a uniform temperature $T_w = 80°C$ placed vertically in quiescent atmospheric air at $T_\infty = 24°C$. Compare the heat transfer rates from the plate when 0.15 and 0.3 m are the vertical heights.

9-11 A circular disk heater of diameter $D = 0.2$ m is exposed to atmospheric air at 25°C. If its surface temperature should not exceed 130°C, estimate the amount of heat it can deliver when it is horizontal and (*a*) faces downward, (*b*) faces upward, and (*c*) vertical.

9-12 The south-facing wall of a building is 7 m high and 10 m wide. The wall absorbs the incident solar energy at a rate of 600 W/m^2. It is assumed that about 200 W/m^2 of the absorbed energy is conducted through the wall into the interior of the wall, and the remaining energy is dissipated by free convection into the surrounding quiescent atmospheric air at 30°C. Calculate the temperature of the outside surface of the wall.

9-13 An electrically heated vertical plate 0.25 m by 0.25 m is insulated on one side and dissipates heat from the other surface at a constant rate of 600 W/m^2 by free convection into quiescent atmospheric air at 30°C. Determine the surface temperature of the plate.
Answer: 120.5°C

9-14 A cube 5 cm by 5 cm by 5 cm is suspended with one of its surfaces in horizontal position in quiescent atmospheric air at 20°C. All surfaces of the cube are maintained at a uniform temperature 100°C. Determine the heat loss by free convection from the cube into the atmosphere.

9-15 A circular hot plate $D = 25$ cm in diameter with both surfaces maintained at a uniform temperature of 100°C is suspended in the horizontal position in atmospheric air at 30°C. Determine the heat transfer rate by free convection from the plate into the atmosphere.

9-16 Consider an electrically heated plate 25 cm by 25 cm in which one surface is thermally insulated and the other is dissipating heat by free convection into atmospheric air at 30°C. The heat flux over the surface of the plate is uniform and results in a mean surface temperature of 50°C. The plate is inclined, making an angle of 50° from the vertical. Determine the heat loss from the plate for (*a*) the heated surface facing up and (*b*) the heated surface facing down.

9-17 A thin electric strip heater of width $H = 20$ cm is placed with its width oriented vertically. It dissipates heat by free convection from both of its surfaces into atmospheric air at $T_\infty = 20°C$. If the surface of the heater should not exceed 225°C, determine the length of the strip in order to dissipate 1000 W of energy into the room.
Answer: $L = 1.73$ m

9-18 A hot iron block at 425°C of sides 10 cm by 15 cm by 20 cm is placed on an asbestos sheet with its 10-cm side oriented vertically. There is negligible heat loss from its surface which is in contact with the asbestos sheet. Calculate the rate of heat loss from its five boundary surfaces by free convection into the surrounding quiescent air at atmospheric pressure and 25°C.

9-19 A plate 0.75 m by 0.75 m is thermally insulated on one side and subjected to a solar radiation flux $q_s = 800$ W/m^2 on the other surface, which has an absorptivity $\alpha_s = 0.9$ for the solar radiation. The plate makes an angle $\theta = -60°$ with the vertical, as illustrated in Fig. 9-4*a*, so that the hot surface is facing upward. If the surface is exposed to quiescent atmospheric air at $T_\infty = 300$ K and the heat transfer is by pure free convection only, calculate the equilibrium temperature of the plate.
Answer: 126°C

Cylinder

9-20 A 5-cm-OD, 1.5-m-long vertical tube at a uniform temperature $T_w = 100°C$ is exposed to quiescent atmospheric air at $T_\infty = 20°C$. Calculate the rate of heat loss from the tube to the surrounding air.
Answer: 101.8 W

9-21 Compare the heat transfer rates by free convection from a 5-cm-OD, 1.5-m-long vertical cylindrical tube maintained at a uniform temperature $T_w = 80°C$ into a quiescent surrounding at $T_\infty = 24°C$ containing (*a*) atmospheric air, (*b*) air at 2 atm, (*c*) carbon dioxide at 1 atm, and (*d*) water.

9-22 An electric heater of outside diameter $D = 2.5$ cm and length $L = 1$ m is immersed horizontally inside a large tank containing engine oil at 20°C. If the surface temperature of the heater is 140°C, determine the rate of heat transfer to the oil.

Answer: 1.614 kW

9-23 A horizontal electric heater of outside diameter $D = 2.5$ cm and length $L = 2$ m dissipates heat by free convection into atmospheric air at $T_\infty = 20°C$. If the surface temperature of the heater is 230°C, calculate the rate of heat transfer from the heater to the air.

9-24 A cylindrical electric heater of outside diameter $D = 2.5$ cm and length $L = 2$ m is immersed horizontally into a pool of mercury at 100°C. If the surface of the heaters is maintained at an average temperature of 300°C, calculate the rate of heat transfer to the mercury.

9-25 A hot gas at 220°C flows through a horizontal pipe of outside diameter $D = 1.5$ cm. The pipe has an uninsulated portion of length $L = 3$ m which is exposed to atmospheric air at temperature $T_\infty =$ 30°C. Assuming the outside surface of the pipe is also at 220°C, determine the rate of heat loss into the atmosphere.

Answer: 296.5 W

9-26 An uninsulated, horizontal duct of diameter $D = 20$ cm carrying cold air at 10°C is exposed to quiescent atmospheric air at 35°C. Determine heat gain by free convection per 1-m length of the duct.

9-27 A 5-cm-diameter horizontal pipe with outer surface at 225°C is exposed to atmospheric air at 25°C. Calculate the heat transfer rate per meter length of the pipe by free convection.

9-28 A 2.5-cm-diameter, 1.5-m-long vertical cylinder maintained at a uniform temperature of 140°C is exposed to atmospheric air at 15°C. Determine the free-convection heat transfer coefficient and the heat transfer rate.

Answer: 6.69 W/(m^2 · °C), 98.5 W

9-29 A tube with $D = 4$ cm OD and $L = 36$ cm, maintained at a uniform temperature $T_w = 390$ K, is placed vertically in quiescent air at atmospheric pressure and $T_\infty = 310$ K. Calculate the average free-convection heat transfer coefficient and the rate of heat loss from the tube to the air.

9-30 Compare the average heat transfer coefficient h_m from an isothermal, horizontal cylinder of diameter $D = 5$ cm at $T_w = 400$ K to quiescent air at atmospheric pressure and $T_\infty = 300$ K obtained from Eqs. (9-85) and (9-87).

Sphere

9-31 A 5-cm-diameter sphere whose surface is maintained at a uniform temperature of 120°C is submerged into quiescent water at 30°C. Determine the heat transfer rate by free convection from the sphere to the water.

Answer: 642.2 W

9-32 Compare the heat transfer rate by free convection from a 20-cm-diameter sphere whose surface is maintained at 140°C with the surrounding fluid at 20°C if the fluid is (*a*) air at 1 atm, (*b*) air at 2 atm, (*c*) carbon dioxide at 1 atm, and (*d*) engine oil.

9-33 A sphere of diameter $D = 0.5$ cm, maintained at a uniform temperature of 60°C, is immersed in water at 20°C. Calculate the rate of heat loss by free convection.

9-34 Compare the heat loss by free convection from a spherical body of diameter $D = 0.2$ m, maintained at a uniform temperature 30°C, to the ambient air at $-10°C$ at $\frac{1}{10}$, $\frac{1}{2}$, 1 and 3 atm.

9-35 A 1.5-cm-diameter, electrically heated sphere is placed in a quiescent body of air at 20°C. Calculate the amount of heat to be supplied by the electric heater in order to keep the surface temperature of the sphere at 100°C.

Answer: 0.74 W

Simplified equations for air

9-36 A vertical plate 0.3 m high and 0.5 m wide at a uniform temperature of 115°C is exposed to atmospheric air at 30°C. Compare the heat transfer coefficient for free convection with that obtainable by the simplified expression for free convection at atmospheric pressure [Eq. (9-70); Eq. (9.69) with Table 9-2; and Table 9-6].

Answer: 5.28, 5.99, 5.83 W/(m^2 · °C), respectively.

9-37 A 0.3-m-high vertical plate maintained at a uniform temperature of 70°C is exposed to cold air at 10°C. Calculate the free-convection heat transfer coefficient, and compare it with that obtainable from the simplified expressions for free convection to air [Eq. (9-70); Eq. (9-69) with Table 9-2; and Table 9-6] for air at (*a*) 0.25, (*b*) 1, and (*c*) 4 atm.

9-38 A square plate 0.2 m by 0.2 m and at a uniform temperature of 140°C is exposed to atmospheric air at 15°C. Calculate the free-convection heat transfer coefficient, and compare the results with those obtainable from the simplified expressions for air at atmospheric pressure for (*a*) the plate vertical, (*b*) the plate horizontal with the heated surface facing up, and (*c*) the plate horizontal with the heated surface facing down.

9-39 A 1-m-high vertical plate at a uniform temperature of 230°C is exposed to atmospheric air at 25°C. Determine the free-convection heat transfer coefficient, and compare it with that obtainable from the simplified expressions for free convection to air at atmospheric pressure [Eqs. (9-71) and (9-69) and Table 9-6].

Answer: 6.87, 5.82, 7.72 W/(m^2 · °C), respectively

9-40 A block 10 cm by 10 cm by 10 cm is suspended with one of its surfaces in horizontal position in atmospheric air at 10°C. All surfaces of the block are maintained at 150°C. Determine the free-convection heat transfer coefficient for all the surfaces of the block, and compare these results with those obtainable from the simplified expressions for air at atmospheric pressure [use Eq. (9-69) with Table 9-3; and Table 9-6].

9-41 A horizontal cylinder 5 cm in diameter and 1 m long, maintained at a uniform temperature 140°C, is exposed to atmospheric air at 10°C. Calculate the free-convection heat transfer coefficient, and compare it with that obtainable from the simplified expression for air at atmospheric pressure [Eqs. (9-85) and (9-87) and Table 9-6].

Answer: 7.24, 7.67, 9.43 W/(m^2 · °C), respectively

9-42 A horizontal electric heater of outside diameter 2.5 cm and length 0.5 m is exposed to atmospheric air at 15°C. If the surface of the heater is at 130°C, determine the heat transfer coefficient for free convection, and compare it with that obtainable from the simplified expression for air at atmospheric pressure [Eqs. (9-85) and (9-87) and Table 9-6].

9-43 Calculate the heat transfer coefficient for free convection from a 2.5-cm-diameter horizontal cylinder maintained at a uniform temperature of 325°C into air at 30°C at (*a*) 0.25, (*b*) 1, and (*c*) 3 atm. Compare these results with those obtainable from the simplified expression for free convection in air [Eqs. (9-85) and (9-87) and Table 9-6].

9-44 A thin, square vertical plate with sides H by H m long, thickness δ m, density ρ kg/m^3, and specific heat c_p J/(kg · °C) is initially at a uniform temperature T_i. At $t = 0$, it is exposed to quiescent atmospheric air at T_∞. Heat transfer from the plate to the air takes place by laminar free convection. Using the simplified correlations for free convection in air and assuming that the lumped-system analysis is applicable, develop an expression for the temperature $T(t)$ of the plate as a function of time.

Enclosure between parallel plates

9-45 Atmospheric air is contained between two large, horizontal parallel plates separated by 5 cm. The lower plate is maintained at 100°C and the upper plate at 30°C. Determine the heat transfer rate between the plates by free convection per square meter of the plate surface. Use Eq. (9-96*a*).

Answer: 233 W/m^2

9-46 Two vertical parallel plates, each 0.5 m by 0.5 m, are separated by 3 cm, and the space is filled with atmospheric air. One of the plates is maintained at a uniform temperature of 250°C and the other at 100°C. Calculate the heat transfer rate by free convection between the plates.

9-47 Two parallel horizontal plates are separated by 2 cm. The lower plate is at a uniform temperature of 220°C and the upper plate at 35°C. Determine the heat transfer rate between the plates per square meter of the plate surface when the fluid between the plates is atmospheric air.

Answer: 1006 W/m^2

9-48 The 2-cm space between two parallel horizontal plates is filled with water. The lower plate is at a uniform temperature of 130°C and the upper plate at 30°C. Determine the heat transfer rate between the plate per meter square of the plate surface.

9-49 Two parallel vertical plates each 2 m high are separated by a 6-cm airspace. One plate is maintained at a uniform temperature of 130°C, and the other at 25°C. Determine the free-convection heat transfer coefficient between the two plates.

9-50 A double-glass window consists of two vertical parallel glasses each 1.6 m by 1.6 m and separated by 1.5-cm airspace at atmospheric pressure. Calculate the free-convection heat transfer coefficient for the airspace for a temperature difference of 35°C. Assume a mean temperature for the air of 27°C. Use Eq. (9-91).

Answer: 2.36 $W/(m^2 \cdot °C)$

9-51 Consider an enclosure formed by two parallel plates, 1.2 m by 1.2 m, and separated by a 3-cm airspace at atmospheric pressure. The lateral boundary of the enclosure is insulated. One of the plates is hot and maintained at a uniform temperature of 120°C; the other plate is at 35°C. Calculate the heat transfer rate across the gap when (*a*) the enclosure is vertical and (*b*) the enclosure is horizontal with the hot plate at the bottom.

9-52 In a horizontal flat-plate solar collector, the absorber plate and the glass cover are separated by an air gap at atmospheric pressure. Estimate the heat transfer coefficient for free convection across the air gap for spacings 1.5 and 3.0 cm, assuming that the absorber plate is at 60°C and the glass cover at 30°C.

9-53 A vertical, double window glass 50 cm by 50 cm is constructed of two glass plates separated by a 2-cm air gap at atmospheric pressure. If the outer and inner glasses are at −15 and 20°C, respectively, calculate the heat transfer rate across the gap.

Enclosures between coaxial horizontal cylinders or concentric spheres

9-54 The annular space between two horizontal, thin-walled coaxial cylinders contains air at atmospheric pressure. The inner cylinder has a diameter $D_1 = 8$ cm and is maintained at a uniform temperature $T_1 = 100°C$, while the outer cylinder has a diameter $D_2 = 12$ cm and is maintained at a uniform temperature $T_2 = 50°C$. Calculate the heat transfer rate by free convection across the airspace per meter length of the cylinders.

Answer: 42.9 W/m

9-55 The annular space between two horizontal, thin-walled coaxial cylinders contains air. The diameters of the inner and outer cylinders are $D_1 = 8$ and $D_2 = 12$ cm, respectively. The temperature difference between the cylinders is 50°C, and the mean temperature of air in the annular space is 27°C. Calculate the heat transfer rate by free convection across the airspace per meter length of the tube for (*a*) air at 1 atm, (*b*) air at 3 atm, and (*c*) air at 5 atm.

9-56 The annular space between two horizontal, thin-walled coaxial cylinders is filled with water. The inner and outer cylinders have diameters $D_1 = 10$ and $D_2 = 13$ cm, respectively. Calculate heat transfer by free convection across the water space per meter length of the cylinder for a temperature difference of 60°C and a mean temperature of 40°C for the water.

9-57 The annular space between two thin-walled, horizontal coaxial cylinders with diameters $D_1 = 7$ and $D_2 = 10$ cm contains water. Calculate the heat transfer rate across the annulus by free convection when the inside and outside cylinders are maintained at 70 and 50°C, respectively.

9-58 The space between two concentric, thin-walled spheres of diameters $D_1 = 10$ and $D_2 = 14$ cm contains air at atmospheric pressure. Calculate the heat transfer rate across the spheres by free convection for a temperature difference of 50°C and a mean air temperature of 27°C.

9-59 Repeat Prob. 9-58 for water at a mean temperature of 40°C contained in the space between the spheres.

9-60 The space between two concentric, thin-walled spheres contains air. The inner and outer spheres have diameters $D_1 = 8$ and $D_2 = 10$ cm, respectively, and the temperature difference between the surfaces is 50°C. Calculate the heat transfer rate across the space by free convection for air at a mean temperature of 27°C and at (*a*) 1, (*b*) 3, and (*c*) 5 atm.

9-61 A spherical storage tank of diameter $D_1 = 1.5$ m contains a cold liquid at $T_1 = 15$°C. To reduce the heat losses, this storage tank is enclosed inside another spherical shell, and the gap spacing is 3 cm. The temperature of the outer sphere is 25°C. Determine the rate of heat loss by free convection across the gap filled with air at (*a*) $\frac{1}{20}$, (*b*) $\frac{1}{10}$, and (*c*) 1 atm.

Combined free and forced convection

9-62 Atmospheric air at 27°C flows upward with a velocity of 2 m/s over a 0.5-m-long vertical plate. Determine the plate temperature for which the effect of free convection on heat transfer will be less than 5 percent.

Answer: 46.6°C

9-63 Air at atmospheric pressure at 77°C flows upwards along a 0.5-m-long vertical plate which is maintained at a uniform temperature of 110°C. Determine the minimum flow velocity below which the effect of free convection on the heat transfer is more than 5 percent.

9-64 Ethylene glycol at 40°C flows along a heated vertical plate $L = 0.5$ m long and is maintained at a uniform temperature of 80°C. Determine the minimum flow velocity below which the effect of free convection on heat transfer becomes more than 5 percent. (For ethylene glycol take Pr = 50.)

REFERENCES

1. Pohlhausen, E.: "Der Wärmeaustrausch zwishen festen Körpen und Flüssigkeiten mit kleiner Reiburg and kleiner Warmeleitung," *Z. Angew. Math. Mech.*, **1**:115 (1921).
2. Schmidt, E., and W. Beckmann: Das Temperatur und Geschwindigkeitsfeld von einer licher Wandtemperatur," *Forsch. Geb. Ingenieures.*, **1**:391 (1930).
3. Schuh, H.: Einige Probleme bei Freirer Strömung Zaher Flüssigkeiten," *Göttinger Monogr. Bd. B., Grenzschichten*, 1946.
4. Eckert, E. R. G., and E. Soehngen: "Interferometric Studies on the Stability and Transition to Turbulence of a Free Convection Boundary Layer," *Proc. Gen. Discuss. Heat Transfer ASME-IME. London, 1951.*
5. Eckert, E. R. G., and T. W. Jackson: "Analysis of Turbulent Free Convection Boundary Layer on a Flat Plate," *NACA Rep.* 1015, 1951.
6. Ostrach, S.: "An Analysis of Laminar Free Convection Flow and Heat Transfer about a Flat Plate Parallel to the Direction of the Generating Body Force," *NACA Rep.* 1111, 1953.
7. Rich, B. R.: "An Investigation of Heat Transfer from an Inclined Flat Plate in Free Convection," *Trans. ASME*, **75**:489–499 (1953).
8. McAdams, W. H.: *Heat Transmission*, 3d ed., McGraw-Hill, New York, 1954.
9. Bayley, F. J.: "An Analysis of Turbulent Free Convection Heat Transfer," *Proc., Inst. Mech. Eng., London*, **169**:361 (1955).
10. LeFevre, E. J.: "Laminar Free Convection from a Vertical Surface," *Mech. Eng. Res. Lab., Heat*, **113** (Gt. Britain), 1956.
11. Sparrow, E. M., and J. L. Gregg: "Laminar Free Convection from a Vertical Plate," *Trans. ASME*, **78**:435–440 (1956).

12. Sparrow, E. M.: *NACA Tech. Note* 3508, 1955; also, E. M. Sparrow and J. L. Gregg, *Trans. ASME*, **78**:435–440 (1956).
13. Sparrow, E. M., and J. L. Gregg: "Details of Exact Low Prandtl Number Boundary Layer Solutions for Forced and Free Convection," *NACA Tech. Memo*. 2-27-59E, 1959.
14. Schlichting, H.: *Boundary Layer Theory*, McGraw-Hill, New York, 1968.
15. MacGregor, R. K., and A. P. Emery: "Free Convection through Vertical Plane Layers: Moderate and High Prandtl Number Fluids," *J. Heat Transfer*, **91**:391 (1969).
16. Cheesewright, R.: "Turbulent Natural Convection from a Vertical Plane Surface," *J. Heat Transfer*, **90**:1 (Feb. 1968).
17. Warner, C. Y., and V. S. Arpaci: "An Investigation of Turbulent Natural Convection in Air at Low Pressure along a Vertical Heated Flat Plate," *Int. J. Heat Mass Transfer*, **11**:397–406 (1968).
18. Warner, C. Y., and V. S. Arpaci: "An Experimental Investigation of Turbulent Natural Convection in Air at Low Pressure for a Vertical Heated Flat Plate," *Int. J. Heat Mass Transfer*, **11**:397 (1968).
19. Vliet, G. C., and C. K. Liu: "An Experimental Study of Natural Convection Boundary Layers," *J. Heat Transfer*, **91C**:517–531 (1969).
20. Vliet, G. C.: "Natural Convection Local Heat Transfer on Constant-Heat-Flux Inclined Surfaces." *J. Heat Transfer* **91C**:511–516 (1969).
21. Muntasser, M. A., and J. C. Mulligan: "A Local Nonsimilarity Analysis of Free Convection from a Horizontal Cylindrical Surface," *J. Heat Transfer*, **100C**: 165–167 (1978).
22. Parsons, J. R., and J. C. Mulligan: "Transient Free Convection from a Suddenly Heated Horizontal Wire," *J. Heat Transfer*, **100C**:423–428, 1978.
23. Fujii, T. and H. Imura: "Natural Convection Heat Transfer from a Plate with Arbitrary Inclination," *Int. J. Heat Mass Transfer*, **15**:755 (1972).
24. Goldstein, R. J., E. M. Sparrow, and D. C. Jones.: "Natural Convection Mass Transfer Adjacent to Horizontal Plates," *Int. J. Heat Mass Transfer*, **16**:1025 (1973).
25. Churchill, S. W. and H. H. S. Chu: "Correlating Equations for Laminar and Turbulent Free Convection from a Vertical Plate," *Int. J. Heat Mass Transfer*, **18**:1323 (1975).
26. Emery, A. F., A. Yang, and J. R. Wilson: "Free Convection Heat Transfer to Newtonian and Non-Newtonian High Prandtl Number Fluids from Vertical and Inclined Surfaces," ASME Prepr. 76-HT-46, 1976.
27. Lloyd, J. R., and W. R. Moran: "Natural Convection Adjacent to Horizontal Surfaces of Various Platforms," ASME Paper 74-WA/HT-66, 1974.
28. Moran, W. R., and J. R. Lloyd: "Natural Convection Mass Transfer Adjacent to Vertical and Downward-Facing Surfaces," *J. Heat Transfer*, **97C**:472 (1975).
29. Sparrow, E. M., and J. L. Gregg: "Laminar-Free-Convection Heat Transfer from the Outer Surface of a Vertical Cylinder," *Trans. ASME*, **78**:1823 (1956).
30. Minkowycz, W. J., and E. M. Sparrow: "Local Nonsimilar Solutions for Natural Convection on a Vertical Cylinder," *J. Heat Transfer*, **96C**:78 (1974).
31. Cebeci, T: "Laminar-Free-Convective-Heat Transfer from the Outer Surface of a Vertical Slender Circular Cylinder," *Fifth Int. Heat Transfer Conf.*, vol. 3, NC 1.4, pp. 15–19, 1974.
32. Churchill, S. W., and H. H. S. Chu: "Correlating Equations for Laminar and Turbulent Free Convection from a Horizontal Cylinder," *Int. J. Heat Mass Transfer*, **18**:1049–1053 (1975).
33. Morgan, V. T.: "The Overall Convective Heat Transfer from Smooth Circular Cylinders," in in T. F. Irvine and J. P. Hartnett (eds), *Advances in Heat Transfer*, vol. 16 Academic, New York, 1975, pp. 199–264.
34. Farouk, B., and S. I. Güçeri: "Natural Convection from a Horizontal Cylinder (a)—Laminar Regime," *J. Heat Transfer*, **103C**:522–527 (1981). (b)—Turbulent Regime," *J. Heat Transfer*, **104C**:228–235 (1982).
35. Yuge, T.: "Experiments on Heat Transfer from Spheres Including Combined Natural and Free Convection," *J. Heat Transfer*, **82C**:214–220 (1960).
36. Kuehn, T. H., and R. J. Goldstein: "Correlating Equations for Natural Convection Heat Transfer Between Circular Cylinders," *Int. J. Heat Mass Transfer*, **19**:1127–1134 (1976).

37. Amato, W. S., and C. Tien: "Free Convection Heat Transfer from Isothermal Spheres in Water," *Int. J. Heat Mass Transfer*, **15**:327–339 (1972).
38. Dropkin, D., and E. Somerscales: "Heat Transfer by Natural Convection in Liquids Confined by Parallel Plates Which Are Inclined at Various Angles with Respect to the Horizontal," *J. Heat Transfer*, **87C**:77–84 (1965).
39. Emery, A., and N. C. Chu: "Heat Transfer across Vertical Layers," *J. Heat Transfer*, **87C**:100–116 (1965).
40. Jacob, M.: "Free Convection through Enclosed Gas Layers," *Trans. ASME*, **68**:189 (1946).
41. Jacob, M.: *Heat Transfer*, vol. 1, Wiley, New York, 1949.
42. Hollands, K. G. T., T. E. Unny, G. D. Raithby, and L. Konicek: "Free Convective Heat Transfer across Inclined Air Layers," *J. Heat Transfer*, **98C**:189–193 (1976).
43. Graff, J. G. A., and E. F. M. Van der Held: "The Relation between the Heat Transfer and Convection Phenomena in Enclosed Air Layers," *Appl. Sci. Res.*, **3**:393 (1952).
44. O'Toole, J., and P. L. Silveston: "Correlation of Convective Heat Transfer in Confined Horizontal Layers," *Chem. Eng. Progr. Symp.*, **57**:81 (196).
45. Schmidt, E.: "Free Convection in Horizontal Fluid Spaces Heated from Below," *Proc. Int. Heat Transfer Conf.*, ASME, Boulder, 1961.
46. Yücel, A., and Y. Bayazitoğlu: "Onset of Convection in Fluid Layers with Nonuniform Volumetric Energy Sources," *J. Heat Transfer*, **101C**:666–671 (1979).
47. Özişik, M. N., and M. A. Hassab: "Effects of Convective Boundary Conditions on the Stability of Conductions Regime in an Inclined Slender Slot," *Int. J. Numer. Heat Transfer*, **2**:251–260 (1979).
48. Globe, S., and D. Dropkin: "Natural-Convection Heat Transfer in Liquids Confined by Two Horizontal Plates and Heated from Below," *J. Heat Transfer*, **81**:24–28 (1959).
49. Goldstein, R. J., and T. Y. Chu: "Thermal Convection in a Horizontal Layer of Air," *Progr. Heat Mass Transfer*, **2**:55 (1969).
50. Ruth, D. W., K. G. T. Holland, and G. D. Raithby: "On Free Convection Experiments in Inclined Air Layers Heated from Below," *J. Fluid Mechanics*, **96**:459–479 (1980).
51. Shaaban, A. H., and M. N. Özişik: "The Effect of Nonlinear Density Stratification on the Stability of a Vertical Water Layer in the Conduction Regime," *J. Heat Transfer*, **105C**:130–137 (1983).
52. Hassab, M. A., and M. N. Özişik: "Effects of Thermal Wall Resistance on the Stability of Conduction Regime in an Inclined Narrow Slot," *Int. J. Heat Mass Transfer*, **24**:739–747 (1981).
53. El Sherbiny, S. M., G. D. Raithby, and K. G. T. Hollands: "Heat Transfer by Natural Convection across Vertical and Inclined Air Layers," *J. Heat Transfer*, **104C**:96–102 (1982).
54. Raithby, G. D., and K. G. T. Hollands: "A General Method of Obtaining Approximate Solutions to Laminar and Turbulent Free Convection Problems," in T. F. Irvine and J. P. Hartnett (eds.), *Advances in Heat Transfer*, vol. 11, Academic, New York, 1975, pp. 265–315.
55. Buchberg, H., I. Catton, and D. K. Edwards: "Natural Convection in Enclosed Spaces: A Review of Application to Solar Energy Collection," ASME Paper 74-WA/HT/12, 1974.
56. Liu, C. Y., W. K. Mueller, and F. Landis: "Natural Convection Heat Transfer in Long Horizontal Annuli," *Int. Dev. Heat Transfer*, **5**:976 (1976).
57. Scanlan, J. A, E. H. Bishop, and R. E. Powe: "Natural Convection Heat Transfer between Concentric Spheres," *Int. J. Heat Mass Transfer*, **13**:1857 (1970).
58. Lloyd, J. R., and E. M. Sparrow: "Combined Forced and Free Convection Flow on Vertical Surfaces, *Int. J. Heat Mass Transfer*, **13**:434–438 (1970).
59. Wilks, G. "Combined Forced and Free Convection Flow on Vertical Surfaces," *Int. J. Heat Mass Transfer*, **16**: 1958–1964 (1973).
60. Mori, Y.: "Buoyancy Effects in Forced Laminar Convection Flow over a Horizontal Flat Plate," *J. Heat Transfer*, **83C**:479–482 (1961).
61. Sparrow, E. M., and W. J. Minkowycz: "Buoyancy Effects on Horizontal Boundary-Layer Flow and Heat Transfer," *Int. J. Heat Mass Transfer*, **5**:505–155 (1962).
62. Hauptmann, E. G.: "Laminar Boundary Layer Flows with Small Buoyancy Effects," *Int. J. Heat Mass Transfer*. **8**:289–295 (1965).

63. Redekopp, L. G., and A. F. Charwatt: "Role of Buoyancy and the Boussinesq Approximation in Horizontal Boundary Layers," *J. Hydronautics*, **6**:34–39 (1972).
64. Leal, L. G.: "Combined Forced and Free Convection Heat Transfer from a Horizontal Flat Plate," *J. Appl. Math. Phys.* (*ZAMP*), **24**:20–42 (1973).
65. Hieber, C. A.: "Mixed Convection above a Heated Horizontal Surface," *Int. J. Heat Mass Transfer*, **16**:769–785 (1973).
66. Robertson, G. E., J. H. Seinfeld, and G. E. Leal: "Combined Forced and Free Convection Flow Passed a Horizontal Flat Plate," *AIChE J.*, **19**:998–1008 (1973).
67. Chen, T. S., E. M. Sparrow, and A. Mucoglu: "Mixed Convection in Boundary Layer Flow on a Horizontal Plate," *J. Heat Transfer*, **99C**:66–71 (1977).
68. Yuge, T.: "Experiments on Heat Transfer from Spheres Including Combined Natural and Free Convection," *J. Heat Transfer*, **82C**:214–220 (1960).
69. Oosthuizen, P. H., and S. Madan: "Combined Convective Heat Transfer from Horizontal Cylinders in Air," *J. Heat Transfer*, **92C**:194–196 (1970).
70. Fand, R. M., and K. K. Keswani: "Combined Natural and Forced Convection Heat Transfer from Horizontal Cylinders to Water," *Int. J. Heat Mass Transfer*, **16**:1175 (1973).
71. Mucoglu, A., and T. S. Chen: "Mixed Convection across a Horizontal Cylinder with Uniform Surface Heat Flux," *J. Heat Transfer*, **99C**:679–682 (1977).
72. Metais, B., and E. R. G. Eckert: "Forced, Mixed and Free Convection Regimes," *J. Heat Transfer*, **86C**:295–296 (1964).
73. Brown, C. K., and W. H. Gauvin: "Combined Free and Forced Convection, Pts. I and II, *Can. J. Chem. Eng.*, **43**:306, 313 (1965).
74. Del Casal, E., and W. N. Gill: "A Note on Natural Convection Effects in Fully Developed Horizontal Tube Flow," *AIChE J.*, **8**:570–574 (1962).
75. Faris, G. N., and R. Viskanta: "An Analysis of Laminar Combined Forced and Free Convection Heat Transfer in a Horizontal Tube," *Int. J. Heat Mass Transfer*, **12**:1295–1309 (1969).
76. Newell, P. H., and A. E. Bergles: "Analysis of Combined Free and Forced Convection for Fully Developed Laminar Flow in Horizontal Tubes," *J. Heat Transfer*, **90C**:83–93 (1970).
77. Bergles, A. E., and R. R. Simonds: "Combined Forced and Free Convection for Laminar Flow in Horizontal Tubes with Uniform Heat Flux," *Int. J. Heat Mass Transfer*, **14**:1989–2000 (1971).
78. Sabbagh, J. A., A. Aziz, A. S. El-Ariny, and G. Hamad: "Combined Free and Forced Convection in Circular Tubes," *J. Heat Transfer*, **98C**:322–324 (1976).
79. Patankar, S. V., S. Ramadhyani, and E. M. Sparrow: "Effect of Circumferentially Nonuniform Heating on Laminar Combined Convection in a Horizontal Tube," *J. Heat Transfer*, **100C**:63–70 (1978).
80. Cheng, K. C., and S. W. Hong: "Combined Free and Forced Laminar Convection in Inclined Tubes," *Appl. Sci. Res.*, **27**:19–38 (1972).
81. Hieber, C. A., and S. K. Sreenivasan: "Mixed Convection in an Isothermally Heated Pipe," *Int. J. Heat Mass Transfer*, **70**:1337–1348 (1974).
82. Depew, C. A., J. L. Franklin, and C. H. Ito: "Combined Free and Forced Convection in Horizontal, Unformly Heated Tubes," ASME Paper 75-HT-19, August 1975.

CHAPTER

TEN

BOILING AND CONDENSATION

Condensers and boilers constitute an important and widely used type of heat exchanger with unique characteristics of heat transfer mechanism on the condensing and boiling side. If a vapor strikes a surface that is at a temperature below the corresponding saturation temperature, the vapor will immediately condense into the liquid phase. If the condensation takes place continuously over the surface which is kept cooled by some cooling process and the condensed liquid is removed from the surface by the motion resulting from gravity, then the condensing surface is usually covered with a thin layer of liquid, and the situation is known as *filmwise condensation.* Under certain conditions, for example, if traces of oil are present during the condensation of steam on a highly polished surface, the film of condensate is broken into droplets, and the situation is known as *dropwise condensation.* The presence of condensate acts as a barrier to heat transfer from the vapor to the metal surface, and the dropwise condensation offers much less resistance to heat flow on the vapor side than the filmwise condensation. If vapor contains some noncondensable gas, this gas will collect on the condensing side while condensation takes place. The presence of noncondensable gas acts as resistance to heat flow on the condensing side because the vapor must diffuse through the noncondensable gas before it comes into contact with the cool surface of the condensate. Therefore, an understanding of the mechanism of heat transfer and an accurate prediction of the heat transfer coefficient for condensing vapors with and without the presence of noncondensable gas are important in the design of condensers.

When a liquid is in contact with a surface maintained at a temperature above the saturation temperature of the liquid, boiling may occur. The phenomenon of heat transfer in boiling is extremely complicated because of a large number of variables involved and very complex hydrodynamic developments occurring in the process. Therefore, considerable work has been directed toward gaining a better understanding of the boiling mechanism, but most of the heat transfer correlations in boiling still remain of empirical and semiempirical nature.

In this chapter we present a simple analysis for the determination of the heat transfer coefficient during filmwise condensation on a plane, vertical surface; we discuss the effects of noncondensable gas on heat transfer; and we give various correlations of the heat transfer coefficient during condensation of vapor. Different boiling regimes and their heat transfer characteristics are described, and pertinent heat transfer relations are presented.

10-1 FILM CONDENSATION THEORY

When the temperature of a vapor is reduced below its saturation temperature, the vapor condenses. In engineering applications, the vapor is condensed by bringing it into contact with a cold surface. The steam condensers for power plants are typical examples of the application of condensing of steam. If the liquid wets the surface, the condensation occurs in the form of a smooth film, which flows down the surface under the action of the gravity. The presence of a liquid film over the surface constitutes a thermal resistance to heat flow. Therefore, numerous experimental and theoretical investigations have been conducted to determine the heat transfer coefficient for film condensation over surfaces [1–34]. The first fundamental analysis leading to the determination of the heat transfer coefficient during filmwise condensation of pure vapors (i.e., without the presence of noncondensable gas) on a flat plate and a circular tube was given by Nusselt [1] in 1916. Over the years, improvements have been made on Nusselt's theory of film condensation. But with the exception of the condensation of liquid metals, Nusselt's original theory has been successful and still is widely used. Here we present Nusselt's theory of film condensation of pure vapors on a vertical plate, because it serves as a basis to better understand heat transfer during condensation.

Condensation on Vertical Surfaces

Consider the condensation of a vapor on a vertical plate, as illustrated in Fig. 10-1. Here x is the axial coordinate, measured downward along the plate, and y is the coordinate normal to the condensing surface. The condensate thickness is represented by $\delta \equiv \delta(x)$. This condensation problem was first analyzed by Nusselt [1] under the following assumptions:

1. The plate is maintained at a uniform temperature T_w that is less than the saturation temperature T_v of the vapor.
2. The vapor is stationary or has low velocity, and so it exerts no drag on the motion of the condensate.
3. The downward flow of condensate under the action of gravity is laminar.
4. The flow velocity associated with the condensate film is low; as a result, the flow acceleration in the condensate layer is negligible.
5. Fluid properties are constant.
6. Heat transfer across the condensate layer is by pure conduction, hence the liquid temperature distribution is linear.

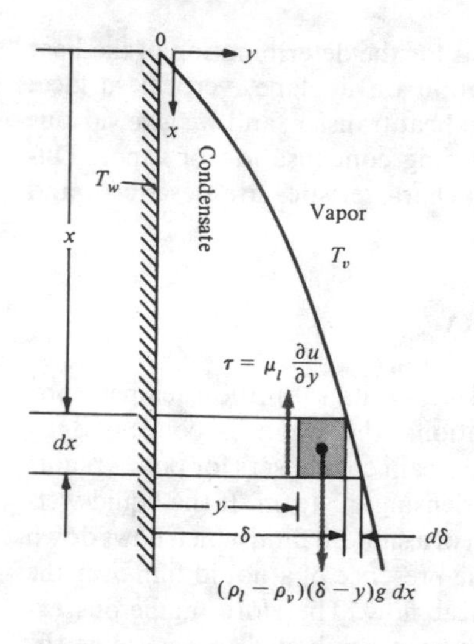

Figure 10-1 Nomenclature for filmwise condensation on a vertical plane surface.

The velocity distribution $u(y)$ at any location x across the condensate layer is determined by writing a force balance on a volume element, shown by the shaded area in Fig. 10-1. Equating the shear force acting upward to the buoyancy force acting downward, we write

$$\mu_l \frac{du}{dy} dx = (\rho_l - \rho_v)(\delta - y)g\,dx$$

or

$$\frac{du}{dy} = \frac{g(\rho_l - \rho_v)}{\mu_l}(\delta - y) \tag{10-1}$$

where $\delta \equiv \delta(x)$ is the thickness of the condensate layer at the position x, μ is viscosity, and the subscripts l and v refer to the liquid and vapor phases, respectively. Here we assume zero shear stress at the liquid-vapor interface, because the vapor is considered almost stationary by assumption 2.

At the wall surface, the liquid velocity is zero:

$$u = 0 \qquad \text{at } y = 0 \tag{10-2}$$

The integration of Eq. (10-1) subject to the boundary condition Eq. (10-2) gives the velocity distribution in the condensate layer:

$$u(y) = \frac{g(\rho_l - \rho_v)}{\mu_l}\left(\delta y - \frac{1}{2}y^2\right) \tag{10-3}$$

The mass flow rate of condensate $m(x)$ through any axial position x per unit width of the plate is given by

$$m(x) = \int_0^\delta \rho_l u \, dy \tag{10-4}$$

Introducing u from Eq. (10-3) into Eq. (10-4) and performing the integration yield

$$m(x) = \frac{g\rho_l(\rho_l - \rho_v)\delta^3}{3\mu_l} \tag{10-5}$$

and differentiation with respect to δ gives

$$dm = \frac{g\rho_l(\rho_l - \rho_v)\delta^2}{\mu_l}\, d\delta \tag{10-6}$$

Here dm represents the rate of condensation over the distance dx per unit width of the plate, since the condensate thickness increases by $d\delta$ over the differential length dx.

The rate of heat released dQ associated with the rate of condensation dm is

$$dQ = h_{fg}\, dm \tag{10-7}$$

where h_{fg} is the latent heat of condensation. The amount of heat released dQ over the area $dx \cdot 1$ must be transferred across the condensate layer of thickness δ by conduction, according to assumption 6. Therefore,

$$dQ = k_l \frac{T_v - T_w}{\delta}\, dx \cdot 1 \tag{10-8}$$

where k_l is the thermal conductivity of liquid and T_v and T_w are the vapor saturation and wall surface temperatures, respectively.

Introducing Eqs. (10-6) and (10-8) into Eq. (10-7), we obtain the following differential equation for the thickness of the condensate layer:

$$\frac{d\delta}{dx} = \frac{\mu_l k_l(T_v - T_w)}{g\rho_l(\rho_l - \rho_v)h_{fg}} \frac{1}{\delta^3} \tag{10-9}$$

The integration of Eq. (10-9) with the condition $\delta = 0$ for $x = 0$ yields the thickness of the condensate layer as a function of the position x along the plate:

$$\boxed{\delta(x) = \left[\frac{4\mu_l k_l(T_v - T_w)x}{g(\rho_l - \rho_v)\rho_l h_{fg}}\right]^{1/4}} \tag{10-10}$$

Since we have established the relation for the thickness of the condensate layer, the local heat transfer coefficient h_x for condensation is determined from the definition

$$h_x(T_v - T_w) = k_l \frac{T_v - T_w}{\delta(x)}$$

or

$$h_x = \frac{k_l}{\delta(x)} \tag{10-11}$$

Introducing $\delta(x)$ from Eq. (10-10) into Eq. (10-11), we obtain

$$h_x = \left[\frac{g\rho_l(\rho_l - \rho_v)h_{fg}k_l^3}{4\mu_l(T_v - T_w)x}\right]^{1/4} \tag{10-12}$$

and the local Nusselt number Nu_x is expressed as

$$\text{Nu}_x \equiv \frac{h_x x}{k_l} = \left[\frac{g\rho_l(\rho_l - \rho_v)h_{fg}k_l^3}{4\mu_l(T_v - T_w)x}\right]^{1/4} \tag{10-13}$$

We note from Eq. (10-12) that the local heat transfer coefficient h_x varies with the distance as $x^{-1/4}$. Then the average heat transfer coefficient h_m over the length $0 \le x \le L$ of the plate is

$$h_m = \frac{1}{L}\int_0^L h_x\,dx = \frac{4}{3}h_x\bigg|_{x=L} \tag{10-14}$$

Or, introducing Eq. (10-12) into Eq. (10-14), we obtain

$$\boxed{h_m = 0.943\left[\frac{g\rho_l(\rho_l - \rho_v)h_{fg}k_l^3}{\mu_l(T_v - T_w)L}\right]^{1/4} \qquad \text{W/(m}^2\cdot{}^\circ\text{C)}} \tag{10-15}$$

The physical properties in Eqs. (10-12) and (10-15), including h_{fg}, should be evaluated at the film temperature

$$T_f = \tfrac{1}{2}(T_w + T_v) \tag{10-16}$$

The additional energy needed to cool the condensate film below saturation temperature is accommodated approximately by evaluating h_{fg} at the film temperature instead of the saturation temperature. The heat transfer coefficient derived above for a vertical plate is also applicable for condensation on the outside or inside surface of a vertical tube, provided that the tube radius is large compared with the thickness of the condensate film.

Condensation on Inclined Surfaces

Nusselt's analysis of filmwise condensation given above for a vertical surface can readily be extended for condensation on an inclined plane surface making an angle φ with the horizontal, as illustrated in Fig. 10-2. The results for the local and the average heat transfer coefficients are given, respectively, as

$$h_x = \left[\frac{g\rho_l(\rho_l - \rho_v)h_{fg}k_l^3}{4\mu_l(T_v - T_w)x}\sin\varphi\right]^{1/4} \tag{10-17}$$

and

$$\boxed{h_m = 0.943\left[\frac{g\rho_l(\rho_l - \rho_v)h_{fg}k_l^3}{\mu_l(T_v - T_w)L}\sin\varphi\right]^{1/4}} \tag{10-18}$$

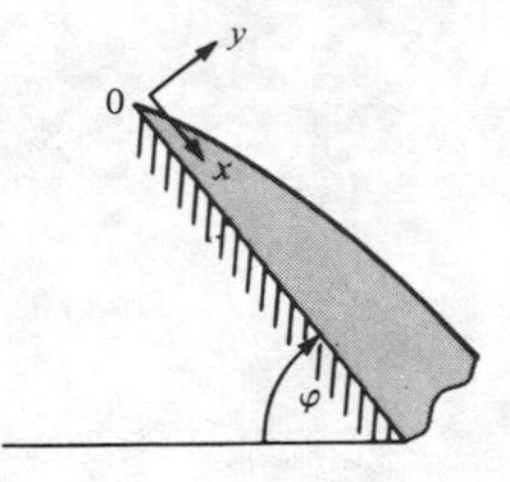

Figure 10-2 Nomenclature for filmwise condensation on an inclined plane surface.

Condensation on a Horizontal Tube

The analysis of heat transfer for condensation on the outside surface of a horizontal tube is more complicated than that for a vertical surface. Nusselt's analysis for laminar filmwise condensation on the surface of a horizontal tube gives the average heat transfer coefficient as

$$\boxed{h_m = 0.725\left[\frac{g\rho_l(\rho_l - \rho_v)h_{fg}k_l^3}{\mu_l(T_v - T_w)D}\right]^{1/4}} \tag{10-19}$$

where D is the outside diameter of the tube.

A comparison of Eqs. (10-15) and (10-19), for condensation on a vertical tube of length L and a horizontal tube of diameter D, yields

$$\frac{h_{m,\,\text{vert}}}{h_{m,\,\text{horz}}} = 1.30\left(\frac{D}{L}\right)^{1/4}$$

This result implies that for a given $T_v - T_w$, the average heat transfer coefficients for a vertical tube of length L and a horizontal tube of diameter D become equal when $L = 2.87D$. For example, when $L = 100D$, theoretically $h_{m,\,\text{horz}}$ would be 2.44 times $h_{m,\,\text{vert}}$. With this consideration, horizontal tube arrangements are generally preferred to vertical tube arrangements in condenser design.

Condensation on Horizontal Tube Banks

Condenser design generally involves horizontal tubes arranged in vertical tiers, as illustrated in Fig. 10-3, in such a way that the condensate from one tube drains onto the tube just below. If it is assumed that the drainage from one tube flows smoothly onto the tube below, then for a vertical tier of N tubes each of diameter D, the average heat transfer coefficient h_m for the N tubes is given by

$$\boxed{h_m\bigg|_{N\ \text{tubes}} = 0.725\left[\frac{g\rho_l(\rho_l - \rho_v)h_{fg}k_l^3}{\mu_l(T_v - T_w)ND}\right]^{1/4} = \frac{1}{N^{1/4}}\,[h_m]_{1\ \text{tube}}} \tag{10-20}$$

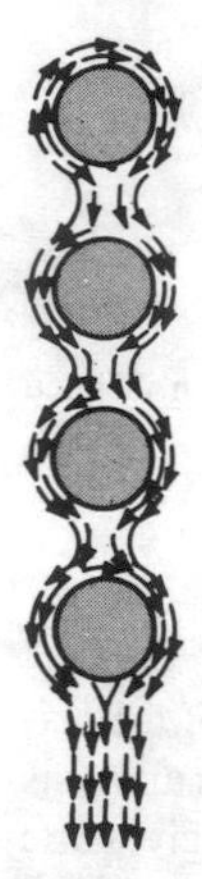

Figure 10-3 Film condensation on horizontal tubes arranged in a vertical tier.

This relation generally yields a conservative result since some turbulence and disturbance of condensate film are unavoidable during drainage and increase the heat transfer coefficient.

Reynolds Number for Condensate Flow

Although the flow hardly changes to turbulent during condensation on a single horizontal tube, turbulence may start at the lower portions of a vertical tube. When turbulence occurs in the condensate film, the average heat transfer coefficient begins to increase with the length of the tube in contrast to its decrease with the length for laminar film condensation. To establish a criterion for transition from laminar to turbulent flow, a *Reynolds number for condensate flow* is defined as

$$\mathrm{Re} = \frac{D_h u_m \rho_l}{\mu_l} \tag{10-21}$$

where u_m is the average velocity of condensate film and D_h is the hydraulic diameter for condensate flow, given by

$$D_h \equiv \frac{4A}{P} = \frac{4 \times \text{cross-sectional area for condensate flow}}{\text{wetted perimeter}} \tag{10-22}$$

From Eqs. (10-21) and (10-22), the Reynolds number for condensate flow becomes

$$\mathrm{Re} = \frac{4(\rho_l u_m A)}{\mu_l P} \tag{10-23}$$

where A = cross-sectional area for condensate flow and P = wetted perimeter.

The Reynolds number at the lowest part of the condensing surface can be expressed in a more convenient form as

$$\mathrm{Re} = \frac{4M}{\mu_l P} \tag{10-24}$$

where M = mass flow rate of condensate at the lowest part of the condensing surface, in kilograms per second. The wetted perimeter depends on the geometry; it is given as

$$P = \begin{cases} \pi D & \text{for vertical tube of outside diameter } D \quad (10\text{-}25a) \\ 2L & \text{for horizontal tube of length } L \quad (10\text{-}25b) \\ w & \text{for vertical or inclined plate of width } w \quad (10\text{-}25c) \end{cases}$$

Experiments have shown that the transition from laminar to turbulent condensation takes place at a Reynolds number of about 1800.

10-2 COMPARISON OF FILM CONDENSATION THEORY WITH EXPERIMENTS

A comparison of the average heat transfer coefficient for vertical surfaces given by Eq. (10-15) with that found by experiments has shown that the measured heat transfer coefficient is about 20 percent higher than theory would suggest. Therefore, McAdams [20] recommends that Nusselt's equation for h_m on a vertical surface be multiplied by a factor of 1.2; hence Eq. (10-15) should be replaced by

$$\boxed{h_m = 1.13\left[\frac{g\rho_l(\rho_l - \rho_v)h_{fg}k_l^3}{\mu_l(T_v - T_w)L}\right]^{1/4}} \tag{10-26}$$

Generally, $\rho_v \ll \rho_l$; hence Eq. (10-26) reduces to

$$h_m = 1.13\left[\frac{g\rho_l^2 h_{fg}k_l^3}{\mu_l(T_v - T_w)L}\right]^{1/4} \tag{10-27a}$$

which can be rearranged in the form

$$\boxed{h_m\left(\frac{\mu_l^2}{k_l^3\rho_l^2 g}\right)^{1/3} = 1.76\ \mathrm{Re}^{-1/3}} \tag{10-27b}$$

which is valid for Re < 1800, and the Reynolds number is defined by Eqs. (10-24) and (10-25).

In the case of condensation on a single horizontal tube, Eq. (10-19) is recommended.

Turbulent Film Condensation

The previous results for film condensation are applicable if condensate flow is laminar. Kirkbride [21] proposed the following empirical correlation for film condensation on a vertical plate after the start of turbulence:

$$h_m\left(\frac{\mu_l^2}{k_l^3 \rho_l^2 g}\right)^{1/3} = 0.0077(\text{Re})^{0.4} \tag{10-28}$$

valid for Re > 1800.

Figure 10-4 shows a plot of Eq. (10-27*b*) and (10-28) as a function of the Reynolds number for condensate flow in the laminar and turbulent regimes, respectively. Included in this figure, shown by the dashed line, is the original coefficient of Nusselt's theory, which lies about 20 percent below the recommended correlation.

In Eq. (10-28), the physical properties of the condensate should be evaluated at $T_f = \frac{1}{2}(T_w + T_v)$.

A more accurate analysis of laminar filmwise condensation of vapors on vertical surfaces has been performed by various investigators [6–10] by using the mathematical techniques of boundary-layer theory. The results show that the Prandtl number of the condensing vapor is a factor that influences the condensation heat transfer coefficient. The Prandtl numbers for steam and other common engineering fluids lie between 1 and 10; in this range, the effect for practical purposes appears to be negligible for laminar film condensation. At low Prandtl numbers in the range of 0.003 to 0.03, which embrace the currently important liquid metals, these analyses show that the heat transfer coefficient drops below the Nusselt prediction as the parameter $c_p(T_v - T_w)/h_{fg}$ increases (i.e., relatively thicker condensate flow). This matter is discussed further later in this chapter.

Example 10-1 Air-free saturated steam at $T_v = 65°\text{C}$ ($P = 25.03$ kPa) condenses on the outer surface of a 2.5-cm-OD, 3-m-long vertical tube maintained at a uniform temperature $T_w = 35°\text{C}$ by the flow of cooling water through the

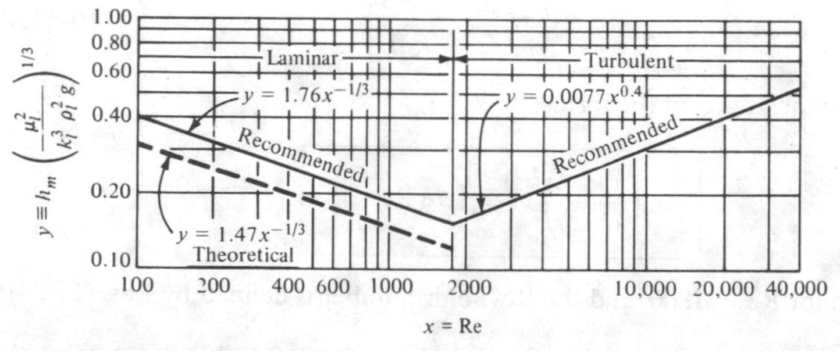

Figure 10-4 Average heat transfer coefficient for filmwise condensation on a vertical surface for laminar and turbulent flow regions.

tube. Assuming film condensation, calculate the average heat transfer coefficient over the entire length of the tube and the rate of condensate flow at the bottom of the tube.

SOLUTION The physical properties of condensate (i.e., water) at the film temperature $T_f = \frac{1}{2}(65 + 35) = 50°C$ are

$$k_l = 0.640 \text{ W/(m}\cdot°\text{C)} \qquad \mu_l = 0.562 \times 10^{-3} \text{ kg/(m}\cdot\text{s)}$$

$$\rho_l = 990 \text{ kg/m}^3 \qquad h_{fg} = 2382 \text{ kJ/kg}$$

and $g = 9.8 \text{ m}^2/\text{s}$ under normal conditions. The average heat transfer coefficient h_m for laminar film condensation on a vertical surface is determined from Eq. (10-26). For $\rho_v \ll \rho_l$, Eq. (10-26) becomes

$$h_m = 1.13\left[\frac{g\rho_l^2 h_{fg} k_l^3}{\mu_l(T_v - T_w)L}\right]^{1/4}$$

The numerical values are now substituted:

$$h_m = 1.13\left[\frac{9.8 \times 990^2 \times (2382 \times 10^3) \times 0.640^3}{(0.562 \times 10^{-3})(65 - 35)(3.0)}\right]^{1/4} = 3729 \text{ W/(m}^2\cdot°\text{C)}$$

The mass flow rate of condensate M at the bottom of the tube is

$$M = \frac{\pi D L h_m(T_v - T_w)}{h_{fg}}$$

$$= \frac{(\pi)(0.025)(3.0)(3729)(65 - 35)}{2346 \times 10^3} = 11.23 \times 10^{-3} \text{ kg/s}$$

In the previous calculations we assumed that the condensate flow was in the laminar regime. To check the validity of this assumption, we calculate the Reynolds number at the bottom of the tube from Eqs. (10-24) and (10-25):

$$\text{Re} = \frac{4M}{\mu_l \pi D} = \frac{4}{0.562 \times 10^{-3}}\,\frac{11.23 \times 10^{-3}}{\pi \times 0.025} = 1018$$

which is less than 1800. Hence the condensate flow is laminar, and the above calculations are valid.

Example 10-2 Determine the average heat transfer coefficient h_m and the total condensation rate for Example 10-1 when the tube is horizontal.

SOLUTION The average heat transfer coefficient for condensation on a horizontal tube is given by Eq. (10-19). For $\rho_v \ll \rho_l$, this equation reduces to

$$h_m = 0.725\left[\frac{g\rho_l^2 h_{fg} k_l^3}{\rho_l(T_v - T_w)D}\right]^{1/4}$$

The numerical values are introduced:

$$h_m = 0.725\left[\frac{9.8 \times 990^2 \times 2382 \times 10^3 \times 0.640^3}{(0.562 \times 10^{-3})(65 - 35)(0.025)}\right]^{1/4} = 7918\ \text{W/(m}^2 \cdot {}^\circ\text{C)}$$

The mass flow rate of condensate M for the tube is

$$M = \frac{\pi D L h_m (T_v - T_w)}{h_{fg}}$$

$$= \frac{(\pi)(0.025)(3.0)(7918)(65 - 35)}{2346 \times 10^3} = 23.86 \times 10^{-3}\ \text{kg/s}$$

The condensate flow rate with the horizontal tube is higher than that with the vertical tube.

Example 10-3 Air-free saturated steam at $T_v = 65°\text{C}$ condenses on the surface of a vertical tube with an OD of $D = 2.5$ cm which is maintained at a uniform temperature $T_w = 35°\text{C}$. Determine the tube length L for a condensate flow rate of $M = 6.0 \times 10^{-3}$ kg/s per tube.

SOLUTION This problem is the inverse of Example 10-1. The physical properties of the condensate are the same as those given previously. We start the calculation by checking the Reynolds number for the condensate:

$$\text{Re} = \frac{4M}{\mu_l \pi D} = \frac{4}{0.562 \times 10^{-3}} \frac{6.0 \times 10^{-3}}{\pi \times 0.025} = 544$$

Hence the condensate flow is in the laminar range. It is more convenient to use Eq. (10-27*b*) to calculate the average heat transfer coefficient:

$$h_m = 1.76\left(\frac{k_l^3 \rho_l^2 g}{\mu_l^2}\right)^{1/3} \text{Re}^{-1/3}$$

Substituting the numerical values, we obtain

$$h_m = 1.76\left[\frac{0.640^3 \times 990^2 \times 9.8}{(0.562 \times 10^{-3})^2}\right]^{1/3}\left(\frac{1}{544}\right)^{1/3} = 4307\ \text{W/(m}^2 \cdot {}^\circ\text{C)}$$

The tube length L is determined from

$$M h_{fg} = (\pi D L) h_m (T_v - T_w)$$

or

$$L = \frac{M h_{fg}}{\pi D h_m (T_v - T_w)}$$

The numerical values are substituted:

$$L = \frac{(6.0 \times 10^{-3})(2382 \times 10^3)}{\pi(0.025)(4307)(65 - 35)} = 1.41\ \text{m}$$

Example 10-4 Air-free saturated steam at $T_v = 85°C$ ($P = 57.83$ kPa) condenses on the outer surface of 225 horizontal tubes of 1.27-cm-OD arranged in a 15-by-15 array. Tube surfaces are maintained at a uniform temperature $T_v = 75°C$. Calculate the total condensation rate per 1-m length of the tube bundle.

SOLUTION The physical properties of water at $T_f = \frac{1}{2}(85 + 75) = 80°C$ are

$$k_l = 0.668 \text{ W/(m} \cdot \text{°C)} \qquad \mu_l = 0.355 \times 10^{-3} \text{ kg/(m} \cdot \text{s)}$$

$$\rho_l = 974 \text{ kg/m}^3 \qquad h_{fg} = 2309 \text{ kJ/kg}$$

The average heat transfer coefficient h_m is determined from Eq. (10-20):

$$h_m = 0.725\left[\frac{g\rho_l(\rho_l - \rho_v)h_{fg}k_l^3}{\mu_l(T_v - T_w)ND}\right]^{1/4}$$

In this equation ρ_v can be neglected in comparison to ρ_l since $\rho_v \ll \rho_l$. The numerical values are introduced:

$$h_m = 0.725\left[\frac{(9.8)(974^2)(2309 \times 10^3)(0.668^3)}{(0.355 \times 10^{-3})(85 - 75)(15)(0.0127)}\right]^{1/4}$$

$$= 7150 \text{ W/(m}^2 \cdot \text{°C)}$$

The total heat transfer surface per meter length of the tube array is

$$\frac{A}{L} = (\text{number of tubes})\pi D = (225)(\pi)(0.0127) = 9.76 \text{ m}^2\text{/m}$$

and the heat flow rate becomes

$$\frac{Q}{L} = h_m\left(\frac{A}{L}\right)(T_v - T_w) = (7150)(9.76)(85 - 75)$$

$$= 697.84 \text{ kW/m}$$

The total mass flow rate of condensate per meter length of the tube bundle is then

$$\frac{M}{L} = \frac{Q/L}{h_{fg}} = \frac{697.84 \times 10^3}{2296 \times 10^3} = 0.3 \text{ kg/(s} \cdot \text{m)}$$

10-3 FILM CONDENSATION INSIDE HORIZONTAL TUBES

In the previous analysis and correlations of film condensation, it is assumed that the vapor is stationary or has a negligible velocity. In practical applications, such as condensers in refrigeration and air conditioning systems, vapor condenses inside the tubes and so has a significant velocity. In such situations the condensation phenomenon is very complicated, and a simple analytical treatment is not possible. Consider, for example, the film condensation on the inside surface of a long vertical tube. The upward flow of vapor retards the condensate flow and causes the thickening of the condensate layer, which in turn decreases the condensation heat transfer coefficient. Conversely, the downward flow of vapor decreases the thickness of the condensate film, hence increases the heat transfer coefficient.

The effect of vapor velocity on the heat transfer coefficient for film condensation inside tubes has been studied [4, 5, 22–27], and more complete information on the subject is given in Ref. 32. Because of the complexity of the problem, we consider here only film condensation on the inside surface of horizontal tubes.

Chato [33] recommends the following correlation for condensation at low vapor velocities inside horizontal tubes:

$$\boxed{h_m = 0.555\left[\frac{g\rho_l(\rho_l - \rho_v)k_l^3 h'_{fg}}{\mu_l(T_{\text{sat}} - T_s)D}\right]^{1/4}} \tag{10-29a}$$

where

$$h'_{fg} \equiv h_{fg} + \tfrac{3}{8}c_{p,l}(T_v - T_w) \tag{10-29b}$$

This result has been developed for the condensation of refrigerants at low-vapor Reynolds numbers such that

$$\text{Re}_v = \frac{\rho_v u_v D}{\mu_v} < 35{,}000 \tag{10-30}$$

where Re_v should be evaluated at the *inlet* conditions and D is the inside diameter of the tube.

For higher flow rates, Akers, Deans, and Crosser [34] propose the following approximate empirical correlation for the average condensation heat transfer coefficient h_m on the inside surface of a horizontal tube of diameter D:

$$\boxed{\frac{h_m D}{k_l} = 0.026\,\text{Pr}_l^{1/3}\left[\text{Re}_l + \text{Re}_v\left(\frac{\rho_l}{\rho_v}\right)^{1/2}\right]^{0.8}} \tag{10-31}$$

where

$$\text{Re}_l = \frac{4M_l}{\pi D\mu_l} \qquad \text{Re}_v = \frac{4M_v}{\pi D\mu_v}$$

Here M_l and M_v are, respectively, the mass flow rate of liquid and vapor, respectively, in kilograms per second. Equation (10-31) correlates experimental data within about 50 percent for the range

$$\text{Re}_v > 20{,}000 \qquad \text{and} \qquad \text{Re}_l > 5000$$

10-4 DROPWISE CONDENSATION

Since the original observation of *dropwise condensation* by Schmidt, Schurig, and Sellschopp [35], numerous investigations of dropwise condensation have been reported [36–50]. In experiments, if traces of oil are present in steam and the condensing surface is highly polished, the condensate film breaks into droplets. This type of condensation is called *dropwise condensation.* Figure 10-5 shows an ideal dropwise condensation of steam on a vertical surface. The droplets grow, coalesce, and run off the surface, leaving a greater portion of the condensing surface freely exposed to incoming steam. Since the entire condensing surface is not covered with a continuous layer of liquid film, the heat transfer for ideal dropwise condensation of steam is much higher than that for filmwise condensation of steam. The heat transfer coefficients may be 5 to 10 times greater, but the overall heat transfer coefficient between the steam and the coolant in a typical surface condenser may be about 2 to 3 times greater for dropwise than for filmwise condensation. Therefore, considerable research has been done with the objective of producing long-lasting dropwise condensation. Various types of promoters such as oleic, stearic, and linoleic acids, benzyl mercaptan, and many other chemicals have been used to promote dropwise condensation. Continuous dropwise conditions obtainable with different promoters vary between 100 and 300 h with pure steam and are shorter with industrial steam or intermittent operations. Failure occurs because of fouling or oxidation of the surface, or by the gradual removal of the promoter on the surface by the flow of condensate, or by a combination of these effects.

To prevent the failure of dropwise condensation due to the oxidation, noble-metal coating of the condensing surface has been tried, and coatings of gold, silver, rhodium, palladium, and platinum have been used. Although some of these coated surfaces could produce dropwise condensation under laboratory

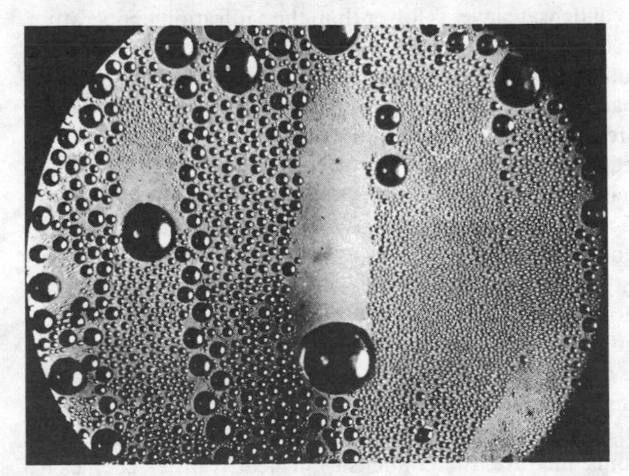

Figure 10-5 Dropwise condensation of steam under ideal conditions. (*From Hampson and Özişik* [*36*].)

conditions for more than 10,000 h of continuous operation, the cost of coating the condensing surface with noble materials is so high that the economics of such an approach for industrial applications has yet to be proved.

It is unlikely that long-lasting dropwise condensation can be produced under practical conditions by a single treatment of any of the promoters currently available. Although it may be possible to produce dropwise condensation for up to a year by the injection of a small quantity of promoter into the steam at regular intervals, the successful operation depends on the amount and the cost of the promoter and to what extent the cumulative effect of the injected promoter can be tolerated in the rest of the plant. Therefore, *in the analysis of a heat exchanger involving the condensation of steam, it is recommended that filmwise condensation be assumed for the condensing surface.*

10-5 CONDENSATION IN THE PRESENCE OF NONCONDENSABLE GAS

Earlier we considered the heat transfer coefficient for condensing vapors that did not contain any noncondensable gas. If noncondensable gas such as air is present in the vapor, even in very small amounts, the heat transfer coefficient for condensation is greatly reduced. The reason is that when a vapor containing noncondensable gas condenses, the noncondensable gas is left at the surface and the incoming condensable vapor must diffuse through this body of vapor-gas mixture collected in the vicinity of the condensate surface before the vapor reaches the cold surface to condense. Therefore, the presence of noncondensable gas adjacent to the condensate surface acts as a thermal resistance to heat transfer. The resistance to this diffusion process causes a drop in the partial pressure of the condensing vapor, which in turn drops the saturation temperature; that is, the temperature of the outside surface of the condensate layer is lower than the saturation temperature at the bulk mixture. A simple condensation theory as discussed earlier is not applicable because the condensate surface temperature is not known; the analysis is complicated in that heat, mass, and momentum transfer problems in the liquid and the vapor-gas mixture should be solved simultaneously.

The prediction of the condensation heat transfer coefficient in the presence of noncondensable gas has been the subject of numerous theoretical and experimental investigations [51–60]. Both analytical and experimental results have shown that the heat transfer coefficient is very much dependent on the vapor flow patterns in the vicinity of the condensing surface. For example, high velocities over the condensing surface tend to reduce the accumulation of the noncondensable gas and to alleviate the adverse effect of noncondensable gas on the heat transfer. If the noncondensable gas can accumulate over the condensing surface, a significant reduction in the heat transfer coefficient can result. Depending on the vapor flow patterns in the vicinity of the condensing surface and the magnitude of the noncondensable gas in the bulk mixture, the condensation heat transfer coefficient can be reduced by many folds. For example, 0.5 percent by mass of air in the steam can reduce the filmwise condensation heat transfer coefficient by a factor of 2, or a 5 percent by mass of air can easily cut h by a factor of 5.

In steam condensers for industrial applications, however, the vapor flow patterns are so complicated that neither an analysis performed under idealized conditions nor the data obtained from simple laboratory experiments can represent the actual situations, but they help to demonstrate the importance of various factors. Therefore, *the general practice in condenser design still is to vent noncondensable gas as much as possible.*

10-6 POOL BOILING REGIMES

Pool boiling provides a convenient starting point for the discussion of the mechanism of heat transfer in boiling systems. Despite the fact that it has been extensively studied and its mechanism is reasonably well understood, it is not possible to predict theoretically the heat transfer characteristics of this apparently most simple boiling system.

Nukiyama [61] was the first investigator who established experimentally the characteristics of pool boiling phenomena. He immersed an electric resistance wire into a body of saturated water and initiated boiling on the surface of the wire by passing current through it. He determined both the heat flux and the temperature from the measurements of current and voltage. Since the original work of Nukiyama, numerous investigations of the pool boiling phenomenon have been reported [62–90]. Figure 10-6 illustrates the characteristics of pool boiling for water at atmospheric pressure. This boiling curve illustrates the variation of the heat transfer coefficient or the heat flux as a function of the temperature difference between the wire and water saturation temperatures.

Free-convection regime In this regime, the energy transfer from the heater surface to the saturated liquid takes place by free convection and can be predicted by the methods discussed in Chap. 9. The surface is only a few degrees above the saturation temperature of the liquid, but the free-convection currents produced in the liquid are sufficient to remove the heat from the surface.

Nucleata boiling regime The nucleata boiling regime in which the bubbles are formed on the surface of the heater can be separated into two distinct regions. In the region designated II, bubbles start to form at the favored sites on the heater surface, but as soon as the bubbles are detached from the surface, they are dissipated in the liquid. In region III, the nucleation sites are numerous and the bubble generation rate is so high that continuous columns of vapor appear. As a result, very high heat fluxes are obtainable in this region. In practical applications, the nucleata boiling regime is most desirable, because large heat fluxes are obtainable with small temperature differences. In the nucleata boiling regime, the heat flux

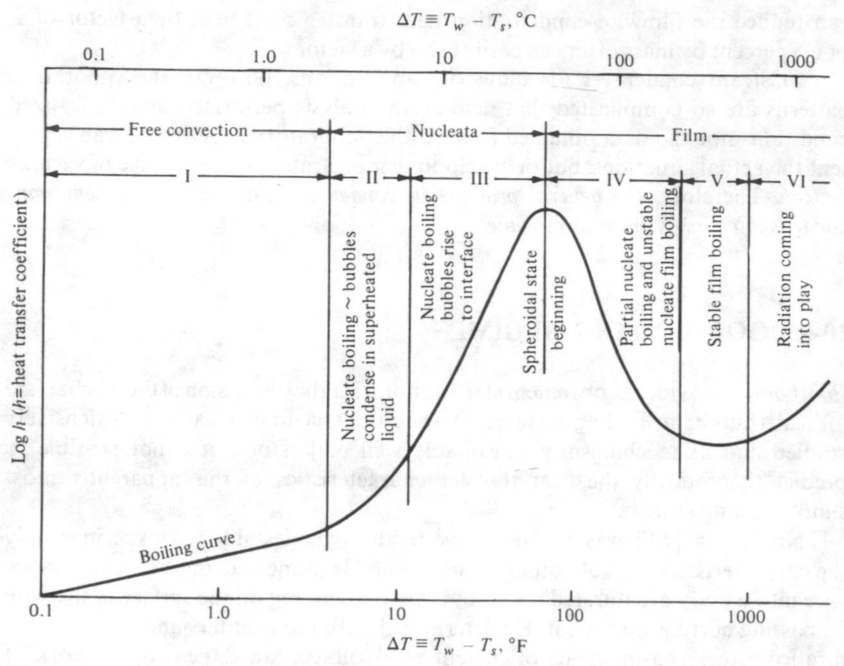

Figure 10-6 Principal boiling regimes in pool boiling of water at atmospheric pressure and saturation temperature T_s from an electrically heated platinum wire. (*From Farber and Scorah* [62].)

increases rapidly with increasing temperature difference until the *peak heat flux* is reached. The location of this peak heat flux is called the *burnout point*, or *departure from nucleata boiling* (DNB), or the *critical heat flux* (CHF). The reason for calling the peak heat flux the burnout point is apparent from Fig. 10-6. As soon as the peak heat flux is exceeded, an extremely large temperature difference is needed to realize the resulting heat flux. Such high temperature differences may cause the burning up, or melting away, of the heating element.

Film boiling regime Figure 10-6 shows that after the peak heat flux is reached, any further increase in the temperature difference causes a reduction in the heat flux. The reason for this curious phenomenon is the blanketing of the heater surface with a vapor film which restricts liquid flow to the surface and has a low thermal conductivity. The film boiling regime can be separated into three distinct regions. The region designated IV is called the *unstable film boiling* region, where the vapor film is unstable, collapsing and reforming under the influence of convective currents and the surface tension. Here the heat flux decreases as the surface temperature increases, because the average wetted area of the heater surface decreases. In region V, the heat flux drops to a minimum, because a continuous vapor film covers the heater surface. This is called the *stable film boiling* region. In region VI, the heat flux begins to increase as the temperature difference increases, because the temperature at the heater surface is sufficiently high for thermal radiation effects to augment heat transfer through the vapor film.

10-7 POOL BOILING CORRELATIONS

In Sec. 10-6 we discussed qualitatively the physical mechanisms characterizing the different regimes in pool boiling. In this section we present some of the correlations for predicting heat transfer in free-convection, nucleata boiling, and film boiling regimes as well as the magnitude of the peak heat flux.

Free-Convection Regime

In this region, the heat transfer takes place by free convection, and heat transfer correlations, as discussed in Chap. 9, are in the form

$$\text{Nu} = f(\text{Gr}, \text{Pr}) \tag{10-32}$$

Once the heat transfer coefficient h is obtained, the heat flux for the free-convection regime is determined from

$$q = h(T_w - T_{\text{sat}}) \tag{10-33}$$

Various correlations are given in Chap. 9 to determine the free-convection heat transfer coefficient h for various geometries, such as vertical, inclined, and horizontal plates; vertical and horizontal cylinders; and many others.

Nucleata Boiling Regime

The nucleata boiling regime involves two separate processes: the formation of bubbles on the surface, which is referred to as the *nucleation*, and the subsequent growth and motion of these bubbles.

The process of nucleation is very complicated. Several theories have been proposed as to exactly how the bubbles first form on the surface. Still there is controversy on this subject. The distribution of active nucleation sites on the surface also must be known for the development of theoretical models of nucleata boiling.

Once the nucleation process is complete, further heat transfer from the surface to the bubble promotes bubble growth. Therefore, considerable effort has been expended to understand the mechanism of bubble growth. The bubble departure size is also another important parameter that affects heat transfer in the nucleata boiling regime.

Clearly the heat transfer in the nucleata boiling regime is affected by the nucleation process, the distribution of active nucleation sites on the surface, and

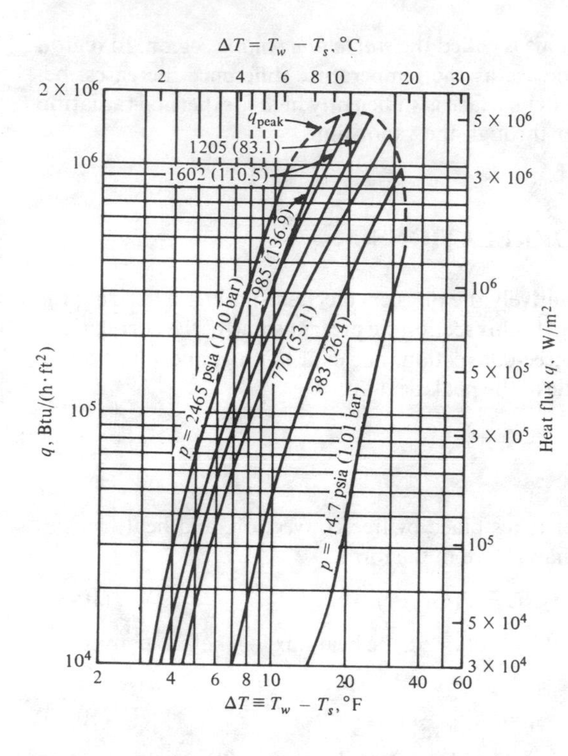

Figure 10-7 Heat flux for boiling of saturated water on a horizontal, 0.61-mm-diameter platinum wire. (*From Addoms* [*71*].)

the growth and departure of bubbles. If the number of active nucleation sites increases, the interaction between the bubbles may become important. In addition to these variables, the state of the fluid (i.e., the fluid properties) and the surface condition (i.e., the mechanical and material properties of the surface) are among the factors that affect heat transfer in the nucleata boiling regime.

Numerous experimental investigations have been reported [89, 90], and various attempts have been made to correlate that portion of the boiling curve characterizing heat transfer in the nucleata boiling regime. For example, Fig. 10-7 shows typical experimental data for the heat flux q plotted as a function of ΔT at various pressure levels for boiling of water on a horizontal, 0.061-cm-diameter, electrically heated platinum wire. The dashed lines denote the location of the peak heat flux in nucleata boiling. Heat flux correlations of the form shown in Fig. 10-7 were used by numerous early investigators for each different liquid by plotting q versus ΔT. Later, attempts were made to obtain more general empirical correlations between q and ΔT that could be applicable to a group of liquids over a wide range of pressure levels. A discussion of such attempts is given in Refs. 66 and 72. The basic concept behind such correlations was the fact that the principal mechanism of heat transfer in nucleate boiling is forced convection caused by the agitation produced by bubbles. The most successful and widely used correlation was developed by Rohsenow [73]. By analyzing the significance of various parameters in relation to forced-convection effects, he proposed the following empirical relation to correlate the heat flux in the entire nucleata boiling regime:

$$\frac{c_{pl}\,\Delta T}{h_{fg}\,\mathrm{Pr}_l^n} = C_{sf}\left[\frac{q}{\mu_l h_{fg}}\sqrt{\frac{g_c \sigma^*}{g(\rho_l - \rho_v)}}\right]^{0.33} \tag{10-34}$$

where c_{pl} = specific heat of saturated liquid, J/(kg · °C) or Btu/(lb · °F)
C_{sf} = constant to be determined from experimental data depending on heating surface–fluid combination
h_{fg} = latent heat of vaporization, J/kg or Btu/lb
g = gravitational acceleration m/s^2 or ft/h^2
g_c = gravitational acceleration conversion factor, 1 kg · m/(N · s^2) or lb · ft/(lbf · h^2); it is not needed in the SI system.
$\mathrm{Pr}_l = c_{pl}\mu_l/k_l$ = Prandtl number of saturated liquid
q = boiling heat flux, W/m^2 or Btu/(h · ft^2)
$\Delta T = T_w - T_s$, temperature difference between wall and saturation temperature, °C or °F
μ_l = viscosity of saturated liquid, kg/(m · s) or lb/(ft · h)
ρ_l, ρ_v = density of liquid and saturated vapor, respectively, kg/m^3 or lb/ft^3
σ^* = surface tension of liquid-vapor interface, N/m or lbf/ft

In Eq. (10-34) the exponent n and the coefficient C_{sf} are the two provisions to adjust the correlation for the liquid-surface combination. Table 10-1 lists the

Table 10-1 Values of the coefficient C_{sf} of Eq. (10-34) for various liquid-surface combinations

Liquid-surface combination	C_{sf}	Reference
Water–copper	0.0130	[75]
Water–scored copper	0.0068	[74]
Water–emergy-polished copper	0.0128	[74]
Water–emery-polished, paraffin-treated copper	0.0147	[74]
Water–chemically etched stainless steel	0.0133	[74]
Water–mechanically polished stainless steel	0.0132	[74]
Water–ground and polished stainless steel	0.0080	[74]
Water–Teflon pitted stainless steel	0.0058	[74]
Water–platinum	0.0130	[71]
Water–brass	0.0060	[74]
Benzene–chromium	0.0100	[70]
Ethyl alcohol–chromium	0.0027	[70]
Carbon tetrachloride–copper	0.0130	[75]
Carbon tetrachloride–emery-polished copper	0.0070	[74]
n-Pentane–emery-polished copper	0.0154	[74]
n-Pentane–emergy-polished nickel	0.0127	[74]
n-Pentane–emery-rubber copper	0.0074	[74]
n-Pentane–lapped copper	0.0049	[74]

Table 10-2 Values of liquid-vapor surface tension σ for various liquids

Liquid	Saturation temperature		Surface tension	
	°F	°C	$\sigma^* \times 10^4$ lbf/ft	$\sigma^* \times 10^3$ N/m
Water	32	0	51.8	75.6
Water	60	15.56	50.2	73.2
Water	100	37.78	47.8	69.7
Water	200	93.34	41.2	60.1
Water	212	100	40.3	58.8
Water	320	160	31.6	46.1
Water	440	226.7	21.9	31.9
Water	560	293.3	11.1	16.2
Water	680	360	1.0	1.46
Water	705.4	374.11	0.0	0
Sodium	1618	881.1	77	11.2
Potassium	1400	760	43	62.7
Rubidium	1270	687.8	30	43.8
Cesium	1260	682.2	20	29.2
Mercury	675	357.2	27	39.4
Benzene (C_6H_6)	176	80	19	27.7
Ethyl alcohol (C_2H_6O)	173	78.3	15	21.9
Freon 11	112	44.4	5.8	8.5

experimentally determined values C_{sf} for a variety of liquid-surface combinations. The value of *n should be taken as 1 for water and 1.7 for all the other liquids* shown in Table 10-1.

Table 10-2 gives the values of the vapor-liquid surface tension for a variety of liquids.

Figure 10-8 shows the correlation of Addoms' experimental data given in Fig. 10-7 with Eq. (10-34) for water with $C_{sf} = 0.013$ over a wide range of conditions. It shows a deviation in heat flux as much as 100 percent, but the typical error in ΔT is about 25 percent. The reason is that according to Eq. (10-34), we have $q \sim \Delta T^3$; as a result, a large error in q corresponds to a small error in ΔT.

A vast amount of experimental data has been accumulated in the literature on heat transfer in the nucleata boiling regime. So one may question whether it is possible to develop a correlation by purely statistical means. In a recent work by Stephan and Abdelsalam [88], this matter was investigated. And for water, for example, their correlation was in excellent agreement with Rohsenow's correlation given by Eq. (10-34).

Example 10-5 Saturated water at $T_v = 100°C$ is boiled with a copper heating element having a heating surface $A = 4 \times 10^{-2}$ m^2 which is maintained at a

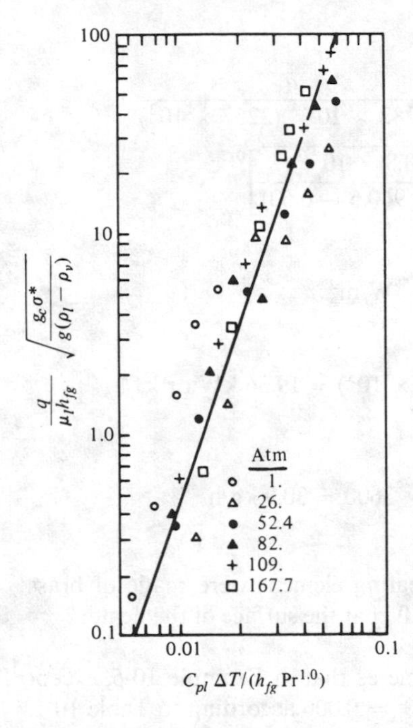

Figure 10-8 Rohsenow's [73] correlation of Addoms' [71] data for boiling of water using Eq. (10-34).

uniform temperature $T_w = 115°C$. Calculate the surface heat flux and the rate of evaporation.

SOLUTION An examination of the heat transfer data in Fig. 10-7 for the boiling of water implies that the boiling is in the nucleata boiling regime. Equation (10-34) can be used to determine the boiling heat flux q:

$$\frac{c_{pl}\,\Delta T}{h_{fg}\,\mathrm{Pr}_l^n} = C_{sf}\left[\frac{q}{\mu_l h_{fg}}\sqrt{\frac{\sigma^*}{g(\rho_l - \rho_v)}}\right]^{0.33}$$

In this equation the gravitational conversion factor g_c is omitted because the SI system of units will be used in the calculations.

For water, the exponent n is taken as 1, and for water–copper the coefficient C_{sf} is obtained from Table 10-1 as 0.013. The physical properties of saturated water and vapor are taken as

$$c_{pl} = 4216\ \mathrm{J/(kg\cdot °C)} \qquad h_{fg} = 2257 \times 10^3\ \mathrm{J/kg}$$

$$\rho_l = 960.6\ \mathrm{kg/m^3} \qquad \rho_v = 0.60\ \mathrm{kg/m^3}$$

$$\mathrm{Pr}_l = 1.74 \qquad \mu_l = 0.282 \times 10^{-3}\ \mathrm{kg/(m\cdot s)} \qquad \sigma^* = 58.8 \times 10^{-3}\ \mathrm{N/m}$$

and

$$\Delta T = T_w - T_v = 15°C$$

These numerical values are substituted:

$$\frac{4216 \times 15}{(2257 \times 10^3)(1.74)} = 0.013\left[\frac{q}{(0.282 \times 10^{-3})(2257 \times 10^3)} \times \sqrt{\frac{58.8 \times 10^{-3}}{(9.8)(960.6 - 0.60)}}\right]^{0.33}$$

Then the heat flux becomes

$$q = 4.84 \times 10^5 \text{ W/m}^2$$

The total rate of heat transfer is

$$Q = \text{area} \times q = (4 \times 10^{-2})(4.84 \times 10^5) = 19.36 \text{ kW or kJ/s}$$

The rate of evaporation becomes

$$M = \frac{Q}{h_{fg}} = \frac{19.36 \times 10^3}{2257 \times 10^3} \times 3600 = 30.9 \text{ kg/h}$$

Example 10-6 In Example 10-5, if the heating element were made of brass instead of copper, what would be the heat flux at the surface of the heater?

SOLUTION The problem is exactly the same as that in Example 10-5, except that $C_{sf} = 0.013$ should be replaced by $C_{sf} = 0.006$ according to Table 10-1. Then from Eq. (10-34) we write

$$\frac{q_{\text{water-brass}}}{q_{\text{water-copper}}} = \left(\frac{C_{sf,\,\text{water-copper}}}{C_{sf,\,\text{water-brass}}}\right)^3$$

Substituting the numerical values yields

$$q_{\text{water-brass}} = (4.84 \times 10^5)\left(\frac{0.013}{0.006}\right)^3 = 4.93 \times 10^6 \text{ W/m}^2$$

Example 10-7 During the boiling of saturated water at $T_v = 100°\text{C}$ with an electric heating element, a heat flux of $q = 5 \times 10^5 \text{ W/m}^2$ is achieved with a temperature difference of $T_w - T_v = 9.3°\text{C}$. What is the value of the coefficient C_{sf} for the heater surface?

SOLUTION The physical properties of saturated water and vapor at 100°C are given in Example 10-5. Introducing these properties into Eq. (10-34) and taking $n = 1$, $\Delta T = 9.3°\text{C}$, and $q = 5 \times 10^5 \text{ W/m}^2$, we obtain

$$\frac{4216 \times 9.3}{(2257 \times 10^3)(1.74)} = C_{sf}\left[\frac{5 \times 10^5}{(0.282 \times 10^{-3})(2257 \times 10^3)} \times \sqrt{\frac{59.8 \times 10^{-3}}{(9.8)(960.6 - 0.6)}}\right]^{0.33}$$

Solving for C_{sf}, we obtain

$$C_{sf} = 0.008$$

Peak Heat Flux

The correlation given by Eq. (10-34) provides information for the heat flux in nucleate boiling, but it cannot predict the *peak heat flux*. The determination of peak heat flux in nucleate boiling is of interest because of burnout considerations; that is, if the applied heat flux exceeds the peak heat flux, the transition takes place from the nucleate to the stable film boiling regime in which, depending on the kind of fluid, boiling may occur at temperature differences well above the melting point of the heating surface. Equations have been developed for the prediction of the peak heat flux [76–79]. Kutateladze [76] treated the failure of nucleate boiling as a purely hydrodynamic problem and developed an expression for q_{max} that is valid for an infinite horizontal plate, but the equation needed a constant. He then suggested the value of the constant as 0.131 on the basis of data for geometries other than an infinite plate.

Zuber [77] and Zuber and Tribus [78] developed a similar expression for the peak heat flux by the stability requirement of the liquid-vapor interface, given in the form

$$\boxed{q_{max} = \frac{\pi}{24}\rho_v h_{fg}\left[\frac{\sigma^* g g_c(\rho_l - \rho_v)}{\rho_v^2}\right]^{1/4}\left(1 + \frac{\rho_v}{\rho_l}\right)^{1/2}} \tag{10-35}$$

where σ^* = surface tension of liquid-vapor interface, N/m or lbf/ft
g = gravitational acceleration, m/s^2 or ft/h^2
g_c = gravitational acceleration conversion factor, 1 $kg \cdot m/(N \cdot s^2)$ or $lb \cdot ft/(lbf \cdot h^2)$; not needed in the SI system
ρ_l, ρ_v = density of liquid and vapor, respectively, kg/m^3 or lb/ft^3
h_{fg} = latent heat of vaporization, J/kg or Btu/lb
q_{max} = peak heat flux, W/m^2 or $Btu/(h \cdot ft^2)$

In Eq. (10-35), the constant $\pi/24 \cong 0.131$ is the same as that proposed by Kutateladze. However, the correlation of Eq. (10-35) with experimental results reveals that $\pi/24$ is low; the value of 0.18 has been recommended by Rohsenow [94]. It is apparent from this equation that large values of h_{fg}, ρ_v, g, and σ^* are desirable for a large value of the peak heat flux. For example, water has a large value of h_{fg}; hence the peak heat flux obtainable with the boiling of water is high. This equation also shows that a reduced gravitational field decreases the peak heat flux.

Although Eq. (10-35) correlates well selected sets of data, it is not so accurate for all systems. For example, in the case of boiling of liquid oxygen, Eq. (10-35) underestimates the peak heat flux [93]. Also the heater geometry is another parameter that affects the peak heat flux.

Table 10-3 Correction factor $F(L')$ for use in Eq. (10-36)

Heater geometry	$F(L')$	Remarks
1. Infinite flat plate facing up	1.14	$L' \geq 2.7$; L is the heater width or diameter
2. Horizontal cylinder	$0.89 + 2.27e^{-3.44\sqrt{L'}}$	$L' \geq 0.15$; L is the cylinder radius
3. Large sphere	0.84	$L' \geq 4.26$; L is the sphere radius
4. Small sphere	$\dfrac{1.734}{(L')^{1/2}}$	$0.15 \leq L' \leq 4.26$; L is the sphere radius
5. Large finite body	~ 0.90	$L' \geq 4$; $L = \dfrac{\text{volume}}{\text{surface area}}$

Based on Refs. 86, 87, 91, and 92.

Lienhard and coworkers [86, 87, 91, 92] reexamined the correlation (10-35) given by Zuber in order to accommodate the effects of heater geometry and size. For the peak heat flux they proposed the following modified form of Eq. (10-35):

$$\boxed{q_{\max} = F(L') \times 0.131\rho_v^{1/2}h_{fg}[\sigma^* g g_c(\rho_l - \rho_v)]^{1/4}} \tag{10-36}$$

where the quantities are defined previously except $F(L')$, which is a correction factor for the effects of heater geometry and size. The factor $F(L')$ depends on the *dimensionless characteristic length* L' of the heater, defined as

$$L' = L\sqrt{\frac{g(\rho_l - \rho_v)}{\sigma^*}} \tag{10-37}$$

where L is the characteristic dimension of the heater and the other quantities are as defined previously.

In Eqs. (10-35) and (10-36), the physical properties of the vapor should be evaluated at

$$T_f = \tfrac{1}{2}(T_w + T_s)$$

The enthalpy of evaporation h_{fg} and the liquid properties should be evaluated at the saturated temperature of the liquid.

We present in Table 10-3 the recommended values of the factor $F(L')$ for heater geometries such as an infinite flat plate, a horizontal cylinder, and a sphere.

Equation (10-36), with the $F(L')$ values as specified in Table 10-3, correlates the experimental data within ± 20 percent. Equation (10-36) may be preferred to Eq. (10-35).

Example 10-8 Water at atmospheric pressure and saturation temperature is boiled in a 25-cm-diameter, electrically heated, mechanically polished, stainless-steel pan. The heated surface of the pan is maintained at a uniform temperature $T_w = 116°C$.

(a) Calculate the surface heat flux.
(b) Calculate the rate of evaporation from the pan.
(c) Calculate the peak heat flux.

SOLUTION The physical properties of saturated water and vapor are taken as

$$c_{pl} = 4216 \text{ J/(kg} \cdot °\text{C)} \qquad h_{fg} = 2257 \times 10^3 \text{ J/kg}$$

$$\rho_l = 960.6 \text{ kg/m}^3 \qquad \rho_v = 0.60 \text{ kg/m}^3$$

$$\Pr_l = 1.74 \qquad \mu_l = 0.282 \times 10^{-3} \text{ kg/(m} \cdot \text{s)}$$

$$\sigma^* = 58.8 \times 10^{-3} \text{ N/m} \qquad \Delta T = T_w - T_v = 16°C$$

(a) Equation (10-34) is used to calculate the surface heat flux;

$$\frac{c_{pl}\,\Delta T}{h_{fg}\,\Pr_l^n} = C_{sf}\left[\frac{q}{\mu_l h_{fg}}\sqrt{\frac{\sigma^*}{g(\rho_l - \rho_v)}}\right]^{0.33}$$

For water, $n = 1$; for water–mechanically polished stainless steel, from Table 10-1 $C_{sf} = 0.0132$. The numerical values of various quantities are introduced into the above equation. We obtain

$$\frac{4216 \times 16}{(2257 \times 10^3)(1.74)} = 0.0132\left[\frac{q}{(0.282 \times 10^{-3})(2257 \times 10^3)} \times \sqrt{\frac{58.8 \times 10^{-3}}{(9.8)(960.6 - 0.60)}}\right]^{0.33}$$

Then the surface heat flux becomes

$$q = 5.61 \times 10^5 \text{ W/m}^2$$

(b) The total rate of heat transfer is

$$Q = \text{area} \times q = \left(\frac{\pi}{4} \times 0.25^2\right)(5.61 \times 10^5)$$

$$= 0.275 \times 10^5 \text{ W or J/s}$$

The rate of evaporation becomes

$$M = \frac{Q}{h_{fg}} = \frac{0.275 \times 10^5}{22.57 \times 10^5} \times 3600 = 43.9 \text{ kg/h}$$

(c) To calculate the peak heat flux, we use Eq. (10-36) with the factor $F(L')$ taken from Table 10-3, case 1. That is,

$$q_{\max} = 1.14 \times 0.131 \rho_v^{1/2} h_{fg}[\sigma^* g(\rho_l - \rho_v)]^{1/4}$$

which is valid for

$$L' \equiv L\sqrt{\frac{g(\rho_l - \rho_v)}{\sigma^*}} \geq 2.7$$

For $L = 0.25$ m and other quantities as given above, we have $L' = 100$, which is larger than the specified lower bound 2.7. Hence, the above equation for q_{max} is applicable. Introducing the numerical values yields the peak heat flux:

$$\begin{aligned} q_{max} &= (1.14)(0.131)(0.6)^{1/2}(2257 \times 10^3) \\ &\quad \times [58.8 \times 10^{-3} \times 9.8(960.6 - 0.6)]^{1/4} \\ &= 1.27 \times 10^6 \text{ W/m}^2 = 1.27 \text{ MW/m}^2 \end{aligned}$$

We note that the surface heat flux $q = 5.61 \times 10^5$ W/m² is well below the peak heat flux $q_{max} = 1.27 \times 10^6$ W/m² or J/(s · m²).

Film Boiling Regime

As illustrated in Fig. 10-6, the nucleate boiling region ends and the unstable film boiling region begins after the peak heat flux is reached. No analysis is available for the prediction of heat flux as a function of the temperature difference $T_w - T_s$ in this unstable region until the minimum point in the boiling curve is reached and the stable film boiling region starts. In stable film boiling regions V and VI, the heating surface is separated from the liquid by a vapor layer across which heat must be transferred. Since the thermal conductivity of the vapor is low, large temperature differences are needed for heat transfer in this region; therefore, heat transfer in this region is generally avoided when high temperatures are involved. However, stable film boiling has numerous applications in the boiling of cryogenic fluids. A theory was developed by Bromley [81] for the prediction of the heat transfer coefficient for stable film boiling on the outside of a horizontal cylinder. The basic approach in the analysis is similar to Nusselt's theory for filmwise condensation on a horizontal tube. The resulting equation for the average heat transfer coefficient h_0 for stable film boiling on the outside of a horizontal cylinder in the *absence of radiation* is given by

$$\boxed{h_0 = 0.62\left[\frac{k_v^3\rho_v(\rho_l - \rho_v)gh_{fg}}{\mu_v D_o\,\Delta T}\left(1 + \frac{0.4c_{pv}\,\Delta T}{h_{fg}}\right)\right]^{1/4}} \qquad (10\text{-}38)$$

where h_0 = average boiling heat transfer coefficient in absence of radiation, W/(m² · °C) or Btu/(h · ft² · °F)

c_{pv} = specific heat of saturated vapor, J/(kg · °C) or Btu/(lb · °F)

D_o = outside diameter of tube, m or ft

g = gravitational acceleration, m/s² or ft/h²

h_{fg} = latent heat of vaporization, J/kg or Btu/lb

k_v = thermal conductivity of saturated vapor, W/(m · °C) or Btu/(h · ft · °F)

$\Delta T = T_w - T_s$, temperature difference between wall and saturation temperatures, °C or °F

In Eq. (10-38), the physical properties of vapor should be evaluated at $T_f = \frac{1}{2}(T_w + T_s)$, and the enthalpy of evaporation h_{fg} and the liquid density ρ_l should be evaluated at the saturation temperature T_s of the liquid.

Equation (10-38) has been derived by assuming that heat transfer across the vapor film is by pure conduction; therefore, it does not include the radiation effects. Bromley [81] suggested that when the surface temperature is sufficiently high for the radiation effects to be important, the average heat transfer coefficient h_m can be determined from the following empirical relation:

$$\boxed{h_m = h_0\left(\frac{h_0}{h_m}\right)^{1/3} + h_r} \tag{10-39a}$$

where h_0 is the boiling heat transfer coefficient given by Eq. (10-38) without the radiation effects and h_r is the radiation heat transfer coefficient which can be estimated approximately from

$$\boxed{h_r = \frac{1}{1/\varepsilon + 1/\alpha - 1}\,\frac{\sigma(T_w^4 - T_s^4)}{T_w - T_s}} \tag{10-39b}$$

where α = absorptivity of liquid
ε = emissivity of hot tube
σ = Stefan-Boltzmann constant
T_w = wall temperature
T_s = saturation temperature of liquid

Equation (10-39) is difficult to use because a trial-and-error approach is needed to determine h_m. When h_r is smaller than h_0, Eq. (10-39*a*) may be replaced by

$$\boxed{h_m = h_0 + \tfrac{3}{4}h_r} \tag{10-40}$$

To check the validity of the above theory, Bromley [81] used Eqs. (10-38) and (10-40) to correlate film boiling data at atmospheric pressure for a number of liquids such as water, benzene, carbon tetrachloride, *n*-pentane, and nitrogen boiling on carbon tubes of various diameters over a wide range of $T_w - T_s$ up to 1400°C. Figure 10-9 shows the comparison of calculated and experimental heat transfer coefficients for nitrogen boiling on a 0.89-cm-diameter tube. The agreement between the theory and experiment is fairly good. However, the correlation of film boiling data for water on very small-diameter wires (i.e., from 0.01 to 0.06 cm) showed that the theory predicted lower values, the deviation ranging from 30 to 100 percent as the wire size decreased from 0.06 to 0.01 cm. Therefore, the above correlation for the effects of radiation should be used with caution for small-diameter wires.

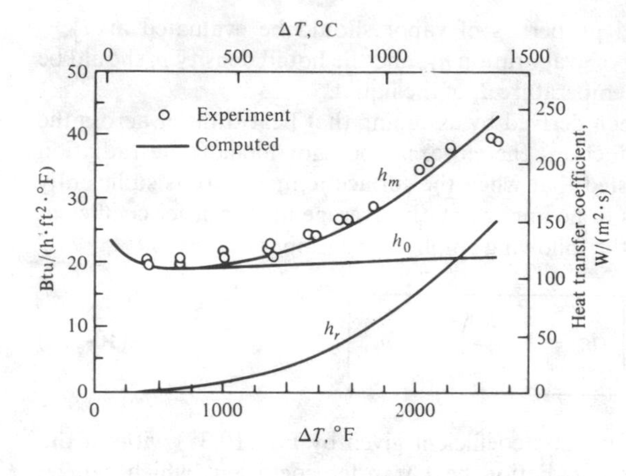

Figure 10-9 Heat transfer coefficient for stable film boiling of liquid nitrogen on an electrically heated, 0.89-cm-diameter carbon tube. (*From Bromley* [*81*].)

Example 10-9 Water at saturation temperature and atmospheric pressure is boiled with an electrically heated, horizontal platinum wire of 0.127-cm diameter. Determine the boiling heat transfer coefficient h_m and the heat flux for a temperature difference $T_w - T_s = 650°\text{C}$.

SOLUTION The heat transfer coefficient for stable film boiling without the radiation effects is given by Eq. (10-38):

$$h_0 = 0.62\left[\frac{k_v^3\rho_v(\rho_l - \rho_v)gh_{fg}}{\mu_v D_o\,\Delta T}\left(1 + \frac{0.4c_{pv}\,\Delta T}{h_{fg}}\right)\right]^{1/4}$$

The physical properties of vapor evaluated at $T_f = \frac{1}{2}(T_w + T_s) = \frac{1}{2}(750 + 100) = 425°\text{C}$ are taken as

$$c_{pv} = 2085\ \text{J/(kg}\cdot°\text{C)} \qquad k_v = 0.0505\ \text{W/(m}\cdot°\text{C)}$$

$$\mu_w = 24.26 \times 10^{-6}\ \text{kg/(m}\cdot\text{s)} \qquad \rho_v = 0.314\ \text{kg/m}^3$$

The liquid density and h_{fg} are evaluated at the saturation temperature T_s:

$$\rho_l = 960.6\ \text{kg/m}^3 \qquad h_{fg} = 2257 \times 10^3\ \text{J/kg}$$

The wire diameter is

$$D_o = 0.127 \times 10^{-2}\ \text{m}$$

and the temperature difference is given by

$$\Delta T = T_w - T_s = 650°\text{C}$$

Substituting these numerical values into the above equation, we find the heat transfer coefficient h_0 as:

$$h_0 = 0.62\left[\frac{(0.0505^3)(0.314)(960.6 - 0.3)(9.8)(2257 \times 10^3)}{(24.26 \times 10^{-6})(0.127 \times 10^{-2})(650)} \times \left(1 + \frac{0.4 \times 2085 \times 650}{2257 \times 10^3}\right)\right]^{1/4}$$

$$= 297.8 \text{ W/(m}^2 \cdot {}^\circ\text{C)}$$

To include the radiation effect, the radiation heat transfer coefficient h_r is determined from Eq. (10-39*b*):

$$h_r = \frac{1}{1/\varepsilon + 1/\alpha - 1}\,\frac{\sigma(T_w^4 - T_s^4)}{T_w - T_s}$$

We assume $\varepsilon = \alpha \cong 1$ and take the Stefan-Boltzmann constant as $\sigma = 5.667 \times 10^{-8}$ W/(m$^2 \cdot$ K^4) and the absolute temperatures as

$$T_w = 750 + 273 = 1023 \text{ K}$$

$$T_s = 100 + 273 = 373 \text{ K}$$

Then h_r becomes

$$h_r = \frac{5.667 \times 10^{-8}(1023^4 - 373^4)}{650} = 93.8 \text{ W/(m}^2 \cdot {}^\circ\text{C)}$$

The linearized equation (10-40) is now used to calculate the total heat transfer coefficient resulting from the combined effect of film boiling and radiation:

$$h_m = h_0 + \tfrac{3}{4}h_r = \underbrace{297.8}_{80.9\%} + \underbrace{\tfrac{3}{4}(93.8)}_{19.1\%} = 368.2 \text{ W/(m}^2 \cdot {}^\circ\text{C)}$$

Then the net heat flux becomes

$$q \doteq h_m\,\Delta T = 368.2 \times 650 = 239.33 \text{ kW/m}^2$$

Thus, the net heat transfer by radiation counts for 19.1 percent of the total heat transfer.

10-8 FORCED-CONVECTION BOILING INSIDE TUBES

In previous sections we considered boiling on a heated surface immersed in a quiescent mass of liquid. If boiling takes place on the inside surface of a heated tube through which the liquid flows with some velocity, boiling is called *forced-convection boiling*. Boiling of liquids in forced flow inside heated tubes has numerous applications in the design of steam generators for nuclear power plants, space power plants, and various advanced power generation systems. Since the velocity inside the tube affects the bubble growth and separation, the mechanism and hydrodynamics of boiling in forced convection are much more complex than in the pool boiling of a quiescent liquid. In addition to the various boiling regimes considered in pool boiling, there are many other boiling regimes in forced-convection boiling inside tubes. The problem is complicated further by the fact that the transition between the regimes should be understood as well. Numerous experimental investigations of this subject, aimed at the understanding of the complex heat transfer mechanism, have been reported [95–106]. An excellent discussion of the problems of forced-convection boiling is given by Collier [97], and an overview of the convective boiling is given in Ref. 106.

To illustrate various flow regimes involved in forced-convection boiling inside tubes, we consider upward flow of water in a long, vertical tube uniformly heated over its length. We examine separately the cases involving a *low heat flux* and a *progressively increasing heat flux*.

Low Heat Flux

Figure 10-10 illustrates, at least qualitatively, various flow patterns and the corresponding heat transfer regimes associated with the upward flow of a subcooled liquid in a vertical tube subjected to a uniform low heat flux over its length.

The subcooled liquid entering the tube is first heated by the mechanism of *forced-convection heat transfer to single-phase liquid*, in region A. In this region, the difference between the wall and the bulk fluid temperatures remains relatively constant.

Region A is followed by the *subcooled nucleata boiling regime*, region B. In the subcooled nucleata boiling regime, the bubbles form and collapse; therefore, the vapor quality remains zero (that is, $x = 0$) until the end of region B. In this region, the temperature difference between the wall and the bulk fluid decreases linearly up to the point where $x = 0$. This implies a gradual increase in the heat transfer coefficient, since the wall heat flux is constant.

In the *saturated nucleata boiling* regime, regions C and D, the temperature difference and therefore the heat transfer coefficient, remains constant.

In the *two-phase forced-convection regime*, regions E and F, the temperature difference decreases and the heat transfer coefficient increases with the distance along the tube. The reason is the reducing thickness of the liquid film as the vapor velocity increases.

At the *dryout* point, there is a sudden rise in the temperature difference with a corresponding decrease in the heat transfer coefficient. Beyond the dryout point, the heat transfer coefficient is equal to that expected for heat transfer by forced convection to saturated steam.

In the *liquid-deficient* regime, region G, the vapor quality continuously increases and the temperature difference decreases with the corresponding increase in the heat transfer coefficient.

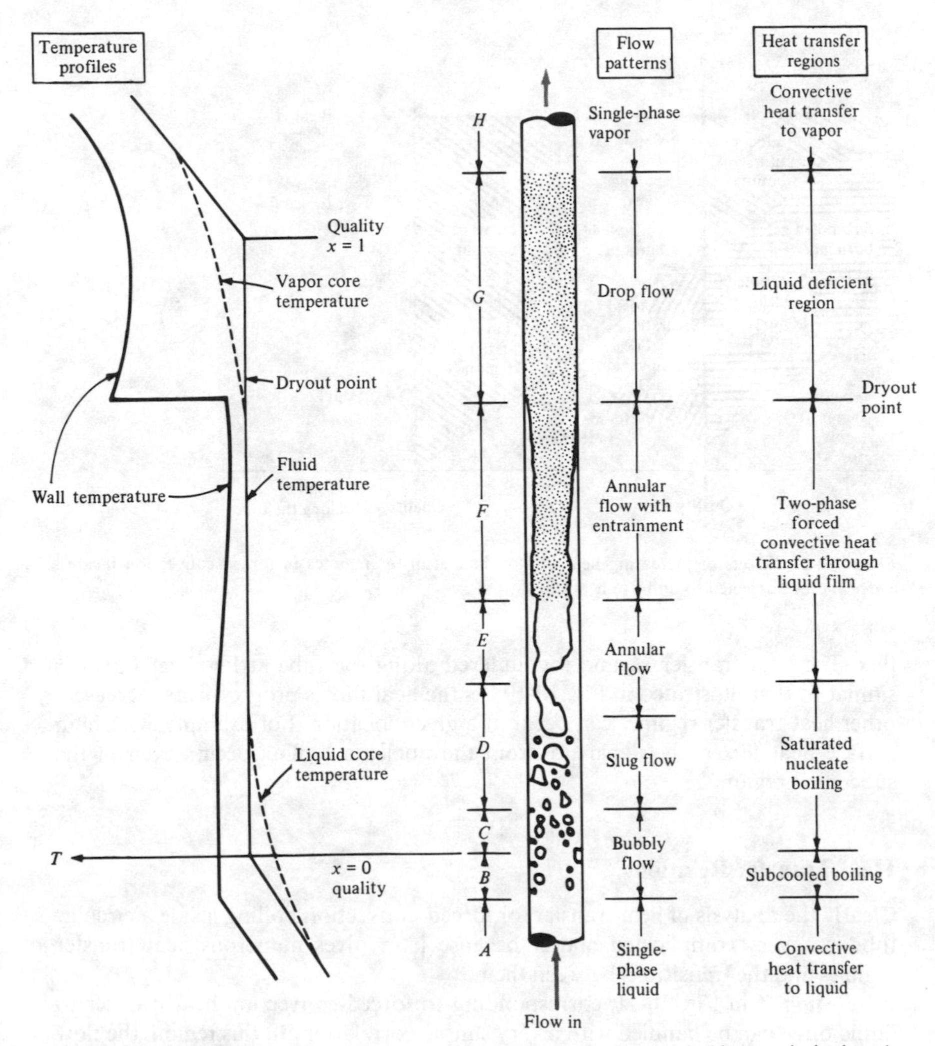

Figure 10-10 Various flow and heat transfer regimes in forced convection inside a vertical tube subjected to uniform heat flux.

High Surface Heat Flux

In the previous discussion, we considered flow in a tube with a relatively low heat flux at the wall. Suppose the flow rate of liquid in the tube remains constant, but higher heat fluxes are applied to the wall surface. Figure 10-11, provided by Collier, illustrates various other heat transfer regimes that are encountered as the surface heat flux is increased progressively. In this figure, the axial coordinate represents the distance along the tube length; and the transverse coordinate, the wall heat

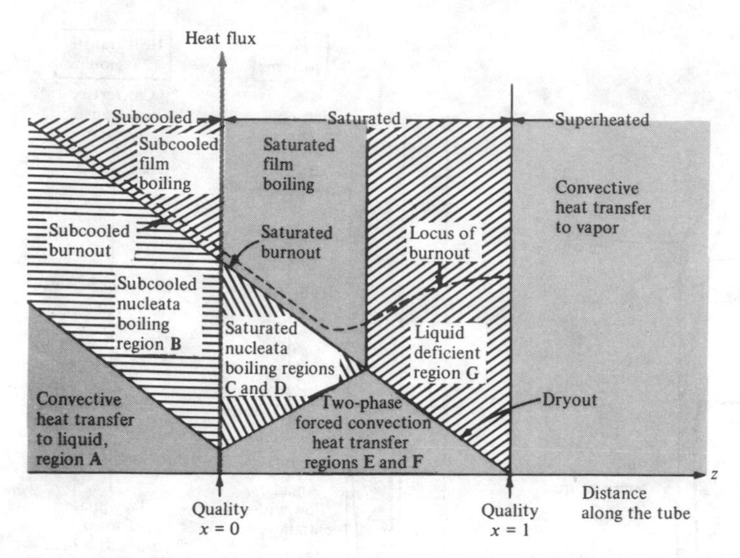

Figure 10-11 Effects of increasing heat flux on heat transfer regimes in forced convection inside a vertical tube subjected to uniform heat flux.

flux. The heat transfer regime encountered along the tube at low heat fluxes is similar to that illustrated in Fig. 10-10. As the heat flux is progressively increased, other heat transfer regimes can occur at a given location. For example, with high surface heat fluxes, the departure from the nucleata boiling occurs even in the subcooled region.

Heat Transfer Relations

Clearly the analysis of heat transfer for forced-convection boiling inside a circular tube is a very complicated matter because it involves numerous heat transfer regimes and the transitions between them.

Region A in Fig. 10-11, corresponding to forced-convection heat transfer to liquid only, can be handled with a very simple correlation. In this region, the flow of subcooled liquid in the tube may be laminar or turbulent. In either case, the heat transfer coefficient can be calculated by the well-established methods discussed in Chap. 7. For example, for fully developed turbulent flow, the Dittus–Boelter equation may be used:

$$\boxed{\mathrm{Nu} \equiv \frac{h_m D}{k} = 0.023\ \mathrm{Re}^{0.8}\ \mathrm{Pr}^{0.4}} \tag{10-41}$$

which is valid for $L/D > 60$ and $\mathrm{Re} > 10{,}000$; various quantities are defined in

Chap. 7. Given the heat transfer coefficient, the heat flux in the tube for the heat transfer regime A is determined from

$$q = h_m(T_w - T_l)$$

where T_w is the tube surface temperature and T_l is the bulk fluid temperature.

After the boiling is initiated, the whole surface eventually becomes covered with bubbles, and the *boiling becomes fully developed*. In the subcooled, fully developed boiling, the surface temperature is essentially a function of the surface heat flux and the system pressure for a given fluid. With this consideration, equations of the same form as that used for pool boiling can be suitable for correlating forced-convection fully developed, subcooled boiling data.

Therefore, heat transfer for the *subcooled boiling regime B*, the forced-convection, fully developed boiling, as discussed by Collier (Chap. 8, Ref. 106), can be correlated by

$$\boxed{\frac{c_{pl}\,\Delta T}{h_{fg}\,\text{Pr}_l^n} = C_{sf}\left[\frac{q}{\mu_l h_{fg}}\sqrt{\frac{g_c\sigma^*}{g(\rho_l - \rho_v)}}\right]^{0.33}} \tag{10-42}$$

where $n = 1$ for water, $n = 1.7$ for other liquids, and all other quantities are as defined for Eq. (10-34) except the coefficient C_{sf}. The recommended values of C_{sf} for forced- (and natural-) convection subcooled boiling in circular tubes, determined by various investigations [107–110], are listed in Table 10-4. The subcooled boiling of water flowing in heated tubes has been investigated by others [111, 112], and alternative correlations have been proposed. However, ΔT calculated from such correlations does not seem to be much different from that obtained by the pool boiling correlation considered above.

Table 10-4 Values of C_{sf} of Eq. (10-42) for forced- (natural-) convection boiling

Geometry	Liquid-surface combination	C_{sf}	Reference
Horizontal tube (14.9-mm ID)	Water–stainless steel	0.015	[107]
Horizontal tube (2.39-mm ID)	Water–stainless steel	0.020	[110]
Vertical tube (4.56-mm ID)	Water–nickel	0.006	[108]
Vertical tube (27.1-mm ID)	Water–copper	0.013	[109]
	Carbon tetrachloride–copper	0.013	
	Isopropyl alcohol–copper	0.0022	
	n-Butyl alcohol–copper	0.0030	
	50 percent K_2CO_3–copper	0.00275	
	35 percent K_2CO_3–copper	0.0054	

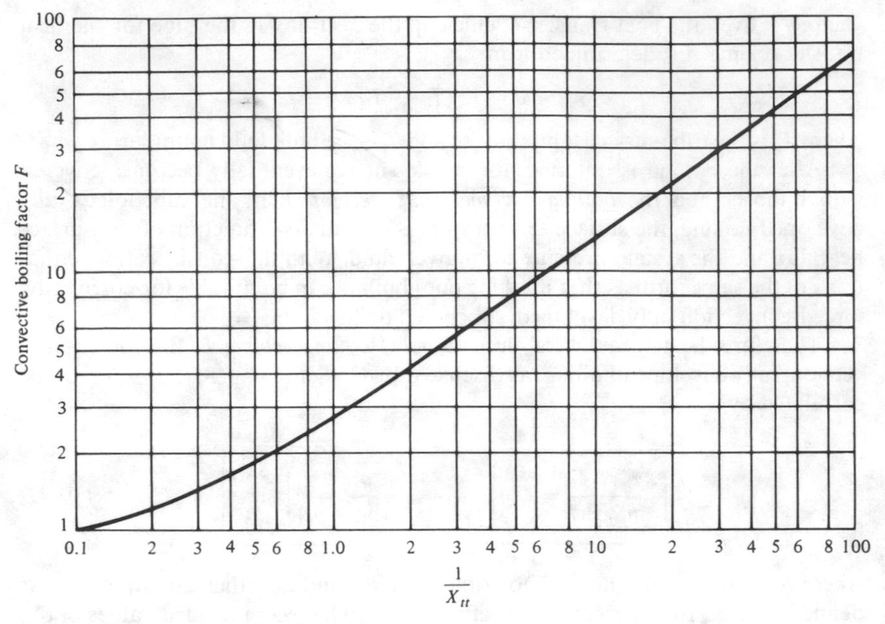

Figure 10-12 Convective boiling factor *F*.

Chen [96] proposed a correlation that covers both the *saturated nucleata boiling regions C and D* and the *two-phase forced-convection regions E and F*, illustrated in Fig. 10-11. The analysis is rather complex. Here we present a summary of Chen's results for the determination of the *two-phase heat transfer coefficient* h_{TP} for these regions.

It is assumed that in these regions heat transfer takes place by both nucleation and convection. Therefore, the two-phase heat transfer coefficient h_{TP} is considered to be composed of the contributions made by these two different mechanisms:

$$h_{TP} = h_{NB} + h_c \tag{10-43}$$

where h_{NB} = contribution from nucleata boiling and h_c = contribution from the forced convection. It is assumed that the forced-convection contribution h_c can be determined from the modified Dittus–Boelter equation, given in the form

$$h_c = 0.023\left(\frac{k_l}{D}\right) \text{Re}_l^{0.8} \text{Pr}_l^{0.4} F \tag{10-44}$$

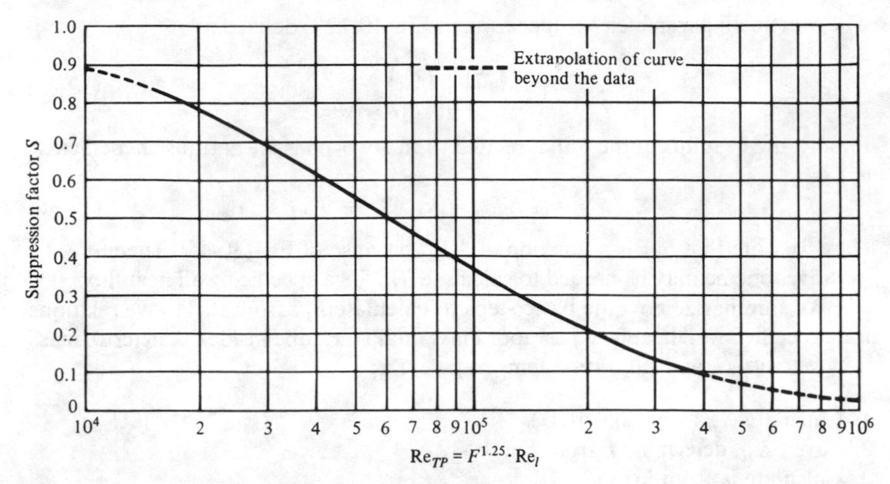

Figure 10-13 Suppression factor S.

where $\mathrm{Re}_l = \dfrac{G(1-x)D}{\mu_l}$

$G = \dfrac{\text{mass flow rate through tube}}{A_{\text{tube}}}$, kg/(m$^2 \cdot$ s)

D = tube inside diameter, m

F = convective boiling factor

x = vapor mass quality

Here the subscript l refers to the liquid phase. Figure 10-12 gives the convective boiling factor F plotted as a function of the Martinelli parameter X_{tt}.

The nucleata boiling heat transfer coefficient h_{NB} is taken as the modified Forster–Zuber [113] equation in the form

$$h_{NB} = 0.00122\left(\frac{k_l^{0.79} c_{pl}^{0.45} \rho_l^{0.49}}{\sigma^{*0.5} \mu_l^{0.29} h_{fg}^{0.24} \rho_v^{0.24}}\right)\Delta T_{\text{sat}}^{0.24} \cdot \Delta P_{\text{sat}}^{0.75} \cdot S \tag{10-45}$$

where $\Delta T_{\text{sat}} = T_w - T_{\text{sat}}$, °C

$\Delta P_{\text{sat}} = P_{\text{sat at } T_w} - P_{\text{sat at } T_{\text{sat}}}$, N/m^2

σ^* = surface tension, N/m

S = suppression factor

The subscripts l and v refer, respectively, to the liquid and the vapor phases.

Figure 10-13 gives the empirical factor S plotted against the two-phase Reynolds number Re_{TP}, defined as

$$\mathrm{Re}_{TP} = F^{1.25}\,\mathrm{Re}_l = F^{1.25}\left[\frac{G(1-x)D}{\mu_l}\right] \tag{10-46}$$

The Martinelli parameter X_{tt} appearing in Fig. 10-12 is defined as

$$X_{tt} = \left(\frac{1-x}{x}\right)^{0.9} \left(\frac{\rho_v}{\rho_l}\right)^{0.5} \left(\frac{\mu_l}{\mu_v}\right)^{0.1} \tag{10-47}$$

Finally, the heat flux at the wall is related to the two-phase heat transfer coefficient h_{TP} by

$$q = h_{TP}\, \Delta T_{sat} = h_{TP}(T_w - T_{sat}) \tag{10-48}$$

Here we note that h_{TP} is a function of ΔT_{sat} because of Eq. (10-45). Therefore, an iterative scheme may be needed to calculate h_{TP} for a specified wall heat flux.

We summarize now the basic steps to calculate h_{TP} from the above relations for a specified wall heat flux q_w, vapor mass quality x, tube inside diameter D, mass flow velocity G, and saturation temperature T_{sat}:

1. Calculate X_{tt} from Eq. (10-47).
2. Given X_{tt}, determine F from Fig. 10-12.
3. Calculate h_c from Eq. (10-44).
4. Calculate Re_{TP} from Eq. (10-46)
5. Determine S from Fig. 10-13.
6. Calculate h_{NB} from Eq. (10-45), and express it in the form

$$h_{NB} = C\, \Delta T_{sat}^{0.24}\, \Delta P_{sat}^{0.75} \tag{10-49}$$

7. Use Eq. (10-43) to determine h_{TP}:

$$h_{TP} = h_{NB} + h_c$$

$$h_{TP} = c\, \Delta T_{sat}^{0.24}\, \Delta P_{sat}^{0.75} + h_c \tag{10-50}$$

8. Use Eq. (10-48) for the wall heat flux q_w:

$$q_w = h_{TP}\, \Delta T_{sat}$$

$$q_w = c\, \Delta T_{sat}^{1.24}\, \Delta P_{sat}^{0.75} + h_c\, \Delta T_{sat} \tag{10-51}$$

where

$$\Delta T_{sat} = T_w - T_{sat} \tag{10-52a}$$

$$\Delta P_{sat} = P_{sat}(\text{ at } T_w) - P_{sat}(\text{at } T_{sat}) \tag{10-52b}$$

When the wall heat flux q_w is specified, an iterative scheme is needed to calculate ΔT_{sat} from Eq. (10-51), because ΔP_{sat} depends on ΔT_{sat}.

One way to perform the iterations is as follows: Make a guess for ΔT_{sat}, and calculate T_w from Eq. (10-52*a*), since T_{sat} is known. Determine P_{sat} at the temperatures T_w and T_{sat} from the physical property table. Then ΔT_{sat} and ΔP_{sat} are available, and the right-hand side of Eq. (10-51) can be computed. If the right-hand side value is different from the specified value of q_w of the left-hand side, repeat the calculations with different values of ΔT_{sat} until the equality is satisfied.

Given ΔT_{sat}, the two-phase heat transfer coefficient h_{TP} can be determined from Eq. (10-48):

$$h_{TP} = \frac{q_w}{\Delta T_{sat}} \tag{10-53}$$

Example 10-10 Saturated water at $T_{sat} = 180°C$ flows with a mass flow rate of $M = 0.1$ kg/s through a 2.5-cm-ID tube, subjected to a uniform wall heat flux of $q_w = 135$ kW/m^2. Calculate the tube wall temperature T_w and the two-phase heat transfer coefficient h_{TP} at the location where the vapor mass quality is $x = 0.25$.

SOLUTION The physical properties of saturated water and steam at $T_{sat} = 180°C$ are taken as

$k_l = 0.675$ W/(m · °C) $\quad \mu_l = 0.154 \times 10^{-3}$ kg/(m · s)

$\mu_v = 0.1525 \times 10^{-4}$ kg/(m · s) $\quad \text{Pr}_l = 1.004$

$c_{pl} = 4417$ J/(kg · °C) $\quad h_{fg} = 2015 \times 10^3$ J/kg

$\rho_l = 889$ kg/m^3 $\quad \rho_v = 5.15$ kg/m^3

$\sigma^* = 41.8 \times 10^{-3}$ N/m $\quad P_{sat}(\text{at } 180°C) = 1.002 \times 10^6$ N/m^2

This problem is now solved by following the calculation procedure just described:

1. Calculate X_{tt} from Eq. (10-47):

$$X_{tt} = \left(\frac{1 - 0.25}{0.25}\right)^{0.9} \left(\frac{5.15}{889}\right)^{0.5} \left(\frac{0.154 \times 10^{-3}}{0.1525 \times 10^{-4}}\right)^{0.1} = 0.258$$

2. Determine F from Fig. 10-12:

$$F = 6.5$$

3. Calculate h_c from Eq. (10-44):

$$G = \frac{M}{A_{tube}} = \frac{0.1}{(\pi/4)(0.025)^2} = 203.7 \text{ kg/(m}^2 \cdot \text{s)}$$

$$\text{Re}_l = \frac{G(1 - x)D}{\mu_l} = \frac{203.7(1 - 0.25)(0.025)}{0.154 \times 10^{-3}} = 24{,}801$$

$$h_c = 0.023\left(\frac{0.675}{0.025}\right)(24{,}801)^{0.8}(1.004)^{0.6}(6.5)$$

$$= 13{,}230$$

4. Calculate Re_{TP} from Eq. (10-46):

$$\text{Re}_{TP} = F^{1.25}\,\text{Re}_l = (6.5)^{1.25}(24{,}801) = 257{,}400$$

5. Determine S from Fig. 10-13:

$$S = 0.17$$

6. Calculate h_{NB} from Eq. (10-45):

$$h_{NB} = 0.00122\left[\frac{(0.675)^{0.79}(4417)^{0.45}(889)^{0.49}}{(0.0418)^{0.5}(0.154 \times 10^{-3})^{0.29}(2015 \times 10^3)^{0.24}(5.15)^{0.24}}\right]$$

$$\times\, \Delta T_{sat}^{0.24}\, \Delta P_{sat}^{0.75}\, 0.17$$

$$= 0.239\, \Delta T_{sat}^{0.24}\, \Delta P_{sat}^{0.75}$$

7. Calculate h_{TP} from Eq. (10-43):

$$h_{TP} = h_{NB} + h_c = 0.239\,\Delta T_{\text{sat}}^{0.24}\,\Delta P_{\text{sat}}^{0.75} + 13{,}230$$

8. Represent q_w according to Eq. (10-48):

$$q_w = h_{TP}\,\Delta T_{\text{sat}}$$

$$135{,}000 = 0.239\,\Delta T_{\text{sat}}^{1.24}\,\Delta P_{\text{sat}}^{0.75} + 13{,}230\,\Delta T_{\text{sat}}$$

We now use this equation to calculate ΔT_{sat} by iteration. Let $\Delta T_{\text{sat}} = 10°\text{C}$. Then

$$T_w = 180 + 10 = 190°\text{C}$$

$$P_{\text{sat}}(\text{at } 190°\text{C}) = 1.2544 \times 10^6\ \text{N/m}^2$$

$$P_{\text{sat}}(\text{at } 180°\text{C}) = 1.0021 \times 10^6\ \text{N/m}^2$$

$$\Delta P_{\text{sat}} = 0.2523 \times 10^6\ \text{N/m}^2$$

These results are introduced into the above expression:

$$135{,}000 \stackrel{?}{=} 0.239(10)^{1.24}(0.2523 \times 10^6)^{0.75} + (13{,}230)(10)$$

$$135{,}000 \neq 179{,}056$$

They are not equal. We try $\Delta T_{\text{sat}} = 8°\text{C}$ and obtain

$$135{,}000 \cong 135{,}390$$

which is sufficiently close. Then

$$\Delta T_{\text{sat}} = 8°\text{C} \qquad T_w = 188°\text{C}$$

and

$$h_{TP} = \frac{135{,}000}{8} = 16{,}875\ \text{W/(m}^2 \cdot °\text{C)}$$

10-9 SUMMARY OF EQUATIONS

In Table 10-5 we summarize the equations given in this chapter.

Table 10-5 Summary of equations

Equation number	Correlation	Remarks
(10-10)	$\delta(x) = \left[\dfrac{4\mu_l k_l(T_v - T_w)x}{g(\rho_l - \rho_v)\rho_l h_{fg}}\right]^{1/4}$	Thickness of condensate film on a vertical plate for laminar film condensation
(10-27a)	$h_m = 1.13\left[\dfrac{g\rho_l^2 h_{fg} k_l^3}{\mu_l(T_v - T_w)L}\right]^{1/4}$	Average heat transfer coefficient for laminar film condensation on a vertical surface
	or	
(10-27b)	$h_m\left(\dfrac{\mu_l^2}{k_l^3\rho_l^2 g}\right)^{1/3} = 1.76\ \text{Re}^{-1/3}$	$\rho_v \ll \rho_l$ $\text{Re} < 1800$

Table 10-5 *(Continued)*

Equation number	Correlation	Remarks
(10-28)	$h_m\left(\frac{\mu_l^2}{k_l^3\rho_l^2 g}\right)^{1/3} = 0.0077\ \mathrm{Re}^{0.4}$	Average heat transfer coefficient on a vertical surface for turbulent film condensation $\rho_v \ll \rho_l \quad \mathrm{Re} > 1800$
(10-29*a*)	$h_m = 0.555\left[\frac{g\rho_l(\rho_l - \rho_v)k_l^3 h'_{fg}}{\mu_l(T_{sat} - T_s)D}\right]^{1/4}$	Film condensation inside horizontal tubes at low vapor velocities
(10-29*b*)	where $h'_{fg} \equiv h_{fg} + \frac{3}{8}c_{pl}(T_v - T_w)$	$\mathrm{Re}_v = \frac{\rho_v u_v D}{\mu_v} < 35{,}000$
(10-31)	$\frac{h_m D}{k_l} = 0.026\ \mathrm{Pr}_l^{1/3}\left[\mathrm{Re}_l + \mathrm{Re}_v\left(\frac{\rho_l}{\rho_v}\right)^{1/2}\right]^{0.8}$	Film condensation inside horizontal tubes at higher flow rates
(10-34)	$\frac{c_{pl}\,\Delta T}{h_{fg}\,\mathrm{Pr}_l^n} = C_{sf}\left[\frac{q}{\mu_l h_{fg}}\sqrt{\frac{g_c\sigma^*}{g(\rho_l - \rho_v)}}\right]^{0.33}$	Heat flux in nucleata pool boiling; coefficient C_{sf} is given in Table 10-1; $n = 1$ for water, $n = 1.7$ for other liquids
(10-35)	$q_{max} = \frac{\pi}{24}\rho_v h_{fg}\left[\frac{\sigma^* g g_c(\rho_l - \rho_v)}{\rho_v^2}\right]^{1/4}\left(1 + \frac{\rho_v}{\rho_l}\right)^{1/2}$	Maximum heat flux in pool boiling
(10-36)	$q_{max} = F(L') \times 0.131\rho_v^{1/2} h_{fg}[\sigma^* g g_c(\rho_l - \rho_v)]^{1/4}$	Maximum heat flux in pool boiling; factor $F(L')$ is for effects of heater geometry and is given in Table 10-3
(10-38)	$h_0 = 0.62\left[\frac{k_v^3\rho_v(\rho_l - \rho_v)g h_{fg}}{\mu_v D_o\,\Delta T}\left(1 + \frac{0.4c_{pv}\,\Delta T}{h_{fg}}\right)\right]^{1/4}$	Heat transfer coefficient for stable film boiling without radiation effects
(10-39*a*)	$h_m = h_0\left(\frac{h_0}{h_m}\right)^{1/3} + h_r$	Heat transfer coefficient for stable film boiling with radiation effects
(10-39*b*)	where $h_r = \frac{1}{1/\varepsilon + 1/\alpha - 1}\,\frac{\sigma(T_w^4 - T_s^4)}{T_w - T_s}$	
(10-42)	$\frac{c_{pl}\,\Delta T}{h_{fg}\,\mathrm{Pr}_l^n} = C_{sf}\left[\frac{q}{\mu_l h_{fg}}\sqrt{\frac{g_c\sigma^*}{g(\rho_l - \rho_v)}}\right]^{0.33}$	Subcooled boiling inside tubes; C_{sf} is given in Table 10-4; $n = 1$ for water, $n = 1.7$ for other liquids
(10-43)	$h_{TP} = = h_{NB} + h_c$	Two-phase heat transfer coefficient h_{TP} for boiling in forced-convection regime inside circular tubes; F and S obtained from Figs. 10-12 and 10-13, respectively
(10-44)	where $h_c = 0.023\left(\frac{k_l}{D}\right)\mathrm{Re}_l^{0.8}\ \mathrm{Pr}^{0.4}\ F$	
(10-45)	$h_{NB} = 0.00122\left(\frac{k_l^{0.79}c_{pl}^{0.45}\rho_l^{0.49}}{\sigma^{*0.5}\mu_l^{0.29}h_{fg}^{0.24}\rho_v^{0.24}}\right)\Delta T_{sat}^{0.24}\,\Delta P_{sat}^{0.75}\ S$	

PROBLEMS

Laminar film condensation

10-1 The condensation heat transfer coefficient for inclined plane surfaces includes sin ϕ, as shown in Eqs. (10-17) and (10-18), where ϕ is the angle with the horizontal. By going through the derivation of the heat transfer coefficient for film condensation, show at which stage of the analysis the term "sin ϕ" enters the problem, and explain the reason.

10-2 Show that the condensation heat transfer coefficient given by Eq. (10-27*a*) also can be arranged in the form given by Eq. (10-27*b*).

10-3 Show that the average of the local heat transfer coefficient h_x given by Eq. (10-17) over the distance $0 \leq x \leq L$ is as given by Eq. (10-18).

10-4 Air-free saturated steam at $T_v = 90°C$ ($P = 70.14$ kPa) condenses on the outer surface of a 1.5-m-long, 2.5-cm-OD vertical tube maintained at a uniform temperature $T_w = 70°C$. Assuming film condensation, calculate (*a*) the local heat transfer coefficient at the bottom of the tube, (*b*) the average heat transfer coefficient over the entire length of the tube, (*c*) the condensate thickness at the bottom of the tube, (*d*) the total rate of condensation at the tube surface, and (*e*) the condensate Reynolds number at the bottom of the tube.

10-5 Air-free saturated steam at $T_v = 50°C$ ($P = 12.35$ kPa) condenses on the outside surface of a 2.5-cm-OD, 2-m-long vertical tube maintained at a uniform temperature $T_w = 30°C$ by the flow of cooling water through the tube. Assuming film condensation, calculate (*a*) the average condensation heat transfer coefficient over the entire length of the tube, (*b*) the rate of condensate flow at the bottom of the tube, and (*c*) the condensate thickness $\delta(x)$ at the bottom of the tube.

Answer: (*a*) 4356 W/(m^2 · °C), (*b*) 5.69×10^{-3} kg/s, (*c*) 0.23 mm

10-6 Calculate the average heat transfer coefficient h_m and the total condensation rate at the tube surface for Prob. 10-5 when the tube is in the horizontal position.

10-7 Repeat Prob. 10-5 for a tube of length $L = 1$ m.

Answer: (*a*) 5180 W/(m^2 · °C), (*b*) 3.383×10^{-3} kg/s, (*c*) 0.19 mm

10-8 Saturated air-free steam at $T_v = 65°C$ ($P = 25.03$ kPa) condenses on the outer surface of a 2.5-cm-OD vertical tube whose surface is maintained at uniform temperature $T_w = 35°C$. Determine the tube length needed to condense 30 kg/h of steam.

10-9 Saturated air-free steam at temperature $T_v = 80°C$ ($P = 47.39$ kPa) condenses on the outer surface of a 1.2-m-long, 0.1-m-diameter vertical tube which is maintained at a uniform temperature $T_w = 40°C$. Calculate (*a*) the average heat transfer coefficient h_m for film condensation over the entire tube length, (*b*) the total rate of steam condensation at the tube surface, and (*c*) the condensate thickness at the bottom of the tube.

10-10 Saturated air-free steam at $T_v = 75°C$ ($P \cong 38.58$ kPa) condenses on a 0.5 m-by-0.5 m vertical plate maintained at a uniform temperature $T_w = 45°C$. Calculate (*a*) the average film condensation heat transfer coefficient h_m over the entire length of the plate, (*b*) the total rate of condensation over the entire surface of the plate, and (*c*) the condensate thickness at the bottom of the plate.

Answer: (*a*) 6142 W/(m^2 · °C), (*b*) 1.953×10^{-2} kg/s, (*c*) 0.17 mm

10-11 Repeat Prob. 10-10 for the cases when the plate makes an angle (*a*) $\phi = 45°$ with the horizontal and (*b*) $\phi = 5°$ with the horizontal.

10-12 Saturated air-free steam at $T_v = 50°C$ ($P = 12.35$ kPa) condenses on the outer surface of a 1-m-long, 2.5-cm-OD vertical tube maintained at a uniform temperature of $T_w = 30°C$. Assuming film condensation, calculate the average condensation heat transfer coefficient h_m over the entire length of the tube and the total rate of condensation at the surface of the tube.

10-13 Air-free saturated steam at $T_v = 90°C$ ($P = 70.14$ kPa) condenses on the outer surface of a 2.54-cm-OD vertical tube maintained at a uniform temperature $T_w = 70°C$. Determine the tube length in order to condense 20 kg/h of steam.

Answer: 1.5 m

10-14 Saturated ammonia vapor at $T_v = -5°C$ ($P = 0.3528$ MPa) condenses on the outer surface of a 0.75-m-long, 1.27-cm-OD vertical tube maintained at a uniform temperature $T_w = -15°C$. Calculate the average condensation heat transfer coefficient h_m and the total rate of condensation of ammonia over the entire length of the tube. (Take $h_{fg} = 1280$ kJ/kg for ammonia.)

10-15 Saturated air-free steam at $T_v = 170°C$ ($P = 0.8$ MPa) condenses on the outer surface of a 2.5-cm-OD 1.2-m-long vertical tube maintained at a uniform temperature of $T_w = 150°C$. Calculate (*a*) the local film condensation heat transfer coefficient at the bottom of the tube, (*b*) the average condensation heat transfer coefficient over the entire length of the tube, and (*c*) the total condensation rate at the tube surface.

10-16 Saturated Freon-12 vapor at $T_v = -5°C$ ($P = 0.26$ MPa) condenses on the outer surface of a 1.2-m-long, 1.27-cm-OD vertical tube maintained at a uniform temperature of $T_w = -15°C$. Calculate (*a*) the average condensation heat transfer coefficient over the entire length of the tube, (*b*) the total heat transfer rate, and (*c*) the mass rate of condensation at the tube surface. (Take $h_{fg} = 154$ kJ/kg for Freon-12.)

Answer: (*a*) 848 W/(m^2 · °C), (*b*) 406 W, (*c*) 2.635×10^{-3} kg/s

Condensation on horizontal tube banks

10-17 Air-free saturated steam at $T_v = 60°C$ ($P = 19.94$ kPa) condenses on the outer surface of 100 horizontal tubes with 2.5-cm OD and 2 m long, arranged in a 10 by 10 square array. The surface of the tubes is maintained at a uniform temperature $T_w = 40°C$. Calculate the average condensation heat transfer coefficient for the entire tube bundle, the total rate of heat transfer, and the rate of condensation at the surface of the tubes in the bundle.

10-18 Consider N horizontal tubes arranged in a vertical tier. What number of tubes will produce an average condensation heat transfer coefficient for the vertical tier equal to half that for the single horizontal tube at the top?

10-19 A steam condenser consists of 625 horizontal tubes, with 1.25-cm OD and 3 m long, arranged in a 25-by-25 array. Saturated steam at $T_v = 50°C$ ($P = 12.35$ kPa) condenses on the outer surface of the tubes which are maintained at a uniform temperature $T_w = 30°C$. Calculate (*a*) the average heat transfer coefficient h_m, (*b*) the total rate of heat transfer, and (*c*) the condensation in the condenser.

Answer: (*a*) 4445 W/(m^2 · °C), (*b*) 6546 kW, (*c*) 2.72 kg/s

10-20 Compare the average condensation heat transfer coefficient for filmwise condensation of air-free steam at atmospheric pressure on a single 1-m-long, 2.5-cm OD vertical tube and on $N = 40$ horizontal tubes with 2.5-cm OD, tubes arranged in a vertical tier.

10-21 Saturated air-free steam at $T_v = 50°C$ ($P = 12.35$ kPa) condenses on the outer surface of a 2.5-cm-OD horizontal tube maintained at a uniform temperature $T_w = 30°C$. Calculate the tube length required to condense $M = 50$ kg/h of steam.

10-22 Air-free saturated steam at $T_v = 85°C$ ($P = 57.83$ kPa) condenses on the outer surface of 196 horizontal tubes with 1.27-cm OD arranged in a 14-by-14 array. The tube surfaces are maintained at a uniform temperature $T_w = 75°C$. Calculate the length of the matrix to condense $M = 0.7$ kg/s of steam.

Answer: 2.84 m

10-23 Some 400 tubes, each with 1.27-cm OD and 2 m long, arranged in a 20-by-20 array, are maintaned at a uniform temperature $T_w = 55°C$. Saturated steam at $T_v = 65°C$ ($P = 25.03$ kPa) condenses at the outer surface of the tubes. Calculate the average film condensation heat transfer coefficient and the total rate of condensation of steam at the surfaces of the tube bank.

Turbulent film condensation

10-24 Air-free saturated steam at $T_v = 70°C$ ($P = 31.19$ kPa) condenses on the outer surface of a 2.5-cm-OD vertical tube maintained at a uniform temperature $T_w = 50°C$. What length of tube would produce a turbulent film condensation?

10-25 Air-free saturated steam at $T_v = 90°C$ ($P = 70.14$ kPa) condenses on the outer surface of a 2.5-cm-OD, 6-m-long vertical tube maintained at a uniform temperature $T_w = 30°C$. Calculate (*a*) the average heat transfer coefficient over the entire length of the tube and (*b*) the total rate of condensation of steam at the tube surface.

Answer: (*a*) 6539 W/(m^2 · °C), (*b*) 0.0784 kg/s

10-26 Air-free steam at $T_v = 100°C$ ($P = 0.1013$ MPa) condenses on a 2-m-long vertical plate. What is the temperature T_w of the plate below which the condensing film at the bottom of the plate will become turbulent?

10-27 Air-free saturated steam at $T_v = 50°C$ ($P = 12.35$ kPa) condenses on the outer surface of a 2.5-cm-OD vertical tube maintained at a uniform temperature $T_w = 30°C$. Calculate the tube length L that would produce turbulent film condensation at the bottom of the tube and the total condensation rate at the tube surface.

Pool boiling

10-28 Saturated water at $T_s = 100°C$ is boiled inside a copper pan having a heating surface $A = 5 \times 10^{-2}$ m^2 which is maintained at a uniform temperature $T_w = 110°C$. Calculate (*a*) the surface heat flux and (*b*) the rate of evaporation.

Answer: (*a*) 144 kW/m^2, (*b*) 11.48 kg/h

10-29 Repeat Prob. 10-28 for a pan made of brass.

10-30 During the boiling of saturated water at $T_s = 100°C$ with an electric heating element, a heat flux of $q = 7 \times 10^5$ W/m^2 is achieved with a temperature difference of $\Delta T = T_w - T_s = 10.4°C$. What is the value of C_{sf} for the heater surface?

10-31 Saturated water at $T_v = 55°C$ ($P = 15.76$ kPa) is boiled with a copper heating element having a heating surface $A = 5 \times 10^{-3}$ m^2 and is maintained at a uniform temperature $T_w = 65°C$. Calculate the rate of evaporation.

Answer: 0.235 kg/h

10-32 A brass heating element of surface area $A = 0.04$ m^2, maintained at a uniform temperature $T_w = 112°C$, is immersed into saturated water at atmospheric pressure at temperature $T_s = 100°C$. Calculate the rate of evaporation.

10-33 Saturated water at $T_s = 100°C$ is boiled in a 20-cm-diameter pan with a temperature difference of $T_w - T_s = 10°C$. Calculate the rate of evaporation for (*a*) copper and (*b*) Teflon-pitted stainless steel.

10-34 Saturated water at $T_s = 100°C$ is boiled with a copper heating element. If the surface heat flux is $q = 400$ kW/m^2, calculate the surface temperature.

Answer: 114.1°C

10-35 An electrically heated copper plate of heating surface $A = 0.2$ m^2, maintained at a uniform temperature $T_w = 108°C$, is immersed in a water tank at $T_s = 100°C$ and atmospheric pressure. Calculate the rate of evaporation.

10-36 Water at a saturation temperature $T_s = 160°C$ ($P = 0.6178$ MPa) is boiled by using an electrically heated copper element with a temperature difference of $T_w - T_s = 10°C$.

(*a*) Calculate the surface heat flux.

(*b*) Compare this heat flux with that obtainable with water at saturation temperature, atmospheric pressure, and with the same temperature difference.

10-37 An electrically heated copper kettle with a flat bottom of diameter $D = 25$ cm is to boil water at atmospheric pressure at a rate of 2.5 kg/h. What is the temperature of the bottom surface of the kettle?

Answer: 106.1°C

10-38 If the coefficient C_{sf} in Eq. (10-34) is increased by a factor of 2, what is the change in the heat flux q while all other quantities remain the same?

Peak heat flux

10-39 Water at atmospheric pressure and saturation temperature is boiled by using an electrically heated, circular disk of diameter $D = 20$ cm with the heated surface facing up. The surface of the element is maintained at a uniform temperature $T_w = 110°C$. Calculate (*a*) the surface heat flux, (*b*) the rate of evaporation, and (*c*) the peak heat flux.

10-40 Saturated water is boiled in the nucleata boiling regime with a large-plate heating element. Calculate the peak heat flux obtainable at (*a*) $P = 1$ and (*b*) $P = 10$ atm.

Answer: (*a*) 1.27, (*b*) 2.98 MW/m^2

10-41 Determine the peak heat flux obtainable with nucleata boiling of saturated water at 1 atm in a gravitational field equal to one-eighth that of the earth. Compare this result with that obtainable at earth's gravitational field. Assume large heating element.

10-42 An electrically heated, copper, spherical heating element of diameter $D = 10$ cm is immersed in water at atmospheric pressure and saturation temperature. The surface of the element is maintained at a uniform temperature $T_w = 115°C$. Calculate (*a*) the surface heat flux (*b*) the rate of evaporation, and (*c*) the peak heat flux.

Answer: (*a*) 484 kW/m^2, (*b*) 24.3 kg/h, (*c*) 0.933 MW/m^2

10-43 Repeat Prob. 10-42 for a horizontal cylindrical heating element of diameter $D = 5$ cm and length $H = 20$ cm.

Film boiling

10-44 Water at saturation temperature and atmospheric pressure is boiled with an electrically heated platinum wire of diameter $D_o = 0.2$ cm, $\varepsilon = 1$ in the stable film boiling regime with a temperature difference $T_w - T_s = 654°C$. Calculate the film boiling heat transfer coefficient h_m and the heat flux.

10-45 Repeat Prob. 10-44 for a heating element of diameter $D_o = 0.6$ cm.

Answer: 272.7 $W/(m^2 \cdot °C)$, 178.4 kW/m^2

10-46 Water at saturation temperature and atmospheric pressure is boiled with an electrically heated, horizontal platinum wire of diameter $D = 0.2$ cm. Boiling takes place with a temperature difference of $T_w - T_s = 454°C$ in the stable film boiling regime. Calculate the film boiling heat transfer coefficient and the heat flux. Assume $\varepsilon = 1$.

10-47 Water at saturation temperature and atmospheric pressure is boiled in the stable film boiling regime with an electrically heated, horizontal platinum wire of diameter $d = 0.127$ cm. Calculate the surface temperature necessary to produce a heat flux $q = 150$ kW/m^2.

Forced-convection boiling

10-48 Saturated water at $T_{sat} = 150°C$ flows with a mass flow rate of $M = 0.05$ kg/s through a 2.5-cm-ID tube subjected to a uniform wall heat flux $q_w = 100$ kW/m^2. Calculate the tube wall temperature T_w and the two-phase heat transfer coefficient h_{TP} at the location where the vapor mass quality is $x = 0.20$.

Answer: 158°C

10-49 Saturated water at $T_{sat} = 120°C$ flows with a mass flow rate of $M = 0.10$ kg/s through a 2.5-cm-ID tube subjected to a uniform wall heat flux $q_w = 120$ kW/m^2. Calculate the tube wall temperature T_w and the two-phase heat transfer coefficient h_{TP} at the location where the vapor mass quality is $x = 0.15$.

REFERENCES

1. Nusselt, W.: "Die Oberflächenkondensation des Wasserdampfes," *Z. Ver. Deut. Ing.* **60**:541–569 (1916).
2. Colburn, A. P.: "The Calculation of Condensation Where a Portion of the Condensate Layer Is the Turbulent Flow," *Trans. AIChE* **30**:187–193 (1933).
3. Kirkbride, C. G.: "Heat Transfer by Condensing Vapors," *Trans. AIChE* **30**:170–186 (1934).
4. Carpenter, E. F., and A. P. Colburn: "The Effect of Vapor Velocity on Condensation inside tubes," *Inst. Mech. Eng.–ASME, Proc. General Discussion Heat Transfer*, pp. 20–26, 1951.
5. Rohsenow, W. M., J. M. Weber, and A. T. Ling: "Effect of Vapor Velocity on Laminar and Turbulent Film Condensation," *Trans. ASME* **78**:1637–1644 (1956).
6. Sparrow, E. M., and J. L. Gregg: "A Boundary Layer Treatment of Laminar Film Condensation," *J. Heat Transfer* **81C**:13–18 (1959).

7. Chen, M. M.: "An Analytical Study of Laminar Film Condensation: Pt. I, Flat Plates," *J. Heat Transfer* **83C**:48–54 (1961).
8. Koh, J. C. Y., E. M. Sparrow, and J. P. Hartnett: "The Two Phase Boundary Layer in Laminar Film Condensation," *Int. J. Heat Mass Transfer* **2**:69–82 (1961).
9. Koh, J. C. Y.: "Film Condensation in a Forced-Convection Boundary Layer Flow," *Int. J. Heat Mass Transfer* **5**:941–954 (1962).
10. Koh, J. C. Y.: "An Integral Treatment of Two-Phase Boundary Layer in Film Condensation," *J. Heat Transfer* **83C**:359–362 (1961).
11. Chung, P. M.: "Unsteady Laminar Film Condensation on Vertical Plate," *J. Heat Transfer* **85C**:63–70 (1963).
12. Madejski, J.: "The Effect of Molecular-Kinetic Resistances on Heat Transfer in Condensation," *Int. J. Heat Mass Transfer* **9**:35–39 (1966).
13. Jones, W. P., and U. Renz: "Condensation from a Turbulent Stream onto a Vertical Surface," *Int. J. Heat Mass Transfer* **17**:1019–1028 (1974).
14. Tamir, A., and I. Rachmilev: "Direct Contact Condensation of an Immiscible Vapor on a Thin Film of Water," *Int. J. Heat Mass Transfer* **17**:1241–1251 (1974).
15. Colburn, A. P., and T. B. Drew: "The Condensation of Mixed Vapors," *Trans. AIChE* **33**:197–208 (1937).
16. Sparrow, E. M., and E. R. G. Eckert: "Effects of Superheated Vapor and Noncondensable Gas on Laminar Film Condensation," *AIChE J.* **7**:473–477 (1961).
17. Sparrow, E. M., and S. H. Lin: "Condensation Heat Transfer in the Presence of Noncondensable Gas," *J. Heat Transfer* **86C**:430–436 (1964).
18. Modlem, D., and S. Siedman: "Theoretical Analysis of Horizontal Condenser-Evaporator Tube," *Int. J. Heat Mass Transfer* **19**: 259–270 (1976).
19. Epstein, M., and D. H. Cho: "Laminar Film Condensation with Heat Generation," *J. Heat Transfer* **97C**, 141–143 (1975).
20. McAdams, W. H.: *Heat Transmission*, 3d ed., McGraw-Hill, New York, 1954.
21. Kirkbride, C. G.: "Heat Transfer by Condensing Vapors," *Trans. AIChE* **30**:170–186 (1934)
22. Colburn, A. P.: "Problems in Design and Research on Condensers of Vapors and Mixtures," *Int. Mech. Eng.–ASME Proc. General Discussions Heat Transfer*, pp. 1–11, 1951.
23. Tepe, J. B., and A. C. Mueller: "Condensation and Subcooling inside an Inclined Tube," *Chem. Eng. Prog.* **43**:267 (1947).
24. Altman, M., F. W. Staub, and R. H. Norris: "Local Heat Transfer and Pressure Drop for Refrigerant-22 Condensing in Horizontal Tubes," *ASME–AIChE* Storrs, Conn., August 1959.
25. Goodykoontz, J. H., and R. G. Dorsch: "Local Heat-Transfer Coefficients and Static Pressures for Condensation of High-Velocity Steam with a Tube," *NASA Tech. Note* D-3953, May 1967.
26. Goodykoontz, J. H., and W. F. Brown: "Local Heat Transfer and Pressure Distribution for Freon-113 Condensing Downward Flow in a Vertical Tube," *NASA Tech. Note* D-3952, May 1967.
27. Soliman, M., J. R. Schuster, and P. J. Berenson: "A General Heat Transfer Correlation for Annular Flow Condensation," *J. Heat Transfer* **90C**:267–276 (1968).
28. Carpenter, F. G.: "Heat Transfer and Pressure Drop for Condensing Pure Vapors inside Vertical Tubes at High Vapor Velocities," Ph.D. dissertation, University of Delaware, Newark, 1948. Also available as American Documentation Institute, Document 3274, Washington.
29. Dhir, V. K.: "Quasi-Steady Laminar Film Condensation of Steam on Copper Spheres," *J. Heat Transfer* **97C**:347–351 (1975).
30. Popiel, Cz. O., and L. Boguslawski: "Heat Transfer by Laminar Film Condensation on Sphere Surfaces," *Int. J. Heat Mass Transfer* **18**:1486–1488 (1975).
31. Karimi, A.: "Laminar Film Condensation on Helical Reflux Condensers and Related Configurations," *Int. J. Heat Mass Transfer* **20**:1137–1144 (1977).
32. Rohsenow, W. M.: *Handbook of Heat Transfer*, McGraw-Hill, New York, 1973, chap. 12.
33. Chato, J. C.: "Laminar Condensation inside Horizontal and Inclined Tubes," *J. Am. Soc. Heating Refrig. Aircond, Engrs.* **4**:52 (1962).

34. Akers, W. W., H. A. Deans, and O. K. Crosser: "Condensing Heat Transfer within Horizontal Tubes," *Chem. Eng. Prog. Symp. Ser.* **55:** 171 (1958).
35. Schmidt, E., W. Schurig, and W. Sellschopp: "Versuche über die Kondensation von Wasserdampf in Film- und Tropfenform," *Tech. Mech. u. Thermodynam.* **1:** 53 (1930).
36. Hampson, H., and N. Özişik: "An Investigation into the Condensation of Steam," *Proc. Inst. Mech. Eng., London,* **1B:** 282–294 (1952).
37. Blackman, L. C. F., and M. J. S. Dewar: "Promoters for Dropwise Condensation of Steam, Pts. I–IV," *J. Chem. Soc.* pp. 162–176, January–March 1957.
38. Blackman, L. C. F., M. J. S. Dewar, and H. Hampson: "An Investigation of Compounds Promoting Dropwise Condensation of Steam," *J. Appl. Chem.* **7:** 160–171 (1957).
39. Osment, B. D. J., D. Tudor, R. M. M. Speirs, and W. Rugman: "Promoters for the Dropwise Condensation of Steam," *Trans. Inst. Chem. Eng.* **40:** 152–160 (1962).
40. Erb. R. A., and E. Thelen: "Promoting Permanent Dropwise Condensation," *Ind. Eng. Chem.* **57:** 49–52 (1965).
41. Umur, A., and P. Griffith: "Mechanism of Dropwise Condensation," ASME Paper 64-WA/HT-3, 1964.
42. Lefevre, E. S., and J. Rose: "A Theory of Dropwise Condensation of Steam," *Proc. ASME–Inst. Mech. Eng. Heat Transfer Conf.,* 1966.
43. Çitakoğlu, E., and J. W. Rose: "Dropwise Condensation—Some Factors Influencing the Validity of Heat Transfer Measurements," *Int. J. Heat Mass Transfer* **11:** 523–537 (1968).
44. Glicksman, L., and A. Hunt: "Numerical Simulation of Dropwise Condensation," *Int. J. Heat Mass Transfer* **15:** 2251–2269 (1972).
45. Tanasawa, I., and J. Ochiai: "Experimental Study of the Dropwise Condensation Process," *Trans. J.S.M.E.* **38:** 3193 (1972).
46. Aksan, S., and J. Rose: "Dropwise Condensation—The Effect of Thermal Properties of the Condenser Material," *Int. J. Heat Mass Transfer* **16:** 461–467 (1973).
47. Graham, C., and P. Griffith: "Dropsize Distribution and Heat Transfer in Dropwise Condensation," *Int. J. Heat Mass Transfer* **16:** 337–346 (1973).
48. Tanasawa, I., F. Tachibana, and J. Ochiai: "A Study of the Process of Drop Growth by Coalescence during Dropwise Condensation," *Bull. J.S.M.E.* **16:** 1367 (1973).
49. Hannemann, R. J., and B. B. Mikic: "An Analysis of the Effect of Surface Thermal Conductivity on the Rate of Heat Transfer in Dropwise Condensation," *Int. J. Heat Mass Transfer* **19:** 1299–1307 (1976).
50. Brodowicz, K., and A. Czaplicki: "Effect of Droplet Coalescence in Dropwise Condensation," *Proc. 7th Int. Heat Transfer Conf.* vol. 5, pp. 3–7, Münich 1982.
51. Othmer, D. F.: "The Condensation of Steam." *Ind. Eng. Chem.* **21:** 576–583 (1929).
52. Colburn, A. P., and T. B. Drew: "The Condensation of Mixed Vapors," *Trans. AIChE* **33:** 197–208 (1937).
53. Meisenburg, S. J., R. M. Boarts, and W. L. Badger: *Trans. AIChE* **31:** 622–637 (1935).
54. Hampson, H.: "The Condensation of Steam on a Metal Surface," *Int. Mech. Eng.–ASME, Proc. General Discussion Heat Transfer,* pp. 58–61, 1951.
55. Sparrow, E. M., and E. R. G. Eckert: "Effects of Superheated Vapor and Noncondensable Gas on Laminar Film Condensation," *AIChE J.* **7:** 473–477 (1961).
56. Sparrow, E. M., and S. H. Lin: "Condensation Heat Transfer in the Presence of Noncondensable Gas," *J. Heat Transfer* **86C:** 430–436 (1964).
57. Minkowycz, W. J., and E. M. Sparrow: "Condensation Heat Transfer in the Presence of Noncondensables, Interfacial Resistance, Superheating, Variable Properties, and Diffusion," *Int. J. Heat Mass Transfer* **9:** 1125–1144 (1966).
58. Mori, Y., and K. Hijikata: "Free Convective Condensation Heat Transfer with Noncondensable Gas on a Vertical Surface," *Int. J. Heat Mass Transfer* **16:** 2229–2240 (1973).
59. Turner, R. H., A. F. Mills, and V. E. Denny: "The Effect of Noncondensable Gas on Laminar Film Condensation of Liquid Metals," *J. Heat Transfer* **95C:** 6–11 (1973).
60. Denny, V. E., and V. J. Jusionis: "Effects of Noncondensable Gas on Forced Flow on Laminar Film Condensation," *Int. J. Heat Mass Transfer* **15:** 315–326 (1972).

61. Nukiyama, S.: "The Maximum and Minimum Values of the Heat *Q* Transmitted from Metal to Boiling Water under Atmospheric Pressure," *J. Jap. Soc. Mech. Eng.* **37:**367–374 (1934). [Transl: *Int. J. Heat Mass Transfer* **9:**1419–1433 (1966).]
62. Farber, E. A., and R. L. Scorah: "Heat Transfer to Water Boiling under Pressure," *Trans. ASME* **79:**369–384 (1948).
63. Jens, W. H., and G. Leppert: "Recent Developments in Boiling Research, Pts, I, II," *J. Am. Soc. Nav. Eng.* **67:**137–155 (1955); **66:**437–456 (1955).
64. Rohsenow, W. M.: "Boiling Heat Transfer," in W. M. Rohsenow (ed.), *Developments in Heat Transfer*," M.I.T. Press, Cambridge, Mass., 1964.
65. Leppert, G., and C. C. Pitt: "Boiling," in T. F. Irvine, Jr., and J. P. Hartnett (eds.), *Advances in Heat Transfer*, vol. 1, Academic, New York, 1964.
66. Tong, L. S.: *Boiling Heat Transfer and Two-Phase Flow*, Wiley, New York, 1966.
67. Jordan, D. P.: "Film and Transition Boiling," in T. F. Irvine, Jr., and J. P. Hartnett (eds.), *Advances in Heat Transfer*, vol. 5, Academic, New York, 1968.
68. Wallis, G. B.: *One-Dimensional Two-Phase Flow*, McGraw-Hill, New York, 1969.
69. Cryder, D. S., and A. C. Finalbargo: "Heat Transmission from Metal Surfaces to Boiling Liquids: Effects of Temperature of the Liquid on Film Coefficient," *Trans. AIChE* **33:**346–362 (1937).
70. Chichelli, M. T., and C. F. Bonilla: "Heat Transfer to Liquids Boiling under Pressure," *Trans. AIChE* **41:**755–787 (1945).
71. Addoms, J. N.: "Heat Transfer at High Rates to Water Boiling outside Cylinders." D. Sc. thesis, M.I.T., Cambridge, Mass., 1948.
72. Rohsenow, W. M., and H. Y. Choi: *Heat, Mass and Momentum Transfer*, Prentice-Hall, Englewood Cliffs, N. J., 1961.
73. Rohsenow, W. M.: "A Method of Correlating Heat Transfer Data for Surface Boiling Liquids," *Trans. ASME* **74:**969–975 (1952).
74. Vahon, R. I., G. H. Nix, and G. E. Tanger: "Evaluation of Constants for the Rohsenow Pool-Boiling Correlation," *J. Heat Transfer* **90C:** 239–247 (1968).
75. Piret, E. L., and H. S. Isbin: "Natural Circulation Evaporation Two-Phase Heat Transfer," *Chem. Eng. Prog.* **50:**305–311 (1954).
76. Kutateladze, S. S.: "A Hydrodynamic Theory of Changes in Boiling Process under Free Convection," *Iz. Akad. Nauk SSSR, Otd. Tekh. Nauk* (4): 524 (1951).
77. Zuber, N.: "On the Stability of Boiling Heat Transfer," *J. Heat Transfer* **80C:**711 (1958).
78. Zuber, N., and M. Tribus: "Further Remarks on the Stability of Boiling Heat Transfer," *Univ. Calif.*, Los Angeles, *Dept. Eng., Rep.* 58-5, 1958.
79. Moissis, R., and P. J. Berenson: "On the Hydrodynamic Transitions in Nucleata Boiling," *J. Heat Transfer* **85C:**221–229 (1963).
80. Usiskin, C. M., and R. Siegel: "An Experimental Study of Boiling in Reduced and Zero Gravity Fields," *Trans. ASME* **83C:**243–253 (1961).
81. Bromley, L. A., "Heat Transfer in Stable Film Boiling," *Chem. Eng. Prog.* **46:**221–227 (1950).
82. Breen, B. P., and J. W. Westwater: "Effect of Diameter of Horizontal Tubes on Film Boiling Heat Transfer," *Chem. Eng. Prog.* **58:**67–72 (1962).
83. Berenson, P. J.: "Film-Boiling Heat Transfer from a Horizontal Surface," *Trans. ASME* **83C:**351–356 (1961).
84. Hsu, Y. Y., and J. W. Westwater: "Approximate Theory for Film Boiling on Vertical Surfaces," *Chem. Eng. Prog., Symp. Ser., AIChE Heat Transfer Conf.*, Storrs, Conn., **56:**15–22 (1959).
85. Bankoff, S. G.: "Discussion of Approximate Theory for Film Boiling on Vertical Surfaces," *Chem. Eng. Prog., Symp. Ser., AIChE Heat Transfer Conf.*, Storrs, Conn., pp. 22–24, 1959.
86. Lienhard, J. H., V. K. Dhir, and D. M. Riherd: "Peak Pool Boiling Heat-Flux Measurements on Finite Horizontal Flat Plates," *J. Heat Transfer* **95C:**152–158 (1973).
87. Sun, K. H., and J. H. Lienhard: "The Peak Boiling Heat Flux on Horizontal Cylinders," *Int. J. Heat Mass Transfer* **13:**1425–1439 (1970).
88. Stephan, K., and M. Abdelsalam: "Heat-Transfer Correlations for Natural Convection Boiling," *Int. J. Heat Mass Transfer* **23:**73–87 (1980).

89. Jakob, M., and W. Fritz: "Versuche über den Verdampfungsvorgang," *Forsch. Geb. Ingenieurs.* **2**:435–447 (1931).
90. Berenson, P. J.: "Experiments on Pool Boiling Heat Transfer," *Int. J. Heat Mass Transfer* **5**:985–999 (1962).
91. Lienhard, J. H., and V. K. Dhir: "Hydrodynamic Prediction of Peak Pool-Boiling Heat Fluxes from Finite Bodies," *J. Heat Transfer* **95C**:152–158 (1973).
92. Ded, J. S., and J. H. Lienhard: "The Peak Boiling Heat Flux from a Sphere," *AIChE J.* **18**:337–342 (1972).
93. Bergles, A. E.: "Burnout in Boiling Heat Transfer, Part 1: Pool-Boiling Systems," *Nuclear Safety* **16**:29–42 (1975).
94. Rohsenow, W. M.: "Boiling," in *Handbook of Heat Transfer*, McGraw-Hill, New York, 1973, pp. 13–28.
95. Chen, J. C.: "A Correlation for Boiling Heat Transfer to Saturated Fluids in Convective Flow," ASME Prepr. 63-HT-34, 6th ASME–AIChE Heat Transfer Conf., Boston, August 1963.
96. Chen, J. C.: "Correlation for Boiling Heat Transfer to Saturated Liquids in Convective Flow," *Int. Eng. Chem. Process Des. Develop.* **5**:322 (1966).
97. Collier, J. G.: *Convective Boiling and Condensation*, McGraw-Hill, New York, 1972.
98. Collier, J. G., and D. J. Pulling: "Heat Transfer to Two-phase Gas-Liquid Systems. Pt. II: Further Data on Steam/Water Mixtures in the Liquid Dispersed Region in an Annulus," AERE-R3809, 1962.
99. Davis, E. J., and G. H. Anderson: "The Incipience of Nucleate Boiling in Forced Convection Flow," *AIChE J.* **12**(4):774–780 (1966).
100. Dengler, C. E., and J. N. Addoms: "Heat Transfer Mechanism for Vaporisation of Water in a Vertical Tube," *Chem. Eng. Prog. Symp. Ser.* **52**:95–103 (1956).
101. Guerrieri, S. A., and R. D. Talty: "A Study of Heat Transfer to Organic Liquids in Single Tube Natural Circulation Vertical Tube Boilers," *Chem. Eng. Prog. Symp. Ser., Heat Transfer, Louisville*, **52**:69–77 (1956).
102. Berenson, P. J., and R. A. Stone: "A Photographic Study of the Mechanism of Forced-Convection Evaporation," Symposium on heat transfer, San Juan, Puerto Rico, *AIChE Reprint* 21, 1963.
103. Konmutsos, K., R. Moissis, and A. Spyridonos: "A Study of Bubble Departure in Forced Convection Boiling," *J. Heat Transfer* **90C**:223–230 (1968).
104. Polomik, E. E., S. Levy, and S. G. Sawochka. "Film Boiling of Steam-Water Mixtures in Annular Flow at 800, 1100, and 1400 Psi," *J. Heat Transfer* **86C**:81–88 (1964).
105. Katto, Y.: "A Generalized Correlation of Critical Heat Flux for the Forced Convection Boiling in Vertical Uniformly Heated Round Tubes," *Int. J. Heat Mass Transfer* **21**:1527–1542 (1978).
106. Bergles, A. E., J. G. Collier, J. M. Delhaye, G. F. Hewitt, and F. Mayinger: *Two-Phase Flow and Heat Transfer in the Power and Process Industries*, Hemisphere, New York, 1981.
107. Krieth, F., and M. Summerfield: "Heat Transfer to Water and High Flux Densities with and without Surface Boiling," *Trans. ASME* **71**:805–815 (1949).
108. Rohsenow, W. M., and J. A. Clarke: "Heat Transfer and Pressure Drop Data for High Heat Flux Densities to Water at High Sub-critical Pressure," 1951 Heat Transfer and Fluid Mechanics Institute, Stanford University Press, Stanford, Calif., 1951.
109. Piret, E. L., and H. S. Isbin: "Two-phase Heat Transfer in Natural Circulation Evaporators," *AIChE Heat Transfer Symp., St. Louis, Chem. Eng. Prog. Symp. Ser.* **50**:305 (1953).
110. Bergles, A. E., and W. M. Rohsenow: "The Determination of Forced Convection Surface Boiling Heat Transfer," *Trans. ASME, J. of Heat Transfer*, **86C**:365 (1964).
111. Jens, W. H., and P. A. Lottes: "Analysis of Heat Transfer Burnout, Pressure Drop and Density Data for High Pressure Water," ANL-4627, 1951.
112. Thom, J. R. S., W. M. Walker, T. A. Fallon, and G. F. S. Reising: "Boiling in Subcooled Water during Flow up Heated Tubes or Annuli," Paper presented at the Symposium on Boiling Heat Transfer in Steam Generating Units and Heat Exchangers, 15–16 September 1965 by Inst. of Mech. Engrs. (London), no. 6, 1965.
113. Forster, H. K. and N. Zuber: "Dynamics of Vapour Bubbles and Boiling Heat Transfer," *AIChE J.* **1**:531–535 (1955).

CHAPTER

ELEVEN

HEAT EXCHANGERS

Heat exchangers are devices that facilitate heat transfer between two or more fluids at different temperatures. Many types of heat exchangers have been developed for use at such varied levels of technological sophistication and sizes as steam power plants, chemical processing plants, building heating and air conditioning, household refrigerators, car radiators, radiators for space vehicles, and so on. In the common types, such as shell-and-tube heat exchangers and car radiators, heat transfer is primarily by conduction and convection from a hot to a cold fluid, which are separated by a metal wall. In boilers and condensers, heat transfer by boiling and condensation is of primary importance. In certain types of heat exchangers, such as cooling towers, hot fluid (i.e., water) is cooled by direct mixing with the cold fluid (i.e., air); that is, the water sprayed or falling down into an induced air draft is cooled by both convection and vaporization. In radiators for space applications, the waste heat carried by the coolant fluid is transported by convection and conduction to the fin surface and from there by thermal radiation into the atmosphere-free space.

The design of heat exchangers is a complicated matter. Heat transfer and pressure drop analysis, sizing and performance estimation, and the economic aspects play important roles in the final design. For example, although the cost considerations are very important for applications in large installations such as power plants and chemical processing plants, the weight and size considerations become the dominant factor in the choice of design for space and aeronautical applications. A comprehensive treatment of heat exchanger design is, therefore, beyond the scope of this chapter. The reader should consult Refs. 1 to 17 for the general theory and design of heat exchangers.

In this chapter we discuss the classification of heat exchangers, the determination of the overall heat transfer coefficient and the logarithmic mean temperature difference, and the methods of rating and sizing of heat exchangers.

11-1 CLASSIFICATION OF HEAT EXCHANGERS

Heat exchangers are made in so many sizes, types, configurations, and flow arrangements that some kind of classification, even though arbitrary, is necessary for the study of heat exchangers. Fraas and Özişik [11], Walker [16], and Kakaç, Shah, and Bergles [15] broadly classify the heat exchangers. In the following discussion we consider classifications based on (1) the transfer process, (2) compactness, (3) construction type, (4) flow arrangement, and (5) heat transfer mechanism.

Classification by Transfer Process

Heat exchangers can be classified as *direct contact* and *indirect contact*. In the direct-contact type, heat transfer takes place between two *inmiscible fluids*, such as a gas and a liquid, coming into direct contact. For example, cooling towers, jet condensers for water vapor, and other vapors utilizing water spray are typical examples of direct-contact exchangers.

Cooling towers have been widely used to dispose of waste heat from industrial processes by rejecting heat into the atmosphere rather than to water in a river, lake, or ocean. The most common types include the *natural-convection* and *forced-convection* cooling towers. In the natural-convection type of cooling tower shown in Fig. 11-1, the water is sprayed directly into the airstream that moves

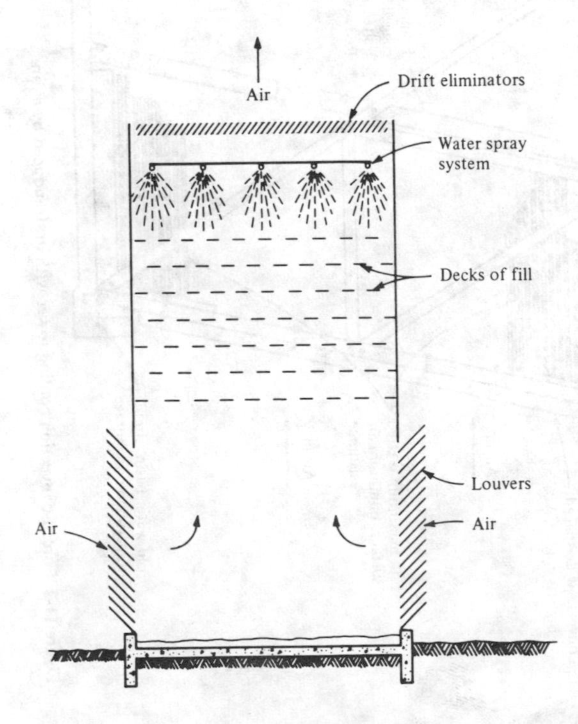

Figure 11-1 Section through a natural-convection cooling tower with "fill" to increase the effective water-droplet surface area by multiple splashing.

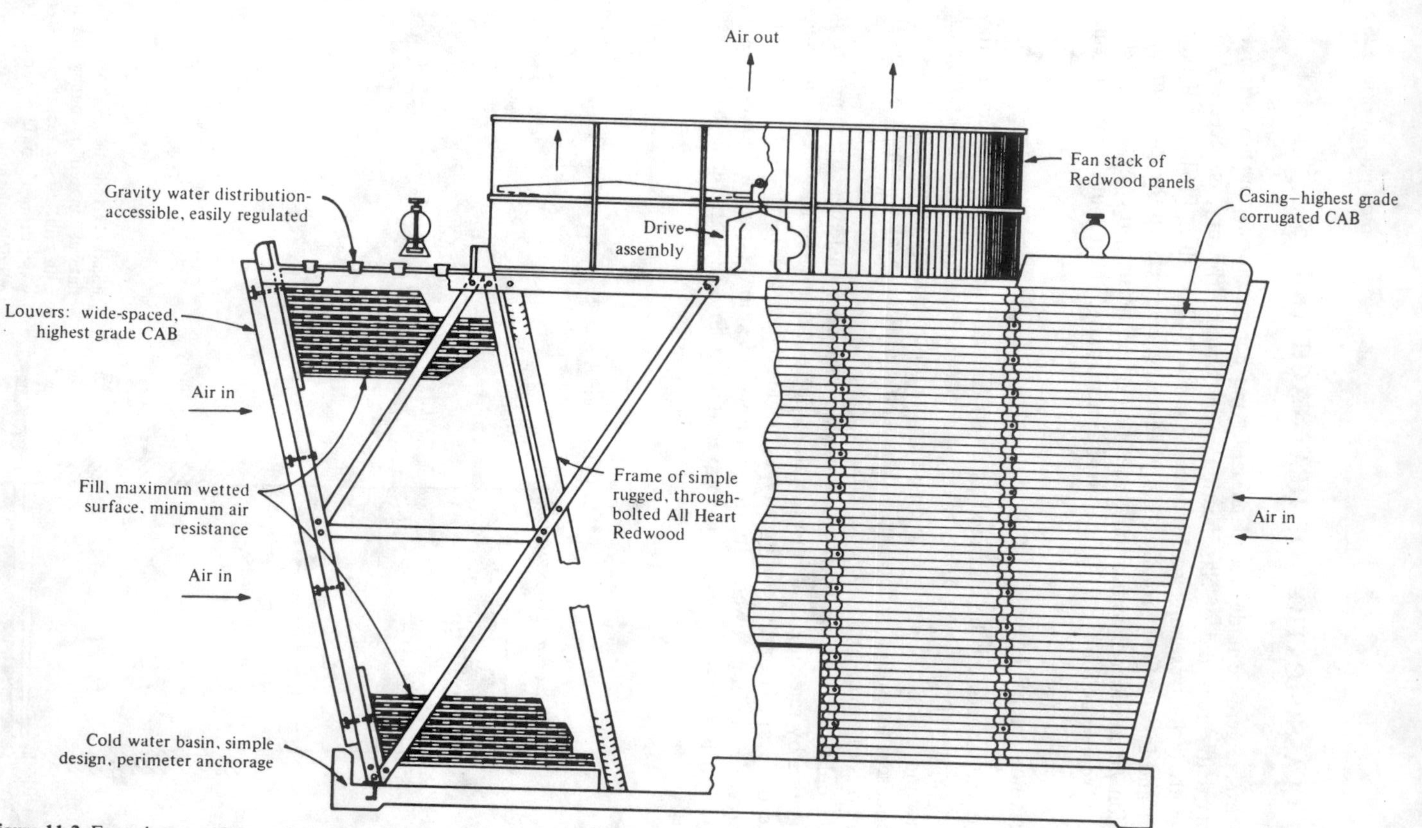

Figure 11-2 Forced-convection cooling tower with draft induced by a fan.

through the cooling tower by thermal convection. The falling water droplets are cooled both by ordinary convection and by the evaporation of water. The deck of *fills* positioned inside the cooling tower reduces the average velocity of the falling droplets and thus increases the time that droplets are exposed to the cooling airstream in falling through the tower. Large natural-convection-type cooling towers over 100 m high have been built to cool waste heat from power plants. In a forced-convection type of cooling tower, the water is sprayed into the airstream circulated through the tower by a fan which can be mounted either on top of the tower so that it draws air upward or just outside the base so that air flows directly inward. Figure 11-2 shows a section through a forced-circulation cooling tower with draft induced by a fan. The increased air circulation increases the heat transfer capacity of the cooling tower. The reader should consult Refs. 18 to 23 for information about the characteristics and performance of cooling towers.

In the indirect-contact type of heat exchangers, such as automobile radiators, the hot and cold fluids are separated by an impervious surface, and they are referred to as *surface heat exchangers*. There is no mixing of the two fluids.

Classification according to Compactness

The definition of compactness is quite an arbitrary matter. The ratio of the heat transfer surface area on one side of the heat exchanger to the volume can be used as a measure of the compactness of heat exchangers. A heat exchanger having a surface area density on any one side greater than about 700 m^2/m^3 quite arbitrarily is referred to as a *compact heat exchanger* regardless of its structural design. For example, automobile radiators having an area density on the order of 1100 m^2/m^3 and the glass ceramic heat exchangers for some vehicular gas-turbine engines having an area density on the order of 6600 m^2/m^3 are compact heat exchangers. The human lungs, with an area density of about 20,000 m^2/m^3, are the most compact heat-and-mass exchanger. The very fine matrix regenerator for the Stirling engine has an area density approaching that of the human lung.

On the other extreme of the compactness scale, plane tubular and shell-and-tube type exchangers, having an area density in the range of 70 to 500 m^2/m^3, are not considered compact.

The incentive for using compact heat exchangers lies in the fact that a high value of compactness reduces the volume for a specified heat exchanger performance. When heat exchangers are to be employed for automobiles, marine uses, aircraft, aerospace vehicles, cryogenic systems, and refrigeration and air conditioning, the weight and size—hence the compactness—become important. To increase the effectiveness or the compactness of heat exchangers, fins are used. In a gas-to-liquid heat exchanger, for example, the heat transfer coefficient on the gas side is an order of magnitude lower than that for the liquid side. Therefore, fins are used on the gas side to obtain a balanced design; the heat transfer surface on the gas side becomes much more compact. Figure 11-3 show a typical automotive radiator.

Figure 11-3 An automobile radiator. (*Courtesy of Harrison Radiator Division of General Motors Corporation.*)

Classification by Construction Type

Heat exchangers also can be classified according to their construction features. For example, there are tubular, plate, plate-fin, tube-fin, and regenerative exchangers.

Tubular heat exchangers Tubular exchangers are widely used, and they are manufactured in many sizes, flow arrangements, and types. They can accommodate a wide range of operating pressures and temperatures. The ease of manufacturing and their relatively low cost have been the principal reason for their widespread use in engineering applications. A commonly used design, called the *shell-and-tube* exchanger, consists of round tubes mounted on a cylindrical shell with their axes parallel to that of the shell. Figure 11-4 illustrates the main features of a shell-and tube exchanger having one fluid flowing inside the tubes and the other flowing outside the tubes. The principal components of this type of heat exchanger are the tube bundle, shell, front and rear end headers, and baffles. The baffles are used to support the tubes, to direct the fluid flow approximately normal to the tubes, and to increase the tubulence of the shell fluid. There are various types of baffles, and the choice of baffle type, spacing, and geometry depends on the flow rate,

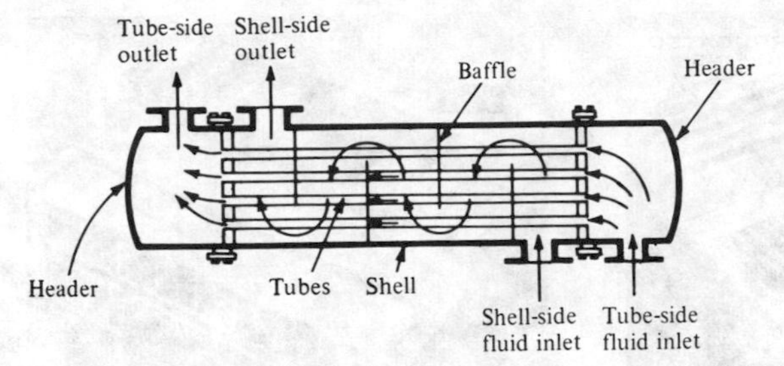

Figure 11-4 A shell-and-tube heat exchanger; one shell pass and one tube pass.

allowable shell-side pressure drop, tube support requirement, and the flow-induced vibrations. Many variations of shell-and-tube exchanger are available; the differences lie in the arrangement of flow configurations and in the details of construction. We discuss this matter further in connection with the classification of heat exchangers by flow arrangement.

The character of the fluids may be *liquid-to-liquid*, *liquid-to-gas*, or *gas-to-gas*. Liquid-to-liquid exchangers have the most common applications. Both fluids are pumped through the exchanger; hence the heat transfer on both the tube side and the shell side is by forced convection. Since the heat transfer coefficient is high with the liquid flow, generally there is no need to use fins.

The liquid-to-gas arrangement is also commonly used; in such cases, the fins usually are added on the gas side of the tubes, where the heat transfer coefficient is low.

Gas-to-gas exchangers are used in the exhaust-gas and air preheating recuperators for gas-turbine systems, cryogenic gas-liquefaction systems, and steel furnaces. Internal and external fins generally are used in the tubes to enhance heat transfer.

Plate heat exchangers As the name implies, plate heat exchangers usually are constructed of thin plates. The plates may be smooth or may have some form of corrugation. Since the plate geometry cannot accommodate as high pressure and/or temperature differentials as a circular tube, they are generally designed for moderate temperature and/or pressure. The compactness factor for plate exchangers ranges from 120 to 230 m^2/m^3.

Plate-fin heat exchangers The compactness factor can be significantly improved (i.e., up to about 6000 m^2/m^3) by using the plate-fin type of heat exchanger. Figure 11-5 illustrates typical plate-fin configurations. Louvered or corrugated fins are separated by flat plates. Cross-flow, counterflow, or parallel-flow arrangements can be obtained readily by properly arranging the fins on each side of the plate. Plate-fin exchangers are generally used for gas-to-gas applications, but they are used for low-pressure applications not exceeding about 10 atm (that is, 1000 kPa). The maximum operating temperatures are limited to about 800°C. Plate-fin heat exchangers also have been used for cryogenic applications.

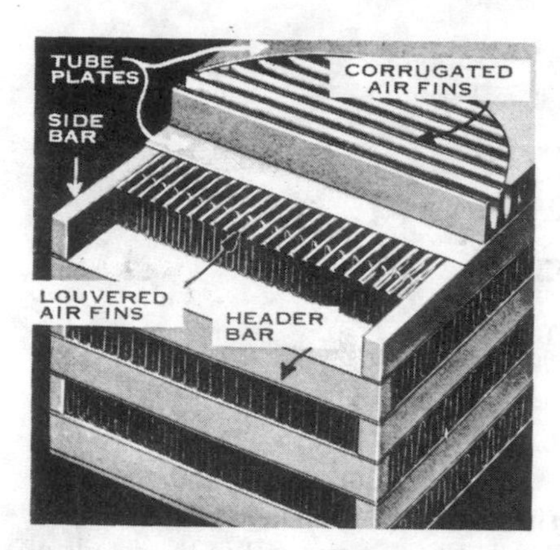

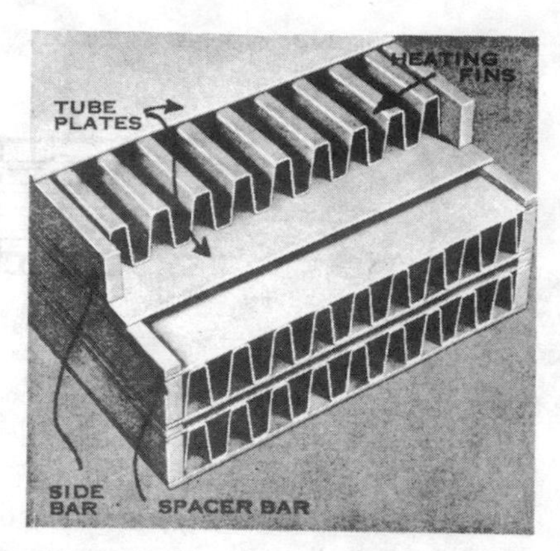

Figure 11-5 Plate-fin heat exchangers. (*Courtesy of Harrison Radiator Division of General Motors Corporation.*)

Tube-fin heat exchangers When a high operating pressure or an extended surface is needed on one side, tube-fin exchangers are used. Figure 11-6 illustrates two typical configurations, one with round tubes and the other with flat tubes. Tube-fin exchangers can be used for a wide range of tube fluid operating pressures not exceeding about 30 atm and operating temperatures from low cryogenic applications to about 870°C. The maximum compactness ratio of about 330 m^2/m^3 is less than that obtainable with plate-fin exchangers.

The tube-fin heat exchangers are used in gas-turbine, nuclear, fuel cell, automobile, airplane, heat pump, refrigeration, electronics, cryogenics, air conditioning, and many other applications.

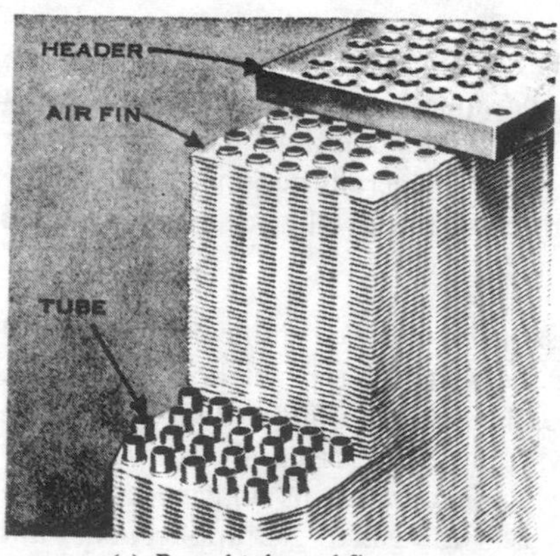

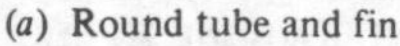
(*a*) Round tube and fin

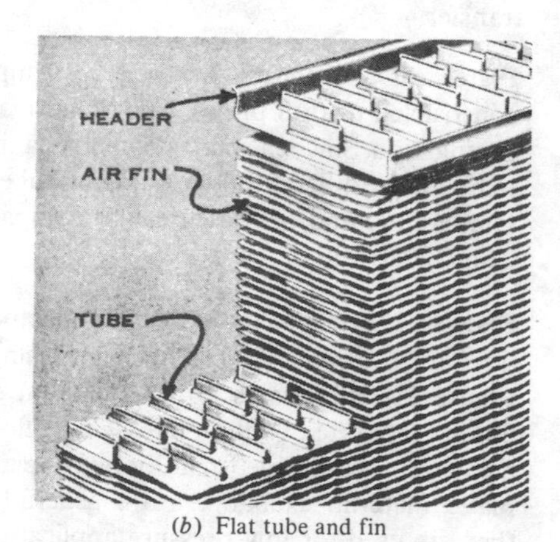

(*b*) Flat tube and fin

Figure 11-6 Tube-fin heat exchangers. (*Courtesy of Harrison Radiator Division of General Motors Corporation.*)

Regenerative heat exchangers Regenerative heat exchangers can be either *static* or *dynamic*. The static type has no moving parts and consists of a porous mass (i.e., balls, pebbles, powders, etc.) through which hot and cold fluids pass alternately. A flow-switching device regulates the periodic flowing of the two fluids. During the flow of the hot fluid, the heat is transferred from the hot fluid to the matrix of the regenerative exchanger. Then the hot fluid flow is switched off, and the cold fluid flow is switched on. During the passage of the cold fluid, heat is transferred from the matrix to the cold fluid. Static-type regenerators can be noncompact for use in high-temperature applications (that is, 900 to 1500°C), such as air preheaters for coke manufacturing and glass melting tanks. Also they can be compact regenerators for use in refrigeration, the Stirling engine.

In dynamic-type regenerators, the matrix is arranged in the form of a drum which rotates about an axis so that a given portion of the matrix passes periodically through the hot stream and then through the cold stream. The heat stored in the matrix during its contact with the hot gas is transferred to the cold gas during its contact with the cold stream. A typical example of the rotary regenerator is the *Ljungstrom regenerative air preheater* shown in Fig. 11-7. The rotary regenerators have been designed for a surface area density up to 6500 m^2/m^3. The metal rotary regenerators can operate at temperatures up to 870°C; ceramic matrices are used for higher temperatures. Rotary-type regenerators are suitable only for gas-to-gas heat exchange, because only for gases is the heat capacity of the heat transfer matrix much greater than the heat capacity of the gas contained in the flow

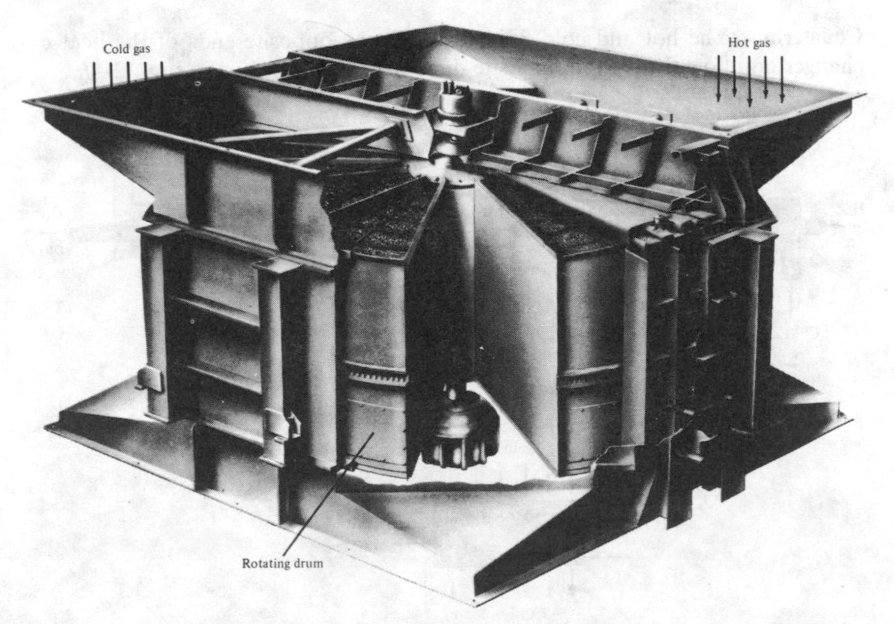

Figure 11-7 Ljungstrom air preheater. (*By permission from C. E. Air Preheater, Combustion Engineering, Inc.*)

passages. It is not suitable for liquid-to-liquid heat exchange, because the heat capacity of the heat transfer matrix is much less than that of the liquid.

Since the heat transfer matrix is rotating, the temperatures of the gases, and the wall depend on space and time; as a result, the heat transfer analysis of regenerators is involved, in that the periodic flow introduces several new variables. For conventional, stationary heat exchangers it is sufficient to define the inlet and the outlet temperatures, the flow rates, the heat transfer coefficients for the two fluids, and the surface areas of the two sides of the heat exchangers. For the rotary heat exchanger, however, it is also necessary to relate the heat capacity of the rotor to that of the fluid streams, the fluid flow rates, and the rotation speed under consideration. The reader should consult Refs. 24 to 32 for detailed discussions of heat transfer and design considerations for regenerative-type heat exchangers.

Classification by Flow Arrangement

Numerous possibilities exist for flow arrangement in heat exchangers. We summarize here the principal ones.

Parallel-flow The hot and cold fluids enter at the same end of the heat exchanger, flow through in the same direction, and leave together at the other end, as illustrated in Fig. 11-8*a*.

Counterflow The hot and cold fluids enter in the opposite ends of the heat exchanger and flow through in opposite directions, as illustrated in Fig. 11-8*b*.

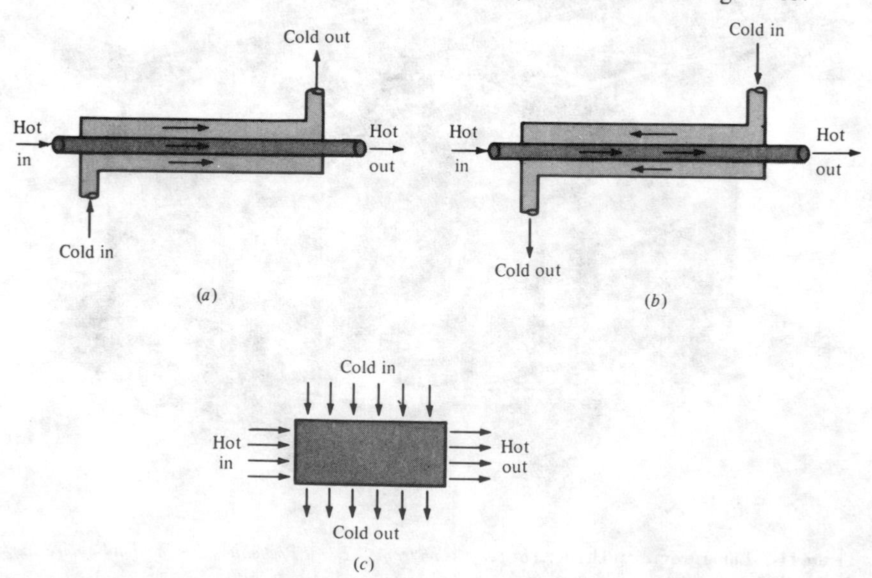

Figure 11-8 (*a*) Parallel-flow, (*b*) counterflow, and (*c*) cross-flow arrangements.

Cross-flow In the cross-flow exchanger, the two fluids usually flow at right angles to each other, as illustrated in Fig. 11-8*c*. In the cross-flow arrangement, the flow may be called *mixed* or *unmixed*, depending on the design.

Figure 11-9*a* shows an arrangement in which both hot and cold fluids flow through individual channels formed by corrugation; therefore the fluids are not free to move in the transverse direction. Then each fluid stream is said to be unmixed. Figure 11-9*b* illustrates a typical temperature profile for the outlet temperatures when both fluids are unmixed, as shown in Fig. 11-9*a*. The inlet temperatures for both fluids are assumed to be uniform, but the outlet temperatures exhibit variation transverse to the flow.

In the flow arrangement shown in Fig. 11-9*c*, the cold fluid flows inside the tubes and so is not free to move in the transverse direction. Therefore, the cold fluid is said to be unmixed. However, the hot fluid flows over the tubes and is free to move in the transverse direction. Therefore, the hot fluid stream is said to be mixed. The mixing tends to make the fluid temperature uniform in the transverse direction; therefore, the exit temperature of a mixed stream exhibits negligible variation in the crosswise direction.

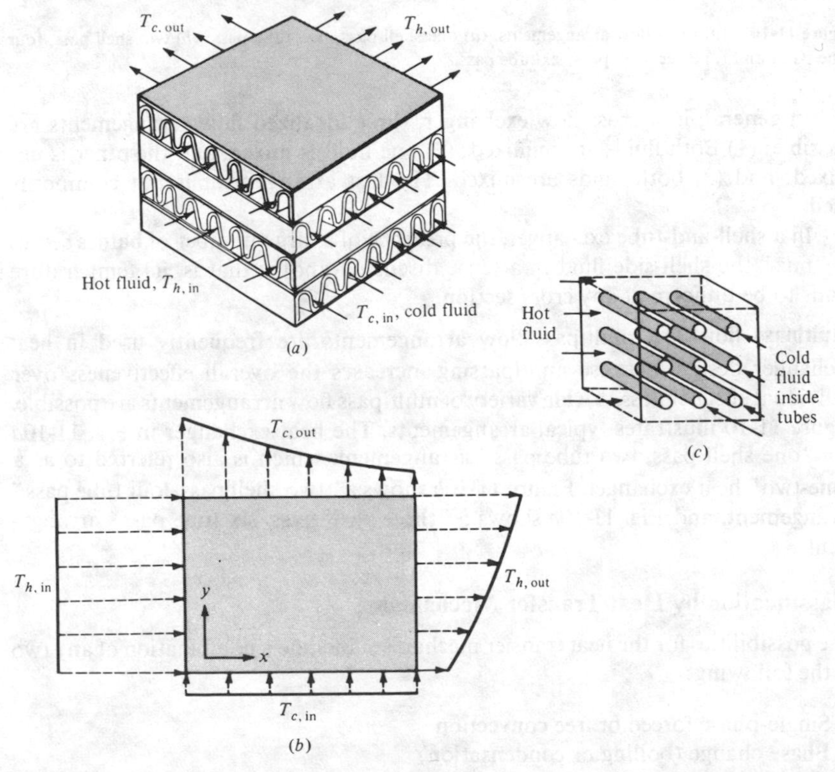

Figure 11-9 Cross-flow arrangements: (*a*) both fluids unmixed; (*b*) temperature profile when both fluids are unmixed; (*c*) cold fluid unmixed, hot fluid mixed.

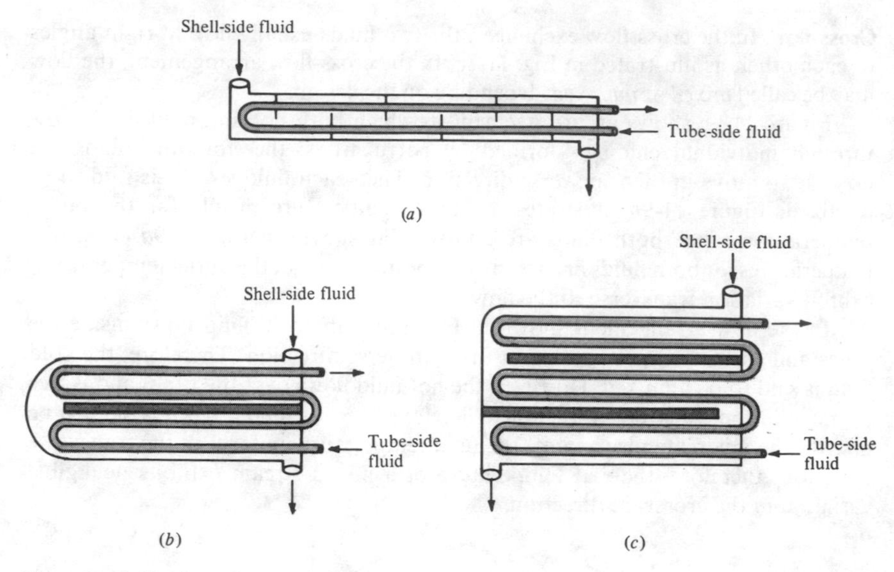

Figure 11-10 Multipass flow arrangements. (*a*) one shell pass, two tube pass; (*b*) two shell pass, four tube pass; and (*c*) three shell pass, six tube pass.

In general, in a cross-flow exchanger, three idealized flow arrangements are possible: (1) Both fluids are unmixed; (2) one fluid is mixed, and the other is unmixed; and (3) both fluids are mixed. The last arrangement is not commonly used.

In a shell-and-tube exchanger, the presence of a large number of baffles serves to "mix" the shell-side fluid in a sense discussed above; that is, its temperature tends to be uniform at any cross section.

Multipass flow The multipass flow arrangements are frequently used in heat exchanger design, because multipassing increases the overall effectiveness over individual effectiveness. A wide variety of multipass flow arrangements are possible. Figure 11-10 illustrates typical arrangements. The heat exchanger in Fig. 11-10*a* is a "one shell pass, two tube pass" arrangement, which is also referred to as a "one-two" heat exchanger. Figure 11-10*b* shows a "two shell pass, four tube pass" arrangement, and Fig. 11-10*c* shows a "three shell pass, six tube pass" arrangement.

Classification by Heat Transfer Mechanism

The possibilities for the heat transfer mechanism include a combination of any two of the following:

1. Single-phase forced or free convection
2. Phase change (boiling or condensation)
3. Radiation or combined convection and radiation

For all the cases discussed earlier we considered single-phase forced convection on both sides of the heat exchanger. Condensers, boilers, and radiators for space power plants include the mechanisms of condensation, boiling, and radiation, respectively, on one of the surfaces of the heat exchanger.

Condensers Condensers are used for such varied applications as steam power plants, chemical processing plants, and nuclear electric plants for space vehicles. The major types include the *surface condensers*, *jet condensers*, and *evaporative condensers*. The most common type is the surface condenser, which has the advantage that the condensate is returned to the boiler through the feedwater system. Figure 11-11 shows a section through a typical two-pass surface condenser for a large steam turbine in a power plant. Since the steam pressure at the turbine exit is only 1.0 to 2.0 inHg abs, the steam density is very low and the volume rate of flow is extremely large. To minimize the pressure loss in transferring steam from the turbine to the condenser, the condenser is normally mounted beneath and attached to the turbine. Cooling water flows horizontally inside the tubes, while the steam flows vertically downward from the large opening at the top and passes transversely over the tubes. Note that provision is made to aspirate cool air from the regions just above the center of the hot well. This is important because the presence of noncondensable gas in the steam reduces the heat transfer coefficient for condensation.

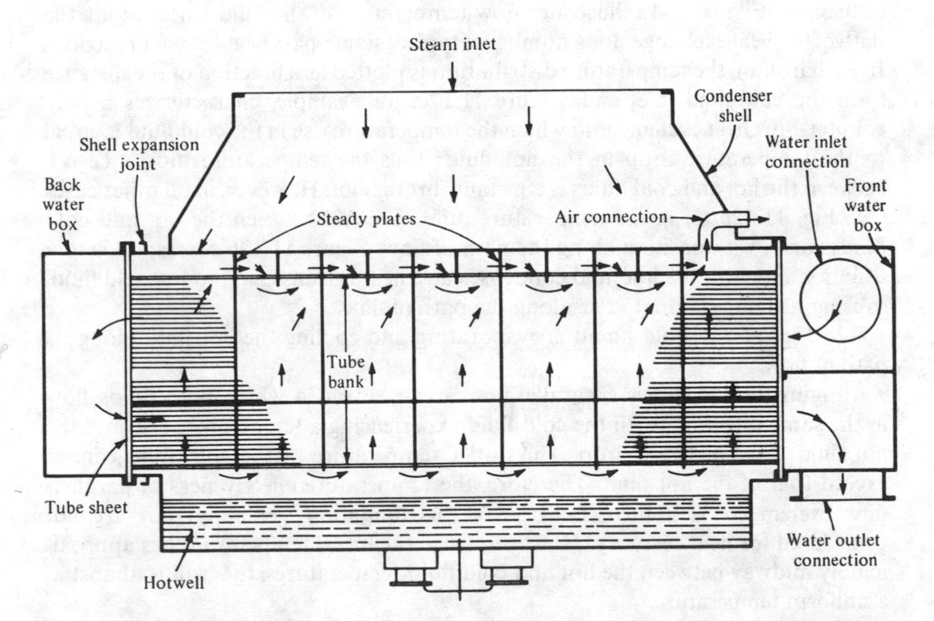

Figure 11-11 Section through a typical two-pass surface condenser for a large steam power plant. (*From Allis-Chalmers Manufacturing Company.*)

Boilers Steam boilers are one of the earliest applications of heat exchangers. The term *steam generator* is often applied to boilers in which the heat source is a hot fluid stream rather than the product of combustion.

An enormous variety of boiler types exist, ranging from small units for house heating applications to huge, complex, expensive units for modern power stations.

Radiators for space power plants The rejection of waste heat from the condenser of a power plant intended to produce electricity for the propulsion, guidance, or the communication equipment of a space vehicle poses serious problems even for a power plant producing only a few kilowatts of electricity. The only way the waste heat can be dissipated from a space vehicle is by thermal radiation by taking advantage of the fourth-power relationship between the absolute temperature of the surface and the radiative-heat flux. Thus, in the operation of some power plants for space vehicles, the thermodynamic cycle is at such high temperatures that the radiator runs red-hot. Even so, it is difficult to keep the radiator size within a reasonable envelope for launch vehicles.

11-2 TEMPERATURE DISTRIBUTION IN HEAT EXCHANGERS

In stationary-type heat exchangers, the heat transfer from the hot to the cold fluid causes a change in temperature of one or both fluids flowing through the heat exchanger. Figure 11-12 illustrates how temperature of the fluid varies along the path of the heat exchanger for a number of typical single-pass heat transfer matrices. In each instant, the temperature distribution is plotted as a function of the distance from the cold-fluid inlet end. Figure 11.12*a*, for example, characterizes a pure counter-flow heat exchanger in which the temperature rise in the cold fluid is equal to the temperature drop in the hot fluid; thus the temperature difference ΔT between the hot and cold fluids is constant throughout. However, in all other cases (i.e., Fig. 11-12*b* to *e*), the temperature difference ΔT between the hot and cold fluids varies with position along the path of flow. Figure 11-12*b* corresponds to a situation in which the hot fluid condenses and heat is transferred to the cold fluid, causing its temperature to rise along the path of flow.

In Fig. 11-12*c*, cold liquid is evaporating and cooling the hot fluid along its path of flow.

Figure 11-12*d* shows a parallel-flow arrangement in which both fluids flow in the same direction, with the cold fluid experiencing a temperature rise and the hot fluid a temperature drop. The outlet temperature of the cold fluid cannot exceed that of the hot fluid. Therefore, the temperature effectiveness of parallel-flow exchangers is limited. Because of this limitation, generally they are not considered for heat recovery. However, since the metal temperature lies approximately midway between the hot and cold fluid temperatures, the wall is almost at a uniform temperature.

Figure 11-12*e* shows a counterflow arrangement in which fluids flow in opposite directions. The exit temperature of the cold fluid can be higher than that of the hot fluid. Theoretically, the exit temperature of one fluid may approach the

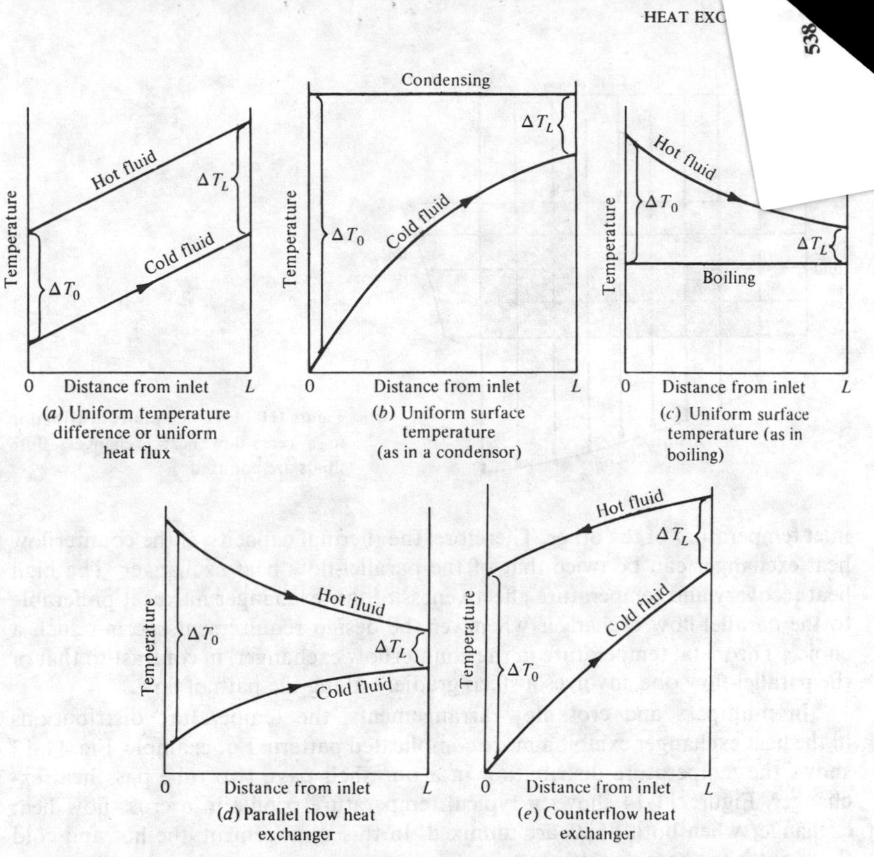

Figure 11-12 Axial temperature distribution in typical single-pass heat transfer matrices.

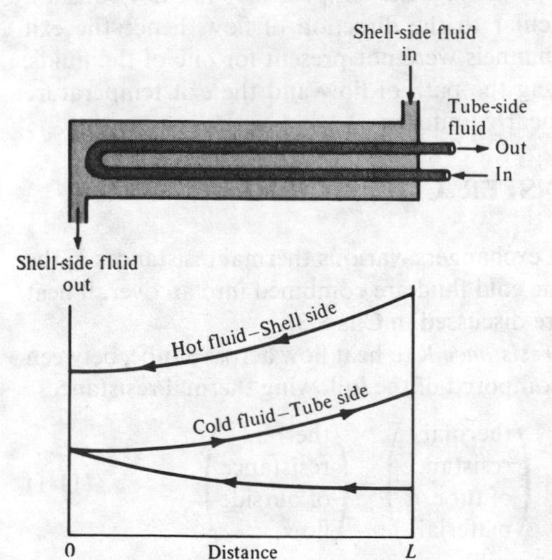

Figure 11-13 Axial temperature distribution in a one shell pass, two tube pass heat exchanger.

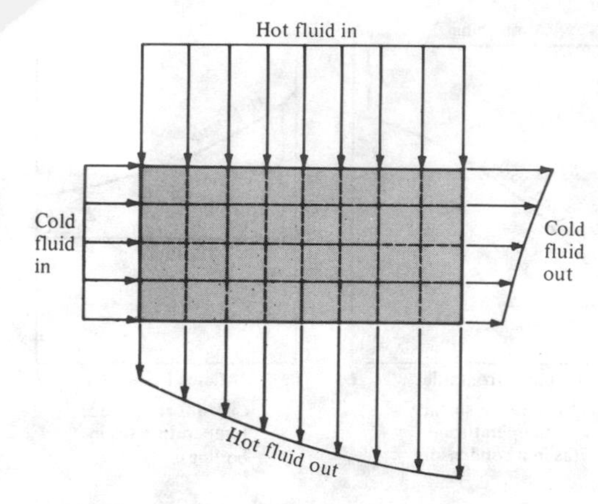

Figure 11-14 Temperature distribution in a cross-flow heat exchanger. Both fluids are unmixed.

inlet temperature of the other. Therefore, the thermal capacity of the counterflow heat exchanger can be twice that of the parallel-flow heat exchanger. The high heat recovery and temperature effectiveness of this exchanger makes it preferable to the parallel-flow exchanger whenever the design requirements permit such a choice. The metal temperature in the counterflow exchanger, in contrast to that of the parallel-flow one, involves a steep gradient along the path of flow.

In multipass and cross-flow arrangements, the temperature distributions in the heat exchanger exhibit a more complicated pattern. For example, Fig. 11-13 shows the temperature distribution in a one shell pass, two tube pass heat exchanger. Figure 11-14 shows a typical temperature profile in a cross-flow heat exchanger when both fluids are unmixed. In this arrangement, the hot and cold fluids enter the heat transfer matrix at uniform temperatures, but since there are channels in the flow path to prevent mixing, the temperatures are not constant across any cross section perpendicular to the direction of flow, hence the exit temperatures are not uniform. If channels were not present for one of the fluids, that fluid could become mixed along the path of flow and the exit temperature for the mixed fluid would become nearly uniform.

11-3 OVERALL HEAT TRANSFER COEFFICIENT

In the heat transfer analysis of heat exchangers, various thermal resistances in the path of heat flow from the hot to the cold fluid are combined into an overall heat transfer coefficient U by a procedure discussed in Chap. 3.

Consider that the total *thermal resistance* R to heat flow across a tube, between the inside and the outside flow, is composed of the following thermal resistances:

$$R = \begin{pmatrix}\text{thermal}\\ \text{resistance}\\ \text{of inside}\\ \text{flow}\end{pmatrix} + \begin{pmatrix}\text{thermal}\\ \text{resistance}\\ \text{of tube}\\ \text{material}\end{pmatrix} + \begin{pmatrix}\text{thermal}\\ \text{resistance}\\ \text{of outside}\\ \text{flow}\end{pmatrix} \tag{11-1}$$

and the various terms are given by

$$R = \frac{1}{A_i h_i} + \frac{t}{kA_m} + \frac{1}{A_o h_o} \tag{11-2}$$

where A_o, A_i = outside and inside surface areas of tube, respectively, m^2

$A_m = \dfrac{A_o - A_i}{\ln(A_o/A_i)}$ = logarithmic mean area, m^2 [see Eq. (3-12*d*)]

h_i, h_o = heat transfer coefficients for inside and outside flow, respectively, $W/(m^2 \cdot {}^\circ C)$

k = thermal conductivity of tube material, $W/(m \cdot {}^\circ C)$

R = total thermal resistance from inside to outside flow °C/W

t = thickness of tube, m

The thermal resistance R given by Eq. (11-2) can be expressed as an overall heat transfer coefficient based on either the inside or the outside surface of the tube. It does not matter on which area it is based as long as it is specified in its definition. For example, the overall heat transfer coefficient U_o based on the *outside surface* of the tube is defined as

$$U_o = \frac{1}{A_o R} = \frac{1}{(A_o/A_i)(1/h_i) + (A_o/A_m)(t/k) + 1/h_o}$$

$$= \frac{1}{(D_o/D_i)(1/h_i) + [1/(2k)]D_o \ln(D_o/D_i) + 1/h_o} \tag{11-3}$$

since

$$\frac{A_o}{A_m} = \frac{D_o}{2t} \ln \frac{D_o}{D_i} \qquad D_o - D_i = 2t \tag{11-4}$$

and D_i and D_o are the inside and outside diameters of the tube, respectively.

Similarly, the overall heat transfer coefficient U_i based on the *inside surface* of the tube is defined as

$$U_i = \frac{1}{A_i R} = \frac{1}{1/h_i + (A_i/A_m)(t/k) + (A_i/A_o)(1/h_o)}$$

$$= \frac{1}{1/h_i + [1/(2k)]D_i \ln(D_o/D_i) + (D_i/D_o)(1/h_o)} \tag{11-5}$$

When the wall thickness is small and its thermal conductivity is high, the tube resistance can be neglected and Eq. (11-5) reduces to

$$U_i = \frac{1}{1/h_i + 1/h_o} \tag{11-5a}$$

In heat exchanger applications, the heat transfer surface is fouled with the accumulation of deposits, which is turn introduces additional thermal resistance in the path of heat flow. The effect of fouling is generally introduced in the form of a *fouling factor* F which has the dimensions $m^2 \cdot °C/W$; this is discussed further later.

We now consider heat transfer across a tube, which is fouled by deposit formation on both the inside and outside surfaces. The thermal resistance R in the path of heat flow for this case is given by

$$R = \frac{1}{A_i h_i} + \frac{F_i}{A_i} + \frac{t}{kA_m} + \frac{F_o}{A_o} + \frac{1}{A_o h_o} \tag{11-6}$$

where F_i and F_o are the fouling factors (i.e., unit fouling resistances) at the inside and outside surfaces of the tube, respectively and the other quantities are as defined previously.

In heat exchanger applications, the overall heat transfer coefficient is usually based on the outer tube surface. Then (11-6) can be represented in terms of the overall heat transfer coefficient based on the outside surface of the tube as

$$U_o = \frac{1}{(D_o/D_i)(1/h_i) + (D_o/D_i)F_i + [D_o/(2k)] \ln (D_o/D_i) + F_o + 1/h_o} \tag{11-7}$$

The values of overall heat transfer coefficients for different types of applications vary widely. Typical ranges of U_o are as follows:

Water-to-oil exchangers:	60 to 350 $W/(m^2 \cdot °C)$
Gas-to-gas exchangers:	60 to 600 $W/(m^2 \cdot °C)$
Air condensers:	350 to 800 $W/(m^2 \cdot °C)$
Ammonia condensers:	800 to 1400 $W/(m^2 \cdot °C)$
Steam condensers:	1500 to 5000 $W/(m^2 \cdot °C)$

It is apparent that U_o is generally low for fluids having low thermal conductivity, such as gases or oils.

Fouling Factor

Considerable effort has been directed during the past decade toward the understanding of fouling [33–44]. During operation, heat exchangers become fouled with an accumulation of deposits of one kind or another on heat transfer surfaces. As a result, the thermal resistance in the path of heat flow increases, which reduces the heat transfer rate. The economic penalty for fouling can be attributed to

1. Higher capital expenditure through oversized units
2. Energy losses due to thermal inefficiencies
3. Costs associated with periodic cleaning of heat exchangers
4. Loss of production during shutdown for cleaning

Epstein [41] has delineated the following six categories of fouling:

1. *Scaling or precipitation fouling*, the crystallization from solution of dissolved substance onto the heat transfer surface
2. *Particulate fouling*, the accumulation of finely divided solids suspended in the process fluid onto the heat transfer surface
3. *Chemical reaction fouling*, the deposit formation on the heat transfer surface by chemical reaction
4. *Corrosion fouling*, the accumulation of corrosion products on the heat transfer surface
5. *Biological fouling*, the attachment of microorganisms to a heat transfer surface
6. *Solidification fouling*, the crystallization of a pure liquid or one component from the liquid phase on a subcooled heat transfer surface

Clearly the mechanism of fouling is very complicated, and no reliable techniques are yet available for its prediction.

When a new heat exchanger is put into service, its performance deteriorates progressively as a result of the buildup of fouling resistance. The fluid velocity and the fluid temperature appear to be among the factors that affect the rate of fouling on a given surface. An increase in the velocity decreases both the rate of deposit and the ultimate amount of deposit on the surface. Increasing the fluid bulk temperature increases both the rate of buildup of fouling and its ultimate stable level.

Based on the experience of manufacturers and users, the Tubular Equipment Manufacturers Association (TEMA) prepared tables of fouling factors as a guide in heat transfer calculations. We present in Table 11-1 some of their results. Fouling is very complicated, and its representation with such a simple listing is highly questionable. But in the absence of anything better, it remains the only reference for estimating the effects of fouling in reducing heat transfer.

Example 11-1 Determine the overall heat transfer coefficient U_o based on the outer surface of a steel pipe with an ID of $D_i = 2.5$ cm and an OD of $D_o = 3.34$ cm $[k = 54\ \mathrm{W/(m \cdot °C)}]$ for the following flow and fouling conditions:

$$h_i = 1800\ \mathrm{W/(m^2 \cdot °C)} \qquad h_o = 1250\ \mathrm{W/(m^2 \cdot °C)}$$

$$F_i = F_o = 0.00018\ \mathrm{m^2 \cdot °C/W}$$

SOLUTION Equation (11-7) can be used to determine U_o:

$$U_o = \frac{1}{(D_o/D_i)(1/h_i) + (D_o/D_i)F_i + [D_o/(2k)]\ln(D_o/D_i) + F_o + 1/h_o}$$

Table 11-1 Unit fouling resistance F for heat transfer equipment

	Water temperature 52°C or less	
	Water velocity 1 m/s and less	Water velocity over 1 m/s
	$m^2 \cdot C/W$	$m^2 \cdot C/W$
Types of Water:		
Seawater	0.000088	0.000088
Distilled	0.000088	0.000088
Treated boiler feedwater	0.00018	0.000088
Engine jacket	0.00018	0.00018
Great Lakes	0.00018	0.00018
Cooling tower and spray pond		
Treated makeup	0.00018	0.00018
Untreated	0.00053	0.00053
Boiler blowdown	0.00035	0.00035
Brackish water	0.00035	0.00018
River water		
Minimum	0.00036	0.00018
Mississippi	0.00053	0.00035
Delaware, Schuylkill	0.00053	0.00035
East River and New York Bay	0.00053	0.00035
Chicago sanitary canal	0.00141	0.00106
Muddy or silty	0.00053	0.00035
Hard (over 15 grains/gal)	0.00053	0.00053
Types of Fluid:		
Industrial oils		
Clean recirculating oil	0.00018	
Machinery and transformer oils	0.00018	
Vegetable oils	0.00053	
Quenching oil	0.00070	
Fuel oil	0.00088	
Industrial gases and vapors		
Organic vapors	0.000088	
Steam (non-oil bearing)	0.000088	
Alcohol vapors	0.000088	
Steam, exhaust	0.00018	
Refrigerating vapors	0.00035	
Air	0.00035	
Industrial liquids		
Organic	0.00018	
Refrigerating liquids	0.00018	
Brine (cooling)	0.00018	

Source: Tubular Exchanger Manufacturers Association [4].

Introducing the numerical values, we obtain

$$U_o = \frac{1}{(0.742 + 0.241 + 0.090 + 0.18 + 0.800) \times 10^{-3}}$$

$$= 487.1 \text{ W/(m}^2 \cdot \text{°C)}$$

Example 11-2 Water at a mean temperature of $T_m = 80°C$ and a mean velocity of $u_m = 0.15$ m/s flows inside a 2.5-cm-ID, thin-walled copper tube. Atmospheric air at $T_\infty = 20°C$ and a velocity of $u_\infty = 10$ m/s flows across the tube. Neglecting the tube wall resistance, calculate the overall heat transfer coefficient and the rate of heat loss per 1-m length of the tube.

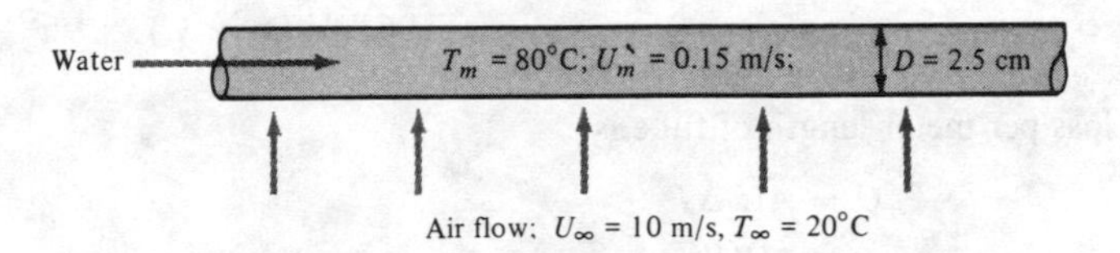

Solution The physical properties of water at $T_m = 80°C$ are

$$\nu = 0.364 \times 10^{-6} \text{ m}^2\text{/s} \qquad k = 0.668 \text{ W/(m} \cdot \text{°C)} \qquad \text{Pr} = 2.22$$

The Reynolds number for water flow is

$$\text{Re} = \frac{u_m D}{\nu} = \frac{0.15 \times 0.025}{0.364 \times 10^{-6}} = 10{,}300$$

We use the Dittus-Boelter equation to determine h_i for water flow:

$$\text{Nu} = 0.023 \text{ Re}^{0.8} \text{ Pr}^{0.3}$$

$$= 0.023(10{,}300)^{0.8}(2.22)^{0.3}$$

$$= 47.4$$

$$h_i = \text{Nu} \frac{k}{D_i} = 47.4 \frac{0.668}{0.025} = 1267 \text{ W/(m}^2 \cdot \text{°C)}$$

To evaluate the physical properties of air at the film temperature, the closest approximation for the film temperature is taken as $T_f \cong (80 + 20)/2 = 50°C$. Then

$$\nu = 18.22 \times 10^{-6} \text{ m}^2\text{/s} \qquad k = 0.0281 \text{ W/(m} \cdot \text{°C)} \qquad \text{Pr} = 0.703$$

The Reynolds number for the air flow becomes

$$\text{Re} = \frac{u_\infty D}{\nu} = \frac{(10)(0.025)}{18.22 \times 10^{-6}} = 13{,}721$$

The Nusselt number for the air flow is determined by Eq. (8-81):

$$\begin{aligned} \mathrm{Nu} &= (0.4\,\mathrm{Re}^{0.5} + 0.06\,\mathrm{Re}^{2/3})\,\mathrm{Pr}^{0.4} \\ &= [0.4(13{,}721)^{0.5} + 0.06(13{,}721)^{2/3}](0.703)^{0.4} \\ &= 70.52 \end{aligned}$$

and

$$h_o = \mathrm{Nu}\,\frac{k}{D_o} = 70.52\,\frac{0.0281}{0.025} = 79.3\ \mathrm{W/(m^2 \cdot {}^\circ C)}$$

By neglecting the tube wall resistance to heat flow and the curvature effect (i.e., assume $D_o \cong D_i$), the overall heat transfer coefficient becomes

$$U = \frac{1}{1/h_i + 1/h_o} = \frac{1}{1/1267 + 1/79.3} = 74.63\ \mathrm{W/(m^2 \cdot {}^\circ C)}$$

The heat loss per meter length of tube is

$$\begin{aligned} Q &= AU\,\Delta T \\ &= \pi D U (T_i - T_o) \\ &= (\pi)(0.025)(74.63)(80 - 20) \\ &= 351.7\ \mathrm{W/m} \end{aligned}$$

Example 11-3 Engine oil at a mean temperature $T_i = 80°C$ and mean velocity $u = 0.1$ m/s flows inside a thin-walled, horizontal copper tube with an ID of $D = 2.5$ cm. The outer surface of the tube dissipates heat by free convection into atmospheric air at $T_\infty = 20°C$. Calculate the temperature of the tube wall, the overall heat transfer coefficient, and the heat loss per meter length of tube.

SOLUTION The physical properties of engine oil at 80°C are

$$\nu = 0.375 \times 10^{-4}\ \mathrm{m^2/s} \qquad k = 0.138\ \mathrm{W/(m \cdot {}^\circ C)}$$

The Reynolds number is

$$\mathrm{Re} = \frac{uD}{\nu} = \frac{0.1 \times 0.025}{0.375 \times 10^{-4}} = 66.7$$

The flow is laminar. By assuming fully developed flow and a constant tube wall temperature, the Nusselt number for oil flow is given by

$$\mathrm{Nu} = 3.66$$

and

$$h_i = 3.66\,\frac{k}{D} = 3.66\,\frac{0.138}{0.025} = 20.2\ \mathrm{W/(m^2 \cdot {}^\circ C)} \qquad (a)$$

We assume that the thermal resistance for a thin-walled copper tube is negligible.

The free-convection heat transfer coefficient h_o at the outer surface of the tube can be calculated from the following simplified expression:

$$h_o = 1.32\left(\frac{T_w - T_\infty}{D}\right)^{0.25} = 1.32\left(\frac{T_w - 20}{0.025}\right)^{0.25}$$

$$= 3.32(T_w - 20)^{0.25} \qquad (b)$$

The tube wall temperature T_w is not known. We consider an overall energy balance equation with the assumption that $A_i \cong A_o$

$$h_i(T_i - T_w) = h_o(T_w - T_\infty) \qquad (c)$$

We introduce the above values of h_i and h_o and T_i and T_∞ into Eq. (c):

$$20.2(80 - T_w) = 3.32(T_w - 20)^{1.25} \qquad (d)$$

The tube wall temperature T_w can now be determined from this equation by iteration. We find

$$T_w = 62.28°\text{C}$$

Then h_o is calculated from Eq. (b):

$$h_o = 8.47 \text{ W/(m}^2 \cdot °\text{C)}$$

and the overall heat transfer coefficient becomes

$$U = \frac{1}{1/h_i + 1/h_o} = \frac{1}{1/20.2 + 1/8.47} = 5.97 \text{ W/(m}^2 \cdot °\text{C)}$$

The heat loss per meter length of tube is

$$Q = \pi D U(T_i - T_\infty)$$

$$= (\pi)(0.025)(5.97)(80 - 20) = 28.13 \text{ W/m}$$

11-4 THE LMTD METHOD FOR HEAT EXCHANGER ANALYSIS

In the thermal analysis of heat exchangers, the total heat transfer rate Q through the heat exchanger is a quantity of primary interest. Here we turn our attention to single-pass heat exchangers having flow arrangements of the type illustrated in Fig. 11-15. It is apparent from this figure that the temperature difference ΔT between the hot and the cold fluids, in general, is not constant; it varies with distance along the heat exchanger.

In the heat transfer analysis of heat exchangers, it is convenient to establish ΔT_m between the hot and cold fluids such that the total heat transfer rate Q between the fluids can be determined from the following simple expression:

$$Q = AU\,\Delta T_m \qquad (11\text{-}8)$$

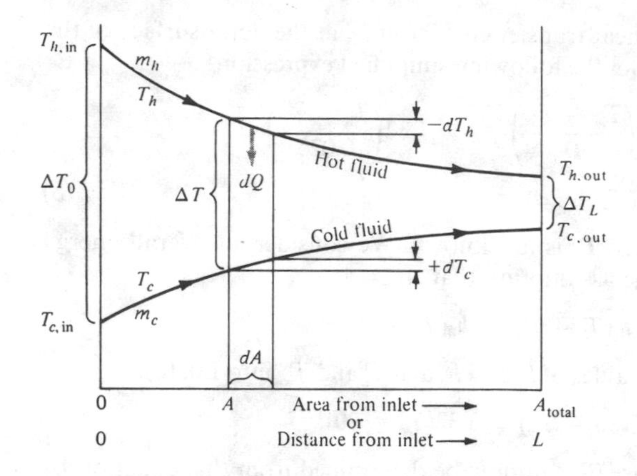

Figure 11-15 Nomenclature for the derivation of logarithmic mean temperature difference $\Delta T_{\ln}$.

where A is the total heat transfer area and U is the average overall heat transfer coefficient based on that area.

In the following analysis we develop an expression for such a mean temperature difference by utilizing the single-pass parallel-flow arrangement shown in Fig. 11-15. But the result developed here is applicable for all the single-flow arrangements shown in Fig. 11-12.

Refer to Fig. 11-15. Let

$A \equiv$ heat transfer area measured from inlet, m^2

$m_c, m_h =$ mass flow rate of cold and hot fluids, respectively, kg/h

$\Delta T = T_h - T_c =$ local temperature difference between hot and cold fluids, °C

$U =$ local overall heat transfer coefficient between two fluids, $W/(m^2 \cdot °C)$

The rate of heat transfer dQ from the hot to the cold fluid through an elemental area dA about the location A is given by

$$dQ = U\, dA\, \Delta T \tag{11-9}$$

However, dQ should equal the heat given up by the hot fluid or gained by the cold fluid flowing from position A to $A + dA$; with this consideration, we write

$$dQ = -m_h c_{ph}\, dT_h \quad \text{(hot fluid)} \tag{11-10a}$$

$$dQ = m_c c_{pc}\, dT_c \quad \text{(cold fluid)} \tag{11-10b}$$

where c_{pc} and c_{ph} are the specific heats and dT_c and dT_h are the changes in the temperatures of the cold and hot fluids, respectively. We note that

$$\Delta T = T_h - T_c \tag{11-11a}$$

or

$$d(\Delta T) = dT_h - dT_c \tag{11-11b}$$

Combining Eqs. (11-10) and utilizing Eq. (11-11*b*), we obtain

$$d(\Delta T) = -\frac{dQ}{m_h c_{ph}} - \frac{dQ}{m_c c_{pc}} = -dQ\left(\frac{1}{m_h c_{ph}} + \frac{1}{m_c c_{pc}}\right) \tag{11-12}$$

which can be written more compactly as

$$d(\Delta T) = -B\, dQ \tag{11-13a}$$

where

$$B = \frac{1}{m_h c_{ph}} + \frac{1}{m_c c_{pc}} \tag{11-13b}$$

The elimination of dQ between Eqs. (11-9) and (11-13*a*) yields

$$\frac{d(\Delta T)}{\Delta T} = -UB\, dA \tag{11-14}$$

The integration of Eq. (11-14) over the entire length of the heat exchanger gives

$$\int_{\Delta T_0}^{\Delta T_L} \frac{d(\Delta T)}{\Delta T} = -B \int_0^{A_t} U\, dA$$

or

$$\int_{\Delta T_0}^{\Delta T_L} \frac{d(\Delta T)}{\Delta T} = -BA_t \frac{\int_0^{A_t} U\, dA}{A_t} \tag{11-15}$$

where A_t is the total heat transfer area of the heat exchanger. We now define the average overall heat transfer coefficient U_m for the entire heat exchanger as

$$U_m = \frac{1}{A_t} \int_0^{A_t} U\, dA \tag{11-16}$$

Then Eq. (11-15) is integrated to yield

$$\ln \frac{\Delta T_0}{\Delta T_L} = BU_m A_t \tag{11-17}$$

The total heat transfer rate Q through the heat exchanger is determined by integrating Eq. (11-13*a*) over the entire length:

$$\int_{\Delta T_0}^{\Delta T_L} d(\Delta T) = -B \int_0^Q dQ$$

$$\Delta T_0 - \Delta T_L = BQ$$

or

$$Q = \frac{\Delta T_0 - \Delta T_L}{B} \tag{11-18}$$

The elimination of B between Eqs. (11-17) and (11-18) results in

$$Q = A_t U_m \frac{\Delta T_0 - \Delta T_L}{\ln (\Delta T_0 / \Delta T_L)} \tag{11-19}$$

Our objective in this analysis was to express the total heat transfer rate through the heat exchanger in terms of a mean temperature difference $\Delta T_{\ln}$ in the form

$$Q = A_t U_m \, \Delta T_{\ln} \tag{11-20}$$

A comparison of the results given by Eqs. (11-19) and (11-20) reveals that the mean temperature difference $\Delta T_{\ln}$ between the hot and cold fluids, over the entire length of the heat exchanger, is given by

$$\Delta T_{\ln} = \frac{\Delta T_0 - \Delta T_L}{\ln (\Delta T_0 / \Delta T_L)} \tag{11-21}$$

The mean temperature difference $\Delta T_{\ln}$, defined by Eq. (11-21), is called the *logarithmic mean temperature difference* (LMTD).

Thus, the total heat transfer rate between the hot and cold fluids for all the single-pass flow arrangements shown in Fig. 11-12 is determined from

$$Q = AU \, \Delta T_{\ln} \tag{11-22}$$

where $\Delta T_{\ln}$ is defined by Eq. (11-21). We note that for the special case of $\Delta T_0 = \Delta T_L$, Eq. (11-21) leads to $\Delta T_{\ln} = 0/0 =$ indeterminate. But by the application of L'Hospital's rule [i.e., by differentiating the numerator and denominator of Eq. (11-21) with respect to ΔT_0], it can be shown that for this particular case $\Delta T_{\ln} = \Delta T_0 = \Delta T_L$.

It is of interest to compare the LMTD of ΔT_0 and ΔT_L with their arithmetic mean:

$$\Delta T_a = \frac{\Delta T_0 + \Delta T_L}{2} \tag{11-23}$$

We present in Table 11-2 a comparison of the logarithmic and the arithmetic means of the two quantities ΔT_0 and ΔT_L. We note that the arithmetic and logarithmic means are equal for $\Delta T_0 = \Delta T_L$. When $\Delta T_0 \neq \Delta T_L$, the LMTD is always *less than* the arithmetic mean; if ΔT_0 is not more than 50 percent greater than ΔT_L, the LMTD can be approximated by the arithmetic mean within about 1.4 percent.

Table 11-2 Comparison of logarithmic and arithmetic means of ΔT_0 and ΔT_L

$\dfrac{\Delta T_0}{\Delta T_L}$	1	1.2	1.5	1.7	2
$\dfrac{\Delta T_a}{\Delta T_{\ln}}$	1	1.0028	1.0137	1.023	1.04

Example 11-4 In a single-pass shell-and-tube heat exchanger, the inlet and outlet temperatures for the hot fluid are, respectively, $T_{h,i} = 260°C$ and $T_{h,o} = 140°C$; for the cold fluid they are $T_{c,i} = 70°C$ and $T_{c,o} = 125°C$. Calculate the logarithmic mean temperature difference for (*a*) counterflow and (*b*) parallel-flow arrangements.

SOLUTION The temperature profiles for the counterflow and parallel-flow arrangements are illustrated in the sketch.

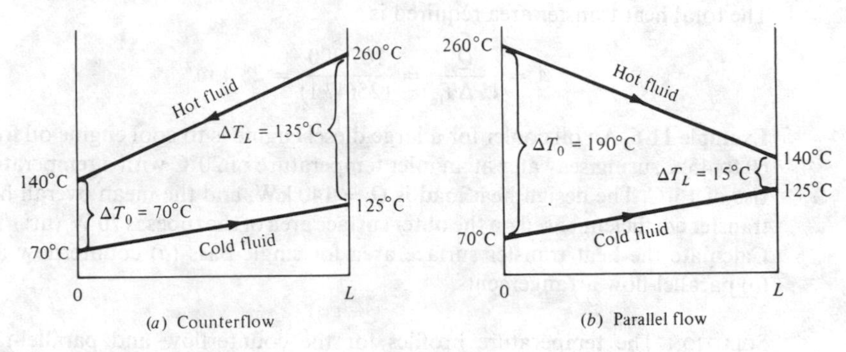

(*a*) Counterflow (*b*) Parallel flow

(*a*) For the counterflow

$$\Delta T_{\ln} = \frac{135 - 70}{\ln \frac{135}{70}} \cong 99$$

(*b*) For the parallel flow:

$$\Delta T_{\ln} = \frac{190 - 15}{\ln \frac{190}{15}} = 68.9$$

Example 11-5 A counterflow shell-and-tube heat exchanger is used to heat water at a rate of $m = 0.8$ kg/s from $T_i = 30°C$ to $T_o = 80°C$, with hot oil entering at 120°C and leaving at 85°C. The overall heat transfer coefficient is $U = 125$ W/(m^2 · °C). Calculate the heat transfer area required.

SOLUTION If we take the specific heat of water as $c_p = 4180$ J/(kg · °C), the total heat load for the exchanger is

$$Q = mc_p(T_o - T_i) = (0.8)(4180)(80 - 30)$$
$$= 167{,}200 \text{ W}$$

The temperature profiles in the exchanger are illustrated in the sketch. The logarithmic mean temperature difference becomes

$$\Delta T_{\ln} = \frac{55 - 40}{\ln \frac{55}{40}} = 47.1°C$$

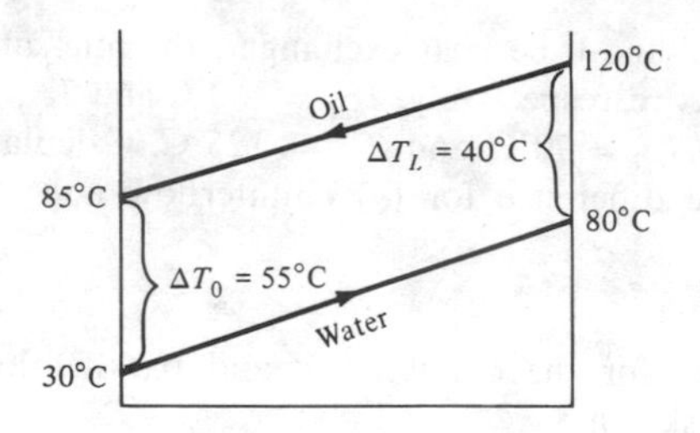

The total heat transfer area required is

$$A = \frac{Q}{U\,\Delta T_{\ln}} = \frac{167{,}200}{125(47.1)} = 28.4\ \text{m}^2$$

Example 11-6 An oil cooler for a large diesel engine is to cool engine oil from 60 to 45°C, using seawater at an inlet temperature of 20°C with a temperature rise of 15°C. The design heat load is $Q = 140$ kW, and the mean overall heat transfer coefficient based on the outer surface area of the tubes is 70 W/(m^2 · °C). Calculate the heat transfer surface area for single-pass (*a*) counterflow and (*b*) parallel-flow arrangements.

SOLUTION The temperature profiles for the counterflow and parallel-flow arrangements are shown in the figure. The heat transfer area for each arrangement is now evaluated.

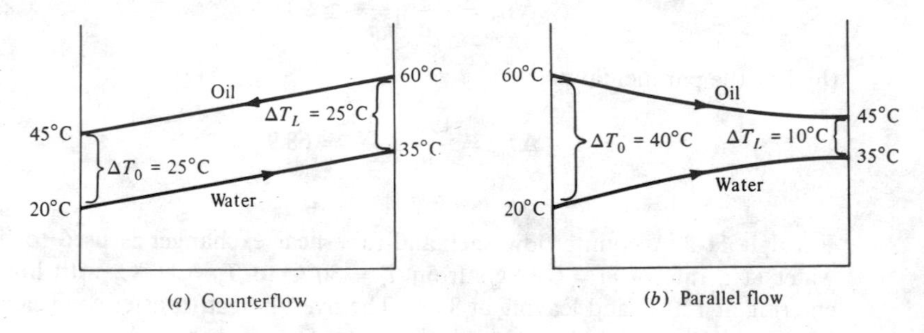

(*a*) Counterflow (*b*) Parallel flow

(*a*) Counterflow:

$$\Delta T_0 = \Delta T_L = 25°\text{C} \quad \therefore \quad \Delta T_{\ln} = 25°\text{C}$$

$$A = \frac{Q}{\Delta T_{\ln}\,U_m} = \frac{140{,}000}{25(70)} = 80\ \text{m}^2$$

(*b*) Parallel flow:

$$\Delta T_0 = 40°\text{C} \qquad \Delta T_L = 10°\text{C} \qquad \therefore \qquad \Delta T_{\ln} = \frac{40 - 10}{\ln \frac{40}{10}} = 21.64°\text{C}$$

$$A = \frac{Q}{\Delta T_{\ln}\,U_m} = \frac{140{,}000}{21.64(70)} = 92.42\ \text{m}^2$$

We note that less area is required with the counterflow arrangement.

Example 11-7 Engine oil is to be cooled from 80 to 50°C by using a single-pass, counterflow, concentric-tube heat exchanger with cooling water available at 20°C. Water flows inside a tube with an ID of $D_i = 2.5$ cm at a rate of $m_w = 0.08$ kg/s, and oil flows through the annulus at a rate of $m_{\text{oil}} = 0.16$ kg/s. The heat transfer coefficients for the water side and oil side are, respectively, $h_w = 1000$ W/(m^2 · °C) and $h_{\text{oil}} = 80$ W/(m^2 · °C); the fouling factors are $F_w = 0.00018$ m^2 · °C/W and $F_{\text{oil}} = 0.00018$ m^2 · °C/W; and the tube wall resistance is negligible. Calculate the tube length required.

SOLUTION The specific heats of water and oil can be taken as

$$c_w = 4180 \text{ J/(kg} \cdot \text{°C)} \qquad c_{\text{oil}} = 2090 \text{ J/(kg} \cdot \text{°C)}$$

The outlet temperature of water can be calculated from an overall energy balance:

$$Q = m_{\text{oil}} c_{\text{oil}}(T_{h,i} - T_{h,o}) = m_w c_w (T_{c,o} - T_{c,i})$$
$$Q = 0.16(2090)(80 - 50) = 0.08(4180)(T_{c,o} - 20)$$

so $$T_{c,o} = 50\text{°C} \qquad \text{and} \qquad Q = 10{,}032 \text{ W}$$

The temperature profiles for the hot and cold fluids are shown in the figure.

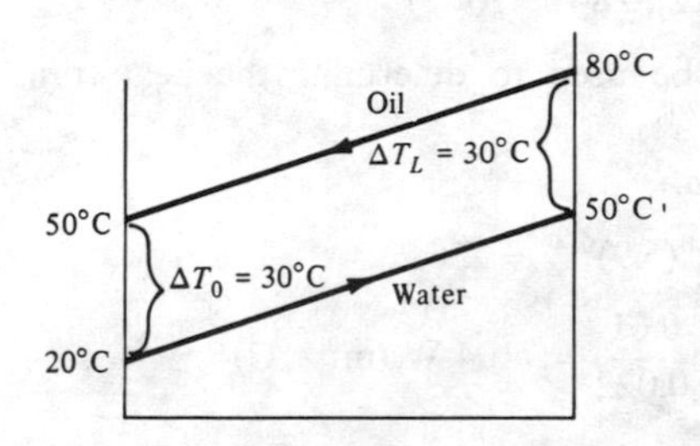

We note that the temperature difference between the hot and cold fluids is constant ($\Delta T = 30$°C) throughout the heat exchanger. The reason is the equal heat capacity rates for the hot and cold fluids, that is $c_w m_w = c_{\text{oil}} m_{\text{oil}}$.

The overall heat transfer coefficient U is determined from Eq. (11-4), by neglecting the thermal resistance of the tube and the curvature effects. We obtain

$$U = \frac{1}{1/h_w + F_w + F_{\text{oil}} + 1/h_{\text{oil}}}$$

$$= \frac{1}{\frac{1}{1000} + 0.00018 + 0.00018 + \frac{1}{80}} = 72.2 \text{ W/(m}^2 \cdot \text{°C)}$$

For $\Delta T_0 = \Delta T_L = 30$°C, we have $\Delta T_{\ln} = 30$°C. The total heat transfer area required becomes

$$A = \frac{Q}{\Delta T_{\ln} U} = \frac{10{,}032}{30(72.2)} = 4.63 \text{ m}^2$$

Then the tube length L

$$L = \frac{A}{\pi D_i} = \frac{4.63}{\pi(0.025)} \cong 60 \text{ m}$$

Example 11-8 A shell-and-tube steam condenser is to be constructed of 2.5-cm-OD, 2.2-cm-ID, single-pass horizontal tubes with steam condensing at $T_s = 54°C$ outside the tubes. The cooling water enters each tube at $T_i = 18°C$, with a flow rate of $m = 0.7$ kg/s per tube and leaves at $T_o = 36°C$. The heat transfer coefficient for the condensation of steam is $h_s = 8000 \text{ W/(m}^2 \cdot °\text{C)}$. Calculate the tube length L. Calculate the condensation rate per tube.

SOLUTION The physical properties of water are taken at $(18 + 36)/2 = 27°C$ as

$$c_p = 4180 \text{ J/(kg} \cdot °\text{C)} \qquad \mu = 0.86 \times 10^{-3} \text{ kg/(m} \cdot \text{s)}$$

$$\text{Pr} = 5.9 \qquad k = 0.61 \text{ W/(m} \cdot °\text{C)}$$

The Reynolds number for flow inside the tube is

$$\text{Re} = \frac{4m}{\pi D \mu} = \frac{(4)(0.7)}{\pi(0.022)(0.86 \times 10^{-3})} = 47{,}107$$

The Dittus-Boelter equation can be used to determine the heat transfer coefficient for the water side:

$$\begin{aligned} \text{Nu} &= 0.023 \text{ Re}^{0.8} \text{ Pr}^{0.4} \\ &= 0.023(47{,}107)^{0.8}(5.9)^{0.4} = 256.2 \end{aligned}$$

$$h_i = \text{Nu} \frac{k}{D} = 256.2 \frac{0.61}{0.022} = 7104 \text{ W/(m}^2 \cdot °\text{C)}$$

The overall heat transfer coefficient based on the outer surface of the tube can be determined from Eq. (11-7). Neglecting the tube wall resistance, we have

$$\begin{aligned} U_o &= \frac{1}{(D_o/D_i)(1/h_i) + 1/h_c} = \frac{1}{(0.025/0.022)(\frac{1}{7104}) + (\frac{1}{8000})} \\ &= 3509 \text{ W/(m}^2 \cdot °\text{C)} \end{aligned}$$

The logarithmic mean temperature difference is

$$\Delta T_{\ln} = \frac{(54 - 18) - (54 - 36)}{\ln [(54 - 18)/(54 - 36)]} = 25.97$$

The tube length L is determined by writing an overall energy balance for a tube:

$$Q_{\text{tube}} = (\pi D_o L) U_o \, \Delta T_{\ln} = m c_p (T_o - T_i)$$

$$Q_{\text{tube}} = \pi(0.025)(L)(3509)(25.97) = 0.7(4180)(36 - 18)$$

$$L = 7.4 \text{ m}$$

The heat transfer rate per tube is

$$Q_{\text{tube}} = 52{,}668 \text{ W}$$

The condensation rate per tube is

$$m_{\text{tube}} = \frac{Q_{\text{tube}}}{h_{fg}} = \frac{52{,}668}{2{,}372{,}400} = 2.22 \times 10^{-2} \text{ kg/s}$$

11-5 CORRECTION FOR LMTD FOR USE WITH CROSS-FLOW AND MULTIPASS EXCHANGERS

The LMTD developed in Sec. 11-4 is not applicable for the heat transfer analysis of cross-flow and multipass exchangers. The effective temperature differences have been determined for cross-flow and multipass arrangements also, but the resulting expressions are very complicated. Therefore, for such situations, it is customary to introduce a *correction factor* F [45–47] so that the simple LMTD can be adjusted to represent the effective temperature difference ΔT_{corr} for the cross-flow and multipass arrangements as

$$\Delta T_{\text{corr}} = F(\Delta T_{\text{ln}} \text{ for counterflow}) \tag{11-24}$$

where ΔT_{ln} should be computed on the basis of counterflow conditions. Namely, ΔT_0 and ΔT_L appearing in the definition of LMTD given by Eq. (11-21) should be taken as (see Fig. 11-12*e*)

$$\Delta T_0 = T_{h,\text{out}} - T_{c,\text{in}} \tag{11-25a}$$

$$\Delta T_L = T_{h,\text{in}} - T_{c,\text{out}} \tag{11-25b}$$

where the subscripts c and h refer, respectively, to the cold and the hot fluids. Figure 11-16 shows the correction factor F for some commonly used heat exchanger configurations. In these figures, the abscissa is a dimensionless ratio P, defined as

$$P = \frac{t_2 - t_1}{T_1 - t_1} \tag{11-26a}$$

where T refers to the *shell-side temperature*, t to the *tube-side temperature*, and subscripts 1 and 2, respectively, to the *inlet* and *outlet* conditions. The parameter R appearing on the curves is defined as

$$R = \frac{T_1 - T_2}{t_2 - t_1} = \frac{(mc_p)_{\text{tube side}}}{(mc_p)_{\text{shell side}}} \tag{11-26b}$$

Note that the correction factors in Fig. 11-16 are applicable whether the hot fluid is in the shell side or the tube side. Correction-factor charts for several other flow arrangements are available in Ref. 45.

Generally F is less than unity for cross-flow and multipass arrangements; it is unity for true counterflow heat exchanger. It represents the degree of departure of the true mean temperature difference from the LMTD for the counterflow.

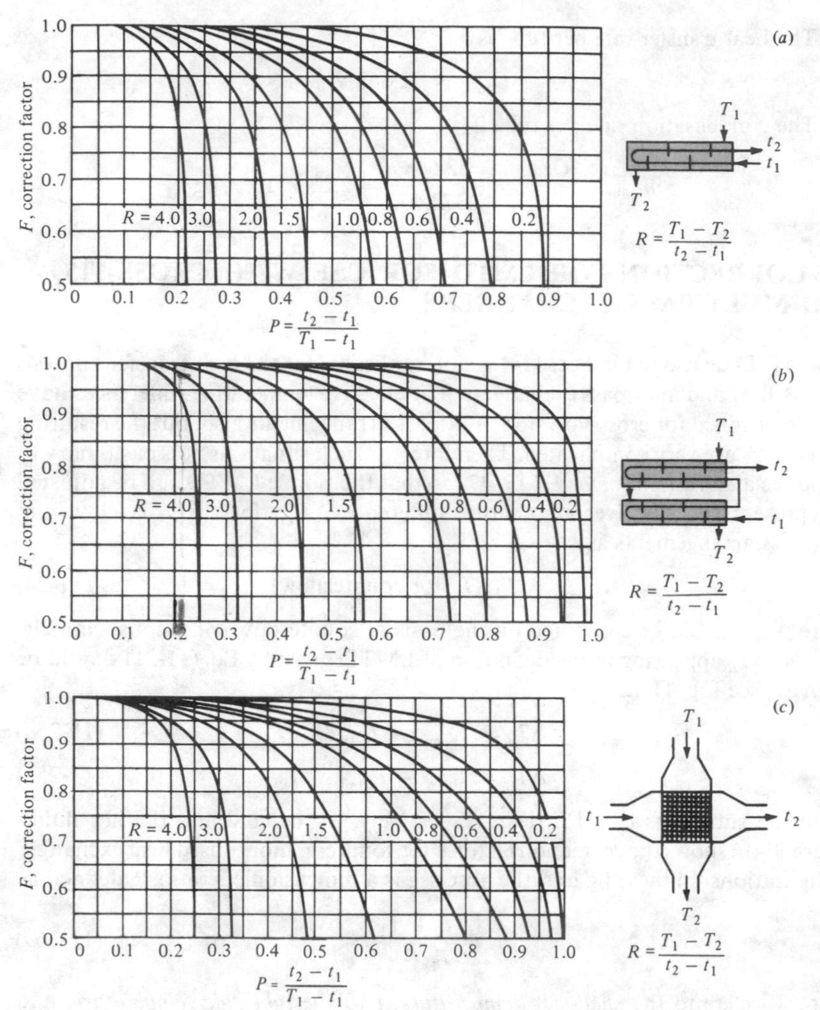

Figure 11-16 Correction factor F for computing $\Delta T_{\text{corrected}}$ for multipass and cross-flow exchangers. (*a*) One shell pass and two tube pass or multiple of two tube pass; (*b*) two shell pass and four tube pass or multiple of four tube pass; (*c*) single-pass, cross-flow, both fluids unmixed. (*From Bowman, Mueller, and Nagle* [45].)

We note from Fig. 11-16 that the value of the parameter P ranges from 0 to 1, and it represents the thermal effectiveness of the tube-side fluid. The value of R ranges from zero to infinity, with zero corresponding to pure vapor condensation on the shell side and infinity to evaporation on the tube side.

Example 11-9 A two shell pass, four tube pass heat exchanger with flow arrangement shown in Fig. 11-16*b* has water on the shell side and brine on the tube side. Water is cooled from $T_1 = 18°C$ to $T_2 = 6°C$ with brine entering at $t_1 = -1°C$ and leaving at $t_2 = 3°C$. The overall heat transfer coefficient is $U = 600\ W/(m^2 \cdot °C)$. Calculate the heat transfer area required for a design heat load of $Q = 24{,}000$ W.

SOLUTION This is a multipass heat exchanger. Therefore, the LMTD based on a *counterflow* arrangement should be corrected by using the correction factor given in Fig. 11-16*b*. The counterflow arrangement is illustrated in the figure.

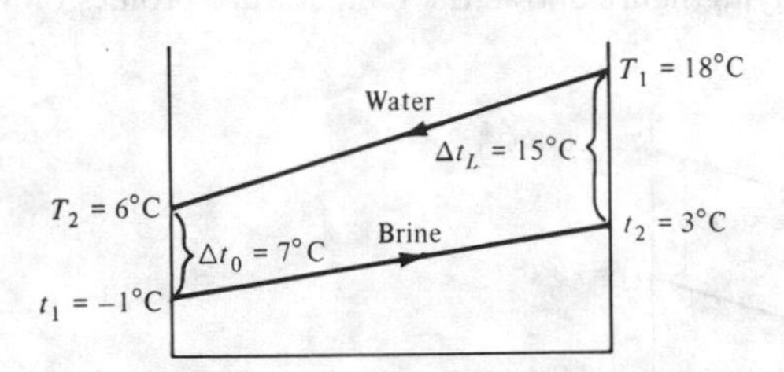

The counterflow LMTD becomes

$$\Delta T_{\ln} = \frac{15 - 7}{\ln \frac{15}{7}} = 10.5°C$$

The parameters P and R are

$$P = \frac{t_2 - t_1}{T_1 - t_1} = \frac{3 - (-1)}{18 - (-1)} = \frac{4}{19} = 0.21$$

$$R = \frac{T_1 - T_2}{t_2 - t_1} = \frac{18 - 6}{3 - (-1)} = \frac{12}{4} = 3$$

Then, from Fig. 11-16*b*, the correction factor F is

$$F = 0.98$$

and the corrected temperature difference becomes

$$\Delta T_{\text{corr}} = 0.98(10.5) = 10.29°C$$

The heat transfer area required is then

$$A = \frac{Q}{U\,\Delta T_{\text{corr}}} = \frac{24{,}000}{600(10.29)} = 3.89\ m^2$$

Example 11-10 A one shell pass, two tube pass heat exchanger with flow arrangement shown in Fig. 11-16*a* has water on the tube side and engine oil on the shell side. It must be designed to heat 1.5 kg/s water from $t_1 = 30°C$ to $t_2 = 80°C$, with hot oil entering at $T_1 = 120°C$ and leaving at $T_2 = 80°C$. The overall heat transfer coefficient is $U = 250\ W/(m^2 \cdot °C)$. Calculate the heat transfer area required.

SOLUTION Taking the specific heat of water as $c_p = 4180$ J/(kg · °C), we find the total heat load:

$$Q = mc_p(t_2 - t_1) = 1.5(4180)(80 - 30)$$
$$= 313{,}500 \text{ W}$$

Since the exchanger is a multipass type, the LMTD should be calculated on the counterflow basis and corrected by utilizing the correction factor from Fig. 11-16*a*. The accompanying figure shows the temperature profiles on the counterflow basis.

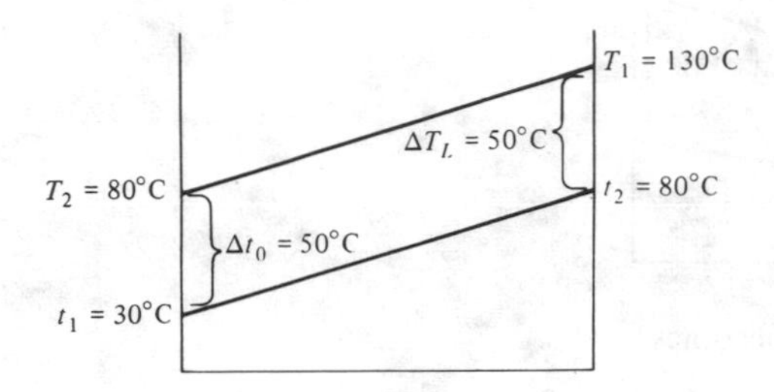

Since $\Delta T_0 = \Delta T_L = 50$°C, we have $\Delta T_{\ln} = 50$°C. Then

$$P = \frac{t_2 - t_1}{T_1 - t_1} = \frac{80 - 30}{130 - 30} = 0.5$$

$$R = \frac{T_1 - T_2}{t_2 - t_1} = \frac{130 - 80}{80 - 30} = 1$$

From Fig. 11-16*a* we find the correction factor F:

$$F = 0.8$$

and the corrected temperature difference becomes

$$\Delta T_{\text{corr}} = 0.8(50) = 40°\text{C}$$

The heat transfer area required is

$$A = \frac{Q}{U\,\Delta T_{\text{corr}}} = \frac{313{,}500}{250(40)} = 31.35 \text{ m}^2$$

Example 11-11 A one shell pass, two tube pass heat exchanger of flow arrangement shown in Fig. 11-16*a* is to be designed to heat $m_c = 2.0$ kg/s of pressurized water from $t_1 = 40$°C to $t_2 = 120$°C flowing on the tube side, by using hot water entering the shell side at $T_1 = 300$°C with a flow rate of $m_h = 1.03$ kg/s. The overall heat transfer coefficient is $U = 1250$ W/(m^2 · °C). Calculate the heat transfer area required.

SOLUTION The outlet temperature of hot water is calculated from an overall heat balance:

$$m_c c_{pc}(t_2 - t_1) = m_h c_{ph}(T_1 - T_2) = Q$$

$$2.0(4200)(120 - 40) = 1.03(4660)(300 - T_2)$$

$$T_2 = 160°\text{C}$$

and the total heat transfer rate Q becomes

$$Q = 672 \text{ kW}$$

The exchanger is multipass, so the mean temperature difference should be evaluated by appropriate correction to the counterflow LMTD. The accompanying figure shows the counterflow temperature profiles.

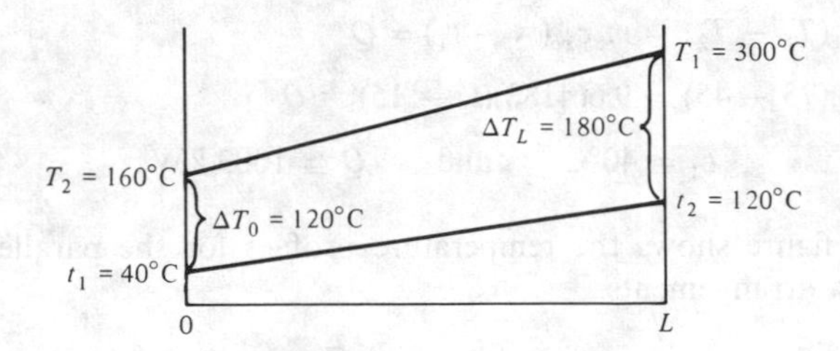

The counterflow LMTD becomes

$$\Delta T_{\ln} = \frac{180 - 120}{\ln \frac{180}{120}} = 148°\text{C}$$

So P and R are

$$P = \frac{t_2 - t_1}{T_1 - t_1} = \frac{120 - 40}{300 - 40} = \frac{80}{260} = 0.31$$

$$R = \frac{T_1 - T_2}{t_2 - t_1} = \frac{300 - 160}{120 - 40} = \frac{140}{80} = 1.75$$

The correction factor F is determined from Fig. 11-16a as

$$F = 0.90$$

Then the corrected temperature difference becomes

$$\Delta T_{\text{corr}} = 0.9(148) = 133.2°\text{C}$$

The heat transfer area required is

$$A = \frac{Q}{U\,\Delta T_{\text{corr}}} = \frac{672{,}000}{1250(133.2)} = 4 \text{ m}^2$$

Example 11-12 A heat exchanger is to be designed to cool $m_h = 8.7$ kg/s an ethyl alcohol solution [$c_{ph} = 3840$ J/(kg · °C)] from $T_1 = 75$°C to $T_2 = 45$°C with cooling water [$c_{pc} = 4180$ J/(kg · °C)] entering the tube side at $t_1 = 15$°C at a rate of $m_c = 9.6$ kg/s. The overall heat transfer coefficient based on the outer tube surface is $U_0 = 500$ W/(m^2 · °C). Calculate the heat transfer area for each of the following flow arrangements:

(*a*) Parallel flow, shell and tube
(*b*) Counterflow, shell and tube
(*c*) One shell pass and two tube pass (Fig. 11-16*a*)
(*d*) Cross-flow, both fluids unmixed (Fig. 11-16*c*)

SOLUTION The outlet temperature t_2 of the cooling water is determined by an overall energy balance

$$m_h c_{ph}(T_1 - T_2) = m_c c_{pc}(t_2 - t_1) = Q$$

$$8.7(3840)(75 - 45) = 9.6(4180)(t_2 - 15) = Q$$

So $$t_2 = 40°\text{C} \quad \text{and} \quad Q \cong 1003 \text{ kW}$$

The accompanying figure shows the temperature profiles for the parallel-flow and counterflow arrangements.

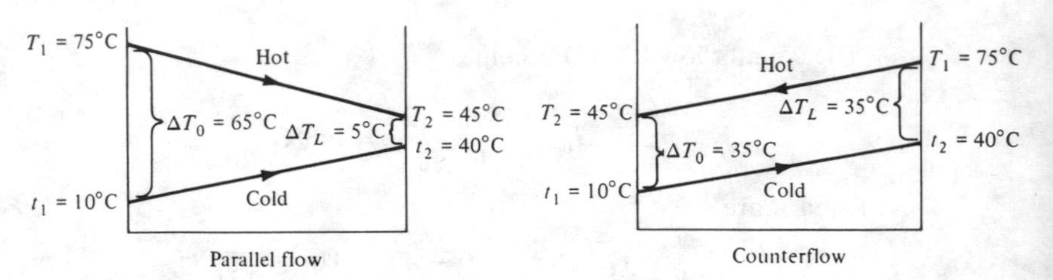

(*a*) Parallel flow: The LMTD becomes

$$\Delta T_{\ln} = \frac{65 - 5}{\ln \frac{65}{5}} = 23.4°\text{C}$$

and the heat transfer area is

$$A = \frac{Q}{U\,\Delta T_{\ln}} = \frac{1{,}003{,}000}{500(23.4)} = 85.73 \text{ m}^2$$

(*b*) Counterflow: The LMTD is

$$\Delta T_{\ln} = 35°\text{C}$$

and the heat transfer area becomes

$$A = \frac{Q}{U\,\Delta T_{\ln}} = \frac{1{,}003{,}000}{500(35)} = 57.3 \text{ m}$$

(*c*) One shell pass, two tube pass: The parameters P and R are

$$P = \frac{t_2 - t_1}{T_1 - t_1} = \frac{40 - 10}{75 - 10} = \frac{30}{65} = 0.46$$

$$R = \frac{T_1 - T_2}{t_2 - t_1} = \frac{75 - 45}{40 - 10} = \frac{30}{30} = 1$$

The correction factor F is found from Fig. 11-16*a*:

$$F = 0.87$$

The corrected temperature difference becomes

$$\Delta T_{corr} = 0.87(35) = 30.45°\text{C}$$

The heat transfer area is

$$A = \frac{1{,}003{,}000}{500(30.45)} = 65.88 \text{ m}^2$$

(*d*) Cross-flow: For $P = 0.46$ and $R = 1$, the correction factor F from Fig. 11.16*c* is

$$F = 0.93$$

The corrected temperature difference becomes

$$\Delta T_{corr} = 0.93(35) = 32.55°\text{C}$$

The heat transfer area is

$$A = \frac{1{,}003{,}000}{500(32.55)} = 61.63 \text{ m}^2$$

11-6 ε-NTU METHOD FOR HEAT EXCHANGER ANALYSIS

The *rating* and *sizing* of heat exchangers are the two important problems encountered in the thermal analysis of heat exchangers. The *rating* problem is concerned with the determination of the heat transfer rate, the fluid outlet temperatures, and the pressure drops for an existing heat exchanger or the one which is already sized; hence the heat transfer surface area and the flow passage dimensions are available. The *sizing* problem is concerned with the determination of the matrix dimensions to meet the specified heat transfer and pressure drop requirements. If we are not concerned with the pressure drop, the rating problem involves the determination of the total heat transfer rate for an existing heat exchanger; and the sizing problem involves the determination of the total heat transfer surface required to meet a specified heat transfer rate.

If the inlet and outlet temperatures of the hot and the cold fluid and the overall heat transfer coefficient are specified, the LMTD method, with or without the correction, can be used to solve the rating or sizing problem.

In some situations only the inlet temperatures and flow rates of the hot and cold fluids are given, and the overall heat transfer coefficient can be estimated. For such cases, the logarithmic mean temperature cannot be determined because the outlet temperatures are not known. Therefore, the use of LMTD method for the thermal analysis of heat exchangers will involve tedious iterations in order to determine the proper value of LMTD which will satisfy the requirement that heat transferred in the heat exchanger be equal to the heat carried out by the fluid.

To illustrate the tedious iteration process involved in such a calculation, we consider a rating problem for the following conditions:

Given: Physical properties of hot and cold fluids
Inlet temperatures $T_{c,\,\mathrm{in}}$ and $T_{h,\,\mathrm{in}}$
Flow rates m_c and m_h, kg/s
Overall heat transfer coefficient U_m
Total heat transfer surface A
Appropriate LMTD correction chart

Determine: Total heat transfer rate Q

The following steps can be followed to solve this problem:

1. Assume an outlet temperature, and determine P and R according to Eqs. (11-26*a*) and (11-26*b*), respectively; also find the LMTD correction factor F from the chart.
2. Calculate ΔT_{ln} for counterflow conditions.
3. Determine Q from

$$Q = AU_m F\,\Delta T_{\mathrm{ln}}$$

4. Calculate the outlet temperatures from the knowledge of Q and the flow rates.
5. Compare the outlet temperatures determined in step 4 with the values assumed step 1.
6. If the assumed and computed values of the outlet temperatures are different, repeat the calculations until a specified convergence is achieved.

Clearly, such a computation is very tedious. The analysis can be simplified significantly by using the *ε-NTU method* or the *effectiveness method*, developed originally by Kays and London [10].

In this method, effectiveness ε is defined as

$$\varepsilon = \frac{Q}{Q_{\max}} = \frac{\text{actual heat transfer rate}}{\text{maximum possible heat transfer rate from one stream to the other}}$$

The maximum possible heat transfer rate $Q_{\max}$ is obtained with a counterflow exchanger if the temperature change of the fluid having the minimum value of mc_p equals the difference in the inlet temperatures of the hot and cold fluids. Here we consider $(mc_p)_{\min}$, because the energy given up by one fluid should equal that received by the other fluid. If we consider $(mc_p)_{\max}$, then the other fluid should undergo a temperature change greater than the maximum available temperature difference; that is, ΔT for the other fluid should be greater than $T_{h,\text{in}} - T_{c,\text{in}}$. This is not possible. With this consideration, $Q_{\max}$ is chosen as

$$Q_{\max} = (mc_p)_{\min}(T_{h,\text{in}} - T_{c,\text{in}}) \tag{11-27}$$

Then, given ε and $Q_{\max}$, the actual heat transfer rate Q is

$$\boxed{Q = \varepsilon(mc_p)_{\min}(T_{h,\text{in}} - T_{c,\text{in}})} \tag{11-28}$$

Here $(mc_p)_{\min}$ is the smaller of $m_h c_{ph}$ and $m_c c_{pc}$ for the hot and cold fluids; $T_{h,\text{in}}$ and $T_{c,\text{in}}$ are the inlet temperatures of the hot and cold fluids, respectively.

Clearly, if the effectiveness ε of the exchanger is known, Eq. (11-28) provides an explicit expression for the determination of Q through the exchanger. We now describe the derivation of the expressions for the effectiveness ε.

Determination of ε The relation for the effectiveness depends on the heat exchanger geometry and the flow arrangement. To illustrate the general procedure for the derivation of ε, we again consider the *parallel-flow* arrangement shown in Fig. 11-15.

From Eq. (11.28) we write

$$\varepsilon = \frac{Q}{(mc_p)_{\min}(T_{h,\text{in}} - T_{c,\text{in}})} \tag{11-29}$$

The actual heat transfer rate Q is given by

$$Q = m_h c_{ph}(T_{h,\text{in}} - T_{h,\text{out}}) = m_c c_{pc}(T_{c,\text{out}} - T_{c,\text{in}}) \tag{11-30}$$

The substitution of Eq. (11-30) into (11-29) yields

$$\varepsilon = \frac{C_h(T_{h,\text{in}} - T_{h,\text{out}})}{C_{\min}(T_{h,\text{in}} - T_{c,\text{in}})} \tag{11-31a}$$

or

$$\varepsilon = \frac{C_c(T_{c,\text{out}} - T_{c,\text{in}})}{C_{\min}(T_{h,\text{in}} - T_{c,\text{in}})} \tag{11-31b}$$

where we defined

$$C_h \equiv m_h c_{ph} \qquad C_c \equiv m_c c_{pc} \tag{11-32}$$

and $C_{\min} \equiv$ smaller of C_h and C_c. Now, our objective is to eliminate the temperature ratio, say, in Eq. (11-31*b*). The procedure is as follows:

We consider Eq. (11-17)

$$\ln \frac{\Delta T_0}{\Delta T_L} = BU_m A \tag{11-33}$$

where, for the parallel-flow arrangement, we have

$$\Delta T_0 = T_{h,\text{in}} - T_{c,\text{in}} \tag{11-34a}$$

$$\Delta T_L = T_{h,\text{out}} - T_{c,\text{out}} \tag{11-34b}$$

Equation (11-33) is exponentiated, and the results of Eq. (11-34) are utilized:

$$\frac{T_{h,\,\text{out}} - T_{c,\,\text{out}}}{T_{h,\,\text{in}} - T_{c,\,\text{in}}} = e^{-BAU_m} \tag{11-35}$$

Equation (11-31) is solved for $T_{h,\,\text{out}}$:

$$T_{h,\,\text{out}} = T_{h,\,\text{in}} - \frac{C_c}{C_h}(T_{c,\,\text{out}} - T_{c,\,\text{in}}) \tag{11-36}$$

This result is introduced into Eq. (11-35) to eliminate $T_{h,\,\text{out}}$:

$$1 - \frac{T_{c,\,\text{out}} - T_{c,\,\text{in}}}{T_{h,\,\text{in}} - T_{c,\,\text{in}}}\left(1 + \frac{C_c}{C_h}\right) = e^{-BAU_m}$$

or

$$\frac{T_{c,\,\text{out}} - T_{c,\,\text{in}}}{T_{h,\,\text{in}} - T_{c,\,\text{in}}} = \frac{1 - e^{-AU_m B}}{1 + C_c/C_h} \tag{11-37}$$

This result is introduced into Eq. (11-31*b*), and the temperature ratio is eliminated. The effectiveness ε is determined as

$$\varepsilon = \frac{1 - e^{-BAU_m}}{C_{\min}/C_c + C_{\min}/C_h} \tag{11-38a}$$

where B is defined by Eq. (11-13*b*) as

$$B = \frac{1}{C_h} + \frac{1}{C_c} \tag{11-38b}$$

Clearly, if we consider a different flow arrangement, we should have a different expression for the effectiveness.

ε-NTU Relation

For convenience in practical applications, a dimensionless parameter called the *number of (heat) transfer units* (NTU) is defined as

$$\text{NTU} = \frac{AU_m}{C_{\min}} \tag{11-39a}$$

For simplicity in the notation, we adopt the following abbreviation:

$$\text{NTU} \equiv N \tag{11-39b}$$

Then Eq. (11-38) is written as

$$\varepsilon = \frac{1 - \exp\left[-N(C_{\min}/C_c + C_{\min}/C_h)\right]}{C_{\min}/C_c + C_{\min}/C_h} \tag{11-40}$$

We now define

$$C \equiv \frac{C_{min}}{C_{max}} \tag{11-41}$$

where C_{min} and C_{max} are, respectively, the smaller and the larger of the two quantities C_h and C_c. Then Eq. (11-40) is written more compactly as

$$\varepsilon = \frac{1 - \exp[-N(1 + C)]}{1 + C} \qquad \text{(for parallel flow)} \tag{11-42}$$

This equation gives the relation between the effectiveness ε and the number of heat transfer units N for a parallel-flow heat exchanger, regardless of whether C_{min} occurs on the hot or the cold side.

Similar calculations can be performed and ε-NTU relations can be developed for heat exchangers having other flow arrangements, such as counterflow, crossflow, multipass, etc. The reader should consult Ref. 10 for the effectiveness charts of various flow arrangements.

In Figs. 11-17 to 11-21 we present some effectiveness charts for typical flow arrangements. Also we list in Table 11-3 some of the functional relationships for ready reference.

Condensers and boilers In the case of condensers and boilers, the fluid temperature on the boiling or the condensing side remains essentially constant. Recall Eqs. (11-31) for the definition of effectiveness. If the effectiveness should remain

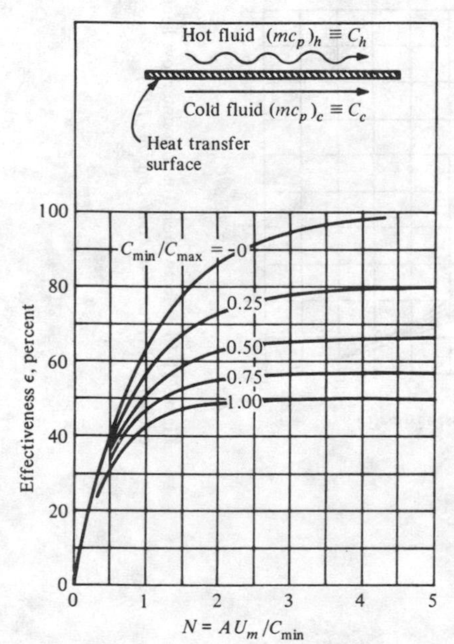

Figure 11-17 Effectiveness for a parallel-flow heat exchanger. (*From Kays and London [10].*)

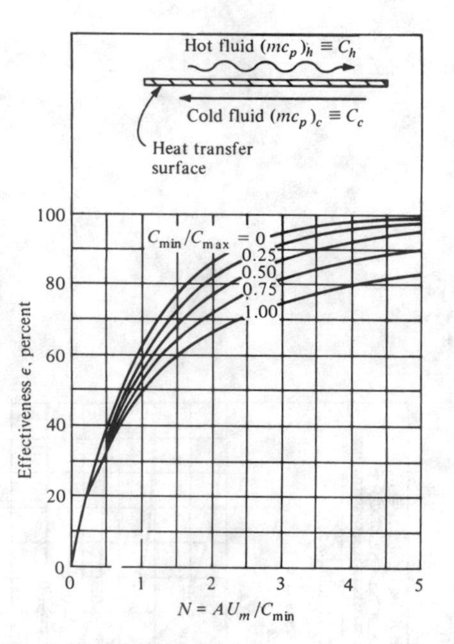

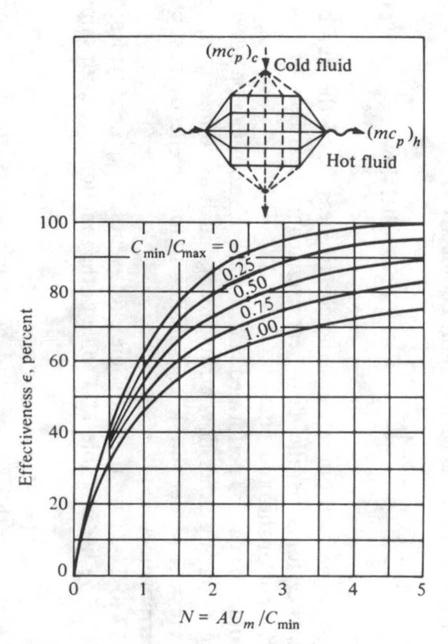

Figure 11-18 Effectiveness for a counterflow heat exchanger. (*From Kays and London* [*10*].)

Figure 11-19 Effectiveness for a cross-flow heat exchanger, both fluids unmixed. (*From Kays and London* [*10*].)

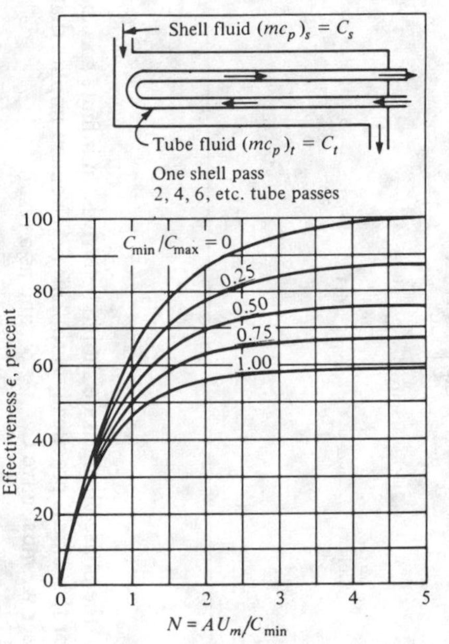

Figure 11-20 Effectiveness for a single shell pass heat exchanger with two, four, six, etc. tube passes. (*From Kays and London* [*10*].)

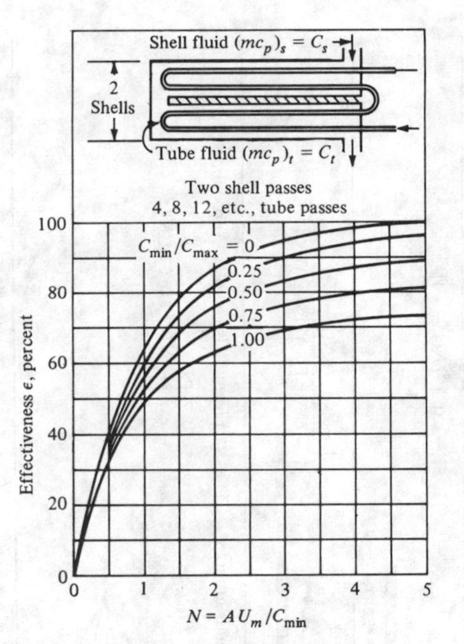

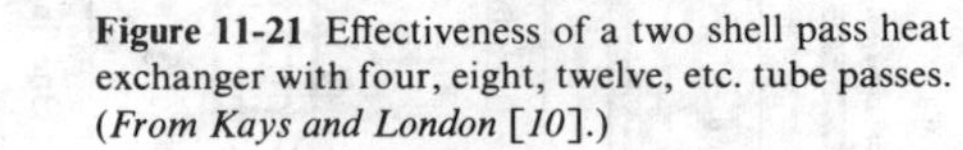

Figure 11-21 Effectiveness of a two shell pass heat exchanger with four, eight, twelve, etc. tube passes. (*From Kays and London* [*10*].)

Table 11-3 Heat exchanger effectiveness formulas;

$$N \equiv NTU = \frac{UA}{C_{\min}} \qquad C = \frac{C_{\min}}{C_{\max}}$$

Flow arrangement	ε formula
Parallel flow	$\varepsilon = \dfrac{1 - \exp[-N(1 + C)]}{1 + C}$
Counterflow 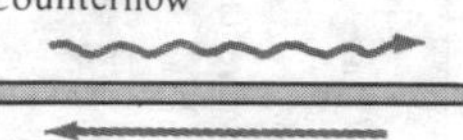	$\varepsilon = \dfrac{1 - \exp[-N(1 - C)]}{1 - C \exp[-N(1 - C)]}$
Cross-flow: Both fluids unmixed 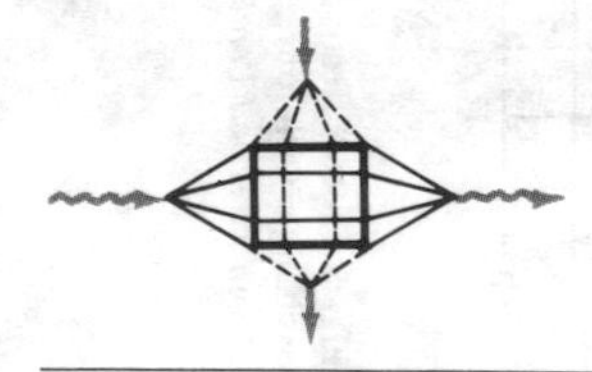	The exact formula is complex. Approximate formula is: $\varepsilon \cong 1 - \exp\left[\dfrac{1}{C} N^{0.22} \exp(-CN^{0.78} - 1)\right]$
Cross-flow: One fluid mixed, other unmixed (Mixed fluid; Unmixed fluid)	(1) $C_{\min}$ mixed, $C_{\max}$ unmixed: 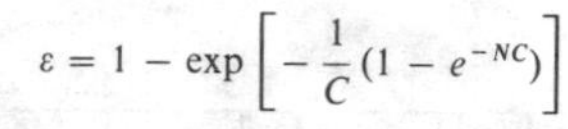 $\varepsilon = 1 - \exp\left[-\dfrac{1}{C}(1 - e^{-NC})\right]$ (2) $C_{\min}$ unmixed, $C_{\max}$ mixed: $\varepsilon = \dfrac{1}{C}\{1 - \exp[-C(1 - e^{-N})]\}$

finite, C_c or C_h on the phase-change side should behave as an infinite specific heat since $T_{\text{in}} - T_{\text{out}}$ for that side is practically zero. This requirement implies that for a boiler or condenser we must have $C_{\max} \to \infty$, and as a result,

$$C = \frac{C_{\min}}{C_{\max}} \to 0 \tag{11-43}$$

And for such situations the expressions given in Table 11-3 simplify to

$$\varepsilon = 1 - e^{-N} \qquad \text{for } C \to 0 \tag{11-44}$$

where $N = AU_m/C_{\min}$.

Physical significance of NTU The physical significance of the dimensionless parameter NTU can be viewed as follows:

$$\text{NTU} = \frac{AU_m}{C_{\min}} = \frac{\text{heat capacity of exchanger, W/°C}}{\text{heat capacity of flow, W/°C}} \tag{11-45}$$

For a specified value of $U_m/C_{\min}$, the NTU is a measure of the actual heat transfer area A, or the "physical size" of the exchanger. The higher the NTU, the larger the physical size.

An examination of the results presented in Figs. 11-17 to 11-21 reveals that for $\varepsilon < 40$ percent, the capacity ratio $C = C_{\min}/C_{\max}$ does not have much effect on the effectiveness ε.

A counterflow exchanger has the highest ε for specified values of NTU and C compared with that for other flow arrangements. Therefore, for a given NTU and C, a counterflow arrangement yields maximum heat transfer performance.

Use of ε-NTU Relations

The ε-NTU relations can be readily used for the solution of both rating and sizing problems.

Rating problem Suppose the inlet temperatures $T_{c,\text{in}}$ and $T_{h,\text{in}}$, the flow rates m_c and m_h, the physical properties of both fluids, the overall heat transfer coefficient U_m, and the total heat transfer area A are all given. The type and the flow arrangement for the exchanger are specified. We wish to determine the total heat flow rate Q and the outlet temperatures $T_{h,\text{out}}$ and $T_{c,\text{out}}$. The calculation is as follows:

1. Calculate $C = C_{\min}/C_{\max}$ and $N \equiv \text{NTU} = U_m A/C_{\min}$ from the specified input data.
2. Knowing N and C, determine ε from the chart or the equation for the specific geometry and flow arrangement.
3. Knowing ε, compute the total heat transfer rate Q from

$$Q = \varepsilon C_{\min}(T_{h,\text{in}} - T_{c,\text{in}})$$

4. Calculate the outlet temperatures from

$$T_{h,\text{out}} = T_{h,\text{in}} - \frac{Q}{C_h}$$

$$T_{c,\text{out}} = T_{c,\text{in}} + \frac{Q}{C_c}$$

The preceding discussion of the ε-NTU method clearly illustrates that the *rating* problem, when the outlet temperatures are not given, can readily be solved with the ε-NTU method, but, a tedious iteration procedure would be required to solve it with the LMTD method, and the convergence might not be easy.

Sizing problem Suppose the inlet and outlet temperatures, the flow rate, the overall heat transfer coefficient, and the total heat transfer rate are given; also the flow arrangement is specified. We wish to determine the total heat transfer surface A.

1. Knowing the outlet and inlet temperatures, calculate ε according to Eqs. (11-31).
2. Calculate $C = C_{\min}/C_{\max}$.
3. Knowing ε and C, determine NTU from the appropriate ε-NTU chart.
4. Knowing NTU, calculate the heat transfer surface A from Eq. (11-39a):

$$A = \frac{(\text{NTU})C_{\min}}{U_m}$$

The use of the ε-NTU method is generally preferred in the design of compact heat exchangers for automotive, aircraft, air-conditioning, and other industrial applications where the inlet temperatures of the hot and cold fluids are specified and the heat transfer rates are to be determined. In the process, power, and petrochemical industries, both the inlet and outlet temperatures of the hot and cold fluids are specified; hence the LMTD method is generally used.

Example 11-13 A counterflow heat exchanger of heat transfer area $A = 12.5\ \text{m}^2$ is to cool oil [$c_{ph} = 2000\ \text{J/(kg}\cdot\text{s)}$] with water [$c_{pc} = 4170\ \text{J/(kg}\cdot\text{s)}$]. The oil enters at $T_{h,\text{in}} = 100°\text{C}$ and $m_h = 2\ \text{kg/s}$, while the water enters at $T_{c,\text{in}} = 20°\text{C}$ and $m_c = 0.48\ \text{kg/s}$. The overall heat transfer coefficient is $U_m = 400\ \text{W/(m}^2\cdot°\text{C)}$. Calculate the exit temperature of water $T_{c,\text{out}}$ and the total heat transfer rate Q.

SOLUTION In this problem the exit temperatures are not known; therefore an iterative solution is required if the LMTD method is to be used. However, the problem can be solved directly without an iteration by using the effectiveness method:

$$C_h = m_h c_{ph} = 2(2000) = 4000\ \text{W/°C}$$

$$C_c = m_c c_{pc} = 0.48(4170) = 2002\ \text{W/°C}$$

Then

$$\frac{C_{\min}}{C_{\max}} = \frac{2002}{4000} = 0.5$$

$$N = \frac{AU_m}{C_{\min}} = \frac{12.5(400)}{2002} = 2.5$$

From Fig. 11-18, we obtain

$$\varepsilon = 0.82$$

From Eq. (11-31*b*), by setting $C_c = C_{\min}$, we obtain

$$\varepsilon = \frac{T_{c,\text{out}} - T_{c,\text{in}}}{T_{h,\text{in}} - T_{c,\text{in}}}$$

or

$$0.82 = \frac{T_{c,\text{out}} - 20}{100 - 20}$$

$$T_{c,\text{out}} = 85.6°\text{C}$$

The total heat transfer rate is now

$$Q = m_c c_{pc}\,\Delta T_c = 0.48(4170)(85.6 - 20) = 131.3\text{ kW}$$

Example 11-14 A cross-flow heat exchanger with the flow arrangement shown in Fig. 11-19 and having a heat transfer area $A = 8.4\text{ m}^2$ is to heat air [$c_{pc} = 1005\text{ J/(kg}\cdot°\text{C)}$] with water [$c_{ph} = 4180\text{ J/(kg}\cdot°\text{C)}$]. Air enters at $T_{c,\text{in}} = 15°\text{C}$ and $m_c = 2.0$ kg/s, while water enters at $T_{h,\text{in}} = 90°\text{C}$ and $m_h = 0.25$ kg/s. The overall heat transfer coefficient is $U_m = 250\text{ W/(m}^2\cdot°\text{C)}$. Calculate the exit temperatures of both air and water as well as the total heat transfer rate Q.

SOLUTION The effectiveness method should be used to solve this problem since the exit temperatures are not known. We have

$$C_c = c_{pc} m_c = 1005(2.0) = 2010\text{ W/°C}$$

$$C_h = c_{ph} m_h = 4180(0.25) = 1045\text{ W/°C}$$

Then

$$\frac{C_{\min}}{C_{\max}} = \frac{1045}{2010} = 0.52$$

and

$$N = \frac{AU_m}{C_{\min}} = \frac{8.4(250)}{1045} = 2.0$$

From Fig. 11-19, we obtain

$$\varepsilon = 0.72$$

From Eq. (11-31*a*), by setting $C_h = C_{\min}$, we obtain

$$\varepsilon = \frac{T_{h,\text{in}} - T_{h,\text{out}}}{T_{h,\text{in}} - T_{c,\text{in}}}$$

$$0.72 = \frac{90 - T_{h,\text{out}}}{90 - 15}$$

$$T_{h,\text{out}} = 36°\text{C}$$

From Eq. (11-31*a*), by setting $C_h = C_{\min}$, we obtain

$$\varepsilon = \frac{C_c(T_{c,\,\text{out}} - T_{c,\,\text{in}})}{C_h(T_{h,\,\text{in}} - T_{c,\,\text{in}})}$$

$$0.72 = \frac{2010(T_{c,\,\text{out}} - 15)}{1045(90 - 15)}$$

$$T_{c,\,\text{out}} = 43.1°\text{C}$$

The total heat transfer rate is

$$Q = m_h c_{ph}\,\Delta T_h = C_h(T_{h,\,\text{in}} - T_{h,\,\text{out}})$$
$$= 1045(90 - 36) = 56.43\ \text{kW}$$

Example 11-15 A two shell pass, four tube pass heat exchanger of flow arrangement shown in Fig. 11-21 is to cool at $m_h = 1.5$ kg/s oil [$c_{ph} = 2100$ J/(kg · °C)] from $T_{h,\,\text{in}} = 90$°C to $T_{h,\,\text{out}} = 40$°C with water [$c_{pc} = 4180$ J/(kg · °C)] entering at $T_{c,\,\text{in}} = 19$°C and $m_c = 1$ kg/s. The overall heat transfer coefficient is $U_m = 250$ W/(m^2 · °C). Calculate the heat transfer area required.

SOLUTION The effectiveness method can be used to solve this problem. We have

$$C_c = m_c c_{pc} = 1(4180) = 4180\ \text{W/°C}$$
$$C_h = m_h c_{ph} = 1.5(2100) = 3150\ \text{W/°C}$$

The effectiveness can be calculated from Eq. (11.31*a*) by setting $C_h = C_{\min}$. We obtain

$$\varepsilon = \frac{T_{h,\,\text{in}} - T_{h,\,\text{out}}}{T_{h,\,\text{in}} - T_{c,\,\text{in}}} = \frac{90 - 40}{90 - 19} = 0.70$$

and

$$\frac{C_{\min}}{C_{\max}} = \frac{3150}{4180} = 0.75$$

Then, from Fig. 11-21, we obtain

$$N = \frac{AU_m}{C_{\min}} = 2.0$$

Therefore,

$$A = \frac{2.0(C_{\min})}{U_m} = \frac{2(3150)}{250} = 25.2\ \text{m}^2$$

Example 11-16 A shell-and-tube steam condenser is constructed with 2.5-cm-OD, single-pass horizontal tubes with steam condensing at $T_h = 54$°C. The cooling water enters the tubes at $T_{c,\,\text{in}} = 18$°C with a flow rate of $m_c = 0.7$

kg/s per tube and leaves at $T_{c,\,\text{out}} = 36°C$. The overall heat transfer coefficient based on the outer surface of the tube is $U_m = 3509$ W/(m$^2 \cdot$ °C). Calculate the tube length L and the heat transfer rate Q.

SOLUTION This is exactly the same problem as that in Example 11-8. Here, we solve it with the effectiveness method.

From Eq. (11-31b), by setting $C_{\min} = C_c$, we obtain

$$\varepsilon = \frac{T_{c,\,\text{out}} - T_{c,\,\text{in}}}{T_{h,\,\text{in}} - T_{c,\,\text{in}}} = \frac{36 - 18}{54 - 18} = 0.5$$

For a condenser, from Eq. (11-44), we have

$$\varepsilon = 1 - e^{-N} \qquad \text{or} \qquad 0.5 = 1 - e^{-N}$$

Then,

$$N = 0.693$$

By definition

$$N = \frac{AU_m}{C_{\min}}$$

where $C_{\min} = m_c c_{pc} = 0.7(4180) = 2926$ W/°C

$$A = \pi DL = \pi(0.025)(L) \qquad \text{m}^2$$

$$U_m = 3509 \text{ W/(m}^2 \cdot \text{°C)}$$

Then

$$0.693 = \frac{(\pi)(0.025L)(3509)}{2926}$$

or

$$L = 7.4 \text{ m}$$

which is the same as that in Example 11-8.

The heat transfer rate per tube is determined by Eq. (11-29) as

$$Q = \varepsilon C_{\min}(T_{h,\,\text{in}} - T_{c,\,\text{in}})$$
$$= (0.5)(2926)(54 - 18) = 52{,}668 \text{ W}$$

11-7 COMPACT HEAT EXCHANGERS

A heat exchanger having a surface area density greater than about 700 m^2/m^3 is quite arbitrarily referred to as a *compact heat exchanger*. These heat exchangers are generally used for applications where gas flows. Hence the heat transfer coefficient is low, and the smallness of weight and size is important. They are available in a wide variety of configurations of the heat transfer matrix, and their heat transfer and pressure drop characteristics have been studied extensively by Kays

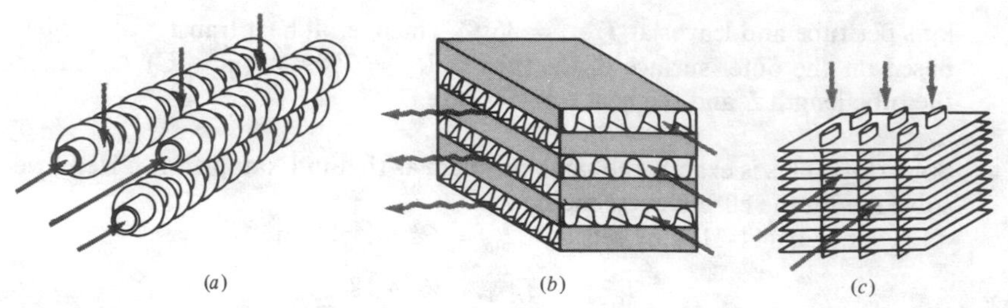

Figure 11-22 Typical heat transfer matrices for compact heat exchangers. (*a*) Circular finned-tube matrix; (*b*) plain plate-fin matrix; (*c*) finned flat-tube matrix.

and London [10]. Figure 11-22 shows typical heat transfer matrices for compact heat exchangers. Figure 11.22*a* shows a *circular finned-tube array* with fins on individual tubes; Fig. 11-22*b* shows a *plain plate-fin matrix* formed by corrugation, and Fig. 11-22*c* shows a *finned flat-tube* matrix.

The heat transfer and pressure drop characteristics of such configurations for use as compact heat exchangers are determined experimentally. For example, Figs. 11-23 to 11-25 show typical heat transfer and friction factor data for three

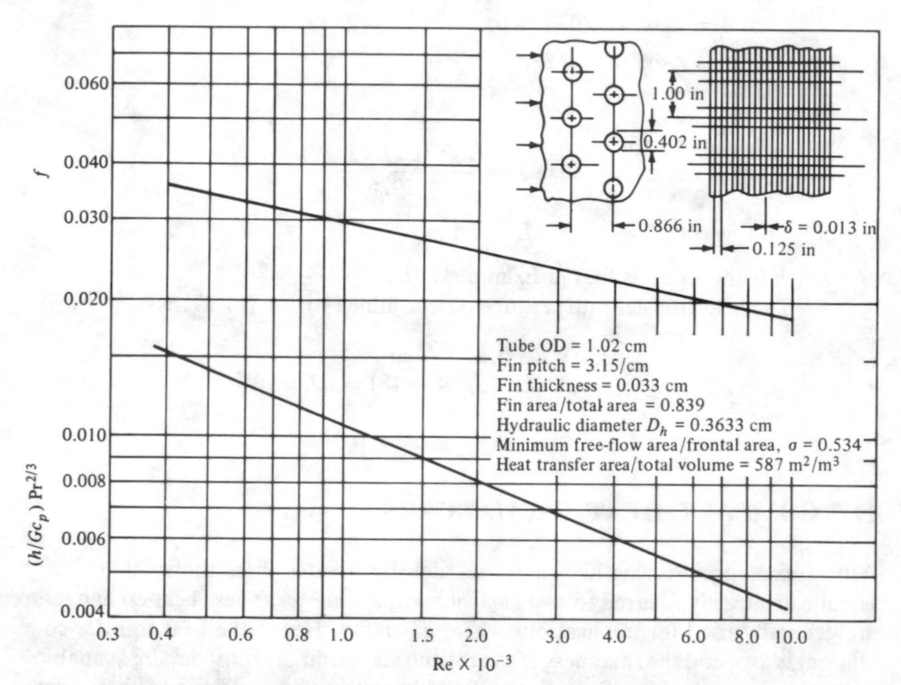

Figure 11-23 Heat transfer and friction factor for flow across plate-finned circular tube matrix. (*From Kays and London* [*10*].)

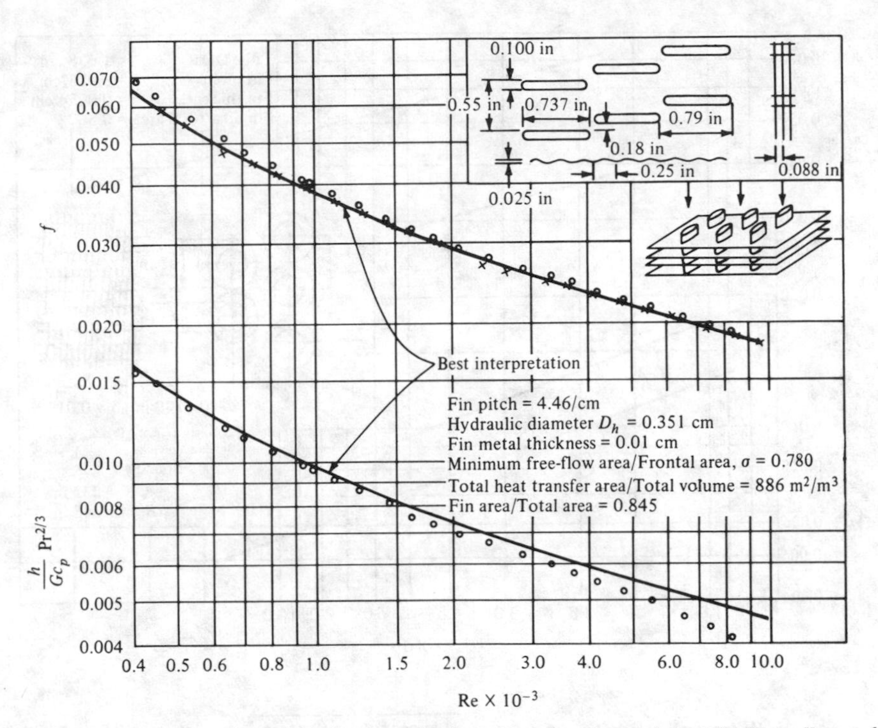

Figure 11-24 Heat transfer and friction factor for flow across finned flat-tube matrix. (*From Kays and London* [*10*].)

different configurations. Note that the principal dimensionless groups governing these correlations include the Stanton, Prandtl, and Reynolds numbers

$$\mathrm{St} = \frac{h}{Gc_p} \qquad \mathrm{Pr} = \frac{c_p \mu}{k} \qquad \mathrm{Re} = \frac{GD_h}{\mu} \tag{11-46}$$

Here G is the *mass velocity* defined as

$$G = \frac{m}{A_{\min}} \qquad \mathrm{kg/(m^2 \cdot s)} \tag{11-47}$$

where m = total mass flow rate of fluid (kg/s) and $A_{\min}$ = minimum free-flow cross-sectional area (m²) regardless of where this minimum occurs.

The magnitude of the *hydraulic diameter* D_h for each configuration is specified on Figs. 11-23 to 11-25. The hydraulic diameter D_h is defined as

$$D_h = 4\frac{LA_{\min}}{A} \tag{11-48}$$

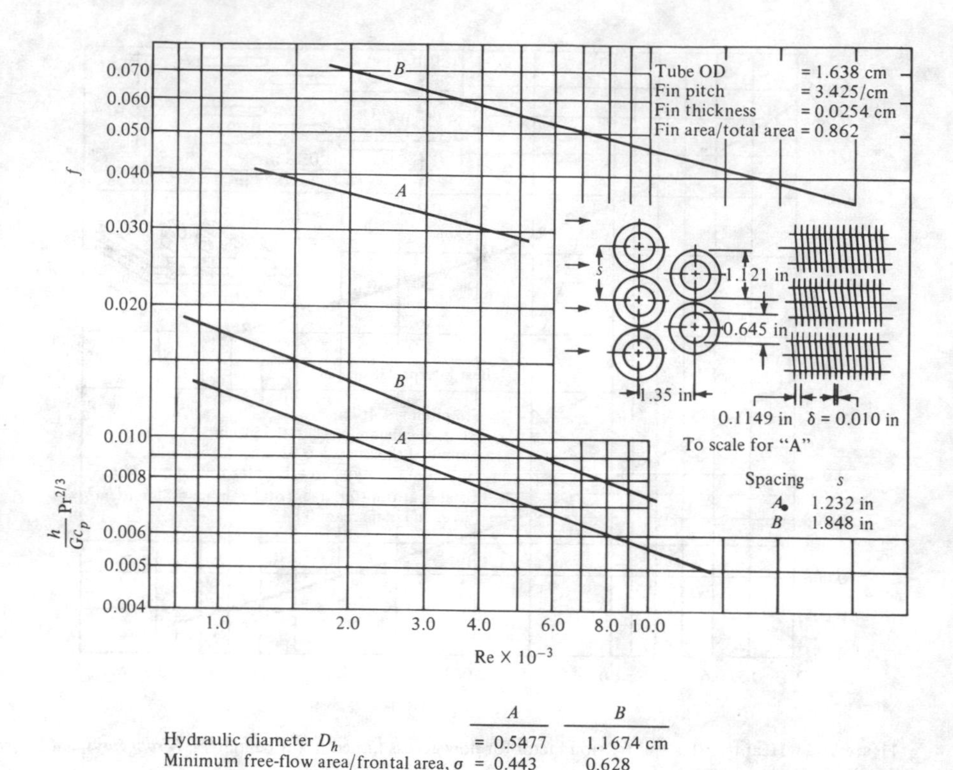

	A	B
Hydraulic diameter D_h	= 0.5477	1.1674 cm
Minimum free-flow area/frontal area, σ	= 0.443	0.628
Heat transfer area/total volume	= 323.8	215.6 m²/m³

Figure 11-25 Heat transfer and friction factor for flow across circular finned-tube matrix. (*From Kays and London* [*10*].)

where A is the total heat transfer area and the quantity $LA_{\min}$ can be regarded as the minimum free-flow passage volume, since L is the flow length of the heat exchanger matrix.

Thus, once the heat transfer and the friction factor charts such as those shown in Fig. 11-23 are available for a specified matrix and the Reynolds number Re for the flow is given, the heat transfer coefficient h and the friction factor f for flow across the matrix can be evaluated. Then the rating and sizing problem associated with the heat exchanger matrix can be performed by utilizing either the LMTD or the effectiveness method of analysis. We now describe the pressure drop analysis for compact heat exchangers.

The pressure drop associated with flow through a compact heat exchanger matrix consists of three components: *the core friction*, the *core acceleration*, and the *entrance and exit losses*.

We now present the pressure drop analysis for *plate-fin* and *tube-fin* heat exchangers.

Pressure Drop for Plate-Fin Exchangers

Consider a *plate-fin* exchanger matrix as illustrated in Fig. 11-22*b*. As the fluid enters the passage, it experiences pressure losses owing to contraction resulting from an area change and an irreversible free expansion after a sudden contraction. As the fluid passes through the heat exchanger matrix (i.e., core), it experiences pressure loss because of fluid friction. Also, depending on whether heating or cooling takes place, a pressure change results from flow acceleration or deceleration. Finally, as the fluid leaves the heat exchanger matrix, there are pressure losses associated with the area change and the flow separation.

Then the total pressure drop for flow of fluid across the heat exchanger matrix is given by [10]

$$\Delta P = \frac{G^2}{2\rho_i}\Bigg[\underbrace{(K_c + 1 - \sigma^2)}_{\substack{\text{entrance}\\ \text{effect}}} + \underbrace{2\left(\frac{\rho_i}{\rho_o} - 1\right)}_{\substack{\text{flow}\\ \text{acceleration}}} + \underbrace{f\frac{A}{A_{\min}}\frac{\rho_i}{\rho_m}}_{\substack{\text{core}\\ \text{friction}}} - \underbrace{(1 - K_e - \sigma^2)\frac{\rho_i}{\rho_o}}_{\substack{\text{exit}\\ \text{effect}}}\Bigg] \tag{11-49}$$

where $\sigma = \dfrac{A_{\min}}{A_{fr}} = \dfrac{\text{minimum free-flow area}}{\text{frontal area}}$

$$\frac{A}{A_{\min}} = \frac{4L}{D_h} = \frac{\text{total heat transfer area}}{\text{minimum free-flow area}}$$

$$G = \frac{\rho u_\infty A_{fr}}{A_{\min}} = \frac{\rho u_\infty}{\sigma} = \text{mass velocity, kg/(m}^2 \cdot \text{s)}$$

K_c, K_e = flow contraction and expansion coefficients, respectively

ρ_i, ρ_o = density at inlet and exit, respectively

$$\frac{1}{\rho_m} = \frac{1}{2}\left(\frac{1}{\rho_i} + \frac{1}{\rho_o}\right)$$

Equation (11-49) is for the pressure drop associated with flow across the heat exchanger matrix. One may consider also the relation for flow inside the tubes of the heat exchanger. Therefore, the total pressure drop through the heat exchanger is equal to the sum of the pressure drops for flow across the tubes and for flow inside the tubes.

In Eq. (11-49), the frictional pressure drop generally dominates and accounts for about 90 percent or more of the total pressure drop across the core. The entrance and exit losses become important for short cores (i.e., small L), small values of σ, high values of the Reynolds number, and gases. For liquids they are negligible. The reader should consult Kays and London [10, pp. 45–46] for typical values of K_c and K_e.

Pressure Drop for Finned-Tube Exchangers

For flow normal to finned-tube banks, such as that illustrated in Fig. 11-22*a*, the entrance and exit losses are generally accounted for in the friction factor, hence $K_c = K_e = 0$. Then, by setting $K_c = K_e = 0$ in Eq. (11-49), the total pressure drop for flow across the tube bank becomes

$$\Delta P = \frac{G^2}{2\rho_i}\Bigg[\underbrace{(1+\sigma^2)\left(\frac{\rho_i}{\rho_o}-1\right)}_{\text{flow acceleration}} + \underbrace{f\,\frac{A}{A_{\min}}\frac{\rho_i}{\rho_m}}_{\text{core friction}}\Bigg] \tag{11-50}$$

Example 11-17 Air at 1 atm and 400 K and with a velocity of $u_\infty = 10$ m/s flows across a compact heat exchanger matrix having the configuration shown in Fig. 11-23.

(*a*) Calculate the heat transfer coefficient.

(*b*) Find the ratio of the frictional pressure drop to the inlet pressure drop for the flow of air across the exchanger. The geometry is shown in the figure

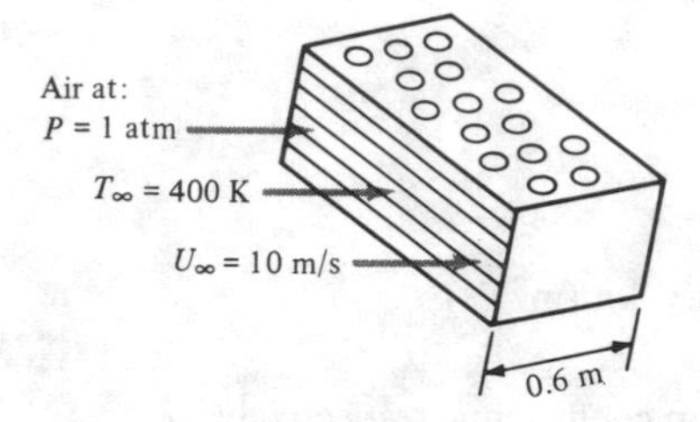

Solution Physical properties of atmospheric air at $T_\infty = 400$ K are

$$\rho = 0.8826 \text{ kg/m}^3 \qquad c_p = 1014 \text{ J(kg} \cdot {}^\circ\text{C)}$$

$$\mu = 2.286 \times 10^{-5} \text{ kg/(m} \cdot \text{s)} \qquad \Pr = 0.689$$

From Fig. 11-23 we have

$$\frac{A_{\min}}{A_{fr}} = \sigma = 0.534 \qquad D_h = 0.3633 \text{ cm}$$

Then

$$G = \frac{\rho u_\infty A_{fr}}{A_{\min}} = \frac{\rho u_\infty}{\sigma} = \frac{(0.8826)(10)}{0.534} = 16.53 \text{ kg/(m}^2 \cdot \text{s)}$$

and

$$\text{Re} = \frac{GD_h}{\mu} = \frac{(16.53)(0.3633 \times 10^{-2})}{2.286 \times 10^{-5}} = 2630$$

(*a*) From Fig. 11-23, for Re = 2630, we obtain

$$\frac{h}{Gc_p} \mathrm{Pr}^{2/3} = 0.0071$$

Then the heat transfer coefficient is

$$h = 0.0071 \frac{Gc_p}{\mathrm{Pr}^{2/3}} = 0.0071 \frac{16.53(1014)}{(0.689)^{2/3}}$$

$$= 152.6 \text{ W/(m}^2 \cdot {}^\circ\text{C)}$$

(*b*) The frictional pressure drop for flow across the heat exchanger matrix is determined from Eq. (11-49):

$$\Delta P_f = f \frac{G^2}{2\rho_a} \frac{A}{A_{\min}}$$

where

$$\frac{A}{A_{\min}} = \frac{4 \times \text{length}}{D_h} = \frac{4 \times 0.6}{0.3633 \times 10^{-2}} = 660.6$$

Also, for Re = 2630, from Fig. 11-23 we have

$$f = 0.025$$

Then

$$\Delta P_f = 0.025 \frac{(16.53)^2}{2(0.8826)} 660.6 = 2556 \text{ N}$$

The inlet pressure is

$$P = 1 \text{ atm} = 1.01132 \times 10^5 \text{ N}$$

Then the ratio of ΔP_f to P becomes

$$\frac{\Delta P_f}{P} = \frac{2556}{1.0132 \times 10^5} = 2.52 \text{ percent}$$

Thus, the frictional pressure drop is 2.5 percent of the inlet pressure.

Example 11-18 This example is intended to illustrate the solution of a complete *rating* problem for a compact heat exchanger.

Hot air at $P_a = 2$ atm and $T_a = 147°\text{C}$ with a flow rate of $m_a = 10$ kg/s flows across a finned flat-tube matrix of configuration shown in Fig. 11-24. Water at $T_w = 15°\text{C}$ and a flow rate of $m_w = 40$ kg/s flows inside the flat tubes. The overall dimensions of the heat exchanger matrix are shown in the figure.

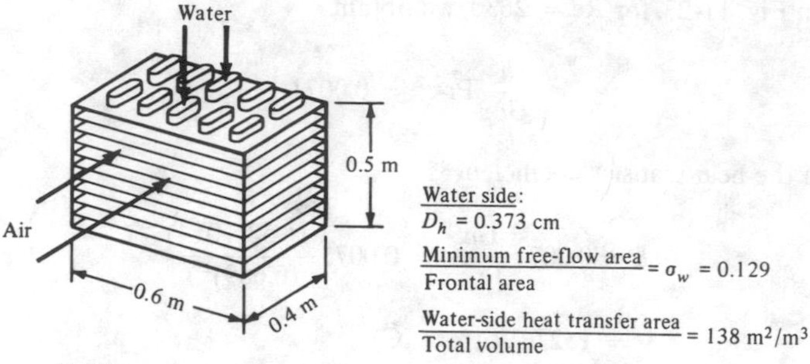

Air side:
Data are given in Fig. 11-24. Additional data are:
t = plate thickness = 0.01 cm
L = effective plate length as a fin = 0.572 cm
$\beta = \dfrac{\text{fin area}}{\text{total area}} = 0.845$

(*a*) Calculate the heat transfer coefficients h_a and h_w for the air and water sides, respectively.
(*b*) Find the overall heat transfer coefficient U_a based on the air-side surface.
(*c*) Determine the total heat transfer rate Q.
(*d*) Calculate the outlet temperature of air and water.

SOLUTION To start the calculations, the physical properties of air and water at their respective mean temperatures are needed. The mean temperatures cannot be calculated because the outlet temperatures are not known. Calculations may be started by evaluating the physical properties at the inlet temperatures, and then the calculations are repeated. We prefer to choose suitable mean temperatures which are to be checked at the end of the calculation.

For the air side we assume a mean temperature 77°C and determine the physical properties of air at $P_a = 2$ atm:

$$\rho_a = 1.996 \text{ kg/m}^3 \qquad c_{pa} = 1009 \text{ J/(kg}\cdot{}^\circ\text{C)}$$
$$\mu_a = 2.075 \times 10^{-5} \text{ kg/(m}\cdot\text{s)} \qquad \text{Pr}_a = 0.697$$

Assuming an average temperature of 20°C for the water, we take the physical properties as

$$\rho_w = 1000 \text{ kg/m}^3 \qquad c_{pw} = 4182 \text{ J/(kg}\cdot{}^\circ\text{C)}$$
$$\nu_w = 1 \times 10^{-6} \text{ m}^2\text{/s} \qquad \text{Pr}_w = 7.02 \qquad k_w = 0.597 \text{ W/(m}\cdot{}^\circ\text{C)}$$

(*a*) *Heat transfer coefficients*:

Air side: $$G_a = \frac{m_a}{A_{\min}} = \frac{m_a}{\sigma_a A_{fr}} \qquad m_a = 10 \text{ kg/s}$$

$$A_{fr} = 0.6(0.5) = 0.3 \text{ m}^2 \qquad \text{(specified on figure)}$$

$$\sigma_a = 0.78 \qquad D_h = 0.351 \text{ cm} \qquad \text{(from Fig. 11-24)}$$

Then, $$G_a = \frac{10}{0.78(0.3)} = 42.74 \text{ kg/(m}^2 \cdot \text{s)}$$

$$\text{Re}_a = \frac{G_a D_h}{\mu_a} = \frac{(42.74)(0.351 \times 10^{-2})}{2.075 \times 10^{-5}}$$

$$= 7230$$

From Fig. 11-24 for $\text{Re}_a = 7230$, we find

$$\frac{h_a}{G_a c_{pa}} \text{Pr}^{2/3} = 0.005$$

Then $$h_a = 0.005 \frac{G_a c_{pa}}{\text{Pr}^{2/3}}$$

$$= 0.005 \frac{42.74(1009)}{(0.697)^{2/3}}$$

$$= 274.3 \text{ W/(m}^2 \cdot {}^\circ\text{C)}$$

Water side: $u_{\max} = \frac{m_w}{\rho_w A_{\min}} = \frac{m_w}{\rho_w \sigma_w A_{fr}} \qquad m_w = 40 \text{ kg/s}$

$\sigma_w = 0.129 \qquad D_h = 0.373 \qquad$ (specified for water side)

$A_{fr} = 0.6(0.4) = 0.24 \text{ m}^2 \qquad$ (from figure)

Then, $$u_{\max} = \frac{40}{10^3(0.129)(0.24)} = 1.292 \text{ m/s}$$

$$\text{Re}_w = \frac{u_{\max} D_h}{\nu_w} = \frac{1.292(0.373 \times 10^{-2})}{1 \times 10^{-6}}$$

$$= 4819$$

The water side can be treated as a fully developed turbulent flow inside a long duct since $L/D_h = 50/0.373 = 134$. Using the Sieder and Tate equation (7-69), we have

$$\text{Nu} = 0.023 \text{ Re}^{0.8} \text{ Pr}^{0.4}$$

so

$$h_w = 0.023 \frac{k}{D_e} \text{Re}^{0.8} \text{ Pr}^{0.4}$$

$$= 0.023 \frac{0.597}{0.373 \times 10^{-2}} (4819)^{0.8} (7.02)^{0.4}$$

$$= 7094 \text{ W/(m}^2 \cdot {}^\circ\text{C)}$$

(*b*) *Overall heat transfer coefficient* U_a. To determine U_a, based on the air-side surface, the fin efficiency η should be determined. The reason is that the effective temperature difference between the fluid and the fin surface is lower than that between the fluid and the fin base. Only the air side has extended surfaces which can be regarded as plate fins; the thermal conductivity and dimensions are given on the figure as

$$k = 170\ \text{W/(m}\cdot{}^\circ\text{C)} \qquad t = \text{thickness} = 0.01\ \text{cm}$$

$$L = \text{effective fin length} = 0.572\ \text{cm}$$

$$\beta = \frac{\text{fin area}}{\text{total area}} = 0.845$$

The fin efficiency η for a plate fin can be calculated, as discussed in Example 3-15, from the relation

$$\eta = \frac{\tanh mL}{mL}$$

where
$$m = \sqrt{\frac{2h_a}{kt}}$$

Then
$$mL = \sqrt{\frac{2(274.3)}{170(1 \times 10^{-4})}}(0.572 \times 10^{-2}) = 1.0275$$

and
$$\eta = \frac{\tanh mL}{mL} = 0.752$$

The area-weighted fin efficiency is determined by Eq. (3-62*b*) as

$$\begin{aligned}\eta' &= \beta\eta + 1 - \beta \\ &= 0.845(0.752) + 1 - 0.845 \\ &= 0.79\end{aligned}$$

The overall heat transfer coefficient U_a, based on the air-side surface, is

$$\frac{1}{U_a} = \frac{1}{\eta' h_a} + \frac{1}{(A_w/A_a)h_w}$$

Here the ratio of the water-side to air-side heat transfer surfaces is found to be

$$\frac{A_w}{A_a} = \frac{\text{water-side heat transfer area/total volume}}{\text{air-side heat transfer area/total volume}} = \frac{138\ \text{m}^2/\text{m}^3}{886\ \text{m}^2/\text{m}^3} = 0.156$$

Then

$$\frac{1}{U_a} = \frac{1}{0.79(274.3)} + \frac{1}{0.156(7094)} = 5.52 \times 10^{-3}$$

(83.6 percent) (16.4 percent)

$$U_a = 181.2\ \text{W/(m}^2\cdot{}^\circ\text{C)}$$

(c) *Total heat transfer rate Q*. We apply the ε-NTU method.

$$\text{Air-side capacity rate} = C_a = m_a c_{pa} = 10{,}090\ \text{W/°C}$$

$$\text{Water-side capacity rate} = C_w = m_w c_{pw} = 167{,}280\ \text{W/°C}$$

$$\text{Total volume of matrix} = V = 0.6(0.5)(0.4) = 0.12\ \text{m}^3$$

$$\frac{A_a}{V} = 886 \qquad \text{(from Fig. 11-24)}$$

Then

$$A_a = 886V = 886(0.12) = 106.32\ \text{m}^2$$

$$N \equiv \text{NTU} = \frac{A_a U_a}{C_{\min}} = \frac{106.32(181.2)}{10{,}090} = 1.91$$

$$\frac{C_{\min}}{C_{\max}} = \frac{10{,}090}{167{,}280} = 0.06$$

From Fig. 11-19, for a cross-flow heat exchanger with both fluids unmixed, for $N = 1.91$ and $C_{\min}/C_{\max} = 0.06$, we obtain

$$\varepsilon = 0.86$$

Then the total heat transfer rate Q, by Eq. (11-29), becomes

$$\begin{aligned} Q &= \varepsilon C_{\min}(T_{h,\text{in}} - T_{c,\text{in}}) \\ &= 0.86(10{,}090)(147 - 15) \\ &= 1145.4\ \text{kW} \end{aligned}$$

(d) *Outlet temperatures.*

$$Q = m_a c_{pa}(T_{a,\text{in}} - T_{a,\text{out}})$$

$$1{,}145{,}400 = 10(1009)(147 - T_{a,\text{out}})$$

So

$$T_{a,\text{out}} = 33.5\text{°C} \qquad \text{for air outlet temperature}$$

$$Q = m_w c_{pw}(T_{w,\text{out}} - T_{w,\text{in}})$$

$$1{,}145{,}400 = 40(4182)(T_{w,\text{out}} - 15)$$

So

$$T_{w,\text{out}} = 21.8\text{°C} \qquad \text{for water outlet temperature}$$

We now check the mean temperatures for the air and water sides:

$$T_{m,\text{air}} = \frac{147 + 33.5}{2} = 90.3\text{°C}$$

$$T_{m,\text{water}} = \frac{15 + 21.8}{2} = 18.4\text{°C}$$

The mean temperatures chosen to evaluate the physical properties of air and water are sufficiently close to these values. If the differences were large, it would be necessary to repeat the calculations by using the above average temperatures to determine the physical properties.

11-8 HEAT EXCHANGER OPTIMIZATION

Although the standard heat exchanger designs may satisfy the need for most small, simple units operating at moderate temperatures and pressures, individually designed units may be needed for numerous special applications.

The heat exchangers are designed for a vast variety of applications; therefore, the criteria for optimization depend on the type of application. For example, the optimization criteria may require minimum weight, minimum volume or heat transfer surface, minimum initial cost, minimum initial and operating costs, maximum heat transfer rate, minimum pressure drop for a specified heat transfer rate, minimum mean temperature difference, and so on.

Therefore, to conduct an optimization study, the thermal design of the heat exchanger should be performed and the calculations should be repeated for each design variable until the optimization criterion is satisfied. Computer programs have been prepared for the thermal design of heat exchangers.

Bell [14, pp. 559–579] describes a computer-aided design procedure for the thermal design of shell-and-tube heat exchangers. Shah [15, pp. 537–566] discusses the basic features of a computer-aided thermal design and optimization process for compact heat exchangers. Spalding [15, pp. 641–655] outlines the general features of a numerical approach for computing fluid dynamics and heat transfer performance of heat exchangers.

To illustrate the basic logical structure of the optimization of heat exchangers, we focus our attention on the compact heat exchangers.

The first step in the optimization process is the solution of the rating and the sizing problems. The *rating* problem is concerned with the determination of the heat transfer rate, the outlet temperatures, and the pressure drop on each side. The following quantities are generally specified in the rating problem: type of heat exchanger, surface geometries, flow arrangement, flow rates, inlet temperatures, and overall dimensions of the matrix.

The sizing problem is concerned with the determination of the matrix dimensions to meet the specified heat transfer and pressure drop requirements. The designer's task is to select the type of construction, flow arrangement, and surface geometries on both sides. The following quantities are generally specified: fluid inlet and outlet temperatures, flow rates, pressure drops, and heat transfer rate.

Shah [15, pp. 537–566] describes the highlight of major computer subroutines needed to perform the sizing and rating calculations. They include the following.

1. *Design specifications.* Complete design specifications should be available as a computer subroutine. The information should include heat exchanger type;

flow arrangement; surface geometries; operating conditions, such as inlet temperatures, pressures, flow rates, types of fluids, etc.; overall dimensions.

2. *Fluid properties.* The physical properties of the fluids, such as specific heat, density, viscosity, thermal conductivity, and the Prandtl number, should be included as a function of temperature in the form of correlations.
3. *Matrix geometry.* The information characterizing the matrix geometry should be provided for each side of the exchanger, including the minimum free flow area, hydraulic diameter, fin dimensions needed for the calculation of fin efficiency, etc.
4. *ε-NTU relation.* Since the ε-NTU method is used in the thermal design of compact heat exchangers, formulas defining the ε-NTU relation should be provided. The relations should be sufficiently general to permit the determination of ε when NTU and $C = C_{min}/C_{max}$ are available, and to calculate NTU when ε and C are available.
5. *h and f relations.* The heat transfer and flow friction characteristics of compact heat exchangers are generally given in the form of j and f charts plotted as a function of the Reynolds number. These data should be provided in the form of correlations.
6. *Fin efficiency.* When extended surfaces are used in the heat transfer matrix, the fin efficiency η and the area-weighted fin efficiency η' are needed in the heat transfer calculations. Therefore the formulas defining the fin efficiency η and the information needed for the computation of η' should be given.
7. *Pressure drop relations.* The pressure drop for flow through the matrix is due to flow friction, acceleration or deceleration resulting from heat transfer, contraction and expansion at the matrix inlet and outlet. Appropriate relations should be given for the calculation of the pressure drop due to such causes. Also, provision must be made for the calculation of pressure drop in bends, turns, headers, manifolds, etc.

Rating problem If the prolem involves the optimization associated with the heat transfer rate or the pressure drop, then the rating problem is solved and the resulting heat transfer rate and the pressure drop are calculated. The solution of the rating problem follows a procedure similar to that illustrated in Example 11-18, except the computations are performed with the computer.

Sizing problem If the problem involves optimization associated with the size, weight, or heat transfer surface, and hence the cost, then the sizing problem is solved and the dimensions of the core matrix and the heat transfer surface are calculated.

Optimization problem As discussed earlier the criterion for optimization depends on the specific application. Therefore, the quantity optimized (i.e., maximized or minimized) should be stated. Also there may be some additional constraints. A variety of approaches may be used to arrive at an optimized design; but whatever approach is adopted, in each case it involves the solution of the rating or the sizing problem.

Suppose the heat exchanger is to be optimized for a minimum total cost. The problem involves explicit constraints, such as a fixed frontal area and the ranges of heat exchanger dimensions, and implicit constraints specifying the minimum heat transfer rate and maximum pressure drops. Once the surface geometry is selected, the designer has the option of imposing additional constraints, such as minimum and maximum values of fin height, fin thickness, fin pitch, fin thermal conductivity, fin length, gas flow rate, etc. Then the problem reduces to that of solving the rating problem within the ranges of variables specified.

PROBLEMS

Classification of heat exchangers

11-1 Illustrate with sketches the flow path arrangement for the following types of shell-and-tube heat exchangers:

(*a*) Single shell pass, two tube pass, counterflow
(*b*) Two shell pass, four tube pass, counterflow
(*c*) Three shell pass, six tube pass counterflow

11-2 Illustrate with sketches the temperature profiles for hot and cold fluids as a function of the distance along the flow path for (*a*) parallel-flow heat exchangers, (*b*) counterflow exchangers, (*c*) condenser, and (*d*) gas-heated boiler.

11-3 Consider a cross-flow heat exchanger with hot and cold fluids entering at uniform temperatures. Illustrate with sketches the exit temperature distribution for the following cases:

(*a*) Both fluids are unmixed.
(*b*) Cold fluid is unmixed, hot fluid is mixed.

Overall heat transfer coefficient

11-4 Water at $T_i = 25°C$ and a velocity of $u_m = 1.5$ m/s enters a brass condenser tube $L = 6$ m long, 1.34-cm ID, 1.58-cm OD, and $k = 110$ W/(m · °C). The heat transfer for condensation at the outer surface of the tube is $h_o = 12{,}000$ W/(m^2 · °C). Calculate the overall heat transfer coefficient U_o based on the outer surface.

Answer: 3600 W/(m^2 · °C)

11-5 Hot water at a mean temperature $T_m = 80°C$ and with a mean velocity $u_m = 0.4$ m/s flows inside a 3.8-cm-ID, 4.8-cm-OD steel tube [$k = 50$ W/(m · °C)]. The flow is considered hydrodynamically and thermally developed. The outside surface is exposed to atmospheric air at $T_\infty = 20°C$, flowing with a velocity of $u_\infty = 3$ m/s normal to the tube. Calculate the overall heat transfer coefficient U_o based on the outer surface of the tube.

11-6 Engine oil at $T_{in} = 50°C$ and a mean velocity of $u_m = 0.25$ m/s enters a brass [$k = 110$ W/(m · °C)] horizontal tube with $D_i = 2.22$ cm ID and $t = 0.17$ cm thick. Heat is dissipated from the outer surface by free convection into an ambient at $T_\infty = 20°C$. Calculate the overall heat transfer coefficient U_o based on the tube's outer surface.

11-7 Determine the overall heat transfer coefficient U_o based on the outer surface of a brass tube with $D_i = 2.5$ cm and $D_o = 3.34$ cm [$k = 110$ W/(m · °C)] for the following conditions: The inside and outside heat transfer coefficients are, respectively, $h_i = 1200$ and $h_o = 2000$ W/(m^2 · °C); the fouling factors for the inside and outside surfaces are $F_i = F_o = 0.00018$ m^2 · °C/W.

Answer: 481.3 W/m^2

LMTD method

11-8 Engine oil at $T_i = 40°C$ at a rate of $m = 0.2$ kg/s enters a tube with $D_i = 2.5$ cm ID which is maintained at a uniform temperature $T_w = 100°C$ by condensing steam outside. Calculate the tube length required to have an outlet temperature $T_o = 80°C$.

11-9 A counterflow heat exchanger is to be used to heat $m_c = 2.5$ kg/s of water from $T_{c,\,in} = 20°C$ to $T_{c,\,out} = 80°C$ by using hot exhaust gas $[c_p = 1000 \text{ J/(kg} \cdot °\text{C)}]$ entering at $T_{h,\,in} = 220°C$ and leaving at $T_{h,\,out} = 90°C$. The overall heat transfer coefficient is $U_m = 250 \text{ W/(m}^2 \cdot °\text{C)}$. Calculate the heat transfer surface required.

Answer: 5.94 m^2

11-10 A single-pass, counterflow, shell-and-tube heat exchanger is used to heat water from $T_{c,\,in} = 15°C$ to $T_{c,\,out} = 80°C$ at a rate of $m_c = 1.5$ kg/s by oil entering the shell side at $T_{h,\,in} = 140°C$ and leaving at $T_{h,\,out} = 90°C$. The overall heat transfer coefficient is $U_m = 250 \text{ W/(m}^2 \cdot °\text{C)}$. Calculate the heat transfer surface required.

11-11 A shell-and-tube heat exchanger is to cool $m = 6$ kg/s of oil $[c_p = 2000 \text{ J/(kg} \cdot °\text{C)}]$ from $T_{h,\,in} = 65°C$ to $T_{h,\,out} = 35°C$ by using $m_c = 10$ kg/s of water at an inlet temperature of $T_{c,\,in} = 20°C$. The average heat transfer coefficient is $U_m = 600 \text{ W/(m}^2 \cdot °\text{C)}$. Calculate the heat transfer surface required for a (*a*) parallel-flow heat exchanger and (*b*) a counterflow heat exchanger. [Water: $C_p = 4200 \text{ J/kg} \cdot °\text{C}$]

11-12 A counterflow heat exchanger is to be designed to cool $m_h = 0.5$ kg/s of oil $[c_p = 2000 \text{ J/(kg} \cdot °\text{C)}]$ from $T_{h,\,in} = 60°C$ to $T_{h,\,out} = 40°C$ with cooling water entering at $T_{c,\,in} = 20°C$ and leaving at $T_{c,\,out} = 30°C$. The overall heat transfer coefficient is $U_m = 200 \text{ W/(m}^2 \cdot °\text{C)}$. Calculate the heat transfer surface required.

11-13 Engine oil at a mean temperature $T_i = 80°C$ and mean velocity $u_m = 0.2$ m/s flows inside a thin-walled, horizontal copper tube with an ID of $D = 1.9$, cm. At the outer surface, atmospheric air at $T_\infty = 15°C$ and a velocity of $u_\infty = 5$ m/s flows across the tube. Neglecting the tube wall resistance, calculate the overall heat transfer coefficient and the rate of heat loss to the air per meter length of tube.

Answer: 18.5 $\text{W/(m}^2 \cdot °\text{C)}$, 71.8 W/m

11-14 Water at a mean temperature $T_i = 80°C$ and a mean velocity $u_m = 0.15$ m/s flows inside a thin-walled, horizontal copper tube with $D = 2.5$ cm ID. At the outer surface of the tube, heat is dissipated by free convection into atmospheric air at $T_\infty = 15°C$. Assuming that the tube wall resistance is negligible, calculate (*a*) the tube wall temperature, (*b*) the overall heat transfer coefficient, and (*c*) the heat loss per meter length of tube.

11-15 The design heat load for an oil cooler is $Q = 500$ kW. Calculate the heat transfer surface A required if the inlet and outlet temperature differences are, respectively, $\Delta T_0 = 40°C$ and $\Delta T_L = 15°C$ and the average value of the overall heat transfer coefficient is $U_m = 500 \text{ W/(m}^2 \cdot °\text{C)}$.

11-16 A counterflow shell-and-tube heat exchanger is to be used to cool water from $T_{h,\,in} = 22°C$ to $T_{h,\,out} = 6°C$ by using brine entering at $T_{c,\,in} = -2°C$ and leaving at $T_{c,\,out} = 3°C$. The overall heat transfer coefficient is estimated to be $U_m = 500 \text{ W/(m}^2 \cdot °\text{C)}$. Calculate the heat transfer surface area for a design heat load of $Q = 10$ kW.

11-17 A counterflow shell-and-tube heat exchanger is to be used to heat water from $T_{c,\,in} = 10°C$ to $T_{c,\,out} = 70°C$ with oil $[c_p = 2100 \text{ J/(kg} \cdot °\text{C)}]$ entering at $T_{h,\,in} = 120°C$ and leaving at $T_{h,\,out} = 60°C$ at a rate of $m_h = 1$ kg/s. The overall heat transfer coefficient is $U_m = 200 \text{ W/(m}^2 \cdot °\text{C)}$. Calculate the heat transfer surface required.

11-18 A counterflow heat exchanger is to cool $m_h = 1$ kg/s of water from 65 to 5°C by using a refrigerant $[c_{pc} = 920 \text{ J/(kg} \cdot °\text{C)}]$ entering at $T_{c,\,in} = -20°C$ with a flow rate of $m_c = 8$ kg/s. The overall heat transfer coefficient is $U_m = 1500 \text{ W/(m}^2 \cdot °\text{C)}$. Calculate the heat transfer surface required. $[c_{pw} = 4180 \text{ J/(kg} \cdot \text{C}°)]$

11-19 Steam condenses at $T_h = 60°C$ on the shell side of a steam condenser while cooling water flows inside the tubes at a rate of $m_c = 3$ kg/s. The inlet and outlet temperatures of the water are $T_{c,\,in} = 20°C$

and $T_{c,out} = 50°C$, respectively. The overall heat transfer coefficient is $U_m = 2000$ W(m^2 · °C). Calculate the surface area required.

Answer: 8.7 m^2

Correction to LMTD

11-20 A one shell pass, two tube pass exchanger is to be designed to heat $m_c = 0.5$ kg/s of water entering the shell side at $T_{c,in} = 10°C$. The hot fluid, oil, enters the tube at $T_{h,in} = 80°C$ at $m_h = 0.3$ kg/s and leaves the exchanger at $T_{h,out} = 30°C$. The overall heat transfer coefficient is $U_m = 250$ W/(m^2 · °C). Calculate the total heat transfer area A required [$c_{ph} = 2000$ J/(kg · °C), $c_{pc} = 4180$ J(kg · °C)].

11-21 Air flowing at a rate of $m_c = 1$ kg/s is to be heated from $T_{c,in} = 20°C$ to $T_{c,out} = 50°C$ with hot water entering at $T_{h,in} = 90°C$ and leaving at $T_{h,out} = 60°C$. The overall heat transfer coefficient is $U_m = 400$ W/(m^2 · °C). Determine the surface area required by using a heat exchanger (*a*) shown in Fig. 11-16*a* and (*b*) shown in Fig. 11-16*b*.

11-22 A two shell pass, four tube pass heat exchanger of flow arrangement shown in Fig. 11-16*b* is to be used to heat water with oil. Water enters the tubes at a flow rate of $m_c = 2$ kg/s and temperature $t_1 = 20°C$ and leaves at $t_2 = 80°C$. Oil enters the shell side at $T_1 = 140°C$ and leaves at $T_2 = 90°C$. Calculate the heat transfer area required for an overall heat transfer coefficient of $U_m = 300$ W/(m^2 · °C) [$c_{pc} = 4180$ J/(kg · °C)].

Answer: 26.6 m^2

11-23 A cross-flow heat exchanger having the flow arrangement shown in Fig. 11-16*c* is to heat $m_c =$ 2 kg/s of air from $T_1 = 10°C$ to $T_2 = 50°C$ flowing on the *shell side*, with hot water entering the tube side at $t_1 = 80°C$ and leaving at $t_2 = 45°C$. The overall heat transfer coefficient is $U_m = 250$ W/(m^2 · °C). Calculate the heat transfer surface A required.

11-24 A one shell pass, two tube pass heat exchanger of flow arrangement shown in Fig. 11-16*d* is used to heat water from $T_1 = 25°C$ to $T_2 = 80°C$ at a rate of $m_c = 1.5$ kg/s with pressurized water entering the tubes at $t_1 = 200°C$ and leaving at $t_2 = 100°C$. The overall heat transfer coefficient is $U_m = 1250$ W/(m^2 · °C). Calculate the heat transfer surface A required [$c_{pc} = 4180$ J/(kg · °C)].

Answer: 7.45 m^2

11-25 A single-pass, cross-flow heat exchanger of flow arrangement shown in Fig. 11-16*c* is used to heat water from $T_1 = 25°C$ to $T_2 = 80°C$ at a rate of $m_c = 0.5$ kg/s by using oil [$c_p = 2100$ J/(kg · °C)]. Oil enters the tubes at $t_1 = 175°C$ at a rate of $m_h = 0.5$ kg/s. The overall heat transfer coefficient is $U_m = 300$ W/(m^2 · °C). Calculate the heat transfer surface A required.

11-26 A one shell pass, two tube pass heat exchanger of flow arrangement shown in Fig. 11-16*a* is to heat water with ethylene glycol. Water enters the shell side at $T_1 = 30°C$ and leaves at $T_2 = 70°C$ with a flow rate of $m_c = 1$ kg/s, while ethylene glycol enters the tube at $t_1 = 100°C$ and leaves at $t_2 = 60°C$. The overall heat transfer coefficient is $U_m = 200$ W/(m^2 · °C). Calculate the heat transfer surface A required [$c_{pc} = 4180$ J/(kg · °C)].

11-27 Repeat Prob. 11-26 with the flow arrangement shown in Fig. 11-16*b*.

11-28 A finned-tube, single-pass, cross-flow heat exchanger, with both fluids unmixed as shown in Fig. 11-16*c*, is to heat air with hot water. The total heat transfer rate is 200 kW. The water enters the tubes at $t_1 = 85°C$ and leaves at $t_2 = 30°C$, while the air enters the shell side at $T_1 = 15°C$ and leaves at $T_2 =$ 50°C. The overall heat transfer coefficient is $U_m = 75$ W/(m^2 · °C). Calculate the heat transfer surface A required.

Answer: 150.7 m^2

11-29 A two shell pass, four tube pass heat exchanger of flow arrangement shown in Fig. 11-16*b* is used to heat water with hot exhaust gases. Water enters the tubes at $t_1 = 50°C$ and leaves at $t_2 = 125°C$ with a flow rate of $m_c = 10$ kg/s, while the hot exhaust gas enters the shell side at $T_1 = 300°C$ and leaves at $T_2 = 125°C$. The total heat transfer surface is $A = 800$ m^2. Calculate the overall heat transfer coefficient [$c_{pc} = 4180$ J/(kg · °C)].

11-30 A cross-flow heat exchanger of flow arrangement shown in Fig. 11-16*c* is to heat water with hot exhaust gas. The exhaust gas enters the exchanger at $T_{h,in} = 250°C$ and leaves at $T_{h,out} = 110°C$. The cold water enters at $t_{c,\,in} = 25°C$ and leaves at $t_{c,\,out} = 100°C$ with a flow rate of $m_c = 2$ kg/s. The overall heat transfer coefficient is estimated to be $U_m = 150$ W/(m^2 · °C). Calculate the heat transfer surface A required.

Answer: 40.2 m^2

ε-NTU method

11-31 A two shell pass, four tube pass heat exchanger of flow arrangement shown in Fig. 11-16*b* is used to cool processed water from $t_1 = 75°C$ to $t_2 = 25°C$ on the tube side at a rate of $m_h = 5$ kg/s, with cold water entering the shell side at $T_1 = 10°C$ at a rate of $m_c = 6$ kg/s. The overall heat transfer coefficient is $U_m = 750$ W/(m^2 · °C). Calculate the heat transfer surface and the outlet temperature of the coolant water.

Answer: 125 m^2, 51.7°C

11-32 A counterflow heat exchanger of flow arrangement shown in Fig. 11-18 is used to cool hot mercury [$c_p = 1370$ J/(kg · °C)] from $T_{h,\,in} = 110°C$ to $T_{h,\,out} = 70°C$ at a rate of $m_h = 1$ kg/s, with water entering at $T_{c,\,in} = 30°C$ at a rate of $m_c = 0.2$ kg/s. The overall heat transfer coefficient is $U_m =$ 250 W/(m^2 · °C). Calculate the heat transfer surface required and the exit temperature of water.

11-33 Hot chemical products [$c_{ph} = 2500$ J/(kg · °C)] at $T_{h,\,in} = 600°C$ and at a flow rate of $m_h =$ 30 kg/s are used to heat chemical products at $m_c = 30$ kg/s [$c_{pc} = 4200$ J/(kg · °C)] at $T_{c,\,in} = 100°C$. The total heat transfer surface is $A = 50$ m^2. Calculate the outlet temperatures of the hot and cold products for the following two cases:

(*a*) A parallel-flow exchanger shown in Fig. 11-17 is used.

(*b*) A counterflow exchanger shown in Fig. 11-18 is used. Take $U_m = 1500$ W/(m^2 · °C).

11-34 A cross-flow heat exchanger of flow arrangement shown in Fig. 11-19 is used to heat water with an engine oil. Water enters at $T_{c,\,in} = 30°C$ and leaves at $T_{c,\,out} = 85°C$ at a rate of $m_c = 1.5$ kg/s, while the engine oil [$c_p = 2300$ J/(kg · °C)] enters at $T_{h,\,in} = 120°C$ at a flow rate of $m_h = 3.5$ kg/s. The heat transfer surface is $A = 30$ m^2. Calculate the overall heat transfer coefficient U_m by using the ε-NTU method.

Answer: 334 W/(m^2 · °C)

11-35 A counterflow shell-and-tube exchanger of flow arrangement shown in Fig. 11-18 is used to heat water with hot exhaust gases. The water [$c_p = 4180$ J/(kg · °C)] enters at $T_{c,\,in} = 25°C$ and leaves at $T_{c,\,out} = 80°C$ at a rate of $m_c = 2$ kg/s. Exhaust gases [$c_p = 1030$ J/(kg · °C)] enters at $T_{h,\,in} = 175°C$ and leaves at $T_{h,\,out} = 90°C$. The overall heat transfer coefficient is $U_m = 200$ W/(m^2 · °C). Calculate the heat transfer surface A required.

11-36 A single shell pass, two tube pass heat exchanger of flow arrangement shown in Fig. 11-20 is used to heat water entering at $T_{c,\,in} = 15°C$ and $m_c = 2$ kg/s with ethylene glycol [$c_p = 2600$ J/(kg · °C)] entering at $T_{h,\,in} = 85°C$ and $m_h = 1$ kg/s. The overall heat transfer coefficient is $U_m = 500$ W/(m^2 · °C). Calculate the rate of heat transfer Q and the outlet temperatures of water and ethylene glycol. The heat transfer area is $A = 10$ m^2.

11-37 A cross-flow heat exchanger of flow arrangement shown in Fig. 11-19 is used to heat pressurized water with a hot exhaust gas. The hot gas [$c_{ph} = 1050$ J/(kg · °C)] enters the exchanger at $T_{h,\,in} =$ 300°C and a flow rate of $m_h = 1$ kg/s, while the pressurized water enters at $T_{c,\,in} = 30°C$ and leaves at $T_{c,\,out} = 130°C$ with a flow rate of $m_c = 0.25$ kg/s. The heat transfer surface area is $A = 3$ m^2. Calculate the overall transfer coefficient U_m.

11-38 A two shell pass, four tube pass heat exchanger shown in Fig. 11-21 is to be used to heat $m_c =$ 1.2 kg/s of water from $T_{c,\,in} = 20°C$ to $T_{c,\,out} = 80°C$ by using $m_h = 2.2$ kg/s of oil entering at $T_{h,\,in} =$ 160°C. The overall heat transfer coefficient is $U_m = 300$ W/(m^2 · °C), and the specific heat of oil is $c_{ph} = 2100$ J/(kg · °C). Determine the heat transfer surface required.

Answer: 13.9 m^2

11-39 A counterflow heat exchanger of flow arrangement shown in Fig. 11-18 is to heat cold fluid entering at $T_{c,\,in} = 30°C$ and $m_c c_{pc} = 15{,}000$ W/°C with the hot fluid entering at $T_{h,\,in} = 120°C$ with

$m_h c_{ph} = 10{,}000$ W/°C. The overall heat transfer coefficient is $U_m = 400$ W/(m^2 · °C), and the total heat transfer surface is $A = 20$ m^2. Calculate the total heat transfer rate Q and the outlet temperatures of the hot and cold fluids.

Answer: 430.2 kW, 77°C, 58.7°C

11-40 A one shell pass, two tube pass heat exchanger of flow arrangement shown in Fig. 11-20 is to heat pressurized cold water from $T_{c,\text{in}} = 40$°C to $T_{c,\text{out}} = 140$°C flowing at $m_c = 2$ kg/s with pressurized hot water entering at $T_{h,\text{in}} = 300$°C with $m_h = 2$ kg/s. The overall heat transfer coefficient is $U_m = 1250$ W/(m^2 · °C). Calculate the heat transfer surface A required and the outlet temperature of the pressurized water.

11-41 A single shell pass, two tube pass condenser of flow arrangement shown in Fig. 11-20 is to condense steam at $T = 55$°C with the cooling water entering at $T_{c,\text{in}} = 20$°C and $m_c = 600$ kg/s. The total heat transfer rate is given by $Q = 14{,}000$ kW. The overall heat transfer coefficient is $U_m = 3500$ W/(m^2 · °C). Calculate the total heat transfer surface A and the outlet temperature $T_{c,\text{out}}$ of the cooling water.

11-42 A cross-flow heat exchanger of flow arrangement shown in Fig. 11-19 is to heat water with hot exhaust gas. The exhaust gas enters at $T_{h,\text{in}} = 200$°C and $m_h = 2.5$ kg/s while the water enters at $T_{c,\text{in}} = 30$°C and $m_c = 1.5$ kg/s. The overall heat transfer coefficient is $U_m = 150$ W/(m^2 · °C). Calculate the total heat transfer rate and the outlet temperatures of water and exhaust gas. Take heat transfer area $A = 17.5$ m^2, $c_{ph} = 1050$ J/(kg · °C), and $c_{pc} = 4180$ J/(kg · °C).

Answer: 214 kW, 118.4°C, 64.2°C

11-43 A two shell pass, four tube pass heat exchanger of flow arrangement shown in Fig. 11-21 is to cool $m_h = 3$ kg/s of oil [$c_{ph} = 2100$ J/(kg · °C)] from $T_{h,\text{in}} = 85$°C to $T_{h,\text{out}} = 35$°C with water [$c_{pc} =$ 4180 J/(kg · °C)] entering the exchanger at $T_{c,\text{in}} = 14$°C and $m_c = 2$ kg/s. The overall heat transfer coefficient is $U_m = 400$ W/(m^2 · °C). Calculate the total heat transfer surface A required.

11-44 A cross-flow heat exchanger of flow arrangement shown in Fig. 11-19 has a heat transfer surface $A = 12$ m^2. It will be used to heat air entering at $T_{c,\text{in}} = 10$°C and $m_c = 3$ kg/s, with hot water entering at $T_{h,\text{in}} = 80$°C and $m_h = 0.4$ kg/s. The overall heat transfer coefficient can be taken as $U_m = 300$ W/(m^2 · °C). Calculate the total heat transfer rate Q and the outlet temperatures of the air and water.

11-45 A counterflow heat exchanger of flow arrangement of Fig. 11-17 is to heat air with the exhaust gas from the turbine. Air enters the exchanger at $T_{c,\text{in}} = 300$°C and leaves at $T_{c,\text{out}} = 500$°C and $m_c =$ 4 kg/s; the exhaust gas enters at $T_{h,\text{in}} = 650$°C and $m_h = 4$ kg/s. The overall heat transfer coefficient is $U_m = 80$ W/(m^2 · °C). The specific heat for both air and the exhaust gas can be taken as $c_{ph} = c_{pc} =$ 1100 J/(kg · °C). Calculate the heat transfer surface A required and the outlet temperature of the exhaust gas.

Answer: 77 m^2, 450.5°C

11-46 Examine the surface area requirement in Prob. 11-45 if the counterflow arrangement is replaced with the cross-flow arrangement in Fig. 11-19.

11-47 A cross-flow heat exchanger of flow arrangement shown in Fig. 11-19 is to heat water with hot exhaust gas. The water enters the tubes at $T_{c,\text{in}} = 25$°C and $m_c = 1$ kg/s, while the exhaust gas enters the exchanger at $T_{h,\text{in}} = 200$°C and $m_h = 2$ kg/s. The total heat transfer surface is $A = 30$ m^2, and the overall heat transfer coefficient is $U_m = 120$ W/(m^2 · °C). The specific heat for the exhaust gas may be taken as $c_{ph} = 1100$ J/(kg · °C). Calculate the total heat transfer rate Q and the outlet temperatures $T_{c,\text{out}}$ and $T_{h,\text{out}}$ of the water and the exhaust gas.

11-48 A counterflow exchanger is to heat air entering at $T_{c,\text{in}} = 400$°C and $m_c = 6$ kg/s with an exhaust gas entering at $T_{h,\text{in}} = 800$°C and $m_h = 4$ kg/s. The overall heat transfer coefficient is $U_m = 100$ W/(m^2 · °C), and the specific heat for both air and the exhaust gas may be taken as $c_{ph} = c_{pc} = 1100$ J/(kg · °C). Calculate the heat transfer surface A required and the outlet temperatures of both the air and the hot exhaust gas, for a heat transfer rate of $Q = 1000$ kW.

11-49 A single-pass, counterflow, tube-and-shell heat exchanger has a heat transfer area $A = 20$ m^2 based on the outer surface of the tubes. Hot oil [$c_p = 2000$ J/(kg · °C)] entering the tubes at $T_{h,\text{in}} =$ 120°C and $m_h = 3$ kg/s is to be cooled with water entering the shell side at $T_{c,\text{in}} = 20$°C and $m_c = 0.75$

kg/s. The overall heat transfer coefficient is $U_m = 350$ W/(m^2 · °C). Calculate the heat transfer rate Q and the exit temperatures of the oil and water.

Answer: 250.8 kW, 78.2°C, 100°C

Compact heat exchangers

11-50 Air at 2 atm and 500 K with a velocity of $u_\infty = 20$ m/s flows across a compact heat exchanger matrix having a configuration shown in Fig. 11-23. Calculate the heat transfer coefficient.

Answer: 330 W/(m^2 · °C)

11-51 Repeat Prob. 11-50 for the heat exchanger matrix configuration shown in Fig. 11-24*a*, and calculate h.

11-52 Repeat Prob. 11-50 for the heat exchanger matrix configuration shown in Fig. 11-25, and calculate h.

11-53 Hot air at $P = 3$ atm and 500 K and at a rate of $m = 20$ kg/s flows across a plain plate-fin matrix configuration shown in Fig. 11-24. The dimensions of the heat exchanger matrix are given in the accompanying figure. Calculate (*a*) the heat transfer coefficient, (*b*) the friction factor, and (*c*) the ratio of the pressure drop due to core friction ΔP_f to the inlet pressure P.

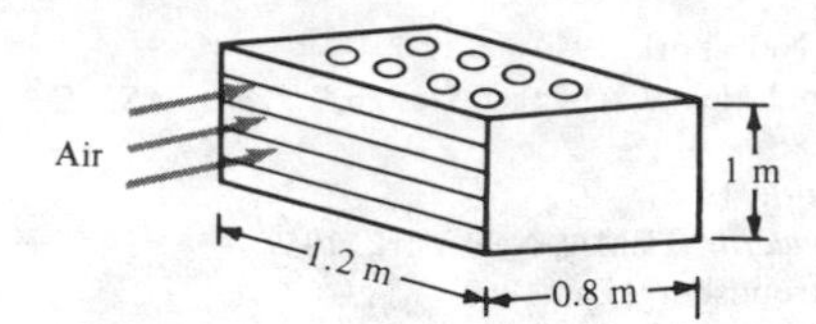

Figure P11-53

11-54 Repeat Prob. 11-53 for a finned circular-tube matrix shown in Fig. 11-23.

Answer: $h = 237$ W/(m^2 · °C), $f = 0.022$, $\Delta P_f/P = 1.5$ percent

11-55 Hot air at $P = 2$ atm and 500 K and at a rate of $m = 12$ kg/s flows across a circular finned-tube matrix configuration shown in Fig. 11-25. The dimensions of the heat exchanger matrix are given in the accompanying figure. Calculate (*a*) the heat transfer coefficient, (*b*) the friction factor, and (*c*) the ratio of the pressure drop due to core friction ΔP_f to the inlet pressure. (Consider case A.)

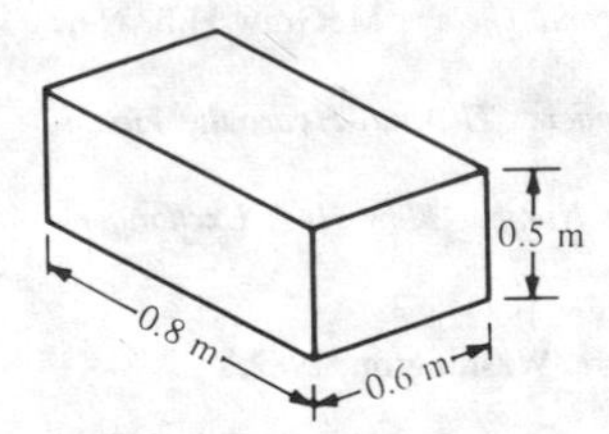

Figure P11-55

11-56 Repeat Prob. 11-55 for a finned flat-tube matrix configuration shown in Fig. 11-24.

Answer: $h = 287$ W/(m^2 · °C), $f = 0.022$, $\Delta P_f/P = 3.9$ percent

11-57 Air at $P = 1$ atm and 500 K and at a rate of $m = 5$ kg/s flows across a plain plate-fin matrix of configuration shown in Fig. 11-24. The dimensions of the heat exchanger matrix are given in the accompanying figure. Calculate (*a*) the heat transfer coefficient, (*b*) the friction factor, and (*c*) the ratio of the pressure drop due to core friction ΔP_f to the inlet pressure.

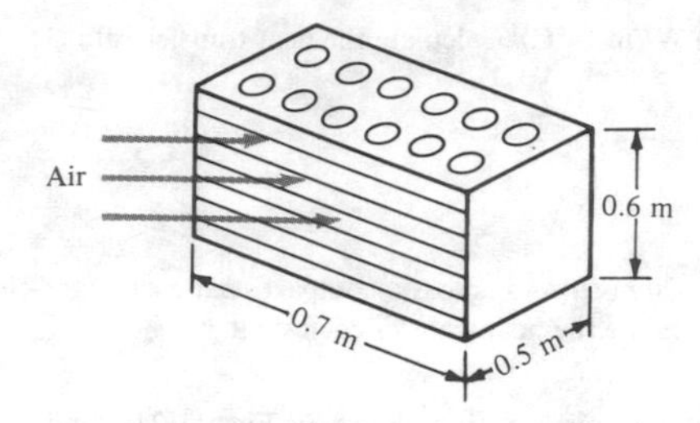

Figure P11-57

11-58 Repeat Example 11-18 for hot air at a flow rate of $m_a = 20$ kg/s, while everything else remains the same.

REFERENCES

General theory and design

1. Kern, D. Q.: *Process Heat Transfer*, McGraw-Hill, New York, 1950.
2. Fax, D. H., and R. R. Mills, Jr.: "General Optimal Heat Exchanger Design," *Trans. ASME* **79**: 653–661 (1957).
3. Hyrnisak, W.: *Heat Exchangers*, Academic, New York, 1958.
4. Tubular Exchanger Manufacturers Association, *Standards*, TEMA, New York, 1959.
5. *Heat Exchangers*, The Patterson-Kelley Co., East Stroudsburg, Pa., 1960.
6. Chilton, C. H. (ed.): *Cost Engineering in the Process Industries*, McGraw-Hill, New York, 1960.
7. Shields, C. D.: *Boilers: Types, Characteristics, and Functions*, McGraw-Hill, New York, 1961.
8. Berman, L. D.: *Evaporative Cooling of Circulating Water*, Pergamon, New York, 1961.
9. Fraas, A. P., and M. N. Özışık: *Steam Generators for High Temperature Gas-Cooled Reactors*, ORNL-3208. Oak Ridge National Laboratory, Oak Ridge, Tenn., April 1963.
10. Kays, W. M., and A. L. London: *Compact Heat Exchangers*, 2d ed., McGraw-Hill, New York, 1964.
11. Fraas, A. P., and M. N. Özışık: *Heat Exchanger Design*, Wiley, New York, 1965.
12. Mueller, A. C.: "Heat Exchangers," W. M. Rohsenow and J. P. Hartnett (eds.), *Handbook of Heat Transfer*, McGraw-Hill, New York, 1973, chap. 18.
13. Afgan, N. H., and E. U. Schlünder: *Heat Exchangers: Design and Theory*, McGraw-Hill, New York, 1974.
14. Kakaç, S., A. E. Bergles, and F. Mayinger (eds.): *Heat Exchangers: Thermal-Hydraulic Fundamentals and Design*, Hemisphere, Washington, 1981.
15. Kakaç, S., R. K. Shah, and A. E. Bergles (eds.): *Low Reynolds Number Flow Heat Exchangers*, Hemisphere, Washington, 1982.
16. Walker, G.: *Industrial Heat Exchangers*, Hemisphere, Washington, 1982.
17. Schlünder, E. U.: *Heat Exchanger Design Handbook*, Hemisphere, Washington, 1982.

Cooling towers

18. Nance, G. R.: "Fundamental Relationships in the Design of Cooling Towers," *Trans. ASME* **61**: 721–725 (1939).
19. London, A. L., W. E. Mason, and L. M. K. Boelter: "Performance Characteristics of Mechanically Induced Draft Counterflow Packed Cooling Towers," *Trans. ASME* **62**: 41–50 (1940).
20. Lichtenstein, J.: "Performance and Selection of Mechanical Draft Cooling Towers," *Trans. ASME* **65**: 779–787 (1943).

21. Kelly, N. W., and L. K. Swenson: "Comparative Performance of Cooling Tower Packing Arrangements," *Chem. Eng. Prog.* **52:**263–268 (1956).
22. Moore, F. K.: "On the Minimum Size of Large Dry Cooling Towers with Combined Mechanical and Natural Draft," *J. Heat Transfer* **95C:**383–389 (1973).
23. Webb, R. L., and R. E. Barry (eds.): *Dry and Wet/Dry Cooling Towers for Power Plants, ASME*, New York, 1973.

Regenerative heat exchangers

24. Hausen, H.: "On the Theory of Heat Exchange in Regenerators," *Z. Angew. Math. Mech.* **9:**193–200 (1929).
25. Karlsson, H., and S. Holm: "Heat Transfer and Fluid Resistance in Ljungstrom-Type Air Preheaters," *Trans. ASME* **65:**61–72 (1943).
26. Coppage, J. E., and A. L. London: "The Periodic Flow Regenerator—A Summary of Design Theory," *Trans. ASME* **75:**779–787 (1953).
27. Harper, D. B., and W. M. Rohsenow: "Effect of Rotary Regenerator Performance on Gas-Turbine-Plant Performance," *Trans. ASME* **75:**759–765 (1953).
28. Lambertson, T. J.: "Performance Factors of a Periodic Flow Heat Exchanger," *Trans. ASME* **80:**586–592 (1953).
29. Harper, D. B.: "Seal Leakage in the Rotary Regenerator and Its Effect on Rotary Regenerator Design for Gas Turbines," *Trans. ASME* **79:**233–245 (1957).
30. Jakob, M.: *Heat Transfer*, vol. 2, Wiley, New York, 1957.
31. Fingold, J. G., and R. H. Sterrett: "Stirling Engine Regenerator Review," *Jet. Prop. Lab. Rept.* 5030-230, California Institute of Technology, Pasadena, 1978.
32. Grossman, D. G., and L. G. Larning: "Aluminuous Keatie Ceramic Regenerators," *Bull. Am. Ceram. Soc.* **56:**474–477 (1977).

Fouling

33. Watkinson, A. P., and N. Epstein: "Gas Oil Fouling in a Sensible Heat Exchanger," *Chem. Eng. Prog. Symp. Ser.* **65**(92)**:**84–90 (1969).
34. Bott, R. R., and R. A. Walker: "Fouling in Heat Transfer Equipment," *Chem. Eng.*, **255:**391–93 (1971).
35. Taborek, J., T. Aoki, R. Ritter, J. Pallen, and J. Knudsen: "Predictive Methods for Fouling Behaviour," *Chem. Eng. Prog.* **68**(7)**:**69 (1972).
36. Watkinson, A. P., and O. Martinez: "Scaling of Heat Exchanger Tubes by Calcium Carbonate," *J. Heat Transfer Trans. ASME*, 76-HT-8 (1976).
37. Suitor, J. W., W. J. Marner, and R. B. Ritter: "The History and Status of Research in Fouling of Heat Exchangers in Cooling Water Service," *Can. J. Chem.* **55:**374–380 (1977).
38. Suitor, J., W. Marner, and R. B. Ritter: "The History and Status of Research in Fouling of Heat Exchangers in Water Service," *Can. J. Chem. Eng.* **35**:347 (1977).
39. Epstein, N.: "Fouling in Heat Exchangers," *Proc. Sixth Int. Heat Transfer Conf.* **6:**235–253 (1978).
40. Pritchard, A. M.: "Heat Exchanger Fouling in British Industry," *Fouling Prevention Res. Dig.* **1**(1–4)**:**iv–vi (1979).
41. Epstein, N.: "Fouling in Heat Exchangers." Keynote paper KS-18 in 6th International Heat Transfer Conference, Toronto, Canada, August 7–11, 1978, vol. 6 of *Proceedings*.
42. Epstein, N.: "Fouling in Heat Exchangers and Fouling: Technical Aspects," in E. F. C. Somerscales and J. G. Knudsen (eds.), *Fouling of Heat Transfer Equipment*, Hemisphere, New York, 1981, pp. 701–734, 31–53.
43. Somerscales, E. F. C., and J. G. Knudsen: *Fouling of Heat Transfer Equipment*, Hemisphere, Washington, 1980.
44. Ma, R. S. T., and N. Epstein: "Optimum Cycles for Falling Rate Processes," *Can. J. Chem. Eng.* in press, 1981.

LMTD correction charts

45. Bowman, R. A., A. C. Mueller, and W. M. Nagle: "Mean Temperature Difference in Design," *Trans. ASME* **62**:283–294 (1940).
46. Gardner, K. A.: "Variable Heat Transfer Rate Correction in Multipass Exchangers, Shell-Side Film Controlling," *Trans. ASME* **67**:31–38 (1945).
47. Stevens, R. A., J. Fernandes, and J. R. Woolf: "Mean Temperature Difference in One, Two and Three-Pass Cross Flow Heat Exchangers," *Trans. ASME* **79**:287–297 (1957).

CHAPTER

TWELVE

RADIATION AMONG SURFACES IN A NONPARTICIPATING MEDIUM

12-1 NATURE OF THERMAL RADIATION

Thermal radiation refers to radiation energy emitted by bodies because of their temperature. All bodies at a temperature above absolute zero emit thermal radiation. Consider, for example, a hot object at temperature T_h placed in a vacuum chamber having cold walls at temperature T_c, as illustrated in Fig. 12-1. Since the hot object is separated from the cold walls by a vacuum, heat transfer by conduction or convection is not possible. Instead, the hot object cools as a result of heat exchange by thermal radiation.

Another example is the transfer of energy from the sun to the earth; the thermal energy emitted from the sun travels through space and reaches the earth's surface. The energy transport by radiation does not require an intervening medium between the hot and cold surfaces. The actual mechanism of radiation propagation is not fully understood, but theories are proposed to explain the propagation process. According to Maxwell's electromagnetic theory, radiation is treated as *electromagnetic waves*, while Max Planck's concept treats radiation as *photons*, or *quanta, of energy*. Both concepts have been utilized to describe the emission and propagation of radiation. For example, the results obtained from the electromagnetic theory have been used to predict radiation properties of materials, while results from Planck's concept have been employed to predict the magnitude of radiation energy emitted by a body at a given temperature.

When radiation is treated as an electromagnetic wave, the radiation from a body at temperature T is considered emitted at all wavelengths from $\lambda = 0$ to $\lambda = \infty$. At temperatures encountered in most engineering applications, the bulk of the thermal energy emitted by a body lies in wavelengths between $\lambda \cong 0.1$ and $\lambda \cong 100\ \mu m$. For this reason, the portion of the wavelength spectrum between $\lambda = 0.1$ and $\lambda = 100\ \mu m$ is generally referred to as the *thermal radiation*. The sun

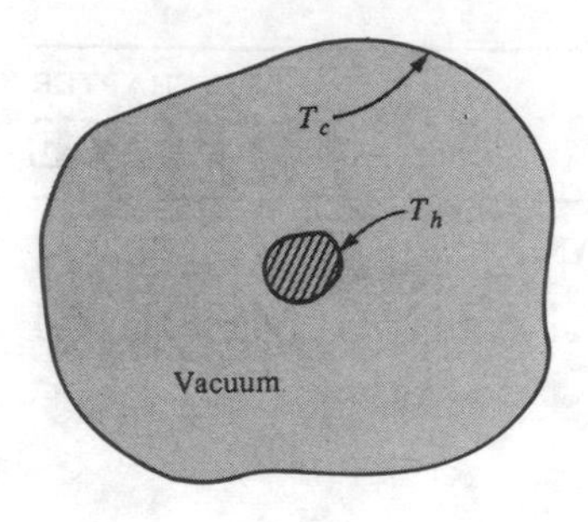

Figure 12-1 Radiation heat exchange.

emits thermal radiation at an effective surface temperature of about 5760 K, and the bulk of this energy lies in wavelengths between $\lambda \cong 0.1$ and $\lambda \cong 3\ \mu m$; therefore, this portion of the spectrum is generally known as the *solar radiation.* The radiation emitted by the sun in wavelengths between $\lambda = 0.4$ and $\lambda = 0.7\ \mu m$ is visible to the eye; therefore, this portion of the spectrum is called *visible radiation* (i.e., light). Figure 12-2 illustrates such subdivisions on the electromagnetic wave spectrum.

The wave nature of thermal radiation implies that the wavelength λ should be associated with the frequency ν of radiation. The relation between λ and ν is given by

$$\lambda = \frac{c}{\nu} \tag{12-1}$$

where c is the speed of propagation in the medium. If the medium in which radiation travels is a vacuum, the speed of propagation is equal to the speed of light, that is,

$$c_0 = 2.9979 \times 10^8 \text{ m/s} \tag{12-2}$$

By utilizing this relationship between λ and ν, we included in Fig. 12-2 the corresponding frequency spectrum.

Other types of radiation, such as x-rays, gamma rays, microwaves, etc., are well known and utilized in various branches of science and engineering. The x-rays are produced by the bombardment of a metal with high-frequency electrons, and the bulk of the energy is in the range between $\lambda \cong 10^{-4}$ and $\lambda \cong 10^{-2}\ \mu m$. Gamma

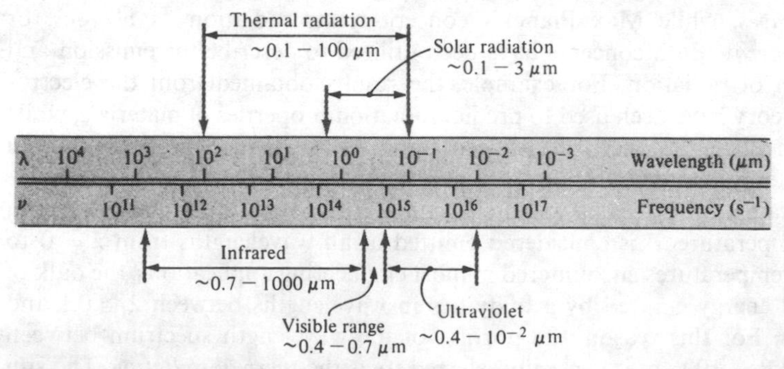

Figure 12-2 Typical spectrum of electromagnetic radiation due to temperature of a body.

rays are produced by the fission of nuclei or by radioactive disintegration, and the bulk of the energy is concentrated in the range of wavelengths shorter than that of x-rays. In this book, we are not concerned with such radiation; our interest is only the thermal radiation as a mechanism of energy transport between objects at different temperatures.

In the study of radiation transfer, a distinction should be made between bodies which are *semitransparent* to radiation and those which are *opaque*. If the material is semitransparent to radiation, such as glass, salt crystals, and gases at elevated temperatures, then the radiation leaving the body from its outer surfaces results from emissions at all depths within the material. The emission of radiation for such cases is a *bulk*, or a *volumetric phenomenon*. If the material is opaque to thermal radiation, such as metals, wood, rocks, etc., then the radiation emitted by the interior regions of the material cannot reach the surface. In such cases, the radiation emitted by the body originates from the material at the immediate vicinity of the surface (i.e., within about 1 μm), and the emission is regarded as a *surface phenomenon*. Also note that a material may behave as a semitransparent medium for certain temperature ranges and as opaque for other temperatures. Glass is a typical example of such behavior; it is semitransparent to thermal radiation at elevated temperatures and opaque at intermediate and low temperatures.

In this chapter we consider radiation transfer only for situations in which radiation can be regarded as a surface phenomenon. Radiation as a bulk or volumetric phenomenon is the subject of the next chapter. The reader interested in a more comprehensive treatment of the subject is referred to Hottel and Sarofim [1], Love [2], Özişik [3], Siegel and Howell [4], and Sparrow and Cess [5].

12-2 BLACKBODY RADIATION

A body at any temperature above absolute zero emits thermal radiation in all wavelengths in all possible directions into space. The concepts of *blackbody* is an idealized situation that serves to compare the emission and absorption characteristics of real bodies.

A blackbody is considered to absorb all incident radiation from all directions at all wavelengths without reflecting, transmitting, or scattering it. For a given temperature and wavelength, no other body at the same temperature can emit more radiation than a blackbody. The radiation emission by a blackbody at any temperature T is the maximum possible emission at that temperature.

The term *black* should be distinguished from its common usage regarding the blackness of a surface to visual observations. The human eye can detect blackness only in the visible range of the spectrum. For example, an object such as ice is bright to the eye but is almost black for long-wave thermal radiation. However, a blackbody is perfectly black to thermal radiation for all wavelengths from $\lambda = 0$ to $\lambda = \infty$.

Radiation is emitted by a body in all directions. It is of interest to know the amount of radiation emitted by a blackbody streaming into a given direction. The fundamental quantity that specifies the magnitude of the radiation energy emitted by a blackbody at an absolute temperature T, at a wavelength λ, in any given direction is called the *spectral blackbody radiation intensity* $I_{b\lambda}(T)$. Here the term *spectral* is used to denote the wavelength dependence of the radiation intensity, and the subscript b refers to the blackbody.

The magnitude of $I_{b\lambda}(T)$ for emission into a vacuum was first determined by Planck [6] and is given by

$$I_{b\lambda}(T) = \frac{2hc^2}{\lambda^5\{\exp[hc/(\lambda kT)] - 1\}} \tag{12-3}$$

where h ($=6.6256 \times 10^{-34}$ J·s) and k ($=1.38054 \times 10^{-23}$ J·K) are the Planck and Boltzmann constants, respectively, c ($=2.9979 \times 10^8$ m/s) is the speed of light in a vacuum, T in kelvins is the absolute temperature, and λ is the wavelength. Here $I_{b\lambda}(T)$ represents *the radiation energy emitted by a blackbody at temperature T, streaming through a unit area perpendicular to the direction of propagation, per unit wavelength about the wavelength λ per unit solid angle about the direction of propagation of the beam.* Based on this definition, the units of $I_{b\lambda}(T)$ can be written as

$$\frac{\text{Energy}}{(\text{Area})(\text{wavelength})(\text{solid angle})} \tag{12-4a}$$

where the area is measured perpendicular to the direction of propagation.

If energy is measured in watts, area in square meters, wavelength in micrometers, and the solid angle in *steradian* (sr), then Eq. (12-4*a*) has the dimension

$$\frac{\text{W}}{\text{m}^2 \cdot \mu\text{m} \cdot \text{sr}} \tag{12-4b}$$

The physical significance of the solid angle is better envisioned by referring to Fig. 12-3. Let $\hat{\Omega}$ be the direction of propagation and O the reference location.

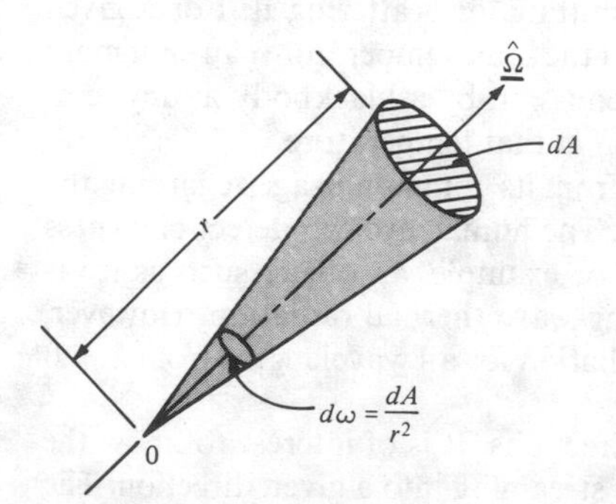

Figure 12-3 Definition of solid angle.

We consider a small area dA at a distance r from O and normal to the direction $\hat{\Omega}$. The solid angle $d\omega$ subtended by dA from O is defined as

$$\boxed{d\omega = \frac{dA}{r^2}} \tag{12-5}$$

Based on this definition, we can readily infer that the solid angle subtended by a hemisphere from its center is 2π (that is, $2\pi r^2/r^2$) and by a full sphere from its center is 4π (that is, $4\pi r^2/r^2$).

In Eq. (12-3), $I_{b\lambda}(T)$ is the blackbody radiation intensity per unit wavelength about the wavelength λ. However, the radiation is emitted at all wavelengths. To determine the *blackbody radiation intensity* $I_b(T)$ emitted by a blackbody at temperature T over all wavelengths, we integrate $I_{b\lambda}(T)$ from $\lambda = 0$ to $\lambda = \infty$:

$$I_b(T) = \int_{\lambda=0}^{\infty} I_{b\lambda}(T)\, d\lambda \qquad \text{W/(m}^2 \cdot \text{sr)} \tag{12-6}$$

Here, $I_b(T)$ is called the *blackbody radiation intensity*.

Example 12-1 Determine the solid angles subtended by the surfaces dA_1 and dA_2, when they are viewed from the point O for the dimensions and the geometric arrangement shown in the accompanying figure.

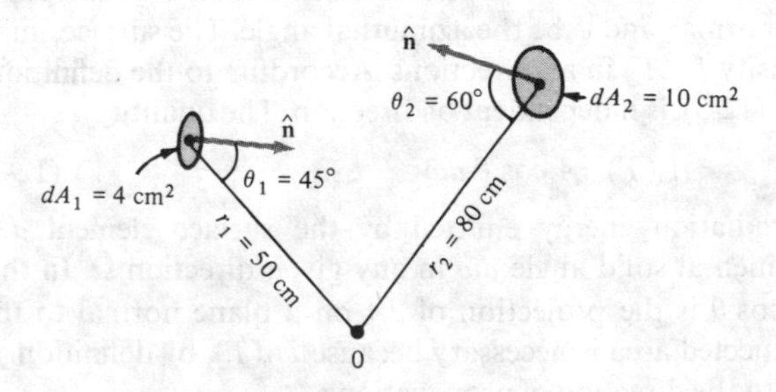

SOLUTION The solid angle $d\omega$ is defined by Eq. (12-5) as

$$d\omega = \frac{dA}{r^2}$$

where dA is the area normal to the direction r. This definition is now applied to the areas dA_1 and dA_2, in order to determine the solid angles.

$$d\omega_1 = \frac{dA_1 \cos\theta_1}{r_1^2} = \frac{4(0.707)}{50^2} = 1.13 \times 10^{-3} \text{ sr}$$

$$d\omega_2 = \frac{dA_2 \cos\theta_2}{r_2^2} = \frac{10(0.5)}{80^2} = 0.78 \times 10^{-3} \text{ sr}$$

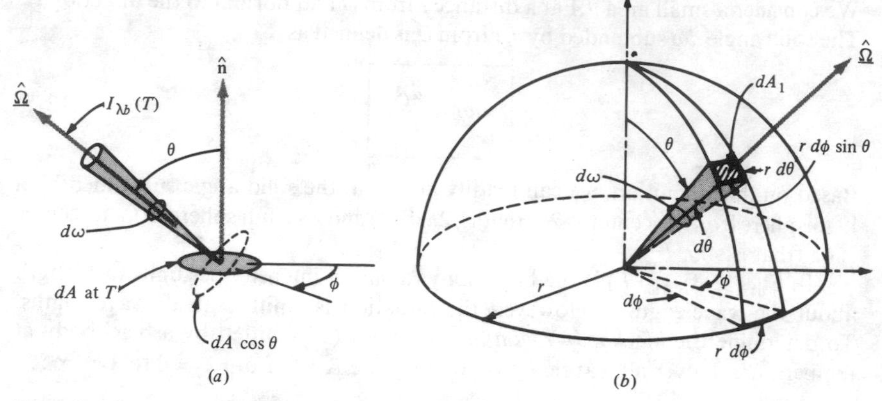

Figure 12-4 Nomenclature for (*a*) emission of radiation from a surface dA, (*b*) definition of the solid angle $d\omega$ in terms of (θ, ϕ).

Blackbody Emissive Power

It is of practical interest to know the amount of radiation energy emitted per unit area of a blackbody at an absolute temperature T in all directions into hemispherical space. To evaluate this quantity, we consider an elemental area dA at temperature T, as illustrated in Fig. 12-4*a*. Let $\hat{n}$ be the normal to this surface, θ be the polar angle measured from this normal, and ϕ be the azimuthal angle. The surface emits radiation of spectral intensity $I_{b\lambda}(T)$ in all directions. According to the definition, this intensity, given by Eq. (12-3), is independent of direction. The quantity

$$I_{b\lambda}(T)\, dA \cos\theta\, d\omega \tag{12-7}$$

represents the spectral radiation energy emitted by the surface element dA, streaming through an elemental solid angle $d\omega$ in any given direction $\hat{\Omega}$. In this expression, the term $dA \cos\theta$ is the projection of dA on a plane normal to the direction $\hat{\Omega}$; the use of projected area is necessary because $I_{\lambda b}(T)$, by definition, is based on the area normal to the direction of propagation.

Dividing Eq. (12-7) by dA, we obtain

$$I_{b\lambda}(T) \cos\theta\, d\omega \tag{12-8}$$

which represents the spectral blackbody radiation energy emitted by a unit surface area, streaming through a differential solid angle $d\omega$ in any direction $\hat{\Omega}$.

Refer to Fig. 12-4*b*. A differential solid angle $d\omega$ can be related to the polar angle θ and the azimuth angle ϕ by

$$d\omega = \frac{dA_1}{r^2} = \frac{(r\, d\theta)(r\, d\phi \sin\theta)}{r^2} = \sin\theta\, d\theta\, d\phi \tag{12-9}$$

Then Eq. (12-8) becomes

$$I_{b\lambda}(T) \cos\theta \sin\theta\, d\theta\, d\phi \tag{12-10}$$

The spectral blackbody radiation emitted per unit surface area in all directions into the hemispherical space is obtained by integrating Eq. (12-10) over $0 \leq \phi \leq 2\pi$ and $0 < \theta \leq \pi/2$. We obtain

$$E_{b\lambda}(T) = I_{b\lambda}(T) \int_{\phi=0}^{2\pi} \int_{\theta=0}^{\pi/2} \cos\theta \sin\theta \, d\theta \, d\phi$$

$$= 2\pi I_{b\lambda}(T) \int_{\theta=0}^{\pi/2} \cos\theta \sin\theta \, d\theta$$

$$= 2\pi I_{b\lambda}(T) [\tfrac{1}{2} \sin^2 \theta]_0^{\pi/2}$$

$$\boxed{E_{b\lambda}(T) = \pi I_{b\lambda}(T)} \qquad \text{W/(m}^2 \cdot \mu\text{m)} \tag{12-11}$$

Here $E_{b\lambda}(T)$ is called the *spectral blackbody emissive power*. It represents the radiation energy emitted by a blackbody at an absolute temperature T per unit area per unit time per unit wavelength about λ in all directions into the hemispherical space. Actually, it is the spectral blackbody radiation flux.

The Planck function defined by Eq. (12-3) is now introduced into Eq. (12-11). We obtain

$$E_{b\lambda}(T) = \frac{c_1}{\lambda^5 \{\exp [c_2/(\lambda T)] - 1\}} \qquad \text{W/(m}^2 \cdot \mu\text{m)} \tag{12-12}$$

where $c_1 = 2\pi hc^2 = 3.743 \times 10^8 \text{ W} \cdot \mu\text{m}^4/\text{m}^2$

$c_2 = \dfrac{hc}{k} = 1.4387 \times 10^4 \ \mu\text{m} \cdot \text{K}$

T = absolute temperature, K

λ = wavelength, μm

Equation (12-12) can be used to compute $E_{b\lambda}(T)$ for any given λ and T. Figure 12-5 shows a plot of $E_{b\lambda}(T)$ as a function of λ at various T. We note from this figure that at any given wavelength, the emitted radiation increases with increasing temperature, and at any given temperature, the emitted radiation varies with wavelength and shows a peak. These peaks tend to shift toward smaller wavelengths as the temperature increases. The locus of these peaks is given by *Wien's displacement law* as

$$\boxed{(\lambda T)_{\text{max}} = 2897.6 \ \mu\text{m} \cdot \text{K}} \tag{12-13}$$

The locations of the peaks are shown in Fig. 12-5 by the dashed line.

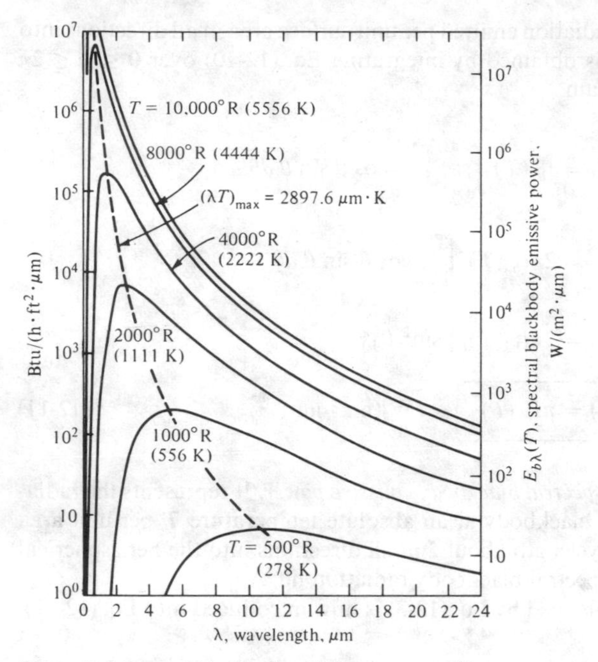

Figure 12-5 Spectral blackbody emissive power at different temperatures.

Stefan-Boltzmann Law

The radiation energy emitted by a blackbody at an absolute temperature T over all wavelengths per unit time per unit area is determined by integrating Eq. (12-12) from $\lambda = 0$ to $\lambda = \infty$:

$$E_b(T) = \int_{\lambda=0}^{\infty} \frac{c_1}{\lambda^5\{\exp\,[c_2/(\lambda T)] - 1\}}\, d\lambda$$

The integration variable is changed from λ to $\lambda T \equiv x$:

$$E_b(T) = T^4 \int_{x=0}^{\infty} \frac{c_1}{x^5[\exp\,(c_2/x) - 1]}\, dx \tag{12-14}$$

The integration can be performed. The result is expressed as

$$\boxed{E_b(T) = \sigma T^4} \qquad \text{W/m}^2 \tag{12-15}$$

where T is in kelvins and σ is called the *Stefan-Boltzmann constant*, which has the numerical value

$$\boxed{\sigma = 5.67 \times 10^{-8} \qquad \text{W/(m}^2 \cdot \text{K}^4)} \tag{12-16}$$

Here $E_b(T)$ is called the *blackbody emissive power*, and the result given by Eq. (12-15) is called the *Stefan-Boltzmann law*. The physical significance of $E_b(T)$ is that it is the blackbody radiation flux emitted from a surface at an absolute temperature T.

A relation can be established between $E_b(T)$ and $I_b(T)$ by integrating Eq. (12-11) over all wavelengths. We obtain

$$\boxed{E_b(T) = \pi I_b(T)} \qquad \text{W/m}^2 \tag{12-17}$$

and from Eqs. (12-15) and (12-17) we write

$$\boxed{I_b(T) = \frac{1}{\pi}\sigma T^4} \qquad \text{W/(m}^2 \cdot \text{sr)} \tag{12-18}$$

Blackbody Radiation Functions

In numerous applications, one is interested in the emission of radiation by a blackbody over the wavelength interval from $\lambda = 0$ to λ as a function of the total emission from $\lambda = 0$ to $\lambda = \infty$. This quantity is determined from its definition, given by

$$f_{0-\lambda}(T) = \frac{\int_0^\lambda E_{b\lambda}(T)\,d\lambda}{\int_0^\infty E_{b\lambda}(T)\,d\lambda} = \frac{\int_0^\lambda E_{b\lambda}(T)\,d\lambda}{\sigma T^4} \tag{12-19}$$

And $E_{b\lambda}(T)$ is introduced from Eq. (12-12) into Eq. (12-19):

$$f_{0-\lambda}(T) = \frac{c_1}{\sigma}\int_{x=0}^{\lambda T} \frac{dx}{x^5[\exp(c_2/x) - 1]} \tag{12-20}$$

where the integration variable is changed from λ to $\lambda T \equiv x$. The integral in Eq. (12-20) can be performed and $f_{0-\lambda}(T)$ calculated for a given λT. Table 12-1 gives the *blackbody radiation function* $f_{0-\lambda}(T)$, as a function of λT, originally computed by Dunkle [7]. In this table, the first and second columns give λT in μm · K and μm · °R, respectively. The third column is useful to compute the spectral blackbody emissive power $E_{b\lambda}(T)$ for a specified temperature and wavelength.

So far we discussed the blackbody radiation intensity and the emissive power, which are useful for the comparison of radiation energy from real surfaces. A perfect blackbody does not exist in reality; however, it can be closely approximated. Consider, for example, a hollow sphere whose interior surface is maintained at a uniform temperature T with a small opening provided at the surface of the sphere. The radiation coming out of such a hole is the closest approximation to blackbody radiation at temperature T.

Table 12-1 Blackbody radiation functions

λT, μm · K	λT, μm · °R	$\frac{E_{b\lambda}}{T^5}$, $\frac{\text{W}}{\text{m}^2 \cdot \text{K}^5 \cdot \mu\text{m}}$ $\times 10^{11}$	$f_{0-\lambda}(T)$	λT, μm · K	λT, μm · °R	$\frac{E_{b\lambda}}{T^5}$, $\frac{\text{W}}{\text{m}^2 \cdot \text{K}^5 \cdot \mu\text{m}}$ + 10^{11}	$f_{0-\lambda}(T)$
555.6	1,000	0.400×10^{-5}	0.00000	5,777.8	10,400	0.52517	0.71806
666.7	1,200	0.120×10^{-3}	0.00000	5,888.9	10,600	0.50261	0.72813
777.8	1,400	0.00122	0.00000	6,000.0	10,800	0.48107	0.73777
888.9	1,600	0.00630	0.00007	6,111.1	11,000	0.46051	0.74700
1,000.0	1,800	0.02111	0.00032	6,222.2	11,200	0.44089	0.75583
1,111.1	2,000	0.05254	0.00101	6,333.3	11.400	0.42218	0.76429
1,222.2	2,200	0.10587	0.00252	6,444.4	11,600	0.40434	0.77238
1,333.3	2,400	0.18275	0.00531	6,555.6	11,800	0.38732	0.78014
1,444.4	2,600	0.28091	0.00983	6,666.7	12,000	0.37111	0.78757
1,555.6	2,800	0.39505	0.01643	6,777.8	12,200	0.35565	0.79469
1,666.7	3,000	0.51841	0.02537	6,888.9	12,400	0.34091	0.80152
1,777.8	3,200	0.64404	0.03677	7,000.0	12,600	0.32687	0.80806
1,888.9	3,400	0.76578	0.05059	7,111.1	12,800	0.31348	0.81433
2,000.0	3,600	0.87878	0.06672	7,222.2	13,000	0.30071	0.82035
2,111.1	3,800	0.97963	0.08496	7,333.3	13,200	0.28855	0.82612
2,222.2	4,000	1.0663	0.10503	7,444.4	13,400	0.27695	0.83166
2,333.3	4,200	1.1378	0.12665	7,555.6	13,600	0.26589	0.83698
2,444.4	4,400	1.1942	0.14953	7,666.7	13,800	0.25534	0.84209
2,555.6	4,600	1.2361	0.17337	7,777.8	14,000	0.24527	0.84699
2,666.7	4,800	1.2645	0.19789	7,888.9	14,200	0.23567	0.85171
2,777.8	5,000	1.2808	0.22285	8,000.0	14,400	0.22651	0.85624
2,888.9	5,200	1.2864	0.24803	8,111.1	14,600	0.21777	0.86059
3,000.0	5,400	1,2827	0.27322	8,222.2	14,800	0.20942	0.86477
3,111.1	5,600	1,2713	0.29825	8,333.3	15,000	0.20145	0.86880
3,222.2	5,800	1.2532	0.32300	8,888.9	16,000	0.16662	0.88677
3,333.3	6,000	1.2299	0.34734	9,444.4	17,000	0.13877	0.90168
3,444.4	6,200	1.2023	0.37118	10,000.0	18,000	0.11635	0.91414
3,555.6	6,400	1.1714	0.39445	10,555.6	19,000	0.09817	0.92462
3,666.7	6,600	1.1380	0.41708	11,111.1	20,000	0.08334	0.93349
3,777.8	6,800	1.1029	0.43905	11,666.7	21,000	0.07116	0.94104
3,888.9	7,000	1.0665	0.46031	12,222.2	22,000	0.06109	0.94751
4,000.0	7,200	1.0295	0.48085	12,777.8	23,000	0.05272	0.95307
4,111.1	7,400	0.99221	0.50066	13,333.3	24,000	0.04572	0.95788
4,222.2	7,600	0.95499	0.51974	13,888.9	25,000	0.03982	0.96207
4,333.3	7,800	0.91813	0.53809	14,444.4	26,000	0.03484	0.96572
4,444.4	8,000	0.88184	0.55573	15,000.0	27,000	0.03061	0.96892
4,555.6	8,200	0.84629	0.57267	15,555.6	28,000	0.02699	0.97174
4,666.7	8,400	0.81163	0.58891	16,111.1	29,000	0.02389	0.97423
4,777.8	8,600	0.77796	0.60449	16,666.7	30,000	0.02122	0.97644
4,888.9	8,800	0.74534	0.61941	22,222.2	40,000	0.00758	0.98915
5,000.0	9,000	0.71383	0.63371	27,777.8	50,000	0.00333	0.99414
5,111.1	9,200	0.68346	0.64740	33,333.3	60,000	0.00168	0.99649
5,222.2	9,400	0.65423	0.66051	38,888.9	70,000	0.940×10^{-3}	0.99773
5,333.3	9,600	0.62617	0.67305	44,444.4	80,000	0.564×10^{-3}	0.99845
5,444.4	9,800	0.59925	0.68506	50,000.0	90.000	0.359×10^{-3}	0.99889
5,555.6	10,000	0.57346	0.69655	55,555.6	100,000	0.239×10^{-3}	0.99918
5,666.7	10,200	0.54877	0.70754	∞	∞	0.	1.00000

Example 12-2 A hole of area $dA = 2\ \text{cm}^2$ is opened on the surface of a large spherical cavity whose inside is maintained at $T = 800$ K.

(*a*) Calculate the radiation energy streaming through the hole in all directions into space.

(*b*) Find the radiation energy streaming per unit solid angle in the direction making a 60° angle with the normal to the surface of the opening.

SOLUTION Radiation from the opening can be approximated as a blackbody radiation at 800 K.

(*a*) Radiation energy streaming through the hole is

$$Q = dA\,\sigma T_1^4 = (2 \times 10^{-4}\ \text{m}^2)\left(5.67 \times 10^{-8}\ \frac{\text{W}}{\text{m}^2\text{K}^4}\right)(800^4\ \text{K}^4)$$
$$= 4.64\ \text{W}$$

(*b*) Radiation energy streaming through a solid angle unity in the direction $\theta = 60°$ with the normal to the surface of the hole is [see Eq. (12-7)]

$$Q = I_b(T)\,dA\cos\theta = \frac{\sigma T^4}{\pi}\,dA\cos\theta$$
$$= \frac{(5.67 \times 10^{-8})(800)^4}{\pi}(2 \times 10^{-4})(\cos 60°) = 0.74\ \text{W}$$

Example 12-3 The emission of radiation from a surface can be approximated as a blackbody radiation at $T = 1000$ K.

(*a*) What fraction of the total energy emitted is below $\lambda = 5\ \mu\text{m}$?

(*b*) What is the wavelength below which the emission is 10.5 percent of the total emission at 1000 K?

(*c*) What is the wavelength at which the maximum spectral emission occurs at $T = 1000$ K?

SOLUTION

(*a*) From Table 12-1, for $\lambda T = 5(1000) = 5000$, we obtain $f_{0-\lambda}(T) = 0.6337$. Then 63.37 percent of the total emission occurs below $\lambda = 5\ \mu\text{m}$.

(*b*) From Table 12-1, for $f_{0-\lambda}(T) = 0.105$, we find $\lambda T = 2222$. Then $\lambda = 2.222\ \mu\text{m}$.

(*c*) From Wien's displacement law, Eq. (12-13), we have

$$(\lambda T)_{max} = 2897.6\ \mu\text{m} \cdot \text{K}$$

Then for $T = 1000$ K, we obtain $\lambda = 2.8976\ \mu\text{m}$.

Example 12-4 A surface with $A = 2\ cm^2$ emits radiation as a blackbody at $T = 1000$ K.

(*a*) Calculate the radiation emitted into a solid angle subtended by $0 \le \phi \le 2\pi$ and $0 \le \theta \le \pi/6$.

(*b*) What fraction is the energy emitted into the above solid angle of that emitted into the entire hemispherical space?

SOLUTION

(*a*) The radiation energy emitted by an area A streaming through a differential solid angle $d\omega = \sin\theta\, d\theta\, d\phi$ in any direction is given by [see Eq. (12-7)]

$$AI_b(T)\cos\theta\sin\theta\, d\theta\, d\phi$$

The energy into the solid angle subtended by the angles $0 \le \phi \le 2\pi$ and $0 \le \theta \le \pi/6$ is obtained by the integration of this quantity:

$$\begin{aligned} Q &= AI_b(T)\int_{\phi=0}^{2\pi}\int_{\theta=0}^{\pi/6}\cos\theta\sin\theta\, d\theta\, d\phi \\ &= 2\pi AI_b(T)\int_{\theta=0}^{\pi/6}\cos\theta\sin\theta\, d\theta \\ &= 2\pi AI_b(T)[\tfrac{1}{2}\sin^2\theta]_0^{\pi/6} \\ &= \frac{\pi}{4}AI_b(T) = \tfrac{1}{4}A\sigma T^4 \\ &= \tfrac{1}{4}(2\times 10^{-4})(5.67\times 10^{-8})(1000)^4 \\ &= 2.835\ \text{W} \end{aligned}$$

(*b*) The energy emitted into the entire hemispherical space is

$$Q_0 = A\sigma T^4$$

Then

$$\frac{Q}{Q_0} = \frac{\frac{1}{4}A\sigma T^4}{A\sigma T^4} = \frac{1}{4}$$

That is, 25 percent of the total energy is emitted into the solid angle $0 \le \phi \le 2\pi$, $0 \le \theta \le \pi/6$.

12-3 RADIATION FROM REAL SURFACES

The spectral radiation intensity emitted by a real surface at temperature T of wavelength λ is always less than that emitted by a blackbody at the same temperature and wavelength. Furthermore, radiation intensity from a real surface depends on direction, whereas the blackbody radiation intensity is independent of direction. To distinguish these two cases, we use the symbol

$$I_\lambda(\theta, \phi) \qquad W/(m^2 \cdot \mu m \cdot sr) \tag{12-21}$$

to denote the *spectral radiation intensity from a real surface*. In fact, this intensity may vary with the position along the surface, but for simplicity of notation we omitted the space variable to appear as a separate variable in the symbol.

The radiation is emitted from a surface at all wavelengths. The integration of $I_\lambda(\theta, \phi)$ from $\lambda = 0$ to $\lambda = \infty$ gives

$$I(\theta, \phi) = \int_{\lambda=0}^{\infty} I_\lambda(\theta, \phi)\, d\lambda \tag{12-22}$$

Here $I(\theta, \phi)$ is called the *radiation intensity*.

In engineering applications, the radiation energy leaving a real surface per unit area per unit time in all directions into the hemispherical space above it is a quantity of practical interest. It is called the *radiation flux* from a surface. An expression defining this quantity in terms of the radiation intensity can be developed by a procedure similar to the derivation of blackbody emissive power.

We refer to the nomenclature in Fig. 12-4. Let $I_\lambda(\theta, \phi)$ be the spectral radiation intensity leaving the surface element in any given direction $\hat{\Omega}$. The spectral radiation energy leaving a unit surface area, streaming through a differential solid angle $d\omega = \sin\theta\, d\theta\, d\phi$ in any direction, is written in analogy to Eq. (12-10) as

$$I_\lambda(\theta, \phi) \cos\theta \sin\theta\, d\theta\, d\phi \tag{12-23}$$

The spectral radiation energy leaving a unit surface area in all directions into the hemispherical space is obtained by integrating Eq. (12-23) over the angles $0 \le \phi \le 2\pi$ and $0 \le \theta \le \pi/2$:

$$\boxed{q_\lambda = \int_{\phi=0}^{2\pi} \int_{\theta=0}^{\pi/2} I_\lambda(\theta, \phi) \cos\theta \sin\theta\, d\theta\, d\phi} \tag{12-24}$$

Here q_λ is called the *spectral radiation flux* from a surface, and it has the dimensions of $W/(m^2 \cdot \mu m)$.

The integration of Eq. (12-24) from $\lambda = 0$ to $\lambda = \infty$ gives

$$q = \int_{\lambda=0}^{\infty} q_\lambda\, d\lambda \qquad W/m^2 \tag{12-25}$$

where q is called the *radiation flux* from a surface.

Radiation Intensity Independent of Direction

When the spectral radiation intensity $I_\lambda(\theta, \phi) \equiv I_\lambda$ is independent of direction, Eq. (12-24) is integrated, to yield

$$q_\lambda = \pi I_\lambda \qquad W/(m^2 \cdot \mu m) \tag{12-26}$$

where q_λ is the spectral radiation flux.

The integration of Eq. (12-26) from $\lambda = 0$ to $\lambda = \infty$ yields

$$q = \pi \int_{\lambda=0}^{\infty} I_\lambda \, d\lambda \qquad \text{W/m}^2 \tag{12-27}$$

Thus, given the spectral distribution of the radiation leaving a surface, the radiation flux q from the surface can be calculated from Eq. (12-27).

12-4 RADIATION INCIDENT ON A SURFACE

In the preceding discussions, we focused our attention on the determination of the radiation energy emitted from a blackbody and that from a real surface. In engineering applications, radiation flux incident on a surface is also of interest.

We refer to the nomenclature shown in Fig. 12-6 for the discussion of radiation flux incident on a surface. Let $I_\lambda^i(\theta, \phi)$ be the spectral radiation intensity incident on a surface element dA in any given direction $\hat{\Omega}$. Here we use the superscript i to distinguish $I_\lambda^i(\theta, \phi)$ from the spectral radiation intensity $I_\lambda(\theta, \phi)$ leaving the surface. Then

$$I_\lambda^i(\theta, \phi) \cos\theta \, d\omega \tag{12-28a}$$

or

$$I_\lambda^i(\theta, \phi) \cos\theta \sin\theta \, d\theta \, d\phi \tag{12-28b}$$

represents the radiation energy incident per unit area of the surface from an irradiation through a differential solid angle $d\omega = \sin\theta \, d\theta \, d\phi$ in a given direction $\hat{\Omega}$. The $\cos\theta$ term appearing in Eq. (12-28) results from the fact that the irradiation energy on the surface is based on the actual area, whereas $I_\lambda^i(\theta, \phi)$ is based on the area normal to the direction of propagation.

The spectral radiation flux incident per unit area of the surface due to irradiation from all directions in the hemispherical space above it is determined by integrating Eq. (12-28b) over the angles $0 \le \phi \le 2\pi$ and $0 < \theta \le \pi/2$. We obtain

$$q_\lambda^i = \int_{\phi=0}^{2\pi} \int_{\theta=0}^{\pi/2} I_\lambda^i(\theta, \phi) \cos\theta \sin\theta \, d\theta \, d\phi \tag{12-29}$$

Here q_λ^i is called the *spectral incident radiation flux* on a surface, and it has the dimensions of W/(m$^2 \cdot \mu$m).

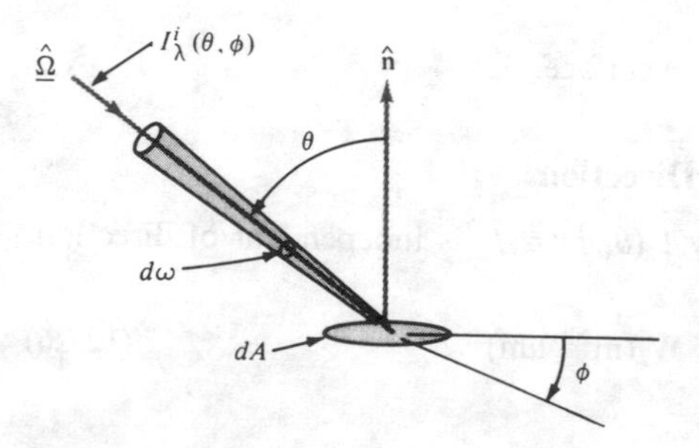

Figure 12-6 Nomenclature for radiation flux incident on a surface.

The integration of Eq. (12-29) from $\lambda = 0$ to $\lambda = \infty$ gives

$$q^i = \int_{\lambda=0}^{\infty} q_\lambda^i \, d\lambda \qquad \text{W/m}^2 \tag{12-30}$$

where q^i is called the *incident radiation flux.*

Radiation Intensity Independent of Direction

When the incident spectral radiation $I_\lambda^i(\theta, \phi) \equiv I_\lambda^i$ is independent of direction, Eqs. (12-29) is integrated to yield

$$q_\lambda^i = \pi I_\lambda^i \qquad \text{W/(m}^2 \cdot \mu\text{m)} \tag{12-31}$$

where q_λ^i is the incident spectral radiation flux.

The integration of Eq. (12-31) from $\lambda = 0$ to $\lambda = \infty$ gives

$$q^i = \pi \int_{\lambda=0}^{\infty} I_\lambda^i \, d\lambda \qquad \text{W/m}^2 \tag{12-32}$$

Given the spectral distribution of the incident radiation, radiation flux q^i incident on a surface can be calculated.

Example 12-5 A small surface of area $A = 5\ \text{cm}^2$ is subjected to irradiation of constant intensity $I^i = 1.8 \times 10^4\ \text{W/(m}^2 \cdot \text{sr)}$ over the solid angle subtended by $0 \le \phi \le 2\pi$, $0 \le \theta \le \pi/6$. Calculate the radiation energy incident on the surface.

SOLUTION The radiation energy incident on the surface through a differential solid angle $d\omega = \sin\theta \, d\theta \, d\phi$ is given by [see Eq. (12-28)]

$$AI^i \cos\theta \sin\theta \, d\theta \, d\phi$$

The total radiation energy Q^i incident on the surface is determined by integrating this quantity over the solid angle of incidence, that is, over the angles $0 \le \phi \le 2\pi$ and $0 \le \theta \le \pi/6$. We obtain

$$Q^i = AI^i \int_{\phi=0}^{2\pi} \int_{\theta=0}^{\pi/6} \cos\theta \sin\theta \, d\theta \, d\phi$$

$$= AI^i(2\pi)\left[\frac{1}{2}\sin^2\theta\right]_0^{\pi/6} = \frac{\pi}{4} AI^i$$

Substituting the numerical values, we obtain

$$Q^i = \frac{\pi}{4}(5 \times 10^{-4})(1.8 \times 10^4) = 7.07\ \text{W}$$

Example 12-6 A surface is irradiated uniformly from all directions in the hemispherical space. The spectral distribution of the incident radiation intensity I_λ^i is

$$
\begin{array}{ll}
0 < \lambda \leq 1\ \mu\text{m} & I_\lambda^i = 0 \\
1 < \lambda \leq 2\ \mu\text{m} & I_\lambda^i = 2000\ \text{W}/(\text{m}^2 \cdot \mu\text{m}) \\
2 < \lambda \leq 4\ \mu\text{m} & I_\lambda^i = 8000\ \text{W}/(\text{m}^2 \cdot \mu\text{m}) \\
4 < \lambda \leq 8\ \mu\text{m} & I_\lambda^i = 4000\ \text{W}/(\text{m}^2 \cdot \mu\text{m}) \\
\lambda > 8\ \mu\text{m} & I_\lambda^i = 0
\end{array}
$$

Calculate the radiation flux q^i incident on the surface.

SOLUTION Since the intensity of the incident radiation is independent of direction, the incident radiation flux q^i can be calculated by Eq. (12-32):

$$q^i = \pi \int_{\lambda=0}^{\infty} I_\lambda^i \, d\lambda \qquad \text{W/m}^2$$

For the stepwise spectral distribution given above, the integral is broken into parts as

$$
\begin{aligned}
q_i &= 2000 \int_1^2 d\lambda + 8000 \int_2^4 d\lambda + 4000 \int_4^8 d\lambda \\
&= 2000(2 - 1) + 8000(4 - 2) + 4000(8 - 4) \\
&= 34{,}000\ \text{W/m}^2
\end{aligned}
$$

12-5 RADIATION PROPERTIES OF SURFACES

The radiation emitted by a real body at a temperature T and wavelength λ is always less than that of the blackbody. Therefore, the blackbody emission is chosen as a reference, and a quantity called the *emissivity* of a surface is defined as the ratio of the energy emitted by a real surface to that by a blackbody at the same temperature; it has a value between zero and unity. Clearly, numerous possibilities exist for making such a comparison; for example, the comparison can be made at a given wavelength, over all wavelengths, for the energy emitted in a specified direction, or for the energy emitted into the hemispherical space. Here we consider the comparison only for the energy emitted into the hemispherical space for both a given wavelength and an average over all wavelengths. With this consideration, the following symbols are used; ε_λ = *spectral hemispherical emissivity* and ε = *hemispherical emissivity*.

A blackbody absorbs all radiation incident upon it at all wavelengths, whereas a real surface absorbs only part of it and the amount of absorption varies with the wavelength of radiation as well as the temperature at which the radiation is emitted. Therefore, a quantity called the *absorptivity* of a surface is defined as the fraction of the incident radiation absorbed by the surface. Clearly, numerous possibilities exist for such a definition; for example, the absorption can be considered at a given wavelength, over all wavelengths, for the energy incident in a given direction, or for the energy incident over all directions in the hemispherical space. Here, we consider only the situation in which the radiation is incident on the surface from all directions in the hemispherical space for a given wavelength and for the average over all wavelengths. With this consideration, the following symbols are used: α_λ = *spectral hemispherical absorptivity* and α = *hemispherical absorptivity*.

When radiation is incident on a real surface, a fraction of it is reflected by the surface. If the surface is perfectly smooth, that is, the roughness of the surface is much smaller than the wavelength of the radiation, the incident and the reflected rays lie symmetric with respect to the normal at the point of incidence, as illustrated in Fig. 12-7*a*. This mirrorlike reflection is called *specular* reflection. If the surface has some roughness, the incident radiation is scattered in all directions. An idealized reflection under such a situation is to assume that the intensity of the reflected radiation is constant for all angles of reflection and independent of the direction of the incident radiation, and it is called *diffuse* reflection. Figure 12-7*b* illustrates a diffuse reflection from a surface. Real surfaces encountered in engineering applications are neither perfectly diffuse nor perfectly specular. However, the concept is useful to study the effects of the two limiting cases on radiation transfer. The *reflectivity* of a surface is defined as the fraction of the incident radiation reflected by a surface. Numerous possibilities exist for the definition of reflectivity; for example, the reflection can be considered at a given wavelength, over all wavelengths, for the energy incident in a given direction, or for the energy incident over all directions in the hemispherical space. There is also the possibility of reflection being specular or diffuse. Here we consider only the diffuse reflection for situations in which the radiation is incident on the surface from all directions in the hemispherical space, for both a given wavelength and the average over all wavelengths. With this consideration, the following symbols are used: ρ_λ = *spectral hemispherical reflectivity* and ρ = *hemispherical reflectivity*.

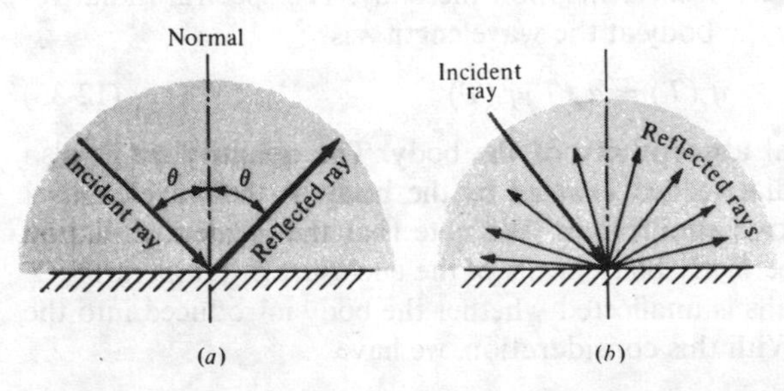

Figure 12-7 Reflection from surfaces. (*a*) Specular reflection, (*b*) diffuse reflection.

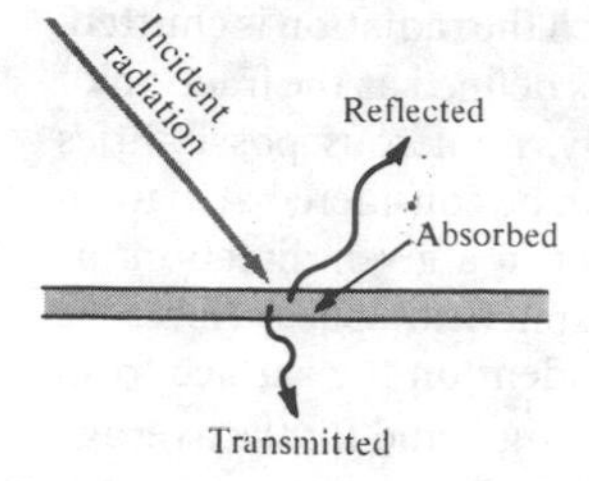

Figure 12-8 Reflection, absorption, and transmission of incident radiation by a semitransparent material.

Finally, if the body is *opaque* to radiation, the sum of the reflectivity and absorptivity of the body should equal unity:

$$\alpha_\lambda + \rho_\lambda = 1 \tag{12-33a}$$

$$\alpha + \rho = 1 \tag{12-33b}$$

If the body is *semitransparent* to radiation, the sum of absorptivity and reflectivity is less than unity, and the difference is called the *transmissivity* of the body. With this consideration we write

$$\alpha_\lambda + \rho_\lambda + \tau_\lambda = 1 \tag{12-34a}$$

$$\alpha + \rho + \tau = 1 \tag{12-34b}$$

where we defined τ_λ = the *spectral transmissivity* and τ = the *transmissivity*.

Figure 12-8 shows that a radiation beam incident on a semitransparent body of finite thickness, such as a glass plate, is partly reflected and absorbed, and the remainder is transmitted through the glass.

Kirchhoff's Law

The absorptivity and the emissivity of a body can be related by Kirchhoff's law of radiation. The reader should consult Ref. 6 for a rigorous derivation of this law.

Consider a body placed inside a perfectly black, closed container whose walls are maintained at a uniform temperature T and allowed to reach an equilibrium with the walls of the container. Let $q_\lambda^i(T)$ be the spectral radiative-heat flux from the walls at temperature T *incident* upon the body. The spectral radiative-heat flux $q_\lambda(T)$ *absorbed* by the body at the wavelength λ is

$$q_\lambda(T) = \alpha_\lambda(T) q_\lambda^i(T) \tag{12-35}$$

where $\alpha_\lambda(T)$ is the spectral absorptivity of the body. The quantity $q_\lambda(T)$ also represents the spectral radiative flux *emitted* by the body at the wavelength λ since the body is in radiative equilibrium. We note that the incident radiation $q_\lambda^i(T)$ is coming from the perfectly black walls of the enclosure at temperature T and the emission by the walls is unaffected whether the body introduced into the enclosure is a blackbody. With this consideration, we have

$$q_{\lambda b}(T) = q_\lambda^i(T) \tag{12-36}$$

where $q_{\lambda b}(T)$ is the spectral blackbody emissive flux at temperature T. From Eqs. (12-35) and (12-36) we write

$$\frac{q_\lambda(T)}{q_{\lambda b}(T)} = \alpha_\lambda(T) \tag{12-37}$$

The spectral emissivity $\varepsilon_\lambda(T)$ of the body for radiation at temperature T is defined as the ratio of the spectral emissive flux $q_\lambda(T)$ of the body to the spectral blackbody emissive flux $q_{\lambda b}(T)$ at the same temperature, that is,

$$\frac{q_\lambda(T)}{q_{\lambda b}(T)} = \varepsilon_\lambda(T) \tag{12-38}$$

From Eqs. (12-37) and (12-38) we obtain

$$\varepsilon_\lambda(T) = \alpha_\lambda(T) \tag{12-39}$$

which is the Kirchhoff law of radiation stating that *spectral emissivity for the emission of radiation at temperature T is equal to the spectral absorptivity for radiation coming from a blackbody at the same temperature T.*

Care must be exercised in the generalization of Eq. (12-39) to the averaged values of α and ε over the entire wavelength, that is, to the case

$$\varepsilon(T) = \alpha(T) \tag{12-40}$$

Equation (12-39) is always valid, but Eq. (12-40) is applicable when the incident and emitted radiation have the same spectral distribution or when the body is *gray*, that is, the radiative properties are independent of wavelength.

The application of Eq. (12-40) greatly simplifies the calculation of radiation heat exchange between the surfaces, as will be apparent later in this chapter.

Graybody

To simplify the analysis of radiative-heat transfer, the *graybody* assumption is frequently made in many applications; that is, the radiative properties α_λ, ε_λ, and ρ_λ are assumed to be uniform over the entire wavelength spectrum. Such bodies are referred to as *graybodies*, and under the graybody assumption the absorptivity and the emissivity are related by the Kirchhoff law as $\alpha = \varepsilon$.

Emissivity

If $q(T)$ is the spectral radiation flux emitted from a real surface at a temperature T and $E_{b\lambda}(T)$ is the blackbody emissive power (i.e., flux) at the same temperature T, then the *spectral hemispherical emissivity* ε_λ of the surface is defined as

$$\boxed{\varepsilon_\lambda = \frac{q_\lambda(T)}{E_{b\lambda}(T)}} \tag{12-41}$$

where $E_{b\lambda}(T)$ is defined by Eq. (12-12) or listed in Table 12-1; the spectral radiation flux emitted by a real surface $q_\lambda(T)$ is given by Eq. (12-24).

The average value of ε_λ over all wavelengths, called the *hemispherical emissivity* ε, is defined as

$$\varepsilon = \frac{\int_0^\infty \varepsilon_\lambda E_{b\lambda}(T)\,d\lambda}{\int_0^\infty E_{b\lambda}(T)\,d\lambda} = \boxed{\frac{\int_0^\infty \varepsilon_\lambda E_{b\lambda}(T)\,d\lambda}{E_b(T)}} \tag{12-42}$$

If ε_λ is known as a function of wavelength, Eq. (12-42) can be used to compute ε. Note that in this averaging process, the spectral blackbody emissive power $E_{b\lambda}(T)$ serves as the weight factor.

Absorptivity

If $q_\lambda^i(T)$ is the spectral radiation flux incident on a surface and $q_\lambda^a(T)$ is the amount of radiation absorbed by the surface, then the spectral hemispherical absorptivity α_λ is defined as

$$\boxed{\alpha_\lambda = \frac{q_\lambda^a(T)}{q_\lambda^i(T)}} \tag{12-43}$$

where $q_\lambda^i(T)$ is given by Eq. (12-29).

The average value of α_λ over all wavelengths, called the *hemispherical absorptivity* α, is defined as

$$\boxed{\alpha = \frac{\int_0^\infty \alpha_\lambda q_\lambda^i(T)\,d\lambda}{\int_0^\infty q_\lambda^i(T)\,d\lambda}} \tag{12-44}$$

Given α_λ as a function of wavelength, Eq. (12-44) can be used to compute α.

We note that the absorptivity α depends on the spectral distribution of the incident radiation $q_\lambda^i(T)$, hence $q_\lambda^i(T)$ is used as the weight factor; but the emissivity depends on the surface temperature, and so the spectral blackbody emissive power $E_{b\lambda}(T)$ at the surface temperature is used as the weight factor in Eq. (12-42).

Reflectivity

If $q_\lambda^i(T)$ is the spectral radiation flux incident on a surface and $q_\lambda^r(T)$ is the amount of radiation reflected by the surface, then the spectral hemispherical reflectivity ρ_λ is defined as

$$\boxed{\rho_\lambda = \frac{q_\lambda^r(T)}{q_\lambda^i(T)}} \tag{12-45}$$

where $q_\lambda^i(T)$ is given by Eq. (12-29).

The average value of ρ_λ over all wavelengths, called the *hemispherical reflectivity* ρ, is defined as

$$\boxed{\rho = \frac{\int_0^\infty \rho_\lambda q_\lambda^i(T)\, d\lambda}{\int_0^\infty q_\lambda^i(T)\, d\lambda}} \tag{12-46}$$

Given ρ_λ as a function of the wavelength, Eq. (12-46) can be used to compute ρ. In this averaging process, the incident spectral radiation flux $q_\lambda^i(T)$ serves as the weight factor.

Transmissivity

The analysis of transmissivity of a semitransparent body is, in general, a complicated matter, because radiation incident upon a semitransparent body penetrates into the depth of the medium where it is attenuated as a result of absorption and in some cases scattering by the material. Therefore, transmissivity depends on the radiation properties of the material, its thickness, and the conditions at the outer surfaces [1–5]. However, in engineering applications there are many situations, such as the transmission of radiation through a sheet of glass, in which the *spectral hemispherical transmissivity* τ_λ is defined as

$$\boxed{\tau_\lambda = \frac{q_\lambda^{tr}(T)}{q_\lambda^i(T)}} \tag{12-47}$$

where $q_\lambda^i(T)$ and $q_\lambda^{tr}(T)$ are the incident and the transmitted radiation fluxes, respectively.

Given the spectral distribution of τ_λ, the *hemispherical transmissivity* τ is determined from

$$\boxed{\tau = \frac{\int_0^\infty \tau_\lambda q_\lambda^i(T)\, d\lambda}{\int_0^\infty q_\lambda^i(T)\, d\lambda}} \tag{12-48}$$

Data on Radiation Properties

There is a vast amount of data on radiation properties of *opaque* surfaces [8–14]. In Appendix Table C-2, we tabulate the hemispherical emissivity ε of various opaque surfaces. To illustrate the effects of various parameters on emissivity, we present some typical experimental data.

Figure 12-9 illustrates the effects of temperature and oxidation on the hemispherical emissivity of opaque surfaces. Clearly, oxidation increases the emissivity. The temperature of the surface also has an effect on the emissivity, but it is less pronounced for highly polished metallic surfaces.

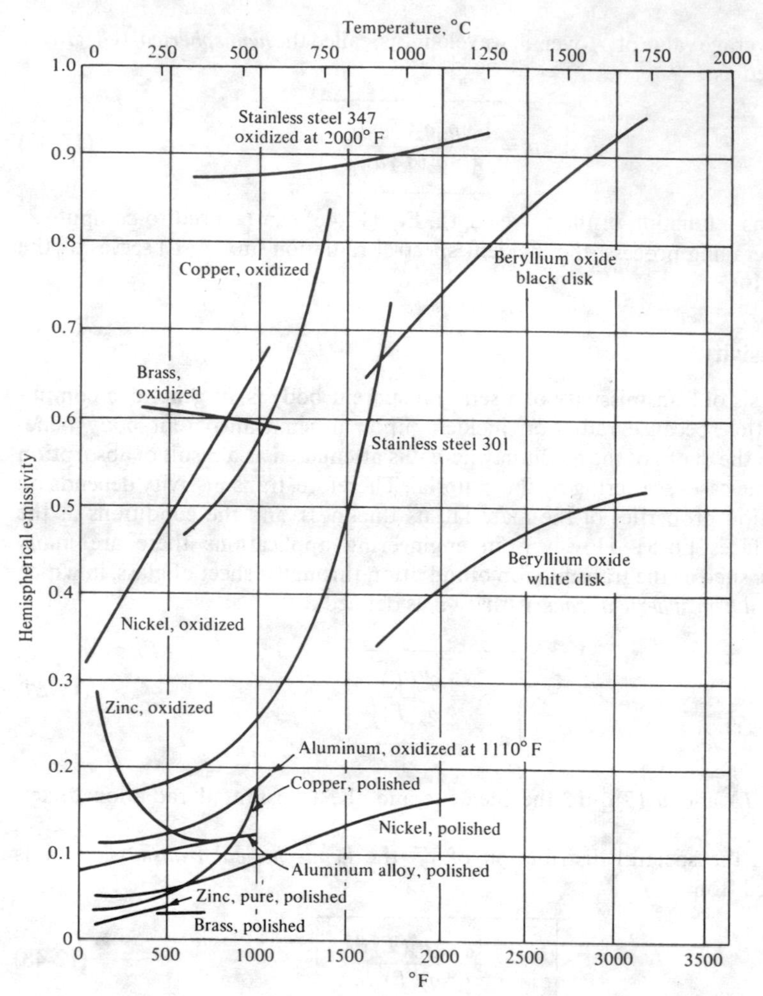

Figure 12-9 Effects of temperature and oxidation on hemispherical emissivity of metals. (*Based on data from Gubareff, Janssen, and Torborg* [8].)

Figure 12-10 shows the effect of the source temperature of the incident radiation (i.e., temperature at which the incident radiation is emitted) on the hemispherical absorptivity of various common materials at room temperature. We note that the absorptivity of nonmetallic substances significantly changes with the source temperature of the incident radiation. Relatively little data are available in the literature on the effects of source temperature on the absorptivity of surfaces, except for the case of solar radiation.

Figure 12-11 shows the spectral emissivity ε_λ as a function of wavelength for several materials.

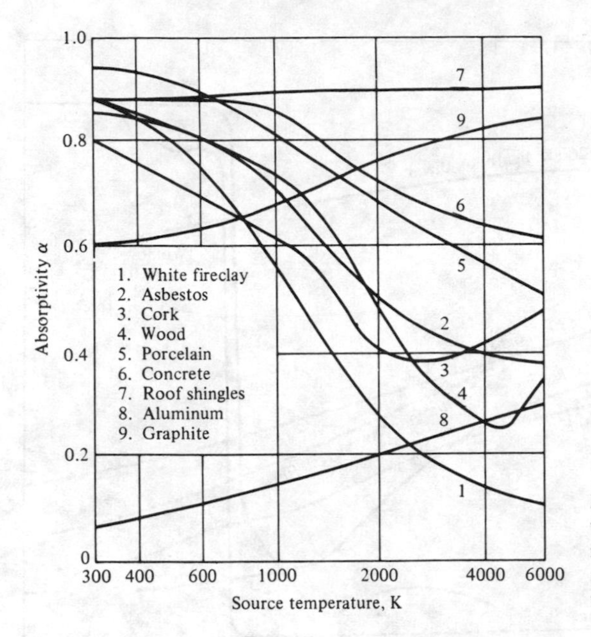

Figure 12-10 Variation of absorptivity with source temperature of incident radiation for various common materials at room temperature. (*From Sieber* [*14*].)

Example 12-7 The spectral hemispherical emissivity of firebrick at $T = 750$ K as a function of wavelength is approximated in stepwise variation as follows:

$$\varepsilon_1 = 0.1 \qquad \text{for } \lambda_0 = 0 \text{ to } \lambda_1 \leq 2\ \mu\text{m}$$

$$\varepsilon_2 = 0.6 \qquad \text{for } \lambda_1 = 2 \text{ to } \lambda_2 \leq 14\ \mu\text{m}$$

$$\varepsilon_3 = 0.8 \qquad \text{for } \lambda_2 = 14 \text{ to } \lambda_3 \to \infty$$

Calculate the hemispherical emissivity ε over all wavelengths.

SOLUTION The spectral distribution of emissivity is given in stepwise variations. Therefore, in applying Eq. (12-42) we break the integral into parts as

$$\varepsilon = \frac{\int_0^\infty \varepsilon_\lambda E_{b\lambda}(T)\, d\lambda}{E_b(T)}$$

$$= \varepsilon_1 \int_0^{\lambda_1} \frac{E_{b\lambda}(T)}{E_b(T)}\, d\lambda + \varepsilon_2 \int_{\lambda_1}^{\lambda_2} \frac{E_{b\lambda}(T)}{E_b(T)}\, d\lambda + \varepsilon_3 \int_{\lambda_2}^{\infty} \frac{E_{b\lambda}(T)}{E_b(T)}\, d\lambda$$

$$= \varepsilon_1 f_{0-\lambda_1} + \varepsilon_2 (f_{0-\lambda_2} - f_{0-\lambda_1}) + \varepsilon_3 (f_{0-\infty} - f_{0-\lambda_2})$$

where $f_{0-\lambda}$ are listed in Table 12-1. We have

$$\lambda_1 T = 2(750) = 1500 \qquad \therefore f_{0-\lambda_1} = 0.013$$

$$\lambda_2 T = 14(750) = 10{,}500 \qquad \therefore f_{0-\lambda_2} = 0.924$$

$$\lambda_3 T \to \infty \qquad \therefore f_{0-\infty} = 1$$

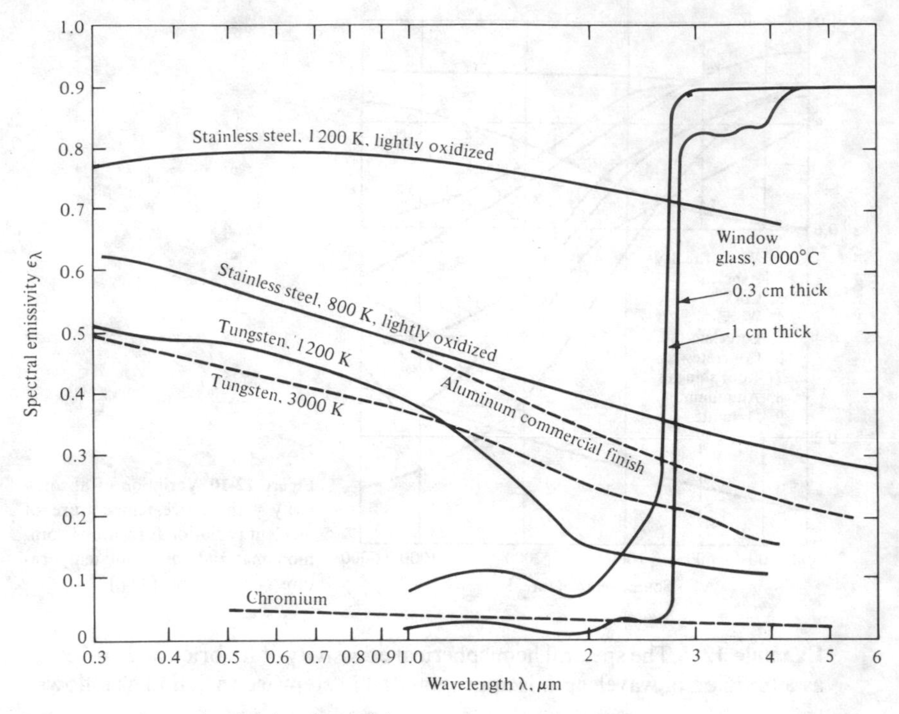

Figure 12-11 Spectral emissivity of materials.

Then

$$\begin{aligned}\varepsilon &= 0.1(0.013) + 0.6(0.924 - 0.013) + 0.8(1 - 0.924)\\ &= 0.609\end{aligned}$$

Example 12-8 The spectral transmissivity of plate glass for incident solar radiation is approximated in stepwise variation as

$$\tau_1 = 0 \qquad \text{for } \lambda_0 = 0 \text{ to } \lambda_1 \leq 0.4\ \mu\text{m}$$

$$\tau_2 = 0.8 \qquad \text{for } \lambda_1 = 0.4 \text{ to } \lambda_2 \leq 3.0\ \mu\text{m}$$

$$\tau_3 = 0 \qquad \text{for } \lambda_2 = 3.0 \text{ to } \lambda_3 \to \infty$$

Calculate the transmissivity of the glass over all wavelengths.

SOLUTION We apply Eq. (12-48) by noting that the incident radiation is a blackbody radiation at the effective surface temperature of the sun (that is, $T \cong 5700$ K). Therefore, Eq. (12-48) is written as

$$\tau = \frac{\int_0^\infty \tau_\lambda E_{\lambda b}(T)\, d\lambda}{E_b(T)}$$

The spectral distribution of τ_λ is given in a stepwise variation, so we break this integral into parts as

$$\tau = \tau_1 \int_0^{\lambda_1} \frac{E_{b\lambda}(T)}{E_b(T)} d\lambda + \tau_2 \int_{\lambda_1}^{\lambda_2} \frac{E_{b\lambda}(T)}{E_b(T)} d\lambda + \tau_3 \int_{\tau_2}^{\infty} \frac{E_{b\lambda}(T)}{E_b(T)} d\lambda$$

For $\tau_1 = \tau_3 = 0$, this relation reduces to

$$\tau = \tau_2(f_{0-\lambda_2} - f_{0-\lambda_1})$$

Taking the sun's surface temperature as $T \cong 5760$ K, we obtain

$$\lambda_1 T = 0.4(5760) = 2304 \qquad \therefore \qquad f_{0-\lambda_1} = 0.125$$

$$\lambda_2 T = 3.0(5760) = 17{,}280 \qquad \therefore \qquad f_{0-\lambda_2} = 0.977$$

Then

$$\tau = 0.8(0.977 - 0.125) \cong 0.68$$

Example 12-9 The filament of a light bulb is assumed to emit radiation as a blackbody at $T = 2400$ K. If the bulb glass has a transmissivity of $\tau = 0.90$ for the radiation emitted by the filament in the visible range, calculate the percentage of the total energy emitted by the filament that reaches the ambient as visible light.

SOLUTION The visible range of the spectrum is taken from $\lambda_1 = 0.38$ to $\lambda_2 = 0.76$ μm. The fraction F of the total energy emitted by the filament that reaches the ambient as light becomes

$$F = \tau \frac{\int_{\lambda_1}^{\lambda_2} E_{b\lambda}(T)\, d\lambda}{E_b(T)} = \tau \int_0^{\lambda_2} \frac{E_{b\lambda}(T)}{E_b(T)} d\lambda - \tau \int_0^{\lambda_1} \frac{E_{b\lambda}(T)}{E_b(T)} d\lambda$$

$$= \tau(f_{0-\lambda_2} - f_{0-\lambda_1})$$

and

$$\lambda_1 T = 0.38(2400) = 912 \qquad \therefore \qquad f_{0-\lambda_1} \cong 0.0002$$

$$\lambda_2 T = 0.76(2400) = 1824 \qquad \therefore \qquad f_{0-\lambda_2} = 0.0436$$

Then

$$F = 0.9(0.0436 - 0.0002) = 0.039$$

Only 3.9 percent of the total energy enters the ambient as light; the remaining energy produces heating.

12-6 SOLAR RADIATION

The sun's energy is created in the interior regions of the sun as a result of a continuous *fusion* reaction. Almost 90 percent of this energy is generated in a region 0.23 times the radius of the sun and then transferred by radiation up to a distance

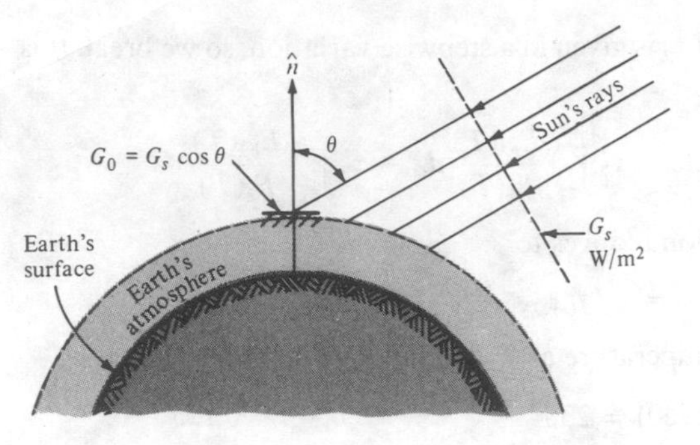

Figure 12-12 Solar constant G_s and extraterrestrial solar irradiation G_o.

about 0.7 times the radius of the sun. Outside this region there is the *convective zone* where the temperature is in the range of 6000 K. The relative coolness of the outer surface of the sun is an indication that the energy created in the interior is dissipated by radiation from the outer surface of the sun. Thus, the sun with its radius $R \cong 6.96 \times 10^5$ km and weight $M \cong 1.99 \times 10^{30}$ kg is almost an inexhaustible source of energy for the earth. Only a small fraction of the sun's energy reaches the earth's surface because of the large distance between them. The intensity of solar radiation reaching the earth's atmosphere has been determined more accurately by a series of high-altitude measurements made by using balloons, aircraft, and spacecraft from 1967 to 1970. The resulting energy, known as the *solar constant* G_s, is established as [15]

$$G_s = 1353 \text{ W/m}^2 \tag{12-49}$$

This quantity represents the *incident solar radiation flux* on a plane normal to the sun's rays, just outside the earth's atmosphere when the earth is at its mean distance from the sun. As the earth moves around the sun in a slightly elliptical orbit, the distance between them varies from 98.3 percent of the mean distance when the earth is closest to the sun to 101.7 percent of the mean distance when the earth-sun distance is maximum. Therefore, the actual value of G_s varies approximately by ± 3.4 percent, that is, from a maximum of 1399 W/m^2 on December 21 to a minimum of 1310 W/m^2 on June 21. However, for practical purposes, such a variation in actual G_s is neglected, and it is taken as a constant equal to 1353 W/m^2. Then the solar energy G_o incident normal to the outer surface of the earth's atmosphere becomes

$$G_o = G_s \cos\theta \qquad \text{W/m}^2 \tag{12-50}$$

where G_o is called the *extraterrestrial* solar irradiation. Figure 12-12 illustrates the physical significance of G_s and G_o in relation to the direction of the solar beam.

The value of G_s can be utilized in the blackbody radiation law to establish an effective temperature T_s for the sun's surface:

$$G_s = \left(\frac{r}{R}\right)^2 \sigma T_s^4 \tag{12-51}$$

where $G_s = 1353\ \text{W/m}^2$
$r = 6.9598 \times 10^8$ m, radius of solar disk
$R = 1.496 \times 10^{11}$ m, mean earth-sun distance
$\sigma = 5.6697 \times 10^{-8}\ \text{W/(m}^2 \cdot \text{K}^4)$ Stefan-Boltzmann constant

Then the effective temperature of the sun's surface is determined as $T_s = 5762$ K.

The solar radiation impinging on the upper surface of the earth's atmosphere must propagate through the earth's atmosphere before reaching the surface. Approximately 99 percent of the atmosphere is contained within a distance of about 30 km from the earth's surface. As the solar radiation passes through the atmosphere, it is *absorbed* and *scattered* by the atmospheric material. Figure 12-13 shows the spectral distribution of the solar radiation $G_{s\lambda}$ just outside the earth's atmosphere and on the ground level under clear atmosphere. We note that the total energy contained under the $G_{s\lambda}$ curve represents the solar radiation flux just outside the earth's atmosphere, that is,

$$\int_0^\infty G_{s\lambda}\, d\lambda = G_s = 1353\ \text{W/m}^2 \tag{12-52}$$

The spectral distribution curve for the solar radiation reaching the earth's surface lies below the $G_{s\lambda}$ curve and exhibits several dips. The reason for this is the absorption of the solar radiation by O_3, O_2, CO_2, and H_2O at various wavelengths. The ozone (O_3), which is concentrated in a layer 10 to 30 km from the earth's surface, strongly absorbs the ultraviolet radiation in the range from $\lambda = 0.2$ to $\lambda = 0.29\ \mu$m and relatively strongly in the range from 0.29 to 0.34 μm. As a result, negligible solar radiation with wavelengths less than about 0.3 μm reaches the earth's surface. Thus the biological systems on the earth are protected from the damaging ultraviolet radiation. The oxygen absorption occurs in a very narrow line centered at $\lambda = 0.76\ \mu$m. The absorption bands due to water vapor are distinctly visible in the range 0.7 to 2.2 μm. Carbon dioxide and water vapor strongly absorb the thermal radiation in wavelengths larger than about 2.2 μm. As a result, the solar radiation reaching the earth's surface is essentially contained in the wavelengths between 0.29 and 2.5 μm. The total energy contained under the curve

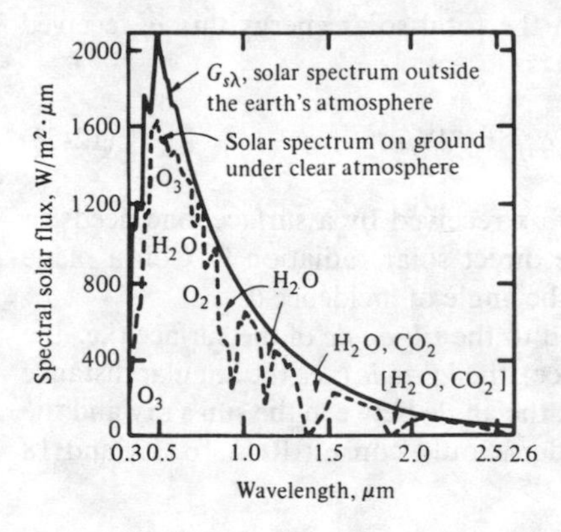

Figure 12-13 Effects of atmospheric attenuation on the spectral distribution of the solar radiation. (*From Thekaekara* [*15*].)

for the solar spectrum on the earth's surface under a clear atmosphere is about 956 W/m^2. This is considerably less than the solar constant 1353 W/m^2 just outside the earth's atmosphere.

In addition to the absorption of the solar radiation, there is the scattering of the solar radiation by air molecules, water droplets contained in the clouds, and aerosols or dust particles as it passes through the atmosphere. The air molecules scatter the solar radiation in very short wavelengths comparable to the size of molecules, and such scattering is called the *Rayleigh scattering.* Water droplets, aerosols, and other atmospheric turbidity scatter radiation in wavelengths comparable to the diameter of such particles.

The part of the solar radiation that is neither scattered nor absorbed by the atmosphere and that reaches the earth's surface as a beam is called *direct solar radiation.* The scattered part of the radiation that reaches the earth's surface from all directions over the sky is called *diffuse solar radiation.* Thus the solar radiation received by the earth's surface is composed of *direct* and *diffuse* components. The diffuse part varies from about 10 percent of the total on a clear day to almost 100 percent on a totally cloudy day.

Solar Radiation Striking the Earth

The amount of solar energy received by a surface on the ground level depends on the orientation of the surface in relation to the sun, the hour of the day, the day of the year, the latitude of the point of observation, and the atmospheric conditions. In the early morning or late afternoon, the solar radiation reaching the earth's surface follows an oblique, longer path through the atmosphere; as a result, the atmospheric attenuation is greater and the intensity is significantly reduced.

The total solar energy flux q_t received per unit area of a surface at ground level consists of direct and diffuse components. Let q_{df} (in watts per square meter) be the *diffuse solar radiation* incident on a horizontal surface due to irradiation from the entire hemispherical space and q_D be the *direct solar radiation* flux per unit area normal to the direction of the solar beam on ground level. Let θ be the *angle of incidence*, that is, the angle between the sun's ray and the normal to a surface, as illustrated in Fig. 12-14. Then the total solar energy flux q_t received per unit area of the surface on ground level is

$$q_t = q_D \cos\theta + q_{df} \qquad \text{W/m}^2 \tag{12-53}$$

Thus, to calculate the total solar energy flux received by a surface, one needs to know the diffuse solar radiation flux, the direct solar radiation flux on a plane normal to the direction of the beam, and the angle of incidence θ.

The angle of incidence θ can be related to the *tilt angle* of the surface (i.e., the angle between the horizontal and the surface), the *latitude* (i.e., the angular distance from the equator), and the *declination* (i.e., the angle between the sun's ray and the equatorial plane at solar noon). The reader should consult Refs. 16, 17, and 18 for a detailed discussion of this subject.

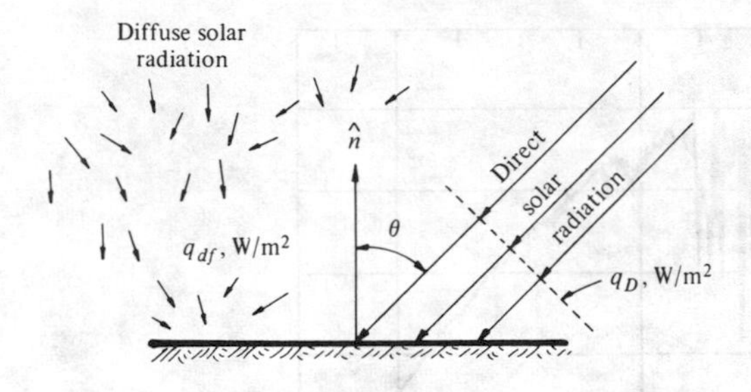

Figure 12-14 Solar radiation received at the earth's surface.

The solar energy striking an opaque surface is partly absorbed by the surface, and the remainder is reflected. In Appendix Table C-2 we present the solar absorptivity of various surfaces.

Solar Radiation Measurements

The daily and hourly records of the amount of solar radiation received at any given location over the earth's surface are essential for the design and optimization of heat transfer systems utilizing solar energy. Such information is also useful for architectural, agricultural, biological, and other purposes. Therefore, solar radiation measurements are continuously made from radiation-monitoring networks located at different parts of the world. Such measurements generally include

1. The direct solar radiation flux q_D at normal incidence
2. The diffuse solar radiation flux q_{df} from the entire sky on a horizontal surface
3. The total (or global) solar radiation flux q_t, which is the sum of direct and diffuse solar radiation received by a horizontal surface
4. The total solar radiation flux on an inclined surface having a specified orientation

In addition, the spectral distribution of solar radiation over certain wavelength bands and the amount of solar radiation reflected from the ground are of interest in certain situations.

Therefore, the solar radiation measurements are made continuously by a system of national and international solar radiation measuring centers at different parts of the world. The reader should consult Refs. 19 to 21 for a discussion of available solar radiation instruments and observing practices.

To illustrate the effects of weather conditions and time of the day on the total (i.e., global) solar radiation flux q_t received by a horizontal surface, we present in Figs. 12-15 and 12-16 the measurements made for a clear and cloudy day, respectively, at Greenbelt, Maryland, reported in Ref. 22. The clear-day record in Fig. 12-15 shows that the peak radiation flux of 1000 W/m^2 occurs near noon, and the sharp dips on the curve are due to occasional clouds blocking the sun. A similar record for a cloudy day is shown in Fig. 12-16. The sharp peaks in mid-afternoon results from the sun coming out briefly. We note that the occasional

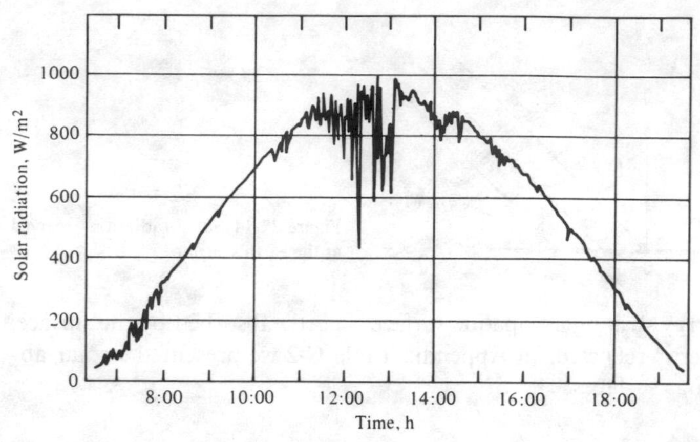

Figure 12-15 Total solar radiation on a horizontal surface, measured at Greenbelt, Maryland, on May 14, 1971. (*From Ref. 22.*)

maximum near 1200 W/m^2, just after 14:00, is 30 percent higher than the clear-sky maximum of 900 W/m^2 for that time of the day. This is due to reflection from the clouds.

Atmospheric Emission

The solar radiation passing through the atmosphere is attenuated owing to absorption by certain atmospheric constituents. Therefore, the atmosphere should emit thermal radiation due to the temperature of such constituents. Also CO_2 and

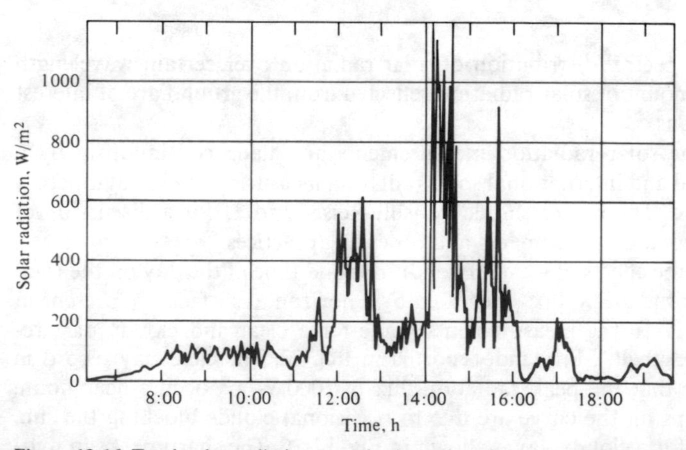

Figure 12-16 Total solar radiation on a horizontal surface, measured at Greenbelt, Maryland, on May 14, 1971. (*From Ref. 22.*)

H_2O are the two principal constituents that give rise to emission over bands in the regions of 5 to 8 μm and above 13 μm. Although such an emission is not a black-body emission, for convenience in the analysis, an *effective sky temperature* T_{sky} has been introduced. The emission of the atmosphere to the earth's surface is represented by

$$q_{sky} = \sigma T_{sky}^4 \qquad \text{W/m}^2 \tag{12-54}$$

The magnitude of this fictitious temperature T_{sky} depends on the atmospheric conditions. The values assigned to it vary from 230 K for cold, clear-sky conditions to about 285 K for warm, cloudy-sky conditions.

Example 12-10 A thin layer of water is exposed to a clear sky in a desert at night. The effective sky temperature is $T_{sky} = 255$ K, the ambient air is at $T_\infty = 278$ K, the heat transfer coefficient for convection is $h = 5$ W/(m$^2 \cdot$ °C), and the emissivity and absorptivity of water to long-wave radiation are $\varepsilon = \alpha = 0.95$. Calculate the equilibrium temperature of water.

SOLUTION An energy balance on the water surface can be stated as

$$\begin{pmatrix}\text{Heat loss}\\ \text{by emission}\end{pmatrix} + \begin{pmatrix}\text{heat loss}\\ \text{by convection}\end{pmatrix} = \begin{pmatrix}\text{heat gain from}\\ \text{sky radiation}\end{pmatrix}$$

$$\varepsilon\sigma T^4 + h(T - T_\infty) = \alpha\sigma T_{sky}^4$$

$$\varepsilon\sigma(T^4 - T_{sky}^4) = h(T_\infty - T)$$

since $\varepsilon = \alpha$. Substituting the numerical values, we obtain

$$0.95 \times 5.67 \times 10^{-8}(T^4 - 255^4) = 5(278 - T)$$

The solution yields the equilibrium temperature of water as

$$T = 268 \text{ K}$$

This result implies that the water will freeze, even though the ambient air temperature T_∞ is above the freezing temperature of water.

Example 12-11 A flat-plate solar collector without a cover receives direct solar radiation flux $q_D = 700$ W/m^2 at an incidence angle $\theta = 30°$ while the diffuse solar radiation flux is $q_{df} = 100$ W/m^2. The collector surface temperature is $T_s = 370$ K, the effective sky temperature is $T_{sky} = 280$ K, and the ambient air temperature is $T_\infty = 300$ K. The collector plate has an absorptivity for solar radiation $\alpha_s = 0.95$, while it has a selective absorber surface with emissivity and absorptivity for the long-wave radiation of $\alpha = \varepsilon = 0.2$. The heat transfer coefficient for convection is $h = 4$ W/(m$^2 \cdot$ °C). Calculate the rate of heat removal per square meter of collector surface.

SOLUTION The heat removal rate Q by the collector is

$$Q = \begin{pmatrix}\text{Solar energy}\\ \text{absorbed}\end{pmatrix} + \begin{pmatrix}\text{sky radiation}\\ \text{absorbed}\end{pmatrix} - \begin{pmatrix}\text{radiation}\\ \text{emitted}\end{pmatrix} - \begin{pmatrix}\text{convection to}\\ \text{ambient air}\end{pmatrix}$$

$$= \alpha_s(q_D \cos\theta + q_{df}) + \alpha\sigma T_{sky}^4 - \varepsilon\sigma T_s^4 - h(T_s - T_\infty)$$

Substituting the numerical values, we obtain

$$Q = 0.95(700 \cos 30^\circ + 100) + 0.2(5.67 \times 10^{-8})(280^4 - 370^4) - 4(370 - 300)$$
$$= 670.9 - 142.8 - 280$$
$$= 248.1 \text{ W/m}^2$$

12-7 CONCEPT OF VIEW FACTOR

So far we discussed radiation to or from a single surface. However, in engineering applications, problems of practical interest involve radiation exchange between two or more surfaces. When the surfaces are separated by a nonparticipating medium that does not absorb, emit, or scatter radiation, then the radiation exchange among the surfaces is unaffected by the medium. A vacuum, for example, is a perfect nonparticipating medium; however, air and many gases closely approximate this condition. For any two given surfaces, the orientation between them affects the fraction of the radiation energy leaving one surface that strikes the other surface directly. Therefore, the orientation of the surfaces plays an important role in radiation heat exchange.

To formalize the effects of orientation in the analysis of radiation heat exchange among the surfaces, the concept of *view factor* has been adopted. The terms *shape factor*, *angle factor*, and *configuration factor* also have been used in the literature. A distinction should be made between a *diffuse view factor* and a *specular view factor*. The former refers to the situation in which the surfaces are diffuse reflectors and diffuse emitters, whereas the latter refers to the situation in which the surfaces are diffuse emitters and specular reflectors. In this book we consider only the cases in which the surfaces are diffuse emitters and diffuse reflectors; therefore, we do not need such a distinction. We use simply the term *view factor*, and it will imply the diffuse view factor.

The physical significance of the view factor between two surfaces is that it represents the *fraction of the radiative energy leaving one surface that strikes the other surface directly*.

View Factor Between Two Elemental Surfaces

To provide some insight to the development of the relations defining the view factors, we derive the expression defining the view factor between two elemental surfaces.

Consider two elemental surfaces dA_1 and dA_2, as illustrated in Fig. 12-17. Let r be the distance between these two surfaces, θ_1 be the polar angle between the

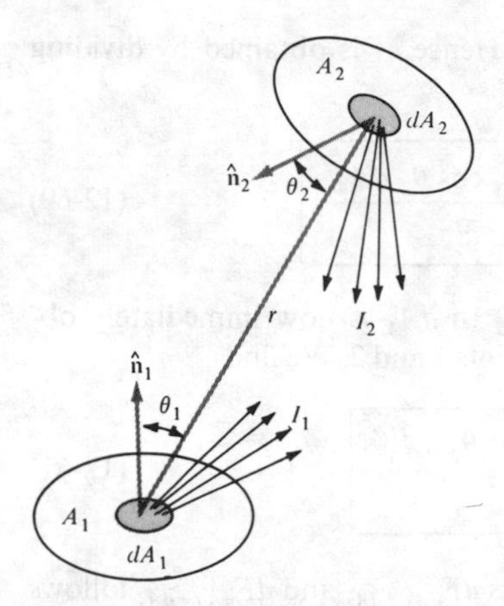

Figure 12-17 Coordinates for the definition of view factor.

normal $\hat{\mathbf{n}}_1$ to the surface element dA_1 and the line r joining dA_1 to dA_2, and θ_2 be the polar angle between the normal $\hat{\mathbf{n}}_2$ to the surface element dA_2 and the line r.

Let $d\omega_{12}$ be the solid angle under which an observer at dA_1 sees the surface element dA_2 and I_1 be the intensity of radiation leaving the surface element diffusely in all directions in hemispherical space. The rate of radiative energy dQ_1 leaving dA_1 that strikes dA_2 is

$$dQ_1 = dA_1\, I_1 \cos\theta_1\, d\omega_{12} \tag{12-55}$$

where the solid angle $d\omega_{12}$ is given by

$$d\omega_{12} = \frac{dA_2 \cos\theta_2}{r^2} \tag{12-56}$$

The substitution of Eq. (12-56) into Eq. (12-55) results in

$$dQ_1 = dA_1\, I_1 \frac{\cos\theta_1 \cos\theta_2\, dA_2}{r^2} \tag{12-57}$$

The rate of radiation energy Q_1 leaving the surface element dA_1 in all directions over hemispherical space is

$$Q_1 = dA_1 \int_{\phi=0}^{2\pi} \int_{\theta_1=0}^{\pi/2} I_1 \cos\theta_1 \sin\theta_1\, d\theta_1\, d\phi \tag{12-58}$$

where ϕ is the azimuthal angle. For a diffusively reflecting and diffusely emitting surface, the radiation intensity leaving the surface is independent of direction. Then, for constant I_1, Eq. (12-58) is integrated to give

$$Q_1 = \pi I_1\, dA_1 \tag{12-59}$$

The elemental view factor $dF_{dA_1-dA_2}$, by definition, is the ratio of the radiative energy leaving dA_1 that strikes dA_2 directly to the radiative energy leaving dA_1

in all directions into the hemispherical space. Hence, it is obtained by dividing Eq. (12-57) by Eq. (12-59) to yield

$$dF_{dA_1-dA_2} = \frac{dQ_1}{Q_1} = \frac{\cos\theta_1 \cos\theta_2 \, dA_2}{\pi r^2} \tag{12-60}$$

The elemental view factor $dF_{dA_2-dA_1}$ from dA_2 to dA_1 is now immediately obtained from Eq. (12-60) by interchanging subscripts 1 and 2. We find

$$dF_{dA_2-dA_1} = \frac{\cos\theta_1 \cos\theta_2 \, dA_1}{\pi r^2} \tag{12-61}$$

The *reciprocity relation* between the view factors $dF_{dA_1-dA_2}$ and $dF_{dA_2-dA_1}$ follows from Eqs. (12-60) and (12-61) as

$$dA_1 \, dF_{dA_1-dA_2} = dA_2 \, dF_{dA_2-dA_1} \tag{12-62}$$

This relation implies that for two elemental surfaces dA_1 and dA_2, when one of the view factors is known, the other is readily computed by the reciprocity relation.

View Factor for Finite Surfaces

We developed the view factor between two elemental surfaces dA_1 and dA_2. These results are now generalized to obtain view factors between an elemental surface dA_1 and a finite surface A_2 or between two finite surfaces A_1 and A_2.

The view factor $F_{dA_1-A_2}$, from dA_1 to A_2, is immediately determined by integrating the elemental view factor $dF_{dA_1-dA_2}$ given by Eq. (12-60) over the area A_2 as

$$F_{dA_1-A_2} = \int_{A_2} \frac{\cos\theta_1 \cos\theta_2}{\pi r^2} \, dA_2 \tag{12-63}$$

The view factor $F_{A_2-dA_1}$ from A_2 to dA_1 is obtained by integrating Eq. (12-61) over the area A_2 and dividing it by A_2:

$$F_{A_2-dA_1} = \frac{dA_1}{A_2} \int_{A_2} \frac{\cos\theta_1 \cos\theta_2}{\pi r^2} \, dA_2 \tag{12-64}$$

The division by A_2 on the right-hand side makes the energy striking dA_1 a fraction of that emitted by A_2 into the entire hemispherical space. From Eqs. (12-63) and

(12-64) we write the *reciprocity relation* between the view factors $F_{dA_1-A_2}$ and $F_{A_2-dA_1}$ as

$$dA_1\, F_{dA_1-A_2} = A_2 F_{A_2-dA_1} \tag{12-65}$$

The view factor from A_2 to A_1 is obtained by integrating Eq. (12-64) over A_1

$$F_{A_2-A_1} = \frac{1}{A_2}\int_{A_2}\int_{A_1} \frac{\cos\theta_1 \cos\theta_2}{\pi r^2}\, dA_1\, dA_2 \tag{12-66}$$

And the view factor from A_1 to A_2 is obtained by integrating Eq. (12-63) over A_1 and dividing by A_1:

$$F_{A_1-A_2} = \frac{1}{A_1}\int_{A_1}\int_{A_2} \frac{\cos\theta_1 \cos\theta_2}{\pi r^2}\, dA_2\, dA_1 \tag{12-67}$$

The division by A_1 on the right-hand side makes the energy striking A_2 a fraction of that emitted by A_1 into the entire hemispherical space.

From Eqs. (12-66) and (12-67), the *reciprocity relation* between the view factors $F_{A_1-A_2}$ and $F_{A_2-A_1}$ is

$$A_1 F_{A_1-A_2} = A_2 F_{A_2-A_1} \tag{12-68}$$

The reciprocity relations given above are useful to determine one of the view factors given knowledge of the other.

Properties of View Factors

We now consider an *enclosure* consisting of N *zones*, each of surface area A_i, $i = 1, 2, \ldots, N$, as illustrated in Fig. 12-18. It is assumed that each zone is isothermal, diffuse emitter, and diffuse reflector. The surface of each zone may be plane, convex, or concave. The view factors between surfaces A_i and A_j of the enclosure obey the following *reciprocity relation*:

$$A_i F_{A_i-A_j} = A_j F_{A_j-A_i} \tag{12-69}$$

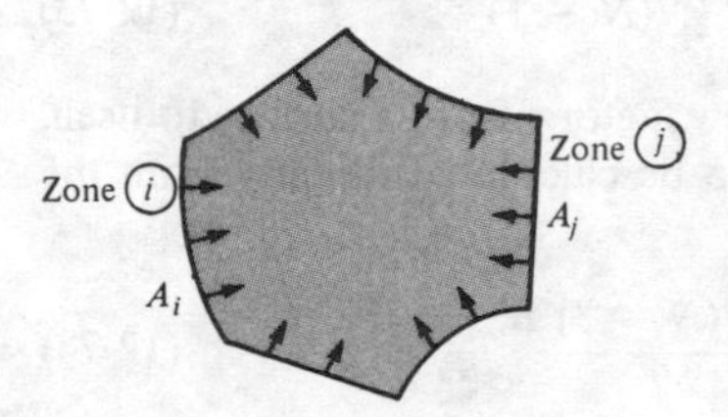

Figure 12-18 An N-zone enclosure.

The view factors from the surface, say A_i, of the enclosure to all surfaces of the enclosure, including to itself, when summed, should be equal to unity by the definition of the view factor. This is called the *summation relation* among the view factors for an enclosure, and it is written

$$\boxed{\sum_{k=1}^{N} F_{A_i - A_k} = 1} \tag{12-70}$$

where N is the number of zones in the enclosure. In this summation the term $F_{A_i - A_i}$ is the view factor from the surface A_i to itself; it represents the fraction of radiative energy leaving the surface A_i that strikes itself directly. Clearly, $F_{A_i - A_i}$ vanishes if A_i is flat or convex, and it is nonzero if A_i is concave; this is stated as

$$F_{A_i - A_i} = 0 \qquad \text{if } A_i \text{ plane or convex} \tag{12-71a}$$

$$F_{A_i - A_i} \neq 0 \qquad \text{if } A_i \text{ concave} \tag{12-71b}$$

The reciprocity and summation rules given are useful in providing additional simple relations to calculate view factors for an enclosure from the knowledge of others. That is, to determine all possible view factors for an enclosure, one need not compute every one of them directly but should make use of the reciprocity and summation relations whenever possible. This situation is envisioned better if all possible view factors for an N-zone enclosure are expressed in matrix notation as

$$F_{ij} \equiv \begin{bmatrix} F_{11} & F_{12} & \cdots & F_{1N} \\ F_{21} & F_{22} & \cdots & F_{2N} \\ \cdots & \cdots & \cdots & \cdots \\ F_{N1} & F_{N2} & \cdots & F_{NN} \end{bmatrix} \tag{12-72}$$

Clearly there are N^2 view factors to be determined for an N-zone enclosure. However, the reciprocity rule provides $N(N-1)/2$ relations, and the summation rule provides N additional relations among the view factors. Then the total number of view factors that are to be calculated for an N-zone enclosure from the view factor expressions becomes

$$N^2 - \tfrac{1}{2}N(N-1) - N = \tfrac{1}{2}N(N-1) \tag{12-73}$$

If the surfaces are convex or flat, N of these view factors, from a surface to itself, vanish and the total number of view factors to be calculated directly from the geometric arrangement of surfaces reduces to

$$\tfrac{1}{2}N(N-1) - N = \frac{N(N-3)}{2} \tag{12-74}$$

For example, for an enclosure with $N = 5$ zones with a flat surface at each zone, out of all the possible $N^2 = 25$ view factors, the number of view factors to be determined from the geometric arrangement of surfaces is only $\frac{1}{2}N(N-3) = 5$.

If the geometry possesses symmetry, some view factors are known from the symmetry condition, thus reducing further the number of view factors to be calculated.

Example 12-12 Two small surfaces $dA_1 = 6\ \text{cm}^2$ and $dA_2 = 12\ \text{cm}^2$ are separated by $r = 60$ cm and oriented as illustrated in the accompanying figure. Calculate the view factors between the surfaces.

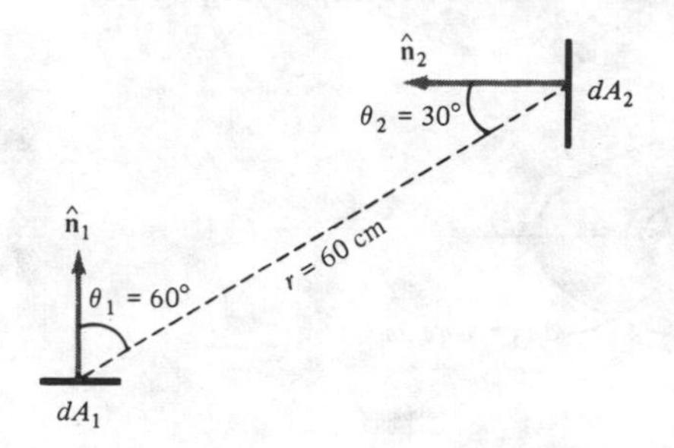

SOLUTION Both surfaces can be approximated as differential surfaces because $dA_i/r^2 \ll 1$ $(i = 1, 2)$. From Eq. (12-60) we have

$$dF_{dA_1-dA_2} = \frac{\cos\theta_1 \cos\theta_2\, dA_2}{\pi r^2}$$

$$= \frac{(\cos 60°)(\cos 30°)(12)}{\pi(60)^2} = 0.46 \times 10^{-3}$$

By the reciprocity relation (12-62) we write

$$dF_{dA_2-dA_1} = \frac{dA_1}{dA_2}\, dF_{dA_1-dA_2} = \tfrac{6}{12} \times 0.46 \times 10^{-3} = 0.23 \times 10^{-3}$$

Example 12-13 View factors are to be determined for all possible combinations of the surfaces of an enclosure consisting of six different zones, each having a flat surface. If the system does not possess symmetry, how many of these view factors are to be computed from the individual view factor relations?

SOLUTION The enclosure contains $N = 6$ zones; hence the maximum possible combination of view factors is $N^2 = 36$. However, the reciprocity relation between view factors provides $\frac{1}{2}N(N-1) = (6 \times 5)/2 = 15$ relations, the summation rule $\sum_{i=1}^{6} F_{i-j} = 1$ provides six additional relations, and the fact that surfaces are flat, hence $F_{i-1} = 0$ $(i = 1$ to $6)$, provides another additional six relations. Then the number of individual view factors to be evaluated from the view factor relations becomes

$$36 - 15 - 6 - 6 = 9$$

This result could also be obtained by Eq. (12-74) as

$$\frac{N(N-3)}{2} = \frac{6 \times 3}{2} = 9$$

Example 12-14 Determine analytically the view factors from an elemental surface dA_1 to a circular disk A_2 of radius R which are parallel to each other and positioned at a distance L, as shown in the accompanying figure.

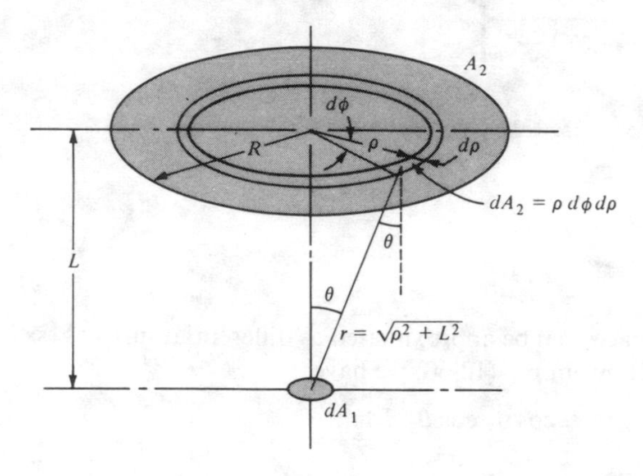

SOLUTION The view factor $F_{dA_1-A_2}$ from an elemental surface dA_1 to a finite surface A_2 is given by Eq. (12-63) as

$$F_{dA_1-A_2} = \int_{A_2} \frac{\cos\theta_1 \cos\theta_2}{\pi r^2}\, dA_2$$

By referring to the geometric arrangement of the accompanying figure, the elemental area dA_1 is expressed in the ρ and ϕ coordinates as

$$dA_2 = \rho\, d\phi\, d\rho$$

where ϕ is the azimuthal angle. The distance r between the elemental surfaces dA_1 and dA_2 is given by

$$r^2 = \rho^2 + L^2$$

and the angles θ_1 and θ_2 are equal to the angle θ shown in the figure and given by

$$\cos\theta_1 = \cos\theta_2 = \cos\theta = \frac{L}{r} = \frac{L}{(\rho^2 + L^2)^{1/2}}$$

Substituting the above expressions for dA_2, r, and θ into the above definition of view factor and performing the integration over the limits $0 \leq \rho \leq R$ and $0 \leq \phi \leq 2\pi$, we obtain the view factor

$$F_{dA_1-A_2} = \frac{1}{\pi} \int_{\rho=0}^{R} \int_{\phi=0}^{2\pi} \frac{L^2}{(\rho^2 + L^2)^2} \rho \, d\rho \, d\phi$$

$$= 2 \int_{\rho=0}^{R} \frac{L^2}{(\rho^2 - L^2)^2} \rho \, d\rho = \frac{R^2}{R^2 + L^2}$$

12-8 METHODS OF DETERMINING VIEW FACTORS

The computation of the view factor between two elemental surfaces, defined by Eqs. (12-60) and (12-61), poses no problem, but the determination of the view factor for finite surfaces involves integration over the surfaces, which is difficult to perform analytically except for simple geometries. Readers should consult Refs. 1 to 5 for a discussion of various analytical techniques for the determination of view factors and Refs. 4 and 23 to 26 for a compilation of view factors. Particularly, Refs. 4 and 23 give comprehensive compilations of the results.

We present in Table 12-2 analytical expressions for view factors for several different simple configurations. Some of the view factor results are presented in graphical form in Figs. 12-19 to 12-23.

We now illustrate with examples the determination of view factors for simple configurations by using both the analytical expressions given in Table 12-2 and the results presented in graphical form. The use of the reciprocity relation and the summation rule also are discussed.

Example 12-15 Determine the view factor $F_{dA_1-A_2}$ from an elemental area $dA_1 = 2 \text{ cm}^2$ to an area A_2 for the configuration shown as case 1 in Table 12-2 by taking $L_1 = L_2 = D = 10$ cm. Compare this result with that presented in graphical form in Fig. 12-19. Also determine the view factor $F_{A_2-dA_1}$ from the surface A_2 to dA_1.

SOLUTION In this example the surface dA_1 can be assumed to be an elemental area because $dA_1/D^2 = 2/10^2 \ll 1$; then the formula given in case 1 of Table 12-2 is applicable. The parameters X and Y become

$$X = \frac{L_1}{D} = \frac{10}{10} = 1 \qquad Y = \frac{L_2}{D} = \frac{10}{10} = 1$$

and the view factor becomes

$$F_{dA_1-A_2} = \frac{1}{2\pi}\left(\frac{1}{\sqrt{2}} \tan^{-1} \frac{1}{\sqrt{2}} + \frac{1}{\sqrt{2}} \tan^{-1} \frac{1}{\sqrt{2}}\right) = \frac{1}{\pi\sqrt{2}} \tan^{-1} \frac{1}{\sqrt{2}}$$

$$= 0.1385$$

Table 12-2 Analytical expressions for view factors for simple geometric arrangements

Geometric arrangement	Analytical expression for view factor
1 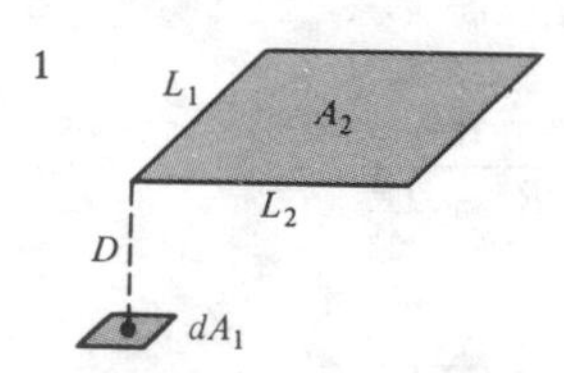 Differential surface parallel to a finite rectangular surface	$F_{dA_1-A_2} = \frac{1}{2\pi}\left(\frac{X}{\sqrt{1+X^2}}\tan^{-1}\frac{Y}{\sqrt{1+X^2}} + \frac{Y}{\sqrt{1+Y^2}}\tan^{-1}\frac{X}{\sqrt{1+Y^2}}\right)$ where $X = L_1/D$ and $Y = L_2/D$ are in radians.
2 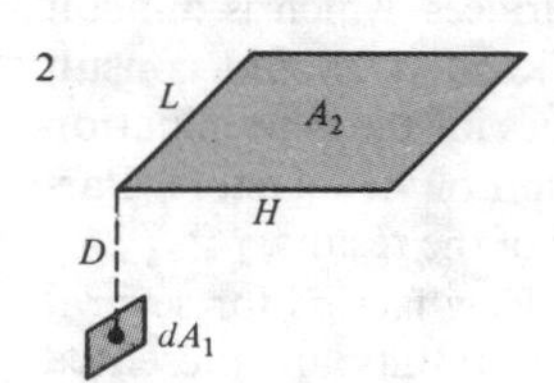 Differential surface perpendicular to a rectangular finite surface	$F_{dA_1-A_2} = \frac{1}{2\pi}\left[\tan^{-1}\frac{1}{X} - \frac{1}{\sqrt{1-(Y/X)^2}}\tan^{-1}\frac{1}{\sqrt{X^2+Y^2}}\right]$ where $X = D/L$ and $Y = H/L$ are in radians.
3 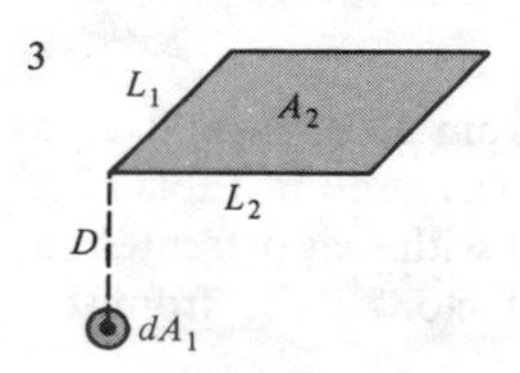 A differential spherical surface and a finite rectangular surface	$F_{dA_1-A_2} = \frac{1}{4\pi}\sin^{-1}\frac{XY}{\sqrt{1+X^2+Y^2+X^2Y^2}}$ where $X = L_1/D$ and $Y = L_2/D$ are in radians.
4 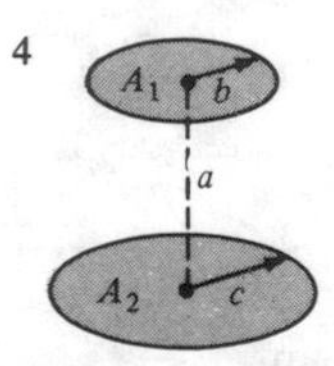 Plane circular surfaces with a common central normal	$F_{A_1-A_2} = \frac{1+B^2+C^3-\sqrt{(1+B^2+C^2)^2-4B^2C^2}}{2B^2}$ where $B = b/a$ and $C = c/a$.

	Geometrical arrangement	Analytical expression for the view factor
5	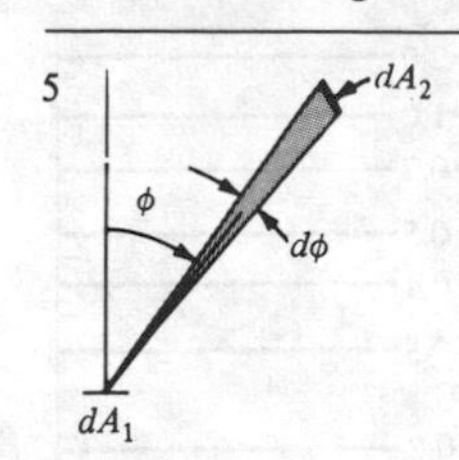	Area dA_1 of differential width and any length, to infinitely long strip dA_2 of differential width and with parallel generating line to dA_1, $dF_{dA_1-dA_2} = \frac{\cos\phi}{2}\, d\phi = \frac{1}{2} d(\sin\phi)$
6	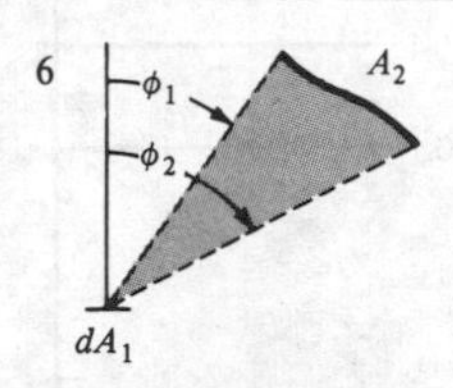	Area dA_1 of differential width and any length to any cylindrical surface A_2 generated by a line of infinite length moving parallel to itself and parallel to the plane of dA_1. $F_{dA_1-A_2} = \frac{1}{2}(\sin\phi_2 - \sin\phi_1)$
7	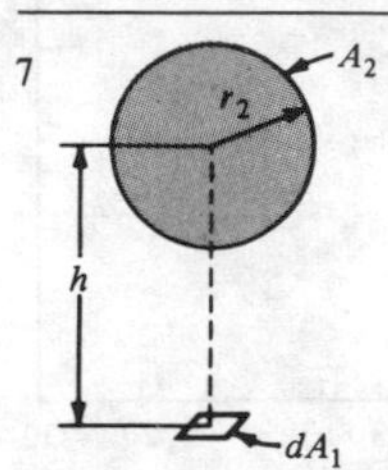	Plane element dA_1 to sphere of radius r_2; normal to center of element passes through center of sphere $F_{dA_1-A_2} = \left(\frac{r_2}{h}\right)^2$
8	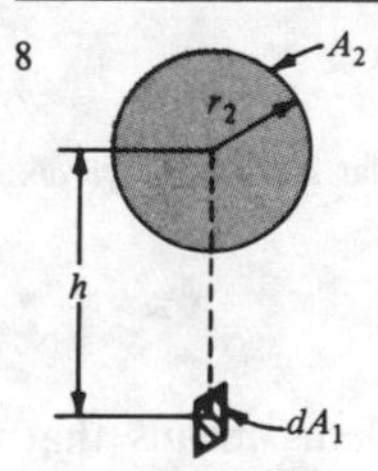	Plane element dA_1 to sphere of radius r_2; tangent to element passes through center of sphere. $F_{dA_1-A_2} = \frac{1}{\pi}\left(\tan^{-1}\frac{1}{\sqrt{H^2-1}} - \frac{\sqrt{H^2-1}}{H^2}\right)$ where $H = h/r_2$.
9	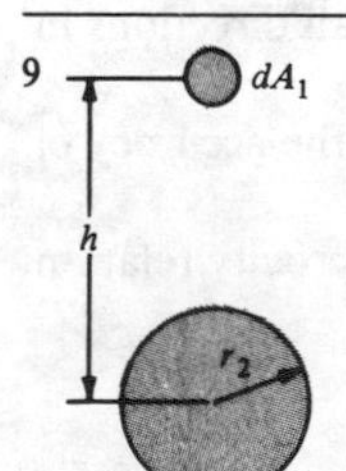	Spherical point source to a sphere of radius r_2. $F_{dA_1-A_2} = \frac{1}{2}(1 - \sqrt{1-R^2})$ where $R = r_2/h$.

From Refs. 4, 25, and 26.

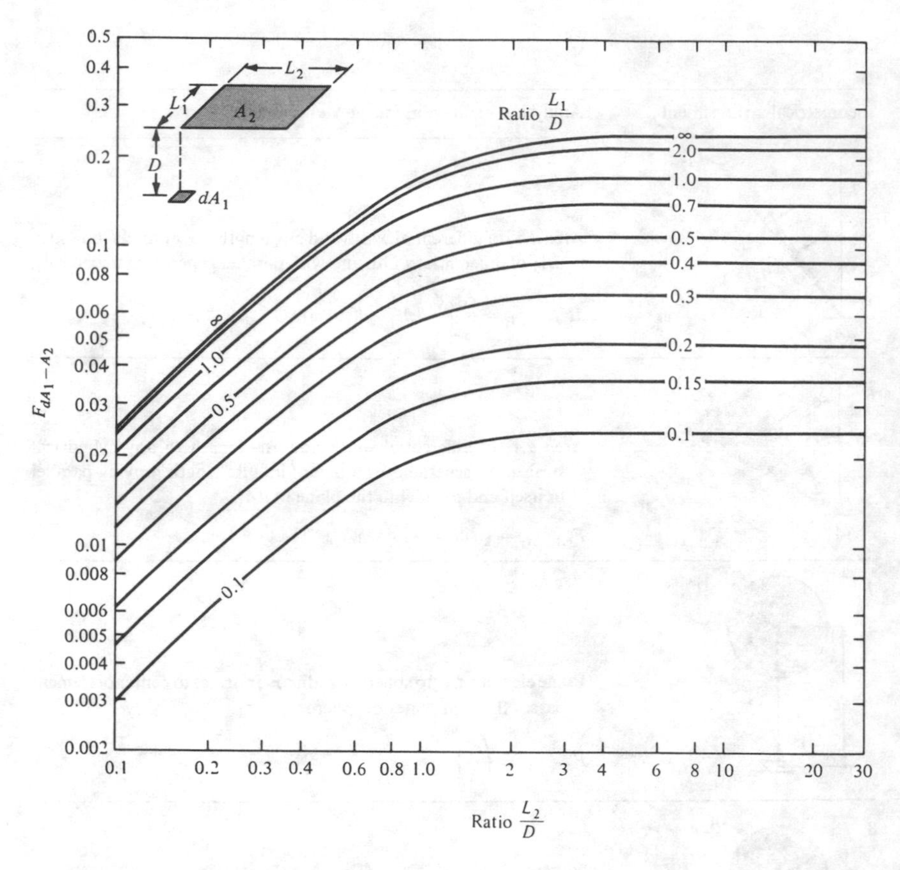

Figure 12-19 View factor $F_{dA_1-A_2}$ from an elemental surface dA_1 to a rectangular surface A_2. (*From Mackey et al.* [25].)

The physical significance of this result in radiation problems means that 13.85 percent of the radiation leaving surface dA_1 diffusely in all directions in hemispherical space strikes surface A_2.

The same result is also obtainable from Fig. 12-19 within the accuracy of readability.

The view factor from A_2 to dA_1 is determined by the reciprocity relation for the view factors:

$$A_2 F_{A_2-dA_1} = dA_1 F_{dA_1-A_2}$$

$$F_{A_2-dA_1} = \frac{dA_1}{A_2} F_{dA_1-A_2} = \frac{2}{10^2}(0.1385) = 0.277 \times 10^{-2}$$

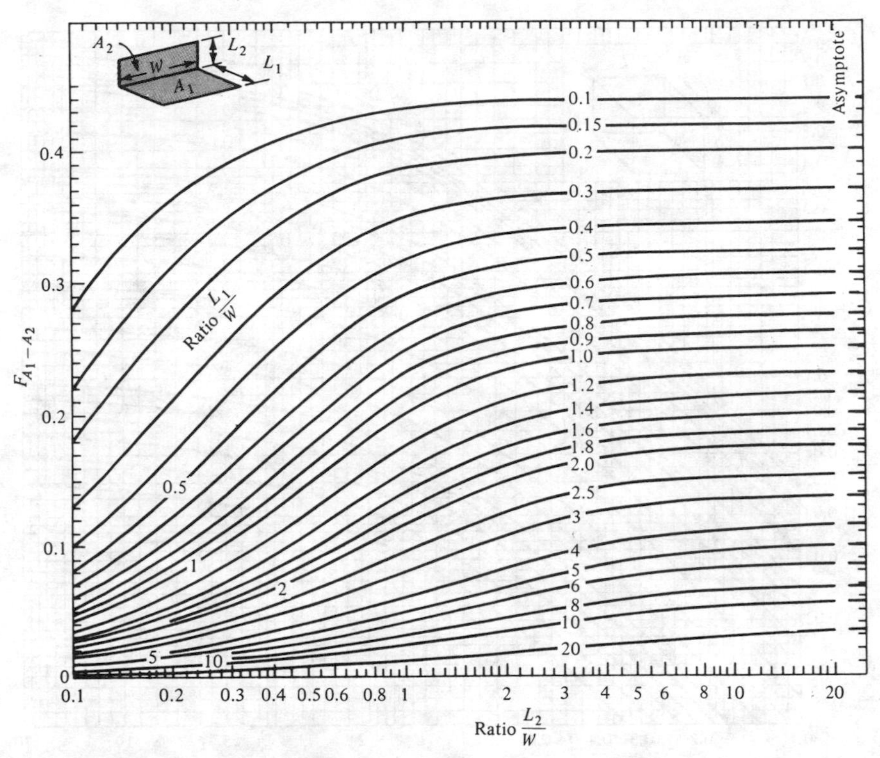

Figure 12-20 View factor $F_{A_1-A_2}$ from a rectangular surface A_1 to a rectangular surface A_2 which are adjacent and in perpendicular planes. (*From Mackey et al.* [25].)

Example 12-16 Determine the following view factors for the geometries shown in the accompanying figures:

(*a*) From the base of a cube to each of its five surfaces

(*b*) From the base of a circular cylinder of radius r and height L to its top surface and the cylindrical surface for the case $r = L$

(*c*) Between the surfaces of two concentric spheres with A_1 and A_2 being the surfaces of the inner and outer spheres

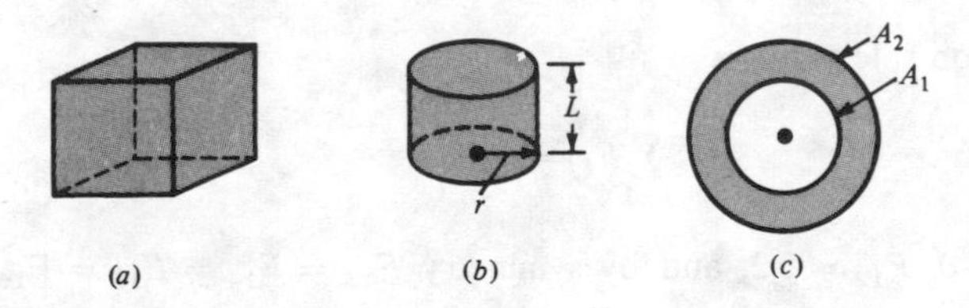

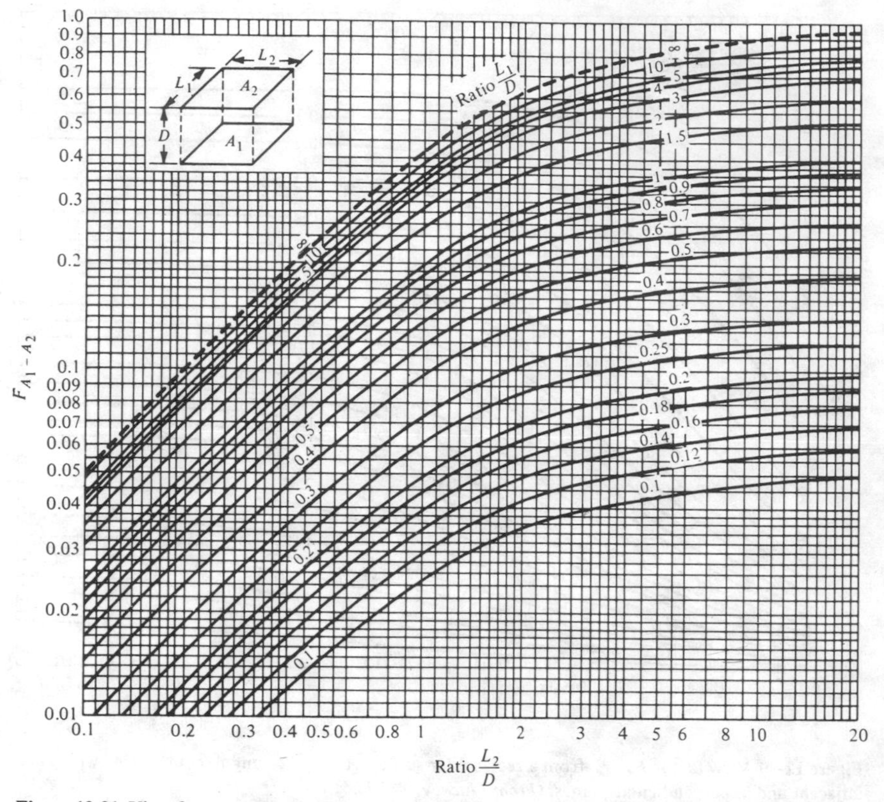

Figure 12-21 View factor $F_{A_1-A_2}$ from a rectangular surface A_1 to a rectangular surface A_2 which are parallel to and directly opposite each other. (*From Mackey et al.* [25].)

SOLUTION For abbreviations, the following notation is used for the view factors:

$$F_{A_i-A_j} \equiv F_{ij}$$

(*a*) For the cube, let 1 and 2 denote the base and top surface, respectively, and 3, 4, 5, and 6 the remaining four lateral surfaces. The view factor F_{12} is determined from Fig. 12-21 as

$$F_{12} = 0.2$$

The summation rule requires

$$\sum_{j=1}^{6} F_{1j} = 1$$

where $F_{11} = 0$, $F_{12} = 0.2$, and by symmetry $F_{13} = F_{14} = F_{15} = F_{16}$.

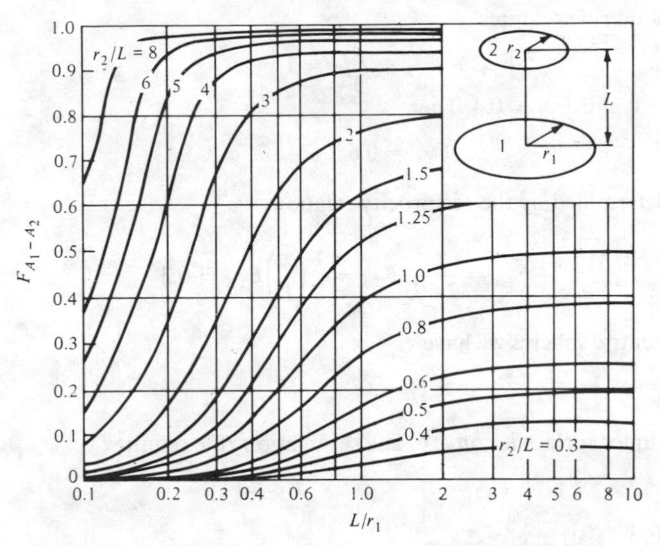

Figure 12-22 View factor $F_{A_1-A_2}$ between two coaxial parallel disks.

Then we obtain

$$F_{1j} = 0.2 \qquad \text{for } j = 2 \text{ to } 6$$

(*b*) For the cylinder, let 1, 2, and 3 denote the base, the top, and the cylindrical surfaces, respectively. For the special case $r = L$, from Fig. 12-22 we obtain

$$F_{12} = 0.4$$

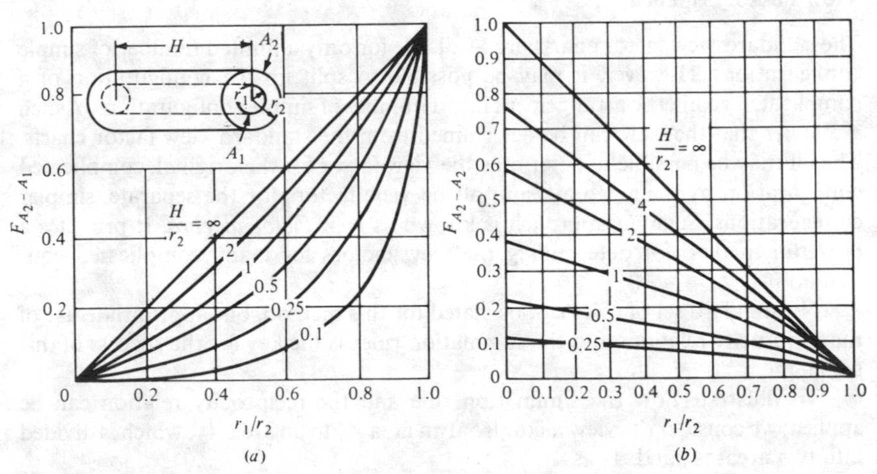

Figure 12-23 View factors $F_{A_2-A_1}$ and $F_{A_2-A_2}$ for concentric cylinders of finite length. (*a*) Outer cylinder to inner cylinder, (*b*) outer cylinder to itself.

The summation rule requires

$$F_{11} + F_{12} + F_{13} = 1$$

where $F_{11} = 0$ and $F_{12} = 0.4$. Then

$$F_{13} = 0.6$$

and F_{31} is determined by the reciprocity relation

$$F_{31} = \frac{A_1}{A_3} F_{13} = \frac{\pi r^2}{2\pi r L} F_{13} = \frac{1}{2}\left(\frac{r}{L}\right) F_{13} = 0.3$$

(*c*) For the concentric spheres we have

$$F_{11} = 0$$

because the inner surface is convex. The summation rule requires

$$F_{11} + F_{12} = 1 \qquad \text{or} \qquad F_{12} = 1$$

The reciprocity relation gives

$$A_1 F_{12} = A_2 F_{21} \qquad \text{or} \qquad F_{21} = \frac{A_1}{A_2}$$

Then F_{22} is determined by the summation rule:

$$F_{21} + F_{22} = 1 \qquad \text{or} \qquad F_{22} = 1 - \frac{A_1}{A_2}$$

View Factor Algebra

The standard view factor charts are available for only a limited number of simple configurations. However, it may be possible to split up the configuration of a complicated geometric arrangement into a number of simple configurations in such a manner that the view can be determined from the standard view factor charts. Then it may be possible to determine the view factor for the original, complicated configuration by the algebraic sum of the view factors for the separate, simpler configurations. Such an approach is known as *view factor algebra.* It provides a powerful method for determining the view factors for many complicated configurations.

No standard set of rules can be stated for this method, but appropriate use of the reciprocity relations and the summation rules is the key for the success of this technique.

To illustrate how the summation rule and the reciprocity relation can be applied, we consider the view factor from an area A_1 to an area A_2, which is divided into two areas A_3 and A_4 as

$$A_2 = A_3 + A_4 \tag{12-75}$$

as illustrated in the accompanying sketch. Then the view factor from A_1 to A_2 can be written as

$$F_{1-2} = F_{1-3} + F_{1-4} \tag{12-76}$$

which is consistent with the definition of the view factor. That is, the fraction of the total energy leaving A_1 that strikes A_3 and A_4 is equal to the fraction that strikes A_2.

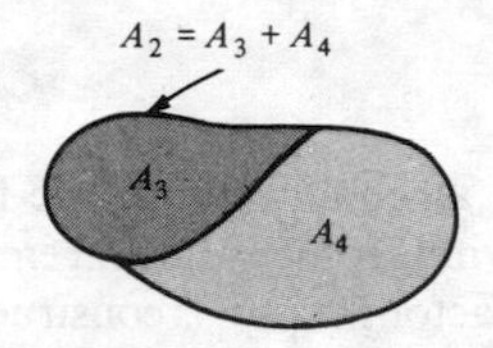

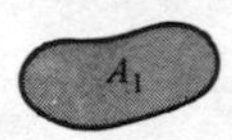

Additional relationships can be obtained among these view factors. For example, both sides of Eq. (12-76) are multiplied by A_1:

$$A_1F_{1-2} = A_1F_{1-3} + A_1F_{1-4}$$

Then the reciprocity relation is applied to each term:

$$A_2F_{2-1} = A_3F_{3-1} + A_4F_{4-1}$$

or

$$F_{2-1} = \frac{A_3F_{3-1} + A_4F_{4-1}}{A_2} = \frac{A_3F_{3-1} + A_4F_{4-1}}{A_3 + A_4} \tag{12-77}$$

Suppose the area A_2 is divided into more parts as

$$A_2 = A_3 + A_4 + \cdots + A_N \tag{12-78}$$

Then the corresponding form of Eq. (12-73) becomes

$$F_{2-1} = \frac{A_3F_{3-1} + A_4F_{4-1} + \cdots + A_NF_{N-1}}{A_3 + A_4 + \cdots + A_N} \tag{12-79}$$

Clearly, similar manipulations can be applied to Eq. (12-77), and other relations can be obtained among the view factors.

Example 12-17 Two parallel surfaces dA_1 and A_2 are arranged as shown in the accompanying figure. Show that by a suitable application of view factor algebra, the view factor for this configuration can be constructed by the algebraic sum of view factors, which can be determined from the standard charts as shown in Fig. 12-19.

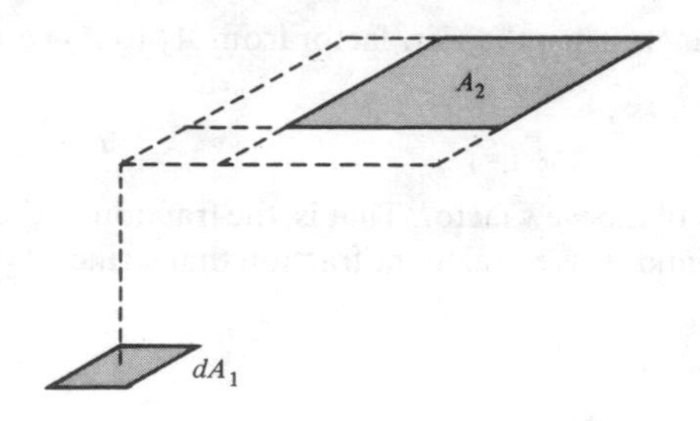

SOLUTION The area A_2 can be expressed as the algebraic sum of the four areas A_3, A_4, A_5, and A_6, as illustrated in the following geometric arrangement, where $A_2 = A_3 - A_4 - A_5 + A_6$. Then the view factor $F_{dA_1-A_2}$ is constructed by the algebraic sum of the view factors for the divided arrangement:

$$F_{dA_1-A_2} = F_{dA_1-A_3} - F_{dA_1-A_4} - F_{dA_1-A_5} + F_{dA_1-A_6}$$

Each view factor on the right-hand side can be obtained from Fig. 12-19; hence $F_{dA_1-A_2}$ can be determined.

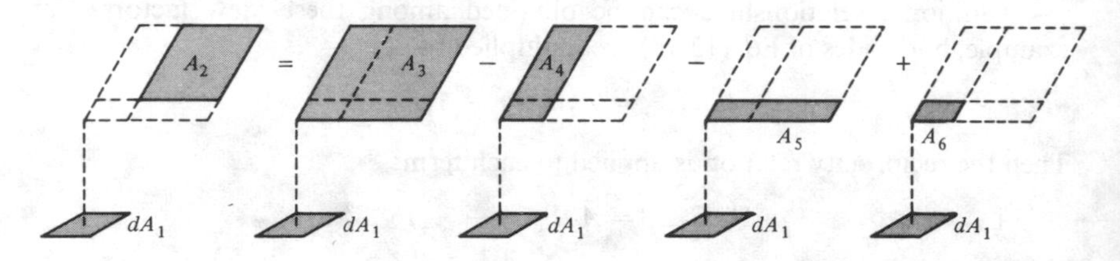

Example 12-18 Determine the view factors F_{1-2} and F_{2-1} between areas A_1 and A_2, which are perpendicular rectangles, as illustrated in the accompanying configuration.

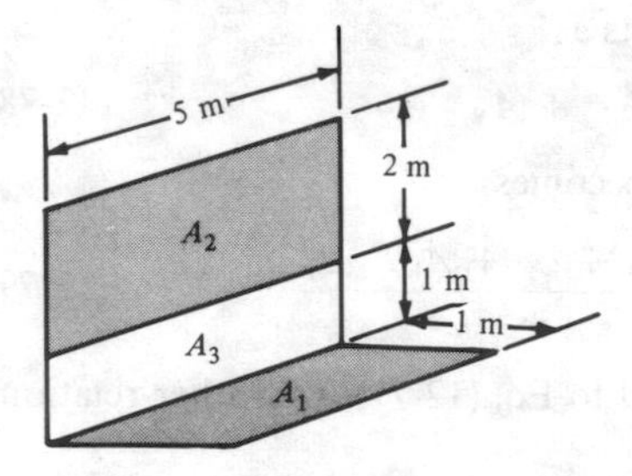

SOLUTION The area A_2 can be written as

$$A_2 = (A_2 + A_3) - A_3$$

Then the view factor from A_1 to A_2 becomes

$$F_{1-2} = F_{1-(2+3)} - F_{1-3}$$

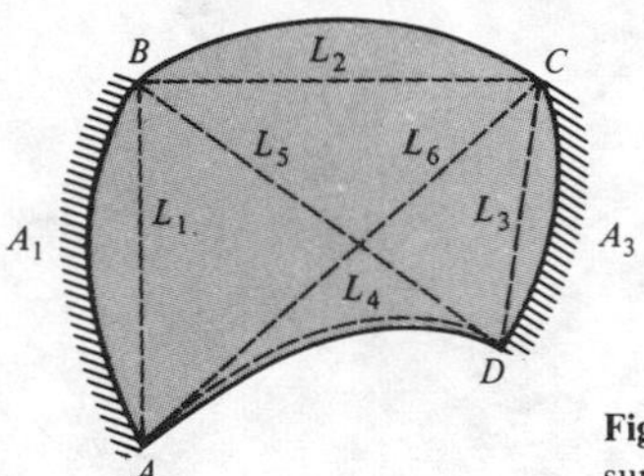

Figure 12-24 Determination of view factor between the surfaces A_1 and A_3 of a long enclosure.

Each of the view factors on the right-hand side can be obtained from Fig. 12-20. That is, $F_{1-(2+3)} = 0.37$ and $F_{1-3} = 0.25$. Then

$$F_{1-2} = 0.37 - 0.27 = 0.10$$

The reciprocity relation is used to determine F_{2-1}:

$$F_{2-1} = \frac{A_1}{A_2} F_{1-2} = \tfrac{1}{2}(0.10) = 0.05$$

Crossed-String Method

Consider an enclosure, as shown in Fig. 12-24, consisting of four surfaces which are very long in the direction perpendicular to the plane of the figure. The surfaces can be flat, convex, or concave. Suppose we wish to find the view factor F_{1-3} between the surfaces A_1 and A_3. We assume imaginary strings, shown by dashed lines in Fig. 12-24, are tightly stretched among the four corners A, B, C, and D of the enclosure. Let L_i ($i = 1, 2, 3, 4, 5, 6$) denote the lengths of the strings joining the corners, as illustrated in this figure. Hottel [27] has shown that the view factor F_{1-3} can be expressed as

$$\boxed{L_1 F_{1-3} = \frac{(L_5 + L_6) - (L_2 + L_4)}{2}} \tag{12-80}$$

Here we note that the term $L_5 + L_6$ is the sum of the lengths of the cross strings, and $L_1 + L_2$ is the sum of the lengths of the uncrossed strings. Equation (12-80) is useful for determining the view factor between the surfaces of a long enclosure, such as a groove, which can be characterized as a two-dimensional geometry of the form illustrated in Fig. 12-24.

Example 12-19 An infinitely long semicylindrical surface A_1 of radius b and an infinitely long flat plate A_3 of half-width c are located a distance d apart, as illustrated in the accompanying figure. Determine the view factor F_{1-3} between surfaces A_1 and A_3 for $b = 5$, $c = 10$, and $d = 8$ cm.

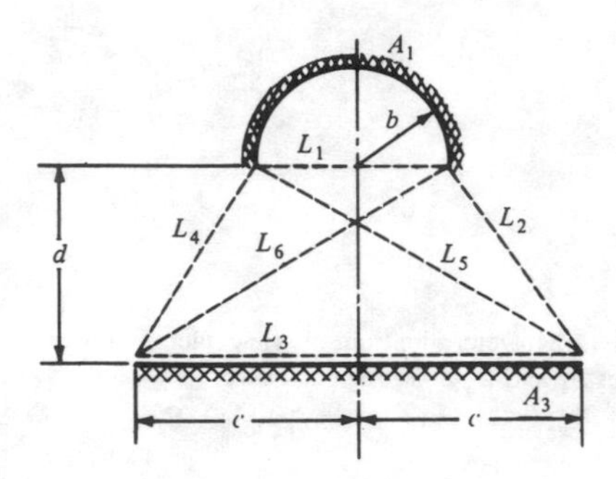

SOLUTION The dashed lines in this figure show the imaginary strings stretched among the four corners of the geometry. The view factor F_{1-3} between surfaces A_1 and A_3 is given by Eq. (12-80) as

$$L_1 F_{1-3} = \frac{(L_5 + L_6) - (L_2 + L_4)}{2}$$

Because of symmetry, we have $L_5 = L_6$ and $L_2 = L_4$; so

$$F_{1-3} = \frac{L_6 - L_4}{L_1}$$

For the geometry of this example, we have

$$L_4 = [(c - b)^2 + d^2]^{1/2}$$
$$L_6 = [(c + b)^2 + d^2]^{1/2}$$
$$L_1 = 2b$$

Substituting these results into the expression for F_{1-3}, we obtain

$$F_{1-3} = \frac{[(c + b)^2 + d^2]^{1/2} - [(c - b)^2 + d^2]^{1/2}}{2b}$$

For the numerical values given, we find

$$[(c + b)^2 + d^2]^{1/2} = [(10 + 5)^2 + 8^2]^{1/2} = 17$$
$$[(c - b)^2 + d^2]^{1/2} = [(10 - 5)^2 + 8^2]^{1/2} = 9.434$$
$$2b = 10$$

Then

$$F_{1-3} = \frac{17 - 9.434}{10} = 0.757$$

12-9 NETWORK METHOD FOR RADIATION EXCHANGE IN AN ENCLOSURE

The analysis of radiation exchange among the surfaces of an enclosure is complicated by the fact that when the surfaces are not black, radiation leaving a surface may be reflected back and forth several times among the surfaces, with partial absorption occurring at each reflection. Therefore, a proper analysis of the problem must include the effects of these multiple reflections. To simplify the analysis, we assume that a given enclosure can be divided into several zones, as illustrated in Fig. 12-25, in such a manner that the following conditions are assumed to hold for each of the zones $i = 1, 2, \ldots, N$:

1. Radiative properties (i.e., reflectivity, emissivity, absorptivity) are uniform and independent of direction and frequency.
2. Surfaces are diffuse emitters and diffuse reflectors.
3. The radiative-heat flux leaving the surface is uniform over the surface of each zone.
4. The irradiation is uniform over the surface of each zone.
5. Surfaces are opaque (that is, $\alpha + \rho = 1$).
6. Either a uniform temperature or a uniform heat flux is prescribed over the surface of each zone.
7. The enclosure is filled with a nonparticipating medium.

Assumptions 3 and 4 are generally not correct, but the analysis becomes very complicated without them.

The objective of the analysis of radiation heat exchange in an enclosure is the *determination of net radiation heat flux* at the zones for which the temperature is prescribed. Several different methods of analysis have been reported for the solution of radiation heat exchange in an enclosure under the simplifying assumptions stated above [27–33]. However, a close scrutiny of all these methods reveals that there is no significant difference among them, since all utilize the same simplifying assumptions. In this section we present the network method originally introduced by Oppenheim [31]. The method is relatively easy to apply for simple problems

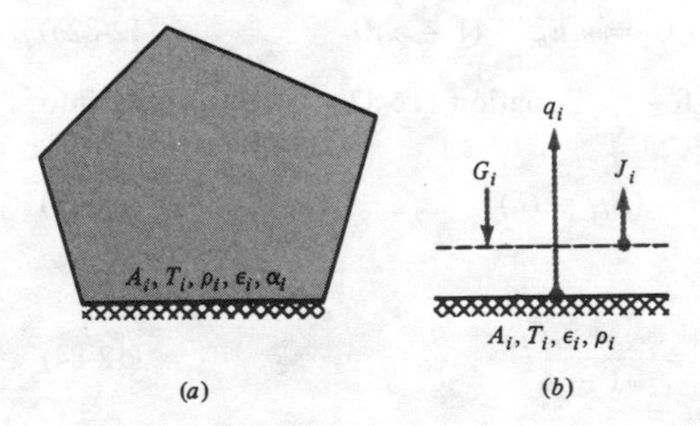

Figure 12-25 (*a*) An enclosure filled with a nonparticipating medium; (*b*) an energy balance per unit area of zone *i*.

which do not involve too many surfaces. Furthermore, it provides good insight to the physical concepts in radiation exchange between surfaces. When many heat transfer surfaces are involved, the method is not so practical. Therefore, in the next section we present the matrix formulation of radiation exchange for enclosures.

The first step in the analysis of radiation exchange by the network method is the development of the concept of *surface resistance to radiation.*

Surface Resistance to Radiation

Consider zone i of an enclosure, as illustrated in Fig. 12-25. We define the following quantities:

G_i = irradiation at zone i; it represents the radiation flux incident on surface A_i, W/m^2

J_i = Radiosity at zone i; it represents the radiation flux leaving surface A_i, W/m^2

q_i = Net radiation flux leaving surface A_i, W/m^2

A distinction should be made between J_i and q_i. The *radiosity* J_i is the radiation energy leaving the surface as observed immediately outside the surface of zone i, at a location illustrated symbolically by the dashed line in Fig. 12-25*b*. The *net radiation heat flux* q_i leaving surface A_i, however, is based on the net energy balance in the interior of surface A_i. Therefore, by definition, q_i is equal to the difference between J_i and G_i:

$$q_i = J_i - G_i \qquad \text{W/m}^2 \tag{12-81}$$

The radiosity, however, is composed of the following components:

$$J_i = \begin{pmatrix} \text{radiation} \\ \text{emitted by} \\ \text{surface} \end{pmatrix} + \begin{pmatrix} \text{radiation} \\ \text{reflected by} \\ \text{surface} \end{pmatrix} \tag{12-82a}$$

Let E_{bi} be the blackbody emissive power, ε_i the emissivity, ρ_i the reflectivity, and G_i the incident radiation flux at zone i. Then Eq. (12-82*a*) becomes

$$J_i = \varepsilon_i E_{bi} + \rho_i G_i = \varepsilon_i E_{bi} + (1 - \varepsilon_i) G_i \tag{12-82b}$$

where we assumed $\rho_i = 1 - \alpha_i = 1 - \varepsilon_i$. Equation (12-82*b*) is substituted into Eq. (12-81) to eliminate J_i:

$$q_i = \varepsilon_i (E_{bi} - G_i) \tag{12-83}$$

Equation (12-82*b*) is solved for G_i:

$$G_i = \frac{J_i - \varepsilon_i E_{bi}}{1 - \varepsilon_i} \tag{12-84}$$

Equation (12-84) is introduced into Eq. (12-83):

$$q_i = \frac{\varepsilon_i}{1 - \varepsilon_i}(E_{bi} - J_i) \qquad \text{W/m}^2 \tag{12-85}$$

The total net radiation heat flow Q_i leaving surface A_i becomes

$$Q_i = A_i q_i = A_i \frac{\varepsilon_i}{1 - \varepsilon_i}(E_{bi} - J_i)$$

which is rearranged as

$$\boxed{Q_i = \frac{E_{bi} - J_i}{R_i}} \qquad \text{W} \tag{12-86a}$$

where

$$\boxed{R_i = \frac{1 - \varepsilon_i}{A_i \varepsilon_i}} \tag{12-86b}$$

Clearly, Eq. (12-86) is analogous to Ohm's law, where R_i represents the surface resistance to radiation. Equation (12-86) is also analogous to the concept of thermal resistance (or the skin resistance) that we discussed in connection with convective heat transfer over a surface. That is, the total heat transfer rate is equal to the potential difference across the surface divided by the thermal resistance to heat flow over the surface.

When the surface is *black*, we have $\varepsilon_i = 1$, which implies that $R_i = 0$. Then Eq. (12-86*a*) reduces to

$$\boxed{J_i = E_{bi} = \sigma T_i^4 \qquad \text{for } \varepsilon_i = 1 \text{ or black surface}} \tag{12-87}$$

Thus, for a black surface, the radiosity is equal to the blackbody emissive power of the surface.

Figure 12-26 illustrates the concept of surface thermal resistance to radiation between the potentials E_{bi} and J_i.

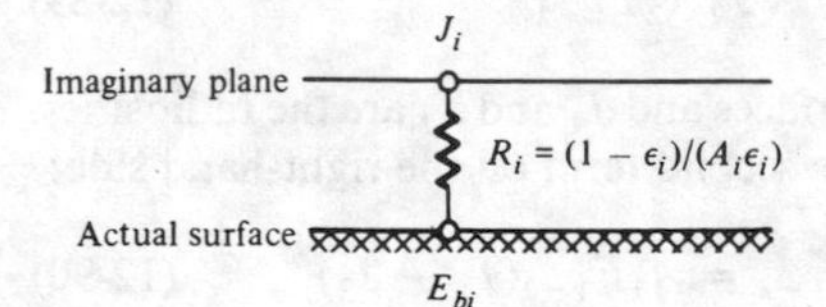

Figure 12-26 Surface resistance to radiation.

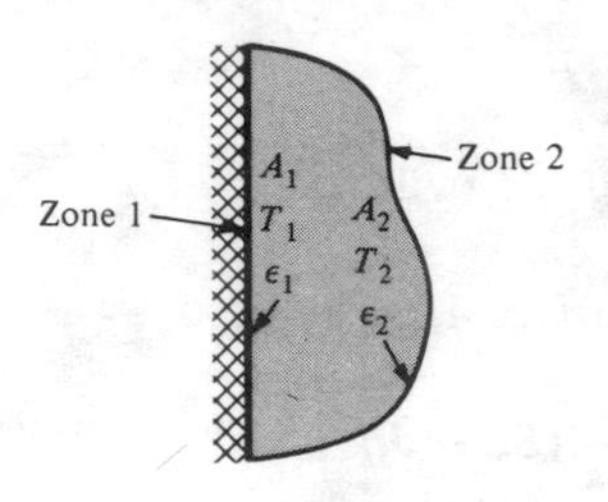

Figure 12-27 Two-zone enclosure and the equivalent radiation network.

Two-Zone Enclosure

Having established the formalism for defining the surface resistance to radiation, we can analyze the simplest enclosure problem, involving heat exchange in an enclosure containing only two zones. Typical examples of such a physical situation include radiation exchange between the surfaces of two large parallel plates, two long coaxial cylinders, or two concentric spheres. For generality, we consider a two-zone enclosure as illustrated in Fig. 12-27. Zone 1 has a surface area A_1 and emissivity ε_1 and is maintained at a uniform temperature T_1. Zone 2 has a surface area A_2 and emissivity ε_2 and is maintained at a uniform temperature T_2. Both surfaces are opaque. Heat exchange takes place between the surfaces because they are at different temperatures. Let

$$Q_{1-2} = \text{net radiation heat transfer from zone 1 to zone 2}$$

Then the energy balance for radiation heat exchange between the two zones can be stated as

$$Q_{1-2} = \begin{pmatrix}\text{radiation energy}\\ \text{leaving } A_1 \text{ that}\\ \text{strikes } A_2\end{pmatrix} - \begin{pmatrix}\text{radiation energy}\\ \text{leaving } A_2 \text{ that}\\ \text{strikes } A_1\end{pmatrix} \qquad (12\text{-}88)$$

The mathematical expressions for each term on the right-hand side are written as

$$Q_{1-2} = J_1 A_1 F_{1-2} - J_2 A_2 F_{2-1} \qquad (12\text{-}89)$$

where F_{i-j} is the view factor between the surfaces and J_1 and J_2 are the radiosities.

We apply the reciprocity relation to the second term on the right-hand side:

$$Q_{1-2} = J_1 A_1 F_{1-2} - J_2 A_1 F_{1-2} = A_1 F_{1-2}(J_1 - J_2) \qquad (12\text{-}90)$$

Equation (12-90) is now rearranged in the form

$$Q_{1-2} = \frac{J_1 - J_2}{R_{1-2}} \tag{12-91}$$

where

$$\boxed{R_{1-2} = \frac{1}{A_1 F_{1-2}}} \tag{12-92}$$

Thus, Eq. (12-92) establishes the resistance to radiation between the potentials J_1 and J_2. Then the thermal radiation network for radiation exchange between zones 1 and 2 is constructed as illustrated in Fig. 12-27. By utilizing this network, the net radiation heat exchange Q_{1-2} between zones 1 and 2 is

$$Q_{1-2} = \frac{E_{b1} - E_{b2}}{R_1 + R_{1-2} + R_2} \tag{12-93}$$

or

$$\boxed{Q_{1-2} = \frac{\sigma T_1^4 - \sigma T_2^4}{(1 - \varepsilon_1)/(A_1\varepsilon_1) + 1/(A_1 F_{1-2}) + (1 - \varepsilon_2)/(A_2\varepsilon_2)}} \tag{12-94}$$

The heat flow rates Q_1, Q_{1-2}, and Q_2 must be related by

$$Q_1 = -Q_2 = Q_{1-2}$$

since there are only two surfaces.

Example 12-20 Consider an enclosure consisting of two parallel, infinite opaque plates. Surfaces 1 and 2 are gray, are kept at uniform temperatures T_1 and T_2, and have emissivities ε_1 and ε_2 and reflectivities ρ_1 and ρ_2, respectively.

(*a*) Develop an expression for radiation exchange between the surfaces.
(*b*) Calculate the net radiation flux q_{1-2} leaving plate 1 for $T_1 = 800$ K, $T_2 = 600$ K, and $\varepsilon_1 = \varepsilon_2 = 0.8$.

SOLUTION This is a problem of radiation exchange in a two-zone enclosure.

(*a*) The expression for net radiation exchange is immediately obtained from Eq. (12-94) by setting $A_1 = A_2 = A$ and $F_{1-2} = 1$. We obtain

$$q_{1-2} = \frac{Q_{1-2}}{A} = \frac{\sigma T_1^4 - \sigma T_2^4}{1/\varepsilon_1 + 1/\varepsilon_2 - 1} \quad \text{W/m}^2$$

(*b*) The net radiation flux q_{1-2} leaving zone 1 for the condition specified becomes

$$q_{1-2} = \frac{(5.67 \times 10^{-8})(800^4 - 600^4)}{1/0.8 + 1/0.8 - 1} = 10{,}453 \text{ W/m}^2$$

Example 12-21 Consider a small convex object of surface area A_1 and emissivity ε_1 that is maintained at a uniform temperature T_1 and placed inside a large cavity of area A_2 with emissivity ε_2 and temperature T_2. The accompanying figure illustrates the configuration.

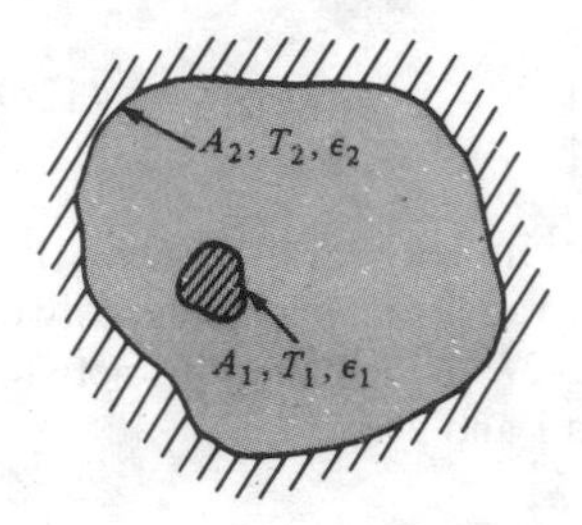

(*a*) Develop an expression for the radiation loss Q_{1-2} from the small object to the surrounding, assuming $A_1/A_2 \to 0$.
(*b*) Calculate the radiation loss Q_{1-2} from the small object for $T_1 = 600$ K, $T_2 = 300$ K, $\varepsilon_1 = 0.9$, and $A_1 = 20\ \text{cm}^2$.

SOLUTION This problem is also a special case of the two-zone analysis.

(*a*) The expression for the heat loss Q_{1-2} is immediately obtained from Eq. (12-94) by setting $A_1/A_2 \to 0$ and $F_{1-2} = 1$ and assuming that $\varepsilon_2 \neq 0$. We find

$$Q_{1-2} = A_1 \varepsilon_1 \sigma (T_1^4 - T_2^4) \qquad \text{W}$$

(*b*) The radiation loss for the conditions stated becomes

$$Q_{1-2} = (20 \times 10^{-4})(0.9)(5.67 \times 10^{-8})(600^4 - 300^4)$$
$$= 12.4\ \text{W}$$

Three-Zone Enclosure

The radiation network approach just described can be readily generalized to enclosures involving three or more zones. However, when the problem involves more than three zones, it is preferable to use the more direct *matrix* approach which lends itself to solution with a digital computer, as will be discussed in the next section. For this reason, we restrict the use of the radiation network method for enclosures up to three zones.

The results given by Eqs. (12-91) and (12-92) can be applied to any two zones *i* and *j* of the enclosure as

$$Q_{i-j} = \frac{J_i - J_j}{R_{i-j}} \tag{12-95a}$$

where

$$R_{i-j} = \frac{1}{A_i F_{i-j}} = \frac{1}{A_j F_{j-i}} \tag{12-95b}$$

and the total net radiation heat flow Q_i leaving the zone i becomes [see Eqs. (12-86)]

$$Q_i = \frac{E_{bi} - J_i}{R_i} \tag{12-96a}$$

where

$$R_i = \frac{1 - \varepsilon_i}{A_i \varepsilon_i} \tag{12-96b}$$

The results given by Eqs. (12-95) and (12-96) when applied to a three-zone enclosure shown in Fig. 12-28*a* yields the radiation network illustrated in Fig. 12-28*b*.

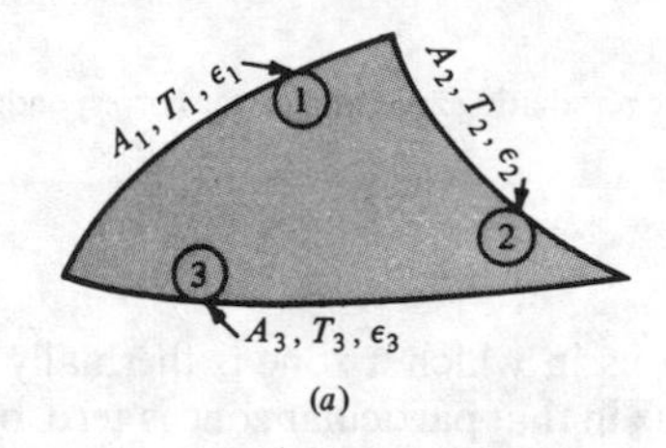

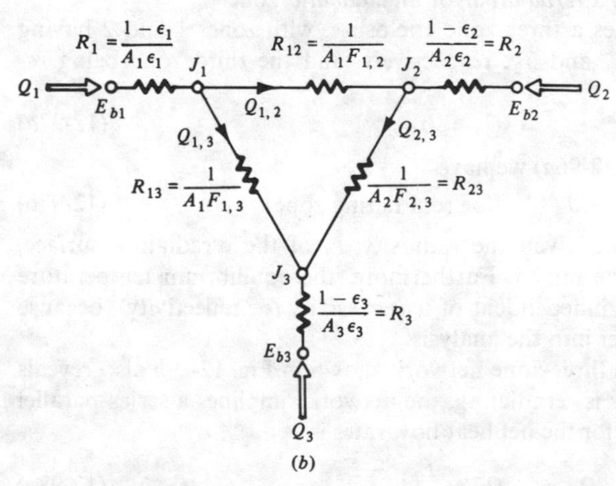

Figure 12-28 (*a*) A three-zone enclosure and (*b*) the corresponding radiation network.

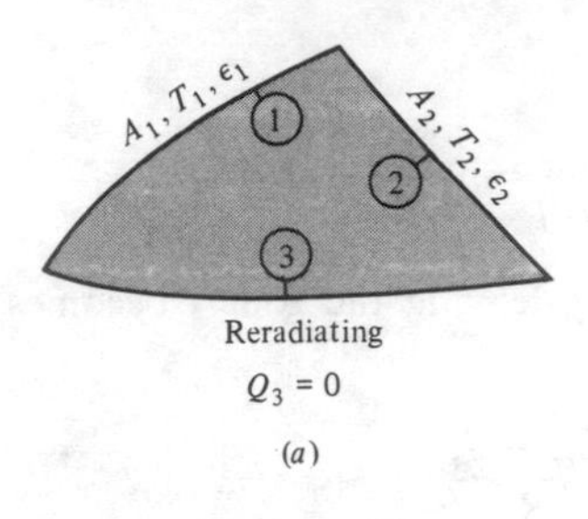

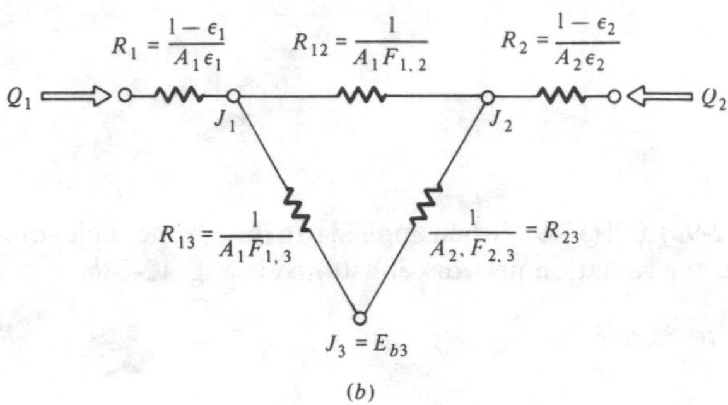

Figure 12-29 (*a*) A three-zone enclosure with one reradiating zone and (*b*) the corresponding radiation network.

Reradiating Surface

There are many engineering applications in which a zone is thermally insulated. In such a case, the net radiation heat flux in that particular zone is zero, because the surface emits as much energy as it receives by radiation from the surrounding zones. Such a zone is called a *reradiating*, or an *adiabatic*, zone.

Figure 12-29*a* illustrates a three-zone enclosure, with zones 1 and 2 having prescribed temperatures T_1 and T_2, respectively, and the third zone being *reradiating*, that is,

$$Q_3 = 0 \tag{12-97a}$$

As a consequence, by Eq. (12-96*a*) we have

$$E_{b3} = J_3 \qquad \text{(for reradiating zone)} \tag{12-97b}$$

This condition implies that, given the radiosity J_3 of the reradiating surface, its temperature can be determined. Furthermore, the equilibrium temperature T_3 of a reradiating zone is independent of its emissivity (or reflectivity), because the emissivity does not enter into the analysis.

An examination of the three-zone network shown in Fig. 12-29*b* also reveals that when one of the zones is reradiating, the network simplifies a series-parallel arrangement. The solution for the net heat flow rates is

$$Q_1 = -Q_2 = \frac{E_{b1} - E_{b2}}{R} \tag{12-98a}$$

where

$$R = R_1 + \left(\frac{1}{R_{12}} + \frac{1}{R_{13} + R_{23}}\right)^{-1} + R_2$$

$$= \frac{1-\varepsilon_1}{A_1\varepsilon_1} + \left[A_1F_{12} + \frac{1}{1/(A_1F_{13}) + 1/(A_2F_{23})}\right]^{-1} + \frac{1-\varepsilon_2}{A_2\varepsilon_2} \quad (12\text{-}98b)$$

Example 12-22 The configuration of a furnace can be approximated as an equilateral triangular duct which is sufficiently long that the end effects are negligible. The hot wall is maintained at $T_1 = 900$ K and has an emissivity $\varepsilon_1 = 0.8$. The cold wall is at $T_2 = 400$ K and has an emissivity $\varepsilon_2 = 0.8$. The third wall is a reradiating zone for which $Q_3 = 0$. The accompanying sketch illustrates the configuration. Calculate the net radiation heat flux leaving the hot wall.

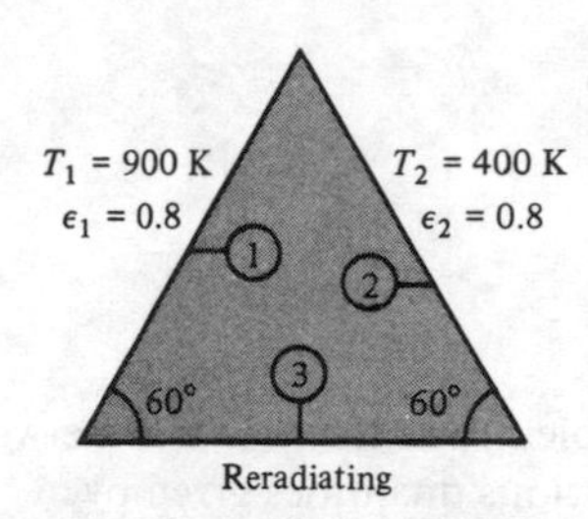

SOLUTION This is a three-zone enclosure problem with one of the zones being a reradiating surface. Therefore, the solution is obtainable from the result given by Eq. (12-98). Various quantities are taken as

$$A_1 = A_2 \equiv A \qquad \varepsilon_1 = \varepsilon_2 = 0.8$$

$$F_{1-2} = F_{1-3} = F_{2-3} = 0.5$$

$$E_{b1} - E_{b2} = \sigma(T_1^4 - T_2^4) = (5.67 \times 10^{-8})(900^4 - 400^4)$$

$$= 35.749 \text{ kW/m}^2$$

Introducing these numerical values into Eq. (12-98*a*), we obtain

$$q_1 = \frac{Q_1}{A} = \frac{\sigma(T_1^4 - T_2^4)}{AR} = \frac{35.749}{AR} \qquad \text{kW/m}^2$$

where AR is determined from Eq. (12-98*b*) as

$$AR = \frac{0.2}{0.8} + \frac{1}{0.5 + (2+2)^{-1}} + \frac{0.2}{0.8} = 1.8333$$

So the net radiation heat flux leaving the hot wall is

$$q_1 = \frac{35.749}{1.8333} = 19.5 \text{ kW/m}^2$$

Example 12-23 Two square plates, each 1 m by 1 m, are parallel to and directly opposite each other at a distance 1 m. The hot plate is at $T_1 = 800$ K and has an emissivity $\varepsilon_1 = 0.8$. The colder plate is at $T_2 = 600$ K and also has an emissivity $\varepsilon_2 = 0.8$. The radiation heat exchange takes place between the plates as well as with a large ambient at $T_3 = 300$ K through the opening between the plates. The configuration is illustrated in the accompanying sketch. Calculate the net heat transfer rate by radiation at each plate and to the ambient.

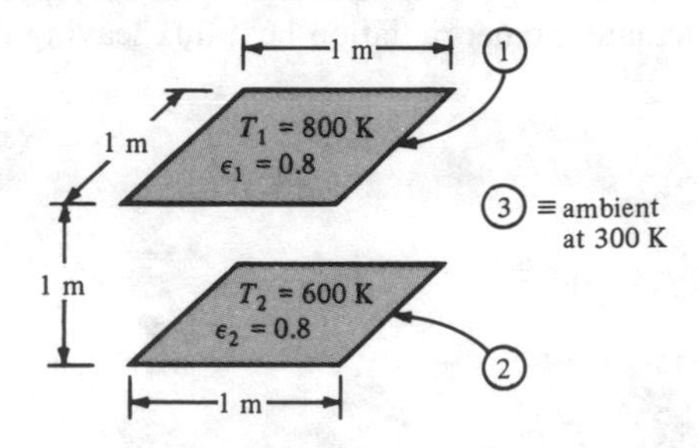

SOLUTION This is a three-zone enclosure problem, and the radiation network is similar to that shown in Fig. 12-28, with various quantities given by

$$A_1 = A_2 = 1 \text{ m}^2 \qquad \frac{A_1}{A_3} \to 0$$

$$\varepsilon_1 = \varepsilon_2 = 0.8$$

$$F_{1-2} = 0.2 \qquad F_{1-3} = F_{2-3} = 0.8$$

$$R_1 = \frac{1-\varepsilon_1}{A_1\varepsilon_1} = \frac{0.2}{1(0.8)} = 0.25 \qquad R_2 = \frac{1-\varepsilon_2}{A_2\varepsilon_2} = \frac{0.2}{1(0.8)} = 0.25$$

$$R_3 = \frac{1-\varepsilon_3}{A_3\varepsilon_3} \to 0 \qquad \therefore J_3 = E_{b3}$$

$$\frac{1}{A_1F_{1-2}} = 5 \qquad \frac{1}{A_1F_{1-3}} = \frac{1}{A_2F_{2-3}} = \frac{1}{0.8} = 1.25$$

$$T_1 = 800 \text{ K} \qquad T_2 = 600 \text{ K} \qquad T_3 = 300 \text{ K}$$

$$E_{b1} = \sigma T_1^4 = (5.67 \times 10^{-8})(800^4) = 23.224 \text{ kW/m}^2$$

$$E_{b2} = \sigma T_2^4 = 7.348 \text{ kW/m}^2$$

$$E_{b3} = \sigma T_3^4 = 0.459 \text{ kW/m}^2$$

These quantities are introduced into the radiation network shown in Fig. 12-28. The results are presented in the accompanying network.

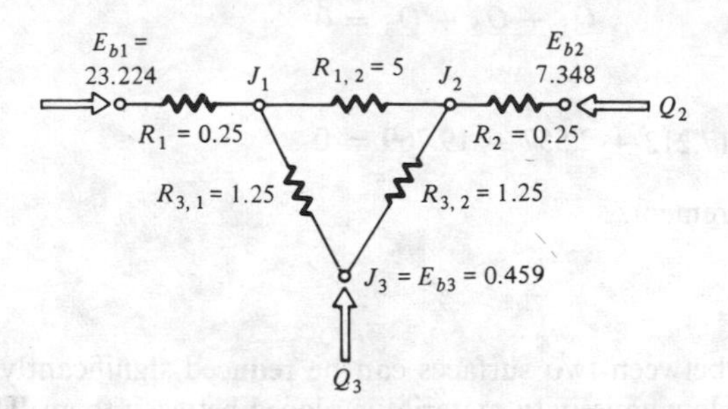

The problem is now reduced to that of determining the radiosities J_1 and J_2. By setting the algebraic sum of the currents at the nodes J_1 and J_2 equal to zero we obtain

$$\text{At } J_1: \qquad \frac{23.224 - J_1}{0.25} + \frac{J_2 - J_1}{5} + \frac{0.459 - J_1}{1.25} = 0$$

$$\text{At } J_2: \qquad \frac{J_1 - J_2}{5} + \frac{7.348 - J_2}{0.25} + \frac{0.459 - J_2}{1.25} = 0$$

A simultaneous solution of these equations yields

$$J_1 = 18.921 \text{ kW/m}^2 \qquad J_2 = 6.709 \text{ kW/m}^2$$

The net rate of heat leaving plate 1 becomes

$$Q_1 = \frac{E_{b1} - J_1}{R_1} = \frac{23.224 - 18.921}{0.25} = 17.212 \text{ kW}$$

The net rate of heat leaving plate 2 is

$$Q_2 = \frac{E_{b2} - J_2}{R_2} = \frac{7.348 - 6.709}{0.25} = 2.557 \text{ kW}$$

The net rate of heat leaving the ambient is

$$\begin{aligned} Q_3 &= \frac{E_{b3} - J_1}{R_{31}} + \frac{E_{b3} - J_2}{R_{32}} \\ &= \frac{0.459 - 18.921}{1.25} + \frac{0.459 - 6.709}{1.25} \\ &= -19.769 \text{ kW} \end{aligned}$$

Clearly, plates 1 and 2 are losing heat, and the ambient is gaining heat; and the algebraic sum of these three quantities should be equal to zero:

$$Q_1 + Q_2 + Q_3 = 0$$

or

$$17.212 + 2.557 - 19.769 = 0$$

which satisfies the requirement.

Radiation Shields

The radiation heat transfer between two surfaces can be reduced significantly if a radiation shield made of a low-emissivity material is placed between them. The reason for this reduction in heat transfer can be envisioned better by recalling the definition of surface resistance to radiation given by Eq. (12-86*b*). That is, the role of the radiation shield is to increase the thermal resistance to radiation, hence to reduce the heat transfer rate. The lower the emissivity of the shield, the higher the thermal resistance.

To illustrate this matter, we first consider radiation heat transfer between two large, opaque parallel plates. Let T_1 and T_2 be the temperatures and ε_1 and ε_2 the emissivities of the surfaces. Then the heat transfer rate Q_0 across an area A through the plates is determined from Example 12-20 as

$$\boxed{Q_0 = \frac{A\sigma(T_1^4 - T_2^4)}{1/\varepsilon_1 + 1/\varepsilon_2 - 1}} \quad \text{W} \qquad (12\text{-}99)$$

We now consider a radiation shield placed between the plates. Let $\varepsilon_{3,1}$ and $\varepsilon_{3,2}$ be the emissivities of the shield at the surfaces facing plates 1 and 2, respectively. The radiation network for the assembly with one shield can be constructed by utilizing Eqs. (12-95) and (12-96). Figure 12-30 shows the radiation network for a parallel-plate assembly with one shield between the plates. By utilizing this network and

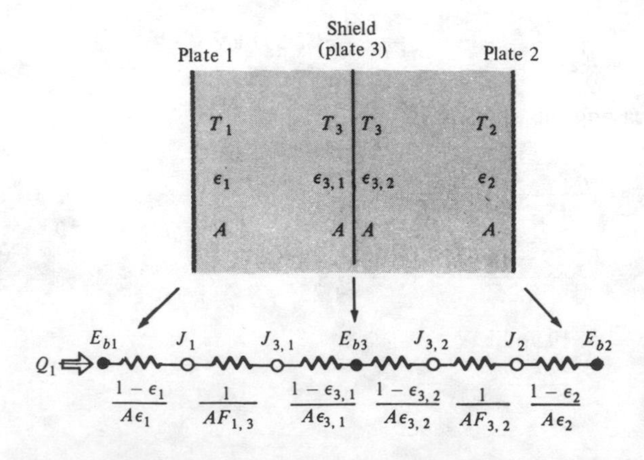

Figure 12-30 Plate-shield assembly and the corresponding radiation network.

noting that $F_{1,3} = F_{3,2} = 1$ for large parallel plates, the heat transfer rate Q_1 across the system with one shield becomes

$$Q_1 = \frac{A\sigma(T_1^4 - T_2^4)}{1/\varepsilon_1 + (1 - \varepsilon_{3,1})/\varepsilon_{3,1} + (1 - \varepsilon_{3,2})/\varepsilon_{3,2} + 1/\varepsilon_2}$$

$$\boxed{Q_1 = \frac{A\sigma(T_1^4 - T_2^4)}{(1/\varepsilon_1 + 1/\varepsilon_2 - 1) + (1/\varepsilon_{3,1} + 1/\varepsilon_{3,2} - 1)}} \quad \text{W} \quad (12\text{-}100)$$

A comparison of Eq. (12-100) with Eq. (12-99) shows that the effect of the radiation shield is to increase the thermal resistance of the system.

If the emissivities of all surfaces are equal, Eq. (12-100) reduces to

$$Q_1 = \frac{A\sigma(T_1^4 - T_2^4)}{2(2/\varepsilon - 1)} \quad \text{W} \quad (12\text{-}101)$$

For a parallel-plate system containing N shields and having the emissivities of all surfaces equal, Eq. (12-101) is generalized as

$$Q_N = \frac{A\sigma(T_1^4 - T_2^4)}{(N + 1)(2/\varepsilon - 1)} \quad \text{W} \quad (12\text{-}102)$$

The ratio of heat transfer rates for parallel-plate systems having N shields and no shield, when all emissivities are equal, is determined from Eq. (12-102):

$$\frac{Q_N}{Q_0} = \frac{1}{N + 1} \quad (12\text{-}103)$$

Concentric Spheres and Long Cylinders

We now consider two concentric spheres or long cylinders with opaque surfaces. Let A_1 and A_2 be the surface areas, T_1 and T_2 the temperatures, and ε_1 and ε_2 the emissivities of the inner and outer surfaces, respectively. The radiation heat transfer across these surfaces can be determined by utilizing Eq. (12-94) and noting that $F_{1-2} = 1$. We obtain

$$\boxed{Q_0 = \frac{A_1\sigma(T_1^4 - T_2^4)}{1/\varepsilon_1 + (A_1/A_2)(1/\varepsilon_2 - 1)}} \quad \text{W} \quad (12\text{-}104)$$

For $A_1 = A_2 \equiv A$, this result reduces to Eq. (12-99) for parallel plates.

We now consider the situation in which a radiation shield is placed between the surfaces. Let $\varepsilon_{3,1}$ and $\varepsilon_{3,2}$ be the emissivities of the shield at the surfaces facing the inner and outer surfaces of the assembly, respectively. The rate of heat transfer across this system containing a shield is determined by utilizing the radiation network shown in Fig. 12-30, using the proper area relations and noting that $F_{1,3} = F_{3,2} = 1$ according to the Example 12-16, case 3. After some manipulation we obtain

$$\boxed{Q_1 = \frac{A_1\sigma(T_1^4 - T_2^4)}{1/\varepsilon_1 + (A_1/A_2)(1/\varepsilon_2 - 1) + (A_1/A_3)(1/\varepsilon_{3,1} + 1/\varepsilon_{3,2} - 1)}} \quad \text{W} \quad (12\text{-}105)$$

A comparison of Eq. (12-105) with Eq. (12-104) clearly shows the contribution of the radiation shield to the thermal resistance to radiation.

For $A_1 = A_2 = A_3 \equiv A$, Eq. (12-105) reduces to Eq. (12-100) for parallel plates.

Example 12-24 Two large parallel plates at $T_1 = 800$ K and $T_2 = 600$ K have emissivities $\varepsilon_1 = 0.5$ and $\varepsilon_2 = 0.8$, respectively. A radiation shield having an emissivity $\varepsilon_{3,1} = 0.1$ on one side and an emissivity $\varepsilon_{3,2} = 0.05$ on the other side is placed between the plates. Calculate the heat transfer rate by radiation per square meter with and without the radiation shield.

SOLUTION The heat transfer rate without the shield is determined by Eq. (12-99):

$$\frac{Q_0}{A} = \frac{\sigma(T_1^4 - T_2^4)}{1/\varepsilon_1 + 1/\varepsilon_2 - 1}$$

The numerical values are introduced:

$$\frac{Q_0}{A} = \frac{(5.67 \times 10^{-8})(800^4 - 600^4)}{1/0.5 + 1/0.8 - 1} = \frac{15{,}876}{2.25} = 7056 \text{ W/m}^2$$

The heat transfer rate with the shield is determined by Eq. (12-100):

$$\frac{Q_1}{A} = \frac{\sigma(T_1^4 - T_2^4)}{(1/\varepsilon_1 + 1/\varepsilon_2 - 1) + (1/\varepsilon_{3,1} + 1/\varepsilon_{3,2} - 1)}$$

The numerical values are introduced:

$$\frac{Q_1}{A} = \frac{(5.67 \times 10^{-8})(800^4 - 600^4)}{(1/0.5 + 1/0.8 - 1) + (1/0.1 + 1/0.05 - 1)} = \frac{15{,}876}{31.24} = 508 \text{ W/m}^2$$

The presence of the radiation shield reduces the heat transfer rate significantly.

Example 12-25 A spherical tank with diameter $D_1 = 40$ cm filled with a cryogenic fluid at $T_1 = 100$ K is placed inside a spherical container of diameter $D_2 = 60$ cm and is maintained $T_2 = 300$ K. The emissivities of the inner and outer tanks are $\varepsilon_1 = 0.15$ and $\varepsilon_2 = 0.2$, respectively. A spherical radiation shield of diameter $D_3 = 50$ cm and having an emissivity $\varepsilon_3 = 0.05$ on both surfaces is placed between the spheres.

Calculate the rate of heat loss from the system by radiation. Then find the rate of evaporation of the cryogenic liquid for $h_{fg} = 2.1 \times 10^5$ W · s/kg.

SOLUTION The rate of heat transfer for the system is determined by Eq. (12-105):

$$Q_1 = \frac{A_1\sigma(T_1^4 - T_2^4)}{1/\varepsilon_1 + (A_1/A_2)(1/\varepsilon_2 - 1) + (A_1/A_3)(1/\varepsilon_{3,1} + 1/\varepsilon_{3,2} - 1)}$$

The numerical values are introduced:

$$Q_1 = \frac{\pi(0.4)^2(5.67 \times 10^{-8})(100^4 - 300^4)}{1/0.15 + (\frac{4}{6})^2(1/0.2 - 1) + (\frac{4}{5})^2(1/0.05 + 1/0.05 - 1)}$$

$$= \frac{-228}{33.4} = -6.83 \text{ W}$$

The evaporation rate M becomes

$$M = \frac{Q_1}{h_{fg}} = \frac{6.83 \text{ W}}{2.1 \times 10^5 \text{ W} \cdot \text{s/kg}} = 3.25 \times 10^{-5} \text{ kg/s}$$

Example 12-26 Two parallel plates are at temperatures T_1 and T_2 and have emissivities $\varepsilon_1 = 0.8$ and $\varepsilon_2 = 0.5$. A radiation shield having the same emissivity ε_3 on both sides is placed between the plates. Calculate the emissivity ε_3 of the shield in order to reduce the radiation loss from the system to one-tenth of that without the shield.

SOLUTION The heat transfer ratio for the cases with and without the shield is obtained from Eqs. (12-99) and (12-100):

$$\frac{Q_1}{Q_0} = \frac{1/\varepsilon_1 + 1/\varepsilon_2 - 1}{(1/\varepsilon_1 + 1/\varepsilon_2 - 1) + (1/\varepsilon_{3,1} + 1/\varepsilon_{3,2} - 1)}$$

The numerical values are introduced:

$$\frac{1}{10} = \frac{1/0.8 + 1/0.5 - 1}{(1/0.8 + 1/0.5 - 1) + (1/\varepsilon_3 + 1/\varepsilon_3 - 1)}$$

The solution is

$$\varepsilon_3 = 0.094$$

12-10 RADIOSITY MATRIX METHOD FOR RADIATION EXCHANGE IN AN ENCLOSURE

The network method is quite easy to apply for determining the radiation exchange in simple enclosures; as the number of zones increases, the manipulations involved to solve the problem becomes enormous, and the method is not so practical. A more direct and simple approach is to transform the problem to the solution of an algebraic matrix equation for the unknown radiosities J_i ($i = 1, 2, \ldots, N$). Once the resulting matrix equation is solved with a digital computer by using any of the standard solution techniques and the radiosities are known, then the net radiation heat flux or the surface temperature at any one of the N zones is immediately determined from its definition.

Figure 12-31 illustrates an enclosure separated into N zones, each having a surface area $A_i, i = 1, 2, 3, \ldots, N$. Let $\alpha_i, \varepsilon_i, \rho_i$, and T_i be the absorptivity, emissivity,

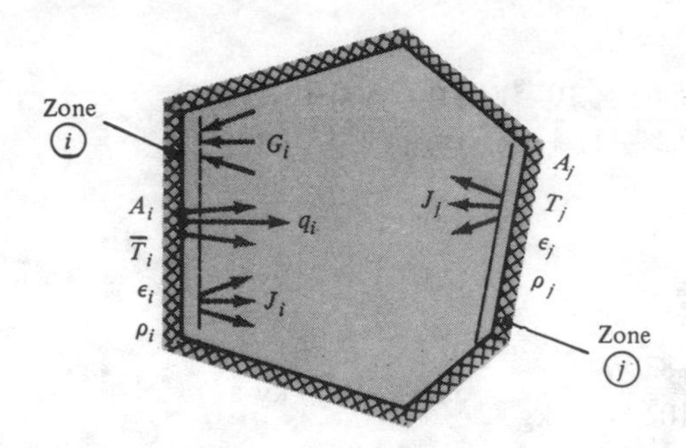

Figure 12-31 N-zone enclosure.

reflectivity, and temperature, respectively, of zone A_i. The assumptions made in connection with the network method of analysis are considered valid. For the graybody assumption made we have $\varepsilon_i = \alpha_i$, and for opaque surfaces we can write $\rho_i = 1 - \alpha_i = 1 - \varepsilon_i$. As discussed previously, we define

G_i = irradiation at zone i, W/m^2

J_i = radiosity at zone i, W/m^2

q_i = net radiation heat flux leaving zone i, W/m^2

As discussed previously, from Eqs. (12-81) and (12-82b), we write respectively,

$$q_i = J_i - G_i \tag{12-106}$$

$$J_i = \varepsilon_i E_{bi} + (1 - \varepsilon_i)G_i \tag{12-107}$$

For an N-zone enclosure illustrated in Fig. 12-31, the incident radiation G_i is related to the radiosities as now described.

The radiation energy leaving the surface A_j that strikes the surface A_i is

$$J_j A_j F_{j-i}$$

or by the reciprocity relation

$$J_j A_i F_{i-j}$$

Then the radiation from all the zones, including i, that strikes surface A_i is determined by summing this expression for $j = 1$ to N. We obtain

$$A_i \sum_{j=1}^{N} J_j F_{i-j}$$

When this expression is divided by A_i, it is equal to G_i; thus we have

$$G_i = \sum_{j=1}^{N} J_j F_{i-j} \tag{12-108}$$

Substituting G_i from Eq. (12-108) into Eqs. (12-106) and (12-107), we find

$$q_i = J_i - \sum_{j=1}^{N} J_j F_{i-j} \qquad i = 1, 2, \ldots, N \tag{12-109}$$

$$E_{bi} = \frac{1}{\varepsilon_i} J_i - \frac{1-\varepsilon_i}{\varepsilon_i} \sum_{j=1}^{N} J_j F_{i-j} \qquad i = 1, 2, \ldots, N \tag{12-110}$$

By eliminating the summation term between these two equations, a simpler expression is obtained for q_i:

$$q_i = \frac{\varepsilon_i}{1-\varepsilon_i}(E_{bi} - J_i) \tag{12-111}$$

Equations (12-109) *and* (12-110) *provide the fundamental relations for obtaining a system of N algebraic equations for the determination of the N unknown radiosities* J_j, $j = 1, 2, \ldots, N$. Once the radiosities are known, the net radiation heat flux q_i at any of zone i is computed either from Eq. (12-109) or from Eq. (12-111).

We now examine the applications of Eqs. (12-109) and (12-110) for the solution of radiation exchange in an N-zone enclosure. Two situations are of particular interest:

1. Temperatures are prescribed for each of the N zones.
2. Temperatures are prescribed for some of the zones, and the net radiation heat fluxes are prescribed for the remaining zones.

We consider these two situations separately.

Temperatures Prescribed for All Zones

Consider an N-zone enclosure in which the temperatures T_i ($i = 1, 2, \ldots, N$) are prescribed for all the surfaces A_i ($i = 1, 2, \ldots, N$). Also, the emissivities ε_i and surface areas A_i are known, and the view factors F_{i-j} between the zones are given. Our objective in this problem is the determination of the net radiation heat fluxes q_i for each zone.

To solve this problem, we consider Eq. (12-110), that is,

$$\frac{1}{\varepsilon_i} J_i - \frac{1-\varepsilon_i}{\varepsilon_i} \sum_{j=1}^{N} J_j F_{i-j} = E_{bi} \qquad i = 1, 2, \ldots, N \tag{12-112}$$

where $E_{bi} = \sigma T_i^4$ is known because the temperatures T_i are prescribed. Then the system (12-112) provides N simultaneous algebraic equations for the N unknown radiosities J_i, $i = 1, 2, \ldots, N$. Once the J_i are obtained from the solution of this system, the net radiation heat flux q_i for any zone i is determined from either Eq. (12-109) or Eq. (12-111).

For convenience in the computations, it is desirable to express Eq. (12-112) in matrix form as

$$[M][J] = [S] \tag{12-113}$$

where $[M]$ is the $N \times N$ coefficient matrix, defined as

$$[M] \equiv \begin{bmatrix} m_{11} & m_{12} & m_{13} & \cdots & m_{1N} \\ m_{21} & m_{22} & m_{23} & \cdots & m_{2N} \\ \cdots & \cdots & \cdots & \cdots & \cdots \\ m_{N1} & m_{N2} & m_{N3} & \cdots & m_{NN} \end{bmatrix} \tag{12-114a}$$

with its elements m_{ij} given by

$$m_{ij} = \frac{\delta_{ij} - (1 - \varepsilon_i)F_{i-j}}{\varepsilon_i} \tag{12-114b}$$

and δ_{ij} is the kronecker delta, defined as

$$\delta_{ij} = \begin{cases} 1 & \text{for } i = j \\ 0 & \text{for } i \neq j \end{cases} \tag{12-114c}$$

The *radiosity* vector $[J]$ and the *surface-input* vector $[S]$ are defined as

$$[J] = \begin{bmatrix} J_1 \\ J_2 \\ \vdots \\ J_N \end{bmatrix} \quad \text{and} \quad [S] = \begin{bmatrix} E_{b1} \\ E_{b2} \\ \vdots \\ E_{bN} \end{bmatrix} = \begin{bmatrix} \sigma T_1^4 \\ \sigma T_2^4 \\ \vdots \\ \sigma T_N^4 \end{bmatrix} \tag{12-115a, b}$$

For this particular case, the vector $[S]$ contains only the surface temperatures, because we consider an enclosure problem in which the temperatures at the zones are specified.

The matrix algebraic equation (12-113) for the unknown radiosities can be solved by a variety of standard computer subroutines available for the solution of a system of algebraic equations. For example, the programmable calculators can readily handle the solution of up to six equations. For a system involving more surfaces, the gaussian elimination procedure discussed in Chap. 5 can be used.

The *matrix inversion* method, well documented in computational practice, also can be used to solve such a system of equations. Suppose the matrix inversion method is used. The solution of the system of Eqs. (12-113) is formally written as

$$[J] = [M]^{-1}[S] \tag{12-116}$$

where $[M]^{-1}$ is the inverse of the coefficient matrix $[M]$. The computer computes $[M]^{-1}$ for a given $[M]$ and stores it. Standard matrix inversion subroutines are available at every computer installation.

Once the radiosities J_i are known from the solution of the system of Eqs. (12-113), the net radiation heat flux q_i, for any zone i for which the temperature is prescribed, is determined from Eq. (12-109) or (12-111). We obtain, respectively,

$$q_i = J_i - \sum_{j=1}^{N} J_i F_{i-j} \tag{12-117}$$

or

$$q_i = \frac{\varepsilon_i}{1 - \varepsilon_i}(E_{bi} - J_i) \tag{12-118}$$

The basic steps in the solution of the enclosure problem by the radiosity matrix method when the temperatures are prescribed for all zones can be summarized as follows:

1. Construct the *radiosity matrix equations* by using either system (12-112) or (12-113). The system (12-113) is more convenient for computational purposes.
2. Evaluate the view factors F_{i-j} and the blackbody emissive power $E_{bi} = \sigma T_i^4$, and specify the emissivities ε_i for all zones of the enclosure.
3. With the information available in step 2, solve the radiosity matrix equations and determine the radiosities J_i for all the zones.
4. Knowing the radiosities, calculate the net radiation heat flux q_i for any of the zones, using Eq. (12-117) or (12-118).

Example 12-27 Two large, opaque, gray, diffusely emitting, diffusely reflecting parallel plates have emissivities $\varepsilon_1 = 0.8$ and $\varepsilon_2 = 0.5$ and are maintained at uniform temperatures $T_1 = 800$ K and $T_2 = 600$ K. Using the radiosity matrix method, calculate the net radiation heat fluxes q_1 and q_2 at the surfaces of the plates.

Solution This is a two-zone enclosure problem with temperatures prescribed for both zones. The radiosity matrix equations are obtained by using Eqs. (12-113) and (12-114). We find

$$\begin{bmatrix} \dfrac{1-(1-\varepsilon_1)F_{11}}{\varepsilon_1} & \dfrac{0-(1-\varepsilon_1)F_{12}}{\varepsilon_1} \\ \dfrac{0-(1-\varepsilon_2)F_{21}}{\varepsilon_2} & \dfrac{1-(1-\varepsilon_2)F_{22}}{\varepsilon_2} \end{bmatrix} \begin{bmatrix} J_1 \\ J_2 \end{bmatrix} = \begin{bmatrix} \sigma T_1^4 \\ \sigma T_2^4 \end{bmatrix}$$

where the view factors are given by

$$F_{11} = F_{22} = 0 \qquad F_{12} = F_{21} = 1$$

Then the matrix equations reduce to

$$\begin{bmatrix} \dfrac{1}{\varepsilon_1} & -\dfrac{1-\varepsilon_1}{\varepsilon_1} \\ -\dfrac{1-\varepsilon_2}{\varepsilon_2} & \dfrac{1}{\varepsilon_2} \end{bmatrix} \begin{bmatrix} J_1 \\ J_2 \end{bmatrix} = \begin{bmatrix} \sigma T_1^4 \\ \sigma T_2^4 \end{bmatrix}$$

where

$$\varepsilon_1 = 0.8 \qquad \varepsilon_2 = 0.5$$

$$E_{b1} \equiv \sigma T_1^4 = (5.67 \times 10^{-8})(800^4) = 23.224 \text{ kW/m}^2$$

$$E_{b2} \equiv \sigma T_2^4 = (5.67 \times 10^{-8})(600^4) = 7.348 \text{ kW/m}^2$$

These numerical values are introduced into the above system. We obtain

$$\begin{bmatrix} 1.25 & -0.25 \\ -1 & 2 \end{bmatrix} \begin{bmatrix} J_1 \\ J_2 \end{bmatrix} = \begin{bmatrix} 23.224 \\ 7.348 \end{bmatrix}$$

and the solution gives the radiosities as

$$J_1 = 21.460 \text{ kW/m}^2 \qquad J_2 = 14.404 \text{ kW/m}^2$$

Equation (12-118) is now used to calculate the net radiation heat fluxes q_1 and q_2 at the surfaces of the plates:

$$q_1 = \frac{\varepsilon_1}{1 - \varepsilon_1}(E_{b1} - J_1) = \frac{0.8}{1 - 0.8}(23.224 - 21.460) = 7.056 \text{ kW/m}^2$$

$$q_2 = \frac{\varepsilon_2}{1 - \varepsilon_2}(E_{b2} - J_2) = \frac{0.5}{1 - 0.5}(7.348 - 14.404) = -7.056 \text{ kW/m}^2$$

As expected, $q_1 = -q_2$.

Example 12-28 An equilateral-triangle enclosure, illustrated in the accompanying figure, is sufficiently long in the direction perpendicular to the

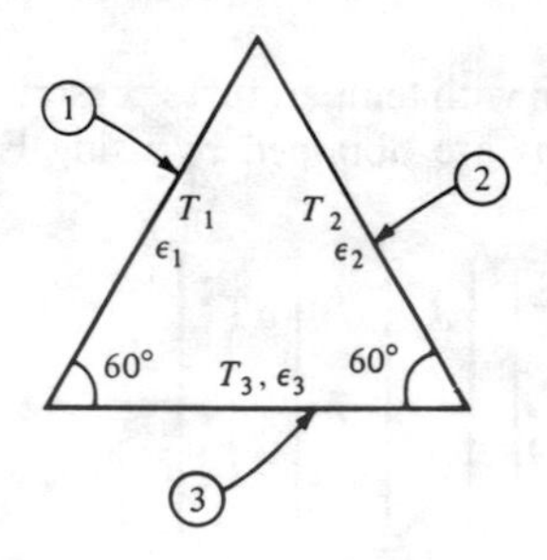

plane of the figure that the end effects can be neglected. Surfaces are opaque, gray, diffuse emitters, and diffuse reflectors and have emissivities

$$\varepsilon_1 = \varepsilon_2 = 0.9 \qquad \varepsilon_3 = 0.5$$

and are maintained at temperatures

$$T_1 = 800 \text{ K} \qquad T_2 = 600 \text{ K} \qquad T_3 = 300 \text{ K}$$

Calculate the net radiation heat fluxes q_i $(i = 1, 2, 3)$ for each of the three zones.

SOLUTION This is a three-zone enclosure problem with prescribed temperature for each zone. The governing matrix equation for the radiosities is written from Eqs. (12-113) and (12-114) as

$$\begin{bmatrix} \dfrac{1-(1-\varepsilon_1)F_{11}}{\varepsilon_1} & \dfrac{0-(1-\varepsilon_1)F_{12}}{\varepsilon_1} & \dfrac{0-(1-\varepsilon_1)F_{13}}{\varepsilon_1} \\ \dfrac{0-(1-\varepsilon_2)F_{21}}{\varepsilon_2} & \dfrac{1-(1-\varepsilon_2)F_{22}}{\varepsilon_2} & \dfrac{0-(1-\varepsilon_2)F_{23}}{\varepsilon_2} \\ \dfrac{0-(1-\varepsilon_3)F_{31}}{\varepsilon_3} & \dfrac{0-(1-\varepsilon_3)F_{32}}{\varepsilon_3} & \dfrac{1-(1-\varepsilon_3)F_{33}}{\varepsilon_3} \end{bmatrix} \begin{bmatrix} J_1 \\ J_2 \\ J_3 \end{bmatrix} = \begin{bmatrix} \sigma T_1^4 \\ \sigma T_2^4 \\ \sigma T_3^4 \end{bmatrix}$$

where

$$F_{11} = F_{22} = F_{33} = 0 \qquad F_{ij} = 0.5 \quad \text{for } i \neq j$$

$$\varepsilon_1 = \varepsilon_2 = 0.8 \qquad \varepsilon_3 = 0.5$$

$$\sigma T_1^4 = 23.224 \qquad \sigma T_2^4 = 7.348 \qquad \sigma T_3^4 = 0.459 \text{ kW/m}^2$$

Introducing these values into the above matrix equation, we obtain

$$\begin{bmatrix} 1.25 & -0.125 & -0.125 \\ -0.125 & 1.25 & -0.125 \\ -0.5 & -0.5 & 2 \end{bmatrix} \begin{bmatrix} J_1 \\ J_2 \\ J_3 \end{bmatrix} = \begin{bmatrix} 23.224 \\ 7.348 \\ 0.459 \end{bmatrix}$$

The solution gives

$$J_1 = 20.187 \qquad J_2 = 8.641 \qquad J_3 = 7.436 \text{ kW/m}^2$$

Equation (12-111) is now used to determine the net radiation heat fluxes:

$$q_1 = \frac{\varepsilon_1}{1-\varepsilon_1}(E_{b1} - J_1) = \frac{0.8}{0.2}(23.224 - 20.187) = 12.148 \text{ kW/m}^2$$

$$q_2 = \frac{\varepsilon_2}{1-\varepsilon_2}(E_{b2} - J_2) = \frac{0.8}{0.2}(7.348 - 8.641) = -5.172 \text{ kW/m}^2$$

$$q_3 = \frac{\varepsilon_3}{1-\varepsilon_3}(E_{b3} - J_3) = \frac{0.5}{0.5}(0.459 - 7.436) = -6.977$$

Since the surface areas for each zone are equal, we should have

$$q_1 + q_2 + q_3 = 0$$

The numerical results check this equality.

Temperature Prescribed for Some Zones and Net Heat Flux Prescribed for Others

In many practical applications, temperatures are prescribed for some of the zones and the net heat fluxes for the remaining zones of an enclosure. In such problems, the interest lies in the determination of net radiation heat fluxes for the surfaces for which temperatures are prescribed and the determination of the temperatures of the surfaces for which the net heat fluxes are specified. This problem also can be solved by utilizing Eqs. (12-109) and (12-110) as now described.

We consider again an N-zone enclosure and assume

1. Temperatures T_i are prescribed for zones $i = 1, 2, 3, \ldots, k$.
2. The net heat fluxes q_i are prescribed for the remaining zones $i = k + 1, k + 2, \ldots, N$.

We need N equations for the determination of N unknown radiosities J_i, $i = 1, 2, \ldots, N$. These equations are obtained from Eqs. (12-109) and (12-110) with the following considerations.

For the zones $i = 1, 2, 3, \ldots, k$ with prescribed surface temperatures T_i, we use Eqs. (12-110), that is,

$$\frac{1}{\varepsilon_i} J_i - \frac{1 - \varepsilon_i}{\varepsilon_i} \sum_{j=1}^{N} J_j F_{i-j} = E_{bi} \qquad i = 1, 2, \ldots, k \tag{12-119}$$

And for zones $i = k + 1, k + 2, \ldots, N$ with prescribed net heat flux q_i, we use Eqs. (12-109), that is,

$$J_i - \sum_{j=1}^{N} J_j F_{i-j} = q_i \qquad i = k + 1, k + 2, \ldots, N \tag{12-120}$$

Equations (12-119) and (12-120) provide N simultaneous algebraic equations for the determination of N unknown radiosities J_i, $i = 1, 2, \ldots, N$.

For computational purposes, it is more convenient to express Eqs. (12-119) and (12-120) in matrix form as

$$[M][J] = [S] \tag{12-121}$$

where $[M]$ is the $N \times N$ coefficient matrix, defined as

$$[M] \equiv \begin{bmatrix} m_{11} & m_{12} & m_{13} & \cdots & m_{1N} \\ m_{21} & m_{22} & m_{23} & \cdots & m_{2N} \\ \cdots & \cdots & \cdots & \cdots & \cdots \\ m_{N1} & m_{N2} & m_{N3} & \cdots & m_{NN} \end{bmatrix} \tag{12-122a}$$

with its elements m_{ij} given by

$$m_{ij} = \frac{\delta_{ij} - (1 - \varepsilon_i) F_{i-j}}{\varepsilon_i} \qquad \text{for } i = 1, 2, \ldots, k \tag{12-122b}$$

$$m_{ij} = \delta_{ij} - F_{i-j} \qquad \text{for } i = k + 1, k + 2, \ldots, N \tag{12-122c}$$

where δ_{ij} is the kronecker delta, defined as

$$\delta_{ij} = \begin{cases} 1 & \text{for } i = j \\ 0 & \text{for } i \neq j \end{cases} \tag{12-122d}$$

The *radiosity* vector $[J]$ and the *surface-input* vector $[S]$ are defined as

$$[J] = \begin{bmatrix} J_1 \\ J_2 \\ \vdots \\ J_N \end{bmatrix} \quad \text{and} \quad [S] = \begin{bmatrix} E_{b1} \\ \vdots \\ E_{bk} \\ q_{k+1} \\ \vdots \\ q_N \end{bmatrix} = \begin{bmatrix} \sigma T_1^4 \\ \vdots \\ \sigma T_k^4 \\ q_{k+1} \\ \vdots \\ q_N \end{bmatrix} \tag{12-122e, f}$$

We note that the elements of the surface input vector contain prescribed surface temperatures and the prescribed net surface heat fluxes. Similarly, the elements m_{ij} of the coefficient matrix contain those appropriate for the prescribed surface temperature for $i = 1, 2, \ldots, k$ and those appropriate for the prescribed net surface heat flux for $i = k + 1, \ldots, N$. Now the problem is reduced to that of solving the matrix equation (12-121) and determining the radiosities J_i ($i = 1, 2, \ldots, N$). The matrix equation can be solved by a variety of methods discussed previously.

Given the radiosities J_i, the net radiation heat flux q_i for any zone with prescribed temperature T_i is determined from Eq. (12-109) or (12-111), that is,

$$q_i = J_i - \sum_{j=1}^{N} J_j F_{i-j} \tag{12-123a}$$

or

$$q_i = \frac{\varepsilon_i}{1 - \varepsilon_i}(E_{bi} - J_i) \tag{12-123b}$$

for $i = 1, 2, \ldots, k$.

The temperature T_i for a zone with a prescribed heat flux q_i is determined from either Eq. (12-110) or Eq. (12-111), that is, from

$$\sigma T_i^4 = \frac{1}{\varepsilon_i} J_i - \frac{1 - \varepsilon_i}{\varepsilon_i} \sum_{j=1}^{N} J_j F_{i-j} \tag{12-124a}$$

or

$$\sigma T_i^4 = J_i + \frac{1 - \varepsilon_i}{\varepsilon_i} q_i \tag{12-124b}$$

for $i = k + 1, k + 2, \ldots, N$.

For a *reradiating* surface, the net heat flux q_i on that surface is zero. Then Eq. (12-124b) requires that

$$\sigma T_i^4 = J_i \qquad \text{(for reradiating surface)} \tag{12-125}$$

We now summarize the computational procedure for the problem in which some surfaces have prescribed temperature and others have prescribed net heat flux.

1. Construct the radiosity matrix equations by using system (12-121).
2. Evaluate the view factors F_{i-j} for all the zones and the blackbody emissive power $E_{bi} = \sigma T_i^4$ for the zones with prescribed temperature. Specify the q_i for the zones with prescribed heat flux and the emissivities ε_i for all the zones.

3. With the information available in step 2, solve the matrix equations (12-121) and determine the radiosities J_i for all the zones.
4. Knowing the radiosities, calculate the net radiation heat fluxes from Eq. (12-123*a*) or Eq. (12-123*b*) for zones $i = 1, 2, \ldots, k$ with prescribed surface temperature. Then calculate the surface temperatures from Eq. (12-124*a*) or Eq. (12-124*b*) for zones $i = k + 1, k + 2, \ldots, N$ with prescribed heat flux.

Example 12-29 In a cubical oven, the top wall is maintained at $T_1 = 800$ K and has an emissivity $\varepsilon_1 = 0.8$, the floor is at $T_2 = 600$ K and has an emissivity $\varepsilon_2 = 0.8$, and the four lateral walls are reradiating surfaces. Calculate the net radiation heat flux leaving the top surface.

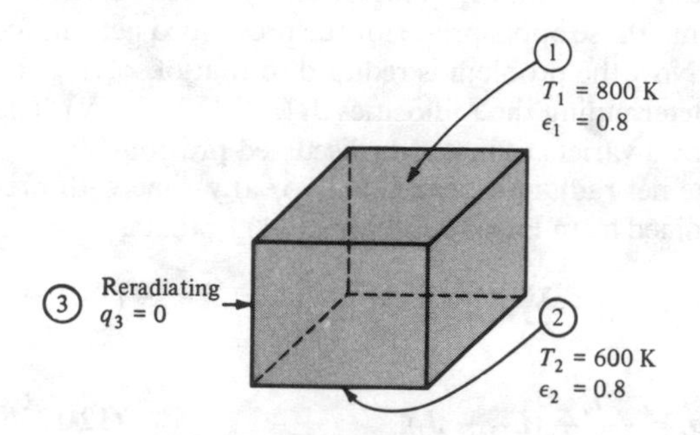

SOLUTION This is a three-zone problem with two of the zones having prescribed surface temperatures and the third zone having a zero net radiation heat flux. The governing equations for the radiosities are obtained from Eq. (12-121) with $k = 2$ and $N = 3$. We find

$$\begin{bmatrix} \dfrac{1-(1-\varepsilon_1)F_{11}}{\varepsilon_1} & \dfrac{0-(1-\varepsilon_1)F_{12}}{\varepsilon_1} & \dfrac{0-(1-\varepsilon_1)F_{13}}{\varepsilon_1} \\ \dfrac{0-(1-\varepsilon_2)F_{21}}{\varepsilon_2} & \dfrac{1-(1-\varepsilon_2)F_{22}}{\varepsilon_2} & \dfrac{0-(1-\varepsilon_2)F_{23}}{\varepsilon_2} \\ 0-F_{31} & 0-F_{32} & 1-F_{33} \end{bmatrix} \begin{bmatrix} J_1 \\ J_2 \\ J_3 \end{bmatrix} = \begin{bmatrix} \sigma T_1^4 \\ \sigma T_2^4 \\ q_3 \end{bmatrix}$$

where

$$\varepsilon_1 = \varepsilon_2 = 0.8$$

$$F_{11} = F_{22} = 0 \qquad F_{12} = F_{21} = 0.2 \qquad F_{13} = F_{23} = 0.8$$

$$F_{31} = F_{32} = \frac{A_2}{A_3} F_{23} = 0.2 \qquad F_{33} = 1 - F_{31} - F_{32} = 0.6$$

$$\sigma T_1^4 = (5.67 \times 10^{-8})(800^4) = 23.224 \text{ kW/m}^2$$

$$\sigma T_2^4 = (5.67 \times 10^{-8})(600^4) = 7.348 \text{ kW/m}^2$$

$$q_3 = 0$$

These numerical values are introduced into the above matrix equations.

$$\begin{bmatrix} 1.25 & -0.05 & -0.2 \\ -0.05 & 1.25 & -0.2 \\ -0.2 & -0.2 & 0.4 \end{bmatrix} \begin{bmatrix} J_1 \\ J_2 \\ J_3 \end{bmatrix} = \begin{bmatrix} 23.224 \\ 7.348 \\ 0 \end{bmatrix}$$

The solution gives

$$J_1 = 21.392 \qquad J_2 = 9.179 \qquad J_3 = 15.286 \text{ kW/m}^2$$

Equation (12-123*b*) is used to determine the net radiation fluxes at surfaces 1 and 2:

$$q_1 = \frac{\varepsilon_1}{1 - \varepsilon_1}(\sigma T_1^4 - J_1) = \frac{0.8}{1 - 0.8}(23.224 - 21.392) = 7.33 \text{ kW/m}^2$$

$$q_2 = \frac{\varepsilon_2}{1 - \varepsilon_2}(\sigma T_2^4 - J_2) = \frac{0.8}{1 - 0.8}(7.348 - 9.179) = -7.33 \text{ kW/m}^2$$

The temperature of the reradiating zone is determined from

$$\sigma T_3^4 = J_3$$

$$T_3 = \left(\frac{15{,}286}{5.67 \times 10^{-8}}\right)^{1/4} = 720 \text{ K}$$

We note that $q_1 = -q_2$, as expected.

12-11 CORRECTION FOR RADIATION EFFECTS IN TEMPERATURE MEASUREMENTS

When a thermometer or thermocouple is used to measure the temperature of a fluid surrounded by an ambient at a temperature different from that of the fluid itself, then the temperature reading is affected by convection to the fluid and radiation to the surrounding ambient. The true temperature of the fluid can be determined only after the measured reading is corrected for the radiation effects with an energy balance at the thermocouple (or thermometer) junction. By assuming that the surrounding is much larger than the size of the measuring device and that thermal equilibrium is established, the energy balance equation can be written as

$$\begin{pmatrix}\text{Rate of heat loss by} \\ \text{convection to fluid}\end{pmatrix} + \begin{pmatrix}\text{rate of heat loss by} \\ \text{radiation into surrounding}\end{pmatrix} = 0 \quad (12\text{-}126a)$$

$$h(T_t - T_\infty) + \varepsilon\sigma(T_t^4 - T_s^4) = 0 \quad (12\text{-}126b)$$

where h = heat transfer coefficient

T_t = thermocouple reading (measured value)

T_∞ = true temperature of fluid

T_s = temperature of surrounding

ε = emissivity of thermocouple

Then the true temperature of the fluid T_∞ can be calculated from Eq. (12-126), since all the other quantities are considered known.

Example 12-30 Hot air flows inside a duct whose walls are maintained at $T_s = 300$ K. The thermocouple inserted into the gas stream reads $T_t = 500$ K. Assuming $\varepsilon = 0.9$ and $h = 300$ W/(m$^2 \cdot$ °C), calculate the true air temperature T_∞.

SOLUTION We apply the energy balance given by Eq. (12-126*b*):

$$h(T_t - T_\infty) + \varepsilon\sigma(T_t^4 - T_s^4) = 0$$

$$300(500 - T_\infty) + (0.9)(5.67 \times 10^{-8})(500^4 - 300^4) = 0$$

Then

$$T_\infty = 509.25 \text{ K}$$

That is, the radiation effect on the thermocouple reading is 9.25°C.

12-12 SUMMARY OF EQUATIONS

We summarize in Table 12-3 the fundamental relations presented in this chapter.

Table 12-3 Summary of equations

Equation number	Description	Equation
(12-5)	Solid angle	$d\omega = \dfrac{dA}{r^2}$
(12-11)	Spectral blackbody emissive power	$E_{b\lambda}(T) = \pi I_{b\lambda}(T)$ W/(m$^2 \cdot \mu$m)
(12-13)	Wien's displacement law	$(\lambda T)_{max} = 2897.6$ μm $\cdot$ K
(12-15)	Blackbody emissive power	$E_b(T) = \sigma T^4$ W/m^2
(12-18)	Blackbody radiation intensity	$I_b(T) = \dfrac{1}{\pi}\sigma T^4$ W/(m$^2 \cdot$ sr)
(12-41)	Spectral hemispherical emissivity	$\varepsilon_\lambda = \dfrac{q_\lambda(T)}{E_{b\lambda}(T)}$
(12-42)	Hemispherical emissivity	$\varepsilon = \dfrac{\int_0^\infty \varepsilon_\lambda E_{b\lambda}(T)\, d\lambda}{E_b(T)}$

Equation number	Description	Equation
(12-43)	Spectral hemispherical absorptivity	$\alpha_\lambda = \dfrac{q_\lambda^a(T)}{q_\lambda^i(T)}$
(12-44)	Hemispherical absorptivity	$\alpha = \dfrac{\int_0^\infty \alpha_\lambda q_\lambda^i(T)\, d\lambda}{\int_0^\infty q_\lambda^i(T)\, d\lambda}$
(12-45)	Spectral hemispherical reflectivity	$\rho_\lambda = \dfrac{q_\lambda^r(T)}{q_\lambda^i(T)}$
(12-46)	Hemispherical reflectivity	$\rho = \dfrac{\int_0^\infty \rho_\lambda q_\lambda^i(T)\, d\lambda}{\int_0^\infty q_\lambda^i(T)\, d\lambda}$
(12-47)	Spectral hemispherical transmissivity	$\tau_\lambda = \dfrac{q_\lambda^{tr}(T)}{q_\lambda^i(T)}$
(12-48)	Hemispherical transmissivity	$\tau = \dfrac{\int_0^\infty \tau_\lambda q_\lambda^i(T)\, d\lambda}{\int_0^\infty q_\lambda^i(T)\, d\lambda}$
(12-60)	Elemental view factor from area dA_1 to dA_2	$dF_{dA_1 - dA_2} = \dfrac{\cos\theta_1 \cos\theta_2\, dA_2}{\pi r^2}$
(12-69)	Reciprocity relation	$A_i F_{A_i - A_j} = A_j F_{A_j - A_i}$
(12-70)	Summation relation	$\sum_{k=1}^{N} F_{A_i - A_k} = 1$
(12-86*a*)	Total net radiation heat flow leaving surface A_i	$Q_i = \dfrac{E_{bi} - J_i}{R_i}$ W
(12-86*b*)		where $R_i = \dfrac{1 - \varepsilon_i}{A_i \varepsilon_i}$
(12-87)	Radiation flux leaving a black surface	$J_i = E_{bi} = \sigma T_i^4$ W/m²
(12-99)	Radiation heat transfer rate Q_0 across an area A through parallel plates	$Q_0 = \dfrac{A\sigma(T_1^4 - T_2^4)}{1/\varepsilon_1 + 1/\varepsilon_2 - 1}$
(12-104)	Radiation heat transfer rate Q_0 across concentric cylinders or spheres	$Q_0 = \dfrac{A_1\sigma(T_1^4 - T_2^4)}{1/\varepsilon_1 + (A_1/A_2)(1/\varepsilon_2 - 1)}$
(12-109)	Fundamental equations of radiosity matrix method of analysis for an N-zone enclosure	$q_i = J_i - \sum_{j=1}^{N} J_j F_{i-j} \qquad i = 1, 2, \ldots, N$
(12-110)		$E_{bi} = \dfrac{1}{\varepsilon_i} J_i - \dfrac{1 - \varepsilon_i}{\varepsilon_i} \sum_{j=1}^{N} J_j F_{i-j} \qquad i = 1, 2, \ldots, N$
(12-111)		$q_i = \dfrac{\varepsilon_i}{1 - \varepsilon_i}(E_{bi} - J_i)$

PROBLEMS

Blackbody radiation

12-1 The sun radiates as a blackbody at an effective surface temperature $T = 5762$ K. What fraction of the total energy is in the (*a*) ultraviolet (that is, $\lambda = 0.01$ to $0.4\ \mu$m), (*b*) visible (that is, $\lambda = 0.4$ to $0.7\ \mu$m), and (*c*) infrared (that is, $\lambda = 0.7$ to $1000\ \mu$m) regions?

Answer: (*a*) 12.11, (*b*) 36.57, (*c*) 51.32 percent

12-2 A tungsten filament is heated to 2500 K. What is the maximum radiative-heat flux from the filament, and what fraction of this energy is in the visible range (that is, $\lambda = 0.4$ to $0.7\ \mu$m)?

12-3 A red-hot surface is at 2000 K. What fraction of the total radiation emitted is in the following wavelength bands? (Assume blackbody radiation.)

$$\Delta\lambda_1 = 1 \text{ to } 5\ \mu\text{m} \qquad \Delta\lambda_3 = 10 \text{ to } 15\ \mu\text{m}$$

$$\Delta\lambda_2 = 5 \text{ to } 10\ \mu\text{m} \qquad \Delta\lambda_4 = 15 \text{ to } 20\ \mu\text{m}$$

12-4 Determine the radiative energy emitted between 2- and 10-μm wavelengths by a 1 m by 1 m gray surface at 600 K which has an emissivity $\varepsilon = 0.8$.

Answer: 4309 W

12-5 Calculate the blackbody emissive power $E_b(T)$ and the blackbody radiation intensity $I_b(T)$ at (*a*) 500, (*b*) 1000, (*c*) 3000, and (*d*) 5762 K.

12-6 The spectral emissive power E_λ for a diffusely emitting surface is

$$E_\lambda = \begin{cases} 0 & \text{for } \lambda < 3\ \mu\text{m} \\ 150\ \text{W/(m}^2\cdot\mu\text{m)} & \text{for } 3 < \lambda < 12\ \mu\text{m} \\ 300\ \text{W/(m}^2\cdot\mu\text{m)} & \text{for } 12 < \lambda < 25\ \mu\text{m} \\ 0 & \text{for } \lambda > 25\ \mu\text{m} \end{cases}$$

(*a*) Calculate the total emissive power of the surface over the entire wavelengths.
(*b*) Calculate the intensity of radiation, assuming I is independent of direction.

12-7 There is a 0.25-cm-diameter hole on a large spherical enclosure whose inner surface is maintained at 600 K. Determine the rate of emission of radiative energy through this opening.

Answer: 3.61×10^{-2} W

12-8 A tungsten filament is heated to 2300 K. What fraction of the total energy is emitted in the wavelength range $\lambda = 0.4$ to $0.8\ \mu$m?

12-9 A laboratory black enclosure at 900 K has a small opening into the atmosphere. Calculate (*a*) the blackbody radiation intensity emerging from the opening and (*b*) the blackbody radiation heat flux from the enclosure.

12-10 A blackbody radiation from a source at T K is incident on a quartz sheet which transmits 90 percent of the incident radiation in the wavelength range $\lambda = 0.2$ to $4\ \mu$m and is opaque for other wavelengths. Calculate the percentages of the incident radiation transmitted through the quartz sheet for each of the blackbody radiation sources at 600, 1200, and 2500 K.

12-11 Calculate the spectral blackbody emissive power $E_{\lambda b}(T)$, in W/(m$^2\cdot\mu$m), at $\lambda = 2$, 5, and 10 μm for a surface at 1000 K.

12-12 Consider a blackbody at 1449 K emitting into air.

(*a*) Determine the wavelength at which the blackbody spectral emissive power $E_{\lambda b}(T)$ is maximum.
(*b*) Calculate the corresponding spectral emissive power and the spectral blackbody radiation intensity.

12-13 A blackbody at 1111 K is emitting into air.

(*a*) Calculate the wavelength at which the blackbody emissive power is maximum.
(*b*) Calculate the energy emitted over the wavelength $\lambda = 1$ to $10\ \mu$m and $\lambda = 10$ to $20\ \mu$m.

Answer: (*a*) 2.61 μm; (*b*) 80,554 and 4810 W/m^2

12-14 A blackbody irradiates such that the maximum spectral emissive power $E_{\lambda b}$ occurs at $\lambda = 3\ \mu m$.

(*a*) What is the temperature of the surface.

(*b*) What is the magnitude of the spectral emissive power $E_{\lambda b}(T)$ at this temperature?

12-15 What is the temperature of a blackbody such that 50 percent of the energy emitted should lie in the wavelength spectrum $\lambda = 0$ to $10\ \mu m$?

12-16 Consider a blackbody emitting at a temperature T K. What fraction of the total hemispherical emissive power is irradiated into a solid angle subtended by $0° \leq \theta \leq 30°$ and $0 \leq \phi \leq 2\pi$?

Answer: 25 percent.

12-17 A large blackbody enclosure has a small opening of area $A_0 = 1\ cm^2$. The radiant energy emitted by the opening is 5.67 W. Determine the temperature of the blackbody enclosure.

12-18 Calculate the wavelengths at which the emission of radiation is maximum by a blackbody at the following temperatures:

(*a*) The effective surface temperature of the sun, 5762 K

(*b*) A tungsten filament at 2300 K

(*c*) A body at room temperature, 300 K

(*d*) A body at 100 K

12-19 Consider a blackbody enclosure at 800 K having an opening $A = 2\ cm^2$. Calculate (*a*) the amount of radiant energy emitted through the opening and (*b*) the intensity of blackbody radiation.

Answer: (*a*) 4.64 W, (*b*) 7392 W/(m^2 · sr)

12-20 Consider a surface emitting as blackbody at 2000 K.

(*a*) Calculate the rate of emission per unit area through a solid angle subtended by $0° \leq \theta \leq 30°$, $0 \leq \phi \leq 2\pi$, over all wavelengths.

(*b*) Find the rate of emission per unit area through a solid angle subtended by $0° \leq \theta \leq 30°$, $0 \leq \phi \leq 2\pi$ over $0 < \lambda < 3\ \mu m$.

12-21 Calculate the fraction of the total radiation energy emitted in the range of wavelengths from $\lambda = 5$ to $20\ \mu m$ by a graybody at 500 K.

12-22 A small surface of area $A = 8\ cm^2$ is subjected to radiation of constant intensity $I = 10^5$ W/(m^2 · sr) over the solid angle subtended by $0 \leq \phi \leq 2\pi, 0 \leq \theta \leq \pi/3$. Calculate the radiation energy received by the surface.

Answer: 188.5 W

12-23 A surface is subjected to irradiation of spectral distribution given by

$$0 < \lambda \leq 1\ \mu m \qquad I_\lambda^i = 0$$

$$1 < \lambda \leq 1.5\ \mu m \qquad I_\lambda^i = 3000\ W/(m^2 \cdot \mu m)$$

$$1.5 < \lambda \leq 5\ \mu m \qquad I_\lambda^i = 8000\ W/(m^2 \cdot \mu m)$$

$$5 < \lambda \leq 10\ \mu m \qquad I_\lambda^i = 2000\ W/(m^2 \cdot \mu m)$$

$$10 < \lambda < \infty\ \mu m \qquad I_\lambda^i = 0$$

Calculate the radiation energy incident on the surface per unit area over a solid angle subtended by $0 \leq \phi \leq 2\pi$ and $0 \leq \theta \leq \pi/3$.

Absorptivity, emissivity, reflectivity, and transmissivity

12-24 The transmissivity of plate glass for incident solar radiation (that is, 5762 K) for various wavelength bands is

$$\tau_1 = 0 \qquad \text{for } \lambda_0 = 0 \text{ to } \lambda_1 = 0.5\ \mu m$$

$$\tau_2 = 0.7 \qquad \text{for } \lambda_1 = 0.5 \text{ to } \lambda_2 = 2.8\ \mu m$$

$$\tau_3 = 0 \qquad \text{for } \lambda_2 = 2.8\ \mu m \text{ to } \lambda_3 \to \infty$$

Determine the average hemispherical transmissivity of the glass over the entire wavelengths.

Answer: 0.51

12-25 Fused quartz transmits 85 percent of thermal radiation at 1700 K in the wavelength band $\lambda_1 = 0.3$ to $\lambda_2 = 3$ μm and is opaque to radiation outside this range. If a blackbody radiation source at 1700 K is placed in front of this fused quartz sheet, what is the rate of energy transmitted through a 0.1-m^2 area of the quartz sheet?

12-26 A plain glass has transmissivity $\tau = 0.90$ in the range $\lambda = 0.2$ to 3 μm and zero transmissivity for other wavelengths. A tinted glass has transmissivity $\tau = 0.90$ in the range $\lambda = 0.5$ to 1 μm and zero transmissivity for all other wavelengths. If the solar energy is incident on both glasses, compare the energy transmitted through each glass.

12-27 The spectral emissivity of a filament at 3000 K is given by

$$\varepsilon_1 = 0.5 \qquad \text{for } \lambda_0 = 0 \text{ to } \lambda_1 = 0.5\ \mu\text{m}$$
$$\varepsilon_2 = 0.1 \qquad \text{for } \lambda_1 = 0.5\ \mu\text{m to } \lambda_2 \to \infty$$

Determine the average emissivity of the filament over the entire wavelengths.

Answer: 0.105

12-28 The spectral emissivity of an opaque surface at 1500 K is

$$\varepsilon_\lambda = \begin{cases} 0.3 & \text{for } \lambda_0 = 0 \text{ to } \lambda_1 = 1.0\ \mu\text{m} \\ 0.8 & \text{for } \lambda_1 = 1.0\ \mu\text{m to } \lambda_2 \to \infty \end{cases}$$

Determine the average emissivity over the entire wavelengths and the emissive power of the material at 1500 K.

12-29 The spectral emissivity of an opaque surface at 1000 K is

$$\varepsilon_\lambda = \begin{cases} 0.1 & \text{for } \lambda_0 = 0 \text{ to } \lambda_1 = 0.5\ \mu\text{m} \\ 0.5 & \text{for } \lambda_1 = 0.5 \text{ to } \lambda_2 = 6\ \mu\text{m} \\ 0.7 & \text{for } \lambda_2 = 6 \text{ to } \lambda_3 = 15\ \mu\text{m} \\ 0.8 & \text{for } \lambda_3 > 15\ \mu\text{m} \end{cases}$$

Determine the average emissivity of the surface over the entire wavelengths.

12-30 The spectral distribution of a radiation flux incident on a surface is

$$G_\lambda = \begin{cases} 500\ \text{W/(m}^2\cdot\mu\text{m)} & \text{for } 0 < \lambda < 1\ \mu\text{m} \\ 2500\ \text{W/(m}^2\cdot\mu\text{m)} & \text{for } 1 < \lambda < 5\ \mu\text{m} \\ 300\ \text{W/m}^2\ \mu\text{m} & \text{for } 5 < \lambda < 10\ \mu\text{m} \\ 0 & \text{for } \lambda > 10\ \mu\text{m} \end{cases}$$

and the spectral absorptivity of the surface for the above incident radiant is

$$\alpha_\lambda = \begin{cases} 0 & \text{for } 0 < \lambda < 1\ \mu\text{m} \\ 0.8 & \text{for } 1 < \lambda < 5\ \mu\text{m} \\ 0.2 & \text{for } \lambda > 5\ \mu\text{m} \end{cases}$$

Determine the average absorptivity α of the surface over the entire wavelength spectrum.

Answer: 0.69

12-31 The spectral hemispherical absorptivity of an opaque surface for solar radiation is

$$\alpha_1 = 0 \qquad \text{for } \lambda_0 = 0 \text{ to } \lambda_1 = 0.2\ \mu\text{m}$$
$$\alpha_2 = 0.85 \qquad \text{for } \lambda_1 = 0.2 \text{ to } \lambda_2 = 2\ \mu\text{m}$$
$$\alpha_3 = 0 \qquad \text{for } \lambda_2 = 2\ \mu\text{m to } \lambda_3 \to \infty$$

Determine the average absorptivity of the surface for solar radiation. What is the rate of absorption of the solar energy by the surface if the incident solar radiation flux is 1000 W/m^2?

12-32 The spectral absorptivity and reflectivity of a surface to solar radiation are

$$\alpha_1 = 0.15 \qquad \rho_1 = 0.1 \qquad \text{for } \lambda_0 = 0 \text{ to } \lambda_1 = 1.2\ \mu\text{m}$$

$$\alpha_2 = 0.7 \qquad \rho_2 = 0.2 \qquad \text{for } \lambda_1 = 1.2\ \mu\text{m to } \lambda_2 \to \infty$$

Determine the average absorptivity, reflectivity, and transmissivity of the surface over the entire wavelengths. Calculate the amount of solar radiation absorbed, reflected, and transmitted for a solar radiation flux of 900 W/m^2.

12-33 The interior walls of a furnace are maintained at 1300 K and can be regarded as black. The furnace has a 10 cm by 10 cm glass window which has a spectral transmissivity to radiation at 1300 K given by

$$\tau_1 = 0.7 \qquad \text{in } 0 < \lambda < 2.5\ \mu\text{m}$$

$$\tau_2 = 0 \qquad \text{in } 2.5 < \lambda < \infty$$

Calculate the average transmissivity of the glass for radiation emitted at 1300 K. Determine the amount of radiant energy transmitted through the window into the surrounding environment.

Answer: 0.23, 373 W

Solar radiation

12-34 The surface of a satellite receives solar radiation at a rate of 1100 W/m^2. The surface has an absorptivity of $\alpha = 0.8$ for solar radiation and an emissivity of $\varepsilon = 0.9$. Assuming no heat losses into the satellite and a heat dissipation by thermal radiation into the space at absolute zero, calculate the equilibrium temperature of the surface.

Answer: 362.4 K

12-35 A surface receives solar radiation at a rate of 900 W/m^2 while the other side is kept insulated. The absorptivity of the surface to solar radiation is $\alpha = 0.9$, and its emissivity is $\varepsilon = 0.5$. Assuming the surface loses heat by radiation into a clear sky at an effective temperature of 40°C, determine the equilibrium temperature of the surface. The absorptivity for long-wave radiation from the sky can be taken as $\alpha_{sky} = 0.5$.

12-36 Repeat Prob. 12-35 for an aluminum surface which has a solar absorptivity $\alpha = 0.15$ and emissivity $\varepsilon = 0.1$.

12-37 A space radiator dissipates 9×10^4 W/m^2 by thermal radiation into an environment at absolute zero temperature. If the surface has an emissivity $\varepsilon = 0.9$, determine the equilibrium temperature of the radiator surface.

12-38 A space radiator is to dissipate heat by thermal radiation into an environment at absolute zero temperature. If the maximum allowable surface temperature is 1500 K, what is the maximum heat transfer rate per square meter of surface fc · a surface emissivity of $\varepsilon = 0.8$?

Answer: 2.3×10^5 W/m^2

12-39 A hot plate 0.3 m by 0.3 m and insulated on the back side dissipates 1.5 kW of heat by convection and radiation from the front surface. The heater is placed outside in a clear night. The effective radiation temperature of the sky on a clear night can be taken as $T_{sky} = 200$ K, the temperature of the air is $T_a =$ 280 K, the convection heat transfer coefficient at the heater surface is $h = 30$ W/(m^2 · °C), and the emissivity of the heater surface is $\varepsilon = 0.9$. What is the surface temperature of the heater? Assume $\alpha = \varepsilon$.

12-40 The outer surface of a spaceship receives radiation heat flux from the sun at a rate of 1380 W/m^2. Assuming the inside surface is perfectly insulated and that the outer surface dissipates heat by radiation into the outer space at 0 K, determine the equilibrium temperature of the surface. Assume $\alpha_s = 0.9$ and $\varepsilon = 0.8$.

12-41 A window has a heat-absorbing glass with an average absorptivity $\alpha = 0.5$ for the solar radiation over the entire wavelength. A solar radiation flux of 800 W/m^2 is incident on the glass. The inside and outside air temperatures are $T_i = 25$°C and $T_o = 20$°C, respectively. The inside and outside heat transfer coefficients for convection are $h_i = 10$ W/(m^2 · °C) and $h_o = 15$ W/(m^2 · °C), respectively. By neglecting the emission of energy by radiation, determine the equilibrium temperature of the glass.

Answer: 38°C

12-42 An opaque surface which is insulated at the back side has an absorptivity of $\alpha_s = 0.8$ for solar radiation and an emissivity $\varepsilon = 0.2$ for long-wave radiation. A solar radiation flux of 800 W/m^2 is incident on this surface. If the surface is exposed to the ambient air at $T_\infty = 300$ K and the convective heat transfer coefficient is $h = 15$ W/(m$^2 \cdot$ °C), determine the equilibrium temperature of the surface. Neglect the sky radiation.

12-43 Radiation from a blackbody source at 3000 K strikes one of the surfaces of a thin, opaque plate at a rate of $q = 3500$ W/m^2. The surface has a spectral absorptivity for this incident radiation given by

$$\alpha_1 = 0.5 \qquad 0 < \lambda < 3\ \mu\text{m}$$

$$\alpha_2 = 0.2 \qquad 3\ \mu\text{m} < \lambda < \infty$$

The plate dissipates heat by convection from both its surfaces into an ambient at $T_\infty = 290$ K with a heat transfer coefficient $h = 20$ W/(m$^2 \cdot$ °C). Calculate the equilibrium temperature of the plate.

Answer: 330.1 K

12-44 Determine the average absorptivities for the solar radiation of two surfaces A and B having the following spectral absorptivities:

$$\alpha_{A_1} = 0.9 \qquad \alpha_{B_1} = 0.1 \qquad \text{for } 0 < \lambda < 3\ \mu\text{m}$$

$$\alpha_{A_2} = 0.1 \qquad \alpha_{B_2} = 0.9 \qquad \text{for } 3\ \mu\text{m} < \lambda\ \infty$$

Explain why the average absorptivity of surface A is higher than that of surface B. Which surface would be desirable for use as an absorber in a flat-plate solar collector?

View factors

12-45 A radiation detector is aimed at a small horizontal surface a distance $L = 5$ m away. The line joining the detector axis to the hot object makes $\theta = 45°$ with the normal to the surface of the hot object. The aperture has an area of 0.1 cm^2.

(*a*) Determine the solid angle subtended by the aperture with respect to the point on the surface of the hot object.

(*b*) If the hot object is a blackbody of area $dA_1 = 100$ cm^2 at temperature 1000 K, determine the energy intercepted by the aperture.

Answer: (a) 4×10^{-7} sr, (*b*) 5.1×10^{-5} W

12-46 Consider three small surfaces each of area $dA_1 = dA_2 = dA_3 = 2$ cm^2 as shown in the accompanying figure. Calculate (*a*) the solid angle $d\omega_{12}$ subtended by area dA_2 with respect to the point on dA_1, (*b*) the solid angle $d\omega_{13}$ subtended by area dA_3 with respect to the point on dA_1, and (*c*) the elemental diffuse view factors $dF_{dA_1-dA_2}$ and $dF_{dA_1-dA_3}$.

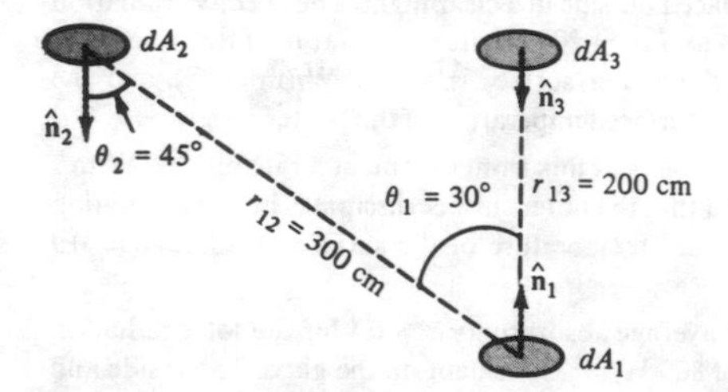

Figure P12-46

12-47 Determine the view factor F_{1-2} between an elemental surface dA_1 and the finite rectangular surface A_2 for the geometric arrangements shown in the accompanying figure.

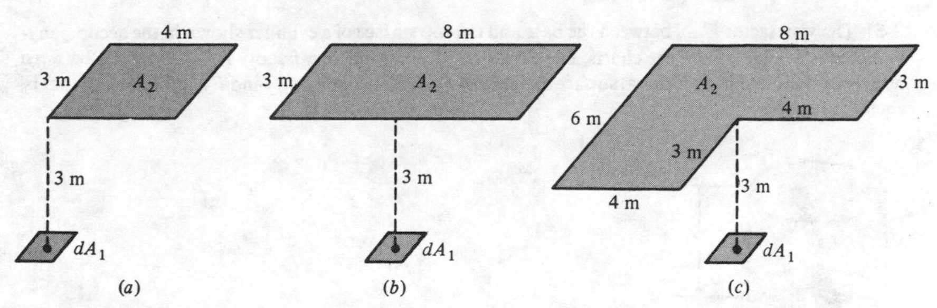

Figure P12-47

12-48 Determine the view factor F_{1-2} between two rectangular surfaces A_1 and A_2 for the geometric arrangement shown in the accompanying figure.

Answer: 0.2275

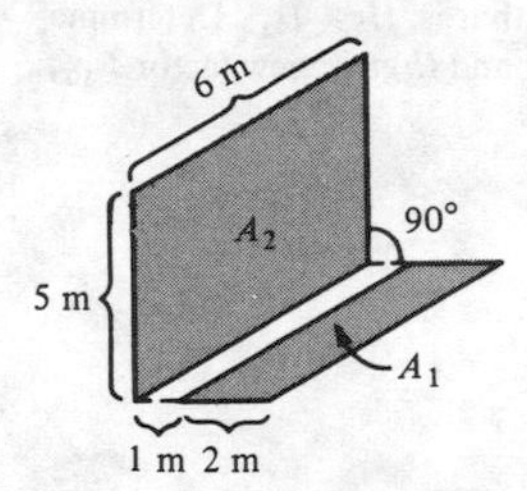

Figure P12-48

12-49 In a cubical enclosure there are 36 individual view factors between the six surfaces of the enclosure. How many different view factors are to be computed?

12-50 Determine the view factors between the surfaces shown below by view factor algebra:

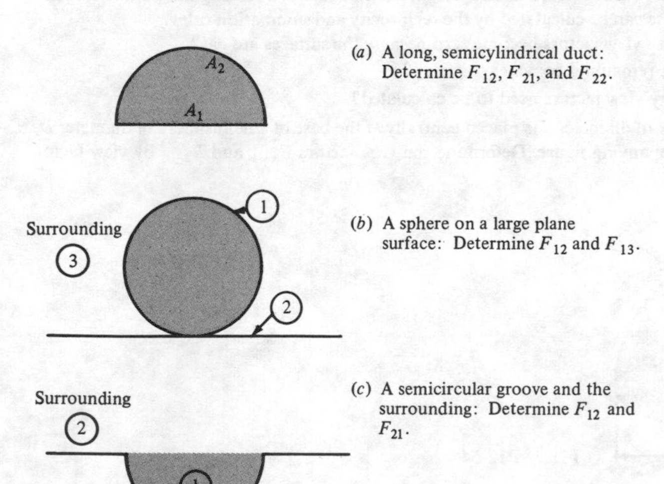

(*a*) A long, semicylindrical duct: Determine F_{12}, F_{21}, and F_{22}.

(*b*) A sphere on a large plane surface: Determine F_{12} and F_{13}.

(*c*) A semicircular groove and the surrounding: Determine F_{12} and F_{21}.

Figure P12-50

12-51 The view factor F_{1-3} between the base and the top surface of a cylinder shown in the accompanying figure is available from the charts. Develop a relation for the view factors F_{1-2} and F_{2-1} between the base and the lateral cylindrical surface in terms of F_{1-3}. Calculate F_{1-2} and F_{2-1} for $H = R$, that is, when $F_{1-3} = 0.39$.

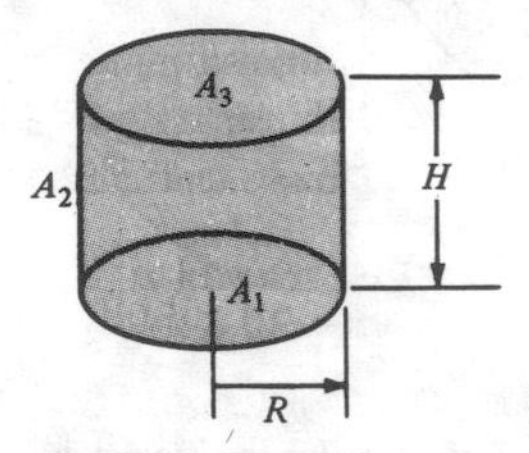

Figure P12-51

12-52 Consider two parallel-plane circular surfaces with a common central normal, as illustrated in the accompanying figure. We assume dA_1 is very small in comparison to the disk A_2. Also the distance between the disks is much larger than the diameter D_2 of the large disk (that is, $H \gg D_2$). Determine (*a*) the solid angle subtended by area A_2 with respect to the point on dA_1 and (*b*) the view factor F_{12}.

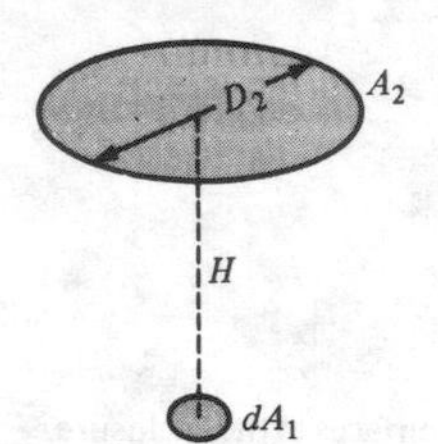

Figure P12-52

12-53 How many view factors are involved in an enclosure in the shape of a rectangular parallelepiped?

(*a*) How many of these can be calculated by the reciprocity and summation rules?
(*b*) How many of these view factors become zero because the surfaces are flat?
(*c*) How many of the remaining possess symmetry?
(*d*) Finally, how many view factors need to be calculated?

12-54 A small circular disk of diameter d is placed centrally at the base of a hemisphere of diameter D, as illustrated in the accompanying figure. Determine the view factors F_{3-1} and F_{3-2} by view factor algebra.

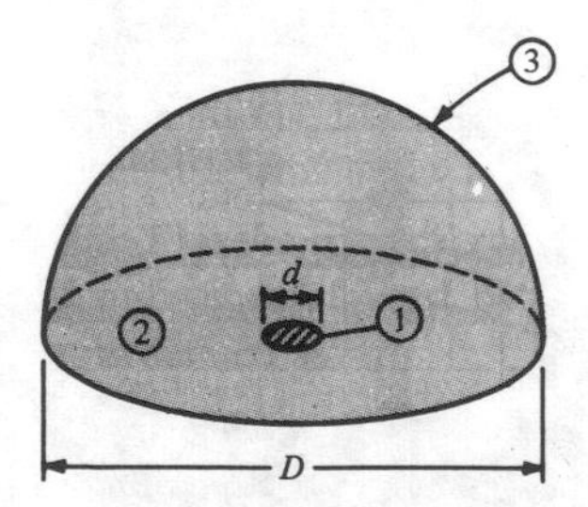

Figure P12-54

Answer: $F_{3-1} = \dfrac{A_1}{A_3} = \dfrac{d^2}{2D^2}$, $F_{3-2} = \dfrac{A_2}{A_3} = \dfrac{D^2 - d^2}{2D^2}$

12-55 Consider two concentric spheres. The inner sphere has a radius $R_1 = 10$ cm. Determine the radius R_2 of the outer sphere such that $F_{21} = 0.64$, where the subscript 1 refers to the inner sphere.

Answer: 12.5 cm

12-56 Consider two very long coaxial cylinders. The outer cylinder has a radius $R_2 = 10$ cm. Determine the radius R_1 of the inner cylinder such that $F_{21} = 0.75$.

12-57 Consider two coaxial cylinders with diameters $D_1 = 0.25$ m and $D_2 = 0.5$ m and equal lengths $L = 0.5$ m. Calculate the view factors between the open ends of annular space between the cylinder.

12-58 The configuration shown in the accompanying figure is very long in the direction perpendicular to the plane of the figure. Calculate the view factor between surfaces A_1 and A_2.

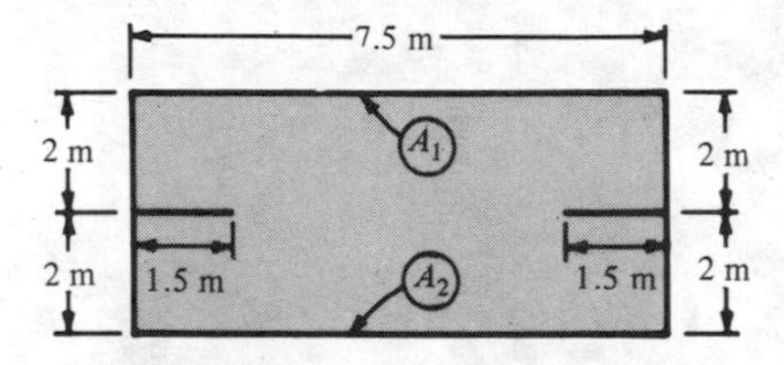

Figure P12-58

Network method

12-59 Calculate the heat dissipated by radiation through a 0.2-m² opening of a furnace at 1100 K into an ambient at 300 K. Assume both the furnace and the ambient are blackbodies.

Answer: 16.5 kW

12-60 Two black rectangular surfaces A_1 and A_2, arranged as shown in the accompanying figure, are located in a large room whose walls are black and kept at 300 K. Determine the net radiative-heat exchange between these two surfaces when A_1 is kept at 1000 K and A_2 at 500 K. (Neglect the radiation from the room.)

Answer: 17 kW

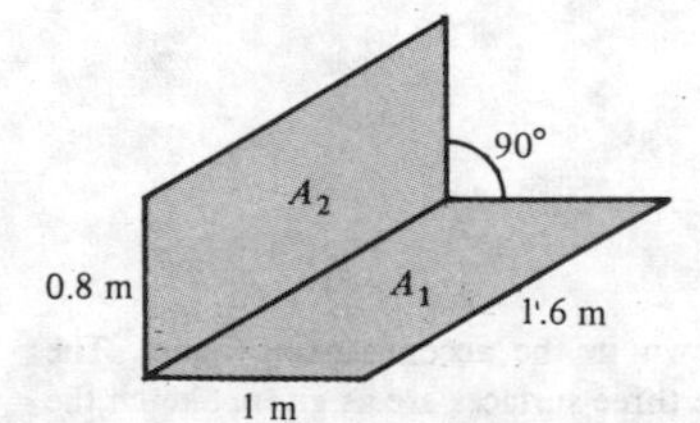

Figure P12-60

12-61 Two aligned, parallel square plates 0.6 m by 0.6 m are separated by $L = 0.3$ m, as illustrated in the accompanying figure. Plate 1 is maintained at $T_1 = 1000$ K and has an emissivity $\varepsilon_1 = 0.7$. Plate 2 is maintained at $T_2 = 500$ K and has an emissivity $\varepsilon_2 = 0.5$. The plates are exposed through the opening between them into an ambient which can be regarded as a black medium at $T_\infty = 300$ K. Sketch the radiation network for the two surfaces and the ambient. Calculate the heat transfer between the plate and the heat loss to the ambient.

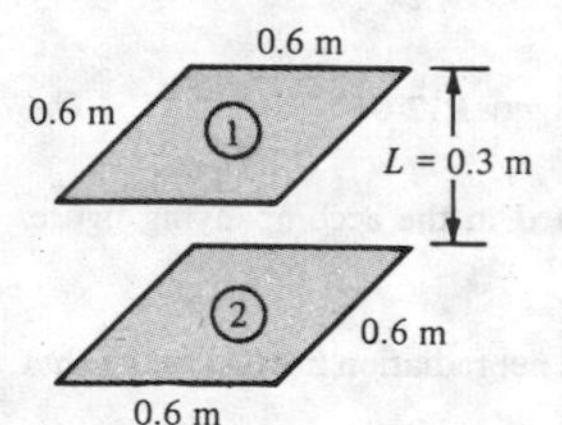

Figure P12-61

12-62 Two rectangular plates 0.5 m by 1.0 m are arranged perpendicular to each other with a common edge, as illustrated in the accompanying figure. Plate 1 is at $T_1 = 1000$ K and has an emissivity $\varepsilon_1 = 0.8$. Plate 2 is at $T_2 = 500$ K and has an emissivity $\varepsilon_2 = 0.6$. The surrounding ambient can be regarded a blackbody at $T_\infty = 300$ K. Sketch the radiation network, and calculate the heat transfer rate between the plates.

Answer: 3.375 kW

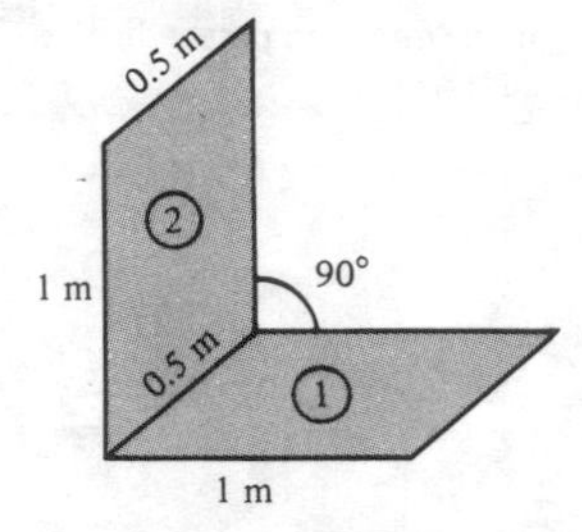

Figure P12-62

12-63 Two parallel circular disks of equal diameter $D = 0.5$ m separated by $L = 0.25$ m have a common central normal, as illustrated in the accompanying figure. One disk is maintained at $T_1 = 800$ K and has an emissivity $\varepsilon_1 = 0.9$. The other is at $T_2 = 500$ K and has an emissivity $\varepsilon_2 = 0.7$. The disks are exposed through the opening between them into an environment which can be regarded as a black medium at $T_\infty = 300$ K. Sketch the radiation network. Calculate the radiation heat transfer between the two disks and the total heat loss into the ambient.

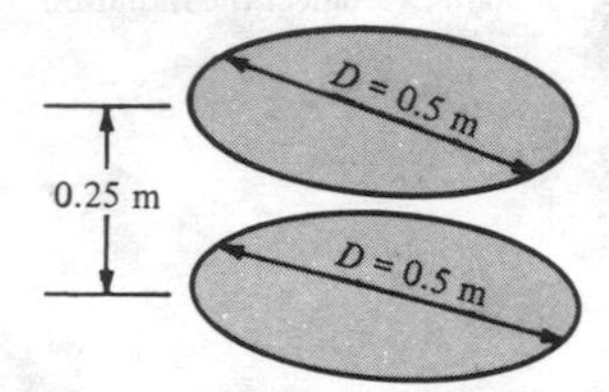

Figure P12-63

12-64 Consider a long equilateral-triangle-shaped duct, shown in the accompanying figure. The radiation properties and the surface conditions for each of the three surfaces are as given. Sketch the radiation network, and calculate the heat transfer rates per unit area at surfaces 2 and 3.

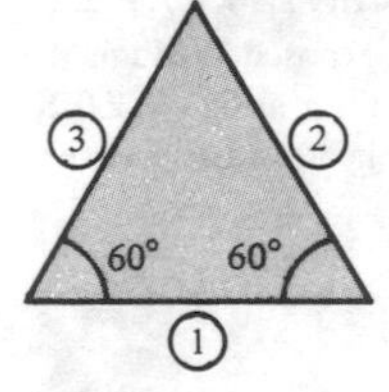

Surface	Surface conditions
1	$T_1 = 300$ K, $\epsilon_1 = 1$
2	$T_2 = 500$ K, $\epsilon_2 = 0.5$
3	$T_3 = 700$ K, $\epsilon_3 = 0.4$

Figure P12-64

12-65 Consider a long, equilateral-triangle-shaped oven, as illustrated in the accompanying figure. The radiation properties and the surface conditions are as given.

(*a*) Sketch the radiation network.

(*b*) Calculate the temperature T_3 of the adiabatic surface and the net radiation heat flux at the hot surface.

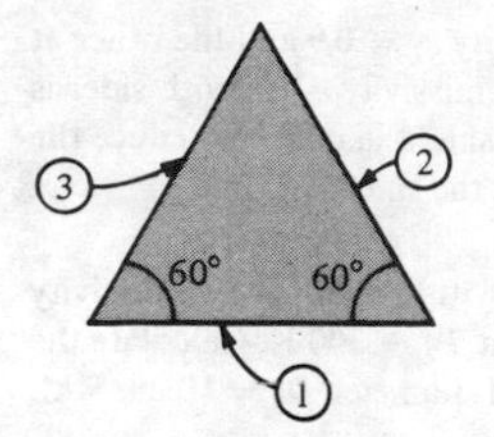

Surface	Surface conditions
1	$T_1 = 500$ K, $\epsilon_1 = 0.7$
2	$T_2 = 1000$ K, $\epsilon_2 = 0.9$
3	Insulated

Figure P12-65

12-66 A 10-cm-OD, 5-m-long steam pipe whose surface is at 110°C passes through a room whose walls are at 10°C. Assuming the emissivity of the pipe is $\varepsilon = 0.9$, determine the rate of heat loss from the pipe by radiation.

Answer: 1.211 kW

12-67 In a test room 3 m by 3 m by 3 m the ceiling is kept at 70°C while the walls and the floor are at 10°C. Assuming that all surfaces have an emissivity $\varepsilon = 0.8$, determine the rate of heat loss from the ceiling by radiation.

12-68 A cubical test room 3 m by 3 m by 3 m is heated through the floor by maintaining it at a uniform temperature $T_1 = 310$ K. Since the side walls are well insulated, the heat loss through them can be considered negligible. The heat loss takes place through the ceiling, which is maintained at 280 K. All surfaces has an emissivity $\varepsilon = 0.85$. Sketch the radiation network, and calculate the rate of heat loss by radiation through the ceiling.

Radiation shields (network method)

12-69 Consider two large parallel plates, one at T_1 K with emissivity $\varepsilon_1 = 0.8$ and the other at T_2 K with emissivity $\varepsilon_2 = 0.4$. An aluminum radiation shield with an emissivity $\varepsilon_3 = 0.05$ is placed between the plates. Sketch the radiation network for the systems with and without the radiation shield. Calculate the percentage reduction in heat transfer rate resulting from the radiation shield.

Answer: 93.4 percent

12-70 A radiation shield (plate 3) is placed between two parallel plates, 1 and 2, as illustrated in the accompanying figure. The radiation properties and the surface conditions for each of these plates are as follows: plate 1: $T_1 = 1000$ K, $\varepsilon_1 = 0.8$; plate 2: $T_2 = 500$ K, $\varepsilon_2 = 0.5$; plate 3: radiation shield, $\varepsilon = 0.05$. Sketch the radiation network. Determine the net radiative-heat transfer per unit area between the plates 1 and 2, and calculate the equilibrium temperature of plate 3.

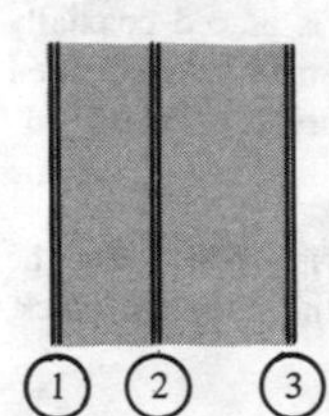

Figure P12-70

12-71 Consider two large parallel plates, one at $T_1 = 1000$ K with emissivity $\varepsilon_1 = 0.8$ and the other at $T_2 = 300$ K with emissivity $\varepsilon_2 = 0.6$. A radiation shield is placed between them. The shield has an emissivity $\varepsilon_{31} = 0.1$ on the side facing the hot plate and an emissivity $\varepsilon_{32} = 0.3$ on the side facing the cold plate. Sketch the radiation network. Calculate the reduction in the heat transfer rate between the hot and cold plates as a result of the radiation shield, and find the equilibrium temperature of the radiation shield.

12-72 Consider two large parallel plates, one at $T_1 = 800$ K with emissivity $\varepsilon_1 = 0.9$ and the other at $T_2 = 300$ K with emissivity $\varepsilon_2 = 0.5$. A radiation shield having an equal emissivity ε_3 on both sides is placed between these two plates. Calculate the emissivity of the radiation shield in order to reduce the radiative heat transfer between the two plates to 10 percent of that without the shield.

Answer: 0.1

12-73 A long tube with an OD of $D = 5$ cm that is maintained at $T_1 = 450$ K and has an emissivity $\varepsilon_1 = 0.9$ is exposed to an ambient which can be regarded as a blackbody at $T_2 = 300$ K. Calculate the reduction in heat loss from the tube by radiation if a radiation shield of diameter $D_3 = 10$ cm with emissivity $\varepsilon_3 = 0.1$ on both surfaces is placed coaxially around the tube.

12-74 In a spaceship, a radiation shield is installed on the exposed surface of a cryogenic tank in order to protect it from the solar radiation flux. The tank surface is at $T_1 = 100$ K and has an emissivity $\varepsilon_1 = 0.15$. The radiation shield has an emissivity of 0.05 for both surfaces. The exposed surface of the radiation shield receives a solar energy flux $G = 1300$ W/m^2 and faces the space surrounding which is regarded a blackbody at 0 K. The tank surface and the radiation shield can be treated as two parallel plates for the analysis. Calculate (*a*) the temperature of the radiation shield and (*b*) the heat flux received by the tank surface.

12-75 A radiation shield having an equal emissivity ε_3 on both its surfaces is placed between two large parallel plates at T_1 and T_2 K with emissivities $\varepsilon_1 = 0.8$ and $\varepsilon_2 = 0.5$. Sketch the radiation network. Calculate the emissivity of the radiation shield in order to reduce the heat transfer between the plates to 8 percent of that without the radiation shield.

Answer 0.0744

12-76 A tube with $D_1 = 3$ cm OD carrying a cryogenic fluid is maintained at $T_1 = 100$ K and has an emissivity $\varepsilon_1 = 0.15$. A radiation shield of diameter $D_3 = 5$ cm with emissivity $\varepsilon_3 = 0.05$ at both its surfaces is installed coaxially over the tube, and the space between them is evacuated. The outer surface of the radiation shield is exposed to an ambient, which can be regarded a blackbody at $T_2 = 300$ K. Sketch the radiation network. Calculate (*a*) the temperature of the radiation shield and (*b*) the reduction in the heat gain by the pipe due to the presence of the radiation shield.

12-77 Consider three large parallel plates. The two outer plates are maintained at $T_1 = 1000$ K and $T_2 = 200$ K and have emissivities $\varepsilon_1 = 0.4$ and $\varepsilon_2 = 0.8$, respectively. Calculate the equilibrium temperature T_3 of the middle plate if it has an emissivity $\varepsilon_3 = 0.1$ on both sides. Determine the heat transfer rate per square meter across the plates. (Neglect heat transfer by convection.)

Answer: 828.9 K, 2.6 kW/m^2

12-78 Consider two coaxial cylinders of radii $r_1 = 5$ cm and $r_2 = 15$ cm which are maintained at $T_1 = 1000$ K and $T_2 = 300$ K, respectively. The emissivity for the surfaces are $\varepsilon_1 = 0.6$ and $\varepsilon_2 = 0.9$, respectively.

(*a*) Calculate the radiation exchange between the cylinders per meter length of the cylinder.

(*b*) A radiation shield of radius $r_3 = 10$ cm having an emissivity $\varepsilon_3 = 0.1$ is placed coaxially between the two cylinders. Sketch the radiation network. Calculate the heat transfer rate between cylinders 1 and 2 with the shield present, and determine the equilibrium temperature T_3 of the shield.

Radiosity matrix method

12-79 In a cubical furnace 1.5 m by 1.5 m by 1.5 m, the ceiling is at 1100 K, the floor is at 550 K, and the walls are refractory (reradiating) surfaces. Assuming that both the ceiling and the floor are black, determine the net radiative-heat exchange between the ceiling and the floor.

Answer: 105 kW

12-80 Repeat Prob. 12-79 for the following surface emissivities:

(*a*) $\varepsilon_{\text{ceiling}} = 1.0$, $\varepsilon_{\text{floor}} = 0.8$
(*b*) $\varepsilon_{\text{ceiling}} = 0.8$, $\varepsilon_{\text{floor}} = 0.8$

12-81 Consider an equilateral-triangle-shaped enclosure which is infinitely long in the direction perpendicular to the plane of the accompanying figure. Surfaces 1 and 2 are kept at prescribed uniform heat fluxes q_1 and q_2, respectively, while surface 3 is kept at a uniform temperature T_3. Write the equations for the determination of the radiosities J_1, J_2, and J_3. (Emissivities are $\varepsilon_1, \varepsilon_2, \varepsilon_3$.)

12-82 The radiation properties and surface conditions for a long, equilateral-triangle-shaped duct, shown in the accompanying figure, are as follows: Surface 1: $q_1 = 600\ \text{W/m}^2$, $\varepsilon_1 = 0.6$; surface 2: $T_2 = 800$ K, $\varepsilon_2 = 0.8$; surface 3: $T_3 = 500$ K, $\varepsilon_3 = 0.7$. Calculate the temperature for surface 1 and the heat fluxes for the surfaces 2 and 3.

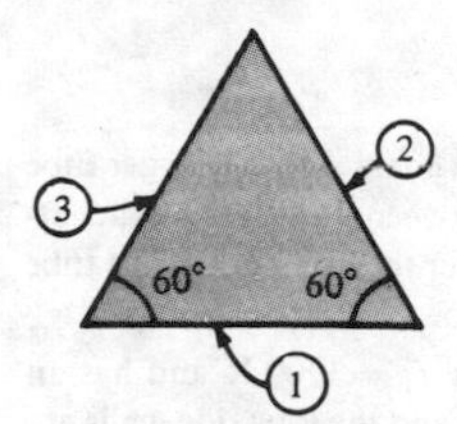

Figure P12-82

12-83 A 2.5-cm-OD hot pipe at $T_1 = 1000$ K is enclosed inside a 7.5-cm-ID pipe at $T_2 = 350$ K. Determine the rate of heat loss by radiation per meter length from the hot to the cold pipe for the following three different surface conditions.

(*a*) $\varepsilon_1 = \varepsilon_2 = 1$
(*b*) $\varepsilon_1 = 1$, $\varepsilon_2 = 0.1$
(*c*) $\varepsilon_1 = 0.1$, $\varepsilon_2 = 0.1$

Answer: (*a*) 4386, (*b*) 1097, (*c*) 337 W/m

12-84 Consider two parallel, circular plate disks of equal diameter $D = 0.6$ m separated by a distance $L = 0.3$ m and having a common central normal. One of the disks is at $T_1 = 600$ K and the other at $T_2 = 400$ K, while their outer surfaces are insulated. The disks are exposed through the opening between them into an environment at $T_3 = 300$ K. Determine the net radiation exchange between the disks as well as heat loss by radiation into the surrounding environment for the following cases:

(*a*) $\varepsilon_1 = \varepsilon_2 = 1$ and the surrounding is black.
(*b*) $\varepsilon_1 = \alpha_1 = \varepsilon_2 = \alpha_2 = 0.8$ and the surrounding is black.

12-85 Two aligned, parallel plates 0.8 m by 0.8 m separated by a distance $L = 0.4$ m are maintained at $T_1 = 800$ K and $T_2 = 400$ K, respectively. The plates are exposed through the opening between them into an ambient which can be regarded as a blackbody at $T_\infty = 300$ K. Calculate the heat transfer rate between the plates and the rate of heat loss through the opening into the surrounding ambient when the emissivities of the plates are (*a*) $\varepsilon_1 = \varepsilon_2 = 1$, (*b*) $\varepsilon_1 = \varepsilon_2 = 0.7$, (*c*) $\varepsilon_1 = \varepsilon_2 = 0.1$.

12-86 Consider two long concentric cylinders. The inner cylinder has radius $R_1 = 8$ cm and emissivity $\varepsilon_1 = 0.8$ and is maintained at a uniform temperature T_1. The outer cylinder has radius $R_2 = 12$ cm and emissivity $\varepsilon_2 = 0.8$ and is maintained at uniform temperature T_2. What change is needed in the emissivity ε_1 of the inner cylinder in order to reduce the heat transfer rate between the cylinders to 70 percent of the original heat transfer rate.

Answer: $\varepsilon_{1,\text{new}} = 0.538$

12-87 Repeat Prob. 12-86 for the case of two concentric spheres.

12-88 A copper rod with $D_1 = 2$ cm diameter dissipates 50 W per meter length as a result of the passage of electric current. The rod is coaxial inside a tube with $D_2 = 5$ cm ID that is maintained at a uniform temperature 300 K. The space inside the tube is evacuated so that the heat transfer is by radiation only. The surfaces have emissivity $\varepsilon_1 = \varepsilon_2 = 0.8$. Calculate the surface temperature of the rod.

Answer: 405.5 K

12-89 A spherical tank with $D_1 = 1$ m diameter containing liquid oxygen is enclosed inside a spherical container of inside diameter $D_2 = 1.25$ m. The space between the tanks is evacuated to reduce the heat transfer to the oxygen. The surfaces have an emissivity $\varepsilon = 0.1$. Calculate the heat transfer rate to the oxygen by radiation when the oxygen tank is at $T_1 = 100$ K and the outer tank at $T_2 = 300$ K. Determine the rate of evaporation of oxygen for $h_{fg} = 2.1 \times 10^5\ \text{W} \cdot \text{s/kg}$.

12-90 Consider two concentric spheres. The inner sphere has diameter $D_1 = 0.4$ m and is maintained at $T_1 = 600$ K; the outer sphere has diameter $D_2 = 0.6$ m and is maintained at $T_2 = 400$ K. The space between the surfaces is evacuated. Determine the total heat transfer rate between the spheres when the emissivities ε_1 and ε_2 for the inner and outer spheres, respectively, are

(*a*) $\varepsilon_1 = \varepsilon_2 = 1$,
(*b*) $\varepsilon_1 = 0.1, \varepsilon_2 = 1$
(*c*) $\varepsilon_1 = 1, \varepsilon_2 = 0.1$
(*d*) $\varepsilon_1 = \varepsilon_2 = 0.1$

12-91 Helium at $T_1 = 80$ K flows inside a tube with $D_1 = 3$ cm OD which is coaxial with a larger tube of inside diameter $D_2 = 6$ cm and is maintained at $T_2 = 300$ K. The space between the tubes is evacuated to reduce the heat transfer. Calculate the heat transfer rate to the helium per meter length of the tube if the emissivities of the inner and outer tubes, respectively, are $\varepsilon_1 = \varepsilon_2 = 0.05$.

12-92 Consider a cubical furnace 0.5 m by 0.5 m by 0.5 m. The ceiling is at $T_1 = 1200$ K and has an emissivity $\varepsilon_1 = 0.9$, the floor is at $T_2 = 600$ K and has an emissivity $\varepsilon_2 = 0.8$, and the four side walls are reradiating (i.e., no net radiation flux) surfaces. Calculate the radiation heat transfer between the ceiling and the floor as well as the equilibrium temperature of the reradiating side walls.

12-93 Consider a cylindrical enclosure of diameter $D = 1$ m and height $L = 1$ m. The top surface is maintained at $T_1 = 700$ K, is opaque, and has an emissivity $\varepsilon_1 = 0.9$. The bottom surface is maintained at $T_2 = 400$ K, is opaque, and has an emissivity $\varepsilon_2 = 0.8$. The lateral cylindrical surface is insulated and so behaves as a reradiating surface. Neglecting the convection losses, determine the rate of heat transfer between the top and bottom surfaces and calculate the equilibrium temperature of the reradiating cylindrical surface. (The view factor between the top and bottom surfaces can be taken as $F_{1-2} = 0.4$.)

12-94 The radiation properties and the surface conditions for a four-zone enclosure with all sides equal, shown in the accompanying figure, are as follows: Surface 1: $T_1 = 800$ K, $\varepsilon_1 = 1$; surface 2: $T_2 = 500$ K, $\varepsilon_2 = 0.8$; surface 3: $T_3 = 400$ K, $\varepsilon_3 = 0.8$; surface 4: insulated (reradiating). Determine the heat fluxes for zones 1, 2, and 3. Calculate the equilibrium temperature for the reradiating zone 4.

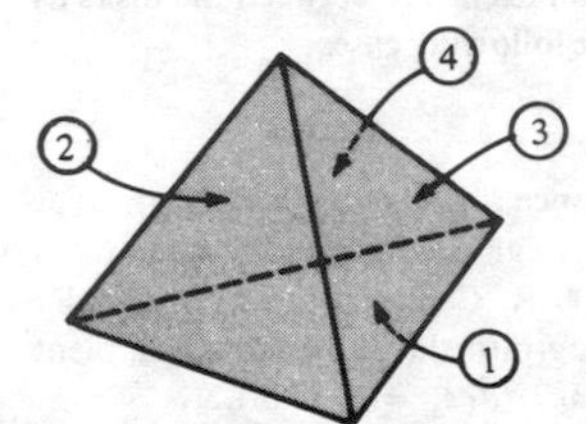

Figure P12-94

12-95 A copper conductor with $D = 1$ cm diameter with surface emissivity $\varepsilon = 0.8$ dissipates 500 W per meter length in an ambient at $T_\infty = 300$ K which can be regarded as black. Assuming that the heat dissipation is by radiation only, determine the surface temperature of the conductor.

Answer: 774 K

12-96 A spherical tank of diameter $D_1 = 0.5$ m containing liquid oxygen is enclosed inside another spherical tank of diameter $D_2 = 0.8$ m, and the space between them is evacuated. The inner and outer spheres are maintained at $T_1 = 90$ K and $T_2 = 280$ K, respectively. Both spheres have an emissivity $\varepsilon = 0.05$. Calculate the rate of heat transfer to the inner sphere and the rate of evaporation of oxygen for $h_{fg} = 2.1 \times 10^5$ W · s/kg for oxygen.

12-97 Consider two concentric spheres 2.5 and 5 cm in diameter. The inner sphere is black and kept at a temperature $T_1 = 1100$ K while the outer sphere has an emissivity $\varepsilon_2 = 0.8$ and is kept at $T_2 = 500$ K. Determine the rate of radiative-heat exchange between the two spheres.

Radiation effect on temperature reading

12-98 A thermocouple is used to measure the temperature of a hot gas flowing in a large duct whose walls are at $T_w = 300$ K. The thermocouple shows a temperature $T_t = 600$ K. Assuming that the emissivity of the thermocouple is $\varepsilon = 0.8$ and the convection heat transfer coefficient is $h = 100$ W/(m$^2 \cdot$ °C), calculate the true air temperature

Answer: 655.1 K

12-99 Air with a true temperature $T_a = 400$ K flows through a large duct whose walls are at $T_w = 300$ K. If a thermocouple having an emissivity $\varepsilon = 0.9$ is inserted into the gas stream and the convection heat transfer coefficient is $h = 150$ W/(m$^2 \cdot$ °C), what temperature will the thermocouple indicate?

12-100 In a room the true air temperature is $T_a = 300$ K, and the walls are at $T_w = 280$ K. A thermocouple having an emissivity $\varepsilon = 0.85$ and subjected to a convective heat transfer coefficient of $h = 15$ W/(m$^2 \cdot$ °C) is placed in the room. What temperature will it show?

Answer: 295.3 K

12-101 A shielded thermocouple shows a temperature $T_t = 600$ K for the hot gas in a large duct whose walls are at $T_w = 400$ K. The shield has an emissivity $\varepsilon_s = 0.2$ for both sides and a diameter 5 times larger than that of the thermocouple wire. The convection heat transfer coefficient for both sides of the shield and for the thermocouple is $h_t = 400$ W/(m$^2 \cdot$ °C). The emissivity of the thermocouple is $\varepsilon_t = 0.9$. Calculate the true temperature of the hot air.

12-102 Hot air is circulated in a room with large glass windows. A thermometer shows 23°C while the average temperature of the walls and the windows is 10°C. The emissivity of the thermometer is $\varepsilon = 0.92$, and the convection heat transfer coefficient for the thermometer is 20 W/(m$^2 \cdot$ °C). Calculate the actual air temperature.

12-103 A thermocouple shows $T_t = 1100$ K in a combustion chamber whose walls are at $T_w = 750$ K. Assuming all black surfaces and a convection heat transfer coefficient $h = 500$ W/(m$^2 \cdot$ °C), calculate the true temperature of the hot gases in the combustion chamber.

Answer: 1230 K

12-104 Find the temperature reading of a thermocouple placed in a large duct with walls at $T_w =$ 600 K and air flowing at $T_a = 1100$ K when the emissivity of the thermocouple is $\varepsilon_t = 0.7$ and the heat transfer coefficient for convection is $h = 120$ W/(m$^2 \cdot$ °C).

12-105 An electrically heated hot plate of diameter $D = 0.3$ m, placed on the floor in a large room, is insulated on the back side and dissipates heat by convection and radiation from the upper surface into the room. The room air is at $T_a = 300$ K, the ceiling and the walls are at $T_w = 290$ K, the convection heat transfer coefficient at the heater surface is $h = 27$ W/(m$^2 \cdot$ °C), and the emissivity of the heater surface is $\varepsilon = 0.9$. If the plate surface temperature should not exceed 600 K, calculate the maximum allowable heat input into the plate.

REFERENCES

1. Hottel, H. C., and A. F. Sarofim: *Radiative Transfer*, McGraw-Hill, New York, 1967.
2. Love, T. J.: *Radiative Heat Transfer*, Merrill, Columbus, Ohio, 1968.
3. Özışık, M. N.: *Radiative Transfer and Interactions with Conduction and Convection*, Wiley, New York, 1973.
4. Siegel, R., and R. Howell: *Thermal Radiation Heat Transfer*, Hemisphere, New York, 1981.
5. Sparrow, E. M., and R. D. Cess: *Radiation Heat Transfer*, Brooks/Cole, Belmont, Calif., 1970.
6. Planck, M.: *The Theory of Heat Radiation*, Dover, New York, 1959.
7. Dunkle, R. V.: "Thermal Radiation Tables and Applications," *Trans. ASME* **76**:549–552 (1954).
8. Gubareff, G. G., J. E. Janssen, and R. H. Torborg: *Thermal Radiation Properties Survey*, Honeywell Research Center, Honeywell Regulator Company, Minneapolis, 1960.
9. Singham, J. R.: "Tables of Emissivity of Surfaces," *Int. J. Heat Mass Transfer* **5**:67–76 (1962).
10. Touloukian, Y. S., and D. P. DeWitt: "Nonmetallic Solids," in *Thermal Radiative Properties*, vol. 8, 1F1/Plenum, New York, 1970.
11. Svet, D. Y.: *Thermal Radiation, Metals, Semiconductors, Ceramics, Partly Transparent Bodies and Films*, Consultants Bureau, New York, 1965.
12. Wood, D. H., H. W. Deem, and C. F. Lucks: *Thermal Radiative Properties*, vol. 3, Plenum, New York, 1964.

13. Touloukian, Y. S., and D. P. DeWitt: "Metallic Elements and Alloys," in *Thermal Radiative Properties*, vol. 7, 1F1/Plenum, New York, 1970.
14. Sieber, W.: *Z. Tech. Phys.* **22**:130–135 (1941).
15. Thekaekara, M. P.: "Data on Incident Solar Radiation," *Suppl. Proc. 20th Ann. Meeting Inst. Environ. Sci.* **21**, 1974.
16. Duffie, J. A., and W. A. Beckman: *Solar Energy Thermal Process*, Wiley, New York, 1974.
17. Meinel, A. B., and M. P. Meinel: *Applied Solar Energy*, Addison-Wesley, Reading, Mass., 1976.
18. Sayigh, A. A. M. (ed.): *Solar Energy Engineering*, Academic, New York, 1977.
19. Hoyt, D. V.: "A Review of Presently Available Solar Radiation Instruments," report and recommendations of *Solar Energy Data Workshop, National Science Foundation* (RANN Grant No. AG-495), Washington, 1974.
20. *Guide to Meteorological Instrumentation and Observing Practices, Secretariat of the World Meteorological Organization*, 4th ed., Geneva, Switzerland, 1971.
21. Coulson, K. L.: *Solar and Terrestrial Radiation: Methods and Measurement*, Academic, New York, 1975.
22. Thekaekara, M. P.: "Solar Radiation Techniques and Instrumentation," *Solar Energy* **18**:309–325 (1976).
23. Hamilton, D. C., and W. R. Morgan: "Radiant Interchange Configuration Factors," *NACA Tech. Note* 2836, 1952.
24. Leuenberger, H., and R. A. Pearson: "Compilation of Radiant Shape Factors for Cylindrical Assemblies," *ASME Paper* 56-A-144, 1956.
25. Mackey, C. O., L. T. Wright, R. E. Clark, and N. R. Gray: "Radiant Heating and Cooling, Pt. I," *Cornell Univ., Eng. Exp. Sta. Bull.* 22, 1943.
26. Jakob, M.: *Heat Transfer* vol. 2, Wiley, New York, 1957.
27. Hottel, H. C.: "Radiant Heat Transmission," in W. H. McAdams (ed.), *Heat Transmission*, 3d ed., McGraw-Hill, New York, 1954, chap. 4.
28. Eckert, E. R. G., and R. M. Drake: *Analysis of Heat and Mass Transfer*, McGraw-Hill, New York, 1972, pp. 619–646.
29. Gebhart, B.: "A New Method for Calculating Radiant Exchanges," *Heat. Piping/Air Cond.* **30**:131–135 (July 1958).
30. Gebhart, B.: "Surface Temperature Calculations in Radiant Surroundings of Arbitrary Complexity for Gray, Diffuse Radiation," *Int. J. Heat Mass Transfer* **3**:341–346 (1961).
31. Oppenheim, A. K.: "Radiation Analysis by the Network Method," *Trans. ASME* **78**:725–735 (1956).
32. Sparrow, E. M.: "Radiation Heat Transfer between Surfaces," in James P. Hartnett and T. F. Irvine, Jr. (eds), *Advances in Heat Transfer*, Academic, New York, 1965, pp. 407–411.
33. Clark, J. A., and E. Korybalski: "Radiation Heat Transfer in an Enclosure Having Surfaces Which Are Adiabatic or of Known Temperature," *First. Natl. Heat Mass Transfer Conf.*, Madras, India, December 1971.

CHAPTER

THIRTEEN

RADIATION IN ABSORBING EMITTING MEDIA

In Chap. 12 we discussed radiation exchange among surfaces when the medium separating them was transparent. Radiation propagating through such a medium remains unchanged. For example, atmospheric air, in general, is transparent to radiation. However, gases such as CO, NO, CO_2, SO_2, H_2O, and various hydrocarbons absorb and emit radiation over certain wavelength regions called *absorption bands*. A body of water and a sheet of glass absorb the solar radiation passing through them. Radiation emitted in the interior regions of molten glass passes through the body of the glass and helps to augment cooling. As a result, during the cooling of molten glass, the temperature distribution within the body is more uniform than that expected from heat transfer by conduction alone.

In numerous other applications, engineers must concern themselves with the absorption and emission of radiation within the body. In general, the analysis of such heat transfer problems is a very complicated matter. In this chapter we present a very simple analysis of radiation exchange in an absorbing and emitting medium, and we discuss the use of some radiation charts for predicting the radiation heat exchange between a body of hot gas and its black enclosure.

The reader should consult Chandrasekhar [1], Hottel and Sarofim [2], Kourganoff [3], Love [4], Özişik [5], Siegel and Howell [6], Sobolev [7], Sparrow and Cess [8], Viskanta [9], and Edwards [10] for detailed discussions of the problems of radiation transfer in participating media.

13-1 EQUATION OF RADIATIVE TRANSFER

As discussed earlier, a beam of radiation traveling through a participating medium is attenuated as a result of absorption. Radiation emitted in the interior of a hot, semitransparent body can pass through the medium and leave the body through the bounding surfaces.

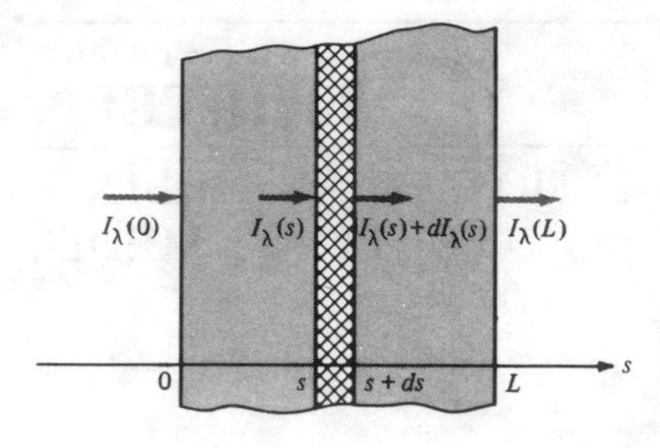

Figure 13-1 Nomenclature for the absorption of radiation in a semitransparent medium.

The distribution of radiation intensity within the body of a semi-infinite medium is governed by the *equation of radiative transfer*. Here we present the development of the equation of radiative transfer for the specific case of an absorbing, emitting medium. The results obtained will help to improve our understanding of the mechanism of radiation transfer in a participating medium.

Figure 13-1 illustrates an absorbing, emitting, semitransparent plate of thickness L that is maintained at a uniform temperature T. A beam of thermal radiation of spectral intensity $I_\lambda(0)$ propagating along a path s normal to the plate is incident on one of the boundary surfaces. When the incident radiation travels through the plate, its intensity is attenuated as a result of absorption by the medium and augmented as a result of emission of radiation owing to the temperature of the plate.

Let $I_\lambda(s)$ be the intensity of radiation along the path s at the location s and $I_\lambda(s) + dI_\lambda(s)$ be the intensity after traveling a distance ds along the path s in the medium. Then the quantity

$$\frac{dI_\lambda(s)}{ds} \tag{13-1}$$

represents the increase in the intensity of radiation per unit length along the direction of propagation. This increase in intensity must be the result of a balance between the augmentation due to emission and the attenuation due to absorption by the medium. Therefore, we state the energy balance in the form

$$\frac{dI_\lambda(s)}{ds} = \begin{pmatrix}\text{emission per}\\ \text{unit volume}\end{pmatrix} - \begin{pmatrix}\text{absorption per}\\ \text{unit volume}\end{pmatrix} \tag{13-2}$$

The absorption per unit volume is given by

$$\begin{pmatrix}\text{Absorption per}\\ \text{unit volume}\end{pmatrix} = \kappa_\lambda I_\lambda(s) \tag{13-3}$$

where κ_λ is called the *spectral absorption coefficient* for the medium and has the dimension m^{-1}.

Assuming that the medium is in local thermodynamic equilibrium and that Kirchhoff's law is valid, κ_λ also represents the emission coefficient for the medium.

Then the emission per unit volume is

$$\begin{pmatrix}\text{Emission per}\\ \text{unit volume}\end{pmatrix} = \kappa_\lambda I_{b\lambda}(T) \tag{13-4}$$

where $I_{b\lambda}(T)$ is the spectral blackbody radiation intensity at the medium temperature T.

Introducing Eqs. (13-3) and (13-4) into Eq. (13-2), we obtain

$$\frac{dI_\lambda(s)}{ds} + \kappa_\lambda I_\lambda(s) = \kappa_\lambda I_{b\lambda}(T) \tag{13-5}$$

which is called the *equation of radiative transfer* for an absorbing, emitting medium. The boundary condition for this equation at $s = 0$ may be specified as

$$I_\lambda(s) = I_\lambda(0) \qquad \text{at } s = 0 \tag{13-6}$$

This boundary condition implies that there is no reflection of radiation at the boundary surface.

13-2 TRANSMISSIVITY, ABSORPTIVITY, AND EMISSIVITY

The spectral radiation intensity $I_\lambda(s)$ along the path s in the absorbing and emitting layer can be determined from the solution of the equation of radiative transfer (13-5) subject to the boundary condition (13-6). Based on the result of such a solution, the relations defining the transmissivity, absorptivity, and emissivity along the path of propagation can be developed.

The solution of Eq. (13-5) subject to the boundary condition (13-6), with the assumption that κ_λ and $I_{b\lambda}(T)$ are constant everywhere in the medium, gives

$$I_\lambda(s) = I_\lambda(0)e^{-\kappa_\lambda s} + (1 - e^{-\kappa_\lambda s})I_{b\lambda}(T) \tag{13-7}$$

The physical significance of various terms in this equation is as follows: The first term on the right-hand side is the contribution to the intensity from the beam entering the medium at the boundary surface $s = 0$. The second term is the contribution to the intensity from the emission along the path $s = 0$ to s. We note that the radiation entering the medium is attenuated exponentially along the path of propagation.

The radiation intensity $I_\lambda(L)$ at the boundary surface $s = L$ is obtained by setting $s = L$ in Eq. (13-7):

$$I_\lambda(L) = I_\lambda(0)e^{-\kappa_\lambda L} + (1 - e^{-\kappa_\lambda L})I_{b\lambda}(T) \tag{13-8}$$

This relation implies that the radiation intensity $I_\lambda(L)$ reaching the boundary surface at $s = L$ is made up of two contributions: The first term on the right-hand side is due to the externally incident radiation, and the second term is due to the emission of radiation from the medium itself because of its temperature.

We now examine two special cases of Eq. (13-8).

Negligible emission If the emission of radiation by the medium is negligible in comparison to the contribution of the externally incident radiation, we set $I_{b\lambda}(T) = 0$, and then Eq. (13-8) reduces to

$$I_\lambda(L) = I_\lambda(0)e^{-\kappa_\lambda L} \tag{13-9}$$

Then the *spectral transmissivity* τ_λ over the path L can be defined as

$$\boxed{\tau_\lambda = \frac{I_\lambda(L)}{I_\lambda(0)} = e^{-\kappa_\lambda L}} \tag{13-10}$$

If the medium is nonreflecting, we must have

$$\tau_\lambda + \alpha_\lambda = 1 \tag{13-11}$$

and the *spectral absorptivity* α_λ over the path L is

$$\boxed{\alpha_\lambda = 1 - e^{-\kappa_\lambda L}} \tag{13-12}$$

When Kirchhoff's law is applicable, the spectral absorptivity α_λ is equal to the *spectral emissivity* ε_λ. Hence we write

$$\boxed{\varepsilon_\lambda = 1 - e^{-\kappa_\lambda L}} \tag{13-13}$$

No externally incident radiation If there is no externally incident radiation but the contribution is due to the emission by the medium only, then Eq. (13-8) reduces to

$$I_\lambda(L) = (1 - e^{-\kappa_\lambda L})I_{b\lambda}(T) \tag{13-14}$$

This equation implies that the term in the parentheses on the right-hand side is like the *spectral emissivity* ε_λ over the path L.

Absorption and Emission Properties of Materials

The absorption and emission characteristics of gases are quite different from those of solids. The absorption (or emission) of radiation by gases does not take place continuously over the entire wavelength spectrum; rather, it occurs over a large number of relatively narrow strips of intense absorption (or emission). Figure 13-2*a* shows the absorption spectrum for carbon dioxide from the data by Edwards [11] for which the density times the thickness (that is, ρL) of the gas layer was $\rho L = 2.44$ kg/m^2. The spectrum is composed of four absorption bands approximately positioned at wavelengths 15, 4.3, 2.7, and 1.9 μm.

In semitransparent solids, the absorption spectrum is not in the form of several discrete absorption bands, but it is more or less continuous. For example, Fig. 13-2*b* shows the spectral absorption coefficient for ordinary window glass as a function of wavelength for several different temperatures as obtained from measurements by Neuroth [12]. It is apparent from this figure that ordinary window glass trans-

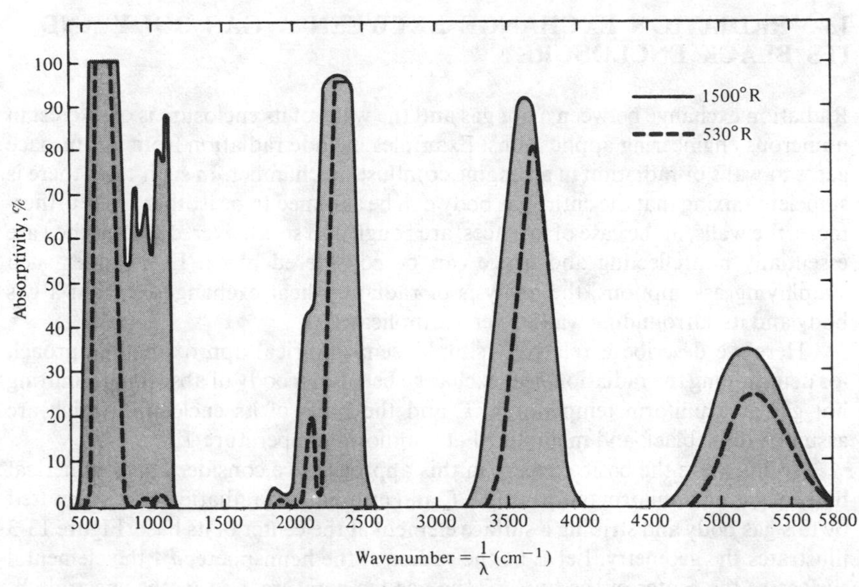

Figure 13-2 (*a*) Spectral absorptivity of CO_2 layer having ρL = 2.441 kg/m² (0.5 lb/ft²). (*From Edwards* [11].)

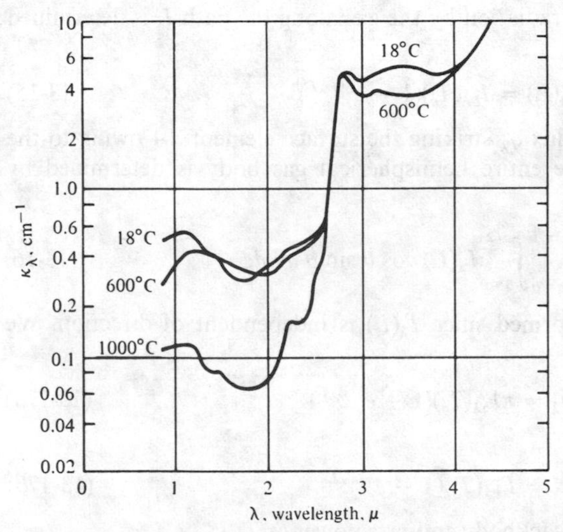

Figure 13-2 (*b*) Spectral absorption coefficient of window glass. (*From Neuroth* [12].)

mits radiation well in the visible range of the spectrum, but in the infrared range (i.e., beyond about 2.6 μm) it is considered opaque to radiation. This behavior of transmittance is the reason for the so-called greenhouse effect, that is, the trapping of solar radiation by a glass enclosure.

13-3 RADIATION EXCHANGE BETWEEN A GAS BODY AND ITS BLACK ENCLOSURE

Radiation exchange between a hot gas and the walls of its enclosure is of interest in numerous engineering applications. Examples include radiation from hot furnace gases to walls or radiation in an engine combustion chamber. In such cases there is sufficient mixing that the entire gas body can be assumed to be isothermal. Furthermore, the walls, in the case of furnaces, are rough and soot-covered so that they are essentially nonreflecting and hence can be considered black. Even under such simplifying assumptions, the analysis of radiation heat exchange between a gas body and its surrounding walls is very complicated.

Here we describe a relatively simple, semiempirical approximate approach for determining the radiation heat exchange between a body of absorbing, emitting hot gas at a uniform temperature T_g and the walls of its enclosure, which are assumed to be black and maintained at a uniform temperature T_w.

To illustrate the basic concept in this approach, we consider a hemispherical body of gas at a uniform temperature T_g and examine the radiation energy emitted by this gas body and striking a surface element at the center of its base. Figure 13-3 illustrates the geometry. Let L be the radius of the hemisphere, dA the elemental surface at the center of the base, T_g the gas temperature, and κ_λ the spectral absorption coefficient of the gas.

The intensity of the spectral radiation $I_\lambda(L)$ striking the surface element dA as a result of the emission of radiation by the gas along the path L is determined from Eq. (13-14):

$$I_\lambda(L) = I_{b\lambda}(T_g)(1 - e^{-\kappa_\lambda L}) \tag{13-15}$$

The spectral radiation heat flux q_λ striking the surface element dA owing to the emission of radiation by the entire hemispherical gas body is determined by integrating Eq. (13-15):

$$q_\lambda = \int_{\phi=0}^{2\pi} \int_{\theta=0}^{\pi/2} I_\lambda(L) \cos\theta \sin\theta \, d\theta \, d\phi \tag{13-16}$$

The integration can be performed since $I_\lambda(L)$ is independent of direction. We obtain

$$q_\lambda = \pi I_{b\lambda}(T_g)(1 - e^{-\kappa_\lambda L}) \tag{13-17a}$$

or

$$q_\lambda = E_{b\lambda}(T_g)(1 - e^{-\kappa_\lambda L}) \tag{13-17b}$$

where $E_{b\lambda}(T_g)$ is the spectral blackbody emissive power at T_g.

From Eq. (13-13), the quantity $(1 - e^{-\kappa_\lambda L})$ is the spectral emissivity of the gas ε_λ for the path length L. Then Eq. (13-17b) is written as

$$q_\lambda = \varepsilon_\lambda E_{b\lambda}(T_g) \tag{13-18a}$$

where

$$\varepsilon_\lambda = 1 - e^{-\kappa_\lambda L} \tag{13-18b}$$

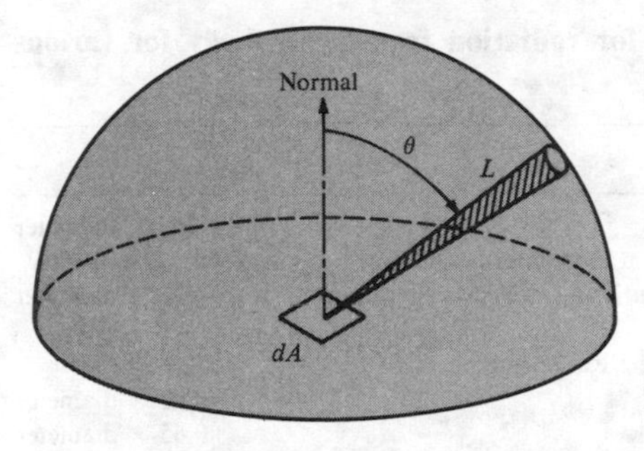

Figure 13-3 Radiation from a hemispherical gas body to a surface element at the center of its base.

Equations (13-18) are a very simple expression for the spectral radiation flux from a hemispherical gas body at uniform temperature T_g of radius L to a surface element at the center of the base. The incident energy depends on the optical radius $\kappa_\lambda L$ of the hemisphere.

The simple form for the incident radiation flux as given by Eqs. (13-18) is not applicable for other geometries. Consider, for example, a body of isothermal gas contained inside a full sphere of diameter D. The radiation energy emitted by the gas striking a spot on the inner surface of the sphere must result from the contributions of beams varying in length from zero to D at normal incidence. If all the contributions due to the varying beam lengths could be approximated by an *equivalent path length* L, related to the diameter D of the sphere, then a simple expression would result for the radiation heat flux from the gas body to the surface.

Similar arguments can be made for the radiation flux from a gas body of any other shape striking all or part of the surface of its enclosure. That is, if one can define an *equivalent path length* L related to the characteristic dimension of the gas volume, then the spectral emissivity of the gas body ε_λ for the radiation flux incident on a particular surface can be expressed in the form given by Eq. (13-18*b*). Table 13-1 lists such equivalent path lengths L for several different geometries. In the absence of information for the equivalent path length, one can estimate it approximately from

$$L \cong 3.5 \frac{V}{A} \tag{13-19}$$

where A = total surface area of the enclosure and V = total volume of the gas.

Emissivity Charts

Hottel [13, 14] and Hottel and Egbert [15] measured gas emissivity ε_g and presented emissivity charts for gases such as CO_2, H_2O, CO, ammonia, SO_2, etc., as a function of temperature and a product term $P_i L$, where P_i is the partial pressure (in atmospheres) of gas i in the gas mass and L is the beam length. Most of the experiments are conducted at a total pressure of 1 atm. However, as the total pressure is increased, the absorption lines are broadened, hence the gas emissivity

Table 13-1 Equivalent path length L for radiation from a gas body for various geometries

Case	Geometry	L
1	Sphere: Radiating to inside surface	0.65 × diameter
2	Hemisphere: Radiating to an element at the center of the base	0.5 × diameter
3	Circular cylinder of infinite height: Radiating to inside cylindrical surface	0.95 × diameter
4	Circular cylinder of semi-infinite height: Radiating to:	
	Element at the center of base	0.90 × diameter
	Entire base	0.65 × diameter
5	Circular cylinder of height equal to two diameters radiating to:	
	Plane end	0.60 × diameter
	Cylindrical surface	0.76 × diameter
	Entire surface	0.73 × diameter
6	Circular cylinder of height equal to $\frac{1}{2}$ diameter radiating to:	
	Plane end	0.43 × diameter
	Cylindrical surface	0.46 × diameter
	Entire surface	0.45 × diameter
7	Volume between two parallel plates radiating to an element on one face	1.8 × spacing between plates
8	Cube radiating to any face	0.6 × edge
9	Gas volume outside infinite bank of tubes radiating to a single tube. Let S = center-to-center spacing between the tubes.	
	Equilateral-triangle array:	
	$S = 2D$	$3.0(S - D)$
	$S = 3D$	$3.8(S - D)$
	Square array: $S = 2D$	$3.5(S - D)$

From [2, 6].

is affected. Therefore, approximate correction factors also are presented to adjust such emissivity data for a total pressure differing from 1 atm. To generalize the utility of these charts for the prediction of gas emissivities for gas masses having shapes other than a hemisphere, corrections are applied by introducing the *equivalent mean path length* which are determined by graphical or analytical means. We now present some of these charts and discuss their application in the prediction of radiative-heat exchange between a gas mass and a surface element.

Figure 13-4*a* gives the emissivity ε_c for carbon dioxide in a gas mass at a total pressure of 1 atm, plotted as a function of gas temperature T_g for several different values of the product P_cL, where P_c is the partial pressure in atmospheres of carbon dioxide and L is the radius in feet or meters of the hemispherical gas mass. Figure 13-4*b* gives an approximate correction factor C_c to adjust the emissivity data of Fig. 13-4*a* for a total pressure of gas mass differing from 1 atm; that is, the emissivity of carbon dioxide ε_c read from Fig. 13-4*a* is multiplied by the factor C_c from the correction chart, Fig. 13-4*b*, to adjust the emissivity for other total pressures.

Figure 13-5*a* shows the emissivity ε_w for water vapor in a gas mass at a total pressure of 1 atm plotted as a function of gas temperature T_g for several different

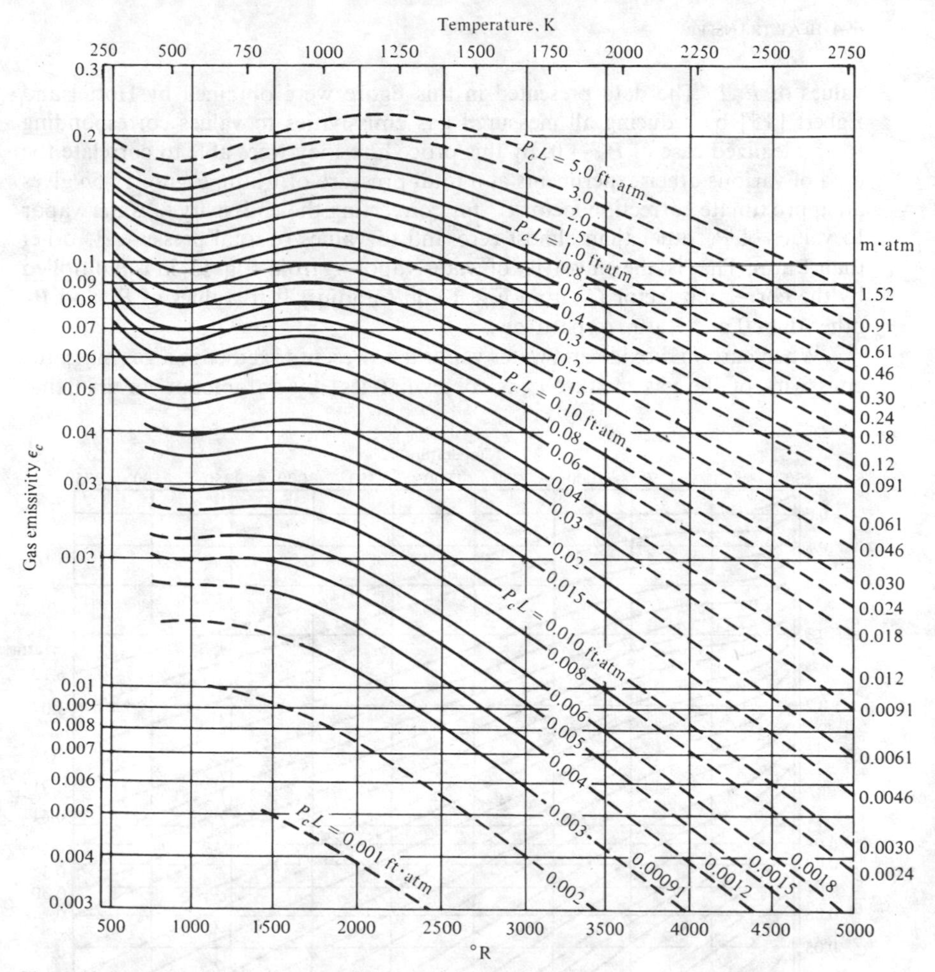

Figure 13-4 (*a*) Emissivity ε_c of carbon dioxide at a total pressure of $P_T = 1$ atm. (*From Hottel* [14].)

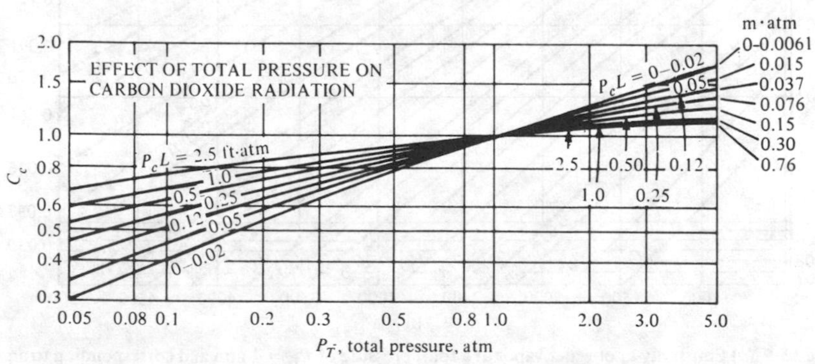

Figure 13-4 (*b*) Correction factor C_c for converting the emissivity of CO_2 at 1 atm to emissivity at P_T atm. (*From Hottel* [14].)

values of $P_w L$. The data presented in this figure were obtained by Hottel and Egbert [15] by reducing all measured gas emissivities to values corresponding to an idealized case of $P_w \to 0$. By this procedure they were able to correlate the data of various other experiments at a total pressure of 1 atm. Figure 13-5*b* gives an approximate correction factor C_w for converting the emissivity of water vapor to values of P_w other than almost zero and to values of total pressure P_T other than 1 atm. That is, the emissivity of water vapor ε_w from Fig. 13-5*a* is multiplied by the correction factor C_w from Fig. 13-5*b* to adjust it to values of P_w and P_T other than 0 and 1 atm, respectively.

When carbon dioxide and water vapor are present together in a gas mass, the emissivity of the gas mixture ε_m is somewhat less by an amount $\Delta\varepsilon$ than that

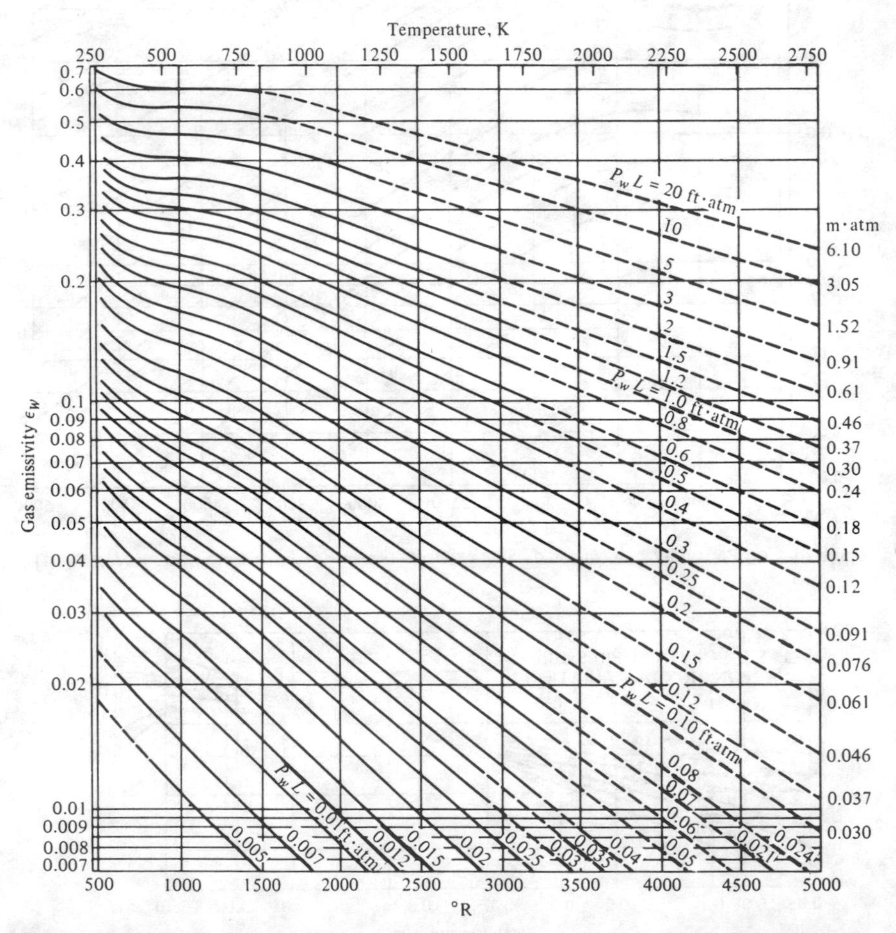

Figure 13-5 (*a*) Emissivity ε_w of water vapor at a total pressure of P_T = 1 atm and corresponding to an idealized case of $P_w \to 0$. (*From Hottel and Egbert* [15].)

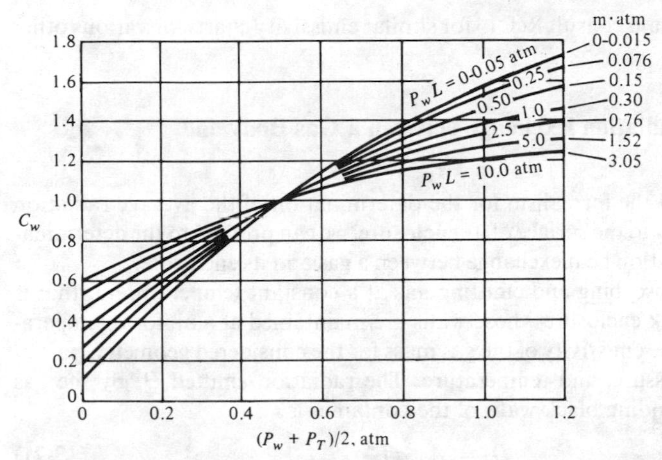

Figure 13-5 (*b*) Correction factor C_w for converting emissivity of H_2O to values of P_w and P_T other than 0 and 1 atm, respectively. (*From Hottel* [14].)

determined by adding separately calculated emissivities ε_c and ε_w for carbon dioxide and water vapor from Figs. 13-4 and 13-5, respectively. The reason is the mutual absorption of radiation. The emissivity ε_m of the mixture of carbon dioxide and water vapor can be determined approximately from

$$\varepsilon_m = \varepsilon_c + \varepsilon_w - \Delta\varepsilon \tag{13-20}$$

where ε_c and ε_w are emissivities of carbon dioxide and water vapor given in Figs. 13-4 and 13-5, respectively, and the emissivity correction $\Delta\varepsilon$ for mutual absorption can be obtained from Fig. 13-6. In this figure P_c and P_w denote, respectively, the partial pressures of carbon dioxide and water vapor.

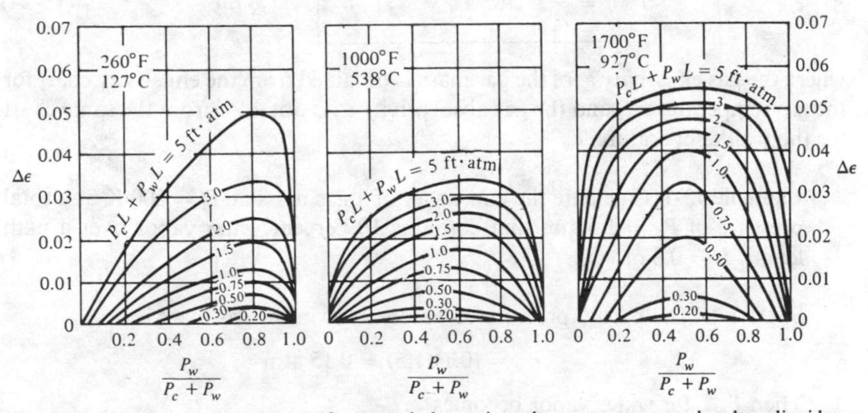

Figure 13-6 Emissivity correction $\Delta\varepsilon$ for mutual absorption when water vapor and carbon dioxide are present in the gas. (*From Hottel* [14].)

The reader should consult Ref. 14 for similar emissivity charts for various other gases.

Calculation of Radiation Exchange between a Gas Body and Its Enclosure

Having established the formalism for the determination of the average radiation flux from a gas body to the surface of its enclosure, we can proceed to the determination of the net radiation heat exchange between a gas and its enclosure.

Consider an absorbing and emitting gas at a constant temperature T_g that is contained in a black enclosure whose walls are maintained at a uniform temperature T_w. Let ε_g be the emissivity of the gas mass for the considered geometry, partial pressure, total pressure, and temperature. The radiation emitted Q_e by the gas mass to the surrounding black walls of the container is

$$Q_e = A\varepsilon_g \sigma T_g^4 \tag{13-21}$$

where A is the surface area of the walls. The radiation emitted by the surrounding black walls of temperature T_w is $A\sigma T_w^4$; if α_g is the absorptivity of the gas for radiation emitted at temperature T_w, then the radiation absorbed Q_a by the gas mass is

$$Q_a = A\alpha_g \sigma T_w^4 \tag{13-22}$$

The net radiative-heat exchange Q between the gas mass at temperature T_g and its black surroundings at temperature T_w is then determined from

$$\boxed{Q = Q_e - Q_a = \sigma A(\varepsilon_g T_g^4 - \alpha_g T_w^4)} \quad \text{W} \tag{13-23}$$

or

$$\boxed{q \equiv \frac{Q}{A} = \sigma(\varepsilon_g T_g^4 - \alpha_g T_w^4)} \quad \text{W/m}^2 \tag{13-24}$$

where the gas emissivity ε_g of the gas mass is obtained from the emissivity chart for the gas temperature T_g and the gas absorptivity α_g is obtained from the same chart for the wall temperature T_w.

Example 13-1 Calculate the emissivity of a gas mass at $T_g = 600$ K at a total pressure of $P_T = 1.5$ atm that contains 10 percent water vapor over a path length $L = 0.8$ m.

SOLUTION The partial pressure of water vapor in the gas mass is

$$P_w = (0.1)(1.5) = 0.15 \text{ atm}$$

Then $P_w L$ for water vapor becomes

$$P_w L = (0.15)(0.8) = 0.12 \text{ m} \cdot \text{atm}$$

From Fig. 13-5*a*, for $T_g = 600$ K and $P_w L = 0.12$, the emissivity of water vapor at a total pressure of 1 atm is $\varepsilon_g = 0.19$. This result must be corrected for the desired total pressure and the partial pressure for the problem by utilizing the correction chart in Fig. 13-5*b*. Thus for

$$\frac{P_w + P_T}{2} = \frac{0.15 + 1.5}{2} = 0.825 \qquad \text{and} \qquad P_w L = 0.12$$

the correction factor is determined from Fig. 13-5*b* as $C_w \cong 1.35$. The corrected emissivity of the gas mass over the path $L = 0.8$ m becomes $\varepsilon_{g,\,\text{corr}} = (0.19)(1.35) = 0.256$.

Example 13-2 A flue gas at $T_g = 1000$ K and total pressure $P_T = 2$ atm containing 10 percent water vapor by volume flows over a tube bank arranged in an equilateral-triangle array, having a tube with diameter $D = 7.6$ cm and a spacing $S = 2D$. The tubes are maintained at a uniform temperature $T_w = 500$ K and considered black. Calculate the net radiation-heat exchange between the gas and the tubes per square meter of tube wall surface.

SOLUTION The equivalent path length L, for the considered gas mass geometry, for radiation to the tube surface is obtained from Table 13-1, case 9:

$$L = 3.0(S - D) = (3.0)(0.076) = 0.228 \text{ m}$$

The partial pressure of water vapor in the gas mass is $P_w = (0.2)(1 \text{ atm}) = 0.2$ atm, and $P_w L = (0.2)(0.228) = 0.0456 \text{ m} \cdot \text{atm}$. Then, from Fig. 13-5*a*, the emissivity of the gas for $T_g = 1000$ K and $P_w L = 0.0456 \text{ m} \cdot \text{atm}$ is $\varepsilon_g = 0.08$.

The absorptivity of the gas also is obtained from Fig. 13-5*a* for $P_w L = 0.0456$, but by using $T_w = 500$ K since the energy absorbed originates from a surface at 500 K. We find $\alpha_g = 0.12$.

Figure 13-5*b* is now used to correct these results for the effects of pressure. We have

$$\frac{P_w + P_T}{2} = \frac{0.2 + 2}{2} = 1.1 \text{ atm} \qquad \text{and} \qquad P_w L = 0.0456$$

and the correction factor is obtained from Fig. 13-5*b* as $C_w = 1.58$. Then

$$\varepsilon_{g,\,\text{corr}} = 0.08(1.58) = 0.126 \qquad \alpha_{g,\,\text{corr}} = 0.12(1.58) = 0.19$$

The net radiation heat flux at the wall is determined from Eq. (13-24):

$$\begin{aligned} q &= \sigma(\varepsilon_{g,\,\text{corr}} T_g^4 - \alpha_{g,\,\text{corr}} T_w^4) \\ &= (5.67 \times 10^{-8})(0.126 \times 1000^4 - 0.19 \times 500^4) \\ &= 6471 \text{ W/m}^2 \end{aligned}$$

13-4 RADIATION FLUX FROM AN ABSORBING, EMITTING SLAB AT UNIFORM TEMPERATURE—AN ANALYTIC SOLUTION

Consider an absorbing, emitting, semitransparent medium confined to a space between two parallel plates at $x = 0$ and $x = L$, as illustrated in Fig. 13-7. The plates are black and are maintained at uniform temperatures T_1 and T_2. The medium between them is gray and has an absorption coefficient κ and at a constant temperature T_0. Problems of this type are encountered in heat transfer from a hot gas or fluid between two parallel plates when the fluid temperature is so high that radiation is the dominant mode of heat transfer. In such problems the radiation heat flux at the walls is of interest, because it governs the amount of cooling needed at the walls to keep them at the specified temperatures.

The net radiation heat flux at the walls can be determined analytically from the equation of radiative transfer, as described in Refs. 3 to 8. Here we present only the resulting expressions for the net radiation heat flux anywhere in the medium and at the boundary surfaces. The reader should consult the original references for the derivation of these results.

In radiation transfer problems of the absorbing, emitting medium, the coordinate is generally measured in the optical variable τ, defined as

$$\tau = \kappa x \tag{13-25}$$

Then the optical thickness of the medium becomes

$$\tau_0 = \kappa L \tag{13-26}$$

The net radiation heat flux $q(\tau)$ at any location τ in the medium is given by

$$\boxed{q(\tau) = 2[\sigma T_1^4 E_3(\tau) - \sigma T_2^4 E_3(\tau_0 - \tau)] + 2\sigma T_0^4[E_3(\tau_0 - \tau) - E_3(\tau)]} \tag{13-27}$$

where T_0 = temperature of medium, K
T_1, T_2 = temperatures of boundary surfaces at $\tau = 0$ and $\tau = \tau_0$, respectively, K
$E_3(\tau)$ = exponential integral function, tabulated in App. D.
σ = Stefan–Boltzmann constant

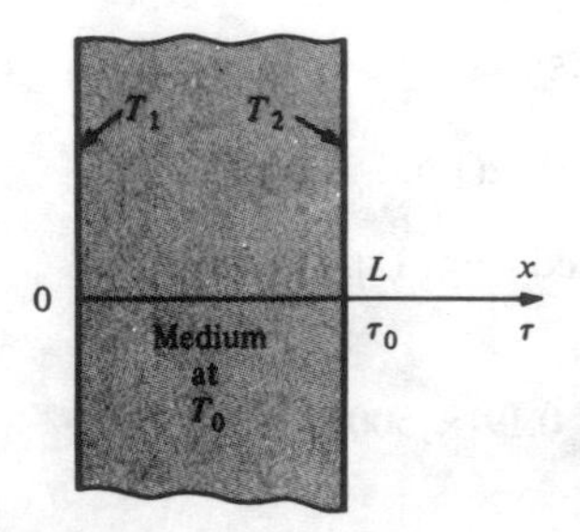

Figure 13-7 Geometry and coordinate.

In Eq. (13-27) the first expression in brackets is the contribution of the boundary surface temperatures to the net radiation flux; the second term in brackets is the contribution of the medium temperature T_0 to the net radiation heat flux.

The net radiation heat flux at the boundary surfaces is immediately obtained from Eq. (13-27) by setting $\tau = 0$ and $\tau = \tau_0$. The net radiation heat flux at the boundary surface $\tau = 0$ becomes

$$q(0) = 2[\tfrac{1}{2}\sigma T_1^4 - \sigma T_2^4 E_3(\tau_0)] + 2\sigma T_0^4[E_3(\tau_0) - \tfrac{1}{2}] \tag{13-28}$$

and the net radiation heat flux at the boundary surface $\tau = \tau_0$ becomes

$$q(\tau_0) = 2[\sigma T_1^4 E_3(\tau_0) - \tfrac{1}{2}\sigma T_2^4] + 2\sigma T_0^4[\tfrac{1}{2} - E_3(\tau_0)] \tag{13-29}$$

To obtain these expressions, we made use of the fact that $E_3(0) = \frac{1}{2}$.

In the above expressions the *heat flow is in the positive τ direction if q is a positive quantity*, whereas in the expressions given in Chap. 12 for the nonparticipating medium, the heat flow is from the surface into the medium (i.e., the enclosure) if q is positive.

If the medium between the plates is transparent to radiation, we have $\kappa = 0$, hence the optical thickness of the medium is $\tau_0 = 0$. For this special case, we set $\tau_0 = 0$ in Eqs. (13-28) and (13-29) and obtain

$$q(0) = \sigma T_1^4 - \sigma T_2^4 \tag{13-30}$$

We note that this result is the same as that given in Chap. 12 for the net radiation heat flux between two black parallel plates at temperatures T_1 and T_2 that are separated by a nonparticipating medium.

Example 13-3 An absorbing, emitting, gray gas at $T_0 = 1000$ K flows between two parallel, black plates maintained at $T_1 = T_2 = 500$ K. Calculate the net radiation heat flux at the walls for an optical thickness $\tau_0 = 0.5$ of the gas between the plates.

SOLUTION The net radiation heat flux at the walls can be evaluated from Eq. (13-28):

$$q(0) = 2[\tfrac{1}{2}\sigma T_1^4 - \sigma T_2^4 E_3(\tau_0)] + 2\sigma T_0^4[E_3(\tau_0) - \tfrac{1}{2}]$$

and because of symmetry, the magnitude of the radiation heat flux at both boundaries is the same.

The numerical values are as follows:

$$\sigma T_1^4 = \sigma T_2^4 = (5.67 \times 10^{-8})(500^4) = 3544 \text{ W/m}^2$$

$$\sigma T_0^4 = (5.67 \times 10^{-8})(1000^4) = 56{,}700 \text{ W/m}^2$$

From Table D-4, $E_3(0.5) = 0.2216$. Introducing these values in the above expression, we obtain

$$q(0) = 2(3544)(\tfrac{1}{2} - 0.2216) + 2(56{,}700)(0.2216 - \tfrac{1}{2})$$
$$= -29.6 \text{ kW/m}^2$$

Here, since $q(0)$ is negative, heat flows from the gas to the plate at $\tau = 0$.

PROBLEMS

Use of emissivity charts

13-1 A gas mass at a uniform temperature $T_g = 1100$ K and a total pressure $P_T = 2$ atm contains 5 percent water vapor by volume. Determine the emissivity of the gas over a path length $L = 1.5$ m.
Answer: 0.235

13-2 Determine the emissivity of a gas mass at $T_g = 800$ K with a total pressure $P_T = 3$ atm that contains 15 percent water vapor by volume over a path length $L = 3$ m.

13-3 A gas body at $T_g = 1100$ K and total pressure $P_T = 2$ atm, containing 15 percent water vapor by volume, flows over a tube bank in a square array. The spacing between the tube centers is $S = 2D$, and the tube diameter is $D = 5$ cm. The tube surface is maintained at a uniform temperature $T_w = 500$ K and can be considered black. Calculate the net radiation heat flux between the gas and the tube surface.

13-4 A gas at a uniform temperature $T_g = 1000$ K and total pressure $P_T = 3$ atm, having 10 percent water vapor by volume, is contained inside a sphere of diameter $D = 0.3$ m. The surface of the sphere is maintained at a uniform temperature $T_w = 500$ K and is considered black. Determine the net radiation heat exchange between the gas and the wall of the sphere.
Answer: 1.85 kW

13-5 A cubical furnace with sides $a = 0.75$ m long has black interior walls which are maintained at a uniform temperature $T_w = 600$ K. It contains hot gases at $T_g = 1400$ K and a total pressure of $P_T = 1$ atm that consist of 22 percent CO_2 and 78 percent nitrogen by volume. Calculate the net radiation heat flux and the total radiation heat transfer rate at the walls.

13-6 A hot gas at $T_g = 1500$ K and total pressure $P_T = 1$ atm, consisting of 20 percent CO_2 and 80 percent nitrogen by volume, is contained in the space between two black, parallel plates maintained at a uniform temperature $T_w = 600$ K. The spacing between the plates is $H = 0.8$ m. Calculate the net radiation heat flux between the hot gas and the wall surface.
Answer: 39.1 kW/m^2

13-7 A cubical furnace of sides $a = 2$ m has interior walls which are maintained at a uniform temperature $T_w = 500$ K and considered black. The furnace is filled with combustion products at $T_g = 1800$ K, at a total pressure $P_T = 2$ atm, which contains 40 percent CO_2 and 60 percent nitrogen by volume. Calculate the net radiation heat flux and the total heat transfer rate at the walls.

13-8 A cylindrical tank with diameter $D = 2$ m and height $H = 4$ m contains a hot gas at $T_g = 1000$ K and a total pressure $P_T = 1$ atm consisting of 25 percent CO_2 and 75 percent nitrogen by volume. The walls of the tank are sufficiently cool so that radiation from them is negligible. Calculate the radiation energy needed to be removed from the tank to keep the walls sufficiently cool (i.e., $T_w \ll T_g$).

13-9 A hot gas at $T_g = 1000$ K and total pressure $P_T = 4$ atm, consisting of 50 percent CO_2 and 50 percent nitrogen by volume, is contained between two parallel plates $H = 15$ cm apart. The plate surfaces are maintained at a uniform temperature $T_w = 500$ K and are considered black. Calculate the net radiation heat flux at the surface of the plates.
Answer: 11.6 kW/m^2

13-10 A cylindrical tank with diameter $D = 4$ m and height $H = 2$ m contains a hot gas at $T_g = 1200$ K and a total pressure $P_T = 1$ atm. The gas is composed of 30 percent CO_2 and 70 percent nitrogen by volume. The walls of the tank are sufficiently cool so that the radiation from the wall is negligible in comparison to that from the hot gas. Calculate the heat removal rate from the tank necessary to keep the walls sufficiently cool.

Analytic solutions

13-11 A gray gas at a uniform temperature $T_0 = 800$ K flows between two black parallel plates, each maintained at a uniform temperature of $T_1 = T_2 = 400$ K. The spacing between the plates is $L = 0.8$ m, and the absorption coefficient of the gas is $\kappa = 0.5\ m^{-1}$. Calculate the net radiation heat flux at the walls.

13-12 An absorbing, emitting, gray gas at a uniform temperature $T_0 = 1000$ K flows between two parallel plates which are regarded as black and are maintained at a uniform temperature $T_1 = T_2 = 400$ K. The spacing between the plates is $L = 1.2$ m. Calculate the absorption coefficient κ for the gas if the amount of cooling needed at each wall surface is $q = 30\ kW/m^2$.

Answer: $0.4\ m^{-1}$

13-13 An absorbing, emitting, gray gas having an absorption coefficient $\kappa = 0.1\ m^{-1}$ that is maintained at a uniform temperature $T_0 = 900$ K flows between two parallel plates separated by a distance $L = 0.6$ m. The plates are considered black and are maintained at uniform, but different, temperatures $T_1 = 400$ K and $T_2 = 600$ K. Calculate the rate of cooling necessary at each wall.

Answer: $9.07\ kW/m^2$ at surface 1; $2.09\ kW/m^2$ at surface 2

13-14 An absorbing, emitting, gray gas at $T_0 = 1200$ K flows between two black parallel plates maintained at $T_1 = 500$ K and $T_2 = 700$ K. Calculate the net radiation heat flux at both surfaces for the following optical thicknesses of the gas: $\tau_0 = 0$, 0.1, and 1.0.

REFERENCES

1. Chandrasekhar, S.: *Radiative Transfer*, Oxford University Press, London, 1950; also, Dover, New York, 1960.
2. Hottel, H. C., and A. F. Sarofim: *Radiative Transfer*, McGraw-Hill, New York, 1967.
3. Kourganoff, V.: *Basic Methods in Transfer Problems*, Dover, New York, 1963.
4. Love, T. J.: *Radiation Heat Transfer*, Merrill, Columbus, Ohio, 1968.
5. Özışık, M. N.: *Radiative Transfer and Interactions with Conduction and Convection*, Wiley, New York, 1973.
6. Siegel, R., and R. Howell: *Thermal Radiation Heat Transfer*, 2d ed., Hemisphere, New York, 1981.
7. Sobolev, V. V.: *A Treatise on Radiative Transfer*, Van Nostrand, Princeton, N.J., 1963.
8. Sparrow, E. M., and R. D. Cess: *Radiation Heat Transfer*, Brooks/Cole, Belmont, Calif., 1970.
9. Viskanta, R.: "Radiation Transfer and Interaction of Convection with Radiation Heat Transfer," in T. F. Irvine and J. P. Hartnett (eds.), *Advances in Heat Transfer*, Academic, New York, 1966.
10. Edwards, D. K.: *Radiation Heat Transfer Notes*, Hemisphere, New York, 1981.
11. Edwards, D. K.: "Radiation Interchange in a Nongray Enclosure Containing an Isothermal Carbon Dioxide–Nitrogen Gas Mixture," *J. Heat Transfer* **84C**:1–11 (1962).
12. Neuroth, N.: "Das Einfluss der Temperatur auf die Spektrale Absorption von Blasern in Ultraroter, I," *Glastech Ber.* **25**:242–249 (1952).
13. Hottel, H. C.: "Heat Transmission by Radiation from Non-luminous Gases," *Trans. AIChE* **19**:173–205 (1927).
14. Hottel, H. C.: "Radiant Heat Transmission," in W. H. McAdams (ed.), *Heat Transmission*, McGraw-Hill, New York, 1954.
15. Hottel, H. C., and R. B. Egbert: "Radiant Heat Transmission from Water Vapor," *Trans. AIChE* **38**:531–565 (1942).

CHAPTER
FOURTEEN
MASS TRANSFER

Mass transfer processes occur in a variety of applications in mechanical, chemical, and aerospace engineering; physics; chemistry; and biology. Typical examples include the transpiration cooling of jet engines and rocket motors, the ablative cooling of space vehicles during reentry into the atmosphere, the mass transfer from laminar and turbulent streams onto the surfaces of a conduit, and evaporation or condensation on the surface of a tube or plate. Processes such as absorption, desorption, distillation, solvent extraction, drying, humidification, sublimation, and many others involve mass transfer. In absorption, a gas is brought into direct contact with a liquid solvent in order to remove the soluble components of the gas. The reverse process occurs in desorption; that is, the transfer of solute takes place from the liquid to the gas. In solvent extraction, one or more components of a liquid mixture are extracted by solution in a selective solvent. In humidification, water is transferred from the liquid to the air. The biological applications include oxygenation of blood, food and drug assimilation, respiration mechanism, and numerous others. The theory and application of mass transfer is such a widespread field that numerous books have been written on the subject [1–16]. Here, in the one chapter devoted to mass transfer we only introduce this subject, with emphasis on the similarity between heat and mass transfer processes. When mass transfer takes place in a fluid at rest, the mass is transferred by purely molecular diffusion resulting from concentration gradients; the process is analogous to heat diffusion resulting from temperature gradients. When the fluid is in motion, mass transfer takes place by both molecular diffusion and convective motion of the bulk fluid; then a knowledge of the velocity field is needed to solve the mass transfer problem. For low concentrations of the mass in the fluid and low mass transfer rates, the convective heat and mass transfer processes are analogous, and many of the results derived in connection with convective heat transfer are applicable to convective mass transfer. Therefore, the mass transfer equations and coefficients presented in this chapter are obtained by analogy directly from the corresponding heat transfer equations. However, under high-mass-flux conditions and with chemical reactions there are significant differences between heat and mass transfer processes; such situations are not considered here. The reader should consult Refs. 7 and 8 for a discussion of mass transfer in chemically reacting systems and Ref. 1 for mass transfer at high mass flux and high concentrations.

14-1 DEFINITIONS OF MASS FLUX

We consider a fluid mixture of two components, say A and B, the composition of which is characterized by the *molal concentration* of the components. The molal concentration c_A of component A is defined as the number of molecules of component A per unit volume of the mixture and may be given in the units $\text{lb} \cdot \text{mol/ft}^3$ or $\text{kg} \cdot \text{mol/m}^3$. Various other definitions are also in use in the literature for expressing the composition. For example, the *mole fraction* χ_A of component A is defined as $\chi_A = c_A/c$, where c is the total molal concentration of the mixture. The *mass concentration* ρ_A of component A is the mass of component A per unit volume and may be given in the units lb/ft^3 or kg/m^3. The *mass fraction* w_A of component A is defined as $w_A = \rho_A/\rho$, where ρ is the total mass density of the mixture.

Various Velocities in the Mixture

Consider a two-component mixture whose concentration varies in the x direction and the fluid undergoes a bulk motion in the same direction. Let u_A and u_B be the *statistical mean velocities* of components A and B, respectively, in the x direction with respect to the stationary coordinate. The *molal average velocity* U of the mixture in the x direction is defined by

$$U = \frac{1}{c}(u_A c_A + u_B c_B) \tag{14-1}$$

where c_A and c_B are the molal concentrations of species A and B, respectively, and c is the *total molal concentration of the mixture.*

The *diffusion velocities* of species A and B with respect to the molal average velocity U are defined as

$$\begin{aligned} u_A - U &= \text{diffusion velocity of species } A \\ u_B - U &= \text{diffusion velocity of species } B \end{aligned} \tag{14-2}$$

Thus, the diffusion velocity indicates the motion of a species relative to the local average motion of the mixture.

Various Fluxes in the Mixture

The *molal fluxes* N_A *and* N_B *of species A and B, respectively, relative to stationary coordinates* in the x direction are given by

$$N_A = c_A u_A \qquad \text{and} \qquad N_B = c_B u_B \tag{14-3}$$

That is, N_A and N_B characterize the moles of species A and B, respectively, that pass through a unit area perpendicular to the x axis per unit time.

The *molal fluxes J_A and J_B of species A and B, respectively, relative to the molal average velocity U* are defined as

$$J_A = c_A(u_A - U) \qquad \text{and} \qquad J_B = c_B(u_B - U) \tag{14-4}$$

That is, the molal fluxes J_A and J_B are the measure of the diffusion rates of species A and B, respectively, in the mixture.

Relationship among Various Fluxes

The relationship among various molal fluxes is now derived by combining the above results. We substitute U from Eq. (14-1) into Eqs. (14-4) to obtain

$$J_A = c_A u_A - \frac{c_A}{c}(u_A c_A + u_B c_B) \tag{14-5a}$$

$$J_B = c_B u_B - \frac{c_B}{c}(u_A c_A + u_B c_B) \tag{14-5b}$$

When the definition of N_A and N_B as given by Eqs. (14-3) is introduced into Eqs. (14-5), we obtain the relations among various fluxes:

$$J_A = N_A - \frac{c_A}{c}(N_A + N_B) \tag{14-6a}$$

$$J_B = N_B - \frac{c_B}{c}(N_A + N_B) \tag{14-6b}$$

These results show that the *molal diffusion flux* J_i of species i is equal to the difference between the molal flux N_i and the bulk flow in the mixture of species i.

It is apparent from Eqs. (14-6) that the sum of J_A and J_B is zero, that is,

$$J_A + J_B = 0 \qquad \text{or} \qquad J_A = -J_B \tag{14-7}$$

since $c = c_A + c_B$. This result implies that in a binary mixture the diffusion fluxes J_A and J_B of the two components are of equal magnitude and in opposite directions.

Fick's First Law

In a binary mixture in which composition varies in the x direction and molecular diffusion occurs within the fluid due to the nonuniformity of composition, the molal fluxes J_A and J_B in the x direction of species A and B are related to the concentration gradients by *Fick's first law*:

$$J_A = -D_{AB}\frac{dc_A}{dx} \tag{14-8a}$$

$$J_B = -D_{BA}\frac{dc_B}{dx} \tag{14-8b}$$

where D_{AB} is the *mass diffusivity* (or *the diffusion coefficient*) of A in B and D_{BA} is the *mass diffusivity of* B in A; they are equal to each other:

$$D_{AB} = D_{BA} \equiv D \tag{14-9}$$

Because mass diffusion takes place in the direction of decreasing concentration, a minus sign is included in Eqs. (14-8) to make the mass flux in the positive x direction a positive quantity when the concentration decreases in the positive x direction. Thus, *when J_A is positive, the mass flux of species A is in the positive x direction, and vice versa.* To give some idea on the units of various quantities in Eqs. (14-8), we list the units:

c_i = concentration of component i in mixture, $i = A$ or B, kg · mol/m^3 or (lb · mol/ft^3)

D = mass diffusivity, or diffusion coefficient, m^2/s or (ft^2/h)

J_i = molal flux of component i in x direction, $i = A$ or B, kg · mol/(m^2 · s) or lb · mol/(ft^2 · h)

x = distance in x direction, m or ft

We note that the mass flux relation given above by Fick's first law is similar to the heat flux expression given by the Fourier law as

$$q = -k\frac{dT}{dx} = -\alpha\frac{d}{dx}(\rho c_p T) \tag{14-10}$$

and to the momentum flux expression given by

$$\tau g = -\mu\frac{du}{dy} = -\nu\frac{d(\rho u)}{dy} \tag{14-11}$$

Clearly, the mass diffusivity D, the heat diffusivity α, and the momentum diffusivity ν have the same units, ft^2/h or m^2/s, and Eqs. (14-8), (14-10), and (14-11) are of the same form.

If the mixture is considered to be a perfect gas, the molal concentrations c_A and c_B are related to the partial pressures p_A and p_B of the species A and B in the mixture by

$$p_i = c_i \mathscr{R} T \qquad i = A \text{ or } B \tag{14-12}$$

where c_i = molal concentration of component i in mixture, kg · mol/m^3 or (lb · mol/ft^3)

p_i = partial pressure of component i in mixture, atm

$\mathscr{R}$ = gas constant
= 0.730 ft^3 · atm/(lb · mol · °R) = 0.08205 m^3 · atm/(kg · mol · K)

Then Eqs. (14-8) can be written as

$$J_i = -\frac{D}{\mathscr{R}T}\frac{dp_i}{dx} \qquad i = A \text{ or } B \qquad \text{kg} \cdot \text{mol/(m}^2 \cdot \text{s)[lb} \cdot \text{mol/(ft}^2 \cdot \text{h)]} \tag{14-13}$$

Various relations given in this section for the definition of mass fluxes are applied in the next sections in the analysis of mass diffusion problems for binary mixtures.

14-2 STEADY-STATE EQUIMOLAL COUNTERDIFFUSION IN GASES

Consideration is now given to the application of the previous relations for flux in the prediction of concentration distribution for the *steady-state equimolal counterdiffusion* in a binary gas mixture composed of components A and B. In this mass transfer process, gases A and B diffuse simultaneously in opposite directions through each other. That is, component A diffuses through component B, and vice versa, and they diffuse at the same molal rate but in opposite directions. This process is approximated in the distillation of a binary system.

Consider that two large vessels containing uniform mixtures of A and B at different concentrations are suddenly connected by a small pipe. It is assumed that both vessels are at the same total pressure p and uniform temperature T. Component A will diffuse from the higher concentration to the lower concentration, and component B will diffuse at the same rate but in the opposite direction through the connecting pipe. Since the vessels are sufficiently large, steady-state equimolal counterdiffusion takes place in the connecting pipe; that is, *the total molal flux with respect to stationary coordinates is zero*, and we have

$$N_A + N_B = 0 \qquad \text{or} \qquad N_A = -N_B \tag{14-14}$$

For this type of mass diffusion process, the molal fluxes of species A and B relative to stationary coordinates are equal and in the opposite directions. The substitution of Eqs. (14-14) into Eqs. (14-6) yields

$$J_A = N_A = -N_B = -J_B \tag{14-15}$$

Substitution of this result into Eq. (14-13) for $i = A$ gives

$$N_A = -\frac{D}{\mathscr{R}T}\frac{dp_A}{dx} \tag{14-16}$$

At steady state N_A and N_B are constant. Then, Eq. (14-16) for constant D implies that the distribution of the partial pressure p_A of component A along the connecting pipe is linear with the distance. A similar conclusion can be drawn for the distribution of p_B either by writing an equation for N_B analogous to Eq. (14-15) or by the fact that the sum of the partial pressures p_A and p_B is equal to the total pressure p, which remains constant, i.e.,

$$p_A + p_B = p = \text{const} \tag{14-17}$$

Let p_{A_1} and p_{A_2} be the partial pressures of component A at the two ends of the connecting pipe $x = x_1$ and $x = x_2$, respectively. The integration of Eq. (14-16) from $x = x_1$ to $x = x_2$ gives

$$N_A = -\frac{D}{\mathscr{R}T}\frac{p_{A_2} - p_{A_1}}{x_2 - x_1} \quad \text{kg}\cdot\text{mol/(m}^2\cdot\text{s) or [lb}\cdot\text{mol/(ft}^2\cdot\text{h)]} \tag{14-18}$$

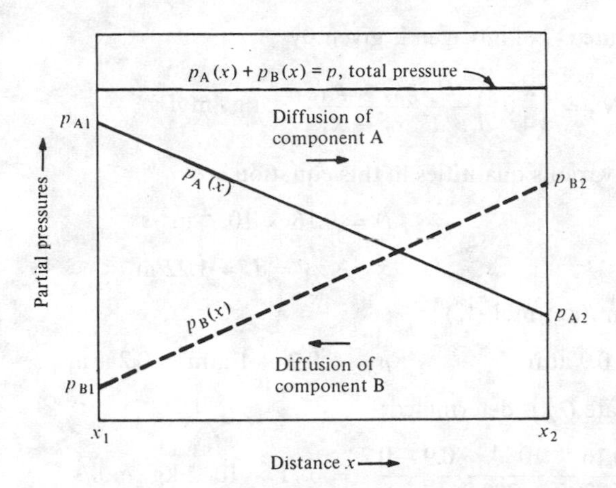

Figure 14-1 Distribution of partial pressures p_A and p_B in equimolal counterdiffusion of a binary gas mixture.

A similar relation can be written for N_B. If it is assumed that $p_{A_1} > p_{A_2}$, then for component B we must have $p_{B_2} > p_{B_1}$, where p_{B_1} and p_{B_2} are the partial pressures of component B at $x = x_1$ and $x = x_2$, respectively. Figure 14-1 shows schematically the distribution of partial pressures of the two components as a function of distance for $p_{A_1} > p_{A_2}$. Clearly, component A diffuses in the direction from x_1 to x_2, and component B in the opposite direction.

Example 14-1 Consider two large vessels, each containing uniform mixtures of nitrogen (i.e., component A) and carbon dioxide (i.e., component B) at 1 atm, $T = 288.9$ K, but at different concentrations. Vessel 1 contains 90 mole percent N_2 and 10 mole percent CO_2, whereas vessel 2 contains 20 mole percent N_2 and 80 mole percent CO_2. The two vessels are connected by a duct of $d =$ 0.1524 m inside diameter and $L = 1.22$ m long. Determine the rate of transfer of nitrogen between the two vessels by assuming that steady-state transfer takes place in view of the large capacity of the two reservoirs. The mass diffusivity for the N_2–CO_2 mixture at 1 atm and 288.9 K can be taken as $D = 0.16 \times 10^{-4}$ m²/s.

SOLUTION In this mass transfer process, nitrogen (component A) is transferred from vessel 1 containing a higher concentration of nitrogen to vessel 2 containing a lower concentration of nitrogen. In the early stages of the mass transfer, the partial pressure of nitrogen in both vessels is considered to remain constant, hence a steady-state transfer can be assumed. Then the mass transfer process can be characterized as an *equimolal counterdiffusion* as described above, and the mass flux of nitrogen through the connecting duct can be determined by using Eq. (14-18):

$$N_A = \frac{D}{\mathscr{R}T} \frac{p_{A_1} - p_{A_2}}{x_2 - x_1} \qquad \text{kg} \cdot \text{mol}/(\text{m}^2 \cdot \text{s})$$

The total mass transfer rate Q_A of nitrogen is given by

$$Q_A = \text{area} \times N_A = \left(\frac{\pi}{4} d^2\right) \frac{D}{\mathscr{R}T} \frac{p_{A_2} - p_{A_1}}{x_2 - x_1} \quad \text{kg} \cdot \text{mol/s}$$

The numerical values of various quantities in this equation are

$d = 0.1524$ m $\qquad D = 0.16 \times 10^{-4}$ m^2/s

$T = 288.9$ K $\qquad x_2 - x_1 = L = 1.22$ m

$\mathscr{R} = 0.08205$ m$^3 \cdot$ atm/(kg $\cdot$ mol $\cdot$ K)

$p_{A_1} = 0.9 \times 1$ atm $= 0.9$ atm $\qquad p_{A_2} = 0.2 \times 1$ atm $= 0.2$ atm

Then the mass transfer rate Q_A is determined:

$$Q_A = \frac{\pi}{4} \times 0.1524^2 \frac{0.16 \times 10^{-4}}{0.08205 \times 288.9} \frac{0.9 - 0.2}{1.22} = 0.71 \times 10^{-8} \text{ kg} \cdot \text{mol/s}$$

14-3 STEADY-STATE UNIDIRECTIONAL DIFFUSION IN GASES

An application of various flux relations discussed previously is now given in relation to the *steady-state unidirectional diffusion* in a binary gas mixture of components A and B. In this process one of the components, say A, diffuses through component B, which remains motionless relative to the stationary coordinates; i.e., the molal flux N_B of component B relative to the stationary coordinates is zero. The situation is better envisioned if we consider a medium containing gas components A and B in which component A is supplied at $x = x_1$ at a steady rate and diffuses in the x direction through gas B to an interface at $x = x_2$ where gas A is absorbed but B is not. This type of diffusion process is approximated in gas absorption, desorption, and adsorption. The determination of the distribution of concentrations of gases A and B in the medium is of interest.

Since in this mass transfer process the net molal flux of component B relative to the stationary coordinates is zero, we set

$$N_B = 0 \tag{14-19}$$

and the molal diffusion flux J_A of component A due to molecular diffusion is given by

$$J_A = -D \frac{dc_A}{dx} \tag{14-20}$$

where D is the mass diffusivity of A in B. The mass flux J_A is now obtained from Eq. (14-6a) as

$$J_A = N_A - \frac{c_A}{c}(N_A + N_B) \tag{14-21}$$

When the results in Eqs. (14-19) and (14-20) are substituted into Eq. (14-21), we obtain

$$-D\frac{dc_A}{dx} = N_A - \frac{c_A}{c}N_A \tag{14-22}$$

Assuming perfect gas, we replace $c_A = p_A/(\mathscr{R}T)$, and Eq. (14-22) becomes

$$-\frac{D}{\mathscr{R}T}\frac{dp_A}{dx} = N_A\left(1 - \frac{p_A}{p}\right)$$

or

$$N_A\,dx = -\frac{Dp}{\mathscr{R}T}\frac{dp_A}{p - p_A} \tag{14-23}$$

where p is the total pressure of the mixture, which is assumed to be constant, and p_A is the partial pressure of A in the mixture. Then

$$p_A + p_B = p = \text{const} \tag{14-24}$$

The integration of Eq. (14-23) for constant D, p, T, N_A with x varying from x_1 to x_2 and p_A varying from p_{A_1} to p_{A_2} gives

$$N_A\int_{x_1}^{x_2} dx = -\frac{Dp}{\mathscr{R}T}\int_{p_{A_1}}^{p_{A_2}}\frac{dp_A}{p - p_A}$$

or

$$N_A = \frac{Dp}{\mathscr{R}T(x_2 - x_1)}\ln\frac{p - p_{A_2}}{p - p_{A_1}} = \frac{Dp}{\mathscr{R}T(x_2 - x_1)}\ln\frac{p_{B_2}}{p_{B_1}} \tag{14-25}$$

since

$$p - p_{A_1} = p_{B_1} \qquad \text{and} \qquad p - p_{A_2} = p_{B_2} \tag{14-26}$$

We note from Eqs. (14-26) that the following relation holds: $p_{B_2} - p_{B_1} = p_{A_1} - p_{A_2}$. Equation (14-25) can be arranged as

$$N_A = \frac{Dp}{\mathscr{R}T(x_2 - x_1)p_{B\ln}}(p_{B_2} - p_{B_1}) = \frac{Dp}{\mathscr{R}T(x_2 - x_1)p_{B\ln}}(p_{A_1} - p_{A_2}) \tag{14-27a}$$

where the *logarithmic mean partial pressure* $p_{B\ln}$ of component B is defined as

$$p_{B\ln} = \frac{p_{B_2} - p_{B_1}}{\ln(p_{B_2}/p_{B_1})} \tag{14-27b}$$

Figure 14-2 illustrates the distribution of partial pressures (or the concentration) of species A and B with distance x in the medium; clearly the distribution is not linear with the position. Also note that there is a diffusion of species B in the medium due to the presence of the concentration gradient set up; but the *net flux of B remains zero* because species B is supplied by the bulk movement at the same ate as it diffuses away.

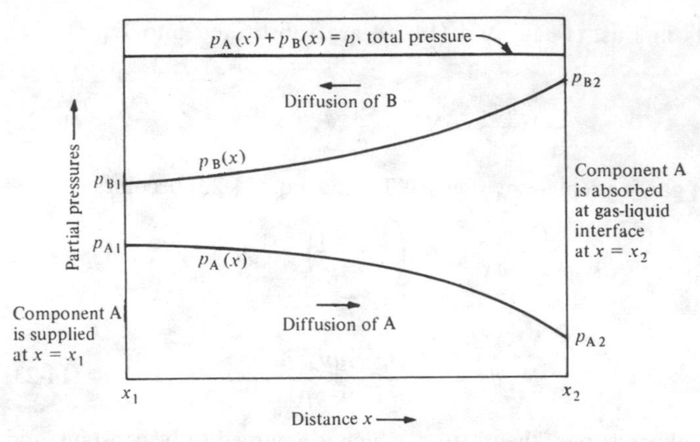

Figure 14-2 Distribution of partial pressures p_A and p_B for unidirectional diffusion of gas A through gas B.

Example 14-2 A deep, narrow cylindrical vessel which is open at the top contains some toluene at the bottom. The air within the vessel is considered motionless, but there is sufficient air current at the top surface of the vessel that any toluene vapor arriving at the top surface is immediately removed to ensure zero toluene concentration at the top surface. The entire system is at atmospheric pressure and 18.7°C. Under these conditions the diffusivity of air-toluene vapor is $D = 0.826 \times 10^{-5}\ m^2/s$ (see Table 14-3), and the saturated vapor pressure of toluene at the liquid surface in the vessel is 0.026 atm. Determine the rate of evaporation of toluene into the air per unit area of the liquid surface if the distance between the liquid toluene surface and the top of the vessel is 1.524 m.

Solution Here toluene vapor is supplied to the air inside the vessel as a result of evaporation from the liquid surface. The toluene vapor diffuses through the stagnant air layer to the top of the vessel where it is immediately removed as a result of air currents, but there is no net removal of air from the vessel. Therefore, the process can be characterized as the steady-state unidirectional diffusion of toluene (component A) through the stationary air (component B) layer of thickness $x_2 - x_1 = 1.524$ m. Equations (14-27) can be used to determine the rate of evaporation of toluene into the air, that is,

$$N_A = \frac{Dp}{\mathscr{R}T(x_2 - x_1)p_{B\ln}}(p_{A_1} - p_{A_2})$$

where

$$p_{B\ln} = \frac{p_{B_2} - p_{B_1}}{\ln(p_{B_2}/p_{B_1})}$$

and the numerical values of various quantities are given as

$$D = 0.826 \times 10^{-5}\ \text{m}^2/\text{s} \qquad T = 291.7\ \text{K}$$

$$x_2 - x_1 = 1.524\ \text{m}$$

$$\mathscr{R} = 0.08205\ \text{m}^3 \cdot \text{atm}/(\text{kg} \cdot \text{mol} \cdot \text{K})$$

$$p = 1\ \text{atm} \qquad p_{A_1} = 0.026\ \text{atm} \qquad p_{A_2} = 0\ \text{atm}$$

$$p_{B_1} = 1 - p_{A_1} = 1 - 0.026 = 0.974\ \text{atm}$$

Then

$$p_{B\,\ln} = \frac{1 - 0.974}{\ln (1/0.974)} = 0.987$$

The rate of evaporation of toluene into the air per unit area of the liquid surface is now determined by substituting these numerical values into the above equation.

$$N_A = \frac{(0.826 \times 10^{-5}) \times 1 \times (0.026 - 0)}{0.08205 \times 291.7 \times 1.524 \times 0.987} = 0.597 \times 10^{-8}\ \text{kg} \cdot \text{mol}/(\text{m}^2 \cdot \text{s})$$

14-4 STEADY-STATE DIFFUSION IN LIQUIDS

In the last two sections, we considered steady-state equimolal counterdiffusion and unidirectional diffusion in binary gas mixtures. We now examine the same diffusion processes for binary mixtures of liquids.

Steady-State Equimolal Counterdiffusion in Liquids

The analysis follows the same approach as that for gases. Equation (14-16) is now written in terms of the molal concentration c_A as

$$J_A = N_A = -D\,\frac{dc_A}{dx} \tag{14-28}$$

The integration of the equation from $x = x_1$ to $x = x_2$ with c_A varying from c_{A_1} to c_{A_2} gives

$$N_A = D\,\frac{c_{A_1} - c_{A_2}}{x_2 - x_1} \tag{14-29}$$

where c_{A_1}, c_{A_2} = molal concentration of species A at x_1 and x_2, respectively, $\text{kg} \cdot \text{mol}/\text{m}^3$

D = mass diffusivity for liquid-to-liquid diffusion in binary liquid mixture, m^2/s

N_A = molal flux of species A, $\text{kg} \cdot \text{mol}/(\text{m}^2 \cdot \text{s})$

x = distance, m

Clearly, Eq. (14-29) is analogous to Eq. (14-18). In the analysis of diffusion in liquids, the mole fraction χ_i is sometimes defined as

$$\chi_i = \frac{c_i}{c} = \text{mole fraction of component } i \qquad i = A \text{ or } B \tag{14-30a}$$

where

$$c_A + c_B = c = \text{total molal concentration of mixture, kg}\cdot\text{mol/m}^3 \tag{14-30b}$$

Then, in terms of the mole fraction, Eq. (14-29) is written

$$N_A = Dc\,\frac{\chi_{A_1} - \chi_{A_2}}{x_2 - x_1} \qquad \text{kg}\cdot\text{mol/(m}^2\cdot\text{s)} \tag{14-31}$$

Steady-State Unidirectional Diffusion in Liquids

The analysis for the determination of the concentration distribution is exactly the same as that for gases. Therefore, Eqs. (14-27) derived previously for gases in terms of partial pressures are now written for liquids in terms of the molal concentrations c_A and c_B:

$$N_A = D\,\frac{c}{x_2 - x_1}\,\frac{c_{B_2} - c_{B_1}}{c_{B\ln}} = D\,\frac{c}{x_2 - x_1}\,\frac{c_{A_1} - c_{A_2}}{c_{B\ln}} \tag{14-32a}$$

where the *logarithmic mean concentration* $c_{B\ln}$ of component B is defined as

$$c_{B\ln} = \frac{c_{B_2} - c_{B_1}}{\ln(c_{B_2}/c_{B_1})} \tag{14-32b}$$

and we note that $c = c_A + c_B$ and $c_{B_2} - c_{B_1} = c_{A_1} - c_{A_2}$.

Equations (14-32) can also be written in terms of mole fractions:

$$N_A = D\,\frac{c}{x_2 - x_1}\,\frac{\chi_{B_2} - \chi_{B_1}}{\chi_{B\ln}} = D\,\frac{c}{x_2 - x_1}\,\frac{\chi_{A_1} - \chi_{A_2}}{\chi_{B\ln}} \tag{14-33a}$$

where the *logarithmic mean mole fraction* $\chi_{B\ln}$ of component B is defined as

$$\chi_{B\ln} = \frac{\chi_{B_2} - \chi_{B_1}}{\ln(\chi_{B_2}/\chi_{B_1})} \tag{14-33b}$$

In Eqs. (14-32) and (14-33), the units of various quantities may be taken as

c_i = molal concentration, $i = A, B$, or $B_{\ln}$, kg · mol/m^3
$\chi_i = c_i/c$ = mole fraction
c = total molal concentration of mixture, kg · mol/m^3
D = mass diffusivity, m^2/s
N_A = molal flux of species A, kg · mol/(m^2 · s)
x = distance, m

14-5 UNSTEADY DIFFUSION

In the previous sections we considered mass transfer problems in which the concentration distribution is a function of one space variable only. In many engineering applications the concentration distribution in the medium varies with both time and position. The mathematical analysis of such unsteady mass transfer problems is complicated because they involve the solution of differential equations with more than one independent variable. However, a large number of unsteady mass transfer problems can be expressed in a form analogous to the one-dimensional, time-dependent heat conduction problems considered in a previous chapter. Therefore, the same mathematical techniques discussed for the solution of a time-dependent heat conduction equation become applicable to the solution of one-dimensional, unsteady mass diffusion problems. In this section we present the mathematical formulation of the one-dimensional, unsteady mass diffusion equation in the rectangular coordinate system and discuss its solution for typical applications.

When there are no mass sources, sinks, or chemical reactions in the medium and the mass transfer takes place as a result of pure molecular diffusion, the mass conservation equation for a given species may be stated as

$$\begin{pmatrix}\text{Net rate of diffusion}\\ \text{of species } A \text{ into}\\ \text{medium}\end{pmatrix} = \begin{pmatrix}\text{rate of increase of}\\ \text{species } A \text{ in}\\ \text{medium}\end{pmatrix} \tag{14-34}$$

We now consider a one-dimensional mass diffusion in the x direction for a differential volume element of thickness Δx and area a, as illustrated in Fig. 14-3, and we write the mathematical expressions for various terms in the mass conservation equation (14-34) as

$$(J_A|_x - J_A|_{x+\Delta x})a = a\,\Delta x\,\frac{\partial c_A}{\partial t} \tag{14-35}$$

where J_A = molal diffusion flux
c_A = molal concentration of component A
t = time

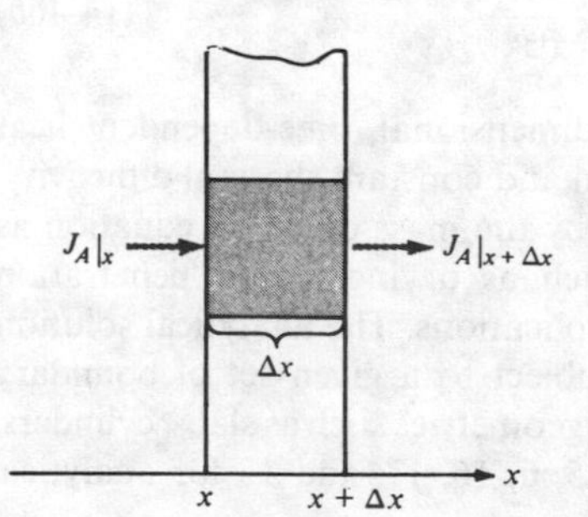

Figure 14-3 Nomenclature for the derivation of the one-dimensional, unsteady mass diffusion equation.

Equation (14-35) is written (as $\Delta x \to 0$)

$$-\frac{\partial J_A}{\partial x} = \frac{\partial c_A}{\partial t} \tag{14-36}$$

By Fick's law of diffusion, the diffusion flux J_A is related to the concentration gradient by [see Eqs. (14-8)]

$$J_A = -D\frac{\partial c_A}{\partial x} \tag{14-37}$$

Substituting Eq. (14-37) into Eq. (14-36) and assuming a constant D, we obtain

$$\frac{\partial^2 c_A}{\partial x^2} = \frac{1}{D}\frac{\partial c_A}{\partial t} \tag{14-38}$$

where c_A = molal concentration of component A, $\text{kg}\cdot\text{mol/m}^3$
D = mass diffusivity for diffusion of component A in medium, m^2/s
t = time, s
x = distance, m

Equation (14-38) is called *the one-dimensional, time-dependent mass diffusion equation* in the rectangular coordinate system for constant mass diffusivity. It is analogous to the heat conduction (or diffusion) equation given previously in the form

$$\frac{\partial^2 T}{\partial x^2} = \frac{1}{\alpha}\frac{\partial T}{\partial t} \tag{14-39}$$

We note that in the mass diffusion equation the mass concentration c_A replaces temperature T, and the mass diffusivity D replaces the heat diffusivity α. Equation (14-38) can readily be generalized to the three-dimensional case:

$$\nabla^2 c_A = \frac{1}{D}\frac{\partial c_A}{\partial t} \tag{14-40a}$$

where ∇^2 is the laplacian operator and in the rectangular coordinate system is given as

$$\nabla^2 \equiv \frac{\partial^2}{\partial x^2} + \frac{\partial^2}{\partial y^2} + \frac{\partial^2}{\partial z^2} \tag{14-40b}$$

Again Eq. (14-40*a*) is analogous to the three-dimensional, time-dependent heat conduction equation, with no energy generation and constant thermal diffusivity.

The mass transfer problems characterized by the mass diffusion equation as given above are encountered in processes such as drying, vapor penetration through porous media, and numerous other applications. The analytical solution of the resulting partial differential equation subject to a given set of boundary and initial conditions can be obtained for simple geometries such as slabs, cylinders, nd spheres. The reader should consult Refs. 5, 6, 16, 17, and 18 for analytical methods of solution of the diffusion equation and for solutions under different boundary conditions. In Example 14-3 we illustrate the solution of the above one-dimensional, unsteady mass diffusion equation for drying of a slab. *Since the mass conservation equation* (14-39) *is of exactly the same form as the one-dimensional, time-dependent heat conduction equation, the same mathematical techniques described previously for the solution of heat conduction problems are applicable.*

Example 14-3 A large slab of clay of thickness L initially contains c_0 mass percent of water distributed uniformly over the volume. The surface of the slab at $x = 0$ is impermeable to moisture flow. The clay is subjected to drying by blowing a stream of low-humidity air over the boundary surface at $x = L$. It is assumed that the drying takes place by the process of moisture diffusion through the clay, and the mass diffusivity D for the diffusion of water vapor through the medium is uniform everywhere. The concentration of water vapor at the boundary surface $x = L$ may be taken as zero since dry air is swept over the surface. Determine a relation for the concentration of water vapor in the clay as a function of time and position. Also give a relation for the concentration of water vapor at the impermeable boundary as a function of time.

SOLUTION The partial differential equation and appropriate boundary and initial conditions for the determination of distribution of concentration $c(x, t) \equiv c$ in mass percent of water in the slab are given as

$$\frac{\partial^2 c}{\partial x^2} = \frac{1}{D}\frac{\partial c}{\partial t} \qquad \text{in } 0 \le x \le L,\ t > 0 \tag{14-41a}$$

$$\frac{\partial c}{\partial x} = 0 \qquad \text{at } x = 0,\ t > 0 \tag{14-41b}$$

$$c = 0 \qquad \text{at } x = L,\ t > 0 \tag{14-41c}$$

$$c = c_0 \qquad \text{in } 0 \le x \le L,\ t = 0 \tag{14-41d}$$

Clearly, boundary condition (14-41*b*) implies that the boundary surface at $x = 0$ is impermeable to moisture, and boundary condition (14-41*c*) states that the vapor concentration is zero at the surface $x = L$. The mass transfer problem described above by Eqs. (14-41) is a special case of the transient heat conduction problem given in Chap. 4. Therefore, the same technique described in Chap. 4 to solve transient heat conduction problems can be used to solve Eqs. (14-41). The solution of Eqs. (14-41) is given as

$$c(x, t) = \sum_{n=1}^{\infty} \frac{1}{N} e^{-D\lambda_n^2 t} \cos \lambda_n x \int_0^L c_0 \cos \lambda_n x' \, dx' \tag{14-42a}$$

where

$$\frac{1}{N} = \frac{2}{L} \tag{14-42b}$$

and the λ_n are the roots of $\cos \lambda L = 0$, or

$$\lambda_n = \frac{2n-1}{2L}\pi \tag{14-42c}$$

The integration in Eq. (14-42*a*) is readily performed since c_0 is constant, and the resulting expression for the distribution of water concentration in the slab becomes

$$\frac{c(x, t)}{c_0} = \frac{2}{L}\sum_{n=1}^{\infty} e^{-D\lambda_n^2 t} \cos \lambda_n x \frac{\sin \lambda_n L}{\lambda_n} \tag{14-43a}$$

where

$$\lambda_n = \frac{2n-1}{2L}\pi \tag{14-43b}$$

The concentration of water at the impermeable boundary $x = 0$ becomes

$$\frac{c(0, t)}{c_0} = \frac{2}{L}\sum_{n=1}^{\infty} e^{-D\lambda_n^2 t} \frac{\sin \lambda_n L}{\lambda_n} \tag{14-44}$$

14-6 MASS DIFFUSIVITY

The prediction of mass diffusivity D has been the subject of extensive investigations, and the mass diffusivity for gases, liquids, and solids requires different considerations. Here we briefly discuss the methods of determination of mass diffusivities.

Mass Diffusivity in Gases

Various models have been proposed for the determination of mass diffusivity $D_{AB} = D_{BA}$ for a binary mixture of gases A and B. The reader should consult Refs. 15, 19, and 20 for detailed discussion of various mathematical models. Here we present the pertinent result for the mass diffusivity of a binary mixture of gas, obtained from the kinetic theory derivation by considering the gas molecules as rigid spheres experiencing elastic collisions and assuming an ideal-gas behavior. According to this model, the mass diffusivity $D_{AB} = D_{BA}$ for a mixture of two gases composed of species A and B, for pressures below 20 atm, can be predicted from

$$D_{AB} = \frac{1.8583 \times 10^{-3}}{p\sigma_{AB}^2 \Omega_{D,AB}} T^{3/2} \sqrt{\frac{1}{M_A} + \frac{1}{M_B}} \tag{14-45}$$

where D_{AB} = mass diffusivity, cm^2/s
p = pressure, atm
T = absolute temperature, K
$, M_B$ = molecular weights of components A and B, respectively
σ_{AB} = *collision diameter*, Å
$_B$ = *collision integral* or a dimensionless function depending on temperature and intermolecular forces

The parameters σ_{AB} and $\Omega_{D,AB}$ for a mixture of two gases composed of components A and B can be determined from Tables 14-1 and 14-2 for the species shown in Table 14-1.

We obtain from Table 14-1 the values of σ_A, $(\varepsilon/k)_A$ and σ_B, $(\varepsilon/k)_B$ for components A and B, respectively. Then σ_{AB} is given by

$$\sigma_{AB} = \tfrac{1}{2}(\sigma_A + \sigma_B) \tag{14-46}$$

Table 14-2 lists the values of $\Omega_{D,AB}$ as a function of kT/ε, which actually implies

$$\frac{kT}{\varepsilon} \equiv \frac{T}{(\varepsilon/k)_{AB}} \tag{14-47a}$$

and the value of $(\varepsilon/k)_{AB}$ for the mixture is taken as

$$\left(\frac{\varepsilon}{k}\right)_{AB} = \sqrt{\left(\frac{\varepsilon}{k}\right)_A \left(\frac{\varepsilon}{k}\right)_B} \tag{14-47b}$$

Table 14-1 Parameter σ and $\frac{\varepsilon}{k}$ for substances*

Substance		Molecular weight M	σ, Å	ε/k, K
Air	Air	28.97	3.617	97.0
Argon	A	39.94	3.418	124.0
Benzene	C_6H_6	78.11	5.270	440.0
Bromine	Br_2	159.83	4.268	520.0
Carbon dioxide	CO_2	44.01	3.996	190.0
Carbon disulfide	CS_2	76.14	4.438	488.0
Carbon monoxide	CO	28.01	3.590	110.0
Carbon tetrachloride	CCl_4	153.84	5.881	327.0
Chlorine	Cl_2	70.91	4.115	357.0
Ethane	C_2H_6	30.07	4.418	230.0
Ethylene	C_2H_4	28.05	4.232	205.0
Fluorine	F_2	38.00	3.653	112.0
Helium	He	4.003	2.576	10.2
Hydrogen	H_2	2.016	2.915	38.0
Iodine	I_2	253.82	4.982	550.0
Krypton	Kr	83.80	3.61	190.0
Methane	CH_4	16.04	3.822	137.0
Methyl chloride	CH_3Cl	50.49	3.375	855.0
Neon	Ne	20.183	2.789	35.7
Nitric oxide	NO	30.01	3.470	119.0
Nitrogen	N_2	28.02	3.681	91.5
Nitrous oxide	N_2O	44.02	3.879	220.0
Oxygen	O_2	32.00	3.433	113.0
Sulfur dioxide	SO_2	64.07	4.290	252.0
Water†	H_2O	18.0	2.641	809.1
Xenon	Xe	131.3	4.055	229.0

* From Hirschfelder, Curtiss, and Bird [20].
† From Svehla [21].

Table 14-2 Values of collision integral $\Omega_{D,AB}$†

kT/ε‡	$\Omega_{D,AB}$‡	kT/ε	$\Omega_{D,AB}$	kT/ε	$\Omega_{D,AB}$
0.30	2.662	1.65	1.153	4.0	0.8836
0.35	2.476	1.70	1.140	4.1	0.8788
0.40	2.318	1.75	1.128	4.2	0.8740
0.45	2.184	1.80	1.116	4.3	0.8694
0.50	2.066	1.85	1.105	4.4	0.8652
0.55	1.966	1.90	1.094	4.5	0.8610
0.60	1.877	1.95	1.084	4.6	0.8568
0.65	1.798	2.00	1.075	4.7	0.8530
0.70	1.729	2.1	1.057	4.8	0.8492
0.75	1.667	2.2	1.041	4.9	0.8456
0.80	1.612	2.3	1.026	5.0	0.8422
0.85	1.562	2.4	1.012	6	0.8124
0.90	1.517	2.5	0.9996	7	0.7896
0.95	1.476	2.6	0.9878	8	0.7712
1.00	1.439	2.7	0.9770	9	0.7556
1.05	1.406	2.8	0.9672	10	0.7424
1.10	1.375	2.9	0.9576	20	0.6640
1.15	1.346	3.0	0.9490	30	0.6232
1.20	1.320	3.1	0.9406	40	0.5960
1.25	1.296	3.2	0.9328	50	0.5756
1.30	1.273	3.3	0.9256	60	0.5596
1.35	1.253	3.4	0.9186	70	0.5464
1.40	1.233	3.5	0.9120	80	0.5352
1.45	1.215	3.6	0.9058	90	0.5256
1.50	1.198	3.7	0.8998	100	0.5130
1.55	1.182	3.8	0.8942	200	0.4644
1.60	1.167	3.9	0.8888	400	0.4170

† From Hirschfelder, Curtiss, and Bird [20].
‡ Reference 20 uses the symbol T^* for kT/ε and $\Omega^{(1,1)*}$ in place of $\Omega_{D,AB}$.

where $(\varepsilon/k)_A$ and $(\varepsilon/k)_B$ for components A and B are readily obtainable from Table 14-1. Thus, given the value of kT/ε, the corresponding value of $\Omega_{D,AB}$ is obtained from Table 14-2.

We present in Table 14-3 measured values of mass diffusivities for typical binary gas systems at 1 atm. Clearly, according to Eq. (14-45), if the experimental value of mass diffusivity D_1 at temperature T_1 is available, the mass diffusivity D_2 at temperature T_2 may be estimated from

$$D_2 = D_1\left(\frac{T_2}{T_1}\right)^{3/2} \frac{(\Omega_{D,AB})_{T_1}}{(\Omega_{D,AB})_{T_2}} \tag{14-48}$$

Example 14-4 Determine the mass diffusivity of the binary mixture of air–CO_2 at 273 K and 1 atm from Eq. (14-45). Compare the result with the experimental value given by Table 14-3.

Table 14-3 Mass diffusivities of binary gas systems at atmospheric pressure (measured values)

System	T, K	Mass diffusivity D	
		cm^2/s	ft^2/h
Air–ammonia	273	0.198	0.769
Air–aniline	298	0.0726	0.282
Air–benzene	298	0.0962	0.374
Air–carbon dioxide	273	0.136	0.528
Air–carbon disulfide	273	0.0883	0.343
Air–chlorine	273	0.124	0.482
Air–ethyl alcohol	298	0.132	0.513
Air–iodine	298	0.0834	0.524
Air–mercury	614	0.473	1.837
Air–naphthalene	298	0.0611	0.237
Air–oxygen	273	0.175	0.680
Air–sulfur dioxide	273	0.122	0.474
Air–toluene	298	0.0844	0.328
Air–water	298	0.260	1.010
CO_2–benzene	318	0.0715	0.278
CO_2–carbon disulfide	318	0.0715	0.278
CO_2–ethyl alcohol	273	0.0693	0.269
CO_2–hydrogen	273	0.550	2.136
CO_2–nitrogen	298	0.158	0.614
CO_2–water	298	0.164	0.637
Oxygen–ammonia	293	0.253	0.983
Oxygen–benzene	296	0.039	0.365

Compiled from Reid and Sherwood [22].

SOLUTION The parameters σ and ε/k and the molecular weights M for air and CO_2 are obtained from Table 14-1:

$$\text{Air}(A): \quad \sigma_A = 3.617 \qquad \left(\frac{\varepsilon}{k}\right)_A = 97.0 \qquad M_A = 28.97$$

$$CO_2(B): \quad \sigma_B = 3.996 \qquad \left(\frac{\varepsilon}{k}\right)_B = 190.0 \qquad M_B = 44.01$$

Then

$$\sigma_{AB} = \tfrac{1}{2}(\sigma_A + \sigma_B) = \tfrac{1}{2}(3.617 + 3.996) = 3.806 \text{ Å}$$

$$\left(\frac{\varepsilon}{k}\right)_{AB} = \sqrt{\left(\frac{\varepsilon}{k}\right)_A\left(\frac{\varepsilon}{k}\right)_B} = \sqrt{97 \times 190} = 135.76 \text{ K}$$

$$\frac{kT}{\varepsilon} = \frac{T}{(\varepsilon/k)_{AB}} = \frac{273}{135.76} = 2.01$$

and the value of the collision integral for $kT/\varepsilon = 2.01$ is obtained from Table 14-2 as $\Omega_{D,AB} = 1.073$

Various quantities as given above, the mass diffusivity for $p = 1$ atm, and $T = 273$ K are determined from Eq. (14-45) as

$$D_{AB} = \frac{1.8583 \times 10^{-3}}{p \times \sigma_{AB}^2 \times \Omega_{D,AB}} T^{3/2} \sqrt{\frac{1}{M_A} + \frac{1}{M_B}}$$

$$= \frac{1.8583 \times 10^{-3}}{1 \times 3.806^2 \times 1.073} 273^{3/2} \sqrt{\frac{1}{28.97} + \frac{1}{44.01}}$$

$$= 0.129 \text{ cm}^2/\text{s}$$

This result should be compared with the measured value $D_{AB} = 0.136$ cm²/s given in Table 14-3.

Mass Diffusivity in Liquids

A rigorous theory is yet to be developed for the prediction of mass diffusivity in liquids. The existing approximate theories lack agreement with experiments; therefore, a number of semiempirical relations have been proposed for the determination of mass diffusivity in liquids. The reader should consult Refs. 1 and 22 for a collection of such expressions. One of these relations, developed by Wilke and Chang [23], gives the mass diffusivity D_{AB} for small concentrations of A in B (that is, A is the solute, B is the solvent) as

$$D_{AB} = (7.4 \times 10^{-8}) \frac{(\xi_B M_B)^{1/2} T}{\mu \tilde{V}_A^{0.6}} \tag{14-49}$$

where D_{AB} = mass diffusivity (A is solute, B is solvent), cm²/s
M_B = molecular weight of solvent B
T = absolute temperature, K
$\tilde{V}_A$ = molal volume of solute A as liquid at its normal boiling point, cm³/(g · mol)
μ = viscosity of solution, cP
ξ_B = "association" factor for solvent B (recommended values are 2.6 for water, 1.9 for methanol, 1.5 for ethanol, and 1.0 for benzene, ether, heptane, and other unassociating solvents)

The liquid molal volumes $\tilde{V}$ at the normal boiling point are given in Table 14-4 for a number of liquids; the reader is referred to Reid and Sherwood [22] for a discussion of various methods of estimating the liquid molal volumes $\tilde{V}$. Equation (14-49) is good within ± 10 percent for the prediction of mass diffusivity of dilute solutions of nondissociating solutes. In Table 14-5 we present the measured values of mass diffusivities for dilute solutions of various liquids in water, methanol, and benzene.

Table 14-4 Liquid molal volumes $\tilde{V}$ at the normal boiling point, calculated by the method of Benson

Compound	Molal volume $\tilde{V}$, cm³/(g · mol)	Compound	Molal volume $\tilde{V}$, cm³/(g · mol)
Acetic acid	62.9	Ethyl mercaptan	76.0
Acetone	81.0	Fluorobenzene	101.0
Acetylene	41.3	Heptane	165.0
Ammonia	25.5	Hydriodic acid	45.9
Benzene	97.0	Hydrochloric acid	31.2
Bromobenzene	115.0	Iodobenzene	131.0
Carbon tetrachloride	103.0	Methane	37.1
Chlorine	44.6	Methanol	42.4
Chlorobenzene	115.0	Methyl chloride	49.8
Diethylamine	113.0	Methyl formate	63.0
Diethyl sulfide	122.0	Propane	75.3
Dimethyl ether	62.7	Sulfur dioxide	44.3
Ethyl acetate	108.0	Water	19.0
Ethylene	45.9		

Compiled from Reid and Sherwood [22].

Table 14-5 Liquid-phase mass diffusivities of dilute solutions in water, methanol, and benzene (measured values)

System		T, °C	$D \times 10^5$ cm²/s	$D \times 10^5$ ft²/h
Bromine	in water	12	0.90	3.50
CO_2	in water	18	1.71	6.64
Chlorine	in water	12	1.40	5.44
Glucose	in water	15	0.52	2.02
Hydrogen	in water	25	3.36	3.05
Iodine	in water	25	1.25	4.85
Methanol	in water	15	1.28	4.97
Nitrogen	in water	22	2.02	7.84
Oxygen	in water	25	2.60	10.10
Aniline	in methanol	15	1.49	5.79
CCl_4	in methanol	15	1.70	6.60
Chloroform	in methanol	15	2.07	8.04
Iodoform	in methanol	15	1.33	5.17
Lactic acid	in methanol	15	1.36	5.28
Acetic acid	in benzene	15	1.92	7.46
Bromine	in benzene	12	2.00	7.77
CCl_4	in benzene	25	2.00	7.77
Chloroform	in benzene	15	2.11	8.19
Iodine	in benzene	20	1.95	7.57

Compiled from Reid and Sherwood [22].

Example 14-5 Determine the mass diffusivity D_{AB} for a dilute solution of CCl_4 (i.e., component A, the solute) in benzene (i.e., component B, the solvent) at 25°C. Compare the predicted value with the measured value given in Table 14-5.

SOLUTION Equation (14-49) can be used to determine the mass diffusivity, and the values of various quantities in this equation are taken as

$$M_B = 78.10 \qquad \tilde{V}_A = 103 \text{ (from Table 14-4)} \qquad T = 298 \text{ K}$$

$$\xi_B = 1 \text{ (for benzene)} \qquad \text{and} \qquad \mu = 0.6 \text{ cP}$$

Then Eq. (14-49) gives

$$D_{AB} = (7.4 \times 10^{-8}) \frac{(\xi_B M_B)^{1/2} T}{\mu \tilde{V}_A^{0.6}} = (7.4 \times 10^{-8}) \frac{(1 \times 78.1)^{1/2} \times 298}{0.6 \times 103^{0.6}}$$

$$= 2.01 \times 10^{-5} \text{ cm}^2/\text{s}$$

which should be compared with the measured value $D_{AB} = 2.00 \times 10^{-5}$ cm^2/s given in Table 14-5.

Mass Diffusivity in Solids

The theoretical prediction of mass diffusivity in solids containing pores and capillaries is an extremely complicated matter, and no fundamental theory is yet available on the subject. When a solid contains capillaries and pores, in some situations the capillary forces are opposed to the concentration gradient; as a result, the mass flux may not even be proportional to the concentration gradient. Describing diffusion inside the tortuous void passages in such a medium by taking into consideration the interaction between different mechanisms of transport appears not to be possible analytically. Therefore, an experimental approach is the only means to determine an *effective mass diffusivity* for a given solid structure and fluid combination. The effective mass diffusivity determined in this manner can be used in the equation of mass conservation (14-38) or (14-40) to predict the mass diffusion in solids. Note that for solids which are not so porous, the mass diffusivity is several orders of magnitude smaller than that for liquids.

14-7 MASS TRANSFER IN LAMINAR AND TURBULENT FLOW

In many engineering applications, mass transfer takes place through a fluid stream that is in laminar or turbulent flow. For example, in transpiration cooling, a cold fluid is injected into the hot gas stream through the perforations in the plate or duct surface to protect the duct surfaces from the hot gas. In high-temperature, gas-cooled nuclear reactors, the fission products that are released into the coolant gas stream are transported by molecular and eddy diffusion from the gas stream to the surfaces of the conduits and the heat exchanger tubes. When the concentration of mass in the fluid stream is very small, the mass transfer to and from the fluid stream is governed by the mass conservation equation, which is similar to the energy equation with no viscous dissipation. For example, the steady-state mass conservation equation for incompressible laminar flow inside a circular tube with cylindrical symmetry in the distribution of mass concentration is given as

$$u\frac{\partial c_A}{\partial x} + v\frac{\partial c_A}{\partial r} = D\left(\frac{\partial^2 c_A}{\partial r^2} + \frac{1}{r}\frac{\partial c_A}{\partial r} + \frac{\partial^2 c_A}{\partial x^2}\right) \tag{14-50}$$

where $c_A = c_A(r, x)$ is the concentration of mass, D is the mass diffusivity, x and r are the axial and radial coordinate, respectively, and u and v are the velocity components in the axial and radial directions, respectively. The velocity components u and v are to be determined from the solution of the continuity and momentum equations.

If the axial diffusion is neglected, Eq. (14-50) simplifies to

$$u\frac{\partial c_A}{\partial x} + v\frac{\partial c_A}{\partial r} = D\left(\frac{\partial^2 c_A}{\partial r^2} + \frac{1}{r}\frac{\partial c_A}{\partial r}\right) \tag{14-51}$$

In the case of fully developed flow, $v = 0$ and Eq. (14-51) is further simplified as

$$u\frac{\partial c_A}{\partial x} = D\left(\frac{\partial^2 c_A}{\partial r^2} + \frac{1}{r}\frac{\partial c_A}{\partial r}\right) \tag{14-52}$$

The solution of these equations subject to appropriate boundary conditions gives the distribution of mass concentration in the flow field. Then the mass flux to and from the conduit surface, in the absence of fluid injection or suction at the wall, is determined from

$$N_A\bigg|_{\text{wall}} = -D\frac{\partial c_A}{\partial r}\bigg|_{\text{wall}} \tag{14-53}$$

(Note that $q = -k\,\partial T/\partial x$ for heat flux.)

Clearly, the mathematical formulation of the mass transfer problem to and from fluid streams is analogous to heat transfer problems to and from fluid streams. In mass transfer, the mass concentration c_A replaces the temperature T, and the mass diffusivity D replaces the heat diffusivity α. The similarity between heat and mass transfer illustrated above for laminar flow is also valid for turbulent flow. Therefore, *when the concentration of mass in the fluid stream is very small*, various expressions given in the previous chapters for heat transfer from laminar and turbulent streams become applicable to mass transfer by proper modification. The reader should consult Ref. 1 for mass transfer at high mass flux and high concentration and Refs. 7 and 8 for mass transfer in chemically reacting systems.

A *mass transfer coefficient* k_m for the transfer of a species A between the fluid stream and the wall surface is defined, in analogy with the heat transfer coefficient, as

$$N_A = k_m(c_{Am} - c_{Aw}) \tag{14-54}$$

where c_{Am} = mean concentration of species A in bulk fluid, $\text{kg} \cdot \text{mol/m}^3$
c_{Aw} = concentration of species A at immediate vicinity of wall surface, $\text{kg} \cdot \text{mol/m}^3$
N_A = mass flux of species A to the wall, $\text{kg} \cdot \text{mol/(m}^2 \cdot \text{s)}$
k_m = mass transfer coefficient for species A, m/s

In heat transfer problems, the *Nusselt number* is defined as

$$\text{Nu} = \frac{hd}{k} \quad \text{for flow inside tube} \tag{14-55a}$$

$$\text{Nu}_x = \frac{hx}{k} \quad \text{for boundary-layer flow} \tag{15-55b}$$

In an analogous manner in mass transfer problems the *Sherwood number* Sh is defined as

$$\text{Sh} = \frac{k_m d}{D} \quad \text{for mass transfer inside tube} \tag{14-56a}$$

$$\text{Sh}_x = \frac{k_m x}{D} \quad \text{for mass transfer in boundary-layer flow} \tag{14-56b}$$

where d = tube diameter
D = mass diffusivity
x = distance along plate

In previous chapters on forced-convection heat transfer, we recall that the Nusselt number for heat transfer in incompressible flows is a function of Reynolds and Prandtl numbers, that is,

$$\text{Nu} = f(\text{Re, Pr}) \tag{14-57}$$

A similar dimensional analysis of the governing equations for mass transfer from an incompressible flow shows that the Sherwood number is a function of Reynolds and Schmidt numbers, that is,

$$\text{Sh} = f(\text{Re, Sc}) \tag{14-58a}$$

Where the Schmidt number is defined as

$$\text{Sc} = \frac{\nu}{D} \tag{14-58b}$$

and the Reynolds and Prandtl numbers have their usual definitions.

We reiterate that *the foregoing similarity between heat and mass transfer processes is valid if the concentration of the species in the fluid stream is small.* For such situations, the mass transfer coefficient k_m is obtained from the analogous heat transfer coefficient by replacing the Nusselt number by the Sherwood number and the Prandtl number by the Schmidt number. We illustrate the procedure for typical cases.

Mass Transfer in Laminar Flow with Fully Developed Velocity and Concentration Distributions inside Ducts

In Chap. 7 we discussed heat transfer for hydrodynamically and thermally developed flow in circular tubes under constant wall temperature and constant wall heat flux boundary conditions. The solution of the mass transfer problem for laminar flow with fully developed velocity and concentration distribution yields analogous results, except the Nusselt number is replaced by the Sherwood number. That is,

$$\text{Sh} = \frac{k_m d}{D} = 3.66 \qquad \text{for uniform wall mass concentration} \qquad (14\text{-}59a)$$

$$= 4.36 \qquad \text{for uniform wall mass flux} \qquad (14\text{-}59b)$$

The Sherwood number for mass transfer in laminar flow inside conduits having noncircular cross section can be obtained from those given in Table 7-1 for heat transfer.

Mass Transfer in Turbulent Flow inside Pipes

The heat transfer relation for turbulent flow inside a pipe was given in Chap. 7 [Eq. (7-68)] as

$$\text{Nu} = 0.023\ \text{Re}^{0.8}\ \text{Pr}^{1/3} \qquad \text{for } 0.7 < \text{Pr} < 100 \qquad \text{Re} > 10{,}000 \quad (14\text{-}60)$$

In the case of mass transfer, experiments performed with a wetted-wall column have shown that the analogous relation is in the form

$$\text{Sh} = 0.023\ \text{Re}^{0.83}\text{Sc}^{1/3} \qquad \text{for } 0.6 < \text{Sc} < 2500 \qquad 2000 < \text{Re} < 35{,}000 \quad (14\text{-}61)$$

The slight increase in the exponent of the Reynolds number may be justified under the experimental conditions leading to this expression. That is, a wetted-wall column is an experimental vertical tube in which liquid flows in a thin film down the inside surface and a gas flows upward in the tube. Mass transfer takes place from the liquid to the gas, or vice versa. Under these conditions, the presence of rippling and wave formation on the liquid surface increases the interface available for mass transfer, which is characterized by slightly stronger dependence of the mass transfer coefficient on Reynolds number. In the case of mass transfer experiments with liquids flowing in turbulent flow inside soluble tubes so that mass is transferred from the inside surface of the tube to the fluid stream, the surface remains smooth and the mass transfer relation takes a form analogous to Eq. (14-60), that is,

$$\mathrm{Sh} = 0.023\ \mathrm{Re}^{0.8}\ \mathrm{Sc}^{1/3} \tag{14-62}$$

Other mass transfer relations are obtainable from the heat transfer relations in a similar manner. For example, the results of analogies between heat and momentum transfer can be recast as analogies between mass and momentum transfer. That is, by replacing the Nusselt number by the Sherwood number and the Prandtl number by the Schmidt number, the Reynolds, Prandtl, and von Kármán analogies become analogies between mass and momentum transfer. Then once the friction factor or the drag coefficient is known, the mass transfer coefficient k_m for turbulent flow inside a pipe or along a flat plate can be determined.

Example 14-6 Air at atmospheric pressure and 25°C containing small quantities of iodine flows with a velocity of $u = 5.18$ m/s inside a tube with an ID of $d = 3.048 \times 10^{-2}$ m. Determine the mass transfer coefficient for iodine transfer from the gas stream to the wall surface. If c_m is the mean concentration of iodine in kg · mol/m^3 in the airstream, determine the rate of deposition of iodine on the tube surface by assuming that the wall surface is a perfect sink for iodine deposition (i.e., the iodine concentration at the immediate vicinity of the wall is assumed to be zero).

SOLUTION The kinematic viscosity of air is $\nu = 1.58 \times 10^{-5}$ m^2/s, and the mass diffusivity for the air-iodine system at 1 atm and 25°C is obtained from Table 14-3 as $D = 0.0834$ cm^2/s. The Reynolds number for the flow is

$$\mathrm{Re} = \frac{ud}{\nu} = \frac{5.18(3.048 \times 10^{-2})}{1.58 \times 10^{-5}} \cong 10^4$$

The flow is turbulent, and the mass transfer coefficient can be determined by Eq. (14-62):

$$\mathrm{Sh} = 0.023\ \mathrm{Re}^{0.8}\ \mathrm{Sc}^{1/3}$$

where

$$\mathrm{Sc} = \frac{\nu}{D} = \frac{1.58 \times 10^{-5}}{0.0834 \times 10^{-4}} = 1.89$$

$$\mathrm{Sh} = \frac{k_m d}{D} = \frac{k_m(3.048 \times 10^{-2})}{0.0384 \times 10^{-4}} = \frac{k_m}{1.26 \times 10^{-4}}$$

$$\frac{k_m}{1.26 \times 10^{-4}} = 0.023(10^4)^{0.8}(1.89)^{1/3}$$

$$k_m = 0.57 \times 10^{-2}\ \mathrm{m/s}$$

The rate of decomposition of iodine on the tube surface is determined from

$$N = k_m(c_m - c_w) \qquad \mathrm{kg \cdot mol/(m^2 \cdot s)}$$

For the perfect sink condition at the wall, $c_w = 0$; then

$$N = k_m c_m = (0.57 \times 10^{-2}) c_m \qquad \text{kg}\cdot\text{mol}/(\text{m}^2\cdot\text{s})$$

where c_m is the mean iodine concentration in $\text{kg}\cdot\text{mol}/\text{m}^3$ in the airstream.

Mass Transfer in Boundary Layer Flow over a Flat Plate

The heat transfer correlations for laminar and turbulent boundary-layer flow over a flat plate presented in Chap. 8 can be recast as mass transfer correlations by changing the Nusselt number to the Sherwood number and the Prandtl number to the Schmidt number.

For example, for mass transfer from laminar boundary-layer flow over a flat plate, the Pohlhausen equation (8-67) can be written as a mass transfer correlation equation in the form

$$\text{Sh}_x = 0.332\ \text{Sc}^{1/3}\ \text{Re}_x^{1/2} \qquad \text{for } \text{Re}_x < 5 \times 10^5 \tag{14-63a}$$

or this expression can be rearranged as

$$\frac{k_m}{u_\infty}\text{Sc}^{2/3} = 0.332\ \text{Re}_x^{-1/2} \qquad \text{for } \text{Re}_x < 5 \times 10^5 \tag{14-63b}$$

where k_m = local mass transfer coefficient at x

$$\text{Sh}_x = \frac{k_m x}{D} = \text{local Sherwood number}$$

$$\text{Sc} = \frac{\nu}{D} = \text{Schmidt number}$$

The average value of the mass transfer coefficient over the distance $0 \le x \le L$ is determined from the definition given by Eq. (8-70) as

$$\bar{k}_m = 2[k_m]_{x=L} \tag{14-64}$$

In mass transfer from turbulent boundary-layer flow over a flat plate, Eq. (8-73a) can be written as mass transfer correlation in the form

$$\frac{k_m}{u_\infty}\text{Sc}^{2/3} = 0.0296\ \text{Re}_x^{-0.2} \qquad \text{for } 5 \times 10^5 < \text{Re}_x < 10^7 \tag{14-65}$$

Since

$$\text{St}_x = \frac{\text{Nu}_x}{\text{Re}_x\,\text{Pr}} \rightarrow \frac{\text{Sh}_x}{\text{Re}_x\,\text{Sc}} = \frac{k_m x}{D}\frac{\nu}{u_\infty x}\frac{D}{\nu} = \frac{k_m}{u_\infty}$$

where k_m is the local mass transfer coefficient.

When heat transfer and mass transfer take place simultaneously, the ratio of the heat transfer coefficient h_x to the mass transfer coefficient k_m can be of interest. Such a relation can be obtained from Eqs. (8-67) and (14-63a) for laminar flow and from Eqs. (8-73*a*) and (14-65) for turbulent flow. We note that for both cases it leads to

$$\frac{h_x}{k_m} = \rho c_p \left(\frac{\text{Sc}}{\text{Pr}}\right)^{2/3} = \rho c_p \,\text{Le}^{2/3} \tag{14-66}$$

where the *Lewis number* Le is defined as

$$\text{Le} = \frac{\text{Sc}}{\text{Pr}} = \frac{\alpha}{D} \tag{14-67}$$

The same relation is also applicable for the ratio of the average values of the heat and mass transfer coefficients, namely,

$$\frac{h_m}{k_m} = \rho c_p \,\text{Le}^{2/3} = \rho c_p \left(\frac{\alpha}{D}\right)^{2/3} \tag{14-68}$$

Example 14-7 Atmospheric air at $T_\infty = 40°\text{C}$ flows over a wet-bulb thermometer (i.e., a thermometer which is wrapped in a damp cloth). The reading of the thermometer, which is called the wet-bulb reading, is $T_w = 20°\text{C}$. Calculate the concentration of water vapor c_∞ in the free stream. Also determine the relative humidity of the airstream (i.e., the ratio of the concentration c_∞ of water vapor in the free stream to the saturation concentration at the free-stream temperature $T_\infty = 40°\text{C}$ obtained from the steam table).

SOLUTION We assume that the steady-state conditions are established so that no net heat exchange takes place at the thermometer. Therefore, heat transferred from the airstream to the wet cloth must be balanced to evaporate the water. Therefore, we state the energy balance as

$$\begin{pmatrix}\text{Rate of heat}\\ \text{transfer from}\\ \text{air to wet}\\ \text{cover}\end{pmatrix} = \begin{pmatrix}\text{heat removed}\\ \text{by evaporation}\\ \text{of water from}\\ \text{cover}\end{pmatrix}$$

or

$$A h_m (T_\infty - T_w) = A k_m (c_w - c_\infty) h_{fg} \tag{a}$$

or

$$\frac{h_m}{k_m}(T_\infty - T_w) = (c_w - c_\infty) h_{fg} \tag{b}$$

The ratio h_m/k_m is obtained from Eq. (14-68):

$$\rho c_p \left(\frac{\alpha}{D}\right)^{2/3} (T_\infty - T_w) = (c_w - c_\infty) h_{fg} \tag{c}$$

The properties are evaluated at the film temperature $T_f = (40 + 20)/2 = 30°C$. We find

Air: $\rho = 1.132 \text{ kg/m}^3$ $\quad c_p = 1.007 \text{ kJ/(kg} \cdot °\text{C)}$

$\alpha = 0.241 \times 10^{-4} \text{ m}^2/\text{s}$

$D = 0.26 \times 10^{-4}$ m/s (from Table 14-3) $\quad \dfrac{\alpha}{D} = 0.927$

Water: $h_{fg} = 2407 \text{ kJ/kg}$

The saturation concentration of water vapor c_w at $T_w = 20°C$ measured by the wet-bulb thermometers is determined from

$$c_w = \frac{P_w M_w}{\mathscr{R} T_w}$$

where $M_w = 18$, molecular weight of water

$T_w = 293$ K, wet-bulb temperature

$P_w = 2.339 \text{ kPa} = 2.339 \text{ kN/m}^2$, from saturated steam tables at $T_w = 20°C$

$\mathscr{R} = 8314 \text{ N} \cdot \text{m/(kg} \cdot \text{mol} \cdot \text{K)}$, universal gas constant

Then C_w is determined as

$$c_w = \frac{(2339)(18)}{(8314)(293)} = 0.01728 \text{ kg/m}^3$$

These quantities are now introduced into Eq. (c):

$$(1.132)(1007)(0.927)^{2/3}(40 - 20) = (0.01728 - c_\infty)(2407 \times 10^3)$$

Then the concentration of water vapor in the free stream is determined as

$$c_\infty = 0.0083 \text{ kg/m}^3$$

The saturation concentration at $T_\infty = 40°C$ from the steam table is

$$c_{\text{sat}} = 0.0512 \text{ kg/m}^3$$

Then the relative humidity becomes

$$\text{RH} = \frac{0.0083}{0.0512} = 16.2\%$$

PROBLEMS

14-1 Two large vessels contain uniform mixtures of air (component A) and sulfur dioxide (component B) at 1 atm and 273 K but at different concentrations. Vessel 1 contains 80 mole percent air and 20 mole percent sulfur dioxide, while vessel 2 contains 30 mole percent air and 70 mole percent sulfur dioxide. The vessels are connected by a 10-cm-ID, 1.8-m-long pipe. Determine the rate of transfer of air between these two vessels by assuming a steady-state transfer takes place.

14-2 Repeat Prob. 14-1 for an air and ammonia mixture.

14-3 An open tank contains benzene at atmospheric pressure and 298 K. The distance from the surface of the benzene layer to the top of the tank is 3 m. It is assumed that the air in the vessel is motionless while there is sufficient air motion outside to remove the benzene vapor arriving at the top surface. Determine the rate of evaporation of benzene per square foot of the benzene surface ($p_{A_1} = 0.01$ atm).

14-4 Repeat Prob. 14-3 for naphthalene inside the tank ($p_{A_1} = 0.001$ atm).

14-5 Calculate the mass diffusivity of the binary gas mixture of air–iodine at 273 K and 1 atm, and compare the result with that given in Table 14-3.

14-6 Repeat Prob. 15-4 for pressures of 5 and 10 atm.

14-7 Repeat Prob. 14-5 for a temperature of 819 K by using Eq. (14-48).

14-8 Calculate the mass diffusivity D_{AB} for a dilute solution of methanol (component A, the solute) in water (component B, the solvent) at 18°C, and compare the result with that given in Table 14-5.

14-9 Determine the mass diffusivity D_{AB} for a dilute solution of acetone (component A, the solute) and benzene (component B, the solvent) at 15°C.

14-10 Repeat Prob. 14-9 for acetic acid (component A) in benzene (component B) at 15°C ($\tilde{V}_A = 64$).

14-11 A slab of clay of thickness $2L$ initially contains c_0 mass percent of water distributed uniformly over the entire volume. The slab is subjected to drying by blowing a stream of low-humidity air over both of its surfaces. The mass diffusivity of water vapor through the clay is D.

(*a*) Give the mathematical formulation of this mass diffusion problem. (Note that this problem can be expressed in the same form as that given by Example 14-3).

(*b*) Derive an expression for the concentration of water vapor as a function of time and position in the slab.

(*c*) Derive an expression for the rate of removal of water vapor per unit area of the surface of the slab.

14-12 Air at 10°C and atmospheric pressure flows over a plane surface covered with naphthalene. The flow velocity is such that the Reynolds number at a distance of 0.6 m from the leading edge of the plate is Re = 9×10^4. Determine the average mass transfer coefficient for the transfer of naphthalene over the 0.6 m length of the surface by making use of the analogy between heat and mass transfer in laminar flow along a flat plate.

14-13 Air at atmospheric pressure and 25°C, containing small quantities of iodine, flows along a flat plate. The flow velocity is such that the Reynolds number at a distance $x = 1.2$ m from the leading edge of the plate is Re = 9×10^4. Determine the average mass transfer coefficient for the transfer of iodine from the airstream into the wall surface over the 1.2-m length of the plate.

14-14 In Prob. 14-13, let c_0 be the mass concentration of iodine in the main airstream. It is assumed that the plate surface is a perfect sink for iodine so that the mass concentration of iodine in the gas at the immediate vicinity of the plate surface is zero. Write the expression for the rate of transfer of iodine from the gas stream to the plate surface per foot width over the 1.2-m length of the plate.

14-15 Air at 25°C and atmospheric pressure flows with a velocity of 7.6 m/s inside a 2.5-cm-ID pipe. The inside surface of the tube contains a deposit of naphthalene. Determine the mass transfer coefficient for the transfer of naphthalene from the pipe surface into the air in regions away from the inlet.

14-16 Dry air at atmospheric pressure and 10°C flows over a flat plate with a velocity of 1 m/s. The plate is covered with a film of water which evaporates into the airstream. Determine the average mass transfer coefficient for the transfer of water vapor over a distance of 0.6 m from the leading edge of the plate.

14-17 Atmospheric air at $T_\infty = 30$°C flows over a wet-bulb thermometer, which reads $T_w = 20$°C. Calculate the concentration of water vapor in the airstream and the relative humidity of the airstream.

REFERENCES

1. Skelland, A. H. P.: *Diffusional Mass Transfer*, Wiley, New York, 1974.
2. Sawistowski, H., and W. Smith: *Mass Transfer Process Calculations*, Interscience, New York, 1963.

3. Hobler, T.: *Mass Transfer and Absorbers* (trans. from Polish by J. Bandrowski), Pergamon, New York, 1966.
4. Spalding, D. B.: *Convective Mass Transfer*, Edward Arnold (Publishers), London, 1963.
5. Jost, W.: *Diffusion in Solids, Liquids, and Gases*, Academic, New York, 1960.
6. Crank, J.: *The Mathematics of Diffusion*, Oxford University Press, London, 1956.
7. Astarita, G.: *Mass Transfer with Chemical Reaction*, Elsevier, Amsterdam, 1967.
8. Danckwerts, P. V.: *Gas-Liquid Reactions*, McGraw-Hill, New York, 1970.
9. Barrer, R. M.: *Diffusion in and through Solids*, Cambridge University Press, New York, 1941.
10. Treybal, R. E.: *Mass Transfer Operations*, McGraw-Hill, New York, 1955.
11. Brian, P. L. T.: *Staged Cascades in Chemical Processing*, Prentice-Hall, Englewood Cliffs, N.J., 1972.
12. Henley, E. J., and H. K. Staffin: *Stagewise Process Design*, Wiley, New York, 1963.
13. Smith, Buford D.: *Design of Equilibrium Stage Processes*, McGraw-Hill, New York, 1963.
14. Sherwood, T. K., and R. L. Pigford: *Absorption and Extraction*, McGraw-Hill, New York, 1952.
15. Bird, R. B., W. E. Stewart, and E. N. Lightfoot: *Transport Phenomena*, Wiley, New York, 1960.
16. Mikhailov, M. D., and M. N. Özısık: *Unified Analysis and Solutions of Heat and Mass Diffusion*, Wiley, New York, 1984.
17. Carslaw, H. S., and J. C. Jaeger: *Conduction of Heat in Solids*, Oxford University Press, London, 1959.
18. Özışık, M. N.: *Heat Conduction*, Wiley, New York, 1980.
19. Present, R. D.: *Kinetic Theory of Gases*, McGraw-Hill, New York, 1958.
20. Hirschfelder, J. O., C. F. Curtiss, and R. B. Bird: *Molecular Theory of Gases and Liquids*, Wiley, New York, 1954.
21. Svehla, R. A.: *NACA Tech. Rep.* R-132, Lewis Research Center, Cleveland, Ohio, 1962.
22. Reid, R. C., and T. K. Sherwood: *The Properties of Gases and Liquids*, McGraw-Hill, New York, 1966, chap. 11.
23. Wilke, C. R., and P. Chang: "Correlations of Diffusion Coefficients in Dilute Solutions," *AIChE J.* **1**:264–270 (1955).

APPENDIX

A
CONVERSION FACTORS

Table A-1 Conversion factors

1. Acceleration
   $1\ ft/s^2 = 0.3048\ m/s^2$
   $1\ m/s^2 = 3.2808\ ft/s^2$

2. Area
   $1\ in^2 = 6.4516\ cm^2$
   $1\ in^2 = 6.4516 \times 10^{-4}\ m^2$
   $1\ ft^2 = 929\ cm^2$
   $1\ ft^2 = 0.0929\ m^2$
   $1\ m^2 = 10.764\ ft^2$

3. Density
   $1\ lb/in^3 = 27.680\ g/cm^3$
   $1\ lb/in^3 = 27.680 \times 10^3\ kg/m^3$
   $1\ lb/ft^3 = 16.019\ kg/m^3$
   $1\ kg/m^3 = 0.06243\ lb/ft^3$
   $1\ slug/ft^3 = 515.38\ kg/m^3$
   $1\ lb \cdot mol/ft^3 = 16.019\ kg \cdot mol/m^3$
   $1\ kg \cdot mol/m^3 = 0.06243\ lb \cdot mol/ft^3$

4. Diffusivity (heat, mass, momentum)
   $1\ ft^2/s = 0.0929\ m^2/s$
   $1\ ft^2/h = 0.2581\ cm^2/s$
   $1\ ft^2/h = 0.2581 \times 10^{-4}\ m^2/s$
   $1\ m^2/s = 10.7639\ ft^2/s$
   $1\ cm^2/s = 3.8745\ ft^2/h$

5. Energy, heat, power
   $1\ J = 1\ W \cdot s = 1\ N \cdot m$
   $1\ J = 10^7\ erg$
   $1\ Btu = 1055.04\ J$
   $1\ Btu = 1055.04\ W \cdot s$
   $1\ Btu = 1055.04\ N \cdot m$
   $1\ Btu = 252\ cal$
   $1\ Btu = 0.252\ kcal$
   $1\ Btu = 778.161\ ft \cdot lbf$
   $1\ Btu/h = 0.2931\ W$
   $1\ Btu/h = 0.2931 \times 10^{-3}\ kW$
   $1\ Btu/h = 3.93 \times 10^{-4}\ hp$
   $1\ cal = 4.1868\ J$ (or $W \cdot s$ or $N \cdot m$)
   $1\ cal = 3.968 \times 10^{-3}\ Btu$
   $1\ kcal = 3.968\ Btu$
   $1\ hp = 550\ ft \cdot lbf/s$
   $1\ hp = 745.7\ W = 745.7\ N \cdot m/s$
   $1\ Wh = 3.413\ Btu$
   $1\ kWh = 3413\ Btu$

6. Heat capacity, heat per unit mass, specific heat
   $1\ Btu/(h \cdot °F) = 0.5274\ W/°C$
   $1\ W/°C = 1.8961\ Btu/(h \cdot °F)$
   $1\ Btu/lb = 2325.9\ J/kg$
   $1\ Btu/lb = 2.3259\ kJ/kg$
   $1\ Btu/(lb \cdot °F) = 4186.69\ J/(kg \cdot °C)$
   $1\ Btu/(lb \cdot °F) = 4.18669\ kJ/(kg \cdot °C)$
   [or $J/(g \cdot °C)$]
   $1\ Btu/(lb \cdot °F) = 1\ cal/(g \cdot °C) = 1\ kcal/(kg \cdot °C)$

7. Heat flux
   $1\ Btu/(h \cdot ft^2) = 3.1537\ W/m^2$
   $1\ Btu/(h \cdot ft^2) = 3.1537 \times 10^{-3}\ kW/m^2$
   $1\ W/m^2 = 0.31709\ Btu/(h \cdot ft^2)$

8. Heat generation rate
   $1\ Btu/(h \cdot ft^3) = 10.35\ W/m^3$
   $1\ Btu/(h \cdot ft^3) = 8.9\ kcal/(h \cdot m^3)$
   $1\ W/m^3 = 0.0966\ Btu/(h \cdot ft^3)$

Table A-1 (*continued*)

9. Heat transfer coefficient
   - 1 Btu/(h·ft^2·°F) = 5.677 W/(m^2·°C)
   - 1 Btu/(h·ft^2·°F) = 5.677 × 10^{-4} W/(cm^2·°C)
   - 1 W/(m^2·°C) = 0.1761 Btu/(h·ft^2·°F)
   - 1 Btu/(h·ft^2·°F) = 4.882 kcal/(h·m^2·°C)
10. Length
    - 1 Å = 10^{-8} cm
    - 1 Å = 10^{-10} m
    - 1 μm = 10^{-3} mm
    - 1 μm = 10^{-4} cm
    - 1 μm = 10^{-6} m
    - 1 in = 2.54 cm
    - 1 in = 2.54 × 10^{-2} m
    - 1 ft = 0.3048 m
    - 1 m = 3.2808 ft
    - 1 mi = 1609.34 m
    - 1 mi = 5280 ft
    - 1 light year = 9.46 × 10^{15} m
11. Mass
    - 1 oz = 28.35 g
    - 1 lb = 16 oz
    - 1 lb = 453.6 g
    - 1 lb = 0.4536 kg
    - 1 kg = 2.2046 lb
    - 1 g = 15.432 g
    - 1 slug = 32.1739 lb
    - 1 t (metric) = 1000 kg
    - 1 t (metric) = 2205 lb
    - 1 ton (short) = 2000 lb
    - 1 ton (long) = 2240 lb
12. Mass flux
    - 1 lb·mol/(ft^2·h) = 1.3563 × 10^{-3} kg·mol/(m^2·s)
    - 1 kg·mol/(m^2·s) = 737.3 lb·mol/(ft^2·h)
    - 1 lb/(ft^2·h) = 1.3563 × 10^{-3} kg/(m^2·s)
    - 1 lb/(ft^2·s) = 4.882 kg/(m^2·s)
    - 1 kg/(m^2·s) = 737.3 lb/(ft^2·h)
    - 1 kg/(m^2·s) = 0.2048 lb/(ft^2·s)
13. Pressure, force
    - 1 N = 1 kg·m/s^2
    - 1 N = 0.22481 lbf
    - 1 N = 7.2333 pdl
    - 1 N = 10^5 dyn
    - 1 lbf = 32.174 ft·lb/s^2
    - 1 lbf = 4.4482 N
    - 1 lbf = 4.4482 kg·m/s^2
    - 1 lbf = 32.1739 pdl
    - 1 lbf/in^2 ≡ 1 psi = 6894.76 N/m^2
    - 1 lbf/ft^2 = 47.880 N/m^2
    - 1 bar = 10^5 N/m^2 = 10^5 Pa
    - 1 atm = 14.696 lbf/in^2
    - 1 atm = 2116.2 lbf/ft^2
    - 1 atm = 1.0132 × 10^5 N/m^2
    - 1 atm = 1.0132 bar
    - 1 Pa = 1 N/m^2
14. Specific heat
    - 1 Btu/(lb·°F) = 1 kcal/(kg·°C) = 1 cal/(g·°C)
    - 1 Btu/(lb·°F) = 4186.69 J/(kg·°C) [or W·s/(kg·°C)]
    - 1 Btu/(lb·°F) = 4.18669 J/(g·K) [or W·s/(g·°C)]
    - 1 J/(g·°C) = 0.23885 Btu/(lb·°F) [cal/(g·°C) or kcal/(kg·°C)]
15. Speed
    - 1 ft/s = 0.3048 m/s
    - 1 m/s = 3.2808 ft/s
    - 1 mi/h = 1.4667 ft/s
    - 1 mi/h = 0.44704 m/s
16. Surface tension
    - 1 lbf/ft = 14.5937 N/m
    - 1 N/m = 0.068529 lbf/ft
17. Temperature
    - 1 K = 1.8°R
    - $T(°\mathrm{F}) = 1.8(\mathrm{K} - 273) + 32$
    - $T(\mathrm{K}) = \frac{1}{1.8}(°\mathrm{F} - 32) + 273$
    - $T(°\mathrm{C}) = \frac{1}{1.8}(°\mathrm{R} - 492)$
    - $\Delta T(°\mathrm{C}) = \Delta T(°\mathrm{F})/1.8$
18. Thermal conductivity
    - 1 Btu/(h·ft·°F) = 1.7303 W/(m·°C)
    - 1 Btu/(h·ft·°F) = 1.7303 × 10^{-2} W/(cm·°C)
    - 1 Btu/(h·ft·°F) = 0.4132 cal/(s·m·°C)
    - 1 W/(m·°C) = 0.5779 Btu/(h·ft·°F)
    - 1 W/(cm·°C) = 57.79 Btu/(h·ft·°F)
19. Thermal resistance
    - 1 h·°F/Btu = 1.896°C/W
    - 1°C/W = 0.528 h·°F/Btu
20. Viscosity
    - 1 P = 1 g/(cm·s)
    - 1 P = 10^2 cP
    - 1 P = 241.9 lb/(ft·h)
    - 1 cP = 2.419 lb/(ft·h)

Table A-1 (*continued*)

1 lb/(ft·s) = 1.4882 kg/(m·s)	1 cm^3 = 0.06102 in^3
1 lb/(ft·s) = 14.882 P	1 oz (U.S. fluid) = 29.573 cm^3
1 lb/(ft·s) = 1488.2 cP	1 ft^3 = 0.0283168 m^3
1 lb/(ft·h) = 0.4134 × 10^{-3} kg/(m·s)	1 ft^3 = 28.3168 liters
1 lb/(ft·h) = 0.4134 × 10^{-2} P	1 ft^3 = 7.4805 gal (U.S.)
1 lb/(ft·h) = 0.4134 cP	1 m^3 = 35.315 ft^3
	1 gal (U.S.) = 3.7854 liters
21. Volume	1 gal (U.S.) = 3.7854 × 10^{-3} m^3
1 in^3 = 16.387 cm^3	1 gal (U.S.) = 0.13368 ft^3

Constants

g_c = gravitational acceleration conversion factor	= 32.1739 ft·lb/(lbf·s^2)
	= 4.1697 × 10^8 ft·lb/(lbf·h^2)
	= 1 g·cm/(dyn·s^2)
	= 1 kg·m/(N·s^2)
	= 1 lb·ft/(pdl·s^2)
	= 1 slug·ft/(lbf·s^2)
J = mechanical equivalent of heat	= 778.16 ft·lbf/Btu
$\mathscr{R}$ = gas constant	= 1544 ft·lbf/(lb·mol·°R)
	= 0.730 ft^3·atm/(lb·mol·°R)
	= 0.08205 m^3·atm/(kg·mol·K)
	= 8.314 J/(g·mol·K)
	= 8.314 N·m/(g·mol·K)
	= 8314 N·m/(kg·mol·K)
	= 1.987 cal/(g·mol·K)
σ = Stefan-Boltzmann constant	= 0.1714 × 10^{-8} Btu/(h·ft^2·°R^4)
	= 5.6697 × 10^{-8} W/(m^2·K^4)

APPENDIX

B

PHYSICAL PROPERTIES

Table B-1 Physical properties of gases at atmospheric pressure

T, K	ρ, $\frac{kg}{m^3}$	c_p, $\frac{kJ}{kg \cdot °C}$	μ, $\frac{kg}{m \cdot s}$	ν, $\frac{m^2}{s} \times 10^6$	k, $\frac{W}{m \cdot K}$	α, $\frac{m^2}{s} \times 10^4$	Pr
Air							
100	3.6010	1.0266	0.6924×10^{-5}	1.923	0.009246	0.02501	0.770
150	2.3675	1.0099	1.0283	4.343	0.013735	0.05745	0.753
200	1.7684	1.0061	1.3289	7.490	0.01809	0.10165	0.739
250	1.4128	1.0053	1.488	9.49	0.02227	0.13161	0.722
300	1.1774	1.0057	1.983	15.68	0.02624	0.22160	0.708
350	0.9980	1.0090	2.075	20.76	0.03003	0.2983	0.697
400	0.8826	1.0140	2.286	25.90	0.03365	0.3760	0.689
450	0.7833	1.0207	2.484	28.86	0.03707	0.4222	0.683
500	0.7048	1.0295	2.671	37.90	0.04038	0.5564	0.680
550	0.6423	1.0392	2.848	44.34	0.04360	0.6532	0.680
600	0.5879	1.0551	3.018	51.34	0.04659	0.7512	0.680
650	0.5430	1.0635	3.177	58.51	0.04953	0.8578	0.682
700	0.5030	1.0752	3.332	66.25	0.05230	0.9672	0.684
750	0.4709	1.0856	3.481	73.91	0.05509	1.0774	0.686
800	0.4405	1.0978	3.625	82.29	0.05779	1.1951	0.689
850	0.4149	1.1095	3.765	90.75	0.06028	1.3097	0.692
900	0.3925	1.1212	3.899	99.3	0.06279	1.4271	0.696
950	0.3716	1.1321	4.023	108.2	0.06525	1.5510	0.699
1000	0.3524	1.1417	4.152	117.8	0.06752	1.6779	0.702
1100	0.3204	1.160	4.44	138.6	0.0732	1.969	0.704
1200	0.2947	1.179	4.69	159.1	0.0782	2.251	0.707
1300	0.2707	1.197	4.93	182.1	0.0837	2.583	0.705
1400	0.2515	1.214	5.17	205.5	0.0891	2.920	0.705
1500	0.2355	1.230	5.40	229.1	0.0946	3.262	0.705
1600	0.2211	1.248	5.63	254.5	0.100	3.609	0.705
1700	0.2082	1.267	5.85	280.5	0.105	3.977	0.705

Table B-1 *(continued)*

T, K	ρ, $\frac{kg}{m^3}$	c_p, $\frac{kJ}{kg \cdot °C}$	μ, $\frac{kg}{m \cdot s}$	ν, $\frac{m^2}{s} \times 10^6$	k, $\frac{W}{m \cdot K}$	α, $\frac{m^2}{s} \times 10^4$	Pr
Air							
1800	0.1970	1.287	6.07	308.1	0.111	4.379	0.704
1900	0.1858	1.309	6.29	338.5	0.117	4.811	0.704
2000	0.1762	1.338	6.50	369.0	0.124	5.260	0.702
2100	0.1682	1.372	6.72	399.6	0.131	5.715	0.700
2200	0.1602	1.419	6.93	432.6	0.139	6.120	0.707
2300	0.1538	1.482	7.14	464.0	0.149	6.540	0.710
2400	0.1458	1.574	7.35	504.0	0.161	7.020	0.718
2500	0.1394	1.688	7.57	543.5	0.175	7.441	0.730
Helium							
3		5.200	8.42×10^{-7}		0.0106		
33	1.4657	5.200	50.2	3.42	0.0353	0.04625	0.74
144	3.3799	5.200	125.5	37.11	0.0928	0.5275	0.70
200	0.2435	5.200	156.6	64.38	0.1177	0.9288	0.694
255	0.1906	5.200	181.7	95.50	0.1357	1.3675	0.70
366	0.13280	5.200	230.5	173.6	0.1691	2.449	0.71
477	0.10204	5.200	275.0	269.3	0.197	3.716	0.72
589	0.08282	5.200	311.3	375.8	0.225	5.215	0.72
700	0.07032	5.200	347.5	494.2	0.251	6.661	0.72
800	0.06023	5.200	381.7	634.1	0.275	8.774	0.72
900	0.05286	5.200	413.6	781.3	0.298	10.834	0.72
Carbon dioxide							
220	2.4733	0.783	11.105×10^{-6}	4.490	0.010805	0.05920	0.818
250	2.1657	0.804	12.590	5.813	0.012884	0.07401	0.793
300	1.7973	0.871	14.958	8.321	0.016572	0.10588	0.770
350	1.5362	0.900	17.205	11.19	0.02047	0.14808	0.755
400	1.3424	0.942	19.32	14.39	0.02461	0.19463	0.738
450	1.1918	0.980	21.34	17.90	0.02897	0.24813	0.721
500	1.0732	1.013	23.26	21.67	0.03352	0.3084	0.702
550	0.9739	1.047	25.08	25.74	0.03821	0.3750	0.685
600	0.8938	1.076	26.83	30.02	0.04311	0.4483	0.668
Carbon monoxide							
220	1.55363	1.0429	13.832×10^{-6}	8.903	0.01906	0.11760	0.758
250	1.3649	1.0425	15.40	11.28	0.02144	0.15063	0.750
300	1.13876	1.0421	17.843	15.67	0.02525	0.21280	0.737
350	0.97425	1.0434	20.09	20.62	0.02883	0.2836	0.728
400	0.85363	1.0484	22.19	25.99	0.03226	0.3605	0.722
450	0.75848	1.0551	24.18	31.88	0.0436	0.4439	0.718

Table B-1 (*continued*)

T, K	ρ, $\frac{\text{kg}}{\text{m}^3}$	c_p, $\frac{\text{kJ}}{\text{kg}\cdot{}^\circ\text{C}}$	μ, $\frac{\text{kg}}{\text{m}\cdot\text{s}}$	ν, $\frac{\text{m}^2}{\text{s}}$ $\times 10^6$	k, $\frac{\text{W}}{\text{m}\cdot\text{K}}$	α, $\frac{\text{m}^2}{\text{s}}$ $\times 10^4$	Pr
Carbon monoxide							
500	0.68223	1.0635	26.06	38.19	0.03863	0.5324	0.718
550	0.62024	1.0756	27.89	44.97	0.04162	0.6240	0.721
600	0.56850	1.0877	29.60	52.06	0.04446	0.7190	0.724
Ammonia, NH_3							
220	0.9304	2.198	7.255×10^{-6}	7.8	0.0171	0.2054	0.93
273	0.7929	2.177	9.353	11.8	0.0220	0.1308	0.90
323	0.6487	2.177	11.035	17.0	0.0270	0.1920	0.88
373	0.5590	2.236	12.886	23.0	0.0327	0.2619	0.87
423	0.4934	2.315	14.672	29.7	0.0391	0.3432	0.87
473	0.4405	2.395	16.49	37.4	0.0467	0.4421	0.84
Steam (H_2O vapor)							
380	0.5863	2.060	12.71×10^{-6}	21.6	0.0246	0.2036	1.060
400	0.5542	2.014	13.44	24.2	0.0261	0.2338	1.040
450	0.4902	1.980	15.25	31.1	0.0299	0.307	1.010
500	0.4405	1.985	17.04	38.6	0.0339	0.387	0.996
550	0.4005	1.997	18.84	47.0	0.0379	0.475	0.991
600	0.3652	2.026	20.67	56.6	0.0422	0.573	0.986
650	0.3380	2.056	22.47	64.4	0.0464	0.666	0.995
700	0.3140	2.085	24.26	77.2	0.0505	0.772	1.000
750	0.2931	2.119	26.04	88.8	0.0549	0.883	1.005
800	0.2739	2.152	27.86	102.0	0.0592	1.001	1.010
850	0.2579	2.186	29.69	115.2	0.0637	1.130	1.019
Hydrogen							
30	0.84722	10.840	1.606×10^{-6}	1.895	0.0228	0.02493	0.759
50	0.50955	10.501	2.516	4.880	0.0362	0.0676	0.721
100	0.24572	11.229	4.212	17.14	0.0665	0.2408	0.712
150	0.16371	12.602	5.595	34.18	0.0981	0.475	0.718
200	0.12270	13.540	6.813	55.53	0.1282	0.772	0.719
250	0.09819	14.059	7.919	80.64	0.1561	1.130	0.713
300	0.08185	14.314	8.963	109.5	0.182	1.554	0.706
350	0.07016	14.436	9.954	141.9	0.206	2.031	0.697
400	0.06135	14.491	10.864	177.1	0.228	2.568	0.690
450	0.05462	14.499	11.779	215.6	0.251	3.164	0.682
500	0.04918	14.507	12.636	257.0	0.272	3.817	0.675
550	0.04469	14.532	13.475	301.6	0.292	4.516	0.668
600	0.04085	14.537	14.285	349.7	0.315	5.306	0.664
700	0.03492	14.574	15.89	455.1	0.351	6.903	0.659
800	0.03060	14.675	17.40	569	0.384	8.563	0.664

Table B-1 *(continued)*

T, K	ρ, $\frac{kg}{m^3}$	c_p, $\frac{kJ}{kg \cdot °C}$	μ, $\frac{kg}{ms}$	ν, $\frac{m^2}{s} \times 10^6$	k, $\frac{W}{m \cdot K}$	α, $\frac{m^2}{s} \times 10^4$	Pr
Hydrogen							
900	0.02723	14.821	18.78	690	0.412	10.217	0.676
1000	0.02451	14.968	20.16	822	0.440	11.997	0.686
1100	0.02227	15.165	21.46	965	0.464	13.726	0.703
1200	0.02050	15.366	22.75	1107	0.488	15.484	0.715
1300	0.01890	15.575	24.08	1273	0.512	17.394	0.733
1333	0.01842	15.638	24.44	1328	0.519	18.013	0.736
Oxygen							
100	3.9918	0.9479	7.768×10^{-6}	1.946	0.00903	0.023876	0.815
150	2.6190	0.9178	11.490	4.387	0.01367	0.05688	0.773
200	1.9559	0.9131	14.850	7.593	0.01824	0.10214	0.745
250	1.5618	0.9157	17.87	11.45	0.02259	0.15794	0.725
300	1.3007	0.9203	20.63	15.86	0.02676	0.22353	0.709
350	1.1133	0.9291	23.16	20.80	0.03070	0.2968	0.702
400	0.9755	0.9420	25.54	26.18	0.03461	0.3768	0.695
450	0.8682	0.9567	27.77	31.99	0.03828	0.4609	0.694
500	0.7801	0.9722	29.91	38.34	0.04173	0.5502	0.697
550	0.7096	0.9881	31.97	45.05	0.04517	0.6441	0.700
600	0.6504	1.0044	33.92	52.15	0.04832	0.7399	0.704
Nitrogen							
100	3.4808	1.0722	6.862×10^{-6}	1.971	0.009450	0.025319	0.786
200	1.7108	1.0429	12.947	7.568	0.01824	0.10224	0.747
300	1.1421	1.0408	17.84	15.63	0.02620	0.22044	0.713
400	0.8538	1.0459	21.98	25.74	0.03335	0.3734	0.691
500	0.6824	1.0555	25.70	37.66	0.03984	0.5530	0.684
600	0.5687	1.0756	29.11	51.19	0.04580	0.7486	0.686
700	0.4934	1.0969	32.13	65.13	0.05123	0.9466	0.691
800	0.4277	1.1225	34.84	81.46	0.05609	1.1685	0.700
900	0.3796	1.1464	37.49	91.06	0.06070	1.3946	0.711
1000	0.3412	1.1677	40.00	117.2	0.06475	1.6250	0.724
1100	0.3108	1.1857	42.28	136.0	0.06850	1.8591	0.736
1200	0.2851	1.2037	44.50	156.1	0.07184	2.0932	0.748

From E. R. G. Eckert and R. M. Drake, *Analysis of Heat Mass Transfer*, McGraw-Hill, New York, 1972.

Table B-2 Physical properties of saturated liquids

t, C	ρ, $\frac{kg}{m^3}$	c_p, $\frac{kJ}{kg \cdot °C}$	ν, $\frac{m^2}{s}$	k, $\frac{W}{m \cdot K}$	α, $\frac{m^2}{s} \times 10^7$	Pr	β, K^{-1}
Ammonia, NH_3							
−50	703.69	4.463	0.435×10^{-6}	0.547	1.742	2.60	
−40	691.68	4.467	0.406	0.547	1.775	2.28	
−30	679.34	4.476	0.387	0.549	1.801	2.15	
−20	666.69	4.509	0.381	0.547	1.819	2.09	
−10	653.55	4.564	0.378	0.543	1.825	2.07	
0	640.10	4.635	0.373	0.540	1.819	2.05	
10	626.16	4.714	0.368	0.531	1.801	2.04	
20	611.75	4.798	0.359	0.521	1.775	2.02	2.45×10^{-3}
30	596.37	4.890	0.349	0.507	1.742	2.01	
40	580.99	4.999	0.340	0.493	1.701	2.00	
50	564.33	5.116	0.330	0.476	1.654	1.99	
Carbon dioxide, CO_2							
−50	1,156.34	1.84	0.119×10^{-6}	0.0855	0.4021	2.96	
−40	1,117.77	1.88	0.118	0.1011	0.4810	2.46	
−30	1,076.76	1.97	0.117	0.1116	0.5272	2.22	
−20	1,032.39	2.05	0.115	0.1151	0.5445	2.12	
−10	983.38	2.18	0.113	0.1099	0.5133	2.20	
0	926.99	2.47	0.108	0.1045	0.4578	2.38	
10	860.03	3.14	0.101	0.0971	0.3608	2.80	
20	772.57	5.0	0.091	0.0872	0.2219	4.10	14.00×10^{-3}
30	597.81	36.4	0.080	0.0703	0.0279	28.7	
Dichlorodifluoromethane (Freon-12), CCl_2F_2							
−50	1,546.75	0.8750	0.310×10^{-6}	0.067	0.501	6.2	2.63×10^{-3}
−40	1,518.71	0.8847	0.279	0.069	0.514	5.4	
−30	1,489.56	0.8956	0.253	0.069	0.526	4.8	
−20	1,460.57	0.9073	0.235	0.071	0.539	4.4	
−10	1,429.49	0.9203	0.221	0.073	0.550	4.0	
0	1,397.45	0.9345	0.214	0.073	0.557	3.8	
10	1,364.30	0.9496	0.203	0.073	0.560	3.6	
20	1,330.18	0.9659	0.198	0.073	0.560	3.5	
30	1,295.10	0.9835	0.194	0.071	0.560	3.5	
40	1,257.13	1.0019	0.191	0.069	0.555	3.5	
50	1,215.96	1.0216	0.190	0.067	0.545	3.5	
Engine oil (unused)							
0	899.12	1.796	0.00428	0.147	0.911	47,100	
20	888.23	1.880	0.00090	0.145	0.872	10,400	0.70×10^{-3}

Table B-2 (*continued*)

t, C	ρ, $\frac{kg}{m^3}$	c_p, $\frac{kJ}{kg \cdot °C}$	ν, $\frac{m^2}{s}$	k, $\frac{W}{m \cdot K}$	α, $\frac{m^2}{s}$ $\times 10^7$	Pr	β, K^{-1}
Engine oil (unused)							
40	876.05	1.964	0.00024	0.144	0.834	2,870	
60	864.04	2.047	0.839×10^{-4}	0.140	0.800	1,050	
80	852.02	2.131	0.375	0.138	0.769	490	
100	840.01	2.219	0.203	0.137	0.738	276	
120	828.96	2.307	0.124	0.135	0.710	175	
140	816.94	2.395	0.080	0.133	0.686	116	
160	805.89	2.483	0.056	0.132	0.663	84	
Ethylene glycol, $C_2H_4(OH_2)$							
0	1,130.75	2.294	57.53×10^{-6}	0.242	0.934	615	
20	1,116.65	2.382	19.18	0.249	0.939	204	0.65×10^{-3}
40	1,101.43	2.474	8.69	0.256	0.939	93	
60	1,087.66	2.562	4.75	0.260	0.932	51	
80	1,077.56	2.650	2.98	0.261	0.921	32.4	
100	1,058.50	2.742	2.03	0.263	0.908	22.4	
Eutectic calcium chloride solution, 29.9% $CaCl_2$							
−50	1,319.76	2.608	36.35×10^{-6}	0.402	1.166	312	
−40	1,314.96	2.6356	24.97	0.415	1.200	208	
−30	1,310.15	2.661,1	17.18	0.429	1.234	139	
−20	1,305.51	2.688	11.04	0.445	1.267	87.1	
−10	1,300.70	2.713	6.96	0.459	1.300	53.6	
0	1,296.06	2.738	4.39	0.472	1.332	33.0	
10	1,291.41	2.763	3.35	0.485	1.363	24.6	
20	1,286.61	2.788	2.72	0.498	1.394	19.6	
30	1,281.96	2.814	2.27	0.511	1.419	16.0	
40	1,277.16	2.839	1.92	0.523	1.445	13.3	
50	1,272.51	2.868	1.65	0.535	1.468	11.3	
Glycerin, $C_3H_5(OH)_3$							
0	1,276.03	2.261	0.00831	0.282	0.983	84.7×10^3	
10	1,270.11	2.319	0.00300	0.284	0.965	31.0	
20	1,264.02	2.386	0.00118	0.286	0.947	12.5	0.50×10^{-3}
30	1,258.09	2.445	0.00050	0.286	0.929	5.38	
40	1,252.01	2.512	0.00022	0.286	0.914	2.45	
50	1,244.96	2.583	0.00015	0.287	0.893	1.63	

Table B-2 (*continued*)

t, C	ρ, $\frac{kg}{m^3}$	c_p, $\frac{kJ}{kg \cdot °C}$	ν, $\frac{m^2}{s}$	k, $\frac{W}{m \cdot K}$	α, $\frac{m^2}{s}$ $\times 10^7$	Pr	β, K^{-1}
Mercury, Hg							
0	13,628.22	0.1403	0.124×10^{-6}	8.20	42.99	0.0288	
20	13,579.04	0.1394	0.114	8.69	46.06	0.0249	1.82×10^{-4}
50	13,505.84	0.1386	0.104	9.40	50.22	0.0207	
100	13,384.58	0.1373	0.0928	10.51	57.16	0.0162	
150	13,264.28	0.1365	0.0853	11.49	63.54	0.0134	
200	13,144.94	0.1360	0.0802	12.34	69.08	0.0116	
250	13,025.60	0.1357	0.0765	13.07	74.06	0.0103	
315.5	12,847	0.134	0.0673	14.02	8.15	0.0083	
Methyl chloride, CH_3Cl							
−50	1,052.58	1.4759	0.320×10^{-6}	0.215	1.388	2.31	
−40	1,033.35	1.4826	0.318	0.209	1.368	2.32	
−30	1,016.53	1.4922	0.314	0.202	1.337	2.35	
−20	999.39	1.5043	0.309	0.196	1.301	2.38	
−10	981.45	1.5194	0.306	0.187	1.257	2.43	
0	962.39	1.5378	0.302	0.178	1.213	2.49	
10	942.36	1.5600	0.297	0.171	1.166	2.55	
20	923.31	1.5860	0.293	0.163	1.112	2.63	
30	903.12	1.6161	0.288	0.154	1.058	2.72	
40	883.10	1.6504	0.281	0.144	0.996	2.83	
50	861.15	1.6890	0.274	0.133	0.921	2.97	
Sulfur dioxide, SO_2							
−50	1,560.84	1.3595	0.484×10^{-6}	0.242	1.141	4.24	
−40	1,536.81	1.3607	0.424	0.235	1.130	3.74	
−30	1,520.64	1.3616	0.371	0.230	1.117	3.31	
−20	1,488.60	1.3624	0.324	0.225	1.107	2.93	
−10	1,463.61	1.3628	0.288	0.218	1.097	2.62	
0	1,438.46	1.3636	0.257	0.211	1.081	2.38	
10	1,412.51	1.3645	0.232	0.204	1.066	2.18	
20	1,386.40	1.3653	0.210	0.199	1.050	2.00	1.94×10^{-3}
30	1,359.33	1.3662	0.190	0.192	1.035	1.83	
40	1,329.22	1.3674	0.173	0.185	1.019	1.70	
50	1,299.10	1.3683	0.162	0.177	0.999	1.61	
Water, H_2O							
0	1,002.28	4.2178	1.788×10^{-6}	0.552	1.308	13.6	
20	1,000.52	4.1818	1.006	0.597	1.430	7.02	0.18×10^{-3}
40	994.59	4.1784	0.658	0.628	1.512	4.34	

Table B-2 *(continued)*

t, C	ρ, $\frac{kg}{m^3}$	c_p, $\frac{kJ}{kg \cdot °C}$	ν, $\frac{m^2}{s}$	k, $\frac{W}{m \cdot K}$	α, $\frac{m^2}{s} \times 10^7$	Pr	β, K^{-1}
Water, H_2O							
60	985.46	4.1843	0.478	0.651	1.554	3.02	
80	974.08	4.1964	0.364	0.668	1.636	2.22	
100	960.63	4.2161	0.294	0.680	1.680	1.74	
120	945.25	4.250	0.247	0.685	1.708	1.446	
140	928.27	4.283	0.214	0.684	1.724	1.241	
160	909.69	4.342	0.190	0.680	1.729	1.099	
180	889.03	4.417	0.173	0.675	1.724	1.004	
200	866.76	4.505	0.160	0.665	1.706	0.937	
220	842.41	4.610	0.150	0.652	1.680	0.891	
240	815.66	4.756	0.143	0.635	1.639	0.871	
260	785.87	4.949	0.137	0.611	1.577	0.874	
280.6	752.55	5.208	0.135	0.580	1.481	0.910	
300	714.26	5.728	0.135	0.540	1.324	1.019	

From E. R. G. Eckert and R. M. Drake, *Analysis of Heat Mass Transfer*, McGraw-Hill, New York, 1972.

Table B-3 Physical properties of liquid metals

Metal	Melting point, °C	Boiling point, °C	T, °C	ρ, $\frac{kg}{m^3}$	c_p, $\frac{kJ}{kg \cdot °C}$	$\mu \times 10^4$, $\frac{kg}{m \cdot s}$	$\nu \times 10^6$, $\frac{m^2}{s}$	k, $\frac{W}{m \cdot °C}$	$\alpha \times 10^6$, $\frac{m^2}{s}$	Pr
Bismuth	271	1477	315	10,011	0.144	16.2	0.160	16.4	11.25	0.0142
			538	9,739	0.155	11.0	0.113	15.6	10.34	0.0110
			760	9,467	0.165	7.9	0.083	15.6	9.98	0.0083
Lead	327	1737	371	10,540	0.159	2.40	0.023	16.1	9.61	0.024
			704	10,140	0.155	1.37	0.014	14.9	9.48	0.0143
Lithium	179	1317	204.4	509.2	4.365	5.416	1.1098	46.37	20.96	0.051
			315.6	498.8	4.270	4.465	0.8982	43.08	20.32	0.0443
			426.7	489.1	4.211	3.927	0.8053	38.24	18.65	0.0432
			537.8	476.3	4.171	3.473	0.7304	30.45	15.40	0.0476
Mercury	−38.9	357	−17.8	13,707.1	0.1415	18.334	0.1342	9.76	5.038	0.0266
			100	13,384.5	0.1373	12.420	0.0928	10.51	5.716	0.0162
			200	13,144.9	0.1570	10.541	0.0802	12.34	6.908	0.0116
Sodium	97.8	883	93.3	931.6	1.384	7.131	0.7689	84.96	56.29	0.0116
			204.4	907.5	1.339	4.521	0.5010	80.81	66.80	0.0075
			315.6	878.5	1.304	3.294	0.3766	75.78	66.47	0.00567
			426.7	852.8	1.277	2.522	0.2968	69.39	64.05	0.00464
			537.8	823.8	1.264	2.315	0.2821	64.37	62.09	0.00455
			648.9	790.0	1.261	1.964	0.2496	60.56	61.10	0.00408
			760.0	767.5	1.270	1.716	0.2245	56.58	58.34	0.00385
Potassium	63.9	760	426.7	741.7	0.766	2.108	0.2839	39.45	69.74	0.0041
			537.8	714.4	0.762	1.711	0.2400	36.51	67.39	0.0036
			648.9	690.3	0.766	1.463	0.2116	33.74	64.10	0.0033
			760.0	667.7	0.783	1.331	0.1987	31.15	59.86	0.0033
NaK (56% Na, 44% K)	−11.1	784	93.3	889.8	1.130	5.622	0.6347	25.78	27.76	0.0246
			204.4	865.6	1.089	3.803	0.4414	26.47	28.23	0.0155
			315.6	838.3	1.068	2.935	0.3515	27.17	30.50	0.0115
			426.7	814.2	1.051	2.150	0.2652	27.68	32.52	0.0081
			537.8	788.4	1.047	2.026	0.2581	27.68	33.71	0.0076
			648.9	759.5	1.051	1.695	0.2240	27.68	34.86	0.0064

Table B-4 Physical properties of metals

Metal	Melting point °C	Properties at 20°C: ρ, $\frac{kg}{m^3}$	c_p, $\frac{kJ}{kg \cdot °C}$	k, $\frac{W}{m \cdot °C}$	α, $\frac{m^2}{s} \times 10^5$	Thermal conductivity k, W/(m · °C): −100°C	0°C	100°C	200°C	300°C	400°C	600°C	800°C	1000°C
Aluminum														
Pure	660	2,707	0.896	204	8.418	215	202	206	215	228	249			
Al-Cu (Duralumin), 94–96% Al, 3–5% Cu, trace Mg		2,787	0.883	164	6.676	126	159	182	194					
Al-Si (Silumin, copper-bearing), 86.5% Al, 1% Cu		2,659	0.867	137	5.933	119	137	144	152	161				
Al-Si (Alusil), 78–80% Al, 20–22% Si		2,627	0.854	161	7.172	144	157	168	175	178				
Al-Mg-Si, 97% Al, 1% Mg, 1% Si, 1% Mn		2,707	0.892	177	7.311		175	189	204					
Beryllium	1277	1,850	1.825	200	5.92									
Bismuth	272	9,780	0.122	7.86	0.66									
Cadmium	321	8,650	0.231	96.8	4.84									
Copper														
Pure	1085	8,954	0.3831	386	11.234	407	386	379	374	369	363	353		
Aluminum bronze 95% Cu, 5% Al		8,666	0.410	83	2.330									
Bronze 75% Cu, 25% Sn		8.666	0.343	26	0.859									
Red brass 85% Cu, 9% Sn, 6% Zn		8.714	0.385	61	1.804		59	71						

Table B-4 *(continued)*

Metal	Melting point °C	Properties at 20°C				Thermal conductivity k, W/m · °C								
		ρ, $\frac{kg}{m^3}$	c_p, $\frac{kJ}{kg \cdot °C}$	k, $\frac{W}{m \cdot °C}$	α, $\frac{m^2}{s} \times 10^5$	−100°C	0°C	100°C	200°C	300°C	400°C	600°C	800°C	1000°C
Brass 70% Cu, 30% Zn		8,522	0.385	111	3.412	88		128	144	147	147			
German silver 62% Cu, 15% Ni, 22% Zn		8,618	0.394	24.9	0.733	19.2		31	40	45	48			
Constantan 60% Cu, 40% Ni		8,922	0.410	22.7	0.612	21		22.2	26					
Iron														
Pure	1537	7,897	0.452	73	2.034	87	73	67	62	55	48	40	36	35
Wrought iron, 0.5% C		7,849	0.46	59	1.626		59	57	52	48	45	36	33	33
Steel (C max ≈ 1.5%):														
Carbon steel														
C ≈ 0.5%		7,833	0.465	54	1.474		55	52	48	45	42	35	31	29
1.0%		7,801	0.473	43	1.172		43	43	42	40	36	33	29	28
1.5%		7,753	0.486	36	0.970		36	36	36	35	33	31	28	28
Nickel steel														
Ni ≈ 0%		7,897	0.452	73	2.026									
20%		7,933	0.46	19	0.526									
40%		8,169	0.46	10	0.279									
80%		8,618	0.46	35	0.872									
Invar 36% Ni		8,137	0.46	10.7	0.286									
Chrome steel														
Cr = 0%		7,897	0.452	73	2.026	87	73	67	62	55	48	40	36	35
1%		7,865	0.46	61	1.665		62	55	52	47	42	36	33	33
5%		7,833	0.46	40	1.110		40	38	36	36	33	29	29	29

20%		7,689	0.46	22	0.635		22	22	22	22	24	24	26	29
Cr-Ni (chrome-nickel):														
15% Cr, 10% Ni		7,865	0.46	19	0.527									
18% Cr, 8% Ni (V2A)		7,817	0.46	16.3	0.444		16.3	17	17	19	19	22	27	31
20% Cr, 15% Ni		7,833	0.46	15.1	0.415									
25% Cr, 20% Ni		7,865	0.46	12.8	0.361									
Tungsten steel														
W = 0%		7,897	0.452	73	2.026									
1%		7,913	0.448	66	1.858									
5%		8,073	0.435	54	1.525									
10%		8,314	0.419	48	1.391									
Lead	328	11,373	0.130	35	2.343	36.9	35.1	33.4	31.5	29.8				
Magnesium														
Pure	650	1,746	1.013	171	9.708	178	171	168	163	157				
Mg-Al (electrolytic)														
6–8% Al, 1–2% Zn		1,810	1.00	66	3.605		52	62	74	83				
Molybdenum	2,621	10,220	0.251	123	4.790	138	125	118	114	111	109	106	102	99
Nickel														
Pure (99.9%)	1,455	8,906	0.4459	90	2.266	104	93	83	73	64	59			
Ni-Cr														
90% Ni, 10% Cr		8,666	0.444	17	0.444		17.1	18.9	20.9	22.8	24.6			
80% Ni, 20% Cr		8,314	0.444	12.6	0.343		12.3	13.8	15.6	17.1	18.0	22.5		
Silver:														
Purest	962	10,524	0.2340	419	17.004	419	417	415	412					
Pure (99.9%)		10,525	0.2340	407	16.563	419	410	415	374	362	360			
Tin, pure	232	7,304	0.2265	64	3.884	74	65.9	59	57					
Tungsten	3,387	19,350	0.1344	163	6.271		166	151	142	133	126	112	76	
Uranium	1,133	19,070	0.116	27.6	1.25									
Zinc, pure	420	7,144	0.3843	112.2	4.106	114	112	109	106	100	93			

From E. R. G. Eckert and R. M. Drake, *Analysis of Heat Mass Transfer*, McGraw-Hill, New York, 1972.

Table B-5 Physical properties of insulating materials

Material	T, °C	k, $\frac{W}{m \cdot °C}$	ρ, $\frac{kg}{m^3}$	c_p, $\frac{kJ}{kg \cdot °C}$	α, $\frac{m^2}{s}$ $\times 10^7$
Asbestos					
Loosely packed	−45	0.149			
	0	0.154	470–570	0.816	3.3–4
	100	0.161			
Asbestos-cement boards	20	0.74			
Sheets	51	0.166			
Felt, 40 laminations/in	38	0.057			
	150	0.069			
	260	0.083			
20 laminations/in	38	0.078			
	150	0.095			
	260	0.112			
Corrugated, 4 plies/in	38	0.087			
	93	0.100			
	150	0.119			
Balsam wool	32	0.04	35		
Board and slab					
Cellular glass	30	0.058	145	1.000	
Glass fiber, organic bonded	30	0.036	105	0.795	
Polystyrene, expanded extruded (R-12)	30	0.027	55	1.210	
Mineral fiberboard; roofing material	30	0.049	265		
Wood, shredded/cemented	30	0.087	350	1.590	
Cardboard, corrugated	—	0.064			
Celotex	32	0.048			
Corkboard	30	0.043	160		
Cork, regranulated	32	0.045	45–120	1.88	2–5.3
Ground	32	0.043	150		
Diatomaceous earth (Sil-o-cel)	0	0.061	320		
Felt, hair	30	0.036	130–200		
Wool	30	0.052	330		
Fiber, insulating board	20	0.048	240		
Glass wool	23	0.038	24	0.7	22.6
Insulex, dry	32	0.064			
Kapok	30	0.035			
Loose fill					
Cork, granulated	30	0.045	160	—	
Diatomaceous silica, coarse powder	30	0.069	350	—	
Diatomaceous silica, fine powder	30	0.091	400	—	
	30	0.052	200	—	
	30	0.061	275	—	

Table B-5 *(continued)*

Material	T, °C	k, $\frac{W}{m \cdot °C}$	ρ, $\frac{kg}{m^3}$	c_p, $\frac{kJ}{kg \cdot °C}$	α, $\frac{m^2}{s} \times 10^7$
Glass fiber, poured or blown	30	0.043	16	0.835	
Vermiculite, flakes	30	0.068	80	0.835	
		0.063	160	1.000	
Formed/Foamed-in-Place					
Mineral wool granules with asbestos/inorganic binders, sprayed	30	0.046	190	—	
Polyvinyl acetate cork mastic; sprayed or troweled	30	0.100	—	—	
Urethane, two-part mixture; rigid foam	30	0.026	70	1.045	
Magnesia, 85%	38	0.067	270		
	93	0.071			
	150	0.074			
	204	0.080			
Rock wool, 10 lb/ft^3	32	0.040	160		
Loosely packed	150	0.067	64		
	260	0.087			
Sawdust	23	0.059			
Silica aerogel	32	0.024	140		
Wood shavings	23	0.059			

From A. I. Brown and S. M. Macro, *Introduction to Heat Transfer*, 3d ed., McGraw-Hill, New York, 1958; *International Critical Tables*, McGraw-Hill, New York, 1926–1930.

Table B-6 Physical properties of nonmetals

Material	T, °C	k, $\frac{W}{m \cdot °C}$	ρ $\frac{kg}{m^3}$	c_p, $\frac{kJ}{kg \cdot °C}$	α, $\frac{m^2}{s}$ $\times 10^7$
Asphalt	20–55	0.74–0.76			
Brick					
Building brick, common	20	0.69	1600	0.84	5.2
Face		1.32	2000		
Carborundum brick	600	18.5			
	1400	11.1			
Chrome brick	200	2.32	3000	0.84	9.2
	550	2.47			9.8
	900	1.99			7.9
Diatomaceous earth, molded and fired	200	0.24			
	870	0.31			
Fireclay brick, burned 1330°C	500	1.04	2000	0.96	5.4
	800	1.07			
	1100	1.09			
Burned 1450°C	500	1.28	2300	0.96	5.8
	800	1.37			
	1100	1.40			
Missouri	200	1.00	2600	0.96	4.0
	600	1.47			
	1400	1.77			
Magnesite	200	3.81		1.13	
	650	2.77			
	1200	1.90			
Clay	30	1.3	1460	0.88	
Cement, portland	23	0.29	1500		
Mortar	23	1.16			
Coal, anthracite	30	0.26	1200–1500	1.26	
Powdered	30	0.116	737	1.30	
Concrete, cinder	23	0.76			
Stone 1-2-4 mix	20	1.37	1900–2300	0.88	8.2–6.8
Cotton	20	0.06	80	1.30	
Glass, window	20	0.78 (avg)	2700	0.84	3.4
Corosilicate	30–75	1.09	2200	–	
Plate (soda lime)	30	1.4	2500	0.75	
Pyrex	30	1.4	2225	0.835	
Paper	30	0.011	930	1.340	
Paraffin	30	0.020	900	2.890	
Plaster, gypsum	20	0.48	1440	0.84	4.0
Metal lath	20	0.47			
Wood lath	20	0.28			

Table B-6 (*continued*)

Material	T, °C	k, $\frac{W}{m \cdot °C}$	ρ, $\frac{kg}{m^3}$	c_p, $\frac{kJ}{kg \cdot °C}$	α, $\frac{m^2}{s} \times 10^7$
Rubber, vulcanized					
Soft	30	0.012	1100	2.010	
Hard	30	0.013	1190	—	
Sand	30	0.027	1515	0.800	
Stone					
Granite		1.73–3.98	2640	0.82	8–18
Limestone	100–300	1.26–1.33	2500	0.90	5.6–5.9
Marble		2.07–2.94	2500–2700	0.80	10–13.6
Sandstone	40	1.83	2160–2300	0.71	11.2–11.9
Teflon	30	0.35	2200	—	
Tissue, human skin	30	0.37	—	—	
Fat layer	30	0.20	—	—	
Muscle	30	0.41	—	—	
Wood (across the grain)					
Balsa	30	0.055	140		
Cypress	30	0.097	460		
Fir	23	0.11	420	2.72	0.96
Maple or oak	30	0.166	540	2.4	1.28
Yellow pine	23	0.147	640	2.8	0.82
White pine	30	0.112	430		

Table B-7 Abbreviated saturated steam table

Temperature T, °C	Pressure P, kPa	Enthalpy of evaporation h_{fg}, kJ/kg	Liquid density ρ_l, kg/m³	Vapor density ρ_v, kg/m³	Temperature T, °C	Pressure P, MPa	Enthalpy of evaporation h_{fg}, kJ/kg	Liquid density ρ_l, kg/m³	Vapor density ρ_v, kg/m³
0.01	0.6113	2501.3	1000.0	0.0049	150	0.4758	2114.3	916.6	2.549
5	0.8721	2489.6	1000.0	0.0068	160	0.6178	2082.6	907.4	3.256
10	1.2276	2477.7	1000.0	0.0094	170	0.7917	2049.5	897.7	4.119
15	1.7051	2465.9	999.0	0.0129	180	1.0021	2015.0	887.3	5.153
20	2.339	2454.1	998.0	0.0173	190	1.2544	1978.8	876.4	6.388
25	3.169	2442.3	997.0	0.0231	200	1.5538	1940.7	864.3	7.852
30	4.246	2430.5	996.0	0.0304	210	1.9062	1900.7	852.5	9.578
35	5.628	2418.6	994.0	0.0397	220	2.318	1858.5	840.3	11.602
40	7.384	2406.7	992.1	0.0512	230	2.795	1813.8	827.1	13.970
45	9.593	2394.8	990.1	0.0655	240	3.344	1766.5	813.7	16.734
50	12.349	2382.7	988.1	0.0831	250	3.973	1716.2	799.4	19.948
55	15.758	2370.7	985.2	0.1045	260	4.688	1662.5	783.7	23.691
60	19.940	2358.5	983.3	0.1304	270	5.499	1605.2	768.1	28.058
65	25.03	2346.2	980.4	0.1614	280	6.412	1543.6	750.8	33.145
70	31.19	2333.8	997.5	0.1983	290	7.436	1477.1	732.1	39.108
75	38.58	2321.4	974.7	0.2421	300	8.581	1404.9	712.3	46.147
80	47.39	2308.8	971.8	0.2935	310	9.856	1326.0	691.1	54.50
85	57.83	2296.0	968.1	0.3536	320	11.274	1238.6	667.1	64.57
90	70.14	2283.2	965.3	0.4235	330	12.845	1140.6	640.6	76.95
95	84.55	2270.2	961.5	0.5045	340	14.586	1027.9	610.5	92.62
100	0.101 35	2257.0	957.9	0.598	350	16.513	893.4	574.7	113.47
110	0.143 27	2230.2	950.6	0.826	360	18.651	720.5	528.3	143.99
120	0.198 53	2202.6	943.4	1.121	370	21.03	441.6	451.9	203.05
130	0.2701	2174.2	934.6	1.496	374.14	22.09	0	316.96	316.96
140	0.3613	2144.7	925.9	1.965					

Adapted from Joseph H. Keenan, Frederick G. Keyes, Philip G. Hill, and Joan G. Moore, *Steam Tables*, Wiley, New York, 1969.

Table B-8 Illustration of physical properties of gases at atmospheric pressure in both Btu and SI units

Temperature, °F	°C	c_p, Btu/lb·°F	$c_p \times 10^{-3}$, W·s/kg·°C	k, Btu/h·ft·°F	k, W/m·°C	μ, lb/ft·h	$\mu \times 10^5$, kg/m·s	ρ, lb/ft³	ρ, kg/m³	ν, ft²/h	$\nu \times 10^4$, m²/s	α, ft²/h	$\alpha \times 10^4$, m²/s	Pr
Air														
−200	−128.9	0.2392	1.001	0.0079	0.0137	0.0252	1.042	0.153	2.462	0.165	0.0426	0.216	0.0557	0.760
0	−17.8	0.2400	1.005	0.014	0.0242	0.0415	1.716	0.0864	1.390	0.480	0.1239	0.675	0.1742	0.711
200	93.3	0.2414	1.011	0.0181	0.0313	0.0519	2.146	0.0602	0.969	0.862	0.2225	1.245	0.3213	0.692
400	204.4	0.2451	1.026	0.0224	0.0388	0.0624	2.580	0.0462	0.743	1.351	0.3487	1.977	0.6103	0.683
N_2														
−200	−128.9	0.252	1.055	0.0079	0.0137	0.0237	0.980	0.148	2.381	0.160	0.0413	0.212	0.0547	0.756
0	−17.8	0.2484	1.040	0.0132	0.0228	0.039	1.612	0.0835	1.344	0.467	0.1205	0.635	0.1639	0.734
200	93.3	0.249	1.042	0.0173	0.0299	0.0498	2.059	0.0582	0.936	0.856	0.2209	1.194	0.3082	0.717
400	204.4	0.2515	1.053	0.021	0.0363	0.0601	2.485	0.0448	0.721	1.342	0.3464	1.864	0.4811	0.719
O_2														
−200	−128.9	0.2175	0.911	0.0079	0.0137	0.0272	1.124	0.169	2.719	0.161	0.0417	0.215	0.0555	0.749
0	−17.8	0.2182	0.914	0.0135	0.0234	0.044	1.819	0.096	1.545	0.458	0.1821	0.644	0.1662	0.711
200	93.4	0.2223	0.931	0.018	0.0311	0.0583	2.410	0.0665	1.070	0.877	0.2264	1.217	0.3141	0.720
400	204.4	0.2305	0.965	0.0233	0.0403	0.0712	2.943	0.0512	0.824	1.391	0.3590	1.973	0.5092	0.704
NH_3														
0	−17.8	0.522	2.185	0.0117	0.0202	0.0213	0.880	0.0441	0.710	0.483	0.1247	0.508	0.1311	0.95
200	93.3	0.532	2.227	0.0192	0.0332	0.0303	1.253	0.0307	0.494	0.999	0.2578	1.173	0.3028	0.84
400	204.4	0.574	2.403	0.0280	0.0484	0.0394	1.629	0.0236	0.380	1.669	0.4308	2.064	0.5327	0.807
Freon-11														
0	−17.8	0.124	0.519	0.00412	0.00713	0.0232	0.959	0.0398	0.640	0.583	0.1505	0.829	0.2140	0.701
100	37.8	0.134	0.561	0.00519	0.00898	0.0274	1.133	0.0322	0.518	0.851	0.2196	1.205	0.3110	0.706
200	93.3	0.145	0.607	0.00627	0.01085	0.0312	1.290	0.0278	0.447	1.122	0.2896	1.555	0.4013	0.722
Steam														
212	100.0	0.451	1.888	0.0145	0.0251	0.0313	1.294	0.0372	0.599	0.842	0.2173	0.864	0.2230	0.96
300	148.9	0.456	1.909	0.0171	0.0296	0.0360	1.488	0.0328	0.528	1.098	0.2834	1.14	0.2942	0.95
400	204.4	0.462	1.934	0.0200	0.0346	0.0407	1.683	0.0288	0.463	1.422	0.3670	1.50	0.3872	0.94

Table B-9 Illustration of physical properties of metals and nonmetals in both Btu and SI units

Material	Temperature, °F	Temperature, °C	c_p, $\frac{Btu}{lb \cdot °F}$	$c_p \times 10^{-3}$, $\frac{W \cdot s}{kg \cdot °C}$	k, $\frac{Btu}{h \cdot ft \cdot °F}$	k, $\frac{W}{m \cdot °C}$	ρ, $\frac{lb}{ft^3}$	ρ, $\frac{kg}{m^3}$	α, $\frac{ft^2}{h}$	$\alpha \times 10^6$, $\frac{m^2}{s}$
Metals										
Aluminum	32	0	0.208	0.871	117	202.4	169	2,719	3.33	85.9
Copper	32	0	0.091	0.381	224	387.6	558	8,978	4.42	114.1
Gold	68	20	0.030	0.126	169	292.4	1204	19,372	4.68	120.8
Iron, pure	32	0	0.104	0.435	36	62.3	491	7,900	0.70	18.1
Cast iron ($c \cong 4\%$)	68	20	0.10	0.417	30	51.9	454	7,304	0.66	17.0
Lead	70	21.1	0.030	0.126	20	34.6	705	11,343	0.95	25.5
Mercury	32	0	0.033	0.138	4.83	8.36	849	13,660	0.172	4.44
Nickel	32	0	0.103	0.431	34.4	59.52	555	8,930	0.60	15.5
Silver	32	0	0.056	0.234	242	418.7	655	10,539	6.60	170.4
Steel, mild	32	0	0.11	0.460	26	45.0	490	7,884	0.48	12.4
Tungsten	32	0	0.032	0.134	92	159.2	1204	19,372	2.39	61.7
Zinc	32	0	0.091	0.381	65	112.5	446	7,176	1.60	41.3
Nonmetals										
Asbestos	32	0	0.25	1.047	0.087	0.151	36	579	0.010	0.258
Brick, fireclay	400	204.4	0.20	0.837	0.58	1.004	144	2,317	0.020	0.516
Cork, ground	100	37.8	0.48	2.010	0.024	0.042	8	128.7	0.006	0.155
Glass, Pyrex			0.20	0.837	0.68	1.177	150	2,413	0.023	0.594
Granite	32	0	0.19	0.796	1.6	2.768	168	2,703	0.050	1.291
Ice	32	0	0.49	2.051	1.28	2.215	57	917	0.046	1.187
Oak, across grain	85	29.4	0.41	1.716	0.111	0.192	44	708	0.0062	0.160
Pine, across grain	85	29.4	0.42	1.758	0.092	0.159	37	595	0.0059	0.152
Quartz sand, dry			0.19	0.796	0.15	0.260	103	1,657	0.008	0.206
Rubber, soft			0.45	1.884	0.10	0.173	69	1,110	0.003	0.077

BIBLIOGRAPHY ON PHYSICAL PROPERTIES OF MATERIALS

1. *International Critical Tables*, McGraw-Hill, New York, 1926–1930.
2. Atomic Energy Commission: *Liquid Metals Handbook*, 2d ed., Department of the Navy, Washington, 1952.
3. Kowalczyk, L. S.: "Thermal Conductivity and Its Variability with Temperature and Pressure," *Trans. ASME* **77**(7): 1021–1036 (1955).
4. Sato, T., and T. Minamiyama: "Viscosity of Steam at High Temperatures and Pressures," *Int. J. Heat Mass Transfer* **7**: 199–209 (1964).
5. Tye, R. R.: *Thermal Conductivity*, vols. 1 and 2, Academic, New York, 1969.
6. Poferl, D. J., R. A. Svehla, and K. Lewandowski: "Thermodynamic and Transport Properties of Air and the Combustion Products of Natural Gas and of ASTM-A-1 Fuel with Air," *NASA Tech. Note* D-5452, 1969.
7. Keenan, Joseph H., Frederick G. Keyes, Philip G. Hill, and Joan G. Moore: *Steam Tables*, Wiley, New York, 1969.
8. Vukalovich, M. P., A. I. Ivanov, L. R. Fokin, and A. T. Yakovelev: *Thermophysical Properties of Mercury*, State Committee on Standards, State Service for Standards and Handbook Data, Monograph Series No. 9, Izd. Standartov, Moscow, 1971.
9. Eckert, E. R. G., and R. M. Drake: *Analysis of Heat and Mass Transfer*, McGraw-Hill, New York, 1972.
10. American Society of Heating, Refrigerating and Air Conditioning Engineers: *ASHRAE Handbook of Fundamentals*, 1972.
11. Touloukian, Y. S., and C. Y. Ho (eds.): *Thermophysical Properties of Matter*, Plenum, New York, vol. 1. *Thermal Conductivity of Metallic Solids*; vol. 2, *Thermal Conductivity of Nonmetallic Solids*; vol. 4, *Specific Heat of Metallic Solids*; vol. 5, *Specific Heat of Nonmetallic Solids*; vol. 7, *Thermal Radiative Properties of Metallic Solids*; vol. 8, *Thermal Radiative Properties of Nonmetallic Solids*; vol. 9, *Thermal Radiative Properties of Coatings*; 1972.
12. Ho, C. Y., R. W. Powell, and P. E. Liley: "Thermal Conductivity of the Elements: A Comprehensive Review," *J. Phys. Chem. Ref. Data* **3**, supp. 1, 1974.
13. Desai, P. D., T. K. Chu, R. H. Bogaard, M. W. Ackermann, and C. Y. Ho: "Part 1: Thermophysical Properties of Carbon Steels, Part II: Thermophysical Properties of Low Chromium Steels, Part III: Thermophysical Properties of Nickel Steels, Part IV: Thermophysical Properties of Stainless Steels," CINDAS Special Report, September 1976.
14. Vargaftik, N. B.: *Tables of Thermophysical Properties of Liquids and Gases*, 2d ed., Hemisphere, New York, 1975.
15. Hanley, E. J., D. P. DeWitt, and R. E. Taylor: "The Thermal Transport Properties at Normal and Elevated Temperature of Eight Representative Rocks," *Proceedings of the Seventh Symposium on Thermophysical Properties*, American Society of Mechanical Engineers, 1977.

APPENDIX

C

RADIATION PROPERTIES

Table C-1 Normal emissivity of surfaces

Surface	T, °C	ε_n
Metals		
Aluminum		
Highly polished, plate	200–600	0.038–0.06
Bright, foil	21	0.04
Heavily oxidized	100–500	0.20–0.33
Antimony, polished	37–260	0.28–0.31
Brass		
Highly polished	250–360	0.028–0.031
Dull plate	50–350	0.22
Oxidized	200–500	0.60
Chromium, polished	37–1100	0.08–0.40
Copper		
Polished, electrolytic	80	0.018
Polished	37–260	0.04–0.05
Calorized	37–260	0.18
Black oxidized	37	0.78
Gold, polished	37–260	0.02
Iron		
Electrolytic, highly polished	175–225	0.052–0.064
Polished	425–1025	0.14–0.38
Freshly emeried	20	0.24
Completely rusted	20	0.69
Oxidized	100	0.74
Rough ingot	925–1100	0.87–0.95
Cast iron, newly turned	22	0.44
Cast iron, oxidized at 600°C	200–600	0.64–0.78
Cast plate, smooth	22	0.80
Cast plate, rough	22	0.82

Table C-1 *(continued)*

Surface	T, °C	ε_n
Metals		
Lead		
Pure, polished	260	0.08
Gray, oxidized	23	0.28
Oxidized at 200°C	200	0.63
Magnesium		
Polished	37–260	0.07–0.13
Oxide	275–825	0.55–0.20
Molybdenum		
Polished	150–480	0.02–0.05
Filament	700–2600	0.10–0.20
Monel		
Polished	37	0.17
Oxidized at 600°C	540	0.45
Nickel		
Electrolytic	37–260	0.04–0.06
Pure, polished	260	0.07
Oxidized at 600°C	260–540	0.37–0.48
Platinum		
Electrolytic	260–540	0.06–0.10
Plate, polished	260–540	0.06–0.10
Oxidized at 600°C	260–540	0.07–0.11
Filament	26–1225	0.04–0.19
Silver		
Polished, pure	225–625	0.02–0.03
Polished	37–370	0.02–0.03
Stainless steel		
Type 301, polished	23	0.16
Type 316, polished	23	0.17
Type 347, polished	23	0.17
Types 301 and 347, after repeated heating and cooling	230	0.57
Type 301, cleaned	23	0.21
Type 316, cleaned	23	0.28
Type 347, cleaned	23	0.39
Tin		
Polished	37	0.05
Bright tinned iron	25	0.04–0.06
Tungsten		
Filament	3300	0.39
Filament, aged	25–3300	0.03–0.35
Zinc		
Polished	225–325	0.05–0.06

Table C-1 (*continued*)

Surface	T, °C	ε_n
Metals		
Oxidized at 400°C	400	0.11
Galvanized sheet iron, bright	27	0.23
Galvanized sheet iron, gray	23	0.28
Nonmetals		
Alumina (85–99.5% Al_2O_3), effect of mean grain size		
10 μm	1000–1560	0.30–0.18
50 μm	1000–1560	0.39–0.28
100 μm	1000–1560	0.50–0.40
Asbestos		
Paper	37	0.93
Board	37	0.96
Brick		
Magnesite, refractory	1000	0.38
Red, rough	21	0.93
Gray, glazed	1100	0.75
Silica	540	0.80
Carbon		
Filament	1050–1400	0.526
Candle soot	95–270	0.952
Lampblack, thin layer	20	0.93
Lampblack, thick layer	20	0.967
Ceramic		
Earthenware, glazed	20	0.90
Earthenware, matte	20	0.93
Porcelain	22	0.92
Refractory, black	93	0.94
Clay, fired	70	0.91
Concrete, rough	37	0.94
Corundum, emery rough	80	0.86
Glass		
Smooth	22	0.94
Pyrex, lead, and soda	260–530	0.95–0.85
Ice		
Smooth	0	0.97
Rough crystals	0	0.985
Marble, light gray, polished	22	0.93
Mica	37	0.75

Table C-1 *(continued)*

Surface	T, °C	ε_r
Nonmetals		
Paints		
Aluminum 10%, lacquer 22%	100	0.52
Aluminum 26%, lacquer 27%	100	0.30
Other aluminum paints	100	0.27
Lacquer, white	100	0.925
Lacquer, black matte	80	0.97
Oil paints, all colors	100	0.92–0.96
Oil paints	20	0.89–0.97
Paper		
Ordinary	20	0.80–0.90
Ordinary	95	0.92
Asbestos	20	0.95
Roofing	20	0.91
Tar	20	0.93
Porcelain, glazed	22	0.92
Quartz		
Fused, rough	21	0.93
Glass, 1.98 mm thick	280	0.90
Glass, 1.98 mm thick	840	0.41
Glass, 6.88 mm thick	280	0.93
Glass, 6.88 mm thick	840	0.47
Rubber		
Hard	23	0.94
Soft, gray	23	0.86
Soil	37	0.93–0.96
Water, deep	0–100	0.96
Wood	20	0.80–0.90

Table C-2 Solar absorptivity of surfaces (receiving surface at room temperature)

Surface	α
Metals	
Aluminum	
Polished	0.10
Anodized	0.14
Foil	0.15

Table C-2 (*continued*)

Surface	α
Metals	
Brass	
Polished	0.3–0.5
Dull	0.4–0.65
Chromium, electroplated	0.41
Copper	
Highly polished	0.18
Clean	0.25
Tarnished by exposure	0.64
Gold	0.21
Iron	
Ground with fine grit	0.36
Galvanized, highly polished	0.34
Galvanized, new	0.64
Matte, oxidized	0.96
Lead roofing, old	0.77
Magnesium, polished	0.19
Nickel	
Highly polished	0.15
Polished	0.36
Oxidized	0.79
Platinum, bright	0.31
Silver	
Highly polished	0.07
Polished	0.13
Stainless steel, type 301	
Polished	0.37
Clean	0.52
Tungsten, highly polished	0.37
Zinc	
Highly polished	0.34
Polished	0.55
Nonmetals	
Asphalt	
Pavement	0.85
Pavement free from dust	0.93
New	0.93

Table C-2 *(continued)*

Surface	α
Nonmetals	
Brick	
White glazed	0.26
Clay, cream glazed	0.36
Red	0.70
Red, darker glazed	0.77
Concrete	
Uncolored	0.65
Brown	0.85
Dark	0.91
Black	0.91
Earth, plowed field	0.75
Granite	0.45
Grass	0.75–0.8
Gravel	0.29
Leaves, green	0.71–0.79
Magnesium oxide (MgO)	0.15
Marble	
White	0.44
Ground, unpolished	0.47
Cleavage	0.60
Paints	
Oil, white lead	0.24–0.26
Oil, light cream	0.30
Oil, light green	0.50
Aluminum	0.55
Oil, light gray	0.75
Oil, black on galvanized iron	0.90
Paper	
Bond	0.25
White	0.28
Sand	0.76
Sawdust	0.75
Slate	
Silver gray	0.79
Blue gray	0.85
Greenish gray	0.88
Dark gray	0.90
Snow, clean	0.2–0.35
Soot, coal	0.95
Whitewash on galvanized iron	0.22
Zinc oxide	0.15

BIBLIOGRAPHY ON RADIATION PROPERTIES

1. Sieber, W.: *Tech. Phys.* **22**:130–135 (1941).
2. Hottel, H.: "Radiant Heat Transmission," in W. H. McAdams (ed.), *Heat Transmission*, 3d ed., McGraw-Hill, New York, 1954.
3. Gubareff, G. G., J. E. Janssen, and R. H. Torborg: *Thermal Radiation Properties Survey*, Honeywell Research Center, Honeywell Regulator Company, Minneapolis, 1960.
4. Singham, J. R.: "Tables of Emissivity of Surfaces," *Int. J. Heat Mass Transfer* **5**:67–76 (1962).
5. Wood, D. H., H. W. Deem, and C. F. Lucks: *Thermal Radiative Properties*, vol. 3, Plenum, New York, 1964.
6. Svet, D. Y.: *Thermal Radiation, Metals, Semiconductors, Ceramics, Partly Transparent Bodies and Films*, Consultants Bureau, Plenum, New York, 1965.
7. Sadykov, B. S.: "Temperature Dependence of the Radiating Power of Metals," *High Temp.* **3**(3): 352–356 (1965).
8. Svet, Darii Ia.: *Thermal Radiation; Metals, Semiconductors, Ceramics, Partly Transparent Bodies, and Films*, Consultants Bureau, Plenum, New York, 1965.
9. Goldstein, R.: "Measurements of Infrared Absorption of Water Vapor at Temperatures to 1000°K," *J. Quant. Spectrosc. Radiat. Transfer* **4**:343–352 (1964).
10. Goldstein, R., and S. S. Penner: "The Near-Infrared Absorption of Liquid Water at Temperatures between 27 and 209°C," *J. Quant. Spectrosc. Radiat. Transfer* **4**:441–451 (1964).
11. Penner, S. S.: *Quantitative Molecular Spectroscopy and Gas Emissivities*, Addison-Wesley, Reading, Mass., 1959.

APPENDIX

D

ERROR FUNCTION, ROOTS OF TRANSCENDENTAL EQUATIONS, AND EXPONENTIAL INTEGRAL FUNCTION

Table D-1 Error function $\mathrm{erf}(\xi)$

$$\mathrm{erf}(\xi) = \frac{2}{\sqrt{\pi}} \int_0^{\xi} e^{-y^2}\,dy$$

$$\mathrm{erf}(\infty) = 1$$

ξ	$\mathrm{erf}(\xi)$	ξ	$\mathrm{erf}(\xi)$	ξ	$\mathrm{erf}(\xi)$
0.00	0.00000	0.76	0.71754	1.52	0.96841
0.02	0.02256	0.78	0.73001	1.54	0.97059
0.04	0.04511	0.80	0.74210	1.56	0.97263
0.06	0.06762	0.82	0.75381	1.58	0.97455
0.08	0.09008	0.84	0.76514	1.60	0.97636
0.10	0.11246	0.86	0.77610	1.62	0.97804
0.12	0.13476	0.88	0.78669	1.64	0.97962
0.14	0.15695	0.90	0.79691	1.66	0.98110
0.16	0.17901	0.92	0.80677	1.68	0.98249
0.18	0.20094	0.94	0.81627	1.70	0.98379
0.20	0.22270	0.96	0.82542	1.72	0.98500
0.22	0.24430	0.98	0.83423	1.74	0.98613
0.24	0.26570	1.00	0.84270	1.76	0.98719
0.26	0.28690	1.02	0.85084	1.78	0.98817
0.28	0.30788	1.04	0.85865	1.80	0.98909

Table D-1 (*continued*)

ξ	erf(ξ)	ξ	erf(ξ)	ξ	erf(ξ)
0.30	0.32863	1.06	0.86614	1.82	0.98994
0.32	0.34913	1.08	0.87333	1.84	0.99074
0.34	0.36936	1.10	0.88020	1.86	0.99147
0.36	0.38933	1.12	0.88079	1.88	0.99216
0.38	0.40901	1.14	0.89308	1.90	0.99279
0.40	0.42839	1.16	0.89910	1.92	0.99338
0.42	0.44749	1.18	0.90484	1.94	0.99392
0.44	0.46622	1.20	0.91031	1.96	0.99443
0.46	0.48466	1.22	0.91553	1.98	0.99489
0.48	0.50275	1.24	0.92050	2.00	0.99532
0.50	0.52050	1.26	0.92524	2.10	0.99702
0.52	0.53790	1.28	0.92973	2.20	0.99813
0.54	0.55494	1.30	0.93401	2.30	0.99885
0.56	0.57162	1.32	0.93806	2.40	0.99931
0.58	0.58792	1.34	0.94191	2.50	0.99959
0.60	0.60386	1.36	0.94556	2.60	0.999764
0.62	0.61941	1.38	0.94902	2.70	0.999866
0.64	0.63459	1.40	0.95228	2.80	0.999925
0.66	0.64938	1.42	0.95538	2.90	0.999959
0.68	0.66278	1.44	0.95830	3.00	0.999978
0.70	0.67780	1.46	0.96105	3.20	0.999994
0.72	0.69143	1.48	0.96365	3.40	0.999998
0.74	0.70468	1.50	0.96610	3.60	1.000000

Table D-2 First roots β_n of $\beta \tan \beta = c$

c	β_1	β_2	β_3	β_4	β_5	β_6
0	0	3.1416	6.2832	9.4248	12.5664	15.7080
0.001	0.0316	3.1419	6.2833	9.4249	12.5665	15.7080
0.002	0.0447	3.1422	6.2835	9.4250	12.5665	15.7081
0.004	0.0632	3.1429	6.2838	9.4252	12.5667	15.7082
0.006	0.0774	3.1435	6.2841	9.4254	12.5668	15.7083
0.008	0.0893	3.1441	6.2845	9.4256	12.5670	15.7085
0.01	0.0998	3.1448	6.2848	9.4258	12.5672	15.7086
0.02	0.1410	3.1479	6.2864	9.4269	12.5680	15.7092
0.04	0.1987	3.1543	6.2895	9.4290	12.5696	15.7105
0.06	0.2425	3.1606	6.2927	9.4311	12.5711	15.7118
0.08	0.2791	3.1668	6.2959	9.4333	12.5727	15.7131
0.1	0.3111	3.1731	6.2991	9.4354	12.5743	15.7143
0.2	0.4328	3.2039	6.3148	9.4459	12.5823	15.7207
0.3	0.5218	3.2341	6.3305	9.4565	12.5902	15.7270
0.4	0.5932	3.2636	6.3461	9.4670	12.5981	15.7334
0.5	0.6533	3.2923	6.3616	9.4775	12.6060	15.7397
0.6	0.7051	3.3204	6.3770	9.4879	12.6139	15.7460
0.7	0.7506	3.3477	6.3923	9.4983	12.6218	15.7524
0.8	0.7910	3.3744	6.4074	9.5087	12.6296	15.7587
0.9	0.8274	3.4003	6.4224	9.5190	12.6375	15.7650
1.0	0.8603	3.4256	6.4373	9.5293	12.6453	15.7713
1.5	0.9882	3.5422	6.5097	9.5801	12.6841	15.8026
2.0	1.0769	3.6436	6.5783	9.6296	12.7223	15.8336
3.0	1.1925	3.8088	6.7040	9.7240	12.7966	15.8945
4.0	1.2646	3.9352	6.8140	9.8119	12.8678	15.9536
5.0	1.3138	4.0336	6.9096	9.8928	12.9352	16.0107
6.0	1.3496	4.1116	6.9924	9.9667	12.9988	16.0654
7.0	1.3766	4.1746	7.0640	10.0339	13.0584	16.1177
8.0	1.3978	4.2264	7.1263	10.0949	13.1141	16.1675
9.0	1.4149	4.2694	7.1806	10.1502	13.1660	16.2147
10.0	1.4289	4.3058	7.2281	10.2003	13.2142	16.2594
15.0	1.4729	4.4255	7.3959	10.3898	13.4078	16.4474
20.0	1.4961	4.4915	7.4954	10.5117	13.5420	16.5864
30.0	1.5202	4.5615	7.6057	10.6543	13.7085	16.7691
40.0	1.5325	4.5979	7.6647	10.7334	13.8048	16.8794
50.0	1.5400	4.6202	7.7012	10.7832	13.8666	16.9519
60.0	1.5451	4.6353	7.7259	10.8172	13.9094	17.0026
80.0	1.5514	4.6543	7.7573	10.8606	13.9644	17.0686
100.0	1.5552	4.6658	7.7764	10.8871	13.9981	17.1093
∞	1.5708	4.7124	7.8540	10.9956	14.1372	17.2788

Roots are all real if $c > 0$.

Table D-3 First six roots β_n of $\beta \cot \beta = -c$

c	β_1	β_2	β_3	β_4	β_5	β_6
−1.0	0	4.4934	7.7253	10.9041	14.0662	17.2208
−0.995	0.1224	4.4945	7.7259	10.9046	14.0666	17.2210
−0.99	0.1730	4.4956	7.7265	10.9050	14.0669	17.2213
−0.98	0.2445	4.4979	7.7278	10.9060	14.0676	17.2219
−0.97	0.2991	4.5001	7.7291	10.9069	14.0683	17.2225
−0.96	0.3450	4.5023	7.7304	10.9078	14.0690	17.2231
−0.95	0.3854	4.5045	7.7317	10.9087	14.0697	17.2237
−0.94	0.4217	4.5068	7.7330	10.9096	14.0705	17.2242
−0.93	0.4551	4.5090	7.7343	10.9105	14.0712	17.2248
−0.92	0.4860	4.5112	7.7356	10.9115	14.0719	17.2254
−0.91	0.5150	4.5134	7.7369	10.9124	14.0726	17.2260
−0.90	0.5423	4.5157	7.7382	10.9133	14.0733	17.2266
−0.85	0.6609	4.5268	7.7447	10.9179	14.0769	17.2295
−0.8	0.7593	4.5379	7.7511	10.9225	14.0804	17.2324
−0.7	0.9208	4.5601	7.7641	10.9316	14.0875	17.2382
−0.6	1.0528	4.5822	7.7770	10.9408	14.0946	17.2440
−0.5	1.1656	4.6042	7.7899	10.9499	14.1017	17.2498
−0.4	1.2644	4.6261	7.8028	10.9591	14.1088	17.2556
−0.3	1.3525	4.6479	7.8156	10.9682	14.1159	17.2614
−0.2	1.4320	4.6696	7.8284	10.9774	14.1230	17.2672
−0.1	1.5044	4.6911	7.8412	10.9865	14.1301	17.2730
0	1.5708	4.7124	7.8540	10.9956	14.1372	17.2788
0.1	1.6320	4.7335	7.8667	11.0047	14.1443	17.2845
0.2	1.6887	4.7544	7.8794	11.0137	14.1513	17.2903
0.3	1.7414	4.7751	7.8920	11.0228	14.1584	17.2961
0.4	1.7906	4.7956	7.9046	11.0318	14.1654	17.3019
0.5	1.8366	4.8158	7.9171	11.0409	14.1724	17.3076
0.6	1.8798	4.8358	7.9295	11.0498	14.1795	17.3134
0.7	1.9203	4.8556	7.9419	11.0588	14.1865	17.3192
0.8	1.9586	4.8751	7.9542	11.0677	14.1935	17.3249
0.9	1.9947	4.8943	7.9665	11.0767	14.2005	17.3306
1.0	2.0288	4.9132	7.9787	11.0856	14.2075	17.3364
1.5	2.1746	5.0037	8.0385	11.1296	14.2421	17.3649
2.0	2.2889	5.0870	8.0962	11.1727	14.2764	17.3932
3.0	2.4557	5.2329	8.2045	11.2560	14.3434	17.4490
4.0	2.5704	5.3540	8.3029	11.3349	14.4080	17.5034
5.0	2.6537	5.4544	8.3914	11.4086	14.4699	17.5562
6.0	2.7165	5.5378	8.4703	11.4773	14.5288	17.6072
7.0	2.7654	5.6078	8.5406	11.5408	14.5847	17.6562
8.0	2.8044	5.6669	8.6031	11.5994	14.6374	17.7032
9.0	2.8363	5.7172	8.6587	11.6532	14.6870	17.7481
10.0	2.8628	5.7606	8.7083	11.7027	14.7335	17.7908
15.0	2.9476	5.9080	8.8898	11.8959	14.9251	17.9742
20.0	2.9930	5.9921	9.0019	12.0250	15.0625	18.1136
30.0	3.0406	6.0831	9.1294	12.1807	15.2380	18.3018
40.0	3.0651	6.1311	9.1987	12.2688	15.3417	18.4180
50.0	3.0801	6.1606	9.2420	12.3247	15.4090	18.4953
60.0	3.0901	6.1805	9.2715	12.3632	15.4559	18.5497
80.0	3.1028	6.2058	9.3089	12.4124	15.5164	18.6209
100.0	3.1105	6.2211	9.3317	12.4426	15.5537	18.6650
∞	3.1416	6.2832	9.4248	12.5664	15.7080	18.8496

Roots are all real if $c > -1$.

Table D-4 Exponential integral functions $E_n(x)$

x	$E_1(x)$	$E_2(x)$	$E_3(x)$	$E_4(x)$
0.00	∞	1.0000	0.5000	0.3333
0.01	4.0379	0.9497	0.4903	0.3284
0.02	3.3547	0.9131	0.4810	0.3235
0.03	2.9591	0.8817	0.4720	0.3188
0.04	2.6813	0.8535	0.4633	0.3141
0.05	2.4679	0.8278	0.4549	0.3095
0.06	2.2953	0.8040	0.4468	0.3050
0.07	2.1508	0.7818	0.4388	0.3006
0.08	2.0269	0.7610	0.4311	0.2962
0.09	1.9187	0.7412	0.4236	0.2919
0.10	1.8229	0.7225	0.4163	0.2877
0.15	1.4645	0.6410	0.3823	0.2678
0.20	1.2227	0.5742	0.3519	0.2494
0.25	1.0443	0.5177	0.3247	0.2325
0.30	0.9057	0.4691	0.3000	0.2169
0.35	0.7942	0.4267	0.2777	0.2025
0.40	0.7024	0.3894	0.2573	0.1891
0.45	0.6253	0.3562	0.2387	0.1767
0.50	0.5598	0.3266	0.2216	0.1652
0.60	0.4544	0.2762	0.1916	0.1446
0.70	0.3738	0.2349	0.1661	0.1268
0.80	0.3106	0.2009	0.1443	0.1113
0.90	0.2602	0.1724	0.1257	0.0978
1.00	0.2194	0.1485	0.1097	0.0861
1.10	0.1860	0.1283	0.0959	0.0758
1.20	0.1584	0.1111	0.0839	0.0668
1.30	0.1355	0.0964	0.0736	0.0590
1.40	0.1162	0.0839	0.0646	0.0521
1.50	0.1000	0.0731	0.0567	0.0460
1.60	0.0863	0.0638	0.0499	0.0407
1.70	0.0747	0.0558	0.0439	0.0360
1.80	0.0647	0.0488	0.0387	0.0319
1.90	0.0562	0.0428	0.0341	0.0282
2.0	4.890×10^{-2}	3.753×10^{-2}	3.013×10^{-2}	2.502×10^{-2}
2.2	3.719	2.898	2.352	1.969
2.4	2.844	2.246	1.841	1.552
2.6	2.185	1.746	1.443	1.225
2.8	1.686	1.362	1.134	0.968
3.0	1.305	1.064	0.893	0.767
3.5	6.970×10^{-3}	5.802×10^{-3}	4.945×10^{-3}	4.296×10^{-3}
4.0	3.779	3.198	2.761	2.423
4.5	2.073	1.779	1.552	1.374
5.0	1.148	0.996	0.878	0.783

The nth exponential integral $E_n(x)$ of the argument x is defined by

$$E_n(x) = \int_1^\infty e^{-xt} t^{-n}\, dt = \int_0^1 e^{-x/\mu} \mu^{n-2}\, d\mu$$

By differentiation

$$\frac{d}{dx} E_n(x) = \begin{cases} -\dfrac{1}{x} e^{-x} & \text{for } n = 1 \\ -E_{n-1}(x) & \text{for } n = 2, 3, 4, \ldots \end{cases}$$

The first four of the functions $E_n(x)$ for values of x from 0 to 5 are given above.

BIBLIOGRAPHY

1. Chandrasekhar, S.: *Radiative Transfer*, Oxford University Press, London, 1950; also Dover, New York, 1960.
2. Kourganoff, V.: *Basic Methods in Transfer Problems*, Dover, New York, 1963.
3. Case, K. M., F. de Hoffmann, and G. Placzek: *Introduction to the Theory of Neutron Diffusion*, Los Alamos Scientific Laboratory, Los Alamos, N. Mex., 1953.
4. Abramowitz, M., and I. A. Stegun (eds.): *Handbook of Mathematical Functions*, Dover, New York, 1965.

APPENDIX

E

DIMENSIONAL DATA FOR TUBES AND STEEL PIPES

Table E-1 Tubes

Outside diameter, in	BWG gauge	Wall thickness, in	Inside diameter, in
$\frac{1}{2}$	16	0.065	0.370
$\frac{1}{2}$	18	0.049	0.402
$\frac{1}{2}$	20	0.035	0.430
$\frac{5}{8}$	14	0.083	0.459
$\frac{5}{8}$	16	0.065	0.495
$\frac{5}{8}$	18	0.049	0.527
$\frac{5}{8}$	20	0.035	0.555
$\frac{3}{4}$	12	0.109	0.532
$\frac{3}{4}$	14	0.083	0.584
$\frac{3}{4}$	16	0.065	0.620
$\frac{3}{4}$	18	0.049	0.652
$\frac{3}{4}$	20	0.035	0.680
1	12	0.109	0.782
1	14	0.083	0.834
1	16	0.065	0.870
1	18	0.049	0.902
1	20	0.035	0.930
$1\frac{1}{4}$	12	0.109	1.032
$1\frac{1}{4}$	14	0.083	1.084
$1\frac{1}{4}$	16	0.065	1.120
$1\frac{1}{4}$	18	0.049	1.152
$1\frac{1}{2}$	12	0.109	1.282
$1\frac{1}{2}$	14	0.083	1.334
$1\frac{1}{2}$	16	0.065	1.370
$1\frac{1}{2}$	18	0.049	1.402
2	12	0.109	1.782
2	14	0.083	1.834
2	16	0.065	1.870

Based on data from Tubular Exchanger Manufacturers Association.

Table E-2 Steel pipes

Nominal pipe size, in	Outside diameter, in	Schedule no.	Wall thickness, in	Inside diameter, in
$\frac{1}{8}$	0.405	40	0.068	0.269
		80	0.095	0.215
$\frac{1}{4}$	0.540	40	0.088	0.364
		80	0.119	0.302
$\frac{3}{8}$	0.675	40	0.091	0.493
		80	0.126	0.423
$\frac{1}{2}$	0.840	40	0.109	0.622
		80	0.147	0.546
$\frac{3}{4}$	1.050	40	0.113	0.824
		80	0.154	0.742
1	1.315	40	0.133	1.049
		80	0.179	0.957
$1\frac{1}{2}$	1.900	40	0.145	1.610
		80	0.200	1.500
2	2.375	40	0.154	2.067
		80	0.218	1.939
3	3.500	40	0.216	3.068
		80	0.300	2.900
4	4.500	40	0.237	4.026
		80	0.337	3.826
5	5.563	40	0.258	5.047
		80	0.375	4.813
6	6.625	40	0.280	6.055
		80	0.432	5.761
10	10.75	40	0.365	10.020
		60	0.500	9.750

Based on ASA Standards B36.10.

NAME INDEX

SUBJECT INDEX